Atomic Collisions
on Solid Surfaces

Atomic Collisions
on Solid Surfaces

E.S. PARILIS

L.M. KISHINEVSKY

N.Yu. TURAEV

B.E. BAKLITZKY

F.F. UMAROV

V.Kh. VERLEGER

S.L. NIZHNAYA

and

I.S. BITENSKY

Arifov Institute of Electronics
Tashkent 700143, Uzbekistan

1993
NORTH-HOLLAND
AMSTERDAM · LONDON · NEW YORK · TOKYO

05187473

PHYSICS

North-Holland

ELSEVIER SCIENCE PUBLISHERS B.V.
Sara Burgerhartstraat 25
P.O. Box 211, 1000 AE Amsterdam, The Netherlands

Library of Congress Cataloging-in-Publication Data

Atomic collisions on solid surfaces / E.S. Parilis ... [et al.].
     p.   cm.
    Translated from Russian.
    Includes bibliographical references and index.
    ISBN 0-444-89105-6
    1. Solids–Surfaces.   2. Surfaces (Physics)   3. Scattering
(Physics)   4. Ion bombardment.   5. Auger effect.   6. Ions–Scattering.   I. Parilis, E.S.
    QC173.4.S94A88   1992
    530.4'17–dc20   92–6503
                                               CIP

ISBN: 0 444 89105 6

Printed in The Netherlands   Printed on acid-free paper

Preface

This book covers the studies which the authors made during the past 30 years. In this period the theory has passed along the way from first using a screened interatomic potential in the problem of atom surface scattering to the simulation of grazing trajectories of atomic and molecular ion scattering in surface semichannels, calculations based on complex programs which are capable of revealing such point defects in the surface structure as vacancies or atomic steps, explanation of the rainbow nature of the *surface peak* in the energy spectrum and orientational effects in the charge state of the scattered atoms, studying both the vibrational and rotational excitation of fast molecules scattered without dissociation by the surface, and describing the *shuttle excitation mechanism* in the cluster propagation through and scattering by the solid.

Starting with the first application of the Firsov mechanism to develop the theory of kinetic electron emission, we proceeded toward predicting the orientational effects in ion-induced Auger electron emission, studying the potential electron emission and *Coulomb explosion sputtering* of insulators under slow multiple-charged ions, proposing the *plane-shaving* and *push-and-stick* models for dimer sputtering and the associative ionization mechanism for their charge-state formation, developing a *shock wave mechanism* of large cluster sputtering and biomolecule desorption under fission fragments.

Our models of atom scattering from solids, and even the associated terminology, such as *single* and *double scattering*, *atomic row effect*, *quasisingle* and *quasidouble* peaks and *surface semichannels* have been accepted by both theorists and experimentalists practically immediately after their appearance in publications and employed actively by many groups both at home and in other countries, thus resulting in copious publications, which presently exceeds, by far, in volume our own papers. Wherever possible, we have tried to sum up the results of these studies, or at least to mention them here, and if we have not succeeded to do this to a full extent, the reason for this can be only lack of information. The authors would be glad to get references to the publications in this area which have somehow eluded their search.

We have limited our consideration to theoretical works and invoke experimental studies only whenever this is needed to compare results with theoretical predictions. For a detailed description of the relevant experiments, the reader is referred to the books by Kaminsky (1965), Arifov (1968), Carter and Colligon (1968), M.W. Thompson (1969), U.A. Arifov

v

and Aliev (1974), Czanderna (1975), Petrov and Abroyan (1977), Lehmann (1977), Mashkova and Molchanov (1985), Ohtsuki (1983), Kurnaev et al. (1985), Ryazanov and Tilinin (1985), Brusilovsky (1990b), and the reviews by Mashkova and Molchanov (1972a,b,c, 1974), Suurmeijer and Boers (1973), Heiland (1973), Boers (1977), Aono (1984), Dodonov et al. (1988), Baranov et al. (1988), Wien (1989), Varga (1989), Hofer (1990), Brusilovsky (1990a), some of which contain also a more or less comprehensive discussion of our theoretical works.

One of the purposes of this edition was to make the English-speaking reader acquainted with our studies published mainly in Russian magazines and partially in two books (Parilis et al. 1987, 1988) issued in Tashkent.

As the theoretical models become increasingly complex, the group of the authors involved expanded in the order as it is shown on the cover. Unfortunately, we have no more with us V. Kivilis who had joined our work in one of its most crucial stages. A. Mukhamedov, Ya. Vinokurov, V. Kitov, Ya. Gilenko, M. Mirakhmedov, A. Goldenberg, I. Wojciechowski and A. Dzhurahalov contributed to the work in various degrees. We are indebted to N. Igamberdieva and G. Skrebtzov for translating the book into English, as well as to Ella Polina for a priceless help in preparing the manuscript.

The authors

Contents

Introduction

This book deals with the theory of collisions of medium-energy atoms on the surface of a solid, and of the accompanying effects, namely, scattering, charge exchange, sputtering and Auger-electron emission. Each of these terms needs a few explanatory comments.

Consider first the term 'scattering'. Since one usually detects only the fraction of the scattered beam which is directed outward with respect to the solid, one should rather consider 'reflection' of atomic articles from its surface. This term is presently widespread and possibly more adequate than 'scattering' since the total flux of the scattered ions contains also particles embedded in the solid. The embedded part of the flux, however, is investigated by totally different methods which do not permit easy extraction of information concerning the scattering at the surface. The effect of scattering from the surface is also difficult to isolate in experiments with fast particles penetrating thin films. In the literature, one employs quite frequently the term 'scattering' also when studying back-scattering from the surface, i.e., ion reflection. Following the well-established terminology, we will use here the terms 'scattering' and 'reflection' to identify the outgoing flux, leaving only the term 'scattering' to describe the result of ion collisions with individual atoms and atomic structures.

The word 'atoms' also requires some explanation. Actually, all the calculations on scattering were carried out here for neutral atoms. The potentials used describe the interaction of neutral atoms with the atoms of a solid, by which one understands both atoms and ions. The use of this terminology is justified by the independence of the scattering process of the charge state of an atom, a fact established sufficiently reliably to hold in the medium-energy range. Note that the reverse certainly is not true; indeed, the charge state of scattered particles depends very strongly on the scattering characteristics. However, since in experiments one employs accelerated ions, one sometimes calls the projectile particles 'ions' to discriminate them from the particles of the target, which are denoted as 'atoms', and we may also follow this convention.

One may find difficulties in a precise definition of the concept 'medium energy'. For specialists in gas dynamics operating with thermal and suprathermal energies, the lower boundary of the energies treated in the present monograph, 0.1 keV, is considered to belong to the high-energy range. At the same time, for the physicists experimenting with modern accelerators, the upper boundary of our range, 100 keV, certainly lies in the region of low or very low energies.

Physically, this energy range is characterized by the dominance of elastic over inelastic energy losses, and by the possibility of considering classical binary collisions using single-center potentials and disregarding the binding energy of the scattering atom in the crystal lattice. The first of these factors determines the upper, and the second, the lower boundary of the energy range. Apart from this, the upper energy limit is historically related to the capability of accelerated technology used in experiments, the results of which were compared with theory. It is usually believed that the main goal of experiments on atom and ion scattering in a solid or on its surface is to determine the potential with which they interact with the atoms of the solid. One can hardly agree with this. In fact, of primary interest are the orientational effects which are typical of the solid state and are determined by the high density and existence of short- and long-range order, as well as by the existence of a sharply defined boundary of the material, the surface of the solid, with its peculiar properties and peculiar structure defects. The effects of correlated collisions (blocking, channeling, and characteristic structure of the energy spectrum of the scattered particles) are closely related with the structure of the solid and its surface and contain very valuable information about them. It is these nontrivial effects, rather than the search for an 'ideal' potential that represent the subject of the studies discussed in the present monograph. As for the interaction potentials, they are derived with a sufficient degree of accuracy from experiments on scattering in gases where one can create conditions for pure binary inter-action and single scattering which will not be distorted by unavoidable action of the neighboring atoms and multiple collisions. They can be used as a good approximation when considering ion scattering in solids. While such a rough approximation as the hard-sphere potential can naturally distort the whole pattern of both angular and energy distributions of the scattered atoms, nevertheless, qualitatively the scattering pattern is about the same for all screened potentials and is determined primarily by multiple collision effects.

In many cases, quantitative results can also be derived. As shown by calculations, the difficulties associated stem in equal degree from the need of taking accurately into account the collision multiplicity, the thermal vibra-tion parameters of the surface atoms, and the shape of the interaction potential. Any single-center potential is nothing more than a more or less good approximation for a solid.

When this research was in its early stage, there was much dispute concerning the influence of atomic bonds in the crystal lattice on the pattern of scattering. This binding was generally believed to be of crucial impor-tance in determining the angular and energy distributions, particularly in the high-energy part of the spectrum. We, however, started from the fact that, first, the binding energy of a scattering atom, which is a few electronvolts, cannot exert a noticeable effect on the scattering of ions having an energy on the order of a kiloelectronvolt, and second, in a collision time of $\approx 10^{-15}$ s the target atom would not move from the equilibrium position to a distance sufficiently large for the binding forces to be felt. Therefore, both in analytical calculations and in the computer simulation programs developed by us we took into consideration the ordered arrangement of atoms in a crystal lattice rather than their bonding.

Later investigations revealed that the bonding effect may be neglected not only for medium energies but in the low-energy range as well, down to a few tens of an electronvolt (Gruich et al. 1975, Tongson and Cooper 1975, Hart and Cooper 1979). The same applies also to the so-called group collisions, i.e., simultaneous collisions of an ion with two, three or four atoms on the surface of a solid. It was assumed that such collisions could account for the presence in the scattered beam of particles with energies in excess of that obtained in single-atom scattering. Attempts were made to describe both the effect of the atomic bonding in the lattice and the group collisions through a formal introduction of an effective scatterer mass exceeding the mass of one free atom in a solid.

From the very beginning, we renounced the consideration of group collisions since in the energy range of interest the distances of closest approach of atoms in binary collisions resulting in large-angle scattering is much smaller than the lattice constant. If we take also into account the atomic displacement due to thermal vibrations, then the situation in which an ion touches symmetrically and simultaneously several surface atoms becomes highly improbable, whereas a nonsymmetrical consecutive contact is indistinguishable from multiple collisions.

The problem of group collisions is closely connected with the binary character of the collisions and the single-center shape of the interaction potential. This is characteristic, however, for much lower energies (0.1–10 eV).

We also renounced the idea of introducing an effective mass, since this technique is of a purely phenomenological nature, contains a certain arbitrariness and is not capable of revealing the microscopic mechanism of scattering. Apart from this, effective mass cannot be a scalar quantity, and introduction of an effective mass tensor for a single crystal is not justified.

Speaking about a surface, one should bear in mind that the bulk of the solid also participates in the process of atom scattering. The scattering depth is greater, the lighter as the atoms are lighter and their energy is higher. In this energy range, heavy atoms are scattered practically by one or two atomic layers. This is a great asset since it offers a possibility of using simple models of single and double scattering, and, under grazing incidence, of calculating scattering produced only by surface atomic rows and the semichannels formed by them.

The possibility of probing only one surface atomic layer by heavy-ion scattering is also unique and does not have analogs in other methods of the surface diagnostics of solids.

Due to the small de Broglie wavelength, which for medium-energy heavy atoms is a few tenths or hundredths of the lattice constant, the amount of information that can be derived from the scattered-beam parameters on the structure of the surface on an atomic level is very large. This is all the more important because the defect density in a real surface, both of natural origin and produced by ion bombardment, exceeds by far that in the bulk. The possibility of establishing correlation between the pattern of the energy spectrum and of the angular distributions of scattered atoms, on the one hand, and individual atomic defects of the surface, on the other, was one of the major points which served to stimulate the research covered in this book.

All the scattering calculations were carried out for metals since ion-scattering experiments are done, as a rule, on high melting point metals whose surface can be cleaned well and which are not charged in the course of ion bombardment. However, the calculations are equally applicable to metals and nonmetals. This gives us grounds to relate the theory to any solid, without specifying its electronic structure. Charge-composition studies showed the main conclusions of the theory to remain also unchanged if the degree of the scattered atom ionization is taken into consideration. A comparison of the energy spectra of scattered heavy atoms for ions and neutrals shows them to retain their characteristic structure of single and double scattering. Actually, introducing a correction for the degree of ionization affects the quantitative relations rather than the qualitative characteristics of a scattered atom beam in the medium-energy range.

Finally, the word 'theory' featuring in the book needs possibly a more precise definition. The bulk of our calculations was carried out from the '60's to the '80's, a period that witnessed an explosive growth of methods of mathematical simulation of the processes involving the passage of fast atomic particles through crystals. It is in these years that the works of M.T. Robinson and Oen (1963), Gibson et al. (1960) and other theorists appeared, who developed a new method of investigation, the so-called mathematical experiment or mathematical simulation, made possible by the advent of fast electronic computers. A substantial part of the calculations presented in this book was made by the computer-simulation technique. This was required by the complexity of the scattering trajectories and by a large number of correlated collision events which prohibit the use of statistical stochastic methods of calculation. Mathematical experiment is similar in some extent to the physical one while permitting us to extract more information from the latter.

One should, however, point out that the principal and most essential results associated with qualitative conclusions from the models used, as well as the radical changes observed when going over from one model to another, were first derived analytically and only thereafter refined, using computer simulation. Even the 'atom row effect' was first shown to exist with a pen, as it were, after which it was obtained routinely with grazing ion scattering programs. In contrast to the channeling which was first observed by M.T. Robinson and Oen (1963) in a computer experiment (although it could have been predicted), the effects obtained in our studies were, as a rule, revealed in a qualitative consideration and subsequently confirmed to exist by computer simulation. Mathematical simulation is certainly an integral part of a theoretical study including the development of an idealized model, a calculation based on physical laws, and a comparison of its results with the data obtained in an experiment, which alone can provide an answer whether the model used is valid or not. As for the problems connected with electronic transitions, they were solved analytically yielding fairly simple final expressions.

One of the major goals of the studies discussed in this book has been the development of a theoretical foundation for the diagnostics of the composition, structure, and state of the surface of solids and thin films, as well as for their quality control needed in present-day microelectronics technology.

CHAPTER 1

General

1. Conditions of applicability of classical mechanics

The conditions of validity of the classical description of scattering has been studied by many authors (Lehmann 1977, Seitz 1949, Bohr 1948, Lindhard 1965, Mott and Massey 1965). Bohr (1948) showed that the elastic scattering of atoms and molecules at medium energies (1–100 keV) through angles greater than the angle determined by the ratio of the de Broglie wavelength of the scattered particle to the double distance of closest approach can be described with good accuracy using classical mechanics. In an elastic collision of two particles, their internal states do not change, thus making it possible to disregard the internal energy of the particles when applying to the collision the energy conservation law.

Classical consideration is possible under the condition that the de Broglie wavelength of a particle of mass m_1 and velocity v_0 is less than the characteristic dimension a of the region within which the scattering field changes substantially, and that the product of this dimension by the scattering-induced change of the particle momentum satisfies Heisenberg's uncertainty principle, i.e., it is required that, first,

$$\hbar / m_1 v_0 \ll a , \tag{1}$$

where $\hbar = h/2\pi$ and h is Planck's constant, and, second, if Δp is the change of momentum in the collision, then

$$a \, \Delta p \gg \hbar . \tag{2}$$

If we consider ion scattering from atoms to be elastic and confine ourselves to scattering angles $\chi > 10^{-2}$ (in the center-of-mass system), then for energies of relative motion, $E_r > 100 \, \text{eV}/m$ (where m is the reduced mass expressed in units of the hydrogen atom mass), the scattering can be described by the laws of classical mechanics. Such estimates show that in the energy range of interest, 0.1–100 keV, and for scattering angles $\chi \geqslant 0.1°$, classical mechanics can be used.

At very high energies, when the relative velocity of colliding atoms exceeds the quantity $Z_1 Z_2 e^2 / \hbar$, classical mechanics is not valid, however, scattering is dominated practically by Coulomb interaction, for which the results of the quantum and classical mechanics coincide. Naturally, the limit in energy shows only that the motion of particles can be approximated by classical trajectory equations and does not mean that quantum-mechanical effects in atomic collisions may be totally neglected. Actually, at energies in excess of ~1 keV, the elastic scattering is accompanied also by processes involving excitation, single- and multiple ionization and charge exchange of colliding atoms (Fedorenko 1959).

Such inelastic collisions are accompanied by a change in the internal energy of individual atoms at the expense of their kinetic energy, which results in an uncertainty in the kinetic energy and interatomic potential of colliding particles. The processes of excitation and ionization are properly described by quantum mechanics.

Firsov (1957, 1958, 1959) showed, however, that irrespective of the inelastic processes occurring in atomic collisions one can calculate approximately the dependence of the scattering angle on the impact parameter using the elastic interaction potential from classical mechanics.

In the medium-energy range, the inelastic energy losses are substantially lower than the elastic ones. They constitute only a few percent of the total energy losses and may be neglected in a first approximation.

2. Laws of momentum and energy conservation

According to Bohr (1948), the interaction of medium-energy particles with the surface of a solid may be considered as a sequence of binary collisions with target atoms which are assumed in a good approximation to be free. The laws of momentum and energy conservation with the inclusion of the assumption of a binary nature of collisions permit one to calculate the momenta and energies of particles after the interaction.

Figure 1a shows the trajectories of particles and their asymptotes in the

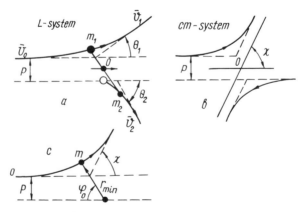

Fig. 1. Scattering schemes in (a) LS and (b) CM, and (c) motion of a particle with reduced mass in a central field (Mashkova and Molchanov 1980).

laboratory system (LS). An ion of mass m_1 and charge Z_1 moves with a velocity $\boldsymbol{v}_0$ (energy E_0) with respect to a fixed (before collision) target atom with an impact parameter p. The impact parameter is understood as the distance from the scattering center to the straight line corresponding to the initial direction of motion of the projectile. As a result of the collision, the moving ion will be deflected from the initial direction by an angle θ_1 and acquire a velocity $\boldsymbol{v}_1$ (and, correspondingly, an energy E_1), while the target particle of mass m_2 (recoil atom), on acquiring a velocity $\boldsymbol{v}_2$ (and, respectively, an energy E_2) will deflect by an angle θ_2 relative to the direction of the initial ion motion. For a given pair of colliding particles the energy E_1 depends on the initial energy E_0, scattering angle θ_1 and mass ratio $\mu = m_2/m_1$. If we consider only elastic collisions, then, starting from the momentum and energy conservation laws, one can derive the unique relation between the quantities E_0, E_1, θ_1, m_1 and m_2. Thus, the LS velocities of the ion and atom after collision will be as follows,

$$\boldsymbol{v}_1 = \left(\frac{1}{1+\mu} \cos\theta_1 \pm \frac{1}{1+\mu} \sqrt{\mu^2 - \sin^2\theta_1} \right) \boldsymbol{v}_0 \,,$$

$$\boldsymbol{v}_2 = \frac{2\cos\theta_2}{1+\mu} \boldsymbol{v}_0 \,; \tag{3}$$

whence for the ion and recoil energies one can write

$$E_1 = \frac{E_0}{(1+\mu)^2} (\cos\theta_1 \pm \sqrt{\mu^2 - \sin^2\theta_1})^2 = \frac{E_0}{(1+\mu)^2} F^2(\theta_1, \mu) \,,$$

$$0 \leqslant \theta_1 \leqslant 180° \,, \tag{4}$$

$$E_2 = \frac{4\mu}{(1+\mu)^2} E_0 \cos^2\theta_2 \,, \quad 0 \leqslant \theta_2 \leqslant 90° \,.$$

For $\mu > 1$, one takes the '+' sign in the expression for E_1, and for $\mu < 1$ (i.e., when the projectile is heavier than the target atom) the expression for E_2 does not change, whereas for E_1 both signs, '+' and '−', are taken, with

$$\theta_1 \leqslant \theta_{\lim} = \arcsin \mu \,, \tag{5}$$

where $\theta_{\lim}$ is the limiting scattering angle.

Thus for $\mu < 1$, single scattering through an angle $\theta_1 > \theta_{\lim}$ is impossible, while for $\theta_1 < \theta_{\lim}$ the energy E_1 becomes a double-valued function of the scattering angle θ_1. In this case, scattering to a given angle $\theta_1 < \theta_{\lim}$ can occur at two different values of the impact parameter p, the '−' sign corresponding to the smaller of them in eq. (4), and the '+' sign to the larger one. Therefore, the recoil angle θ_2 and, hence, the energy E_2 turn out to be also double-valued functions of the angle θ_1. While this was noted as far back as 1951 (A.B. Brown et al. 1951), no experimental verification of such

double-valued behavior in the scattering of atoms on the surface of a solid
could be obtained for a long time.

In a head-on collision, where $p = 0$, the projectile is scattered either
through an angle $\theta_1 = 180°$ (for $\mu > 1$), or through $\theta_1 = 0°$ (for $\mu < 1$), which
correspond to a CM (centre-of-mass system) scattering angle $\chi = 180°$.
Therefore, the particle initially at rest will be imparted the maximum
energy,

$$E_2 = \frac{4\mu}{(1 + \mu)^2} E_0 .\tag{6}$$

On expressing, in terms of θ_1, the particle velocities and energies after
collision, one can find the relation between the energy transferred in the
collision, E_2, and the ion scattering angle,

$$\cos \theta_1 = \left(1 - \frac{1 + \mu}{2} \frac{E_2}{E_0}\right)\Big/ \sqrt{1 - (E_2/E_0)} .\tag{7}$$

The angles of scattering, θ_1, and of recoil, θ_2, in the LS and the CM
scattering angle χ (fig. 1b) are related in the following way;

$$\tan \theta_1 = \frac{\sin \chi}{\dfrac{1}{\mu} + \cos \chi} , \qquad \theta_2 = \tfrac{1}{2}(\pi - \chi) ,$$

$$\chi = \theta_1 + (-1)^n \arcsin \frac{\sin \theta_1}{\mu} + n\pi ,\tag{8}$$

$$\theta_2 = \tfrac{1}{2}\pi(1 - n) - \tfrac{1}{2}\theta_1 - \tfrac{1}{2}(-1)^n \arcsin \frac{\sin \theta_1}{\mu} ,$$

where

$$n = 0 \qquad \text{for } \mu \geqslant 1 ,$$

$$n = 0; 1 \quad \text{for } \mu < 1 .$$

Note that for $\mu < 1$, the values of expressions (8) for χ and θ_2 at $n = 0$
correspond to the larger, and for $n = 1$, to the smaller of the impact
parameters. Thus, the laws of the energy and momentum conservation for
an elastic collision of two particles yield relations between the initial energy
E_0, scattering angles of the ion, θ_1, and recoil atom, θ_2, masses m_1 and m_2,
and energies of the colliding particles after the collision, E_1 and E_2, which
do not depend on the interaction potential. The directions of particle motion
after the collision, i.e., the angles θ_1 and θ_2, depend on the actual law
governing the interaction of the particles, and on their separation in the
collision.

3. Deflection function

If we choose an interaction potential $V(r)$, then the problem of an elastic collision of two particles in the CM system (Fig. 1b,c) reduces to that of the scattering of one particle of reduced mass $m = m_1 m_2/(m_1 + m_2)$ in the field $V(r)$ of a fixed force center located at the CM of the particles (Landau and Lifshitz 1960).

In classical mechanics, the angle χ of deflection of a particle trajectory in a potential field $V(r)$ as a function of the impact parameter p and relative energy $E_r = m_2 E_0/(m_1 + m_2)$ is called the deflection function and determined by the scattering integral,

$$\chi = \pi - 2\varphi_0, \quad \varphi_0 = p \int_{r_0}^{\infty} \frac{dr}{r^2 \left(1 - \frac{V(r)}{E_r} - \frac{p^2}{r^2}\right)^{1/2}} ; \tag{9}$$

where r_0 is the distance of closest approch of the particles in the collision corresponding to the point where the radial particle velocity vanishes; r_0 is the root of the expression under the radical in eq. (9). The distance of the point with reduced mass to the origin as a function of time is determined by the integral

$$t = 2 \int_{r_0}^{r} v^{-1} \left(1 - \frac{V(r)}{E_r} - \frac{p^2}{r^2}\right)^{-1/2} dr . \tag{10}$$

It should be pointed out that the integrals in eqs. (9) and (10) can be calculated analytically only for certain potentials $V(r)$ (Goldstein 1959); for real potentials one usually applies approximate or numerical methods of integrations. For single scattering, the deflection function $\chi(p)$ decreases monotonically to 0 for $p \to \infty$. As will be shown later for double scattering, the function $\chi(p)$ may be nonmonotonic, and the function $p(\chi)$, inverse to it, double valued.

When simulating the scattering of atoms in matter, one uses the deflection function in the LS, $\theta(p)$.

In contrast to the energy of a scattered particle, in order to calculate the deflection function, one has to know the interaction potential.

4. Interaction potentials

The most important problem in the study of scattering is the determination of the interaction forces between atoms (Torrens 1972, Kumakhov and

Komarov 1979). The interaction forces are expressed in terms of the potential energy $V(r)$ due to many-particle interactions of electrons and nuclei. In the medium-energy region, the Coulomb interaction of nuclei is screened by electronic shells. In this case, the expression for the Coulomb potential is complemented by a screening function.

There is, unfortunately, at present no potential which would be valid in the entire range of internuclear separations, and, therefore, one uses approximate-potential functions, each of them being applicable only within a particular region of internuclear separation. The most well-known potentials were introduced by Bohr (1948), L.H. Thomas (1927) and Fermi (1928), Born and Mayer (1932), Molière (1949), Firsov (1957), Abrahamson (1969), Lindhard et al. (1963), O'Connor and Biersack (1986), Goodstein et al. (1989).

An empirical potential with an exponential screening function was proposed by Bohr (1948),

$$V(r) = \frac{Z_1 Z_2 e^2}{r} \exp\left(-\frac{r}{a_B}\right),$$ (11)

where a_B is a quantity called the screening parameter, which is constant for the given pair, $a_B = a_0(Z_1^{2/3} + Z_2^{2/3})^{-1/2}$; here $a_0 = 0.5292$ Å is the radius of the first Bohr orbit of the hydrogen atom.

This potential falls off rapidly with increasing interatomic distance r, which produces a strong underestimation of the interaction energy, and, therefore, it is applicable only for $r < a_0$. In the region $r > a_0$, one uses ordinarily the semiempirical potential function of Born–Mayer,

$$V(r) = A \exp\left(-\frac{r}{b}\right),$$ (12)

where A and b are constants whose magnitude depends on the nature of the colliding particles. There are several methods of determining these constants (Abrahamson 1963, 1969, Gaydaenko and Nikulin 1970, Günther 1964, Andersen and Sigmund 1965).

For interatomic distances less than ~1 Å, the most widely used function is the Thomas–Fermi–Firsov (TFF) screened Coulomb potential,

$$V_F(r) = \frac{Z_1 Z_2 e^2}{r} \chi(r/a).$$ (13)

The screening function $\chi(r/a)$ was calculated by Firsov basing on the statistical model of Thomas–Fermi, its values being tabulated (Torrens 1972, Kumakhov and Komarov 1979).

The Firsov potential falls off too slowly with increasing interatomic separation, thus resulting in an overevaluation of the scattering angles and cross sections for $r \geqslant 1$ Å. A number of analytical approximations of the screening function of the TFF potential were proposed (Torrens 1972), the most well-known of them being the approximation of Molière (1949),

$$V_M(r) = \frac{Z_1 Z_2 e^2}{r} \left[0.35 \exp\left(-\frac{0.3r}{a}\right) + 0.55 \exp\left(-\frac{1.2r}{a}\right) \right.$$

$$\left. + 0.1 \exp\left(-\frac{6.0r}{a}\right) \right]. \tag{14}$$

Calculations using this potential agree well with experiment up to $r \leqslant 1.5$ Å (Torrens 1972, Oen and Robinson 1976a, Hou and Robinson 1976).

Firsov (1957) and Lindhard et al. (1963) proposed expressions for the screening parameter a which are close to one another,

$$a_F = 0.8853 a_0 (Z_1^{1/2} + Z_2^{1/2})^{-2/3},$$

$$a_L = 0.8853 a_0 (Z_1^{2/3} + Z_2^{2/3})^{-1/2}. \tag{15}$$

Boers (1977) and Poelsema et al. (1977a) represent the screening parameter in the Thomas–Fermi potential in the form

$$a = 0.8853 c a_0 (Z_1^{1/2} + Z_2^{1/2})^{-2/3},$$

where c (a fitting parameter less than unity) is chosen from a comparison of the theory with experiment. The approximation of O'Connor and Biersack (1986) for the screening function in the Thomas–Fermi potential takes into account the exchange and correlation energies, the so-called 'universal' potential obtained in this way shows good agreement with experiment,

$$V_{BZ}(r) = \frac{Z_1 Z_2 e^2}{r} \left[0.1818 \exp\left(-\frac{3.2}{a} r\right) + 0.5099 \exp\left(-\frac{0.9423}{a} r\right) \right.$$

$$+ 0.2802 \exp\left(-\frac{0.4029}{a} r\right)$$

$$\left. + 0.02817 \exp\left(-\frac{0.2016}{a} r\right) \right], \tag{16}$$

where

$$a = \frac{0.8853 a_0}{(Z_1^{0.23} + Z_2^{0.23})}.$$

For $r \leqslant 2a_0$, the Firsov potential is well approximated by an inverse square potential of the form

$$V(r) = C/r^2, \quad C = \frac{3.05 Z_1 Z_2}{(Z_1^{1/2} + Z_2^{1/2})^{2/3}} \text{ eV·Å}^2, \tag{17}$$

or (Karpuzov et al. 1969a)

$$V(r) = \frac{C}{r^2} - \Delta V ,$$

where

$$\Delta V = 3.5 \times 10^{-3} \frac{Z_1 Z_2 e^2 (Z_1^{1/2} + Z_2^{1/2})^{2/3}}{a_{TF}} , \quad a_{TF} = 0.468 \text{ Å} . \tag{18}$$

Potential (12) is approximated by a model potential of the form (Karpuzov et al. 1969a, Karpuzov and Yurasova 1971a, Karpuzov et al. 1966, Sigmund and Vajda 1964)

$$V(r) = \frac{\alpha}{r^2} \left(1 - \frac{r}{r_c}\right)^2 \quad \text{for} \quad r < r_c ,$$

$$= 0 \qquad\qquad \text{for} \quad r \geqslant r_c , \tag{19}$$

where r_c is the cut-off radius, α is a constant chosen in such a way that for $r < a_0$ the interaction potential (19) is close to the potential of Firsov, and for $r > a_0$, to that of Born–Mayer. The model potential describes particle interaction over a wide range of interatomic separations and yields an analytic dependence between the scattering angle and the impact parameter.

In numerical simulation of the scattering of medium-energy ions the TFF potential is usually matched with some other potential which describes well the interaction of particles at distances $r \geqslant a_0$. Thus, Shulga (1980) used a potential of the form

$$V(r) = C/r^2 \qquad\qquad \text{for} \quad r \leqslant r_c ,$$

$$= A \exp(-r/b) \quad \text{for} \quad r > r_c , \tag{20}$$

where $b = 0.196$ Å; $r_c = 2b$; $A = C(e/2b)^2$; $e = 2.7182 \dots$. Potential (20) was obtained by matching the inverse square approximation of the Firsov potential with the potential of Born–Mayer for a given b. The expressions for A and r_c were obtained from the condition that at $r = r_c$ both the potentials and their first derivatives are joined.

5. Differential scattering cross sections

The differential scattering cross section is determined by the type of interaction potential and represents the most essential characteristic of the scattering process.

Let the number of particles deflected through an angle from χ to $\chi + d\chi$ be dN. The cross section of scattering is the ratio $d\sigma = dN/n$, where n is the incident beam density. If the scattering function $\chi(p)$ is single valued, then the particles falling into a given angle interval $[\chi, \chi + d\chi]$ will be those with impact parameters lying in the range $[p(\chi), p(\chi) + dp(\chi)]$, i.e., all the particles which pass through a ring with radii p and $p + dp$ near a spherically symmetric scatterer. The scattering cross section in the angular interval from χ to $\chi + d\chi$,

$$d\sigma = 2\pi p \, dp \, , \tag{21}$$

as a function of scattering angle can be written as

$$d\sigma = 2\pi p(\chi) \left| \frac{dp(\chi)}{d\chi} \right| d\chi \, . \tag{22}$$

The use of the absolute value of the derivative is explained by the fact that $d\sigma$ should always be positive by its very meaning. One can relate $d\sigma$ with an element not of the plane angle $d\chi$, but rather of the solid angle $d\Omega$ formed by the cones with vertex angles χ and $\chi + d\chi$, which is equal to $d\Omega = 2\pi \sin \chi \, d\chi$. In this case, eq. (22) yields

$$d\sigma(\chi) = \frac{p(\chi)}{\sin \chi} \left| \frac{dp(\chi)}{d\chi} \right| d\Omega \, . \tag{23}$$

The quantity $d\sigma/d\Omega$ has the dimension of area per unit solid angle. It is called the differential scattering cross section.

Thus, the differential scattering cross section, eq. (23), is directly expressed in terms of the deflection function $\chi(p, E_r)$. In a general case, the calculation of $\chi(p, E_r)$ and, hence, of $d\sigma/d\Omega$ requires carrying out a numerical integration which was done for the screened Coulomb potential by Everhart et al. (1955) and for Born–Mayer potential by M.T. Robinson (1963), or the use of approximate methods, proposed for potentials (12) and (13) by Sigmund and Vajda (1964) and Firsov (1958), correspondingly. The inverse problem, the reconstruction of the interatomic potential from experimental values of $d\sigma(\theta, E)$, was also solved (Firsov 1953).

The impact parameter in the barycentric energy range 1–100 keV corresponding to potential (17) is

$$p(E_0, \chi) = \left(\frac{C}{E_r} \right)^{1/2} \frac{\pi - \chi}{[(2\pi - \chi)\chi]^{1/2}} \, . \tag{24}$$

In this case,

$$\chi = \pi \left[1 - \frac{p}{\left(p^2 + \dfrac{C}{E_r} \right)^{1/2}} \right]. \tag{25}$$

Using eq. (23), one can determine by means of eq. (24) the differential scattering cross section for an angle χ,

$$\sigma(E_0, \chi) = \frac{\pi^2(m_1 + m_2)a_{TF}^2 Z_1 Z_2}{m_2(Z_1^{1/2} + Z_2^{1/2})^{2/3}} \frac{13.68}{E_0} \frac{\pi - \chi}{\chi^2(2\pi - \chi)^2 \sin \chi} . \tag{26}$$

For the model potential, eq. (19), the $\chi(E_0, p)$ relation can be written in the form

$$\chi = \pi - 2\left[\frac{p}{\left(p^2 + \dfrac{\alpha_0}{E_r}\right)^{1/2}} \left(\frac{\pi}{2} - \arcsin \frac{p^2}{r_c\left(p^2 + \dfrac{\alpha_0}{E_r} - \dfrac{p^2}{r_c^2} \dfrac{\alpha_0}{E_r}\right)^{1/2}}\right) \right.$$
$$\left. + \arcsin \frac{p}{r_c} \right], \tag{27}$$

where $E_r = E_0 m_2/(m_1 + m_2)$ is the energy of relative motion.

For low projectile energies (0.1–1 keV), one can use the Born–Mayer potential for which the dependences of the scattering angle χ on the impact parameter can be approximated in the form (Sigmund and Vajda 1964, Andersen and Sigmund 1966)

(a) for $\chi \leqslant \frac{1}{2}\pi$

$$\sin^2 \frac{\chi}{2} = \frac{\dfrac{2p}{b} - 1}{\left(\dfrac{p}{b} + 2\dfrac{E_r}{A} e^{p/b}\right)^2 - 4\dfrac{E_r}{A} e^{p/b}} ; \tag{28a}$$

(b) for $\chi > \frac{1}{2}\pi$

$$\sin^2 \frac{\chi}{2} = \frac{1 - \left(\dfrac{p}{R_0}\right)^2\left(1 - \dfrac{b}{R_0}\right)^2}{1 + 4\dfrac{b}{R_0}\left(\dfrac{p}{R_0}\right)^2} , \tag{28b}$$

where R_0 is the head-on collision radius derived from

$$R_0 = b \ln \frac{A}{E_r} ,$$

(c) for $\chi = \pi$

$$\cos \frac{\chi}{2} = \frac{p}{R_0}\left(1 + \frac{b}{R_0}\right), \tag{28c}$$

where A and b are the constants in potential (12).

By means of these expressions, one can construct with a high accuracy the deflection function $\chi(p)$ for a given relative energy E_r in the entire range of variation of the impact parameters $0 \leqslant p \leqslant \infty$.

Note that the result obtained from eq. (28a) is approximately 5% less than the exact value of χ, and the from expression (28b), based on the method of Lehmann (1977), is greater by about the same amount. The differential scattering cross sections in the approximation of Andersen and Sigmund (1966) can be written in the form

(a)　　$\sigma(E_0, \chi)$

$$= \frac{pb\left[\left(\frac{p}{b} + 2\frac{E_r}{A}e^{p/b}\right)^2 - 4\frac{E_r}{A}e^{p/b}\right]^2}{4\left\{\left(\frac{p}{b} + 2\frac{E_r}{A}e^{p/b}\right)^2 - 4\frac{E_r}{A}e^{p/b} - \left(2\frac{p}{b} - 1\right)\left[2\frac{E_r}{A}e^{p/b}\left(\frac{p}{b} + 2\frac{E_r}{A}e^{p/b}\right) + \frac{p}{b}\right]\right\}};$$

(b)　　$\sigma(E_0, \chi) = \dfrac{\left(b \ln \dfrac{A}{E_r}\right)^2 \left(3 + \ln \dfrac{A}{E_r}\right)}{4 \ln \dfrac{A}{E_r}\left(\dfrac{4\sin^2\frac{1}{2}\chi - 1}{\ln \dfrac{A}{E_r}} + 1\right)^2};$　　　(29)

(c)　　$\sigma(E_0, \chi) = \dfrac{1}{4} \dfrac{b^2\left(\ln \dfrac{A}{E_r}\right)^4}{\left(\ln \dfrac{A}{E_r} + 1\right)^2}.$

In the range of large impact parameters, i.e., small scattering angles, $\theta(p)$ is calculated in the impulse approximation directly in the CM system (Landau and Lifshitz 1960, Goldstein 1959),

$$\theta_1 = -\frac{1}{E_0} \int_0^\infty \frac{dV}{dr} \frac{p}{(r^2 - p^2)^{1/2}} \, dr. \qquad (30)$$

In the case of a power law potential, $V(r) = A_n/r^n$, $n > 0$, calculation of integral (30) yields

$$\theta = \frac{1}{E_0} \frac{A_n}{p^n} \frac{\Gamma(\frac{1}{2})\Gamma(\frac{1}{2}n + 1))}{\Gamma(\frac{1}{2}n)}, \qquad (31)$$

where $\Gamma(x)$ is the gamma function.

For the Molière potential, eq. (14), the $\theta(p)$ dependence in the impulse approximation can be written in the form (Erginsoy 1965)

$$\theta = \frac{Z_1 Z_2 e^2}{E_0 a}\left[0.1K_1\left(\frac{6p}{a}\right) + 0.55K_1\left(\frac{1.2p}{a}\right) + 0.35K_1\left(\frac{0.3p}{a}\right)\right], \qquad (32)$$

where K_1 is the modified Bessel function.

6. Inelastic energy losses

A certain amount of kinetic energy is expended in the excitation and ionization of electronic shells ,of the colliding atoms. Although in the medium-energy range they make up only a small fraction (5–10%) of the total losses, they play an essential role in the ionization and electron-emission processes accompanying the ion scattering, and determine to a considerable extent their charge composition (Fedorenko 1959, Morgan and Everhart 1962, Lindhard and Scharff 1961, Lindhard and Winter 1964).

The expressions for the ion and recoil energies taking into account inelastic losses $\mathscr{E}(E_0, p)$ can be written in the form

$$E_1 = (1 + \mu)^{-2} E_0 \left(\cos \theta_1 \pm \sqrt{(\mu f)^2 - \sin^2\theta_1}\right)^2 ,$$

$$E_2 = \mu(1 + \mu)^{-2} E_0 \left(\cos \theta_2 \pm \sqrt{f^2 - \sin^2\theta_2}\right)^2 ,$$

$$(33)$$

where

$$f = \left[1 - \frac{1 + \mu}{\mu} \frac{\mathscr{E}(E_0, p)}{E_0}\right]^{1/2} .$$

The relations between the angles θ_1, θ_2 and the CM scattering angle χ, eq. (8) in this case, take the form (Lehmann 1977)

$$\tan \theta_1 = \frac{\sin \chi}{(\mu f)^{-1} + \cos \chi} ,$$

$$\tan \theta_2 = \frac{\sin \chi}{f^{-1} - \cos \chi} .$$

$$(34)$$

As follows from eq. (34), for a given impact parameter the angles of scattering θ_1 and recoil θ_2 are smaller than is the case with the elastic process (Lehmann 1977), which means that inelastic collisions result in a change of the ion and recoil directions of motion. This leads to an interesting consequence associated with the fact that all specific features of an elementary collision event originating from the inverse mass ratio of the colliding particles ($\mu < 1$), will also be observed in the practically essential case of $\mu = 1$ (e.g., for collisions of crystal atoms with one another), and even for μ slightly in excess of unity. The inclusion of inelastic energy losses results for $\mu < 1$ in a reduction of the value of θ_{lim} compared with that calculated for the elastic process. In this case

$$\theta_1 \leq \theta_{1 \text{ lim}} = \arcsin(f\mu) . \tag{35}$$

As follows from eq. (35), there may exist $\theta_{1 \text{ lim}}$ for $\mu = 1$ as well as for μ slightly more than unity, provided $f\mu < 1$. Another essential consequence of the inclusion of inelastic energy losses is the existence (irrespective of the value of μ) of a limiting angle θ_2 for the recoil atom,

$$\theta_2 \leqslant \theta_{2\,\text{lim}} = \arcsin f .$$ (36)

Figure 2 shows schematically the dependences of the angles θ_1 and θ_2 on the impact parameter p for the case $\mu = 1$ with inclusion (full curves) and without inclusion (broken curves) of inelastic energy losses. We see that when the inelastic energy losses are included, the angles θ_1 and θ_2 reach maximum values less than 90° for certain p, with $\theta_1 + \theta_2 \neq 90°$. In head-on collisions ($p = 0$), the angle $\theta_1 = 0°$ rather than 90° as is the case with the elastic process. The quantities $\theta_{1\,\text{lim}}$ and $\theta_{2\,\text{lim}}$ are determined by the nature of the particles and by the magnitude of the inelastic energy losses.

Several relations are presently available to evaluate the kinetic energy lost in inelastic collisions (Firsov 1959, Lindhard and Scharff 1961, Lindhard and Winter 1964, Russek and Thomas 1958, Kishinevsky 1962, Kishinevsky and Parilis 1962, Oen and Robinson 1976b).

Firsov (1959) and Lindhard and Scharff (1961) developed theories, based on the Thomas–Fermi model, of the electronic stopping power (inelastic energy losses) which yield the same linear dependence of specific inelastic energy losses on ion velocity. The fraction of the kinetic energy lost in inelastic processes is proportional to the velocity of the atomic particle and is a monotonically growing function of the atomic numbers Z_1 and Z_2 of the collision partners.

If we limit ourselves to a consideration of rectilinear motion of the nuclei of colliding atoms, i.e., to collisions with large impact parameters resulting in scattering by small angles, then for the energy transfer one can obtain a simple formula (Firsov 1959),

$$\mathscr{E}(E_0, p) = \frac{4.3 \times 10^{-8}(Z_1 + Z_2)^{5/3} v_0}{[1 + 3.1 \times 10^7 (Z_1 + Z_2)^{1/3} r_0]^5} ,$$ (37)

where $\mathscr{E}$ is the energy in electronvolts, v_0 is the velocity in centimeters per second, and r_0 is the distance of closest approach in centimeters.

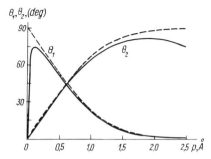

Fig. 2. Scattering angles θ_1 and θ_2 versus the impact parameter p for the case $\mu = 1$ with inclusion (full curves) and without inclusion (broken curves) of inelastic energy losses.

The expression of Firsov directly relates the inelastic energy loss and the impact parameter and, therefore, is widely used in processing experimental data on the energy losses of scattered ions. This expression is sufficiently accurate when the atomic numbers of colliding particles differ by no more than a factor four. Kishinevsky (1962) and Kishinevsky and Parilis (1962) derived an expression for the inelastic energy transfer for the case of two arbitrary atoms and for any impact parameter, without invoking the simplifying assumption of rectilinear and uniform motion of the nuclei,

$$
\mathscr{E}(E_0, p) = \frac{0.3 \times 10^{-7} v Z_1 (Z_1^{1/2} + Z_2^{1/2})(Z_1^{1/6} + Z_2^{1/6})}{\left[1 + \frac{0.67\sqrt{Z_1} r_0}{a_{TF}(Z_1^{1/6} + Z_2^{1/6})} \right]^3} \left[1 - 0.68 \frac{V(r_0)}{E_r} \right],
$$

(38)

where v and E_r are the velocity and energy of relative atomic motion, Z_1 is the greater, and Z_2 the smaller of the atomic numbers, and r_0 is in units of Å.

We really see that eq. (38) reduces to eq. (37) in the limiting case of small scattering angles. To estimate the inelastic energy losses occurring in the scattering of light ions (H^+, He^+) from solids, Oen and Robinson (1976b) proposed a relation

$$
\mathscr{E}(E_0, p) = \left(\frac{0.045}{\pi a_F^2 N} \right) k E^{1/2} \exp\left(-0.3 \frac{r_0}{a_F} \right),
$$

(39)

where N is the target atom density, and k is an electronic stopping power parameter (Lindhard and Scharff 1961).

Lindhard and Scharff (1961) proposed an expression for the calculation of specific inelastic energy losses,

$$
\left(-\frac{dE}{dx} \right)^{LS}_{inel} = \frac{8\pi e^2 N a_0 Z_1 Z_2}{(Z_1^{2/3} + Z_2^{2/3})^{3/2}} \frac{v}{v_B} \xi_\ell,
$$

(40)

where $\xi_\ell \approx (1 - 2)$ scales as $Z_1^{1/6}$, and v_B is the velocity of an electron in the first Bohr orbit of the hydrogen atom.

Figure 3 shows the dependences of the inelastic energy losses of Ne^+ ions with $E_0 = 5$ keV scattered from a Ni atom, on the impact parameter calculated with three expressions, eqs. (37)–(39). The $r_0(p)$ relation was calculated with the Biersack–Ziegler potential. The values of $\mathscr{E}$ are seen to agree well with one another in the region $p \geq 0.5$ Å. For smaller p, the expression of Firsov becomes invalid, the values of $\mathscr{E}(p)$ calculated with it reaching a maximum at $p = 0$. The relation of Oen and Robinson (1976b) behaves in a similar way, although there are quantitative discrepancies. At the same time, the behavior of $\mathscr{E}(p)$ predicted by the expression of Kishinevsky

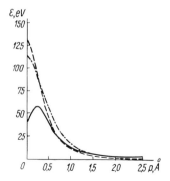

Fig. 3. Inelastic energy losses of Ne$^+$ ions with $E_0 = 5$ keV, scattered from a Ni atom, versus the impact parameter calculated with three expressions, eqs. (37–39); Firsov (1959) (broken curve), Oen and Robinson (1976b) (chain curve) and Kishinevsky (1962) (full curve).

(1962) in the region of small p differs strongly from the two preceding cases. The curve passes through a maximum at $p \neq 0$ (in our case at $p \approx 0.25$ Å). Next, as p falls off to 0, $\mathscr{E}(p)$ decreases too. Note that the values of $\mathscr{E}(p)$ calculated with this formula for $p < 0.5$ Å are substantially smaller than those obtained with eqs. (37) and (39). Thus, in the region $p < 0.5$ Å, in accordance with the relation of Kishinevsky, a double-valued behavior of the inelastic energy losses is observed, namely, the same values of $\mathscr{E}$ correspond to two different impact parameters. This behavior of $\mathscr{E}(p)$ is accounted for by the corresponding dependences of r_0 and v on p in the expression of Kishinevsky.

The theories of Firsov and Lindhard were checked more than once (Teplova et al. 1962, Mayer et al. 1970, O'Connor and MacDonald 1980a,b, Bertrand 1980, Xu and Sullivan 1990, Kasi et al. 1989). On the whole, experimental data agree well with these theories, however, in some cases the calculated values exhibit discrepancies from the measurement which reach 30–40% (Snowdon et al. 1989b, Monreal et al. 1989). The inelastic energy losses observed in the single scattering peak are explained in terms of the existing theories, whereas those found in the region of multiple scattering cannot be calculated without knowing exactly the scattered ion trajectories. To determine the latter, one has to perform computer simulation of the ion scattering from the surface of crystal.

7. Binary scattering model

Atomic collisions on the surface of a solid and in a gas have much in common (Kaminsky 1965, U.A. Arifov 1968, Mashkova and Molchanov 1985, Bohr 1948). The difference lies in the fact that in gases collisions are primarily single, whereas in the scattering of ions from a solid there always is a certain contribution of multiple interaction. In connection with this, a so-called 'gas-scattering model' was proposed to explain the main features of medium-energy ion scattering from the surface of a solid (Parilis 1965), where in the initial approximation the well-studied features of ion scattering in gases were used (Fedorenko 1959, Morgan and Everhart 1962).

For not too low energies the ion–atom interaction on the surface of a solid may be considered as isolated binary collisions and their sequences. This conclusion is based on a consideration of the times and energies characteristic of such interaction events.

Since the repulsive part of the potential falls off rapidly with increasing interatomic separation, the interaction in such collisions lasts on the order of 10^{-15} s, which is much less than the lattice vibration period which is approximately 10^{-13}–10^{-12} s. Therefore, if the energy transferred in a collision exceeds the binding energy of the target atom (about 10–25 eV), one may disregard its interaction with the lattice. Simultaneous interaction of an ion with several target atoms results in an increase of the scattered ion energy compared with the value calculated by eq. (4). One may ask at what minimum energy of the projectile ion the binary collision approximation still holds, i.e., when expression (4) may still be used to calculate the particle energies after collision. There is a wealth of experimental and theoretical data on this problem. As follows from literature (Kaminsky 1965, U.A. Arifov 1968, Mashkova and Molchanov 1985, Veksler 1978), using the binary collision approximation to describe the ion interaction with a target atom in the energy range 1–100 keV is valid and yields results in good accord with experiment. There is, however, disagreement concerning the possibility of employing this approximation at energies below 1 keV (Veksler 1978). Gay and Harrison (1964) considered the possibility of using the binary model in the case of collisions of a copper ion projectile with a copper atom embedded in the lattice. The ion energy was 1 keV and less. A comparison of the computer calculations of the ion collision with an isolated copper atom and with an atom incorporated in the lattice leads to a conclusion of the inapplicability of the binary collision model for energies below 500 eV. This conclusion, however, as pointed out by M.T. Robinson (1981), is not well grounded and is a consequence of the shortcoming of the technique used by Gay and Harrison to compare the binary collision approximation with a dynamic calculation. Karpuzov and Yurasova (1971a), starting with different models of the ion collision with a lattice atom, obtained trajectories of argon ions of energy 50–500 eV impinging on a copper single crystal at normal incidence and at 70° (measured from the surface normal). It was shown that at particle energies ~100 eV taking into account the binding energy and nonbinary character of collisions in a crystal does not affect essentially the trajectories of the reflected ions and acts primarily on those of the ions penetrating into the crystal. Andersen and

Sigmund (1966), in their study of the mechanism of passage of slow ions through a crystal lattice, showed that the binding forces of atoms in the lattice start to influence the scattering process only at energies on the order of 10 eV or less.

Mashkova and Molchanov (1985), U.A. Arifov and Aliev (1974) and U.A. Arifov and Aliev (1968a) measured the energy distributions of reflected ions and established that the position of the single-scattering peak, rather than always coinciding with the theoretical prediction, is shifted towards higher energies. This, however, can be explained as due not only to simultaneous interaction with several atoms in the lattice but also to the effect of neighbor atoms in a lattice row on 'single' scattering (Kivilis et al. 1967). The possibility of interaction of a ≈ 100 eV projectile ion simultaneously with several target atoms was studied experimentally (Veksler 1962, 1978; Veksler and Evstifeev 1973, Petrov 1960a). The multipeak structure of the spectrum obtained in these experiments is described in the hard-sphere approximation using the concept of the target atom 'effective mass' m_{eff} which is slightly greater than m_2.

The possibility of such a scattering mechanism was suggested also by Shkarban (1978), who measured the energy accommodation coefficients for the scattering of Ar^+, Kr^+ and Xe^+ ions on polycrystalline copper and polycrystalline and single-crystal Mo. The decrease of the accommodation coefficients at low energies turned out to be disproportionately large. The 'effective mass' concept implies that m_{eff} should depend not only on energy but on the direction of collision as well, i.e., it should be a tensor. This leads to an unjustified complication of the model and of the calculations based on it.

Tongson and Cooper (1975) made a precise measurement of the single-scattering peak position for the scattering of 20–1000 eV He^+ and Ne^+ ions from polycrystalline copper. The theoretical and experimental positions of the single-scattering peak were found to practically coincide (with a discrepancy less than 1%).

Note, however, that both Tongson and Cooper (1975) and a number of other experimenters used as condition of validity of the binary collision approximation the position of the single-scattering peak. The trajectories of the particles responsible for this peak lies mainly outside the crystal, so that simultaneous ion interaction with several target atoms should not affect them markedly. This interaction should influence most strongly the ions suffering multiple collisions in the bulk of the crystal (Hart and Cooper 1979, Karpuzov and Yurasova 1971a).

CHAPTER 2

Computer Simulation of Surface Collisions

1. Simulation techniques

The difficulties entailed in describing many-particle interaction make theoretical consideration of multiple ion-scattering from atoms on a solid a difficult problem (Sigmund 1989). This has stimulated interest in applying computer-simulation techniques to the scattering process. Computer simulation (Agranovich and Kirsanov 1976) resembles to a certain extent an experimental study and, thus, can be very useful in the analysis of models of processes and phenomena. After the fundamental work of Gibson et al. (1960), which was a pioneering attempt at simulating radiation damage, computer modeling was used to advantage in studying a broad range of physical phenomena, among them the scattering of ions from the surface of a solid (Agranovich and Kirsanov 1976). Basically, there are two methods to simulate ion scattering from a solid, namely,

(1) solving coupled equations of motion of the ion and of target atoms (molecular-dynamics method), and

(2) constructing an ion trajectory in a solid in the approximation of sequential isolated binary collisions.

Direct integration of the equations of motion permits a description of the interaction of an ion or displaced atom simultaneously with several target atoms, involving either multistep (method of Adams) or single-step (Runge–Kutta method) difference techniques (Mosunov 1983, Forsythe et al. 1977, Harrison et al. 1969). Among the numerical techniques suitable for integration of the equations of motion of a particle in a crystal, the most frequently used is the 'mean force' approach (Harrison et al. 1969), falling in the family of the Runge–Kutta methods. However, because of the considerable computer time involved, the methods based on direct integration of the equations of motion are not very popular.

The binary collision method has found much broader recognition, particularly in the medium-energy range (Agranovich and Kirsanov 1976, M.T. Robinson and Torrens 1974). This technique was first used by Beeler (1966, 1970) and Beeler and Besco (1963) in radiation damage calculations for crystals, as well as in the simulation of energetic atom ranges in solids (Jackson 1975, M.T. Robinson and Oen 1963) and of channeling (D.V. Morgan and Van Vliet 1968, Barrett 1971, Oen et al. 1963). To the simulation of medium-energy ion scattering from solids, it was first applied by Karpuzov et al. (1966) and Parilis and Turaev (1966).

Besides the methods of sequential trajectory simulation, one can use the Monte Carlo technique to calculate the probability distribution functions for the quantities describing the scattering process (Strizhenov et al. 1971, Kalmykov et al. 1972, Pugachova and Khakimov 1979, Pugachova et al. 1975, Konoplev 1986, Akkerman et al. 1972). In computer simulation, both techniques are frequently employed (Mosunov 1983, D.V. Morgan and Van Vliet 1968, Barrett 1971, Preuss 1978, Hou 1989).

2. Particle dynamics in the binary collision model

In the binary collision model, at each given moment of time the ion interacts only with one target atom. As for the interaction of the scattering atom with the crystal, during the collision time it is considered to be insignificant. Indeed, the displacement of a target atom during the collision is $\delta \sim 2(m_1/m_2)\theta$ (D.V. Morgan and Van Vliet 1968). For the change of the scattering angle $\delta\theta$ caused by a change in the impact parameter δp, one can write

$$\delta\theta \sim \frac{\partial\theta}{\partial p}\,\delta p \sim 2\,\frac{m_1}{m_2}\,\theta^2\,. \tag{1}$$

For ions of medium mass and energy, usually $\delta\theta \ll \theta$, so that the effect of the remainder of the lattice on the ion–atom scattering process is indeed very small. In the binary collision model, particles move along straight-line segments representing asymptotes to their LS trajectories, and one determines not a particle trajectory but rather the difference between the angles characterizing the initial and final directions of motion (Parilis 1965, Mashkova and Molchanov 1985, M.T. Robinson and Torrens 1974, Barrett 1971). While this approach permits one to cut the required computer time (compared with direct integration of the equations of motion), it also entails a systematic error due to the fact that over short segments of path, the real ion trajectory differs from the asymptotes used to replace the former. This error was estimated by M.T. Robinson and Torrens (1974) for the Cu–Cu pair, a number of potentials (Molière, hard sphere, Coulomb and inverse square), and three values of energy (3.3 eV, 1.6 and 328 keV).

It was established that the deviation of an asymptote from the real trajectory is essential only for head-on collision and high energies; however, because of the smallness of the closest approach distance, particles reach their asymptotic trajectories before the next collision also in these cases. The projectile scattering angle and the coordinates of the injection point of the trajectory (the point of intersection of the asymptotes) are found by calculating the integrals given in eqs. (9) and (10) of chapter 1. In the program developed by Beeler Jr (1970) and Beeler Jr and Besco (1963), the target atom is fixed during the projectile approach and starts to move as a result of the collision directly from the lattice site it occupied before.

However, already in the very first attempts at computer simulation of low-energy atomic displacement cascades (Gibson et al. 1960), this approximation was shown to be inaccurate. A combined analysis of the motion of the projectile and of the lattice atom during the interaction permits one to find the LS position of the intersection of the projectile trajectory asymptotes, which lies at a distance Δ from the corresponding point in the case of a fixed target atom (Mashkova and Molchanov 1985, Lehmann 1977, M.T. Robinson 1963, 1981). The quantity Δ determines the displacement of the projectile during the collision and can be written as

$$\Delta = 2\tau \frac{m_1}{m_1 + m_2} + p \frac{m_2 - m_1}{m_1 + m_2} \tan \tfrac{1}{2}\chi , \tag{2}$$

where τ is the time integral. For the hard-sphere potential, $\tau = p \tan \tfrac{1}{2}\chi$ (M.T. Robinson and Torrens 1974), and $\Delta = p \tan \tfrac{1}{2}\chi$.

M.T. Robinson (1963, 1981) evaluated Δ for sequential atomic collisions in crystals. It was found that for Cu and Au, Δ becomes comparable with the interatomic separations in the lattice only for fairly low energies (~ 10 eV). In other cases, the fixed target approximation yields quite satisfactory results.

3. Crystal models; inclusion of thermal vibrations

Algorithms use for the description of the target a series of increasingly more complex models of the surface of a single crystal which approximate its real properties with a progressively improving accuracy, namely, the two atom (Parilis 1965, Mashkova and Molchanov 1985) and atomic row (Kivilis et al. 1967) models (fig. 1), semichannel and crystallite models including two or more atomic layers (Kivilis et al. 1967, 1970, Karpuzov et al. 1969a,b) (fig. 2), models of a nonperfect defected surface allowing for the presence of potential barriers, adsorbed atoms, atomic steps, vacancies and their clusters (fig. 3).

A good agreement between the results of the ion-scattering simulation with these models of the surface and the real process was ensured by their good fit to experimental data, a thorough testing of the computer modeling technique, and a thorough analysis of the validity of the approximations used.

A large number of calculations was performed on crystallites comprising up to ten or more atomic layers. The constitution of the lattice is the simplest for surface models since only one layer of atoms is used in this case. As the number of the layers increases, the atoms in the single-crystal lattice become arranged in coordination spheres in accordance with the actual crystal structure, face, plane and direction chosen. In the case of a polycryst-

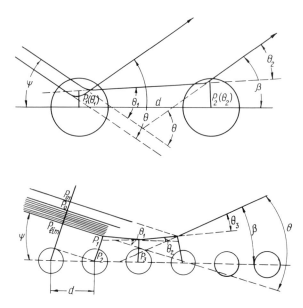

Fig. 1. The two atom and atomic row scattering models.

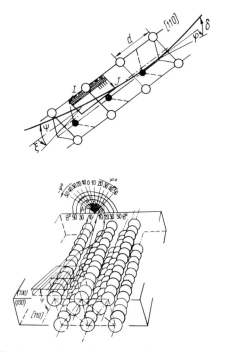

Fig. 2. The semichannel and crystallite scattering models.

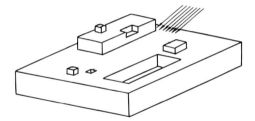

Fig. 3. The models of a nonperfect defected surface.

al, the atoms take up their positions in the coordination spheres in a random way, the anisotropic distribution of the target atoms over the coordination spheres being replaced by an isotropic one (Mosunov 1983, Forsythe et al. 1977). In such a model of the polycrystal, the density and short-range order of the atoms are retained while the long-range order brakes down. Mosunov (1983) proposed a model in which the ordered structure characteristic of polycrystal grains is retained. One chooses a certain number of coordination spheres in an ideal fcc crystal. Next, this crystallite is turned in space in a random way. For each projectile ion the target is constructed anew. In Robinson's program, a similar procedure, associated with a random rotation of the crystal structure prior to each collision, was employed to simulate the scattering of ions from the surface of an amorphous solid (Hou and Robinson 1976). While the density of the crystal was retained in this model, no correlations related to the direction were present.

As a rule, one enters into the computer memory not the entire crystallite of atoms but rather a limited number of nearest neighbors in the first two or three coordination spheres. This permits one to avoid overloading the computer memory and large increases in computing time. However, more efficient has proved to be the procedure in which for each atom a list of its nearest neighbors feeling the interaction potential is drawn, and only the interaction with the atoms in the list is taken into account. Rather than being stored in computer memory, this look-up table is generated rapidly for each atom near the trajectory of the scattered ion (M.T. Robinson and Torrens 1974) in such a way that the target atom closest to the ion always resides at the center of two coordination spheres. The trajectories of the moving ions are computed in the field of two coordination spheres. The surface is simulated by excluding from the calculation the atoms on the coordination spheres which happen to be beyond the prescribed boundary of the ideal single crystal. When moving in the bulk of the crystal, situations may arise where the distances to several nearest neighbors will be less than the potential cut-off radius r_c and the inclusion of simultaneous interaction of the ion with these atoms becomes necessary (M.T. Robinson and Torrens 1974, Hou and Robinson 1976).

To ensure the binary nature of interaction, Hou and Robinson (1976) and Shulga and Yurasova (1971) chose the potential in such a way that midway between atoms in the lattice both the potential and its derivative vanished.

Thermal vibrations of atoms in a crystal affect substantially the dynamics of the ion-scattering process (Parilis 1965, Mashkova and Molchanov 1985, Mosunov 1983, M.T. Robinson and Torrens 1974, Parilis and Turaev 1966, Preuss 1978, Shulga and Yurasova 1971, Hutchence and Hontzeas 1974, Parilis et al. 1967, Nelson et al. 1962). There are two principal methods of describing thermal vibrations of atoms on the surface of a solid. In the first of them, the atoms are assumed to vibrate independently of one another, and in the second, correlations are included. The first simple model of thermal vibrations was proposed by Parilis et al. (1967), who considered interaction of a projectile ion with a frozen atomic row bent by thermal vibrations. The correlations were included in such a way that the atoms were in opposite phases. This approach was based on the fact that the most intense frequency in the phonon spectrum of a row corresponds to the minimum wavelength $\lambda = 2d$, where d is the distance between the atoms in the row.

The instantaneous local configuration of a row propagates along it with a sonic velocity $u_s \simeq 3 \times 10^5$ cm/s. At the same time, a fast ion moving with a velocity $v \simeq 10^7$–10^8 cm/s approaches and leaves the bent frozen row in a time 10^{-14}–10^{-15} s. Hence, the row configuration cannot change substantially during the ion scattering time.

The mean square displacement of an atom from its equilibrium position is

$$\frac{\Delta s^2}{d^2} \simeq \frac{1.5 \times 10^{-2}}{T_m} T \quad \text{for } T \geqslant \theta_D ,$$

$$\simeq \frac{1.5 \times 10^{-2}}{T_m} \frac{\theta_D}{4} \quad \text{for } T = 0 ,$$

(3)

where θ_D is the Debye temperature, and T the absolute temperature.

Thus, the displacement of an atom $\Delta s \sim \sqrt{T}$ and is $\sim 0.1d$ close to the melting point of a metal, and $0.05d$ for $T = \frac{1}{4} T_m$ (Maradudin et al. 1963).

In most of the simulation studies, the atoms are assumed to vibrate independently of one another (Agranovich and Kirsanov 1976, Mosunov 1983, M.T. Robinson and Torrens 1974, Preuss 1978, Shulga and Yurasova 1971, Hutchence and Hontzeas 1974), and the displacements from the equilibrium positions of the lattice atoms to obey the Gaussian distribution (Landau and Lifshitz 1964),

$$\rho(x) = \frac{1}{(2\pi\sigma^2)^{1/2}} e^{-\Delta x^2/2\sigma^2} ,$$

(4)

where σ^2 is the variance, and Δx the displacement from the lattice site. This assumption proves to be valid for cubic crystals. In a cubic lattice, the displacements Δx, Δy, Δz do not depend on one another, their rms values being equal $(\overline{\Delta x^2} = \overline{\Delta y^2} = \overline{\Delta z^2} = \frac{1}{3} \overline{\Delta r^2})$. The variance of the distribution was calculated in the Debye–Waller approximation (Maradudin et al. 1963),

$$\sigma^2 \equiv \overline{\Delta x^2} = \frac{3\hbar^2}{m_2 k \theta_D} \left[\frac{1}{4} + \frac{1}{x^2} \varphi(x) \right], \quad x = \frac{\theta_D}{T}, \quad \varphi(x) = \int_0^x \frac{t \, dt}{e^t - 1}, \quad (5)$$

where φ is the Debye function. D.V. Morgan and Van Vliet (1968) approximated the Gaussian distribution of atomic displacements with a triangular function. Barrett and Jackson (1980) studied the effect of correlations in the lattice thermal atom vibrations on the channeling process. It turned out that the correlations come from effects qualitatively similar to those which appear when the vibration amplitude decreases, and that their inclusion is essential for trajectories directed along an atomic row and less important for a randomly oriented trajectory. Relations (3)–(5) describe the bulk of the crystal. On the surface, however, the atomic vibration amplitude is substantially different (Jackson 1974, Black et al. 1980, Landman et al. 1988). Indeed, atoms on the surface have a number of nearest neighbors and distances to them different from those in the bulk. The associated anisotropy of thermal vibrations on the surface of a crystal can be fairly large. For instance, in the bulk of single crystal W, $\overline{\Delta z^2} = 2.62 \times 10^{-19}$ cm^2, while on the (100) plane, $\overline{\Delta z^2} = 5.643 \times 10^{-19}$ cm^2, and $\overline{\Delta u^2} = 4.33 \times 10^{-19}$ cm^2, where Δz is the displacement normal to the surface, and Δu is that in the target [100] plane.

4. Algorithm and program of calculation

The algorithm used in our calculations to construct the trajectories of the ions scattered by atoms on the surface of a solid is based on two assumptions, namely,

(1) only binary ion collisions with target atoms are considered; and

(2) the path which an ion goes between collisions is represented by straight-line segments.

A parallel, uniform, monoenergetic ion beam impinges on a target area on the surface of a crystal (fig. 4). It is assumed that the ion beam has a sufficiently low density that the ions do not strike twice the same point. The shape of the target area is chosen such that by translating it one could cover the entire surface of the crystal. The planes of incidence and exit of the ions are under angles ξ and φ with the chosen crystallographic direction, respectively; the incoming and outgoing asymptotes of the ion trajectory form angles ψ and β with the surface. The method used here permits one to establish a one-to-one correspondence between the scattering direction of

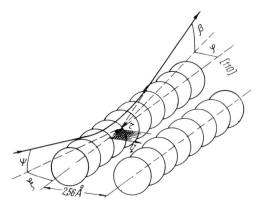

Fig. 4. Scheme of scattering of ions by a three-dimensional atomic chain.

the ion and its energy, on the one hand, and the aiming point on the crystal surface, the incident direction of the ion, and its energy, on the other.

The initial direction of the ion beam incident on the surface was defined through the projections of the unit vector $L_0(l_0, m_0, n_0)$ on the coordinate axes,

$$l_0 = \cos \psi \cos \xi, \qquad m_0 = \cos \psi \sin \xi, \qquad n_0 = -\sin \psi. \qquad (6)$$

The potential target atom on the surface for the first collision with an incident ion was determined in the following way. First, the point on the surface is determined whose distance from the initial ion direction is p_{lim}, and an atom with integral coordinates is chosen in the vicinity of this point. It is with this atom that the incident ion is considered to interact.

The particle path in the crystal is followed by means of a special procedure depending on the method used to construct the numerical target crystallite. The scattering angles on the ion θ_1 and of the recoil atom θ_2, as well as the coordinates of the asymptote intersection point X_1 are determined by calculating the scattering integral, eq. (9) of chapter 1, and time integral, eq. (10) of chapter 1 (fig. 5). Figure 5 illustrates schematically the classical scattering of a projectile particle from an initially fixed target atom in the laboratory (L) system.

The unit vector $L_0(l_0, m_0, n_0)$ defines the initial direction of the ion and passes through the aiming point $X(x, y, z)$. The distance of closest approach (impact parameter) of the projectile ion to the target atom with coordinates $X_0(x_0, y_0, z_0)$ is given by

$$p = \frac{\sqrt{\left| \begin{matrix} y_0 - y & z_0 - z \\ m_0 & n_0 \end{matrix} \right|^2 + \left| \begin{matrix} z_0 - z & x_0 - x \\ n_0 & l_0 \end{matrix} \right|^2 + \left| \begin{matrix} x_0 - x & y_0 - y \\ l_0 & m_0 \end{matrix} \right|^2}}{\sqrt{l_0^2 + m_0^2 + n_0^2}}.$$

$$(7)$$

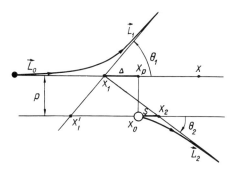

Fig. 5. Scheme of classical scattering of a particle colliding with an initially immovable target atom in the laboratory system.

To determine the trajectories of the scattered ion, $L_1(l_1, m_1, n_1)$, and of the recoil, $L_2(l_2, m_2, n_2)$, one has to know, besides the angles θ_1 and θ_2, also the position of the point X_1 where the asymptotes of the trajectories of the interacting particles intersect. As shown in fig. 5, the intersection point of the scattered ion trajectory asymptotes $X_1(x_1, y_1, z_1)$ is at a distance Δ from the projection $X_p(x_p, y_p, z_p)$ of the original position of the target atom in the initial direction of ion motion,

$$\Delta = [(1 + f)\tau + (f\mu - 1)p \tan \tfrac{1}{2}\chi]/f(1 + \mu) , \tag{8}$$

with $f = 1$ in the case of elastic scattering. In the hard-sphere model, $\Delta = p \tan \tfrac{1}{2}\chi$. The position of the point $X_p(x_p, y_p, z_p)$ is determined by means of the expressions

$$x_p = [l_0 t + m_0(m_0 x - l_0 y) + n_0(n_0 x - l_0 z)]/u ,$$
$$y_p = [l_0(l_0 y - m_0 x) + m_0 t + n_0(n_0 y - m_0 z)]/u , \tag{9}$$
$$z_p = [l_0(l_0 z - n_0 x) + m_0(m_0 z - n_0 y) + n_0 t]/u ,$$

where $t = l_0 x_0 + m_0 y_0 + n_0 z_0$, $u = l_0^2 + m_0^2 + n_0^2$. Then, for $X_1(x_1, y_1, z_1)$ we can write

$$x_1 = x_p - l_0 \Delta/\sqrt{u} ,$$
$$y_1 = y_p - m_0 \Delta/\sqrt{u} , \tag{10}$$
$$z_1 = z_p - n_0 \Delta/\sqrt{u} .$$

As follows from the spherical symmetry of the binary interaction potential, the initial, L_0, and final, L_1, directions of the scattered ion, the original coordinates of the target atom $X_0(x_0, y_0, z_0)$ and its final direction L_2 lie in one plane. Knowing the coordinates of the points X_1, X_1' and X_2, one can now determine the final directions of the ion and of the recoil using the equation of a straight line passing through two points. As seen from fig. 5, the point $X_2(x_2, y_2, z_2)$ lies at a distance s from $X_0(x_0, y_0, z_0)$:

$$s = \frac{p}{\tan \theta_2} - \Delta . \tag{11}$$

In the hard sphere approximation, $s = 0$. Thus

$$x_2 = x_0 + l_0 s/\sqrt{u} ,$$
$$y_2 = y_0 + m_0 s/\sqrt{u} , \tag{12}$$
$$z_2 = z_0 + n_0 s/\sqrt{u} .$$

As for the point $X_1'(x_1', y_1', z_1')$, its coordinates are found from the relations

$$x_1' = x_0 - \left(\Delta + \frac{p}{\tan \theta_1}\right) l_0/\sqrt{u} ,$$

$$y_1' = y_0 - \left(\Delta + \frac{p}{\tan \theta_1}\right) m_0/\sqrt{u} , \tag{13}$$

$$z_1' = z_0 - \left(\Delta + \frac{p}{\tan \theta_1}\right) n_0/\sqrt{u} .$$

The final directions of the ion and recoil can be written as

$$l_1 = (x_1 - x_1') \sin \theta_1/p ,$$
$$m_1 = (y_1 - y_1') \sin \theta_1/p , \tag{14}$$
$$n_1 = (z_1 - z_1') \sin \theta_1/p ;$$

$$l_2 = (x_2 - x_1) \sin \theta_2/p ,$$
$$m_2 = (y_2 - y_1) \sin \theta_2/p , \tag{15}$$
$$n_2 = (z_2 - z_1) \sin \theta_2/p .$$

Thus, solving eqs. (14) and (15) yields the final directions of motion for the ion, L_1, and for the recoil atom, L_2. As the trajectories are generated further, the quantities E_0, L_0, X are replaced by E_1, L_1, X_1 (for the ion) and by E_2, L_2, X_1 (for the recoil).

The computer program for the calculation of the ion and recoil trajectories in sequential correlated ion collisions with atoms of the solid took into consideration the displacements Δ and s determined by the time integral τ. The accuracy of computation of these quantities for a given interaction potential dominates the accuracy of the computer simulation. Figure 6 shows the calculated values of Δ and s reduced to the screening factor a versus the impact parameter p for an Ne$^+$–Ni pair and $E_0 = 5$ keV. Here curves 2 and 3 relate to the quantities Δ and s calculated with the Biersack–Ziegler potential with time integral included, and curve 1 to the

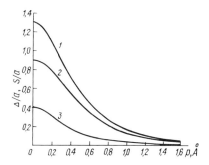

Fig. 6. Calculated values of Δ and s reduced to the screening factor a versus the impact parameter p for an Ne$^+$–Ni pair and $E_0 = 5$ keV.

quantity Δ calculated with the same potential but in the hard-sphere approximation. The quantity s in the hard-sphere model is zero. The inclusion of Δ is seen to be essential in the region of small parameters ($p < 0.5$ Å), curve 1 lying substantially higher than curve 2. At the same time, for $p \geqslant 1$ Å, they practically merge. The inclusion of s is also essential only for small parameters ($p < 0.5$ Å), while for $p \geqslant 1$ Å the approximation of a fixed target atom is valid.

The trajectories of moving ions were calculated in the field of two coordination spheres. To simulate the surface, the atoms in the coordination spheres found to lie beyond the specified boundary of an ideal single crystal were excluded from calculation. For each ion trajectory passing through the given aiming point, the impact parameters p_i for the atoms within two coordination spheres are computed by expression (7) which are then compared with the limiting impact parameter $p_{\lim}$. Values of p_i smaller than $p_{\lim}$ are chosen, and the changes in the ion trajectory after interaction with these atoms are calculated consecutively by expressions (8)–(15) in accordance with the corresponding collision times, i.e., the scattering angles, energies and new directions of ion motion are determined. Next, the interaction of the ion with energy E_{1i} and a new direction of motion l_{1i}, m_{1i}, n_{1i} and a new $p_{\lim}(E_{1i})$ with the other atoms in the crystallite is followed. Apart from this, the energies E_{2i} and recoil directions $\boldsymbol{L}_{2i}(l_{2i}, m_{2i}, n_{2i})$ are calculated and entered into computer memory. This procedure is repeated until the search through the entire crystallite of atoms has been computed. Next, the total scattering angle θ (the angle between the incoming and outgoing ion directions), the escape angle β (the angle between the scattered ion direction and its projection on the surface plane), and the azimuthal scattering angle φ are computed,

$$\theta = \arccos \frac{l_0 l_k + m_0 m_k + n_0 n_k}{\sqrt{u}\sqrt{w}},$$

$$\beta = \arcsin n_k \,,$$

(16)

$$\varphi = \arctan(m_k/l_k) \,,$$

where $w = l_k^2 + m_k^2 + n_k^2$. The particle is considered to be reflected if the outgoing part of its trajectory is directed away from the crystal surface and it does not interact with the atoms in the surface layer.

After this, the trajectory for the next aiming point on the surface is chosen. The target area is searched until all aiming points have been analyzed.

By successively using the various scattering models in the program, namely, two-atom, atomic row, semichannel, crystallite of several atomic layers, as well as nonideal models including the presence on the surface of adsorbed atoms, atomic steps, vacancies and their clusters, one can steady by computer simulation the effect of target size on the spatial, angular, and energy distributions and trajectories of scattered ions and recoils.

In contrast to scattering from an atomic row with the ion moving in space between two neighboring atomic rows in the first layer (fig. 4) or in the bulk of the crystal, one may envisage situations where the distances to several nearest neighbors will be less than $p_{\lim}$, thus requiring the inclusion of simultaneous interaction of the ion with several target atoms. In the binary collision approximation, taking into account simultaneous interaction of an ion with several lattice atoms is a fairly complex problem. In some publications, it is solved just by truncating the interaction potential (i.e., by equating it to zero) at distances $p_{\lim} = \frac{1}{2}a$, where a is the distance between the neighboring atomic rows, which corresponds to scattering from atoms of 'isolated' atomic rows. Thus, for $p < p_{\lim}$, only interactions with atoms of the nearest atomic row occur. The shortcomings of such an approach in a description of correlated collision events, e.g., in channeling, are obvious, namely, it introduces an instability into the motion of a particle and increases its energy losses. The simultaneous character of interaction must also be included in studies of the so-called zigzag collisions which occur as an ion moves between neighboring atomic rows on the surface of a single crystal.

Our algorithm makes use of the procedure which includes simultaneous interaction within the binary collision model proposed by M.T. Robinson and Torrens (1974) (fig. 7). When more than one target atom satisfies the conditions for interaction, the program starts with selecting the first of those lying along the projectile trajectory, say T_1 in fig. 7a. Next, for each of the other target atoms the following quantities are computed:

$$\Delta \zeta_{1i} = \zeta_i - \zeta_1 \,,$$

(17)

$$d_{1i}^2 = p_1^2 + (\Delta \zeta_{1i})^2 \,; \qquad d_{i1}^2 = p_i^2 + (\Delta \zeta_{1i})^2 \,.$$

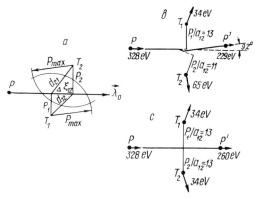

Fig. 7. (a, c) Scheme which includes simultaneous interaction within the binary collision model proposed by M.T. Robinson and Torrens (1974) and (b) a comparison with the sequential treatment of collisions.

Here, p_1 and p_i are the impact parameters for the first and ith target atom, respectively, and $\zeta = \lambda_0 \cdot \Delta x_i$. If $\Delta\zeta_{1i} < \zeta_m$, $d_{1i} < p_{lim}$, and $d_{i1} < p_{lim}$, then the second and first collisions are taken to occur simultaneously. Each collision of the set of simultaneous events is computed as if it were the only one to occur. Next, the new direction of ion motion is determined by vector addition of the scattering angles thus obtained in the LS. The ion undergoes each collision with the same initial energy E_0. Next, the energies transferred to the various target atoms are added, and the energy losses thus found are subtracted from E_0. In fig. 7b this procedure is compared in a simple example with the sequential treatment of collisions. The presence of simultaneous collisions was checked in the program by comparing the values of ζ_m and p_{lim}.

To compare experimental energy distributions of scattered ions with the results of a computer simulation, one has to know the intensity I of the particles scattered in a given direction θ and φ with a given energy E.

Using analyzers discriminating the particles in solid angle Ω and energy E, one measures the double differential cross section with respect to angle and energy, i.e.,

$$I = \frac{d^2\sigma}{d\Omega \, dE} \, . \tag{18}$$

In the case of single scattering,

$$\frac{d^2\sigma}{d\Omega \, dE} = \frac{d\sigma}{dE} \, \delta[E - E(\theta)] \, . \tag{19}$$

Replacing the delta function with the analyzer response function, we obtain

$$\frac{d^2\sigma}{d\Omega \, dE} = \frac{d\sigma}{dE} \, f[E - E(\theta)] \, ,$$

where

$$f[E - E(\theta)] = 0 \qquad \text{for} \quad [E - E(\theta)] > \Delta E ,$$

$$= \frac{1}{\Delta E} \quad \text{for} \quad [E - E(\theta)] < \Delta E , \tag{20}$$

where ΔE is the analyzer resolution.

The differential cross section of scattering at a given angle is determined by the ratio of the unit target area ds to the unit solid angle $d\Omega = \sin \theta \, d\theta \, d\varphi$,

$$\frac{d\sigma}{d\Omega} = \frac{ds \sin \psi}{\sin \theta \, d\theta \, d\varphi} . \tag{21}$$

The change of the unit target area entailed in transition from the rq- to $\theta\varphi$-coordinates is given by the Jacobian D,

$$D = \frac{\partial(r, q)}{\partial(\theta, \varphi)} = \begin{vmatrix} \dfrac{\partial\theta}{\partial r} & \dfrac{\partial\theta}{\partial q} \\[2mm] \dfrac{\partial\varphi}{\partial r} & \dfrac{\partial\varphi}{\partial q} \end{vmatrix} = \frac{\partial\theta}{\partial r}\frac{\partial\varphi}{\partial q} - \frac{\partial\theta}{\partial q}\frac{\partial\varphi}{\partial r} . \tag{22}$$

Substituting eq. (22) in eq. (20), we obtain

$$I = \frac{1}{\Delta E}\frac{d\sigma}{d\Omega} = \frac{\sin \psi \, dr \, dq}{\Delta E \sin \theta (d\theta_r \, d\varphi_q - d\theta_q \, d\varphi_r)}$$

$$\approx \frac{\sin \psi \, \Delta r \, \Delta q}{\Delta E \sin \theta_{r,q}(\Delta\theta_r \, \Delta\varphi_q - \Delta\theta_q \, \Delta\varphi_r)} , \tag{23}$$

where $\theta_{r,q}$ is the scattering angle corresponding to the given aiming point with integral coordinates r and q,

$$\Delta\theta_r = (\theta_{r+1} - \theta_r)_{q=\text{const}} ; \qquad \Delta\theta_q = (\theta_{q+1} - \theta_q)_{r=\text{const}} ;$$

$$\Delta\varphi_q = (\varphi_{q+1} - \varphi_q)_{r=\text{const}} ; \qquad \Delta\varphi_r = (\varphi_{r+1} - \varphi_r)_{q=\text{const}} ;$$

The algorithm is constructed in such a way that the program considers first the scattering for the coordinate r from 0 to r_{max} for $q = 0$. The values of $\Delta\theta_r$ and $\Delta\varphi_r$ are computed and stored in memory. Next, for $q = 1$, the values of $\Delta\varphi_q$ and $\Delta\theta_q$ corresponding to the same value of r are calculated. After this, the intensity in the energy distribution of scattered ions is determined by eq. (23).

Apart from this, spatial angular and energy distributions of the scattered ions and recoil atoms were constructed by the more accurate histogram technique based on counting the trajectories of particles which enter the detector with a given angular and energy spread.

CHAPTER 3

Single and Double Scattering

1. Single-scattering effect

The theory of medium-energy ion scattering from solids has evolved through a number of successive approximations. In the earlier stages, the dominant concept considered the scatterer as a whole and assumed an analogy with the scattering of hard spheres from a solid wall, smooth or rough. It soon, however, became clear that the dominant role in this phenomenon is played by binary collisions of the ion with individual atoms on the surface of the solid.

Experimental studies of ion scattering from solids were started fairly long ago by Gurney (1928), Eremeev and Zubchaninov (1942), U.A. Arifov and Ayukhanov (1951), and Mashkova and Molchanov (1980). These and other studies (U.A. Arifov 1961, Mashkova and Molchanov 1962, U.A. Arifov et al. 1962a, Gruich et al. 1964) succeeded in reliably isolating the directly reflected particles from the total emission accompanying the bombardment of a solid with ion beams, as well as in investigating some of their properties. It turned out that the scattered ions make up a small fraction of the particles emitted by the surface and possess considerable energies, comparable with that of the incident ion beam. Among them one detected beam and target ions with energies characteristic for a single binary collision of an ion with the target atom.

U.A. Arifov and Ayukhanov (1951) and Eremeev (1951) established experimentally that the maximum energy of the ions with $E_0 = 0.1$–3.0 keV reflected at large angles by a polycrystalline surface corresponds to a single elastic scattering of an ion from a free atom in the solid.

A.B. Brown et al. (1951) observed in the energy distribution of 1.2 MeV protons reflected from a lithium surface peaks corresponding to single scattering from carbon and oxygen atoms on the target surface. This has been essentially the first observation of ion scattering from atoms on the surface of a solid.

In the medium-energy range, Panin (1962) was the first to study the energy spectra of the ions reflected in a given direction and showed that they contain a narrow peak at the energy which an ion retains after a single elastic collision with the atom of a solid.

Datz and Snoek (1964) also pointed out that a sizable fraction of the reflected ions possess energies characteristic for a single collision of two particles. The geometry of the experiment (relatively large incidence and scattering angles) was here an essential factor, since at small angles multiple ion collisions with target atoms will dominate.

Consider a single reflection of an ion from a target atom. The single scattering model permits one to evaluate fairly simply the coefficient of an ion scattering from a solid. Earlier, U.A. Arifov et al. (1961b) studied the reflection of alkali-metal ions from the surface of a polycrystal at normal incidence. For qualitative evaluation of K_N, the authors propose the expression

$$K_N = \sigma_0 N \lambda_0 , \tag{1}$$

where σ_0 is the backscattering cross section in elastic collision, λ_0 is the effective depth for the ion backscattering, and N is the number of atoms in 1 cm^3 of the solid.

The total backscattering cross section in elastic collision can be written as

$$\sigma_0 = \pi p_0^2 , \tag{2}$$

where p_0 is the impact parameter corresponding to the scattering angle $\theta = 90°$. The calculation was made with the Firsov potential.

Figure 1 compares the calculated (full curves) and experimental (broken curves) $K_N(E_0)$ values for Na$^+$ and K$^+$ ions reflected from Mo. We see them to disagree, the faster fall-off of the calculations with increasing energy compared with experimental data being accounted for by the accepted constant value of λ_0 (although λ_0, naturally, also grows with E_0) and neglect of multiple scattering. For the single scattering probability including possible reflection from deeper layers in the solid, one can write (Parilis 1965)

$$K_1(E_1, \theta_1) = \frac{N}{\sin \psi} \int_0^\infty \sigma(E_0, \theta_1) \exp\left(- \frac{x}{\lambda(E_0) \sin \psi}\right)$$

$$\times \exp\left(- \frac{x}{\lambda(E_1) \sin \beta}\right) dx$$

$$= \sigma(E_0, \theta_1) c(\theta_1, \psi, \beta) \lambda(E_0) N , \tag{3}$$

where

$$c(\theta_1, \psi, \beta) = \frac{\sin \beta}{\sin \beta + \dfrac{\lambda(E_0)}{\lambda(E_1)} \sin \psi} = \frac{\sin \psi}{\sin \beta + \dfrac{(1 + \mu)^2}{F^2(\theta_1)} \sin \psi} ; \tag{4}$$

$\lambda(E_0)$ and $\lambda(E_1)$ are the ranges of the projectile and reflected atom, respectively.

The quantities $\lambda(E_0)$ and $\lambda(E_1)$ require some comment. Since the Firsov potential is long ranged, the range of the ion of any energy is, in principle, zero. We had in mind, however, a comparison with experiment where the

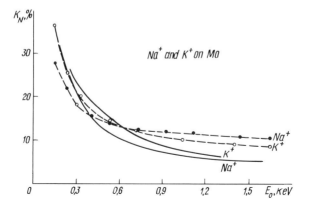

Fig. 1. Scattering coefficient of Na$^+$ and K$^+$ ions falling on an Mo surface.

single scattering peak has a finite width, therefore, one can determine these quantities in such a way that they will correspond to the ion energy being conserved within the peak width. In other words, one estimates the probability of the reflected ion to contribute to the single scattering peak. In this case, the ranges λ will be determined by the cross section of scattering through a small but finite angle $\Delta\theta$ associated with an energy loss which does not exceed the instrumental width of the principal maximum E. Such an approximation can be used to evaluate the relative intensity of the single scattering peaks. The factor $c(\theta, \psi, \beta)$ takes into account the transmission of atomic layers for the incident and reflected beams and yields a correct dependence of the single scattering probability on the grazing and escape angles, ψ and β, respectively, which agrees with experiment. Figure 2a presents the single scattering probability ratio $K(\psi_1)/K(\psi_2)$ for the Ar$^+$–Cu pair and $E_0 = 25$ keV, and grazing angles $\psi_1 = 4°$, $\psi_2 = 10°$. The triangles denote the experimental data obtained by Molchanov and Soszka (1964). The dependences are seen to agree qualitatively, although the calculated curve lies below the experimental points. This is due to the fact that the fraction of multiply scattered particles, other conditions being equal, is substantially larger at $\psi_1 = 4°$ than at $\psi_2 = 10°$.

Figure 2b gives an idea of the effect of the cross section $\sigma(\theta)$ and of the factor $c(\theta, \psi, \beta)$ on the angular dependence of the single scattering probability $K_1(\theta_1)$. Shown for comparison is curve 3 obtained in the hard-sphere approximation. The invalidity of this model was established already by U.A. Arifov et al. (1961b) and Parilis (1965) in calculations of the energy dependence of the total reflection coefficient. As is well-known, the hard-sphere approximation results in the independence of the reflection coefficient on ion energy which is in sharp contrast with experiment showing K_N

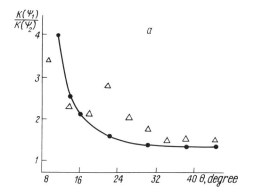

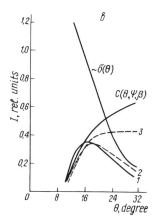

Fig. 2. (a) $K(\psi_1)/K(\psi_2)$ ratio for $\psi_{1,2} = 4°$ and $10°$ [triangles: experiment (Molchanov and Soszka 1964); full line: calculation] and (b) angular distribution of 25 keV Ar^+ ions reflected from a Cu surface [1: theory (Parilis 1965), 2: experiment (Mashkova and Molchanov 1962), 3: hard-sphere model].

to fall-off with increasing energy in full accord with the scattering law obtained with a screened Coulomb potential.

Experiments (Kaminsky 1965, U.A. Arifov 1968, Mashkova and Molchanov 1985) reveal, in the energy distribution of ions scattered in a given direction, a fairly sharp maximum at an energy $E = E_1(E_0, \theta_1)$. An exclusion are the cases with $m_1 > m_2$, where for $\theta_1 > \arcsin \mu$ single scattering is impossible, i.e., there exists a limiting angle $\theta_{\lim} = \arcsin \mu$ beyond which single scattering cannot occur. For $\theta_1 < \arcsin \mu$, one should observe two maxima corresponding to the two signs in eq. (5) of chapter 1. By elementary theory of scattering, their intensities should be proportional to

$\sigma(\chi_1)$ and $\sigma(\chi_2)$, where χ_1 and χ_2 relate to the two signs in the expression

$$\cos \chi_{1,2} = -\frac{1}{\mu}\left(\sin^2\theta_1 \pm \cos\theta_1 \sqrt{\mu^2 - \sin^2\theta_1}\right). \tag{5}$$

The double-valued nature of the impact parameter revealed in scattering at a given angle is demonstrated in fig. 3a. We see that as the scattering angle grows, the impact parameters $p(\chi_1)$ and $p(\chi_2)$ corresponding to the two branches draw nearer, which makes the peaks on the energy scale corresponding to scattering at angles χ_1 and χ_2 come closer to one another. As the scattering angle increases, the separation between the maxima decreases until at $\theta_1 = \arcsin\mu$, they finally merge. In fig. 3b, one can follow the evolution of the position and height of these peaks as a function of θ for the case of a graphite surface bombarded by Ar^+ ions, and of a molybdenum surface with Cs^+ ions of an energy of 30 keV. In the first case, $\theta_{lim} = 17.5°$, and as θ_1 approaches θ_{lim}, the maxima come closer to one another, becoming equal in intensity at $\theta_1 = \theta_{lim}$. As already pointed out, this doubling of the single-scattering peak near θ_{lim} has long defied experimental verification (Shoji et al. 1989, Bastasz et al. 1989). Eckstein and Matschke (1976) were the first to observe two single-scattering peaks in the energy spectrum of Ne^+ ions backscattered from a graphite target. Later, Algra et al. (1980a) experimentally revealed the double-peak structure in the single scattering of $^{84}Kr^+$ ions from a Cu (410) surface close to the maximum scattering angles $(\theta = 40–50°)$. The surface of this high-indexed face was stepped, thus providing favorable conditions for the observation of single scattering from

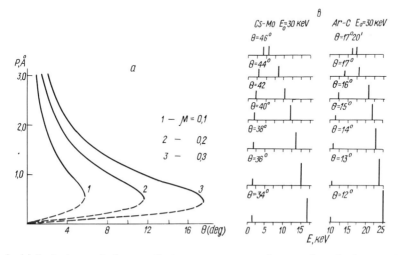

Fig. 3. (a) Scattering angle θ versus the impact parameter p for a number of values $\mu < 1$, and (b) position of single-scattering peaks in the spectrum for $\mu < 1$.

step edge atoms. In full agreement with theory, and also due to the stronger neutralization of the ions, the low-energy peak was found to be nearly an order of magnitude lower than the high-energy one, becoming equal to it in intensity only close to the limiting angle.

Another exception where single-scattering peaks cannot be observed in the energy distribution is associated with the screening action of neighbor atoms at glancing incidence of the ions on the surface. However, because of the presence of defects on the surface, the single-scattering peak was observed in a number of cases also for small grazing and escape angles (Van der Weg and Bierman 1969, Dahl and Sandager 1969).

2. Double-scattering effect

As more and more experimental and theoretical data became available, it was recognized that one cannot obtain a comprehensive description of the process of ion scattering from atoms on the surface of a solid without the inclusion of multiple scattering (Panin 1962, Datz and Snoek 1964, Parilis 1965, Mashkova et al. 1965, Parilis and Turaev 1964, 1965). When the single-scattering peak was observed, it was pointed out that the part of the energy distribution lying to the left of it should be due to multiple scattering (Parilis 1965). It soon turned out that this relates not only to the low- but also to the high-energy part of the distribution, since, as follows from elementary considerations, the energy of the ion scattered at an angle θ may, as a result of several collisions, become not only less but even greater than $E_1(E_0, \theta)$ (Datz and Snoek 1964, Parilis 1965, Mashkova et al. 1965, Parilis and Turaev 1964, 1965). If the grazing and scattering angles are not too small, the probability of scattering in a double-collision event will be next in decreasing order after the single-scattering probability and, when added to it, will make up a larger part of the total probability of scattering at a given angle θ (Parilis and Turaev 1964, 1966, Parilis 1965).

An ion colliding with a pair of closely situated atoms can become scattered at an angle θ at least in two ways, namely, as a result of single scattering with one of the atoms, and because of scatterings at angles θ_1 and θ_2 in two sequential collisions with the atoms (fig. 4). In the first case, its

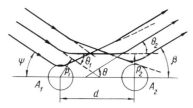

Fig. 4. Ion scattering from two surface atoms.

energy in the first elastic collision will be $E = E_1(E_0, \theta)$, and in the second, $E_2(E_0, \theta) \lesseqgtr E_1(E_0, \theta)$.

The equation $E_2(E_0, \theta, \theta_1, \varphi_1) = E_1(E_0, \theta)$ defines the cone of directions of the first scattering events (when ion scattering from a diatomic molecule is considered), for which the energy of double scattering does not differ from that of the single-scattering event,

$$
\cos \varphi_1 = \frac{1}{2 \sin \theta_1 \sin \theta} \left[(1 + \mu) \frac{F(\theta)}{F(\theta_1)} + \frac{F(\theta_1)(1 - \mu)}{F(\theta)} - 2 \cos \theta_1 \cos \theta \right],
$$

(6)

where φ_1 is the azimuthal angle of the direction of ion motion after the first scattering. Outside the cone, $E_2 < E_1(E_0, \theta)$, whereas inside it $E_2 > E_1(E_0, \theta)$. This means that double-scattering events also contribute to reflection with an energy $E_1(E_0, \theta)$, the 'single'-collision peak containing also multiply scattered particles. As for the energy retained by the ion scattered at an angle θ after two sequential collisions with the two atoms, it will be

$$
E_2(E_0, \theta_1, \theta_2) = \frac{E_0}{(1 + \mu)^4} (\cos \theta_1 \pm \sqrt{\mu^2 - \sin^2\theta_1})^2 (\cos \theta_2 \pm \sqrt{\mu^2 - \sin^2\theta_2})^2
$$

$$
= \frac{E_0}{(1 + \mu)^4} F^2(\theta_1) F^2(\theta_2),
$$

(7)

and will depend on the interaction potential, since the angles θ_1 and θ_2 are determined (in the plane case) by the equations

$$
\sin(\psi - \theta) = \frac{p_1(E_0, \theta_1) - p_2[E(\theta_1), \theta_2]}{d}.
$$

(8)

Here, p_1 and p_2 are the impact parameters for the first and second collision, respectively, ψ is the grazing angle, and d is the atomic separation. In the general case, eq. (8) contains also the azimuthal angle φ_1, the angle θ_2 being defined as

$$
\cos \theta_2 = \cos \theta_1 \cos \theta + \sin \theta_1 \sin \theta \cos \varphi_1.
$$

(9)

In the approximation used to calculate the single-scattering probability, the probability of double scattering will be expressed as

$$
K_2(E_2, \theta) = \sum \sigma(E_0, \theta_1) \sigma[E_1(\theta_1), \theta_2] c(\theta_1, \psi, \theta_1 - \psi)
$$

$$
\times c(\theta_2, \psi - \theta_1, \beta) \lambda(E_0) \lambda(E_1) N^2.
$$

(10)

The summation in eq. (10) should be carried out over the equal-energy cones, $E_2(E_0, \theta) = \chi^2 E_0$,

$$\cos \varphi_1 = \frac{1}{2 \sin \theta_1 \sin \theta} \left[\frac{\chi(1+\mu)^2}{F(\theta_1)} + \frac{(1-\mu)F(\theta_1)}{(1+\mu)\chi} - 2 \cos \theta_1 \cos \theta \right],$$

(11)

relating θ_1 to $\varphi_1 [\chi = (E_2/E_0)^{1/2}]$.

For multiple scattering, the summation of the probabilities $K_i(E, \theta)$ should be performed also over all combinations of intermediate angles which yield in the end the given energy.

3. Double scattering from a single crystal

In the energy distribution of the ions scattered from single crystals, one can reveal a structure due to double collision events (Parilis and Turaev 1964).

In contrast to a polycrystal where every intermediate direction of scattering is equally probable, in a single crystal the intermediate scattering angles θ_1, φ_1, and θ_2 can take on only certain discrete values. For double collisions, this leads to a discrete pattern of the spectrum which, as predicted by Parilis and Turaev (1964), can be observed experimentally against a general background caused by multiple scattering. It was assumed that an ion can enter the analyzer in two ways, namely, as a result of single scattering from an atom located at the origin [001], or after two sequential collisions with the [001] atom and an [ijk] atom, following which the ion is scattered in the direction of the analyzer. Higher multiplicity scattering was not considered (fig. 5a).

An estimate made by Parilis and Turaev (1964) for Ar^+ ions back-scattered from Cu(100) at $E_0 = 25$ keV, and for K^+ ions from W(100) at $E_0 = 3$ keV for different ψ and β, and including only the elastic energy losses showed the double-scattering peaks to be of noticeable intensity, the separation between them being fairly large. The very first attempt to observe this experiment (Mashkova et al. 1965) resulted in a reliable isolation of at least one peak corresponding to double collisions along $\langle 110 \rangle$. Its position (fig. 5b) agrees well (if inelastic energy losses are included) with the calculations and is not related to the isotopic effect (broken curve in the spectrum).

Further calculations were carried out with a program for an M-20 computer, where one entered the crystallographic coordinates of the atoms in a crystallite with cubic symmetry containing about 60 atoms surrounding the target atom (fig. 5a) involved in the first collision and assumed to be at rest at the origin (Parilis and Turaev 1964). The computer simulation of this crystallite of atoms combined into a cubic lattice permitted the revealing of a fine structure in the spectrum which depends on the composition and dimensions of the crystal lattice.

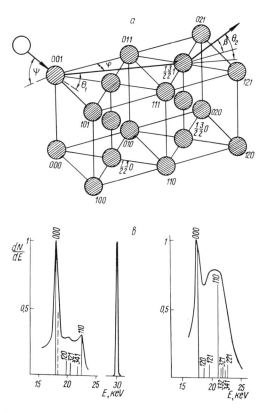

Fig. 5. Ion reflection from a single crystal. (a) Atoms on side planes not shown, with central atoms specified for the bcc lattice; (b) energy distributions of Ar ions backscattered from single crystalline Cu ($E_0 = 30$ keV, (110) reflection plane) for the faces (100) (left) and (114) (right).

Particular attention in writing the program was focussed not on increasing the crystallite size but on the possibility of varying Z_1, Z_2, E_0, μ, ψ; type and parameters of the crystal lattice, and the directions of incidence and reflection. The energy spectra calculated in this way showed that in the case of a complex compound, e.g., KCl, one may observe secondary peaks of the same index corresponding to the K–K, K–Cl, Cl–K, Cl–Cl combinations. This permits one to investigate the chemical structure of the surfaces of crystals or of crystallite (e.g., epitaxial) films from the energy spectrum of reflected ions. The advantage of this method over the others consists in that here only the upper atomic layers of a solid are analyzed. The similar quadrupling of the peaks corresponding to two isotopes (e.g., [63]Cu and [65]Cu) can hardly be resolved experimentally, although this was done by Datz and Snoek (1964) who established the existence of an isotopic splitting of the main peak.

The possibility of experimental observation of a peak in the spectrum is determined by its FWHM. The spread of the secondary peaks due to thermal vibrations of both atoms perpendicular to the connecting line is characterized by the FWHM (Parilis 1965),

$$\Delta E_2(\theta_1, \theta_2) = \frac{4E_2}{d} \sqrt{\frac{\ln 4}{\gamma}} \left[\frac{\mathrm{d}}{\mathrm{d}\theta_1} \ln F(\theta_1) - \frac{\mathrm{d}}{\mathrm{d}\theta_2} \ln F(\theta_2) \right], \tag{12}$$

where γ is the Gaussian parameter of isotropic atom vibrations at a lattice site. Leibfried (1955) found $1/\gamma \simeq 0.010 a_{min}^2 T/T_m$ (a_{min} is the distance between the neighbor lattice atoms, T is the absolute temperature, T_m is the melting temperature).

Because θ_1 and θ_2 vary in opposite ways in the course of atomic vibrations, the full width at half maximum of the peaks, eq. (12) is not large, only 5–15% of their energy. The most narrow peaks correspond to close values of θ_1 and θ_2.

Lattice defects, dislocations, etc. also contribute to the peak width. The increase in peak intensity caused by the vibration of atoms along the connecting axis,

$$\Delta K_2 = K_2 \frac{3}{\gamma d^2} = 0.03 \frac{a_0^2}{d^2} \frac{T}{T_m} K_2 ,$$

does not exceed 1–2% at room temperature. Here, K_2 is the peak intensity at absolute zero.

Thus, it was shown that the above-mentioned fine structure in the energy spectrum should be looked for in the high-energy part, for close values of θ_1 and θ_2, large values of μ and moderate temperatures.

The prediction of the double-scattering effect on single crystals initiated an intense experimental investigation of the structure in the energy distributions of scattered ions over a wide range of their initial energies, grazing and scattering angles, and for different directions in single crystals. The energy range 10–60 keV was covered by Mashkova and Molchanov (1966, 1972a, 1980), Chicherov (1968), Van der Weg and Bierman (1968); the 1–4 keV range by U.A. Arifov and Aliev (1967, 1968a, 1968b, 1974), Suurmeijer and Boers (1973), Boers (1977), Heiland et al. (1973a, 1976); and at energies below 1 keV, by U.A. Arifov et al. (1972). In all these studies the double-scattering effect was confirmed experimentally.

The fine structure in the energy spectrum of reflected ions originates from the ordered arrangement of atoms on the surface of a solid and, hence, depends on the degree of this ordering. A decrease in order should result in a smoothening of the energy structure, and its total disappearance in the formation of the continuous spectrum characteristic of polycrystals.

Revealing the discrete nature of the energy spectrum of scattered atoms and its relation to the crystalline structure of the surface has opened the way to diagnostics of the surface structure at an atomic level by spectroscopy of the scattered heavy-particle energy.

4. Screening and blocking in two-atom scattering

In the case of sequential scattering from two atoms, the scattering function (i.e., dependence of the scattering angle θ on the impact parameter p) is nonmonotonic because of the screening action of atoms located in the incidence and reflection directions. As shown as far back as 1965 (Parilis 1965), already the first atom in the direction of reflection produces an additional scattering to an angle corresponding to the parameter

$$p_2[E_1(\theta_1), \theta_2] = p_1(E_0, \theta_1) + d \sin(\theta_1 - \psi) , \qquad (13)$$

where θ_1, θ_2 are the angles of scattering from the first and second atom, respectively, with $\theta = \theta_1 + \theta_2$ in the plane case ($\varphi_1 = -\varphi_2 = 0$); E_1 is the ion energy after scattering at an angle θ_1, and ψ is the grazing angle. Because of this, the escape angle $\beta = \theta_1 + \theta_2 - \psi$ not only does not vanish for $\theta_1 \to 0$, but starts to grow reaching a maximum θ_{max} (at sufficiently small ψ). As a result, the $\theta(p_1)$ dependence, rather than being a monotonically decreasing function as is the case with scattering from an atom, represents a curve with two extrema whose height and position depend on ψ, E_0, d, the reflected beam being limited by the minimum and maximum scattering angles.

Thus, in sequential scattering from two atoms there are screening cones (i.e., regions forbidden for the screened ions to enter) behind both the first and the second atom.

One should discriminate the shadow cones formed behind the first cone from the blocking cones behind the second one (Mashkova and Molchanov 1985, Turkenburg et al. 1976). The difference consists in that in the latter case the 'point source' of the primary particles lies at a certain distance from the blocking atom close to the nucleus of the first atom, whereas in the first case a parallel beam falls on the shadowing atom.

The blocking effect was revealed by Domeij and Bjorquist (1965) and Tulinov et al. (1965a,b), and Oen (1965) interpreted it. The radius of the screening cone R_s was calculated by Martynenko (1964), who studied the sputtering of single crystals. He showed that in the region $R_s < p_1 < 1.05 R_s$ behind the shadow there is a narrow, brightly illuminated ring, the intensity of the flux throughout the ring's width exceeding that of the undistorted flux by $\sim 10\%$.

The extremal values of the scattering angle (θ_{min} and θ_{max}) in the $\theta(p_1)$ dependence originating from the shadowing and blocking effects cause the appearance of the so-called rainbow effect in the differential cross section of double scattering, i.e., ion focusing near the shadow boundaries.

The angle θ_{min} can be found by solving eq. (13) together with the condition of the minimum angle (Parilis 1965),

$$\frac{dp_1}{d\theta_1} + d\cos(\theta_1 - \psi) = -\frac{dp_2}{d\theta_2} . \tag{14}$$

The minimum angle decreases with increasing initial energy E_0 and interatomic distance d. Figure 6 presents a $\theta_{min}(\psi)$ plot for the Ar^+–Cu pair calculated with the Firsov potential.

Limitation of reflection at large angles is likewise connected with the screening action of surface atoms. As the grazing angle ψ is reduced continuously to small values, the surface atoms enter gradually the shadow formed by the neighbor atoms. It thus becomes impossible to reach first the large impact parameters corresponding to scattering at small angles, and then the small parameters (large reflection angles). The angular distribution is cut off at the angle θ_{max} determined by the equation

$$p_2[E_1(\theta_1), \theta_{max}] = (R_s - d\sin\psi)\cos\psi , \tag{15}$$

where $R_s = 2.2[db^2/(1+\mu)]^{1/3}$ is the shadow radius at a distance d from the atom (Parilis and Turaev 1965); θ_1 is the scattering angle corresponding to R_s which is determined by the parameter

$$p_1(E_0, \theta_1) = \left(\frac{\pi d \cos\psi \, b^2}{1+\mu}\right)^{1/3} , \tag{16}$$

and b is the distance of closest approach in a head-on collision with an energy E_0.

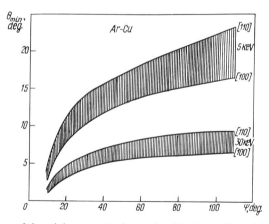

Fig. 6. Dependence of the minimum scattering angle of Ar ions reflected from a Cu surface on the grazing angle.

Thus, the reflected beam is bounded by the escape angles $\beta_{\min} < \beta < \theta_{\max} - \psi$. The upper boundary decreases rapidly with decreasing ψ and the reflected beam becomes narrower.

Equations (13) and (14) do not, however, have an analytic solution in a general form. The solution of the double-scattering problem in the impulse approximation for the plane case, as well as expressions for the minimum and maximum scattering angles, scattered-ion intensity and relative intensity of double scattering were obtained by Martynenko (1973). The interaction potential used was an inverse power law, $V = V_0 r^{-n}$. In this case,

$$\theta_1 = \frac{C}{p_1^n}, \qquad p_2 = p_1 + d\left(\frac{C}{p_1^n} - \psi\right), \qquad \theta_2 = \frac{C}{\left[p_1 + d\left(\frac{C}{p_1^n} - \psi\right)\right]^n},$$

(17)

$$C = \frac{\sqrt{\pi}\Gamma[\frac{1}{2}(n+1)]V_0}{\Gamma(\frac{1}{2}n)E_0}, \qquad \theta = \theta_1 + \theta_2.$$

As follows from eq. (17) and fig. 7a, $\theta(p_1)$ has a minimum, $\theta_{\min}$, at

$$\psi > \psi_{\mathrm{cr}} = \frac{R_s}{d} = \frac{n+1}{n}\frac{p_1^s}{d},$$

(18)

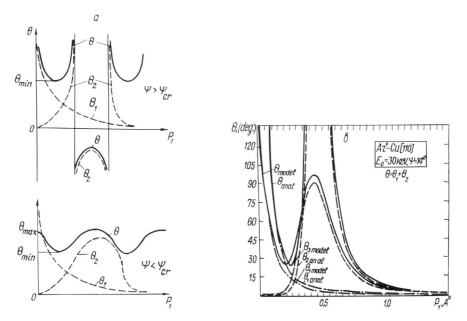

Fig. 7. (a) $\theta_1(p_1)$, $\theta_2(p_1)$, $\theta(p_1)$ dependences for $\psi < \psi_{\mathrm{cr}}$ and $\psi > \psi_{\mathrm{cr}}$, and (b) comparison with computer calculation.

where $R_s = [(n+1)/n](ndC)^{1/n+1}$ is the radius of the shadow produced by the first atom, $p_1^s = (ndC)^{1/n+1}$ is the impact parameter corresponding to scattering at the edge of the shadow. The ratio of the shadow radius to the interatomic separation (R_s/d) defines the angles for which screening should be taken into consideration.

For $\psi > \psi_{cr}$, the $\theta(p_1)$ curve does not have the upper extremum and is discontinuous. For $\psi < \psi_{cr}$, an upper limit θ_{max} appears which is due to the fact that the impact parameters smaller than $R_s - \psi d$ cannot be reached since they are in the shadow of the neighbor atom. The falling off part of the $\theta(p_1)$ curve corresponds to quasisingle scattering, and the rising one, to the quasidouble process. The region of possible values of θ becomes more narrow as the grazing angle decreases, and for $\psi = \psi_{min}$, $\theta_{min} = \theta_{max} = 2\psi$. The values of θ_{min} and θ_{max} and the corresponding impact parameters p_{1min} and p_{1max} can be found from the following equations:

For $\psi > \psi_{cr}$,

$$p_{1min} = p_1^s\left[n\left(1 + \frac{\psi d}{p_1^s}\right)\right]^{-1/n}, \qquad \theta_{min} = \psi + \psi_{cr} - \left(\frac{p_1^s}{n\psi d}\right)^{1/n}; \qquad (19)$$

for $\psi \rightarrow \psi_{cr}$,

$$p_{1max} = p_1^s\left[1 - \frac{(n+1)^n}{n^{n+1}}\left(1 - \frac{\psi}{\psi_{cr}}\right)^{n+1}\right],$$

$$\theta_{max} = \frac{n^n}{(n+1)^{n+1}}\frac{\psi_{cr}}{\left(1 - \frac{\psi}{\psi_{cr}}\right)^n}.$$

For $\psi \leqslant \psi_{min}$, interaction with many atoms becomes more essential, and the atomic row model more acceptable. Figure 7b presents dependences $\theta_1(p_1)$, $\theta_2(p_1)$ and $\theta(p_1)$ obtained analytically and by computer simulation for the case of scattering from two atoms. The rising branch for increasing p_1 is due to double scattering, and the falling off branches to single scattering from the first (for small p_1) and second (for large p_1) atoms, respectively. We readily see that already in scattering from two atoms their mutual screening action results in the reflected beam being limited by a minimum, β_{min}, and a maximum, β_{max}, escape angle, their values depending on the initial ion energy E_0, interatomic separation d and beam orientation relative to the crystallographic axes ψ.

As shown by a comparison of the analytical calculations with computer simulations made with the same interaction potential, in both cases the interval of possible double-scattering angles becomes ever more narrow with decreasing ψ, with only specular scattering remaining at $\psi = \psi_{min}$ when $\theta_{min} = \theta_{max} = 2\psi$. The values of θ_{min} and θ_{max} in the two cases are, however, different. The values of θ_{max} calculated analytically for each given ψ exceed by far those obtained by computer simulation. In the region $\psi = 8-13°$, the values of θ_{min} are about the same, but as ψ increases still more the angles θ_{min} derived by computer become greater than those obtained analytically. Thus, because of the mutual screening action of neighbor atoms the limitation of the reflected beam by a minimum and a maximum escape angle obtained by computer simulation turns out to be stronger than in the case of analytical calculations.

Note also that the steepness of the $\theta(p_1)$ curves in the region of double scattering is different, namely, analytical calculation yields a steeper curve, and since the scattering cross section is proportional to $d\theta/dp_1$, this leads naturally to different values for the double-scattering cross section.

Figure 8a presents $\theta(p_1)$ dependences for the scattering from two atoms

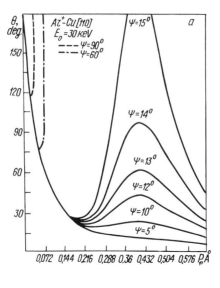

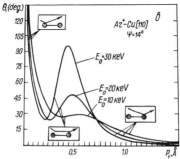

Fig. 8. $\theta(p_1)$ plots for (a) different grazing angles, and (b) different initial energies.

obtained for different ψ. We see that for sufficiently large ψ the ions leaving the surface are limited only on the side of small escape angles, the $\theta(p_1)$ curve exhibiting a discontinuity. The region of impact parameters p_1 corresponding to this discontinuity determines the trajectories of the particles penetrating beyond the line connecting two atoms. As the angle ψ decreases a constraint appears also on the side of large escape angles (Martynenko 1973), the interval of allowed double-scattering angles, $\theta_{min} \leq \theta \leq \theta_{max}$, contracting with decreasing ψ, E_0 (fig. 8b) and interatomic separation d (Mukhamedov et al. 1973).

The above estimates can obviously be no more than approximate, since each of the two atoms is screened by the preceding one and, in its turn, screens the subsequent atoms in the reflection plane.

The above-mentioned $\theta(p_1)$ dependences imply also that scattering to a given angle θ in the interval $\theta_{min} \leq \theta \leq \theta_{max}$ can occur along three different trajectories which, in general, yield different ion energy losses [see Mashkova and Molchanov (1966, 1972c, 1980), Muller-Jahreis (1970), where the two-atomic scattering model is studied]. In the approximation of correlated scattering by a pair of atoms through small angles using eq. (13) and the inverse square approximation of the Firsov potential, the following expression was obtained,

$$\left(\frac{\pi C}{2d^2 E_0}\right)^{1/2} \left(\frac{1}{\sqrt{\theta - \theta_1}} - \frac{1}{\sqrt{\theta_1}}\right) = (\theta_1 - \psi). \tag{20}$$

In deriving this equation, the relation between the scattering angles θ_1 and θ_2 and the impact parameter p_1 was determined from eq. (17). When eq. (20) is solved graphically, the intersection points of the curve described by the left-hand part of the equation with a $y = \theta_1 - \psi$ straight line determine the number of possible different trajectories which, in general, yield different energy losses for the scattered ions [fig. 7.5 in Mashkova and Molchanov (1980)]. For large enough θ and $\psi \approx \frac{1}{2}\theta$, there are three intersection points, two of them corresponding to the small first (θ_1) or second ($\theta - \theta_1$) scattering angles, and the third to approximately symmetric scattering where $\theta_2 = \theta - \theta_1 = \frac{1}{2}\theta$. Therefore, the energy distribution of scattered ions should always contain three peaks (or two, when $\psi = \frac{1}{2}\theta$). For ψ other than $\frac{1}{2}\theta$ the energy distribution exhibits only one peak corresponding to ions which have suffered either strong scattering from the first atom and weak scattering from the second (for $\psi < \frac{1}{2}\theta$), or conversely, weak scattering from the first atom followed by a strong interaction with the second one (for $\psi > \frac{1}{2}\theta$). As the scattering angle θ decreases, the interval of the grazing angles ψ where all three intersection points are possible becomes increasingly narrow, so that when θ drops below θ_{lim}, only one point of intersection is possible for any grazing angle, in which case only one peak should be observed in the energy distribution. The value of θ_{lim} can be obtained from the condition

$$\frac{d}{d\theta_1} \left\{ \frac{1}{d} \left[p_2(\theta - \theta_1) - p_1(\theta_1) \right] \right\} \bigg|_{\theta_1 = \theta_{\text{lim}}/2} = \frac{d}{d\theta_1} (\theta_1 - \psi) . \tag{21}$$

An experimental check of the predictions of the two-atom scattering model showed (Mashkova and Molchanov 1966, 1975, 1985) that the measured energy distributions of 30 keV Ar^+ ions scattered from the [110] planes of copper and silicon always contain two peaks both in the regions of the angles ψ and θ where the model predicts the existence of one peak only, and in the regions where three peaks should exist. The observation of two instead of three peaks can be accounted for by a low analyzer resolution, whereas the presence of two peaks in the regions of small, either grazing or scattering, angles (Mashkova and Molchanov 1972c, 1985, Müller-Jahreis 1970) where the model predicts one peak only, suggests that the two-atom model is invalid for the description of ion scattering by the lattice atoms. The possible combinations of the first, θ_1, and second, θ_2, scattering angles shown in fig. 7.5 in Mashkova and Molchanov (1985) which result in scattering through a given angle θ were calculated in the impulse approximation.

Figure 9 presents in a graphical form the relations between θ_1, θ_2 and θ which are not limited by the small-angle approximation for the case of

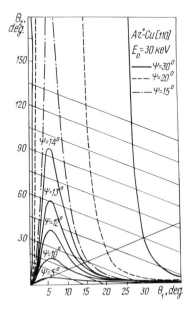

Fig. 9. Relation between the first (θ_1) and second (θ_2) scattering angles as a function of grazing angle ψ.

reflection at the ion incidence plane and are derived by computer simulation
of ion scattering from two atoms (Mukhamedov et al. 1973, Parilis et al.
1975). In the region of small θ angles, the values of θ_1 and θ_2 obtained with a
computer agree well with the results produced by the graphical method. As
seen from the figure, there are three scattering branches, two of them
corresponding to quasisingle scattering characterized by a small value of one
of the angles (θ_1 and θ_2) and, accordingly, a large value of the other angle,
and one branch for double scattering. The intersections of the curves with
inclined straight lines $\theta = \theta_1 + \theta_2 = \text{const}$ define the three possible combinations of the θ_1 and θ_2 angles, whose values at $\theta = \text{const}$ vary with the
variation of ψ, E_0 and d.

According to the calculated relation between the first, θ_1, and total, θ,
scattering angles, the condition $\theta_1 = \psi$ is satisfied only in the case corresponding to specular scattering, i.e., when $\psi = \frac{1}{2}\theta$. In all other cases, θ_1
differs from ψ, which should be taken into account when determining the
positions of the single- and double-scattering peaks in the energy distribution of scattered ions.

5. Energy distribution of scattered projectiles

In the two-atom model, scattering at a given angle may occur as a result of
one or two collisions, which yields different energies for the reflected ions.
The energy E_0 of ions which are double scattered to an angle θ may be
either greater or smaller than the energy E_1 of ions which are single
scattered to the same angle, depending on θ_1, with $E_2 > E_1$ for $0 < \theta_1 < \theta$,
and $E_{2\text{max}}$ corresponding to the angle $\theta_1 = \frac{1}{2}\theta$ (Parilis 1965).

The family of double-scattering trajectories with the same final energy E_2
forms in the space of first scattering directions (θ_1, φ_1) a conical surface
described by eq. (11). The $\varphi_1(\theta_1)$ dependence (fig. 10) shows that as E_2

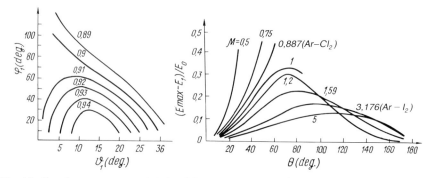

Fig. 10. Equal energy cones $\varphi_1(\theta_1)$ for different E_2/E_0 for $\text{Ar}^+ - \text{I}_2$ ($E_0 = 30$ keV, $\theta = 30°$) (left)
and $(E_{\text{max}} - E_1)/E_0$ versus θ plot for different μ (right).

increases, the equienergetic cones $E_2(\theta) = $ const become narrower and merge into a point for $E_2 \rightarrow E_{2\max}(\theta)$ (Mukhamedov and Parilis 1981). One can calculate, taking into account only the elastic losses, the maximum energy of the ions scattered to an angle θ,

$$E_{2\max} = E_2(E_0, \mu, \tfrac{1}{2}\theta) = \frac{E_0}{(1 + \mu)^4} F^4(\mu, \tfrac{1}{2}\theta)$$

$$= \frac{E_0}{(1 + \mu)^4} \left(\cos\frac{\theta}{2} + \sqrt{\mu^2 - \sin^2\frac{\theta}{2}} \right)^4 , \qquad (22)$$

as well as their minimum energy,

$$E_{2\min} = E_2(E_0, \mu, \pi - \tfrac{1}{2}\theta) = \frac{E_0}{(1 + \mu)^4} F^4(\mu, \pi - \tfrac{1}{2}\theta)$$

$$= \frac{E_0}{(1 + \mu)^4} \left(\cos\frac{\theta}{2} - \sqrt{\mu^2 - \sin^2\frac{\theta}{2}} \right)^4 . \qquad (23)$$

The energy spectrum is confined between these two boundaries which shift only slightly when the inelastic energy losses are included. The width of the spectrum,

$$\Delta E(\theta) = \frac{E_0}{(1 + \mu)^4} [F^4(\tfrac{1}{2}\theta) - F^4(\pi - \tfrac{1}{2}\theta)]$$

$$= \frac{4E_0}{(1 + \mu)^4} (\mu^2 + \cos\theta)[(\mu^2 + \cos\theta)^2 - (\mu^2 - 1)^2]^{1/2} , \qquad (24)$$

depends on the scattering angle θ and mass ratio μ.

The energy $E_1(\theta)$ corresponding to single scattering lies at a distance from the high-energy edge,

$$\Delta E_{\max}(\theta) = \frac{E_0}{(1 + \mu)^2} \left[\frac{F^4(\tfrac{1}{2}\theta)}{(1 + \mu)^2} - F^2(\theta) \right] , \qquad (25)$$

which is much less than the distance to the low-energy edge,

$$\Delta E_{\min}(\theta) = \frac{E_0}{(1 + \mu)^2} \left[F^2(\theta) - \frac{F^4(\pi - \tfrac{1}{2}\theta)}{(1 + \mu)^2} \right] , \qquad (26)$$

while being still sufficiently large to permit a reliable experimental study of this part of the energy spectrum which is of major interest. The relative width of this region of spectrum, $\Delta E_{\max}(\theta)/E_0$, passes through a maximum as the scattering angle is varied and depends on μ. For the $Ar^+ - I_2$ pair in the angular range 60–120° (fig. 10) the width $\Delta E_{\max}(\theta)/E_0$ grows with decreasing μ for all angles θ lying before the maximum that corresponds to $\theta \approx \tfrac{1}{2}\pi$. For larger angles, however, a deviation from this relation is observed. For the reverse mass ratio, the curve exhibits a cut-off at the maximum single-scattering angle θ_{lim}, and one can readily see that double scattering can increase this angle by a factor two (Mukhamedov and Parilis 1982). As for the double-scattering energy $E_2(\theta)$, it turns out to be a four-valued function in the region $0 < \theta < \theta_{lim}$, and a two-valued one for $\theta_{lim} < \theta < 2\theta_{lim}$.

Only elastic energy losses were included in this analysis. Figure 11a presents a scattering-angle dependence of the energy of the ions reflected in one and two collisions computed with inclusion of inelastic energy losses (full curve) as well as obtained analytically (Parilis et al. 1975) without their inclusion (broken curve). Shown in the same figure is the $E(\theta)$ dependence for the scattering from one atom. The ion energies resulting from quasisingle scattering at a given angle are seen to be somewhat greater than those for the single-scattering event. The slight difference in the energies for

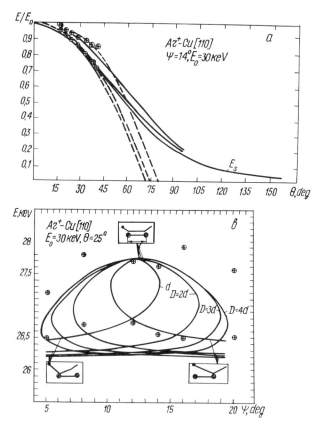

Fig. 11. (a) E/E_0 versus scattering angle θ for double scattering, and (b) scattered ion energy versus grazing angle; the crosses denote experimental values (Mashkova and Molchanov 1980).

quasisingle scattering from the first and second atoms originates from differences in the combinations of the first, θ_1, and second, θ_2, scattering angle. A comparison shows that in the region of small scattering angles, up to $\theta = 45°$, all calculations yield close results. The cut-off of the scattered beam on the small scattering angle side is confirmed both analytically and by computer calculations.

Figure 11b shows the dependence of the energy of the ions scattered at an angle $\theta = 25°$ on the grazing angle ψ for various interatomic distances d. The crosses relate to experimental data (Mashkova and Molchanov 1975). The calculated curve $E(\psi)$ represent actually loops, their different branches corresponding to quasisingle and double collisions. As follows from the calculations, for $D = d$ double scattering exists only in the region $11° \leqslant \psi \leqslant 14°$. At the same time, experiment reveals the existence of double scattering peaks throughout the whole ψ range of 2 to 22°. As seen from the figure, increasing the separation between atoms permits one to broaden the range where double scattering exists.

Figure 12a presents a scattering angle dependence of inelastic energy losses for $E_0 = 30$, 20, 10 keV computed by means of Firsov's expression (Parilis et al. 1975, Umarov 1975). The upper branches of the ovals relate to losses in double scattering, the lower ones to losses in quasisingle scattering from the first atom, and the branch between then to those suffered in quasisingle scattering from the second atom. In accordance with experiment (Snoek et al. 1966), the inelastic energy losses are seen to grow with increasing E_0 and θ.

While the total energy losses resulting from double scattering at a given angle are smaller than those suffered in a single-scattering event, in the case of inelastic losses the reverse is true, namely, the losses grow with increasing collision multiplicity. As E_0 decreases, however, the collision multiplicity affects only weakly the magnitude of the inelastic energy losses, and the separation between the branches of the ovals decreases. Shown in fig. 12b is the dependence of inelastic energy losses suffered in scattering at a given angle $\theta = 25°$ on the grazing angle ψ, i.e., on the orientation of the beam relative to the axis of the two-atom pair. The calculated curves represent loops, their branches corresponding to quasisingle and double collisions. The calculated inelastic energy losses correlate in absolute magnitude with the experimental data (Snoek et al. 1966) obtained at higher energies.

Figure 13 presents a dependence of scattered ion energy E on scattering angle θ. One can compare the experimental data [full curve, Mashkova and Molchanov (1974)] with the analytical calculation with a two-atom model [chain curve, Martynenko (1973)], and results of a computer simulation [broken curve, Parilis et al. (1975)]. Analytical calculations are seen to yield a much higher energy separation between the peaks than that which follows from experiment and computer simulation. This is due primarily to position-

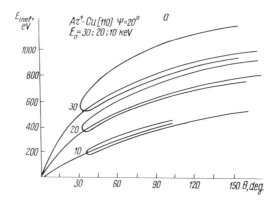

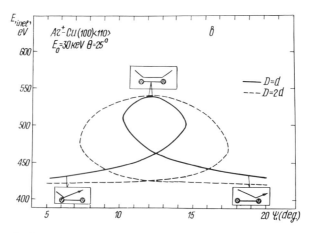

Fig. 12. (a) Inelastic energy losses versus scattering angle for different E_0, and (b) relation between inelastic energy losses and grazing angle ψ.

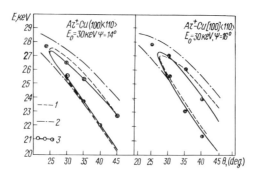

Fig. 13. Scattered ion energy versus scattering angle θ; (1) results from computer simulation, (2) analytical calculations, (3) experiment.

ing the double-scattering peak. The position of the double-scattering peak depends essentially on the scattering angles θ_1 and θ_2, resulting in scattering at a given angle $\theta = \theta_1 + \theta_2$. If for any θ one chooses a combination of the angles θ_1 and θ_2 where $\theta_1 = \psi$, and $\theta_2 = \theta - \psi$, i.e., if one assumes double scattering to be symmetrical, then the double-scattering peak will move in energy to the right with respect to the peak whose position was calculated based on a correct combination of the angles (see fig. 13). The additional shift of the double-scattering peak to the right originates from the exclusion of inelastic energy losses in the analytical calculations. The experimentally measured and calculated double-scattering peak positions do not coincide and differ by $\approx 10\%$.

Mashkova and Molchanov (1966, 1967, 1972c, 1973, 1975) and Molchanov et al. (1969) carried out a careful experimental check of the predictions of the two-atom scattering model. Figure 5.15 from Mashkova and Molchanov (1980) presents the energy distribution of Ar^+ ions scattered from a copper crystal. The scattering angle was constant and equal to $\theta = 34.5°$, while the grazing angle was varied from 5 to 24°. Also shown are the peak positions of single and double scattering calculated with the two-atom model. In accordance with the predictions of the model, the energy distributions reveal a structure in their high-energy part not in all cases but only at grazing angles in the range 14–21°. However, the three-peak structure predicted by the model has never been observed experimentally.

The two-peak structure was found to exist also in the scattering of Ar^+ ions from a silicon crystal (Mashkova and Molchanov 1972c).

Experimental energy distributions of 30 keV Ar^+ ions scattered from Cu and Si [110] planes always contain two peaks both in the region of ψ and θ angles where the model predicts the existence of only one peak and in the regions where three peaks should be observed.

Parilis (1965) discussed the effect of thermal vibrations of lattice atoms on the energy distribution of ions scattered from a pair of atoms, as well as on the region where double scattering exists, as a function of θ and ψ. It was established that the spread of the peaks in the energy distribution caused by thermal vibrations of both atoms perpendicular to the line connecting them is small, namely, 5–15% of the ion energy, because θ_1 and θ_2 vary in opposite directions. The vibrations perpendicular to the crystal surface affect the grazing angle while not acting on the scattering angle θ.

The effect of these vibrations at a constant θ and variable ψ was considered by Müller-Jahreis (1970). The effect of thermal vibrations is included by introducing $\pm\Delta\psi$, which corresponds to a deflection of surface atoms by ≈ 0.2 Å. A strong manifestation of the thermal vibrations is observed to exist only near the limiting angles ψ of double scattering. In the symmetric case, $\psi = \frac{1}{2}\theta$, both peaks broaden slightly, the double-scattering peak shifting insignificantly toward lower energies. The inclusion of thermal vibrations permits one to account for the experimentally observed structure in the distributions at small scattering angles.

Verhey et al. (1975) calculated and experimentally measured the energy spectra of Kr^+ ions with $E_0 = 10\,keV$ scattered from a Cu (100) plane at various temperatures. At low temperatures, the conventional double-peak energy pattern is observed. However, as the temperature increases, a quasitriple scattering peak appears in the high-energy part of the distribution, which is due to scattering from three successively located surface atoms. An analysis of the behavior of this new peak permitted the determination of the Debye temperature for the surface.

Later, the same authors (Poelsema et al. 1975, 1976a, 1976b, 1977b, Boers and Verhey 1973) made a comprehensive study of the effect of thermal vibrations on the angular and energy distributions of scattered ions. The results were discussed in terms of the atomic row model (we will dwell on them in more detail in the next chapter).

In the evaluation of the dependence of the mean (or most probable) particle energy loss ΔE on the scattering angle θ, an interesting modification of single scattering was studied which takes into account the effect of the target atoms lying outside the scattering plane (Evdokimov 1976). It is known that usually ΔE grows monotonically with increasing θ. However, at very small grazing angles ψ one observes an anomalous drop of ΔE with increasing θ (Evdokimov et al. 1969) so that the $\Delta E(\theta)$ dependence may cease to be single valued, resulting in the appearance in the energy distribution of scattered particles of two peaks of comparable intensity.

6. Relative probability of double scattering

The relative height of the double-scattering peak represents a convenient parameter for describing the shape of the energy distribution. It coincides with the relative probability of double scattering only in the cases where the peaks are well resolved (Mashkova and Molchanov 1985, Williams et al. 1990).

Parilis and Turaev (1966) used the following expression to calculate the relative probability of double scattering,

$$K_{1,2} = \frac{K_2(E_2, \theta_1, \theta_2)}{K_1(E_1, \theta)} = \frac{\sigma(\theta_1)\sigma(\theta_2)}{\sigma(\theta)} \frac{c(\theta_1, \psi, \theta_1 - \psi)c(\theta_2, \psi - \theta_1, \beta)}{c(\theta, \psi, \beta)}$$

$$= c(Z_1) \frac{R(Z_1, Z_2)\phi(\theta_1, m_1)\phi(\theta_2, m_2)}{d^2 E_0 F^2(\theta_1, m_1)\phi(\theta, m_1)} \frac{c(\theta_1, \psi, \theta_1 - \psi)c(\theta_2, \psi - \theta_1, \beta)}{c(\theta, \psi, \beta)},$$

$$(27)$$

where

$$\phi(\theta, \mu) = \frac{\left[\pi - \theta - \arcsin \dfrac{\sin \theta}{\mu}\right] \operatorname{cosec}\left[\theta + \arcsin \dfrac{\sin \theta}{\mu}\right]}{\left[2\pi - \theta - \arcsin \dfrac{\sin \theta}{\mu}\right]^2 \left[\theta + \arcsin \dfrac{\sin \theta}{\mu}\right]^2} \ ;$$

$$R(Z_1, Z_2) = \frac{Z_1 Z_2}{(Z_1^{1/2} + Z_2^{1/2})^{2/3}} \ ;$$

$$c(Z_1) = 0.45 \pi^2 30.4 \frac{m_1 + m_2}{m_1} \left(\frac{a_{TF}}{a}\right)^2 .$$

Here, a is the lattice constant.

The double-scattering intensity depends on the intermediate scattering angles θ_1 and θ_2, the interatomic separation d, the initial energy E_0, the mass ratio of the colliding atoms μ, and such factors as the type and composition of the crystal lattice, isotopic abundance, etc. For a fixed scattering angle, a decrease in the interatomic distance d and of the energy E_0 produces an increase in the relative intensity of double scattering. The relative intensities of the double-scattering peaks grow also with increasing μ, however, this is accompanied by a narrowing of the spectrum, i.e., a decrease in the separation between the single and double-scattering peaks.

Martynenko (1973) derived expressions for the probabilities of single and double scattering in the approximation of independent scattering from a pair of atoms.

The probability of scattering at an elementary solid angle $d\Omega = d\theta \, d\varphi$ can be written as

$$dI = p_1 \, dp_1 \, d\varphi = p_1 \left| \frac{\partial p_1}{\partial \theta} \frac{\partial \varphi_0}{\partial \varphi} \right| d\Omega = \frac{p_1 \, d\Omega}{\dfrac{d\theta}{dp_1}\left(\theta + \psi\theta_2 \dfrac{d}{p_2}\right)}, \qquad (28)$$

where φ_0 is the azimuthal angle of the target point of the first atom. We see that

$$\frac{dI}{d\Omega} = \frac{1}{d^2} \frac{d\sigma(\theta_1)}{d\Omega} \frac{d\sigma(\theta_2)}{d\Omega} ,$$

only for p_1 and $p_2 \ll \theta_1 d$ and $\theta_2 d$, i.e., only in the case where the angles of the first and second scattering are sufficiently large. This condition can obviously not be met if one of the two scatterings occurs at a small angle. To obtain more detailed information concerning the peak heights, one has to derive the function $p_1(\theta)$ by means of eqs. (24) and (28) of chapter 1, and substitute the result in eq. (28).

Figure 14 presents on a semilog scale the scattering angle dependence of the probabilities of single and double scattering obtained with a computer (Parilis et al. 1975). To facilitate our analysis, we will follow also the $\theta(p_1)$ plot (see fig. 7b). We readily see that the left-hand part of this relation extending up to $\theta_{min} = 25°$ is due to quasisingle scattering from the first atom, the curve becoming steeper with decreasing θ. In fig. 14, to this branch corresponds the section of the curve labelled 1–2 (I_1). Just as the differential single-scattering cross section (chain curve), this branch rises with decreasing θ. However, at $\theta_{min} = 25°$, curve 1–2 undergoes a discontinuity (here, $d\theta/dp_1 = 0$), after which it reverses its course (curve 3–4). Curve 3–4 on the $\theta(p_1)$ diagram corresponds to the middle rising branch due to double scattering, the $\theta(p_1)$ relation reversing the sign. Curve 3–4 corresponding to double scattering (I_2) extends to the maximum scattering angle $\theta_{max} = 96°$, after which it suffers a discontinuity (here also $d\theta/dp_1 = 0$), and curve 5–6 due to quasisingle scattering from the second atom (I_1') begins. Curve 5–6 in the two-atom model extends down to $\theta = 0$. The $\theta(p_1)$ relation corresponds to the right-hand dropping part of the curve.

An analysis of the general pattern of the relationship reveals that the probability of double scattering at a given angle θ exceeds that of single scattering to the same angle near the points where the $\theta(p_1)$ relation undergoes a turn, namely, at $\theta < 26°$ and at $\theta > 72°$. Accordingly, the slope of the double-scattering branch in these regions of θ exceeds that of the quasisingle scattering.

The approximation of independent scattering by a pair of atoms disre-

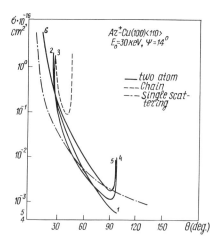

Fig. 14. Scattering angle dependence of the single- and double-scattering probabilities: curve 1–2 denotes quasisingle scattering from the first atom (I_1); curve 3–4 double scattering (I_2); and curve 5–6 quasisingle scattering from the second atom (I_i').

gards the nonuniformity in flux of ions scattered by a first atom near a second atom. If the number of ions entering the analyzer after being scattered from the first atom through an angle θ is $I_1 = I_0(d\sigma/d\Omega)(\theta, E_0)$, where I_0 is the primary ion beam intensity, and $d\sigma/d\Omega$ is the differential scattering cross section in LS; the number of ions entering the analyzer after double scattering is $I_2 = I(\theta_1, E_0, d)(d\sigma(\theta_2, E_1)/d\Omega)\,d\Omega$, where E_1 is the energy of the ions scattered from the first atom at an angle θ_1, and $I(\theta_1, E_0, d) = I_0(d\sigma/d\Omega)(\theta_1, E_0)/d^2$ is the flux intensity of ions scattered from the first atom onto the second atom, then for the ratio $R = I_2/I_1$ we will have (Mashkova and Molchanov 1980)

$$R = \frac{\dfrac{d\sigma}{d\Omega}(\theta_1, E_0)\,\dfrac{d\sigma}{d\Omega}(\theta_2, E_1)}{d^2\,\dfrac{d\sigma}{d\Omega}(\theta, E_0)}. \tag{29}$$

Using Firsov's potential in the small-angle approximation for symmetric scattering in the beam incidence plane ($\theta \approx \theta_1 + \theta_2$), eq. (29) yields (Mashkova and Molchanov 1966, 1980, Shulga 1990a)

$$R = \frac{I_2}{I_1} \sim \left(\frac{Z_1 Z_2}{E_0}\right)\left(\frac{1}{d^2}\right)\left(\frac{1}{\theta^3}\right). \tag{30}$$

Earlier it has been shown, however, that there is no purely single scattering (for sufficiently small ψ and θ) in the two-atom model; actually, it is quasisingle, i.e., represents a combination of large and small (θ_1' and θ_2') scattering angles, with $\theta = \theta_1' + \theta_2'$. It thus follows that eq. (29) should be replaced by the relation

$$R = \frac{\dfrac{d\sigma}{d\Omega}(\theta_1 E_0)\,\dfrac{d\sigma}{d\Omega}(\theta_2 E_1)}{\dfrac{d\sigma}{d\Omega}(\theta_1')\,\dfrac{d\sigma}{d\Omega}(\theta_2')d^2}, \tag{31}$$

where θ_1' and θ_2' are the first and second angle in quasisingle scattering.

In our computer simulation the differential single- and double-scattering cross sections were obtained with eq. (23) of chapter 2.

Figure 15 presents scattering-angle dependences of R for the Ar^+–Cu pair calculated using eqs. (30) and (31) (the broken and the chain curve, respectively) and a computer simulation with the two-atom (full curves for different d) and atomic row (dotted curve) models (Parilis et al. 1975). In all cases the quantity R is seen to decrease with increasing d. Calculations using eq. (31) yield very small (10^{-2}–10^{-4}) values for R, differing from the values obtained by eq. (30) and by computer simulation.

The difference between the values of $R(\theta)$ derived by eq. (30) and by

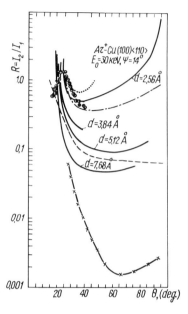

Fig. 15. $R = I_2/I_1$ versus θ plots calculated by eqs. (30) and (31) (broken and chain curves, respectively); full curves are from a computer simulation based on the two-atom model; dotted curve based on the atom row model; circles are from experiment.

computer simulation in the region of small angles θ (25–40°) is small, however, it grows with increasing θ. The experimental values of R (circles) lie closer to the curve obtained by simulation. One could not expect them to be in full agreement since the computations were carried out for all scattered particles whereas in experiment one detects the charged component only.

Figure 16 presents in the same notation the dependence of R on the energy E_0. Also shown are the experimental data obtained by Mashkova and Molchanov (1980). While the calculation using eq. (30) (curve 2) yields a qualitatively correct $R(E_0)$ relation, the values of R disagree with experiment. One of the reasons for this discrepancy lies in that the expressions for the differential cross sections of scattering from a parallel beam cannot be applied to intermediate collisions in multiple scattering.

Experimental studies were made on the dependence of the relative double-scattering intensity R on the atomic number of the ion and target atom (Mashkova and Molchanov 1967), scattering angle θ and interatomic separation d (U.A. Arifov and Aliev 1974, Mashkova and Molchanov 1967, 1975), initial energy E_0 (Van der Weg and Bierman 1968, Molchanov et al. 1969, Algra et al. 1982a), angle of incidence ψ (Mashkova et al. 1966,

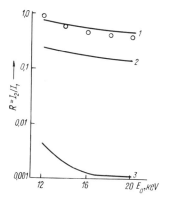

Fig. 16. R versus E_0 plots. Curve 1 is obtained using a computer simulation; curve 2 calculation using eq. (30); curve 3 calculation using eq. (31); circles experimental values.

Mashkova and Molchanov 1972c), and target temperature (Chicherov 1968, Mashkova et al. 1967, Evdokimov et al. 1967a).

Algra et al. (1982a) measured experimentally for Ne$^+$, Ar$^+$ and K$^+$ scattered from Cu (100) the dependences of the relative single-scattering intensity, I_1/I_2, on E_0 and θ, both for the total number of scattered particles, $R = I_1/I_2$, and for the ionic component, $R^+ = R\eta_1^+/\eta_2^+$, where η_1^+ and η_2^+ are the fractions of charged particles undergoing single and double scattering, respectively. The experimental data were compared with calculations based on the two-atom model using the Molière potential and a semianalytic three-atom model with the same potential (Poelsema et al. 1977b). Note that Algra et al. (1982a) denote by R the single-to-double scattering intensity ratio, i.e., R^{-1} in our nomenclature. By Algra et al. (1982b), for alkali-metal ions $\eta_1^+ = \eta_2^+$ and, hence, $R^+ = R$. This, however, does not apply to noble gases (Luitjens et al. 1980a) where $\eta_2^+ < \eta_1^+$. It is very essential that Luitjens et al. (1980a) measured with the TOF technique the dependences of R^{-1} on E_0 and θ both for the total scattered flux (R) of Ne, Ar and K, and for its ionic component (R^+). This made it possible to compare the results obtained with calculations. One may write for the quantity R^{-1} for the total scattered flux

$$R^{-1} = \frac{I_1}{I_2} = \frac{\sigma(\theta, E_0)}{\sigma(\theta_1, E_0)\sigma(\theta_2, E_1)}\, d^2 . \tag{32}$$

As follows from eq. (32), the ratio R^{-1} for two directions on a surface with different interatomic distances d_1 and d_2 is

$$R_1^{-1}/R_2^{-1} = d_1^2/d_2^2 . \tag{33}$$

The measured values of R^+ are much larger than the corresponding values of R because of the charge exchange and neutralization processes occurring in the interaction of an ion with the surface (Van Zoest et al. 1984). As already pointed out, in the case of Ne^+ and Ar^+ ions scattered from copper the values for η_2^+ are much smaller than those for η_1^+, and this accounts for the large values of R^{-1}. Apart from this, η_2^+ and η_1^+ exhibit different dependences on E_0. For argon, η_2^+ grows with E_0 faster than η_1^+ does (Luitjens et al. 1980a), which manifests itself in a decrease of R^+ with increasing E_0, whereas in the case of Ne, conversely, R^+ grows with increasing E_0 (Luitjens et al. 1980b). The experimental and calculated values of R^{-1} for argon are approximately the same as those for potassium, since the scattering cross sections of ^{39}K and ^{40}Ar on copper are about equal. The values of $(R^+)^{-1}$ for potassium calculated in terms of the two-atom model as a function of E_0 for $\theta = 60°$ and $\psi = \frac{1}{2}\theta$ correlate well with one another, except for low energies ($E_0 < 5$ keV). As E_0 decreases, the screening of the neighbor atoms can no more be neglected, thus making the two-atom model inapplicable. On the other hand, the probability of the so-called zigzag collisions with atoms located in parallel rows on the surface grows with decreasing E_0. While for low E_0 the zigzag collisions have substantially greater cross sections, and the ions involved in them have a final energy close to E_1, the values of R^{-1} become greater than those calculated by eq. (32).

An analysis of the data obtained by various authors reveals a qualitative agreement between the experimental dependences of R^{-1} on E_0, θ and d, and expression (30), although quantitatively there are discrepancies.

7. Double scattering in nuclei

There is a far-reaching analogy between the scattering of heavy atoms from a two-atom molecule and double scattering of nucleons from a deuterium atom (Allaby et al. 1969) which originates from a similarity in the short-range nature between the internuclear and interatomic potentials, and from close relations between the corresponding scattering lengths and dimensions of the deuteron and a molecule.

When the de Broglie wavelength λ of a particle involved in an interaction is small enough compared with the mean separation between nucleons in a nucleus, $l \sim 10^{-13}$ cm, the situation becomes quasiclassical, thus justifying the approximations of particle trajectories and two-particle interactions inside a nucleus. For this to be true, the energy of the primary particle should exceed a few tens of MeV.

For 15–26 GeV protons backscattered at small angles from carbon, tantalum, and beryllium, Karplus and Yamaguchi (1967) discovered a structure in the energy spectrum near the upper boundary, which they attributed to single collisions with one of the target nucleons. They also studied the possibility of scattering in two sequential noncorrelated collisions with nucleons in a nucleus. The calculations showed that a single scattering at a given angle may be less probable than two sequential scatterings, each to an angle one half of the former. It turned out that multiple scattering dominates over single quasielastic interactions in the momentum transfer range 0.75–1.5 GeV/s. For lower momentum transfers, multiple scattering becomes less probable. Moreover, for large nuclei triple collisions become noticeable in the range 1.3–1.7 GeV/s, which results in a broadening of the momentum loss spectrum.

Cocconi et al. (1962) were the first to separate experimentally the single- and double quasielastic scattering peaks in the momentum transfer spectrum. A detailed study of double proton scattering from deuterium at 19.2 GeV/s was later carried out on a proton synchrotron (Allaby et al. 1969). The experiments showed the proton momentum spectra to have a two-peak structure (fig. 17). An increase of the scattering angle was found to produce a better peak separation and an increase in their relative intensity. The position of the low-momentum peak corresponds to proton–proton scattering in a single collision for which one can write in LS

$$(\Delta P)_1 \approx \frac{q^2}{2M} = \frac{P^2\theta^2}{2M} , \qquad (34)$$

where q is the momentum transfer, P and θ are the scattered proton

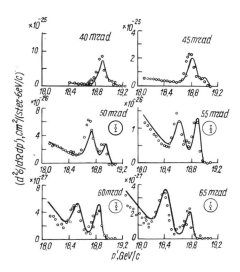

Fig. 17. Momentum spectra of protons backscattered from deuterium at angles of 40–65 mrad for an initial momentum of 19.2 GeV/s (Allaby et al. 1969).

momentum and angle, respectively, and M is the nucleon mass. From the standpoint of kinematics, the greater momentum peak corresponds to elastic scattering of a proton from deuterium involving a momentum loss

$$(\Delta P)_2 = \frac{P^2\theta^2}{2M_d} = \frac{P^2\theta^2}{4M} = \frac{1}{2}(\Delta P)_1 , \tag{35}$$

where M_d is the deuteron mass.

Using qualitative arguments, the authors succeeded in explaining the appearance of a sharp peak in the momentum loss spectrum at $\frac{1}{2}(\Delta P)_1$. The deuterium target was chosen because of the weak binding energy of the proton and neutron in the deuteron, as well as because the nuclear effects in fast-particle scattering can be satisfactorily explained by means of Glauber's theory (Glauber 1971). By this theory, scattering from deuterium represents a superposition of direct scattering on one of the nucleons and of a process in which the projectile interacts sequentially with both nucleons. The structure of deuterium (primarily the mean separation between the proton and the neutron) affects substantially the phenomena associated with double rescattering. In Glauber's theory, the total cross section of interaction of a particle with the deuteron can be written as (Kolybasov and Marinov 1973):

$$\sigma_d = \sigma_p + \sigma_n - \Delta , \tag{36}$$

where Δ is the cross section defect (a quantity which is positive at fairly high energies and corresponds to the concept of nucleon screening).

In a rough approximation, Δ can be presented in the form (Glauber 1955)

$$\Delta = \frac{1}{4\pi R^2} \sigma_p \sigma_n , \tag{37}$$

where R is the distance between the proton and the neutron.

In Glauber's theory, the total cross section is determined by separating the cross section mathematically into three parts, each of them depending on the amplitude of elastic scattering from a nucleon and on the deuteron wave function,

$$\frac{d\sigma}{d\Omega} = \frac{d\sigma^{(1)}}{d\Omega} + \frac{d\sigma^{(2)}}{d\Omega} + \frac{d\sigma^{(1,2)}}{d\Omega} . \tag{38}$$

The first term describes the contribution of single scattering, the second that of double collisions, and the third that of their interference. To determine the double-scattering contribution, the angular distributions were measured and compared with theoretical predictions (Kolybasov and Marinov 1973). It was found that for momentum transfers in excess of 0.6–0.7 GeV/s, the differential cross section is totally determined by double scattering.

8. Ion scattering from a diatomic molecule

Double scattering of fast ions from neighbor atoms turned out to be a convenient, while simplified, description of scattering from a surface of a solid. However, in scattering from solids the characteristic double-peak structure of the energy spectrum of the ions scattered at a given angle is invariably observed against the background of multiple scattering. In scattering from a diatomic molecule this background is excluded, so that the double-scattering structure can be observed in its pure form.

Double scattering of fast atoms from a molecule has not yet been observed experimentally; one has not undertaken by attempts to do this, although theoretical considerations of a general nature have already been put forward (Parilis 1965) and the probability of the process has been directly evaluated (Mukhamedov and Parilis 1981, 1975, Parilis 1969c). At the same time, the structure of the energy spectrum of heavy ions scattered from a molecule is very convenient for experimental investigation and should be quite diverse because of the rich multiplicity of the projectile and target atomic mass ratios, as well as of the energy losses expended in electron shell excitation. It may serve, e.g., to study electron processes in 'carambole' collisions (Foster and Saris 1973).

Mukhamedov and Parilis (1982) studied theoretically the double scattering of ions of medium mass and energy from a diatomic molecule and carried out a computer simulation for the cases Ar^+-I_2 and Ar^+-Cl_2 differing in the mass ratio μ.

Multiple scattering from the atoms of a molecule is of interest in studies of density effects, i.e., of nonadditive processes occurring in the electron shells of ions in sequential collisions in a solid and separated by intervals of time shorter than the relaxation time. In this case, one can calculate the kinematics of both collisions and relate the elements of the ion trajectory with measurables, namely, the scattering angles and energy of the scattered particles. The double-scattering trajectory represents a three-dimensional figure with the polar (θ_1 and θ_2) and azimuthal (φ_1 and φ_2) scattering angles related by eq. (11).

After double scattering from a diatomic molecule, the ion energy may be both greater and less than $E_1(\theta)$, i.e., the single-scattering energy, as well as equal to it. The maximum, $E_{2max}(\theta)$, and minimum, $E_{2min}(\theta)$, values of the double-scattered ion energy are found from eqs. (22) and (23).

The energy imparted to a molecule in its collision with a fast ion is high enough to cause its break up, the scattering being accompanied by the production of recoils. The recoil atoms can form in several ways (fig. 5 from Mukhamedov and Parilis 1982; Shulga et al. 1989, Dodonov et al. 1989). In the first case, single scattering at an angle θ results in an ejection of one recoil of energy E at an angle δ_1,

$$\delta_1 = \tfrac{1}{2}(\pi - \chi) ; \qquad E(\delta_1) = KE_0 \sin^2 \tfrac{1}{2}\chi , \qquad (39)$$

where

$$K = \frac{4m_1 m_2}{(m_1 + m_2)^2} .$$

In the second case, scattering at an angle θ is accompanied by the production of two recoils ejected at the angles

$$\delta_1 = \tfrac{1}{2}[\pi - \chi_1(\theta_1)] , \qquad \delta_2 = \tfrac{1}{2}[\pi - \chi_2(\theta_2)] , \qquad (40)$$

and with energy

$$E(\delta_1) = KE_0 \sin^2 \tfrac{1}{2}\chi_1 , \qquad E(\delta_2) = KE_1(\theta_1) \sin^2 \tfrac{1}{2}\chi_2 , \qquad (41)$$

where θ_1 and θ_2 are the angles of ion scattering from the first and second atom.

Since the energy losses suffered in sequential collisions may not be additive, an experiment can be proposed where simultaneous measurement of the energy of all the particles formed in the collision of an ion with a molecule could permit studying the inelastic processes accompanying scattering.

As already mentioned, the analytical expressions for the double-scattering cross section, eqs. (27) and (28), expressed in terms of the product of scattering cross sections for intermediate angles θ_1 and θ_2 are not well grounded, thus calling for computer simulation of the trajectories of the ions scattered from a diatomic molecule (Mukhamedov and Parilis 1981, 1982). The differential cross section for a molecule which orientation is defined by the polar, ψ, and azimuthal, ξ, angles were calculated by the expression

$$\left.\frac{\mathrm{d}\sigma}{\mathrm{d}\Omega}(\theta)\right|_{\psi,\xi} = |D \sin \theta|^{-1} , \qquad (42)$$

where D,

$$D = \left| \frac{\partial\theta}{\partial r} \frac{\partial\varphi}{\partial q} - \frac{\partial\theta}{\partial q} \frac{\partial\varphi}{\partial r} \right| ,$$

is the Jacobian for the transformation of the Cartesian coordinates of the target point r, q into the angular coordinates of the scattering direction θ, φ. In double scattering at a given angle θ, to each energy $E_2(\theta)$ corresponds a set of trajectories forming a cone in the direction space θ_1, φ_1. Each such cone can be identified in a single-valued way with a cone in the ψ, ξ space. Therefore, in order to calculate the energy distribution, one has to carry out integrations over equal energy cones in the ψ, ξ space of molecule orientations,

$$\frac{d^2\sigma}{d\Omega \, dE_2}(\theta, E_2) = \frac{1}{2\pi} \int_0^{\xi_{max}} \frac{d\sigma}{d\Omega}(\theta, \psi) \sin \psi \, \frac{d\psi}{dE_2} \, d\xi , \tag{43}$$

where ξ_{max} is the maximum value of ξ on the $E_2(\psi, \xi) = $ const cone.

Since eq. (43) determines the relative probability of scattering at an angle θ with energy E_2, the shape of the double spectrum should depend also on the probability distribution of molecule orientations in ψ, ξ space. However, for a free molecule, all its orientations are equally probable. The differential cross section of scattering at a given angle θ_{min} is obtained by integration in energy $E_2(\theta)$ over the interval $E_{2min}(\theta) \leqslant E_2 \leqslant E_{2max}(\theta)$,

$$\frac{d\sigma}{d\Omega}(\theta) = \int_{E_{2min}(\theta)}^{E_{2max}(\theta)} \frac{d^2\sigma}{d\Omega \, dE_2}(\theta, E_2) \, dE_2 . \tag{44}$$

Note that because of the complexity of the expressions relating to the trajectory parameters $\theta_1, \theta_2, \varphi_1, p_1, p_2$ with the molecule orientation ψ, ξ, no analytic solution of the Jacobian in the integrand has yet been obtained. Therefore, in order to derive the energy and angular distributions, one has to carry out either a numerical calculation of the Jacobian or a computer simulation of a large number of trajectories of scattered particles.

Figure 18a shows the results of a computer simulation for the relation between the angles θ_1 and θ_2 as a function of grazing angle ψ for double scattering of 30 keV Ar$^+$ ions from a molecule of iodine at an angle $\theta = 30°$. Depending on the actual conditions of collisions, doubly scattered particles can be divided into particles with grazing and penetrating (i.e., traversing the line connecting the atoms) trajectories which contribute to the energy region $E_{2min} \leqslant E \leqslant E_{2max}$. The $E_2/E_0(\psi)$ dependence for these trajectories consists of two separate curves lying, respectively, to the right and left of the $E_2 = E_1(\theta)$ point (fig. 18b).

The energy spectrum of the atoms scattered from a diatomic molecule consists of an intense peak whose position on the energy scale corresponds to the single-scattering energy $E_1(\theta)$, and of a small high-energy double-scattering peak (fig. 18c), separated by 0.6 keV from the main peak. The area under the double-scattering peak is $\approx 2\%$ of that bounded by the main peak. The separation of the double- from the single-scattering peak and its intensity are sufficient to ensure reliable identification and experimental study.

To explain the shape of the spectrum, one has to analyze the behavior of the integrands $(d\sigma/d\Omega)(\theta, E_2)$ and $(d\psi/dE_2)(\theta, E_2)$ in eq. (43) which represent the differential double-scattering cross section and orientational energy density function, respectively, within the range of E_2 energies occupied by atoms scattered at an angle θ.

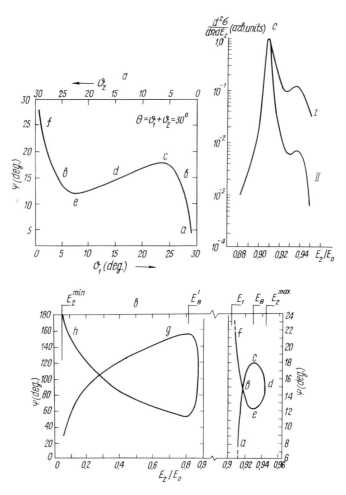

Fig. 18. Dependence of (a) contributions of θ_1, θ_2 angles and (b) E_2/E_0 ratio on grazing angle ψ, and (c) energy distribution of Ar atoms scattered from the two-atom molecule I_2. I denotes plane case; II the spatial case; a–b–c quasisingle scattering from the first atom of the molecule; e–b–f quasisingle scattering from the second atom; c–d–e double scattering; g and h double scattering sections for penetrating trajectories.

As seen from fig. 19a, the differential double-scattering cross section $(d\sigma/d\Omega)(\theta, E_2)$ on each cone $E_2(\theta, \psi, \xi) = \text{const}$ falls off dramatically with increasing azimuthal angle ξ. This means that the largest contribution to double scattering comes from trajectories passing near the analyzer plane ($\xi = 0°$). One can perform a qualitative analysis of the mechanism of energy spectrum formation within the plane case ($\varphi_1 = \xi = 0°$). A comparison of the energy distribution of Ar$^+$ ions scattered from I_2 molecules in the plane (I)

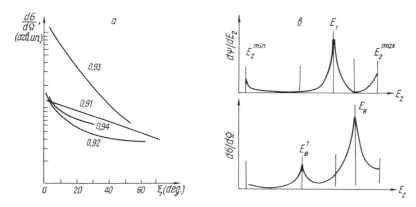

Fig. 19. (a) Double-scattering cross section versus azimuthal angle of the two-atom molecule axis and (b) behavior of the integrands $d\psi/dE$ and $d\sigma/d\Omega$ in the energy range studied.

and spatial (II) cases shows that inclusion of the spatial factor results in a sharp decrease of the relative height of the double-scattering peak, its shape being affected only insignificantly (fig. 18c). At two extreme points of the energy scale, $E_{2min}(\theta)$, $E_{2max}(\theta)$, the derivative $d\psi/dE_2$ tends to infinity (fig. 19b). However, due to the fact that for $E_2(\theta) \rightarrow E_{2max}(\theta)$ the solid angle $d\Omega_{\psi,\xi}$ of the cone vanishes at a faster rate, these two infinities do not produce high peaks at the extreme points of the distribution function. As $E_2(\theta) \rightarrow E_1(\theta)$, i.e., to the energy of single scattering, the derivative $d\psi/dE_2$ also tends to infinity at a rate depending on the actual interaction potential chosen. The behavior of the derivative close to $E_2 \approx E_1(\theta)$ accounts for the appearance of the quasisingle scattering peak, since the corresponding trajectories which are close in energy to $E_1(\theta)$ have a much greater statistical weight.

The differential cross section $(d\sigma/d\Omega)(\theta, E_2)$ is also singular at the two points of the energy scale (fig. 19b) corresponding to the molecule orientations (ψ, ξ) at which the derivative $d\theta/dp_1$, entering implicitly the Jacobian, eq. (42), and, hence, the Jacobian itself, vanish. On the other hand, it is at these orientations that the derivative $d\psi/dE_2$ is zero (fig. 19b). Nevertheless, calculations revealed the possible appearance of a peak at the energies $E_2(\theta) = E_B$ (where E_B is the energy of the atoms that underwent double scattering from molecules in ψ_B^{min} and ψ_B^{max} orientations), since near E_B the differential cross section tends to infinity at a faster rate than the derivative $d\psi/dE_2$ vanishes.

Thus, the appearance of the high-energy peak in the energy distribution originates from double scattering at the p_1 and ψ values corresponding to the rainbow effect, which influence on the shape of the spectrum continues to be felt also after averaging over all possible molecule orientations. A comparison of the energy distributions of Ar atoms scattered from I_2 and Cl_2 molecules at different angles θ reveals (fig. 20) that the shape of the spectrum depends both on the scattering angle and the mass ratio μ. An increase of θ and a decrease of μ bring about the same result, namely, the spectra shift as a whole towards lower energies, the relative height of the double-scattering peak decreasing, and the peak separation increasing.

It is interesting to compare the differential cross section of scattering from a two-atom molecule with that from a single atom (Mukhamedov and Parilis 1982). For certain orientations of the molecule relative to the beam direction, one of its atoms can be shadowed by the other atom. As a result, the differential single-scattering cross section for a molecule will decrease compared with that for a single atom, in a first approximation, by an amount $\omega_s/4\pi$ becoming

$$\frac{d\sigma}{d\Omega}(\theta) = 2 \frac{d\sigma}{d\Omega}(\theta)_{at}\left(1 - \frac{\omega_s}{4\pi}\right),\tag{45}$$

where ω_s,

$$\omega_s = \frac{\pi R_s^2(E_0)}{d^2},$$

is the solid angle of the shadow cone, and R_s is the shadow radius at distance d from the atom.

As follows from eq. (45), the decrease of the differential single-scattering cross section caused by shadowing in this approximation does not depend on the scattering angle θ. Therefore, the integral scattering cross section also reduces by a factor $(1 - \omega_s/4\pi)$. Indeed, when double scattering is included, the ratio $(d\sigma/d\Omega)/2(d\sigma/d\Omega)_{at}$ depends on the scattering angle θ.

The quantity $(d\sigma/d\Omega)(\theta)$ was determined by numerical integration over

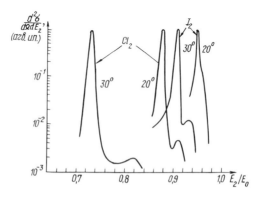

Fig. 20. Energy spectra of Ar atoms scattered from I_2 and Cl_2 molecules for different θ angles.

all molecule orientations,

$$\frac{d\sigma}{d\Omega}(\theta) = \frac{1}{4\pi} \int\limits_0^{2\pi} \int\limits_0^{\pi} \frac{d\sigma}{d\Omega}(\theta, \psi, \xi) \sin \psi \, d\psi \, d\xi .$$ (46)

As shown by calculation (fig. 21a), the scattering cross section for a molecule for large angles θ is small compared with $2(d\sigma/d\Omega)_{at}$. This can be accounted for by the decrease in the differential single-scattering cross section for a molecule being substantially greater than the double-scattering contribution due to shadowing. However, as θ decreases, the double-scattering probability starts to grow, so that for certain θ the ratio $(d\sigma/d\Omega)_M/2(d\sigma/d\Omega)_{at}$ becomes greater than unity (in the case considered, for $\theta \leqslant 35°$). A particular role in the formation of the double-scattered component is played by the trajectories close to the values of p_1, ψ at which the Jacobian, eq. (42), becomes a singularity. To reveal the role of this effect in the angular distribution, we analyzed the dependence of the double-scattering cross section on θ for each of the scattering branches. Figure 21b presents two of the differential scattering cross section contributions for a molecule: $(d\sigma/d\Omega)(\theta)_{II}$, the cross section for glancing double scattering (II), and $(d\sigma/d\Omega)(\theta)_{III}$, cross section for scattering with penetration (III).

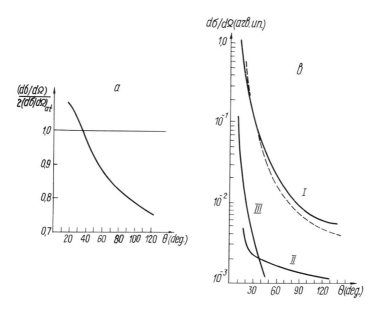

Fig. 21. (a) Relation between double scattering from a molecule and doubled atomic cross section for different θ, and (b) differential cross section versus θ plot (broken line); I denotes doubled atomic cross section; II grazing double scattering; III penetrating scattering.

As seen from the figure, the first dominates at large, and the second, at small angles.

It is known that the double scattering due to the first mechanism can be observed only for angles $\theta \leqslant \theta^*_{min}$. The cross section of this component of double scattering will be the largest at $\theta = \theta^*_{min}$, with only the second double-scattering component remaining for $\theta < \theta^*_{min}$. Apparently, if one chooses conditions excluding the possibility of double scattering involving 'penetrating' trajectories (e.g., by reducing E_0 until the penetrating trajectories have become impossible), one may expect the appearance of a halo effect for the scattering from molecules similar to that observed by Demkov (1980, 1981).

By placing scattered and recoil atom detectors at proper angles and connecting them in coincidence, one can identify the double-scattering events of interest here, the coincidence technique permitting reconstruction of the trajectory of each defected particle (Mukhamedov and Parilis 1981, 1982). While the calculation of double scattering of atoms from molecules is interesting in itself, it can also be used to advantage as a model of atom scattering from polycrystals.

9. Double scattering from polycrystals

The experimental energy distributions of medium-energy heavy ions scattered from the surface of a polycrystal differ from those observed in single crystals, namely, these spectra exhibit a clearly pronounced single-scattering peak and a shoulder on the high-energy side of the peak (U.A. Arifov 1968, Mashkova and Molchanov 1985, Datz and Snoek 1964, Molchanov and Soszka 1964, Sharma and Buck 1975, Sharma 1975). The presence of this shoulder is attributed to the effect of multiple collisions. Datz and Snoek (1964), who revealed the shoulder to the right of the single-scattering peak in the spectrum of the ions scattered from a polycrystalline target interpreted it in terms of the double-scattering model. Initially, it is for polycrystals that the double-scattering model was developed (Parilis 1965), and only later it was shown that in the case of single crystals the ordered atomic arrangement results in the appearance of a separate maximum (Mashkova et al. 1965, Parilis and Turaev 1964).

Experiments on the scattering of inert gas ions of energy up to 30 keV from polycrystalline copper and gold targets through large angles ($\theta > 90°$) showed, however (Sharma and Buck 1975, Sharma 1975, Baun 1978, Ball et al. 1972), that the shoulder in the high-energy part of the distribution can transform into a distinct peak (fig. 22a). The presence of this peak was associated with the effect of double scattering in polycrystals (Mashkova and Molchanov 1985, Mukhamedov et al. 1976). Sharma and Buck (1975),

Sharma (1975), Balashova et al. (1977a,b, 1978) carried out systematic studies of double scattering from polycrystals. The energy spectra of Ne^+, Ar^+ and Kr^+ ions of energy $E_0 = 4$–30 keV scattered at angles $\theta = 90$–$150°$ from polycrystalline gold and copper targets, and estimates made basing on the single- and double-scattering model showed (Sharma and Buck 1975) the width and relative height of the high-energy shoulder to depend on the ion–target atom mass ratio and scattering angle. The width of the shoulder was found to decrease in all cases with increasing scattering angle and pass through a maximum at $\theta = 90$–$110°$. This is in accord with the theoretical estimate (Mukhamedov et al. 1976), where for the shoulder width one takes the distance from the main peak to the point of maximum energy in double scattering. However, the ratio of the intensity at the double-scattering shoulder to that of the main peak did not agree with the value calculated by eq. (30).

Balashova et al. (1977a, 1977b, 1978) measured the energy distributions of Ar^+ ions scattered from polycrystalline copper. A clearly pronounced double-peak structure depending on the angles ψ and θ was observed in certain intervals of E_0 (fig. 22b) and shifted towards lower E_0 with increasing scattering angle. Terzic et al. (1976, 1979a, 1979b) studied the effect of double scattering from polycrystalline silver for low-energy (0.4–3.2 keV) sodium ions. The major fraction of the reflected beam resides in this case in the ionized state. The measured energy distributions of Na^+ ions scattered at angles $\theta = 50$–$130°$ for $\psi = 45°$ and $E_0 = 0.4$–3.2 keV also revealed an intense hump in the high-energy part which, for certain E_0 and θ, transformed into a peak. The analytical method of ion-scattering calculations based on the two-atomic model which was developed by Parilis (1965) and Parilis and Turaev (1964) was used by Sharma (1975) to analyze the effect of double scattering from a polycrystal. To explain the shape of the energy spectrum, integration over the cones, eq. (11), was performed in an explicit way, and the calculated intensities were compared with the observed spectra. The probability of single scattering in an energy interval between E and $E + \delta E$ was assumed to be equal to

$$I(E_0, E)\, \delta E = -Bp\, \mathrm{d}p = BN\lambda(E_0)\sigma(\theta) \sin\theta\, \frac{\mathrm{d}\theta}{\mathrm{d}E}\, \delta E , \tag{47}$$

where $\lambda(E_0)$ is the effective range of the incident ion, N is the number of scattering centers in $1\,\mathrm{cm}^3$ and B is a constant. The double-scattering probability was evaluated by the expression

$$I_2(E_0, E_2)\, \delta E_2 = B^2 N^2 \lambda(E_0)\bar{\lambda}\, \delta E_2 \int_{\theta_1} \sigma(\theta_1) \sin\theta_1\, \sigma(\theta_2) \frac{\sqrt{\mu^2 - \sin^2\theta_2}}{2E_2}\, \mathrm{d}\theta_1 , \tag{48}$$

where $\bar{\lambda} = \lambda(E_1)$ is the mean effective range.

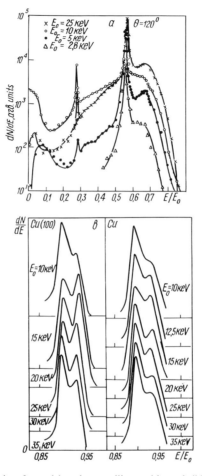

Fig. 22. Double scattering from (a) polycrystalline gold, and (b) polycrystalline and single crystal copper for incident Ar ions.

From the energy spectra shown in fig. 23a, for different θ, it is seen that the high-energy branches of the calculated spectra (broken lines) fall off in all cases faster than the experimental ones do. It was suggested (Sharma 1975) that this discrepancy comes from an inaccuracy of the cross section calculated with the Bohr potential. It is more probable, however, that the reason for this disagreement lies in inadequacy of relation (28), which is based on a description of two sequential collisions as independent scattering events. As already mentioned, relation (28) indeed cannot be considered as well grounded, since, as follows from the definition of the scattering cross section (Landau and Lifshitz 1960), the scattering center should intercept a

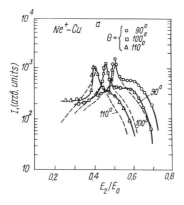

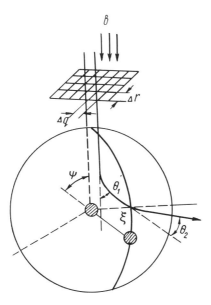

Fig. 23. (a) Energy spectra of Ne ions backscattered from polycrystalline copper, and (b) two-atom ion scattering from a polycrystalline surface.

parallel beam whereas after the first collision the trajectory asymptotes become in a general case intersecting straight lines.

Thus, the analytical expressions for the relative double-scattering intensity, eqs. (27) and (30), regrettably do not permit a reliable calculation of the structure of the energy spectrum. Therefore, a comprehensive calculation of a scattered ion energy spectrum requires a computer simulation of all possible trajectories for various mutual orientations of the ion beam relative to the axis connecting the two atoms (Umarov 1975, Mukhamedov et al.

1976). Mosunov et al. (1978) carried out a computer simulation of the reflection of Na^+ ions from polycrystalline silver using three volume models of the polycrystal. The calculated energy spectrum consisted of the principal peak and a less pronounced high-energy peak to the right of it. A trajectory analysis showed the high-energy peak to be due to double scattering. However, the extent to which the calculations were in agreement with experiment (Tersic et al. 1976, 1979a, 1979b) was found to be about the same for all models, making it purely qualitative. Mukhamedov et al. (1976, 1979) studied double scattering from a polycrystal based on a model of a free-rotating two-atom pair, all its positions being assumed equally probable in a first approximation (fig. 23b). An essential point in the calculation was the choice of the number of coordination spheres in the bulk of the target within which the scattering is considered to occur. An evaluation of the shadow cone radius by expression (15) revealed at $E_0 = 30$ keV about 20% of the area of the second coordination sphere, the total shadowed area on the surface of the third sphere being 40%. An increase of the shadowed area makes direct collision of single-scattered particles with atoms of the sub-sequent screened spheres less probable. Mukhamedov et al. (1979) took this point into account when they considered double scattering within only the first two coordination spheres with the radii $R_1 = d_{\langle 110 \rangle} = 2.56$ Å, and $R_2 = d_{\langle 100 \rangle} = 3.62$ Å, respectively. For the two-atomic model, there are, in general, no forbidden regions in the orientation (ψ, ξ) or scattering direction (θ, φ) spaces. This means that for any arbitrarily chosen direction (θ, φ), one can find a corresponding position of the two-atom pair providing single or double scattering. However, as was shown earlier, the mutual screening action of the neighbor atoms places a constraint on the scattered flux on both the large and small sides of the escape angle. One readily sees a similar limitation to exist also on the grazing angles, namely, each fixed scattering angle θ can become active only within a certain orientation interval, $\psi_{min}(\theta) < \psi < \psi_{max}(\theta)$ (fig. 23b). For not too small scattering angles, the dependence of the limiting angles on θ for the potential C/r^2 can be written as

$$\psi_{min}(\theta) = \arcsin\left[\frac{p'_{lim} - p(\theta)}{d_i} \right], \qquad \psi_{max}(\theta) = \arcsin\left[\frac{p''_{lim} - p(\theta)}{d_i} \right],$$

(49)

where

$$p'_{lim} = \frac{C}{E_0} f(\alpha^*); \qquad p''_{lim} = \frac{C}{E_1(\theta)} f(\alpha^*); \qquad f(\alpha^*) = \frac{\pi - \alpha^*}{(2\pi - \alpha^*)\alpha^*};$$

α^* is the limiting angle of additional deflection from the atoms in the row closest to the pair under consideration, which does not result in a substantial distortion of the particle trajectories. The agreement with experiment

(Mashkova and Molchanov 1980) and with the computer simulations of scattering from an atomic row (Kivilis et al. 1970, Nizhnaya et al. 1979a,b,c,d) turns out to be good if $\alpha^* \approx 1°$. Thus, the screening action of an atomic row limiting the region of existence of double scattering can be taken into account by reducing the interval of the two-atom pair orientations over which integration is performed.

We have until now been discussing double scattering on the surface. However, medium-energy projectiles are capable of penetrating into the bulk of the solid. Therefore, double collisions with atoms in the bulk of the target should likewise be included in the analysis. To take into consideration the attenuation of the flux of atoms which are double-scattered in the bulk, U.A. Arifov (1968) and Parilis (1965) proposed to introduce factors of the type $\exp(-l/\lambda)$, where l is the particle path in the material, and λ is the mean free path. Let α^* be the minimum angle of additional deflection above which the particle may be considered as lost from the beam. Then the mean free path to such a collision can be evaluated as

$$\lambda(\alpha^*) = \frac{1}{n_0 \int\limits_{\alpha^*}^{\pi} \sigma(\theta)\, d\Omega} , \tag{50}$$

where n_0 is the material's density.

Using the expression for the differential scattering cross section with a power-law potential, eq. (13) of chapter 1,

$$\sigma(\theta) = \frac{B}{\sin^3 \frac{1}{2}\theta} , \tag{51}$$

we obtain

$$\lambda^{-1} = n_0 \int\limits_{\alpha^*}^{\pi} \sigma(\theta)\, d\Omega = 2\pi n_0 \int\limits_{\alpha^*}^{\pi} \frac{B}{\sin^3 \frac{1}{2}\theta} \sin\theta\, d\theta = 8\pi n_0 K , \tag{52}$$

where

$$B = 4.1\, e^2 a_{TF} Z_1 Z_2 \frac{m_1 + m_2}{32\pi m_2 E_0} ; \qquad K = \frac{1}{\sin \frac{1}{2}\alpha^*} - 1 .$$

Assuming the flux attenuation along a path l to follow an exponential law, $P(l) = \exp(-l/\lambda)$, the total attenuation of the flux of double-scattered particles can be determined in terms of the product of attenuations along each rectilinear segment of the trajectory between collisions,

$$P(l) = \exp\left[-\left(\frac{l_1}{\lambda_1} + \frac{l_2}{\lambda_2} + \frac{l_3}{\lambda_3} \right) \right]. \tag{53}$$

In the case of scattering from a pair of atoms at fixed angles ψ and θ,

$$l_1 = x \sin \psi \; ; \qquad l_2 = D \; ; \qquad l_3 = x \sin(\theta - \psi) \; , \tag{54}$$

where x is the distance from the surface to the point in the bulk where the first hard-sphere collision took place.

Thus, the intensity of double scattering in the bulk of the target is given by the expression

$$\left(\frac{d^2\sigma}{d\Omega \, dE} \right)_{\!\! \Sigma} (\theta, E_2) = \int_0^{x_{max}} n_0^{1/3} \frac{d^2\sigma}{d\Omega \, dE_2} (\theta, E_2, x) P(E_2, x) \, dx \; , \tag{55}$$

x_{max} is the maximum penetration depth from which a scattered particle still can escape outward. Integral (55) can be readily calculated if one disregards the continuous energy losses and assumes the distribution in depth to be isotropic. Then

$$\left(\frac{d^2\sigma}{d\Omega \, dE_2} \right)_{\!\! \Sigma} (\theta, E_2) = n_0^{1/3} \frac{d^2\sigma}{d\Omega \, dE_2} (\theta, E_2) \int_0^{x_{max}} \exp[-(t_1 + t_2 + t_3)x] \, dx$$

$$= \frac{n_0^{1/3}}{T} \frac{d^2\sigma}{d\Omega \, dE_2} (\theta, E_2)(1 - e^{-Tx_{max}}) \; , \tag{56}$$

where

$$T = t_1 + t_2 + t_3 \; ; \quad t_1 = \frac{\sin \alpha}{\lambda_1} \; ; \quad t_2 = \frac{D}{\lambda_2} \; ; \quad t_3 = \frac{\sin(\theta - \alpha)}{\lambda_3} \; .$$

For $E_0 = 30$ keV, $\theta = 30°$ and $\alpha = 15°$ (the angle between the beam direction and the surface), $T \cong 0.1$, i.e., the total contribution to scattering from a semiinfinite ($x_{max} = \infty$) target is equivalent to an unattenuated contribution of ten atomic layers. The major part of this contribution builds up over distances about 10–15 Å, i.e., actually a few atomic layers. This statement is in accord with the data of Pugachova (1978), who showed by simulating ion scattering from a polycrystal on a computer-created crystallite of atoms that, in the angular range $10° < \theta < 50°$, backscattering occurs primarily in the first two layers, the dominant contribution to the reflected flux coming from particles undergoing single and double collisions.

As already pointed out, the orientation cones corresponding to different energies in scattering at an angle θ are symmetric with respect to the $\frac{1}{2}\theta$ direction. Hence, their position relative to the surface can be uniquely determined if one specifies the incidence angle ψ. When the scattering takes place in the volume confined within the first two layers, a certain part of the cone, corresponding to the angles $\psi > \alpha$, will extend above the surface. For instance, if the angle of incidence is one half that of scattering, then the double-scattering cone will be submerged only for one half into the material, its upper part not contributing under these conditions. The second layer in the fcc lattice of copper is formed by atoms lying in the center of the (010) and (001) faces. To simulate scattering from this layer, the center of the coordination sphere should be positioned at a distance $\frac{1}{2}a$ from the surface. Then over a broad range of θ angles $[\theta < 2\arcsin(a/2d_i)]$ the double-scattering cones will be completely submerged into the solid. Since the ranges of the particles corresponding to different $\Omega(\theta, E_2)$ cones are different, different parts of the double-scattering spectrum may be expected to be attenuated in a different degree. Calculations showed that the behavior of the attenuation along the $E_2/E_0(\psi)$ oval for the first layer represents a double-valued function of $p(E_2)$. Figure 24 presents curves for the case of $Ar^+–Cu$, $E_0 = 30$ keV, $\theta = 30°$ and $50°$, and $\alpha = \frac{1}{2}\theta$. Curves II corresponding to scattering from the second layer are seen to pass far below curves I, thus evidencing a substantial attenuation of the reflected flux starting from the second and third layers. The degree of attenuation varies also with variation of the conditions of scattered particle detection. As the scattering angle θ and, accordingly, the grazing angle $\alpha = \frac{1}{2}\theta$ increase, the scattering takes on bulk features, and the influence of the attenuation function on the escape of the particles increases.

The computer simulation of double scattering from a polycrystal was based on the algorithm used to calculate the scattering from a two-atomic molecule and including the above-mentioned features of scattering in a solid (Mukhamedov and Parilis 1982, Mukhamedov et al. 1976). The position of the two-atom pair was specified by the angles ψ and ξ, all its orientations being considered equally probable. The resultant energy spectrum was obtained by averaging the spectra over all orientations. Mukhamedov et al.

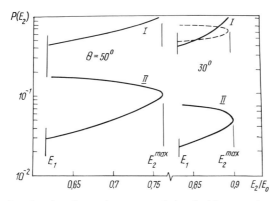

Fig. 24. Attenuation function for various parts of the double-scattering cones in the high-energy region: curves I denote scattering from the first layer; curves II the same from the second layer; broken curve scattering from a nonideal surface (Ar–Cu; $\theta = 30°$, $50°$; $E_0 = 30$ keV).

(1976) carried out a numerical computation for Ar^+ ions with $E_0 = 16\,\mathrm{keV}$ scattered at $\theta = 40°$ from polycrystalline copper.

Figure 25 presents the scattering cones in θ_1 and φ_1 coordinates corresponding to different E_2/E_0 ratios, and the probability distributions for such trajectories over different parts of the cones, calculated analytically and with a computer. As shown by a comparison of panels a and b, the probability of scattering to a given angle falls off with increasing φ_1, the highest probability corresponding to the values of φ_1 closest to zero. This supports our earlier conclusion that the bulk of the ions is scattered close to the ridge of the atom pair. The $(dI/d\Omega)(\theta)$ curves obtained either analytically or by computer simulation are shaped as concave arcs, however, whereas the ends of these arcs corresponding to quasisingle scattering tend strongly upwards if derived analytically, computer simulation yields for their height a finite value. As already pointed out, this discrepancy comes from the inadequacy of expression (28). Computations based on the plane model (Mukhamedov et al. 1973) showed a strong increase in intensity in the impact parameter range corresponding to the transition from single to double scattering. Analytical calculations (Martynenko 1973) also suggest the presence of this

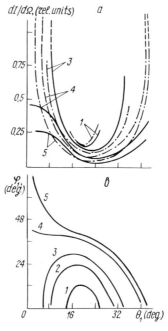

Fig. 25. (a) Distribution of scattering probabilities over equal-energy cones (full curves are from computer simulation, chain curves from analytical calculation), and (b) equal-energy cones in scattering at $\theta = 40°$; (1) $E_2/E_0 = 0.85$; (2) $E_2/E_0 = 0.81$; (3) $E_2/E_0 = 0.79$; (4) $E_2/E_0 = 0.73$; (5) $E_2/E_0 = 0.71$.

feature close to the shadow boundary. At $\theta = 40°$, the intensity grows dramatically if $\psi = 14$ and $25°$. As seen from fig. 25a, the high probabilities for $E_2/E_0 = 0.79–0.81$ are obtained exactly for such orientations of the two-atomic pair. Figure 26 presents $(E_2/E_0)(\psi)$ dependence. The numbered labelled parts of the curve correspond to the various types of scattering: 1–2 quasisingle from the first atom, 2–3 double, 3–4 quasisingle from the second atom. The parts 5–6 and 6–7 relate to quasisingle and double scattering where the ion trajectory passes between the two atoms. As the azimuthal grazing angle varies, this curve traces a surface, so that to each energy E_2 there will now correspond a set of values of ψ and ξ, and, hence, a set of possible trajectories making up a conical surface in space. As seen from fig. 26, double scattering with an energy in excess of E_1 occurs within a narrow interval in the orientation space of the atomic pair. All the other orientations will yield quasisingle and double scatterings with energies less than E_1, which are characterized by large θ_1 and θ_2 angles and a low probability. The computer calculation of the energy spectrum of scattered ions included integration over the equal energy cones, eq. (43). To compare the calculated spectrum with the experimental, one has to take into account the

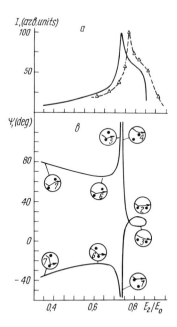

Fig. 26. (a) Energy spectrum of 16 keV Ar^+ ions scattered at $\theta = 40°$ from polycrystalline copper [full curve: computer simulation, broken curve: experiment (Balashova et al. 1978)], and (b) scattered ion energy E_2 versus grazing angle ψ.

analyzer resolution ΔE. Then

$$I = \frac{1}{\Delta E} \int\limits_{E_2}^{E_2 + \Delta E} \int\limits_{\psi_{min}}^{\psi_{max}} \frac{\mathrm{d}I}{\mathrm{d}\Omega} (\psi) c(\psi) \sin \theta_1(\psi) \begin{vmatrix} \dfrac{\partial \theta_1}{\partial \psi} & \dfrac{\partial \theta_1}{\partial \xi} \\[2mm] \dfrac{\partial \varphi_1}{\partial \psi} & \dfrac{\partial \varphi_1}{\partial \xi} \end{vmatrix} \frac{\mathrm{d}\xi}{\mathrm{d}E_2} \, \mathrm{d}\psi \, \mathrm{d}E_2 \,.$$

(57)

Since this calculation does not include the charge state of the scattered particles, its results should lie closer to the experimental data corresponding to the neutral component; note also that for the inert gases neutral atoms make up the bulk of the scattered beam. As seen from fig. 26a, for the conditions chosen in the calculation, averaging over orientations does not result in the appearance of an individual peak, which is also confirmed by an experiment carried out under the same conditions (Chicherov 1972).

Mukhamedov et al. (1979) performed computer simulations of Ar^+ ions scattering from polycrystalline copper over a wide range of E_0 (12.5–30 keV) and θ (20–150°). The double differential coefficient of double scattering was determined as

$$I = \frac{\mathrm{d}^2 N}{\mathrm{d}E \, \mathrm{d}\Omega} = \frac{N}{4\pi \sin \theta} \int\limits_0^{\xi_{max}} \frac{\mathrm{d}I}{\mathrm{d}\Omega} (\psi) \sin \psi \, \frac{\mathrm{d}\psi}{\mathrm{d}E_2} \, \mathrm{d}\xi \,,$$

(58)

where $\mathrm{d}I/\mathrm{d}\Omega$ is the double-scattering cross section. The integration was carried out over the cone of possible orientations of the two-atomic pair which yield scattering at a given angle θ with a given energy $E_2 = \text{const}$ (Parilis 1965). The derivative $\mathrm{d}\psi/\mathrm{d}E_2$ determines the number of pair orientations per given energy interval, the derivative $\mathrm{d}I/\mathrm{d}\Omega$ being found by simulating the ion trajectories using eq. (58). The function $\mathrm{d}I/\mathrm{d}\Omega$ was calculated within the energy interval $E_{min} \leqslant E_2 \leqslant E_{max}$. E_{min} is the minimum energy when double scattering at a given angle; E_{max} is the maximum possible energy in the double-scattering model. The calculations showed that, in all the cases considered, $\mathrm{d}I/\mathrm{d}\Omega$ has a singularity at a point E_B within the interval $E_1 < E_B < E_{max}$ (E_1 is the energy of single scattering to an angle θ). The discontinuity of $\mathrm{d}I/\mathrm{d}\Omega$ is due to the rainbow effect (Martynenko 1973, Mukhamedov et al. 1976) occurring at a certain orientation of the pair relative to the beam. Under blocking conditions, there always exists a region of $E \approx E_B$ where the scattering intensity grows dramatically for each θ at the corresponding ψ_B^{min} and ψ_B^{max} (fig. 27a). Estimates showed that $\mathrm{d}I/\mathrm{d}\Omega$ tends to infinity in accordance with

$$\frac{\mathrm{d}I}{\mathrm{d}\Omega}\bigg|_{E \approx E_B} \sim [E(\psi) - E_B]^{-k}, \quad k < 1 \,.$$

(59)

In the remainder of the region, $\mathrm{d}I/\mathrm{d}\Omega$ falls off monotonically. Figure 27b

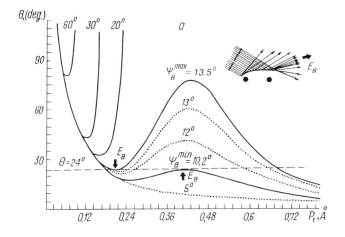

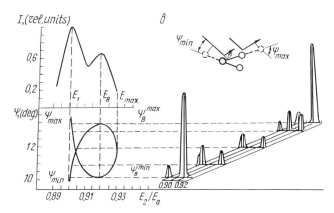

Fig. 27. (a) Double-collision scheme and $\theta(p_1)$ dependence, and (b) loop relation between ψ and E_2/E_0, and the energy spectrum (the full curves are for ψ_B^{min} and ψ_B^{max} taking on extremal values for $\theta = 24°$ and corresponding to $E = E_B$).

presents the $E_2/E_0(\psi)$ relation for $\theta = 24°$ which reveals the following properties of the function $d\psi/dE_2$:

(1) a sharp maximum in the region of quasisingle scattering ($E \approx E_1$), which is due to the large number of orientations contributing to the relatively narrow energy region of quasisingle scattering;

(2) the vanishing of the derivative $d\psi/dE_2$ for $E_2 \to E_B$;

(3) tending of $d\psi/dE_2$ to infinity for $E \to E_{max}$.

A calculation using eq. (58) showed that despite these features of the integrand, the integral converges throughout the energy range considered. Figure 27b presents also the results of a calculation of the double differential scattering cross section. The energy spectrum is seen to exhibit a double-peak structure. The reason for this is obvious, namely, the first peak of quasisingle scattering is due to the large number of pair orientations contributing to this energy interval, and the second to a focussing effect associated with the large cross section of double scattering where $dI/d\Omega \to \infty$ for the angles ψ_{B}^{min}, ψ_{B}^{max} and for $E = E_{B}$. As for the extreme right-hand part of the energy spectrum, the peak here is not always resolvable despite the fact that $d\psi/dE_{2}$ tends to infinity, since for $E \to E_{max}$ the solid angle of the cone vanishes quite rapidly. As follows from fig. 27b, as the scattering angle θ grows, the interval of the angles ψ where double scattering is possible also increases, in other words, the number of the two-atom orientations, including the deeper lying layers of the target which contribute to the double-scattering process becomes increasingly greater. For very large θ, the energy separation between the extremal points and the quasisingle scattering branches decreases to the extent where they become indistinguishable. On the other hand, the double-scattering cross section falls off fairly rapidly with increasing θ while behaving like a delta function close to the extremal points. Therefore, the singularity disappears nearly completely when averaged over all possible orientations, no noticeable maximum being visible in the high-energy region. The double-scattering shoulder represents in this case a curve which falls off steeply towards increasing E_{2}, the height of the highest energy point in the spectrum becoming for $\theta > 90°$ an order of magnitude lower than that of the single-scattering peak (see fig. 27). As already pointed out, the calculations were carried out separately for two-atomic pairs with $d = d_{\langle 100 \rangle}$, and $d = d_{\langle 110 \rangle}$, d is the interatomic separation in the pair.

The results of calculations for different initial energies E_{0} and for $\theta = 30°$ are presented in fig. 28a. Also shown are averaged spectra which are compared with experiment (Balashova et al. 1977b). In the averaging over the orientations, the following factors were taken into consideration:

(1) crystal symmetry, due to which averaging should be performed only over the solid angle per one atom in the corresponding coordination sphere;

(2) short-range order in the crystal, due to which a third atom responsible for the blocking of the scattering lies at the continuation of the axis connecting the atoms in the pair (the corresponding minimum, ψ_{min}, and maximum, ψ_{max}, angles provide an additional narrowing of the region of orientations over which the averaging is done; this cut-off is different for different axes, thus producing a shift of the quasisingle scattering peaks for different d and a broadening of the spectrum in the averaging);

(3) equal probability of all microcrystal orientations.

However, the relative concentration of microcrystals of different orientations may vary in the course of intense ion bombardment and, hence, the spectrum of scattered ions may also vary as a manifestation of this process.

As seen from fig. 28a, the shape of the double scattering loops in the

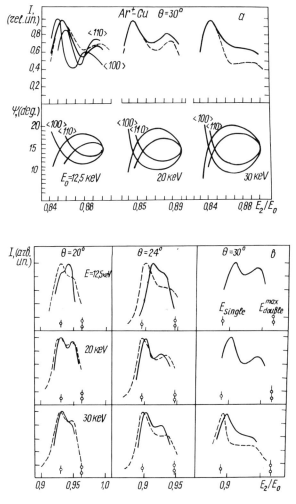

Fig. 28. (a) E_2/E_0 versus ψ plots for different E_0 and interatomic spacing d with the corresponding spectra, and (b) calculated energy distributions of Ar^+ ions backscattered from polycrystalline copper for different θ and E_0 [full curves: averaged spectrum, broken curves: experiment (Balashova et al. 1977b)].

diagrams is very sensitive to the value of d, the extremal points of $E_2/E_0(\psi)$ shifting strongly toward higher energies with decreasing d. As a result, the double-scattering maximum in the energy spectrum broadens, its position being dependent on the relative concentration of microcrystals of various orientations in the polycrystal. Interestingly, the region of quasisingle scattering also shifts to the right with decreasing d, however, the separation between the single- and double-scattering peaks changes very little. In the averaging this results in a broadening of both peaks compared with the peaks obtained under the same conditions from single crystals (Parilis and Turaev 1964). Shown in fig. 28b is a series of experimentally measured (Balashova et al. 1977a,b, 1978) and calculated energy distributions. The calculated distributions bear an overall similarity with the experimental ones. Both spectra transform in the same way on variation of θ and E_0. The reduction in height of the double-scattering peak and its gradual transformation into a shoulder with decreasing E_0 can be accounted for by the narrowing of the region of atomic pair orientations allowable for double scattering as a result of increasing shadow size. Conversely, as the energy increases, the relative height of the double-scattering peak grows too. Further increase of E_0, however, again results in a decrease and, eventually, disappearance altogether of the high-energy peak, since the differential double-scattering cross section falls off with increasing E_0 and θ much faster than the single-scattering cross section does (U.A. Arifov 1968, Parilis and Turaev 1966, Parilis 1965).

At the same time it should be pointed out that for small scattering angles the energy distributions obtained by computer simulation extend over narrower energy range than the experimental ones. Apart from this, in this region of θ angles the relative height of the double-scattering peak in the computed spectra is higher than that seen in experiment. Within the model of a perfect (not disordered) surface, one cannot explain the shift toward lower energies of the measured single-scattering peak and its relatively large height. The long tail of the energy distribution extending into the high-energy region which is predicted by the model can originate from the presence in the reflected flux of ions that suffered multiple collisions.

The agreement with experiment can be improved by using an interaction potential falling off at large distances ($r > 1$ Å) faster than the potential, given in eq. (13) of chapter 1 does.

The conditions of quasisingle and double scattering change dramatically if one assumes that the surface structure of the target, rather than being perfect, contains various point defects. Indeed, as will be shown later, ion bombardment of a solid produces on its surface a large number of vacancy-type defects and their clusters, isolated atoms and atom pairs, and steps, the degree of surface disorder varying from one layer to another.

To include the nonideal surface structure into the computational procedure, a layer-by-layer distribution of vacancy concentration was introduced in accordance with Bitensky and Parilis (1978b). It turned out that the behavior of the flux attenuation function is very sensitive to the atom density distribution in the various layers. A change in the attenuation function results in a substantial change in the shape of the spectrum. A better agreement with experiment was obtained under the assumption that the uppermost layer is made up of individual isolated atoms (fig. 29a). In the region of large θ angles, the double-scattering shoulder in calculated spectra lies substantially below its experimental position (fig. 29b). This discrepancy evidences the inapplicability of the two-collision approximation to the description of large-angle scattering.

Thus, the shape of the energy spectrum of heavy ions backscattered from a polycrystal in the double-scattering model is a result of averaging over the various orientations of the atom pair, as well as over the interatomic separations. An isolated maximum appears in the cases where the spectra to be averaged are displaced somewhat with respect to one another. The position of the maximum corresponds to the atomic pair orientation at which the density of the scattering trajectories corresponding to a narrow energy interval increases dramatically due to the rainbow effect. The shape of the energy spectrum of scattered ions may be useful in detecting changes in the polycrystalline surface structure of metals.

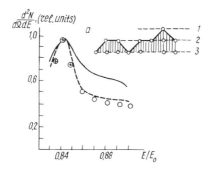

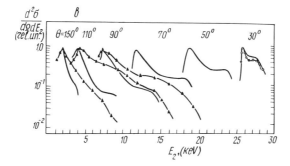

Fig. 29. Energy distributions of 30 keV Ar ions (a) double-scattered at $\theta = 30°$ from nonideal polycrystalline Cu surfaces, and (b) scattered at larger angles. Full curve: ideal model calculation; broken curve: nonideal model calculation; crosses: experiment (Balashova et al. 1977b).

10. Enhancement of 180° backscattering

Aono et al. (1981) proposed a method of surface structure analysis of single crystals involving the measurement of the intensity of low-energy ions backscattered at large angles, including 180°. Yamamura and Takeuchi (1984), and Takeuchi and Yamamura (1985) used the two-atomic model to calculate the large-angle backscattering intensity. Following Oen (1982), who used the blocking and focusing phenomena in this model to account for the so-called Pronko effect (Pronko et al. 1979), i.e., an enhancement of reflection at 180° in the region of Rutherford scattering, they were able to show that large-angle (including 180°) reflection enhancement can occur also at low energies under the conditions of blocking and focusing.

An enhancement of 180° reflection, which manifests itself in a growth of the single-scattering peak in the energy spectrum, can be observed, however, not under the blocking and focusing conditions corresponding to a particular geometry of scattering. Dzhurakhalov et al. (1987a) showed that the reflection can be enhanced also in the absence of blocking and focusing, as a result of addition of intensities of single- and double-scattering at 180°. The possibility of such enhancement of 180° reflection was later supported by Demkov (1988), who used for this purpose the angular diagram technique. Consider the results obtained by Dzhurakhalov et al. (1987a). It is well-known that the energy of an ion which suffered an elastic single scattering from a target atom at an angle θ is governed by the laws of conservation irrespective of the actual kind of the interaction potential and can be calculated using eq. (4) of chapter 1. In backscattering at 180°, this relation takes on the form

$$\frac{E_1(E_0, 180°)}{E_0} = \frac{(m_2 - m_1)^2}{(m_2 + m_1)^2} . \tag{60}$$

When scattered at a given angle θ in two sequential collisions with two atoms whose position is fixed, the expression for the ion energy can be written as

$$E_2(E_0, \theta_1, \theta_2) = \frac{E_0}{(1 + \mu)^4} (\cos \theta_1 + \sqrt{\mu^2 - \sin^2\theta_1})^2$$

$$\times (\cos \theta_2 + \sqrt{\mu^2 - \sin^2\theta_2})^2 , \tag{61}$$

and depends on the interaction potential (Parilis 1965) since it affects the combination of the θ_1 and θ_2 angles.

This expression is valid for all scattering angles except $\theta = 180°$, for which, in the case of double scattering, eq. (61) reduces to eq. (60) for any combination of θ_1 and θ_2. It thus follows that the energy E_2 of the double-scattered ions at 180° is exactly equal to the energy E_1 of the single-scattered ions at the same angle, and does not depend on the type of the potential used. Hence, one should observe an enhancement of the single-scattering peak in the energy spectrum in the region corresponding to single scattering, E_1.

Figure 30a presents the dependences of the single-scattering energy (I), as well as of the maximum and minimum possible double-scattering energies (II) an θ for the Na^+–Mo case. The maximum and minimum energies correspond to double-scattering trajectories of different type, 2, 2*, and are given by the expressions (Parilis 1965)

$$E_{2max}(E_0, \theta, \mu, \tfrac{1}{2}\theta, \tfrac{1}{2}\theta) = \frac{E_0}{(1+\mu)^4} (\cos \tfrac{1}{2}\theta + \sqrt{\mu^2 - \sin^2 \tfrac{1}{2}\theta})^4 ; \qquad (62)$$

$$E_{2min}(E_0, \theta, \mu, \pi - \tfrac{1}{2}\theta, \pi - \tfrac{1}{2}\theta) = \frac{E_0}{(1+\mu)^4} (\cos \tfrac{1}{2}\theta - \sqrt{\mu^2 - \sin^2 \tfrac{1}{2}\theta})^4 . \qquad (63)$$

E_{2max} corresponds to the usual double forward scattering at angles $\theta_1 = \theta_2 \approx \tfrac{1}{2}\theta$ sequentially from the first (A) and second (B) atoms, whereas in the case of E_{2min} an ion is first scattered by the second atom in the direction of the first and, by the latter, toward the analyzer (Mashkova and Molchanov

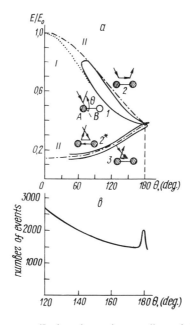

Fig. 30. Single-scattering energy (I, dotted curve), as well as of the maximum and minimum possible double scattering energies (II, chain curve) versus θ for the Na^+–Mo case (a) and angular dependence of the intensity of singly-scattered Na^+-ions with the contribution of double-scattered ions (b).

1980). This case includes two sequential scatterings at large angles $\theta_1 = \theta_2 \simeq \pi - \frac{1}{2}\theta$. We readily see that for $\theta = 180°$ expressions (62) and (63) reduce to eq. (60), the single- and double-scattering branches merging. Figure 30a presents also the results of a calculation for $\psi = 45°$, $E_0 = 250$ eV and interatomic distance in Mo, $d = 3.14$ Å. The interaction was approximately by the Molière potential with Firsov's screening parameter multiplied by a coefficient 0.8. We see that, in agreement with Takeuchi and Yamamura (1985), scattering at $\theta > 180°$ is possible, and that there is also a triple scattering branch 3 which, however, does not intersect with branches 1, 2, and 2^* at $\theta = 180°$ and passes below them for all θ. Thus, the $E/E_0(\theta)$ dependence consists of four branches. Branches 1, 2, 2^* correspond to the quasisingle, quasidouble forward and quasidouble backscattering, and are closed ovals bounded by the values θ_{min} and θ_{max}; branch 3 corresponds to triple scattering, it lies separated from the others and below E_{2min}. Hence, the energy distribution of reflected ions should contain, besides the three peaks (single, double forward and triple) pointed out by Takeuchi and Yamamura (1985), also a fourth peak of double backscattering located between the single and triple peaks. The difference in the nature and relative intensity R of trajectories of Na^+ ions with $E_0 = 250$ eV scattered from a pair of Mo atoms at 180° is illustrated in table 1.

Figure 30b presents a computed angular dependence of the intensity of singly-scattered Na^+ ions, with the contribution of double-scattered ions included for 180°. The angular dimensions of the 'analyzer' accepting the scattered particles were taken equal to $\Delta\theta = \Delta\varphi = \pm 0.5°$, where φ is the azimuthal scattering angle measured from the axis connecting the atoms. The incident beam density was 4×10^7 part./Å^2. We see that the summation of the intensities of the single and both types of double scattering at $\theta = 180 \pm 0.5°$ produces enhanced reflection in the region of the single-scattering peak. At $\psi = 45°$, the total reflected particle intensity increases by 40%. It should be pointed out that this enhancement is due primarily to the summation of the single- and double-scattering intensities and only partially to blocking at θ_{max} which in this case is 189°. The intensities are summed also under the blocking conditions. The angle θ_{max} grows with increasing grazing

Table 1
The characteristics of different trajectories.

Trajectory type (i)	θ_1 deg.	θ_2 deg.	θ_3 deg.	E eV	$R = I_i/I_1$
Single (1)	0.27	179.50	0.76	94.06	1
Double forward (2)	31.16	150.12	1.30	93.54	0.15
Double backscattering (2^*)	0.40	154.91	25.46	93.93	0.25
Triple scattering (3)	29.00	172.21	21.21	86.10	0.05

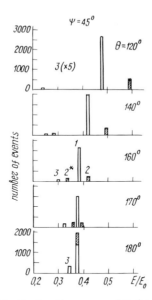

Fig. 31. Calculated energy-distribution histograms of Na$^+$ ions, scattered from Mo.

angle ψ, and the enhancement of 180° reflection as a result of intensity summation can be directly observed.

The energy distributions (histograms) in fig. 31 specify the positions and relative intensities of the calculated single, double forward and back, and triple scattering peaks as functions of scattering angle θ at 120°, 140°, 160°, 170°, and 180°. We see that the merging of the single- and double-scattering peaks at $\theta = 180°$ brings about an enhancement of the single-reflection peak to the extent where it becomes about equal to the peak of single-scattering at $\theta = 140°$.

The enhancement of the single 180° reflection peak caused by the summation of the intensities of single and double scattering can be revealed experimentally by comparing the intensity of this peak with those of the peaks of single scattering at angles close to 180° under conditions far from blocking.

This effect may be an instrumental effect in the experimental determination of the inelastic energy losses in double scattering at 180° as a function of E_0. Based on the fact that the energies of the ions suffering elastic single and double scattering at 180° should be equal, the inelastic energy losses in double scattering can be found from the shift of the single- and double-scattering peaks with respect to the energy conserved in the elastic process.

CHAPTER 4

Atomic Row Model

At small grazing angles, scattering does not lend itself for a description in terms of the single- and double-collision models which are applicable at large grazing angles. At glancing incidence, the projectile interacts with a row of atoms lying in a given crystallographic direction.

There are two major methods of describing the interaction with an atomic row:

(1) the model of Lindhard (1965), or the string model, making use of a continuous potential; and

(2) the atomic row model (Kivilis et al. 1967) where scattering is visualized as a sequence of correlated binary collisions with individual atoms in a row.

1. String model

Lindhard (1965) proposed an approximation for the consideration of orientational effects caused by the regular arrangement of the target atoms. This method is based on four principal assumptions permitting one to determine and explain particle motion on a purely theoretical basis.

(1) One uses the classical concept of interaction which is valid because the wavelength of the projectile particle is small compared with the characteristic scatterer dimensions.

(2) One considers an idealized perfect lattice and a perfect atomic row. The thermal and zero lattice vibrations are deviations from the perfect conditions.

(3) One assumes particle scattering to be caused by nuclear collisions, these collisions being nearly elastic, and the particle scattering angles small.

(4) One studies only small grazing angles relative to the axis of the atomic row, since only in this case the collision with an atomic row, in a first approximation, be considered as involving a continuous scattering potential (the string model).

To be able to consider scattering from an atomic row as that from a string, Lindhard (1965) introduced the concept of continuous potential at a distance R from the axis of the atomic row,

$$U(r) = \int_{-\infty}^{+\infty} V(R = \sqrt{z^2 + r^2}) \, \frac{\mathrm{d}z}{d} , \tag{1}$$

where z is the axis of the row, and $V(R)$ is the ion–atom potential. He studied ion–atom interaction described by the Thomas–Fermi potential with different screening functions,

$$V(R) = \frac{Z_1 Z_2 e^2}{R} \varphi_0\left(\frac{R}{a}\right),$$ (2)

where a is the screening parameter defined in eq. (15) of chapter 1, and $\varphi_0(R/a)$ is the screening function for an isolated atom.

As seen from eq. (1), in a general case the dependence of U on $1/r$ is one power weaker than that of V on $1/r$,

$$U(r) = \frac{Z_1 Z_2 e^2}{d} \xi(r/a).$$ (3)

Having studied the behavior of the $\xi(r/a)$ function, Lindhard showed that, for small r,

$$\xi(r/a) = 2 \ln\left(\frac{ca}{r}\right) \quad \text{for} \quad r < ca$$ (4)

(here $2 \ln c$ is a constant of integration).

A somewhat better approximation, valid for any value of r, is given by the expression which is called the standard Lindhard potential,

$$\xi(r/a) = \ln\left[\left(\frac{ca}{r}\right)^2 + 1\right].$$ (5)

One usually takes $c = \sqrt{3}$, while for small r/a a better approximation is obtained with a smaller c, and for large r/a with a somewhat larger value of c.

According to eq. (5), for $r > ca$, $\xi = ca/r^2$, in the range $r \sim a$, the potential scales as $\sim R^{-2}$, and $\xi(r/a) \sim \pi a/2r$.

At high energies, where $r_{min} \to 0$, and the potential, eq. (3), is valid, it is required that

$$\frac{ca}{\psi d} \exp\left(-\frac{\psi^2 d}{2b}\right) > 1,$$ (6)

where $b = Z_1 Z_2 e^2/E$ is the collision diameter in LS.

Condition (6) is met if the exponent is less than unity, and $ca/\psi d > 1$, i.e., if

$$\psi < \psi_1 = \sqrt{\frac{2b}{d}} = \sqrt{\frac{E_1}{E}}; \quad E_1 = \frac{2 Z_1 Z_2 e^2}{d};$$

$$\psi_1 < \frac{a}{d} \quad \text{or} \quad E > E' = 2 Z_1 Z_2 e^2 d/a^2.$$ (7)

For low energies, where eq. (7) is already inapplicable, the approximate value for the critical angle is

$$\psi < \psi_2 = \left(\frac{ca}{d\sqrt{2}}\psi_1\right)^{1/2} \approx \left(\frac{a\psi_1}{d}\right)^{1/2} ; \tag{8}$$

since

$$c/\sqrt{2} = \frac{\sqrt{3}}{\sqrt{2}} \approx 1 , \tag{9}$$

the critical angle ψ_2 can be found for $\psi_1 > a/d$ or $E > E'$.

The results described by eqs. (7) and (8) give an idea of the behavior of a particle beam propagating through the lattice. If the initial angle ψ is less than $\psi_{cr}(\psi_1, \psi_2)$, then the concept of a continuous atomic row holds, so that the particles undergo repulsive collisions with the row and escape from it at the same angle ψ they had before collision.

Consider the trajectory of a particle in a field with a continuous potential $U(r)$ created by an atomic row lying along the z axis (fig. 1a,b).

The angle of scattering from an atomic row in the transverse plane is

$$\theta_\perp(p) = \pi - 2p \int_{r_{min}}^{\infty} \frac{dr}{r^2\left[1 - \frac{U(r)}{E_\perp} - \frac{p^2}{r^2}\right]^{1/2}} . \tag{10}$$

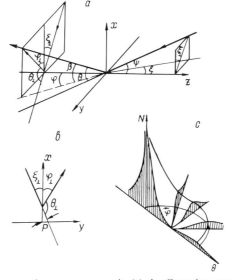

Fig. 1. Scattering from a continuous atom row in (a) the Cartesian system and (b) transverse plane, and (c) the general angular distribution of protons backscattered from a continuous atomic row for an incidence angle $\psi < \Psi_{cr}$.

The scattering angle θ (i.e., the angle between the initial and final directions of particle motion) is related to the angle $\theta_\perp$ through

$$\theta(p) = 2\psi \sin \tfrac{1}{2}\theta_\perp(p) . \tag{11}$$

The azimuthal angle of scattering can be written as

$$\varphi = \arcsin(\tan \varphi_1 \tan \beta) , \tag{12}$$

where $\varphi_1 = \pi - (\theta_\perp + \xi_\perp)$. If $\theta_\perp(p) = 180°$, then $\theta = 2\psi$, i.e., if the scattering plane passes the axis of the atomic row ($p = 0$), then the angle of incidence onto the row coincides with that of reflection.

Consider the cross section of scattering from a continuous potential of a string. The number of particles scattered into an angular interval $(\varphi_\perp ; \varphi_\perp + d\varphi_\perp)$ is equal to that of particles with impact parameters $(p, p + dp)$, $dN = N_0 \, dp$, where N_0 is the number of particles per unit length in the transverse plane.

In this case, the scattering cross section will have the dimension of length

$$d\sigma = \frac{dN}{N_0} = dp = \frac{dp}{d\varphi_\perp} \, d\varphi_\perp . \tag{13}$$

For the inverse square interaction potential

$$\theta_\perp(p) = \pi - 2\varphi_{0\perp} = \pi - \frac{\pi p}{\sqrt{p^2 + \dfrac{C}{E_\perp}}} , \tag{14}$$

and

$$d\sigma = \frac{dp}{d\xi} = \sqrt{\frac{C}{E_\perp}} \frac{\pi^2}{[\pi^2 - (\pi - \theta_\perp)^2]^{3/2}} . \tag{15}$$

Figure 1c presents the dependence of the scattering cross section on the angles θ and φ (Iferov et al. 1972).

The continuous potential approximation used widely in the description of light ion channeling deep in the bulk of a crystal and for energies in the range of millions and billions of electronvolts (Lindhard 1965, Kumakhov and Shirmer 1980, Gemmel 1974) is presently finding application also in the consideration of glancing scattering of medium-energy heavy ions in surface layers of single crystals (semi- and hyperchanneling), i.e., in all those cases where the grazing angle is so small (less than the Lindhard angle) that the atomic structure of the row may be neglected.

2. Scattering from a discrete atomic row

In the model of a continuous string only scattering in the atomic row plane through an angle $\theta = 2\psi$, i.e., specular reflection, can occur. Apart from this, only one kind of scattering trajectory for a given angle θ is possible. Therefore, the energy distribution of scattered particles represents a spectrum with one maximum. However, as shown in the preceding chapter, the energy spectrum of ions reflected from a single crystal should contain a structure due to the existence of different types of scattering trajectories (single and double scattering). Such structure of the energy spectrum under glancing incidence was obtained with the discrete atomic row model proposed by Kivilis et al. (1967) and subsequently revealed experimentally (U.A. Arifov 1968, Mashkova and Molchanov 1985).

In the atomic row model, scattering is considered as a sequence of correlated binary collisions with individual atoms in the row (fig. 2). The first collision is assumed to occur with the atom for which the impact parameter does not exceed a certain limiting value p_{lim} corresponding to scattering at a given small angle, e.g., $\theta_{lim} \approx 0.5°$. For the Firsov potential $E_0 = 30$, 10 and 5 keV, and the Ar^+–Cu pair, $p_{lim} = 1.5$, 2.0 and 2.5 Å, respectively.

The scattering angle in each collision, $\theta_i(p_i)$, is determined by the interaction potential for each projectile energy E_i, which is calculated using eqs. (4) and (37) of chapter 1 with due inclusion of elastic and inelastic energy losses. The impact parameter for each subsequent collision, p_{i+1}, is determined by the preceding impact parameter p_i and the scattering angles θ_i,

$$p_{i+1}[E_i(\theta_i), \theta_{i+1}] = p_i[E_{i-1}(\theta_{i-1}, \theta_i), \theta_i] - d \sin\left(\psi - \sum_{i=1}^{n} \theta_i\right). \tag{16}$$

The interaction with the row stops when $p > p_{lim}$ and the ion escapes from the row on scattering at the angle $\theta = \Sigma\, \theta_i$. This usually occurs after three to ten collisions. For a given incidence angle ψ, the escape angle β and the scattering angle $\theta = \psi + \beta$ are determined unambiguously by the first impact parameter p_1. Obviously enough, $\theta(p_1)$ is a periodic function of p_1 with a period $a = d \sin \psi$.

Figure 3a presents the $\theta(p_1)$ dependence for different values of ψ. We readily see that for all ψ the scattering angle is limited by the minimum and maximum scattering angles θ_{min} and θ_{max} (and, accordingly, by the minimum

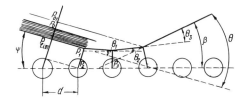

Fig. 2. Scattering of a particle from an atom row.

and maximum escape angles). This phenomenon was called 'blocking' and originates from the mutual screening action of the atoms in the row. The limitation on the side of large escape angles comes from the fact that an ion cannot approach an atomic row close enough because of the incoming blocking. The limitation on the small escape angles comes from the outgoing blocking. At scattering angles less than a certain θ_{min}; the ion approaches sufficiently close the row and suffers a number of additional interactions until the distance to the row determined by the scattering angles has become greater than p_{lim}, and $\theta = \Sigma \, \theta_i > \theta_{min}$, while $\beta > \beta_{min}$.

Figure 3b shows the boundaries of possible escape angles for the case of scattering of Ar^+ ions with initial energy $E_0 = 30$, 10 and 5 keV from a Cu [110] row versus grazing angle ψ. For small ψ, the reflection is close to specular (broken curve in fig. 3b), this effect being stronger pronounced for higher initial energies. In this region, scattering can be satisfactorily described in terms of the continuous atomic row model. As ψ grows, the number of collisions with atoms in the row decreases, multiple collisions transferring gradually to single and double scattering. The screening effect disappears, and θ_{max} may attain any large value, up to 180°.

Consider now the energy distributions of the ions scattered from an atomic row. In the two-atom model, scattering at a given angle occurs as a result of one or two collisions, which yields different energies of the reflected ions and corresponds to two peaks in the ion reflection spectrum. Such double-valued pattern of the dependence of the energy on scattering angle, $E(\theta)$, maintains also in scattering from an atomic row. As seen from fig. 3a, each $\theta(p_1)$ curve passes twice within its period at a given value of θ, at two different values of p_1. Since to each p_1 corresponds its own value of E, the energy of the ions reflected in a given direction will be a double-valued function.

Figure 4a presents a characteristic oval in the $E(\theta)$ dependence. In the plane case, the cross section of scattering from an atomic row $d\sigma$ is

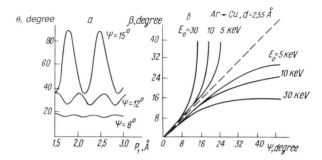

Fig. 3. Periodic dependence of (a) scattering angle θ on the first impact parameter p_1, and (b) maximum and minimum escape angle β on grazing angle ψ for different initial energy E_0.

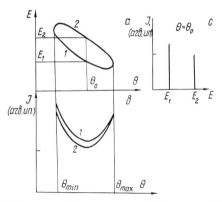

Fig. 4. Dependence of (a) scattered ion energy and (b) scattered ion intensity on scattering angle θ, and (c) spectrum of ions scattered from a one-dimensional row.

proportional to $\mathrm{d}p_1/\mathrm{d}\theta$. Since scattering at a given angle θ from an ideal row is possible only for two types of trajectories each corresponding to a definite energy, the energy spectrum should consist of two infinitely narrow peaks of height $\mathrm{d}p_1/\mathrm{d}\theta$. Figure 4c shows a spectrum of ions scattered from a one-dimensional atomic row, and fig. 4b the scattering-angle dependence of the scattered particle intensity $I \sim \mathrm{d}\sigma/\mathrm{d}\theta$ (Yurasova et al. 1968).

As seen from fig. 4a, at the points $\theta_{\max}$ and $\theta_{\min}$,

$$\frac{\mathrm{d}\sigma}{\mathrm{d}\theta} \approx \frac{\mathrm{d}p_1}{\mathrm{d}\theta} = \frac{1}{\mathrm{d}\theta/\mathrm{d}p_1},$$

tends to infinity, the different bending of curves 1 and 2 corresponding to different types of trajectories being accounted for by the different slopes of the curve $\theta(p_1)$ in regions 1 and 2.

Kivilis et al. (1970) studied the effect of various parameters on the characteristics of ion scattering. Figure 5 presents the dependences of the

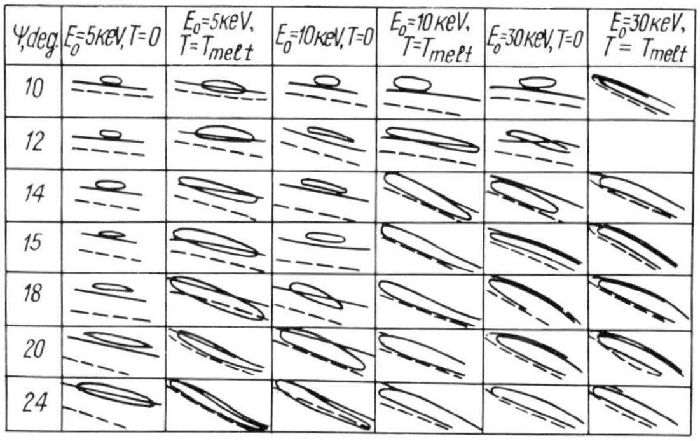

ψ,deg	E_0=5keV, T=0	E_0=5keV, T=T_{melt}	E_0=10keV, T=0	E_0=10 keV, T=T_{melt}	E_0=30keV, T=0	E_0=30keV, T = T_{melt}
10						
12						
14						
15						
18						
20						
24						

Fig. 5. Dependence of the $E(\theta)$ function on E_0, T, and ψ.

reflected ion energy on scattering angle, $E(\theta)$, for different initial energies E_0, grazing angles ψ and target temperature T. The simulation of thermal vibrations and the effect of temperature on the ion scattering characteristics will be discussed later. The ovals correspond to scattering from a row. The broken curves identify the $E(\theta)$ function for single scattering, and the full curves passing higher, the same dependence for double scattering. The left-hand edge of each oval corresponds to θ_{min}, and the right-hand one to θ_{max}. The ovals fit into single and double-scattering curves only for large ψ. As ψ decreases, the width of the oval becomes less than the distance between the $E(\theta)$ plots for single and double scattering. Apart from this, the entire oval rises with decreasing ψ, so that its lower arc may become at the level of, or even above, the full curve. This calls for caution when identifying experimentally observed peaks as 'single' or 'double' from their position on the energy scale. The apparent disappearance of the single-scattering peak may sometimes be only a consequence of such a shift, and that of the double-scattering one of difficulties encountered in the resolution of two close peaks.

Actually, at glancing incidence angles the concepts of single- and double-scattering peaks lose their original meaning. The peaks of the lower and upper oval arcs correspond to a number of sequential collisions, one or, accordingly, two of them being related to large-angle scattering, and the others to small-angle scattering involving low-energy losses. For instance, for $\psi = 12°$, $\theta = 25°30'$ the peak of the lower arc corresponds to scattering at angles $0°54'$, $3°$, $18°40'$, $2°11'$ and $0°45'$, and that of the upper arc, to $0°44'$, $1°38'$, $7°23'$, $13°12'$, $1°50'$ and $0°45'$.

Thus, the atomic row model developed by Kivilis et al. (1967, 1970) yields the following main results called the atomic row effects:

(1) mutual screening action of neighbor atoms confines the scattered beam within a minimum and a maximum escape angle, i.e., results in blocking;

(2) the double-valued pattern of the dependence of particle energy on scattering angle, $E(\theta)$, is retained, i.e., the energy spectrum of scattering contains two peaks, namely, a quasisingle (QS) and a quasidouble (QD) peak;

(3) both peaks shift towards higher energies relative to the purely single and double interactions, this shift increasing with decreasing grazing angle.

The smaller the grazing angle ψ is, the stronger is the atomic row effect. For a given ψ, the same trend is observed to exist with decreasing temperature T, energy E_0 and interatomic separation d in the row. The ovals become more narrow, shrink in size and rise above the single- and double-scattering curves. Despite the fact that the calculations of Kivilis et al. (1967, 1970) were numerical, the origin of the above-mentioned relations can be pointed out. The sequence of the correlated ion collisions with atoms in the row is determined by the angles β_i formed by the straight segments of the trajectory with the axis of the row. This angle can be roughly estimated from

$$\sin \beta_i = [p(\theta_{i+1}) - p(\theta_i) - \delta(T)]/d . \tag{17}$$

Since the thermal displacement of an atom $\delta(T) \sim \sqrt{T}$, and the impact parameters $p(\theta) \sim E_0^{-1/2}$ and grow with decreasing θ (for specular reflection $\theta \approx 2\psi$), similar trajectories and, hence, similar ovals $E(\theta)$ are obtained for d, E_0, ψ and T change in the same direction.

The double-valued pattern of the energy dependence $E(\theta)$ and the blocking effect predicted by Kivilis et al. (1967) were confirmed experimentally by Evdokimov et al. (1969). The characteristic ovals were obtained for the first time, however, they were found to be displaced relative to the theoretical predictions. Besides, no confinement of the scattered beam was observed at small grazing angles, although the single- and double-scattering peaks turned out to be so close (this follows also from theory) as to be unresolvable. The peaks were observed to approach one another also when crossing over from less to more closely packed atomic rows. A certain discrepancy between experiment and calculation (Kivilis et al. 1967) can be traced to the fact that Evdokimov et al. (1969) compared the experimental data obtained at $T \approx 300$ K with the calculation (Kivilis et al. 1967) made for $T = 0$ K. Indeed, an atomic row calculation including thermal vibrations (Kivilis et al. 1970) revealed a better agreement with experiment. Heiland et al. (1976), Taglauer and Heiland (1972, 1976), Heiland et al. (1973a,b) carried out a comprehensive computer simulation study of the energy dependences of scattered particles. They confirmed the existence of an oval in the $E(\theta)$ dependence and the confinement of the scattered beam by a maximum and a minimum escape angle. A change in the screening parameter of the interaction potential was shown to affect these angles.

Verhey et al. (1975) developed the atomic row model proposed by us (Kivilis et al. 1967, 1970) by including thermal displacements of atoms. The broadening of the single- and double-scattering peaks was confirmed, and a possibility of appearance of a third peak at large vibration amplitudes was pointed out. Taglauer and Heiland (1972) and Heiland et al. (1973b) measured the angular and energy distributions of Ne^+ ions with $E_0 = 200$ eV–2 keV scattered from Ni (110) and Ni (111), and compared them with atomic row calculations. Characteristic ovals in the $E(\theta)$ were revealed, and the calculated values of θ_{min} and θ_{max} were identified with measurements. While the atomic row model was confirmed qualitatively, the measured values of scattered particle energy turned out to lie somewhat lower, and the values of θ_{max} somewhat higher than the theoretical predictions. An investigation of the dependence of scattering on d for the $\langle 110 \rangle$ and $\langle 112 \rangle$ directions also revealed a good agreement with the atomic row model. By comparing the experimental and calculated angular distributions and the blocking angles, the same researchers determined the screening parameter in the Thomas–Fermi potential. The values thus obtained were found to lie 30% below the Lindhard parameter, and 15% below the Firsov screening parameter. This discrepancy stems apparently from the fact that the model used considered an ideal atomic row without the inclusion of thermal vibrations and defects. De Wit et al. (1975) made an experimental and a theoretical study of Ne^+ ion scattering with $E_0 = 1$ keV from Cu (100) and (110). The measured energy spectra were used to derive the minimum and maximum scattering angles. These angles were also calculated by computer simulation of ion trajectories in the crystal using the Born–Mayer potential. It was shown that the left-hand part of the $E(\theta)$ oval (small scattering angles) is more sensitive to a change of the constant b, while its right-hand part is equally sensitive to the variation of both constants, A and b, in the potential. A comparison of the values of θ_{min} and θ_{max} calculated with the constants A and b found by Yurasova et al. (1968) and by Abrahamson (1963) with experiment is in favor for the latter. However, the values of θ_{max} calculated with Abrahamson's constants turned out to be less than the experimental figures, which is apparently due to the neglect of thermal vibrations of the lattice atoms.

The validity of the atomic row model was also discussed by Preuss (1978), who compared computer simulation data with experiment. The size of the crystallite was varied depending on the actual scattering geometry, and uncorrelated thermal vibrations, inelastic energy losses and various surface defects were included. The calculations were carried out with the Born–Mayer and Molière potentials for $T = 0$ and 300 K. A comparison of the calculations with experiment confirmed the existence of the atomic row effect, the best agreement for $T = 300$ K being obtained with the Molière potential.

A scattering study from a discrete atomic row was found also useful in the investigation of the channeling processes. The limited scope of the existing analytical theories of the channeling effect at small depths calls for the need of mathematical simulation of particle trajectories in a crystal. Indeed, using the continuous potential model in describing the oscillations in the depth dependence of the ion backscattering yield (Abel et al. 1976, Agranovich and Ryabov 1970) does not permit a correct approximation of the quasichanneling and obtaining agreement with experimental data. The theory of the channeling effect at small depths based on the Fokker–Planck equations yields good results, but only for very small angles between the initial beam's incidence direction and the crystallographic axes (or planes).

When describing quasichanneling and orientational effects at small depths, one should take into account that quasichanneling particles penetrate in the inner regions of the atomic rows and planes, thus requiring an analysis of individual collisions of these particles with atoms in the crystal. The binary collision model permits the inclusion of the projectile interaction with individual atoms in the crystal, a consideration of particle scattering with any values of transverse energy, investigation of crystals of any chemical composition taking into account thermal vibrations and atomic displacements, as well as various radiation-induced defects.

The model of scattering from a discrete atomic row, first developed by Kivilis et al. (1967, 1970), was later used in numerous studies of channeling (Kadmensky et al. 1974, Kadmensky and Samarin 1982, 1983, Kadmensky and Tulinov 1973) where scattering from a row is considered as a fundamental interaction event in axial channeling and in the investigation of the shadowing effect (Ryabov 1968). Kovalyova and Shipatov (1979, 1980a, 1980b, 1983) used a computer simulation to study scattering from binary atomic rows, i.e., rows consisting of atoms of two species. It was shown (Kovalyova and Shipatov 1983) that since the parameters characterizing the motion of quasichanneled particles are determined by the size of the shadow and by the density of trajectories at its boundary, as well as by the cross section of scattering from atoms in the row, the structure and composition of the rows affect strongly the escape of backscattered quasichanneled particles. In particular, there may appear two groups of particles quasichanneled relative to the binary row. An analysis was made of the backscattering of ions from a K–F atomic row oriented along the ⟨100⟩ direction in KF (Kovalyova and Shipatov 1979, 1980b), from a Pb–O row lying in the ⟨110⟩ direction in the cubic phase of $PbTiO_3$, as well as scattering of He^+ ions from ⟨001⟩, ⟨110⟩, and ⟨111⟩ rows in GaP and GaAs crystals (Kovalyova and Shipatov 1980a). A study was made of the yield of scattered ions as a function of their atomic number and initial energy. For binary atomic rows consisting of atoms of various species, the dominant contribution to backscattering comes from atoms with a large atomic number Z. This accounts for the increase of the backscattering cross section with atomic number Z. A comparison of backscattering probabilities for monoatomic rows also shows that a high yield is observed for rows consisting of atoms with a large atomic number. The cross section of large-angle scattering and the angular width of the shadow decrease with increasing E_0. Therefore, an increase in the energy results in a decrease in the peak height and in the angles at which this peak is observed.

The peak region in the angular dependence of the scattering intensity is sensitive to atomic row structure. The dependence of the angular width of the shadow on Z produces a fine peak structure in the case of binary atomic rows. Thermal vibrations of atoms broaden the peaks, reduce their height and smear out the fine structure in their distributions.

3. Spatial distribution of scattered ions

Establishing a mutual one-to-one correspondence between the direction of
the scattered ion and its energy on the one hand, and the target point at the
atomic row, the incidence direction of the ion, and its initial energy on the
other, provides a clue to the interpretation of experimentally measured
angular and energy distributions of scattered ions and offers a possibility of
studying the atomic structure of crystal surface.

Nizhnaya et al. (1979a) and Marchenko et al. (1972, 1973) used the binary
collision model to calculate scattering from a three-dimensional atomic row.
Figure 4 of chapter 2 shows schematically ion scattering in a three-dimensional atomic row. The computer simulation technique was employed to
study Ar^+ ion scattering with $E_0 = 30$ keV under glancing incidence on a Cu
[110] row. A square target area of 2.55×2.55 Å^2 with sides parallel to the
$\langle 110 \rangle$ direction was chosen on the (100) plane. The sides of the square were
divided into 100 (r-coordinate) and 200 (q-coordinate) segments. The nodes
of the grid thus formed served as target points. The incidence and escape
planes made angles ξ and φ with the axis of the rows, the initial and final
asymptotes of the ion trajectory forming grazing, ψ, and escape, β, angles
with the (100) plane, respectively. One calculated the scattered ion energy E
and angles φ and β as functions of the angles ξ and ψ and target point
coordinates, r and q. Figure 6 shows typical angular and energy distributions
of 30 keV Ar^+ ions scattered from a $\langle 100 \rangle$ row in Cu for $\psi = 13°$ and $\psi = 14°$
at $\xi = 0$. As seen from the figure, the ions with the target points lying on
$q = $ const straight lines are scattered into ovals, the positions of the points on
an oval being determined by the values of the coordinate r (specified next to
the oval). The oval corresponding to aiming exactly along the ridge of the
row, $q = 0$, degenerated into a straight line. The inner and outer boundaries
of the scattered beam are clearly distinguished, the oval envelopes defining
the minimum and maximum scattering angles. As the azimuthal angle φ
increases, the beam narrows down. Note that blocking on the side of large
escape angles increases, and on the small angles decreases. Figure 6 presents
the dependence of scattered particle energy on azimuthal escape angle for
different distances from the axis of the row, i.e., for different q. We see that
the angular size of the ovals decreases with increasing q, the ovals themself
shifting towards smaller escape angles. The boundaries of the scattered
beam drift apart as ψ increases.

$E(\theta)$ plots for different ψ are presented in fig. 7 for a plane problem.
Calculations show that $E(\theta)$ ovals grow with increasing ψ and shift toward
larger scattering angles and, hence, lower energies. As ψ increases, the $E(\theta)$
ovals suffer a break up in their right-hand part. This break up is connected
with the projectile ions becoming incorporated into the atomic row. Shown
on the left in fig. 7 is the $E(\theta)$ dependence for $\langle 100 \rangle$ and $\langle 110 \rangle$ rows.

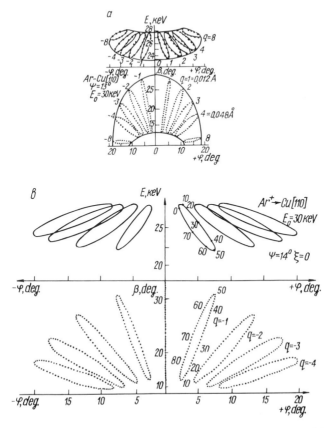

Fig. 6. Angular and energy distributions of scattered ions for grazing angles (a) $\psi = 13°$, and (b) $\psi = 14°$. Direction of incidence coincides with the row axis ($\xi = 0°$).

Scattering from a less closely packed $\langle 100 \rangle$ row yields a broader spatial and energy distribution of the scattered ions than that from a $\langle 110 \rangle$ row. The lower part of the $E(\psi)$ ovals corresponds to quasisingle scattering, where the energy losses are larger, and the upper oval part to quasidouble scattering with lower energy losses.

In the three-dimensional case, the right-hand parts of the ovals correspond to trajectories involving sequential scattering at two comparatively lage angles and a number of small angles and, in this sense, represent an analog of double scattering, whereas the left-hand parts relate to single scattering, i.e., to trajectories involving one large and a number of small scattering angles. Just as in the plane case, the right- and left-hand parts of the ovals are separated in energy; however, as a result of the quasisingle- and quasidouble-scattering trajectories being bent differently they are sepa-

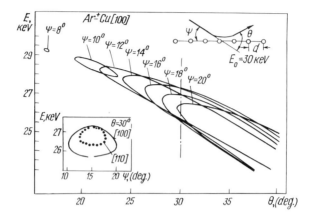

Fig. 7. Energy versus scattering angle plot for different ψ. The inset shows $E(\psi)$ dependence for the $\langle 100 \rangle$ and $\langle 110 \rangle$ rows.

rated in space as well, and this is what results in the appearance of characteristic ovals in the $E(\theta)$ dependence. Shown in the upper part of fig. 6 is the $E(\varphi)$ dependence. To each point on the oval in the energy distribution corresponds a point on the oval in the angular distribution and a point in the target area, r and q. The high-energy part of the spectrum corresponds to the inner boundary of the reflected beam (small β and φ), and the low-energy part to the other boundary. The angular distribution of reflected particles does not have ovals (for $\beta > 0$) extending beyond $q = 10$. For $q > 10$, the scattered ions escape under the plane ($\beta < 0$). This means that the only contribution to ion reflection in this model comes from a narrow band exactly above the ridge, its width being $\frac{20}{200}$ of the interatomic separation along the $\{110\}$ axis (0.25 Å). The closer the target points are to the ridge of the atomic row, the closer to the incidence plane is the ion escape trajectory, in other words, the smaller is q, the smaller will be φ. For $q \neq 0$, no ions will be reflected in the plane of incidence altogether.

4. Effect of thermal vibrations and interaction potential

Parilis et al. (1967) were the first to propose a model for scattering from an atomic row including thermal vibrations. This model was based on correlated thermal vibrations. They were introduced as waves of wavelength $2d$ propagating along a row, which by Debye's theory correspond to the lowest and most intense frequency. Kivilis et al. (1970) and Parilis et al. (1967) considered interaction of a projectile with an atomic row bent by thermal vibrations and frozen in its extreme position. The rms displacement of each atom from the corresponding equilibrium position is a fraction of the mean radius of the cell, r_s (Maradudin et al. 1963),

$$\sigma^2 = \hbar^2 T / m_2 k \theta_D^2 . \tag{18}$$

Only the atomic displacements in the direction perpendicular to the row were taken into account. Kivilis et al. (1970) calculated the energies of ions with $E_0 = 5$, 10, 30 keV scattered from a Cu [110] row for two values of the target temperature (fig. 5). The first temperature corresponded to the fixed ideal lattice, and the second to an amplitude of atomic vibrations of $0.1d$ (d is the interatomic separation in the row). As seen from fig. 5, the ovals in the $E(\theta)$ dependence grow with increasing temperature, which means that

(1) the interval of possible scattering angles increases for given ψ and E_0;

(2) the distance between the peaks in the energy spectrum grows, the peaks themself shifting towards higher energies.

It was shown that the $E(\theta)$ ovals change in a similar way when E_0, ψ, d and T change in the same direction. The thermal vibration model proposed by Kivilis et al. (1970) was considered in more detail by Nizhnaya et al. (1976, 1979a). The calculations were carried out for rows with different interatomic separations (d, for an ideal row, and $2d$, for a one half as closely packed row) and for different thermal displacement amplitudes of surface atoms.

Figure 8 shows $E(\theta)$ plots for Ar$^+$ ions with $E_0 = 30$ keV scattered from Cu (100) in the [100] direction. Presented on the left is an $E(\theta)$ plot for scattering from a row with interatomic separation d, and on the right the

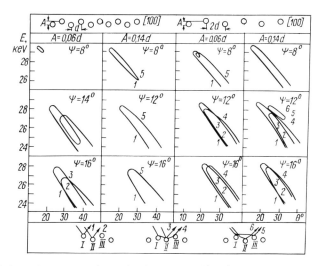

Fig. 8. $E(\theta)$ dependence for Ar$^+$–Cu (100) scattering ($\langle 110 \rangle$ direction, $E_0 = 30$ keV, $A = 2\sigma$).

same dependence for a row with separation $2d$, for thermal displacement amplitudes $\sigma = 0.03d$, $\sigma = 0.07d$, and difference $\psi(A = 2\sigma)$. The types of the trajectories involved are shown schematically below. We see that as the thermal vibration amplitude increases, the ovals not only grow and shift toward lower energies but also reveal in some cases a more complex spectrum structure. Peaks 1, 3 correspond to single and double scattering from a thermal 'downward' step, and peaks 2, 4 to that from an 'upward' step. The difference in their position originates from different blocking actions of the ion-facing atom. For smaller ψ and a larger vibration amplitude the ion interacts, as it were, with a row which is only one half as closely packed (peaks 1, 5). For a thermal vibration amplitude of $\sigma = 0.07d$, and $\psi = 12-14°$, a six-peak structure appears in the case of the row which is one half as dense, where peaks 1–5 are the same as before, and peak 6 comes from quasitriple collisions similar to those revealed by Verhey et al. (1975).

Nizhnaya et al. (1979a) studied the effect of the interaction potential on the angular and energy distributions of backscattered ions for different thermal displacement amplitudes of atoms in a row.

Figure 9 presents the results of $E(\theta)$ calculations made with two potentials: (1) the Firsov potential, (2) and (3) the potential of Sigmund and Vajda (1964) with two sets of constants:

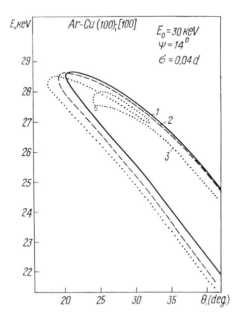

Fig. 9. $E(\theta)$ dependence calculated for three potentials on a row with thermal vibration amplitude $\sigma = 0.04d$.

(a) $r_c = 3$ Å, $B_2 = 447$ Å^2 eV, $B_1 = -278$ Å eV, $B_3 = 43$ eV;

(b) $r_c = 1.752$ Å, $B_2 = 350$ Å^2 eV, $B_1 = -389.453$ Å eV, $B_3 = 114.025$ eV for a row with a thermal vibration amplitude $\sigma = 0.04d$. It was shown that the application of a shorter-range (harder) potential broadens the $E(\theta)$ oval and lowers it down into the region of lower energies, i.e., acts in the same way as an increase of interatomic separation in a surface atomic row, or of the thermal displacement amplitude. This may be accounted for by the fact that the harder the potential is, the stronger it is cut off, the less will be the number of atoms in the row that participate in the scattering, and the less will be, accordingly, the blocking by the neighboring atoms. The same weakening of blocking occurs in scattering from a cut-off row or with increasing interatomic separation in the row. When calculated with potential 2, the scattering spectrum reveals a more complex than double-peak structure, in particular, a peak appears that corresponds to quasitriple collisions and that is absent when the other potential is used, since under strong blocking (potentials 1, 3) the lower-lying atom in the bent row is screened by the neighboring atoms, so that scattering occurs, as it were, from a row with twice the initial separation, the use of shorter-range potentials weakening the blocking. However, on the energy scale the quasitriple peak lies below the double collision one which appears when other potentials are employed. Hence, when comparing calculations with experiment, the major attention should be focused on the shape of the energy spectrum, which calls for a sufficiently high-energy resolution. Apart from this, it was shown that similar $E(\theta)$ ovals can be obtained also with different interaction potentials and different thermal displacement amplitudes. This creates additional difficulties in the choice of the interaction potential or in the determination of the thermal displacement amplitude σ from a comparison of calculations with experiment. Since σ depends on the target temperature T and surface Debye temperature θ_D, one can, knowing T and finding θ_D from other sources, choose properly the interaction potential governing the ion scattering by an atomic row.

Nizhanya et al. (1979a) and Marchenko et al. (1973) considered a model of thermal vibrations by which the projectile ion interacts with an atomic row frozen in a prescribed position. A comprehensive analysis of other models was carried out by Poelsema et al. (1975, 1976a,b, 1977a,b), as well as by Walker and Martin (1982), who studied uncorrelated thermal vibrations with a Gaussian distribution and correlated vibrations with different correlation coefficients.

Poelsema et al. (1975) investigated qualitatively the behavior of the scattered ion energy spectra with increasing T. Figure 10 illustrates the effect of thermal vibrations on the shape of the energy distributions of the ions backscattered from an atomic row, and the characteristic ion trajectories contributing to different energy regions. Figure 10a presents the

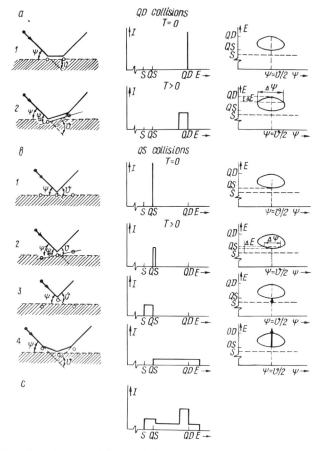

Fig. 10. Trajectories of scattered ions and formation of energy spectra in the presence of thermal vibrations of the atoms in a row.

formation of the QD peak. Shown above (1) are (left-to-right) the trajectory of the QD peak for a non-vibrating row $(T = 0)$, the energy distribution for this type of trajectory, and an $E(\Psi)$ plot with the corresponding value of QD energy for $\psi = \frac{1}{2}\theta$. In fig. 10a (2), the atoms are displaced from their equilibrium positions because of thermal vibrations $(T > 0)$. In this case, the grazing angle ψ may differ from the macroscopic angle ψ. Because of thermal vibrations, the angle ψ_a lies in the interval $\Delta\psi$ around ψ. This interval is specified on the $E(\psi)$ plot. It corresponds to the energy interval ΔE for QD scattering presented in the same plot. Thus, thermal vibrations of the atoms in the row result in a broadening of the QD peak in the energy distribution; obviously, this broadening leads to a reduced height of the peak.

Figure 10b illustrates the effect of thermal vibrations on QS scattering. The process of QS scattering from a fixed row is presented in fig. 10b (1). For QS scattering (for $T > 0$), one can also envisage configurations of the row where the true grazing angle ψ_a will lie in an interval $\Delta\psi$ around the macroscopic angle ψ, which will result, just as in the case of QD collisions, in a certain broadening of the QS peak and a reduction of its intensity [fig. 10b (2)]. However, in the case of QS scattering the interval of allowed angles $\Delta\psi$ is only one half of that for QD collisions because of the doubling of the separation between the atoms in the row that determine the ion trajectory. However, three atoms do not necessarily lie on one straight line, and in the case where scattering occurs primarily from the atom located above its neighbors, QS scattering approaches the single collision (S). This is shown in fig. 10b (3).

Finally, the diagram in fig. 10b (4) illustrates the situation where the central atom lies below the two neighboring atoms. In this case the scattering reveals more pronounced multiple features than that from a fixed row. Naturally, if a scattering angle θ is actually a sum of three angles θ_i, where $\theta_i = \frac{1}{3}\theta$, the energy losses will be smaller than those under QD scattering at the same angle. This collision was called quasitriple (QT), and the corresponding peak the quasitriple peak. This peak appears in the case where thermal displacements are sufficiently large, i.e., for sufficiently high T. Poelsema et al. (1975) investigated changes in the spatial distribution of scattered ions induced by increasing T. A study was made of the energy $E(\psi)$ (fig. 11a) and of the total scattered ion yield $I(\psi)$ (fig. 11b) for a fixed scattering angle θ, as well as of $I(\theta)$ (fig. 12c) and $E(\theta)$ (fig. 12a) for a fixed ψ. For $T = 0\,\mathrm{K}$ and $\psi = \psi_{\min}$ or $\psi = \psi_{\max}$, the scattered ion intensity is infinite. This property is associated with a singularity in the behavior of the derivative $\mathrm{d}\theta/\mathrm{d}p_1$ in the region of external angles (Martynenko 1973, Parilis et al. 1975), $\mathrm{d}\theta/\mathrm{d}p_1 = 0$ (see Fig. 3a). When thermal atomic displacements are included, the 'effective' grazing angle ψ_a differs from ψ by an amount $\Delta\psi$ which grows with temperature.

Thus, for $T \neq 0\,\mathrm{K}$, scattering at an angle θ is possible for ψ angles lying in the interval $\psi_{\min} - \Delta\psi_a \leqslant \psi \leqslant \psi_{\max} + \Delta\psi_a$. The range of allowable ψ angles expands with increasing temperature, its boundaries becoming diffuse. As a result, $I(\psi)$ will no more be infinite for $\psi = \psi_{\max}$ and $\psi_{\min}$. The behavior of $I(\psi)$ for different temperatures is shown schematically in fig. 11b.

In a similar way one can explain the behavior of the $E(\theta)$ and $I(\theta)$ dependences (fig. 12a,b,c). As seen from fig. 12b, the interval of allowed scattering angles $\theta_{\min} \leqslant \theta \leqslant \theta_{\max}$ grows with increasing ψ in such a way that $\Delta\theta_{\max} = 2\Delta\psi + \Delta\theta_+$, and $\Delta\theta_{\min} = -2\Delta\psi + \Delta\theta_-$. Since $\Delta\theta_+$ has the same sign as $\Delta\psi$, and $\Delta\theta_-$ the opposite one, the expansion of the angular interval $\Delta\theta_{\max}$ towards larger angles is greater than that towards smaller θ.

Poelsema et al. (1976b) considered the possibility of using a two-dimen-

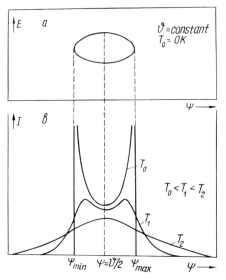

Fig. 11. (a) $E(\psi)$ dependence for $T = 0\,\mathrm{K}$, and (b) schematic representation of the spatial distribution of ions scattered at an angle θ.

sional atomic row when calculating scattering along low-index directions. To do this, scattering from a three-dimensional atomic row was calculated with the inclusion of thermal vibrations which were assumed to be Gaussian and uncorrelated. Account was taken of the thermal displacements occurring both in the plane of incidence passing through the atomic row and in the plane perpendicular to it.

Figure 13 compares energy distributions calculated in terms of the three- and two-dimensional models of a vibrating atomic row. The shape of the energy spectrum depends only weakly on the actual type of the model used. This means that application of the two-dimensional thermal displacement model to an atomic row yields sufficiently good results.

Boers (1977), Verhey et al. (1975), Poelsema et al. (1975, 1976a, 1976b, 1977b) studied the combined effect of the various parameters, such as the correlation coefficients, potential parameters and surface Debye temperature, on the shape of the energy spectrum. For this purpose, the scattering of Ar^+ ions with $E_0 = 6\,\mathrm{keV}$ from a Cu(100) row was calculated for different values of the surface Debye temperature (θ_D) and different screening parameters in the Molière potential. The screening potential was varied by multiplying a_F by c, where c was chosen to be equal to 1 (M1 potential), 0.875 (M2), 0.75 (M3) and 0.5 (M4). Figure 14 shows these potentials, as well as the potentials of Firsov (F) and Born–Mayer with Abrahamson's (A) (Abrahamson 1969) and Andersen and Sigmund's (AS) (Andersen and Sigmund 1965) constants.

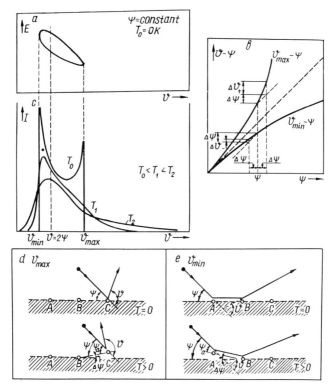

Fig. 12. (a) $E(\theta)$ relations for $T = 0$ K, (b) dependence of the boundary scattering angles θ_{max} and θ_{min} on ψ; and effect of thermal vibrations of the atom row on (c) the spatial distribution of scattered ions for a given ψ and different T, and on the quantities (d) θ_{max}, and (e) θ_{min}.

The results of the calculation are presented in fig. 15 for (a) $\theta = 30°$ and (b) 60°. A change of the screening parameter in a potential produces a stronger effect at $\theta = 30°$. As one crosses over from the M4 to M1 potential, the spectrum takes on an increasingly more quasisingle pattern, in other words, the relative intensity of single scattering grows compared to the double scattering. The two-peak structure of the spectrum becomes washed out by thermal vibrations. The effect of temperature is stronger when a potential with a large screening parameter is used. Figure 15 (a2 and b2) illustrates the influence of θ_D on the shape of the energy spectra. As θ_D decreases, the rms thermal displacement of atoms increases and the spectra becomes washed out just as this occurs in a transition from the M4 potential to M1. This complicates the proper choice of the potential parameters on the one hand, and of the surface Debye temperature on the other. Boers (1977) suggests a possibility for determining these parameters. θ_D can be derived from the appearance of the QT peak in the high-energy part of the

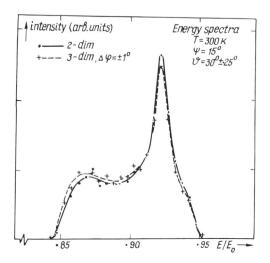

Fig. 13. Comparison of the energy distributions of 6 keV Ar^+ ions scattered from a (1) two-dimensional, and (2) three-dimensional vibrating Cu (100) row for $T = 300$ K and $\Delta\varphi = \pm 1°$.

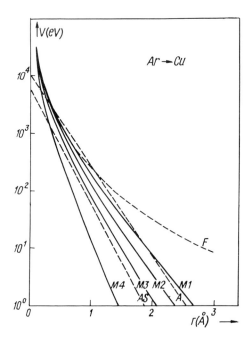

Fig. 14. Dependence of various interaction potentials on separation between colliding particles.

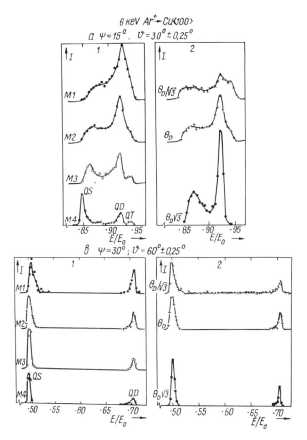

Fig. 15. Effect of (1) potential parameters, and (2) Debye temperature on the shape of the energy spectrum of 6 keV Ar$^+$ ions scattered from Cu (100) for $T = 300$ K. (a) $\psi = 15°$, $\theta = 30° \pm 0.25$, (b) $\psi = 30°$, $\theta = 60° \pm 0.25°$.

scattered ion spectrum at a sufficiently high target temperature. Calculations show that the QT peak intensity depends on the choice of the interaction potential, while the appearance of the peak, i.e., the temperature cut-off at which it appears (fig. 16) is practically independent of the choice of the potential and is rather determined primarily by the thermal displacement of surface atoms which at a given temperature depends on θ_D.

In our experiments, the QT peak is observed if the target temperature is high enough. By comparing the measured temperature cut-off with the calculation, one can derive the Debye temperature of the vibrations perpendicular to the surface. Knowing θ_D, a comparison of experimental with calculated spectra can yield the parameters of the potential. For Cu it was found that $\theta_D = 147$ K $\pm 6\%$, to be compared with the experimental value

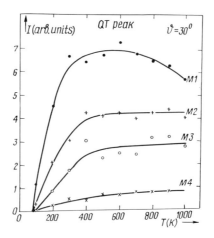

Fig. 16. A QT-peak intensity versus temperature plot for different parameters of the potential.

$\theta_D = 133$ K $\pm 3\%$ (Jackson 1974). The discrepancy provides information on the degree of correlation between the surface atoms.

In order to investigate the effect of correlation between thermal vibrations on the energy spectra of scattered ions, a calculation was made of Ar^+ ions scattering from a Cu atomic row for various values of the correlation coefficients, including $K = 0$, i.e., for uncorrelated vibrations. The calculation was performed for equal mean-squared differences between the thermal displacements of neighboring atoms. It was found that the QS and QD scattering intensities are governed primarily by this rms quantity, and only the QT intensity is lower, as the correlation coefficient K is greater (fig. 17a). This can apparently be accounted for by the fact that the conditions for QT collisions become unfavorable with increasing K, since the 'wells' in the atomic row required for scattering at an angle $\theta = 30°$ are smoothed out. Having prescribed the value of θ_D and calculated the QS peak intensity versus ψ, one can derive the screening parameter by varying it properly. The same goal can be reached by studying the intensity of the main peak as a function of θ for various E_0. The screening parameter is here determined from a comparison of measured angles θ_{max} with those calculated for different screening parameters in the Molière potential.

Heiland et al. (1976) compared the results of calculations within the atomic row model with a computer simulation using the MARLOWE program and showed that the atomic row model can be used in determining the blocking angles of the backscattered beam, the screening parameter in the Thomas–Fermi potential, and the surface Debye temperature. In table 1, we present experimental and calculated values of the screening parameter in the Molière potential.

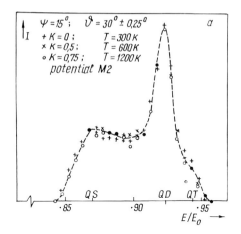

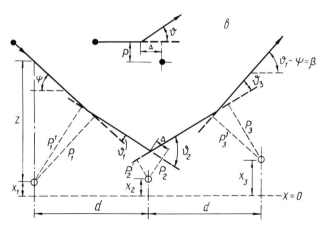

Fig. 17. (a) Effect of correlation between thermal atom displacements on spectral shape, and (b) three-atom scattering model.

Table 1
The values of screening parameters.

Pair	Experiment (Heiland et al. 1976)	Lindhard (1969)	Firsov (1957)
Ne–Ni	0.0945 ± 0.002	0.1258	0.1129
Ne–Ar	$0.105 \ \pm 0.005$	0.1115	0.1008
Ne–W	$0.094 \ \pm 0.005$	0.0993	0.0905

The values of θ_D obtained by Heiland et al. (1976) are close to experiment (Jackson 1974). The quantitative discrepancies apparently originate from the simplified model of thermal vibrations (uncorrelated vibrations with a Gaussian distribution) accepted by Jackson (1974).

Thus, a correct inclusion of thermal vibrations is not less essential for the description of scattering than the knowledge of the parameters of the interaction potential. These parameters cannot be chosen from a comparison with experiment without proper inclusion of thermal vibrations. A further improvement of the model is based on inclusion of surface defects on the one hand, and of the charge exchange effects on the scattered ions on the other.

5. Three-atom scattering model

Poelsema and Boers (1978) proposed a simplified semianalytical model where the interaction of a projectile with an atomic row is described as a sequence of collisions with three atoms. The model includes correlated thermal vibrations of atoms and permits obtaining the energy and angular distributions of scattered particles by calculation of one integral only.

Instantaneous displacements of atoms x_1, x_2 and x_3 from equilibrium positions are prescribed (fig. 17b). The trajectory asymptotes are determined by the collision parameters, LS scattering angle θ and the quantity Δ (M.T. Robinson and Torrens 1974).

Scattering at an angle θ from three atoms is possible if

$$u \equiv x_1 - x_2 = [p_2(\theta_2) - p_1(\theta_1) + d \sin(\psi - \theta_1)] \cos^{-1}(\psi - \theta_1) , \qquad (19)$$

$$v \equiv x_2 - x_3 = [p_3(\theta_3) - p_2'(\theta_2) + d \sin(\psi - \theta_1 - \theta_2)] \cos^{-1}(\psi - \theta_1 - \theta_2) , \qquad (20)$$

where $\theta_1 + \theta_2 + \theta_3 = \theta$; and $p' = p \cos \theta + \Delta \sin \theta$ is the impact parameter after interaction.

Thermal displacements have a Gaussian distribution with a standard deviation σ. For the probability density function for the displacement matrix $\mathbf{X} = (x_1, x_2, x_3)$ we can write

$$g(\mathbf{X}) = (2\pi)^{-3/2} |\mathbf{V}|^{-1/2} \exp(-\tfrac{1}{2}\mathbf{X}^T\mathbf{V}^{-1}\mathbf{X}) , \qquad (21)$$

where $\mathbf{V}$ is a covariance matrix.

The yield of ions backscattered with a given relative energy E at a given angle θ is a product of the probability of existence of a given atomic structure, of the collision sequence and of the scattering probability. The density of the scattered ion distribution probability on the axis is

$$f(x_1, x_2, x_3, z) = N_0 \tan^{-1} \psi \, g(\mathbf{X}) , \tag{22}$$

where N_0 is the number of incident ions per unit length.

The probability density function for the variables u, v, z is found from the expression

$$f_2(u, v, z) = \frac{N_0}{2\pi\sigma^2 N^{1/2} \tan \psi}$$

$$\times \exp\left[-\frac{(1 - p_{23})u^2 + (1 - p_{12} + p_{13} - p_{23})uv + (1 - p_{12})v^2}{N\sigma^2} \right],$$

$$\tag{23}$$

where

$$N = 1 - p_{12}^2 - p_{13}^2 - p_{23}^2 + 2p_{12}p_{13}p_{23} + 2(1 - p_{12})(1 - p_{13})(1 - p_{23}) , \tag{24}$$

and p_{ij} are the correlation coefficients.

For uncorrelated thermal displacements,

$$f_2(u, v, z) = \frac{N_0}{2\pi\sigma^2\sqrt{3} \tan \psi} \exp\left[-\frac{u^2 + uv + v^2}{3\sigma^2} \right]. \tag{25}$$

The energy of the ion scattered at an angle θ with inclusion of elastic losses is determined by the relation

$$E = E_0\{[\cos\theta + (\mu^2 - sin^2\theta)^{1/2}](1 + \mu)^{-1}\}^2 , \tag{26}$$

which for small θ can be replaced by

$$E = E_0 \exp(-\theta^2\mu^{-1}) . \tag{26a}$$

The Taylor expansions eqs. (26) and (26a), coincide down to fourth-order terms. Equation (26a) predicts for the Ar–Cu pair the energy to within 1% for scattering angles up to $\theta = 85°$. For scattering from three atoms,

$$E = E_0 \exp[-(\theta_1^2 + \theta_2^2 + \theta_3^2)\mu^{-1}] . \tag{27}$$

Thus, for the probability density function for the variables (E, θ, θ_1) one can write

$$f_2(E_1, \theta, \theta_1) = f_2(u, v, z)|D| , \tag{28}$$

where D is a functional determinant:

$$D = \frac{\delta z}{\delta\theta_1} \frac{\delta u}{\delta\theta_2} \frac{\delta v}{\delta\theta_3} \left(\frac{\delta E}{\delta\theta_3} - \frac{\delta E}{\delta\theta_2} \right)^{-1} . \tag{29}$$

From eqs. (19), (20), (27) and (28), we obtain

$$f'_2(E, \theta, \theta_1) =$$

$$f_2(u, v, z) \left| \frac{\dfrac{dp_1}{d\theta_1} \dfrac{dp_2}{d\theta_2} \dfrac{dp_3}{d\theta_3}}{\cos \psi \cos(\psi - \theta_1) \cos(\psi - \theta_1 - \theta_2)} \frac{\mu}{2E(\theta_3 - \theta_2)} \right|. \tag{30}$$

The yield of the ions scattered with an energy E at an angle θ is obtained by integrating eq. (30) over all possible angles,

$$f_3(E, \theta) = \int_{\theta_{1\,min}}^{\theta_{1\,max}} f'_2(E, \theta, \theta_1)\, d\theta_1 . \tag{31}$$

Thus, the energy distribution of the ions scattered in the three-atom model is derived by calculation of a simple integral eq. (31), for various energies E.

Three-atom model calculations were carried out with the Molière potential using different screening parameters. The angular and energy distributions obtained in this way correlate with the Monte Carlo simulation studies. The proposed calculation describes well the thermal vibrations in the angular distribution of scattered ions, however, the three-atom approximation fails for small ψ as the temperature increases (at $t = 700$ K).

The energy distribution is also well approximated by this model, particularly in the low-energy range, since the region between the quasisingle and quasidouble peaks is not affected by inclusion of interaction with the central atom, as this is required by the model. The difference between the results obtained with the three-atom model and by Monte Carlo simulation decreases for potentials with a smaller screening parameter. Note that while the three-atom model describes well the energy and angular distributions, it is inapplicable to the simulation of channeling and blocking.

On the whole, the atomic row model has played a significant role in the development of the theory of medium-energy atom scattering from the surface of single crystals. It should be stressed that its success originates not only from a correct description of the correlated nature of collisions under glancing incidence of atoms onto a surface, but also from the fact that the atomic row is a real structure on the surface of a single crystal which provides a major contribution to scattering under glancing incidence.

6. Atomic row effect for an inverse mass ratio

The previous description of the atomic row effect and of its experimental verification corresponds to the direct mass ratio of colliding particles, i.e., to cases where the mass of the projectile m_1 is less than that of an atom in the row m_2, so that $\mu = m_2/m_1 > 1$. At the same time, in experimental studies of small-angle ion scattering from the surface of crystals, in particular, of alloys, a situation may arise where an alloy component will be lighter than the projectile ion. It would thus be of interest to investigate the atomic row effect for the inverse mass ration of collision partners, i.e., where the mass of the ion is greater than that of the atom in the row, and $\mu = m_2/m_1 < 1$ (Parilis et al. 1989a).

The atomic row effect manifests itself in this case in a number of features associated with the existence of a limiting scattering angle $\theta_{lim} = \arcsin(f\mu)$ in a single collision, as well as with the fact that scattering at a given angle $\theta < \theta_{lim}$ is possible for two values of the impact parameter p. The scattered ion energy determined by eq. (33) of chapter 1 is a double-valued function of the scattering angle, with the '+' sign in the expression corresponding to the larger ($p > p_{lim}$), and the '−' sign to the smaller ($p < p_{lim}$) of the two impact parameters. Here, $p_{lim}(E_0)$ is the impact parameter corresponding to scattering at an angle $\theta = \theta_{lim}$.

Computer simulation was used to study in the binary collision approximation the trajectories of $E_0 = 15$ keV Kr^+, Xe^+ and Rn^+ ions scattered from atomic row ridges on Cu (100) in the $\langle 110 \rangle$ direction with a grazing angle interval $\psi = 5$–$25°$. The calculations were carried out with two interaction potentials, namely, that of Biersack–Ziegler, eq. (16) of chapter 1 and a matched one, eq. (20) of chapter 1, with inclusion of elastic and inelastic energy losses. Figure 18 presents dependences of the energy E retained by the Kr^+ ions scattered from copper atomic rows on the scattering angle θ for $\psi = 17.5$ and $22.5°$. The broken curve shows a similar

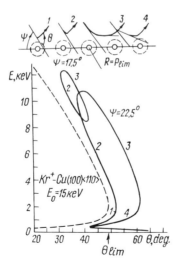

Fig. 18. Energy E retained by the Kr^+ ions scattered from copper atom rows versus the scattering angle θ for $\psi = 17.5°$ and $22.5°$.

dependence for single scattering. In this case, $\mu = 0.758$, $\theta_{\lim} \simeq 49.3°$, and $p_{\lim} \simeq 0.10$ Å. A trajectory analysis revealed that for small grazing angles (in the Kr$^+$–Cu case, up to $\psi = 20°$) and for $\mu < 1$, the usual atomic row effect is observed, just as in the case $\mu > 1$. Indeed, for $\psi = 17.5°$ one observes a double-valued dependence $E(\theta)$ caused by the conventional quasisingle (branch 2) and quasidouble (branch 3) scattering, as well as blocking on the sides of small and large scattering angles. Note that for small ψ the scattering angles in the case of an atomic row are smaller than $\theta_{\lim}$ for a single collision, the ions do not approach close enough the atoms in the row, and the impact parameters $p_i > p_{\lim}$. As shown on the right of fig. 18, one observes in this case the usual behavior of $\theta(p)$, with the plus sign taken in the expression for energy, eq. (33) of chapter 1. As the grazing angle increases ($\psi = 22.5°$), small impact parameters $p_i < p_{\lim}$ for which the $\theta(p)$ dependence reverses its course become accessible, adding new types of trajectories $(1, 4, 5)$ to the conventional ones, QS and QD $(2, 3)$, which results in a substantial change of the $E(\theta)$ behavior.

Trajectory 1 corresponds to QS scattering with an impact parameter in the principal collision $p < p_{\lim}$, trajectory 4 to QD scattering with an impact parameter in one or two strong collisions $p < p_{\lim}$, and trajectory 5 to multiple scattering with $p < p_{\lim}$ in one or several collisions. A specific feature of the atomic row effect in this case is the appearance of additional branches $(1, 4, 5)$ in the $E(\theta)$ dependence, which should result in the formation, in the energy spectrum of scattering at angles θ close to $\theta_{\lim}$ of three peaks in the low-energy part, as well as in scattering at angles θ considerably in excess of $\theta_{\lim}$. Indeed, for Rn$^+$ ions and for $\psi = 20°$, scattering angles are possible that exceed by a factor three $\theta_{\lim} \simeq 16.6°$ for a single collision.

The lower panel of fig. 19 shows the angular distribution of $E_0 = 15$ keV Kr$^+$ ions scattered from a copper atom row at a grazing angle $\psi = 20°$. In contrast to the case of direct mass ratio, one observes here additional maxima caused by the rainbow effect in QS (1), QD (4) and multiple (5) scattering. Taking into consideration the thermal vibrations of the atomic row results in a blurring of the peaks rather than their merging since the separations between them are fairly large.

The pattern of the scattering of heavy ions (Kr$^+$, Xe$^+$, Rn$^+$) from a row of light atoms (Cu) exhibits the following variation as a function of the grazing angle: for $\psi \leqslant 10°$, the scattering is specular in all cases, i.e., the escape angle $\delta = \psi$; for $\psi > 10°$ the region of possible scattering angles grows with increasing ψ, and the atomic row effect makes scattering at angles greater than $\theta_{\lim}$ possible with a factor three excess reached, as already mentioned, with Rn$^+$ ions at $\psi = 20°$. Interestingly, not only the maximum but also the minimum escape angles exceed in this case the specular angle, so that the scattering becomes supraspecular. As ψ increases still more, collisions

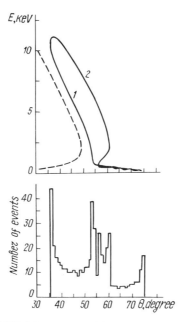

Fig. 19. $E(\theta)$ oval for 15 keV Kr$^+$ ions scattered from copper atom rows (top panel), and the corresponding angular distribution (bottom panel).

become essentially multiple (while for a direct mass ratio the number of collisions, in this case, conversely, decreases), with a fraction of the ions beginning to penetrate through the atomic row, just as in the $\mu > 1$ case.

Thus, the specific features of the atomic row effect observed with the inverse mass ratio of colliding particles result in more complex angular and energy distributions of scattered ions which makes studying the latter an interesting experimental problem. Note that the experimental studies of small-angle scattering in the case of inverse mass ratio for the pairs Kr$^+$–Cu (100), Ar$^+$–Si (100) (Mashkova and Molchanov 1980) and Kr$^+$–V (Eckstein et al. 1974) were carried out for θ smaller than the limiting scattering angle θ_{lim}, whereas more interesting from the standpoint of the results obtained here is the measurement of scattering energy spectra for $\theta > \theta_{\mathrm{lim}}$.

7. Ion scattering from complex composition surfaces

The atomic row effect exhibits remarkable features in small-angle ion scattering from a single-crystal surface of complex composition when the rows in certain crystallographic directions consist of alternating atoms of different species. It was shown, e.g. (U.A. Arifov 1968), that in the case of a KCl compound the energy distributions of scattered particles reveal distinct peaks corresponding to single- and double scattering from the atomic combinations K–K, Cl–Cl, K–Cl, and Cl–K. The possibility of using double scattering to study the degree of disorder in the Cu_3Au alloy was demonstrated by Nizhnaya et al. (1979d). Edge et al. (1974) used computer simulation to investigate glancing scattering of $E_0 = 334$ keV protons from primitive and nonprimitive atomic rows, i.e., rows consisting of atoms of one or two species, respectively. The proton scattering studied was from Al atoms and from rows of alternating atoms, such as K–Cl, Na–Cl, K–I. It was shown that in the last case, besides the peaks corresponding to the minimum and maximum escape angles, the angular distributions may have also additional peaks between them.

Dzhurakhalov et al. (1989b, 1989c, 1990b) used the model of scattering from rows of alloy component atoms to investigate the dynamics of structural changes on the alloy surface in the course of ordering. In contrast to Nizhnaya et al. (1979d), a study was made not only of atom rows with an alternating sequence of the alloy species, but also of rows where an arbitrary number of atoms of one species was succeeded by an arbitrary number of atoms of another species. The trajectories of the ions scattered from atomic row ridges were computer simulated in the binary collision approximation, with inclusion of elastic and inelastic energy loses and using a matched interaction potential, eq. (20) of chapter 1. Calculations were performed for the alloy Cu_3Au (100) where the atoms of Cu and Au form a continuous sequence of substitutional solid solutions. The temperature of the transition from the disordered to ordered phase is in the case ~400°C; at higher temperatures all lattice sites are populated uniformly by atoms of both species, while at lower temperatures nonuniform occupation of sites by atoms of different species is typical. Note that $\langle 100 \rangle$ rows contain either only Cu or only Au atoms, whereas in the $\langle 110 \rangle$ rows the Cu and Au atoms alternate. Figure 20a presents ovals depicting the dependence of the energy of scattered Ar^+ ions with initial energy $E_0 = 5$ keV on the scattering angle θ, which consists of QS (1) and QD (2) scattering branches. Shown in the bottom panel are the corresponding angular distributions of ions scattered from $\langle 110 \rangle$ rows in the Au (100), Cu (100), and Cu_3Au (100) planes. The dashed oval corresponds to the Au atoms, the chain one to the Cu atoms, and the solid oval with a characteristic break (3) to a row of alternating Cu and Au atoms. In the case of a mixed row the oval has an unusual shape in that the QD branch is replaced by quasitriple scattering (3) from a sequence of Au–Cu–Au atoms with the principal deflection from the Cu atom which results in considerable energy losses. The QS and QD scatterings occur in this case primarily from Au atoms. The angular distribution contains in the region corresponding to the break in the oval an additional maximum and

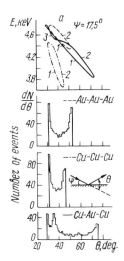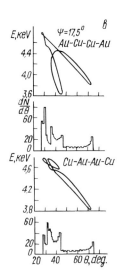

Fig. 20. (a) Ovals of Ar⁺ ion energy dependence on scattering angle θ and angular distributions in scattering from ⟨110⟩ atom rows on Cu(100) (chain curve), Au(100) (broken curve), and Cu₃Au(100) (full curve); (b) same dependence for cases where one of the alloy atoms repeats twice in the row.

exhibits a dramatic broadening. This maximum appears as a shoulder at $\psi = 15°$, grows with increasing grazing angle and exceeds in height at $\psi = 19°$ the maxima at the limiting scattering angles. As follows from the calculation, the additional maximum comes from the rainbow effect associated with the enhanced scattering close to the shadow-cone boundary of the Au atom toward the nearest Cu atom in a narrow impact parameter region along the row.

Figure 20b shows also the scattered ion energy versus the scattering angle ovals and the corresponding angular distributions, however, in this case, in the alternating sequence of atoms one of them, either Au or Cu, is repeated twice in succession. Both the energy ovals and the angular distributions are seen to vary dramatically. A comparison with fig. 20a reveals that when two neighbor atoms belong to the same species, the characteristic break (3) transforms into an additional oval which lies lower than the principal one and corresponds to QS and QD scattering from a pair of Cu atoms. To this oval corresponds two additional maxima in the angular distribution. A similar pattern is also observed in the case when two Au atoms adjoin one another in a mixed row. The relative height of the additional maxima was found to depend on the number of neighbor atoms of the same species in a row.

Increasing the number of neighboring identical atoms up to three or more results in an increase of the corresponding additional maxima in the angular distribution. Note that the shape of the energy oval and the interval of possible scattering angles are retained, whereas the relative heights of the maxima at the limiting scattering angles decrease. The characteristic structure of the angular distribution with the number of maxima is retained also when thermal vibrations of atoms in the alloy are included. Thus, a comparison of the angular and energy distributions of ions scattered from the surface of an alloy in the process of ordering with similar distributions for pure targets made up of the alloy components permits a conclusion that two or more neighboring atoms in an alternating sequence are of the same species. The results obtained can be used to study short-range order in alloys undergoing ordering.

CHAPTER 5

Ion Scattering in Semichannels

The next step in the development of model concepts concerning ion scattering from a solid consists in taking into account the interaction not with one but rather with several neighboring atomic rows which form on the surface of a single crystal a channel which is open at the top, i.e., a semichannel. Kivilis et al. (1967) were the first to introduce this term. Atomic rows arranged along the various crystallographic directions make up semichannels of different kinds and size (fig. 1). Two types of semichannels are usually considered, namely, open (frequently called surface channels) in which the atomic rows of the second layer lie under those of the top layer (fig. 1a), and the semichannel itself with a bottom row where the atomic row of the second layer is aligned between those of the first layer (fig. 1b). The size and shape of a semichannel are determined by its depth h (i.e., depth of the second layer), distance between atomic rows on the surface, a, and separation between atoms in a row, d.

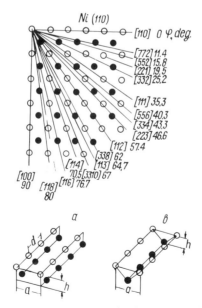

Fig. 1. Atom rows in the first two layers on the (100) face of an fcc crystal, and (a,b) two typical cases of their arrangement; open circles denote atoms in the topmost layer, and full circles second layer atoms.

Depending on the actual crystallographic orientation, atoms in the rows making up the walls of a semichannel can be arranged symmetrically (e.g., the [110], [100] directions in an fcc lattice on the (110) plane) or nonsymmetrically (the [112] direction in an fcc lattice on the (110) plane) with respect to the semichannel axis.

1. Spatial and energy distribution

The first publications dealing with medium-energy heavy-ion scattering from semichannels on the surface of single crystals appeared in the period 1967–1970 (Kivilis et al. 1967, Yurasova et al. 1967, 1968, Parilis et al. 1969, 1970, Karpuzov et al. 1969b). They reported on computer simulation studies of the spatial and energy relations of scattering as a function of grazing angle, target orientation and temperature, as well as on the primary beam energy. The calculations were carried out in the binary collision approximation, by following the trajectories of the ions incident on the surface of a unit cell until their escape out of the crystal. The direction of ion escape and the ion energy as a function of the aiming point were given.

Parilis et al. (1970) carried out a calculation of backscattering of Ar^+ ions with $E_0 = 30$ keV from the Cu (100) plane for $8° \leqslant \psi \leqslant 20°$.

Figure 2 presents the angular distributions of the rejected ions and their energy spectrum. Two principal spots, A and B, stand out against the scattering diagram. Spot A is close to the specular reflection angle; in the target area it is related with the points lying on the atomic row ridges. The energy spectrum of group A is in agreement with results obtained with the atomic row model. One can distinguish in it QS and QD scattering peaks. Spot B accommodates particles with aiming points at the bottom of the semichannels and close to the ridges of surface atomic rows, their trajectories being zigzag shaped. These are ions trapped in semichannels, suffering up to 20–30 collisions and scattered at small angles. If one follows the $\theta = $ const arc in the region of spot A, it will have a maximum at $\varphi = 0$, whereas in the region of spot B the curve will be double-humped with a minimum at $\varphi = 0$.

Spot B vanishes with increasing ψ, which means a decrease of the number of ions trapped in semichannels. The calculations of Karpuzov et al. (1969b), Yurasova et al. (1968), Yurasova and Karpuzov (1967), Shulga et al. (1971), and Yurasova et al. (1972) performed in a similar way reveal in some cases an intensity maximum at $\varphi = 0$ for ions of group B (fig. 3). The same results were obtained also by Yamamura and Takeuchi (1979), who drew attention to discrepancy between the results of Parilis et al. (1970) on one hand, and Shulga et al. (1971) and Yurasova et al. (1972) on the other.

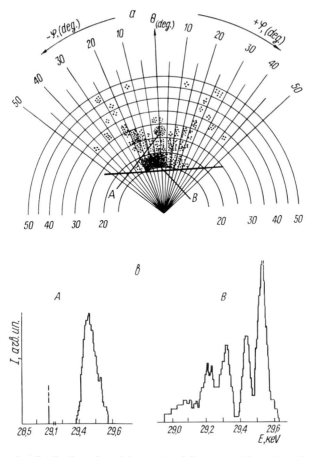

Fig. 2. (a) Angular distribution of particles scattered from a semichannel on Cu (100), and (b) their energy spectrum.

This apparently can be attributed to the effect of ion focusing in semichannels, to be discussed in section 5.2.

Nizhnaya et al. (1979a) compared the scattering on an ideal atomic row with that on a semichannel. The conditions and method of the calculations coincided with those employed by Parilis et al. (1970). It was found that the bulk of the ions is scattered from a semichannel at angular ranges $\varphi = 0-25°$ and β (escape angle) $= 0-30°$, which corresponds to a target area close to the atomic-row ridge; in this case the spatial distribution of the scattered ions coincides with that for an isolated atomic row.

Another, smaller fraction of the scattered ions, which originates from particles penetrating down to the second and third layers in a single crystal and suffering multiple collisions with the walls of a semichannel, has a

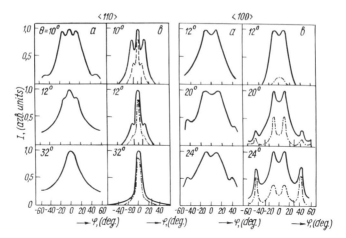

Fig. 3. (a) Experimental, and (b) calculated spatial distributions of Ar ions scattered from Cu (001) for two directions, $\langle 110 \rangle$ and $\langle 100 \rangle$ ($E_0 = 4.5$ keV, $\psi = 15°$; broken curve denotes the contribution of the second layer).

broader angular extent. It corresponds to aiming points remote from the ridge. In scattering from a semichannel, there is a certain probability for a particle to return into the initial plane even for aiming points outside the ridge of the atomic row.

Nizhnaya et al. (1979a), Parilis et al. (1969, 1970), and Parilis and Turaev (1972) conclude that the major contribution to scattering in the incidence plane comes from particles falling along the ridges.

Karpuzov et al. (1969a), Yurasova et al. (1968), Yurasova and Karpuzov (1967), Shulga et al. (1971), and Yurasova et al. (1972) studied the contribution of the various trajectories of particles backscattered from the surface of a single crystal to the angular and energy distributions. In the energy and angular ranges considered ($E_0 = 2.2$–4.5 keV, $\psi > 20°$), the scattering was shown to occur mainly in the first two surface layers. In the plane of incidence it is observed to take place not only from the ridges but in semichannels as well, the relative contribution coming from the atomic row and semichannel trajectories depending on the actual conditions of bombardment and scattering geometry.

In some cases, one observed experimentally in the energy spectra, besides the quasisingle- and quasidouble-scattering peaks predicted by the one-dimensional atomic row model (Kivilis et al. 1967), also additional peaks (Karpuzov et al. 1983, Niehus and Preuss 1982, Von dem Hagen et al. 1982, Taglauer et al. 1980, Jackson et al. 1981, Helbig et al. 1982). Computer simulation using a three-dimensional model of the crystal showed the appearance of these peaks to be due to the so-called zigzag collisions, where the ion is scattered sequentially by two neighboring rows. The cross section of such zigzag collisions increases with decreasing E_0 faster than that of

single and double scattering in the scattering plane (Algra et al. 1982a).
Depending on the actual type of the zigzag trajectories, they can contribute
to various parts of the energy spectrum, including the region to the left of
the single-scattering peak (Von dem Hagen et al. 1982, Taglauer et al. 1980,
Jackson et al. 1981, Helbig et al. 1982).

Thus, the above-mentioned studies dealing with calculation of ion scattering in surface layers of a single crystal showed that at sufficiently small
grazing angles a considerable fraction of the reflected beam comes from ions
undergoing quasichanneling in semichannels on the crystal surface. In later
investigations, the ion motion in semichannels was studied in great detail
both theoretically and by computer simulation, which resulted in the
interpretation of a number of experimentally observed effects.

In particular, one used computer simulation to study the trajectories of
ions scattered from an atomic row, two neighboring rows in the top layer,
and a semichannel on the crystal surface, and to evaluate their contributions
to the energy and angular distributions (Dzhurakhalov et al. 1987b). Figure
4a shows schematically a semichannel on the Ni (100) face in the $\langle 110 \rangle$
direction and the target area on it, and identifies the angles used in the
computation. The aiming points on the crystal surface filled a rectangle,
whose sides were divided into 500 segments in the beam incidence plane (I
coordinate) and in the perpendicular direction (J coordinate). The binary
collision approximation was employed to consider the scattering of $E_0 = 5$,
30 and 35.8 keV Ne^+ ions from the $\langle 110 \rangle$ atomic rows on the Ni (100)
plane.

The scattered particle trajectories were analyzed in a grazing angle
interval $\psi = 5\text{--}15°$, i.e., both below and above Lindhard's critical angle
which for $E_0 = 5$ and 30 keV is, accordingly, 12.8 and 8.2°. The Molière

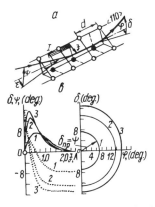

Fig. 4. (a) Schematic of a semichannel; (b) dependence of the δ and φ angles of scattered ions
on coordinate J, and the relation between these angles for one isolated atomic row, with 5 keV,
$\xi = 0°$ Ne^+ ions scattered from Ni (100) in the $\langle 110 \rangle$ direction.

potential with the screening lengths defined by expression (15) of chapter 1 was used. The quantities determined for each computed trajectory included the scattered-ion energy, the inelastic energy losses, the number of collisions involved, the impact parameters of the first and last collisions and the angles δ and φ as functions of E_0, angles ψ and ξ, and the aiming point coordinates I and J. Wherever required, the ion trajectories were followed to the end. Figure 4b shows the dependence of the scattered ion angles δ (broken curves) and φ (full curves) on coordinate J for $I = 250$, and the relation between the angles δ and φ for one isolated atomic row and three values of the angle ψ. The circle at the origin specifies the atom in the row corresponding to the aiming point coordinates $I = 250$, $J = 0$. We see that to the ridge of the row ($J = 0$) corresponds the maximum values of the escape angle δ_{max}, the azimuthal scattering angles φ being zero. In scattering from a ridge, the number of collisions varies as a function of screening length a in $V_M(r)$ from 9 to 12. As the aiming point moves away from the ridge, the escape angle decreases, and the azimuthal angle increases. The corresponding ion trajectories are approximately symmetrical about the row axis and consist of two parts, a descending and an ascending one, thus implying that the atomic row is still capable of raising an ion trajectory above its axis. The number of collisions suffered by ions at $\psi = 7.5°$ before reflection is approximately the same as in scattering from the ridge. For a given J (which is the smaller, the larger is ψ), the angle φ reaches its maximum value φ_{max}, the corresponding angle becoming zero. As ψ increases, the value of φ_{max} grows too, the maximum shifting closer to the ridge, and the half-width of the ion distribution in δ and φ decreasing. The contribution to ion reflection from an atomic row is determined by the width of the strip above the ridge within which the angle δ varies from δ_{max} to zero, and the angle φ from zero to φ_{max}. The width of this strip decreases with growing E_0, grazing angle ψ, and interatomic separation in the row d. However, as the screening length a in $V_M(r)$ is increased ($0.8a_F$; a_F; a_L) which results in a hardening of the potential, the strip width also increases. For a harder potential, to the same impact parameter corresponds a larger scattering angle. As a result of different variation rates of the angles δ and φ with coordinate J (the angle φ varying faster than δ), to the chosen angular size of the detector window $\Delta\delta$ in the incidence plane corresponds a larger angular dimension $\Delta\varphi$ in the transverse direction. Indeed, in the case of $\psi = 7.5°$, to $\Delta\delta = \pm0.5°$ corresponds $\Delta\varphi \cong \pm2.7°$. Thus, if the experimental angular spread both in $\Delta\delta$ and in $\Delta\varphi$ is $\pm0.5°$, then only a fraction of the ions backscattered at an angle δ will be analyzed, another part with $0.5° < \Delta\varphi < 2.7°$ remaining undetected. In scattering from one row, the value of φ_{max} decreases only slightly with growing θ_{lim} while increasing with decreasing potential hardness. The relation between the angles δ and φ for $\psi = 7.5°$ (fig. 4b, curve 1) is described by a circle. In the small scattering angle approximation, the sum of the squares of these angles yields the squared spatial escape angle (Varelas et al. 1977),

$$\delta_{lim}^2 = \delta^2 + \varphi^2 . \tag{1}$$

Under specular reflection, $\delta_{lim} \cong \psi$, and the radius of curve 1 is δ_{lim} which means that the angle δ_{lim} is no more a function of δ and φ, and the relation between the latter can be represented in the form

$$\delta^2 \cong \psi^2 - \varphi^2 . \tag{2}$$

As the angle ψ and, accordingly, the angles δ and φ increase, the relation between them is described by curves 2 and 3 which differ from curve 1. This is a result of the possibility of scattering from a row at angles other than specular, which appears as ψ increases (Kivilis et al. 1967). In this case, the angle δ_{lim} becomes dependent on δ and φ and may take on values not equal to ψ. In connection with this, relation (2) fails, so that one can no more obtain an analytic dependence between the angles δ and φ. The values of δ_{lim} determined by eq. (1) will now be only approximate. It can be shown that in the general case the true value of δ_{lim} can be found from the expression

$$\cos \delta_{lim} = \cos \delta \cos \varphi . \tag{3}$$

As follows from a comparison of eqs. (1) and (3), the results obtained with them disagree when the angles δ and φ are both large enough, and are equal when either one of the angles δ and φ, or both of them, are small. The disagreement grows with increasing ψ and E_0. Figure 5 presents the magnitude of δ and φ as a function of coordinate J for ions scattered in the field of two neighboring surface atomic rows (curves 2) and in a semichannel formed by three rows with one of them lying in the second layer (curves 3); curves 1 for one atomic row are shown for comparison. One can see that the effect of a neighboring row on the ion trajectory starts to manifest itself only at a certain value of coordinate J, which depends on ψ, E_0, d, θ_{lim} and the screening length a in $V_M(r)$ (the point where curves 1 and 2 begin to diverge). To the angle φ_{max} corresponds the angle δ_{min} rather than $\delta = 0°$ as is the case with a single row; note that there is more than one maximum in the angle φ, and they are different in magnitude and sign. The angles φ take on both positive and negative values, thus evidencing an oscillatory (zigzag shaped) form of ion trajectories between two rows. The number of collisions suffered by ions before reflection increases and reaches 22–24, depending on the magnitude of a in $V_M(r)$. The number of collisions grows with increasing hardness of the potential. The effect of the second row on the ion trajectories increases with increasing coordinate J. However, at a certain value of J the ions penetrate the plane connecting both rows, the angle δ acquiring negative values. As coordinate J increases still more, up to midway between

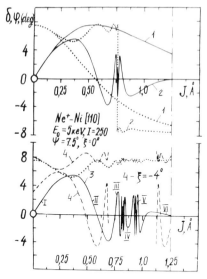

Fig. 5. Dependence of the δ and φ angles of 5 keV Ne$^+$ ions scattered from Ni(100) in the $\langle 110 \rangle$ direction for $\psi = 7.5°$ and $\xi = 0°$, on coordinate J for $I = 250$ for one row (curves 1), two adjacent rows with first layer (curves 2), and semichannel (curves 3).

the rows, the angle φ, on passing through a small maximum, vanishes, the angle $\delta = -\psi$, and entrance focusing appears (Shulga 1983a). Thus, the contribution of two neighboring rows to scattering is determined by the region of J from the point where curves 1 and 2 start to diverge to the dip ($\delta = -\psi$). As ψ decreases down to values smaller than ψ_{cr} for surface hyperchanneling (Evdokimov et al. 1979), which in the present case is 5.7°, the ions will not penetrate the plane connecting both rows, the scattering being determined by the mechanism of surface hyperchanneling. The presence of a row in the second layer results in reflection from the surface also of those ions that penetrated the plane connecting the two neighboring rows in the first layer ($\delta \cong \psi$); note that a very small change of coordinate J in the region IV–V leads to sharp changes both in magnitude and sign, and depending on ψ, E_0 and $V(r)$ one may expect both reflection of ions from the surface and their transfer from a semichannel into the bulk of the target or into neighboring semichannels. Regions I and II of curve 3 correspond to the motion of particles in the field of one and two neighboring rows, respectively, the effect of the bottom row being practically not felt here at all. The number of collisions with atoms of three rows in the region III–V is the largest and reaches 85–200. In region VI, the angle φ, on passing through a small maximum, vanishes as one approaches the center of the semichannel, the angle $\delta \cong \psi$, and exit focusing is observed (Shulga 1983a). As the potential hardens, when $a = a_L$, the focusing action of the semichannel in region VI on the escaping particles increases. Conversely, when $a = 0.8a_F$, which corresponds to a softer potential, the ions entering a semichannel in the interval III–V of J penetrate from the semichannel into the bulk of the target, so that no exit focusing in region VI is observed. Such dependence of the angles δ and φ of the scattered ions on coordinate J in the course of motion in the semichannel was obtained by Evdokimov et al. (1981), while Karpuzov et al. (1978) identified ion trajectories in semichannels on the crystal surface. The calculations were performed both with a three-dimensional program (Karpuzov and Yurasova 1971b) and in the string approximation (Lindhard 1965).

Throughout the entire range of variation of coordinate J, the scattered ions escape at an angle which close or equal to specular angle ($\delta \cong \psi$), and which decreases substantially only in the case of scattering from two rows in the region $0.25 < J < 0.5$ Å ($\delta \cong 5$–$6°$). Scattering at such small angles occurs outside the plane of incidence, the corresponding azimuthal angles being $\varphi \cong \varphi_{max}$. Thus, in the impact parameter range from the ridge to the center of a semichannel one can distinguish regions in coordinate J where the ions interact only with one row, two neighboring rows in the first layer, and a semichannel, and find their contributions to the energy and angular ion distributions. The distributions shown in fig. 6 were obtained for $a = a_F$. Figure 6a presents the contributions due to one row (I), two neighboring rows in the top layer (II, III), and the second layer (IV–VI) to the energy distribution of ions scattered in the interval $\Delta\varphi = \pm4°$. One clearly discriminates the peaks due to scattering from the ridge of an atomic row (I) and in a semichannel (VI). The peaks of the ions scattered in the field of two rows in the first layer (II, III) lie on the low-energy side of peak I. The separation between peaks I and VI (175 eV) permits their reliable experimental identification. The shaded part of the histogram in fig. 6b shows the contribution of one row, the broken line that of two neighboring rows in the top layer, and the full line the contribution from the second layer to the spatial angular distribution of scattered ions. The major contribution to the distribution in

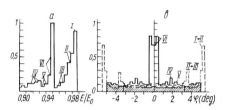

Fig. 6. (a) Energy, and (b) spatial angular distribution of 5 keV Ne$^+$ ions scattered from Ni (100) in the $\langle 110 \rangle$ direction for $\psi = 7.5°$ and $\xi = 0°$.

the plane of incidence ($\varphi = 0$) is seen to come from particles scattered close to the center of a semichannel (VI), while the particles scattered from the top layer (I, II, III) contribute primarily to the distribution outside the plane of incidence ($\varphi = 3.5$ and $5°$). Variation of the screening length a in the interaction potential affects markedly the shape of the distributions. Indeed, for $a = a_L$ the height of peak VI increases, peak II→III is observed at a smaller value, $\varphi \cong 2.5°$, while peaks IV and V are absent altogether. For $a = 0.8a_F$, peak VI decreases dramatically because of the absence of focusing in the semichannel, whereas peaks I and II are seen to exist at larger azimuthal angles φ.

Shulga (1983b) followed the pattern of variations in the spatial distribution of Ar ions caused by $\langle 110 \rangle$ semichannels on Cu (100) as a function of E_0. The ion trajectories were found by computer integration of the equations of ion motion in the field of all neighboring atomic rows, which is described by an averaged potential $U(r)$ corresponding to the Molière ion–atom potential. It was found that close to $E_0 = E_f$ (E_f is the focusing energy) the relative contribution of the second layer increases sharply. The spatial distributions calculated by us for $E_0 = 5$ keV (fig. 6b) and $E_0 = E_f = 35.8$ keV turned out to resemble those obtained by Shulga (1983b). This supports the possibility of using the binary collision technique involving individual free target atoms in the approximation of straight asymptotes for describing ion scattering from semichannels on the crystal surface. As shown by an analysis of the formation of the multipeak structure in the spatial distribution of scattered ions, the maxima in the distributions correspond to the regions of coordinate J where the azimuthal scattering angle φ reaches a maximum and then reverses its course (fig. 5). These regions exhibit features associated with the fact that the derivative $dJ/d\varphi$ characterizing the scattering probability tends to infinity. A small local maximum at $\varphi = 0°$, which was observed also by Shulga (1983b) and Evdokimov (1983), comes from the fact that in region VI this feature is seen to exist at $\varphi \neq 0$. This finds its explanation in that for $E_0 < E_f$ the entrance focusing (fig. 5) for a substantial part of the ions occurs at nonzero angles φ, in other words, the ions impinge on the ridge of the row in the second layer at small angles φ. The exit focusing for these ions likewise occurs at small angles φ to the axis of the row. The multipeak structure of the distribution in escape angle also originates from the features associated with the fact that in certain regions of J (fig. 5) the derivative $dJ/d\delta$ tends to infinity.

Thus, in the spatial angular and energy distributions one can clearly distinguish the regions where scattering from one row, two rows in the top layer, and the semichannel provide dominant contributions. The multipeak structure in the angular distributions is due to the rainbow effect in φ and δ angles.

2. Ion focusing effect

A study of scattering from a single crystal at grazing incidence angles revealed the existence of effects (ion focusing, surface channeling, hyperchanneling) which can be interpreted only in terms of the semichannel model.

The ion-focusing effect manifests itself in a growth of intensity and decrease of the FWHM of the scattered flux at certain target orientations relative to the incident beam. Actually, the ions passing between two atomic rows on the surface are focused (in the plane perpendicular to the channel) onto the bottom of the semichannel. This situation is illustrated schematically in fig. 7.

In the impulse approximation, the total scattering angle of an ion with the aiming point between two surface rows will simply be the difference between the scattering angles for each of them,

$$\theta(\rho) = \frac{1}{E} \left[f(a - \rho) - f(a + \rho) \right], \tag{4}$$

where $(a - \rho)$ and $(a + \rho)$ are the impact parameters of a collision with surface rows in the transverse plane, $f(\rho)$ being determined by the actual potential used [see eq. (30) of chapter 1].

For a continuous row (string) potential, $U(r) = \pi C/dr$,

$$\theta = \frac{\pi C}{E_\perp d} \left(\frac{1}{a - \rho} - \frac{1}{a + \rho} \right), \tag{5}$$

and for the paraxial part of the beam $(\rho \ll a)$,

$$\theta = \frac{2\pi C \rho}{E_\perp a^2 d}. \tag{6}$$

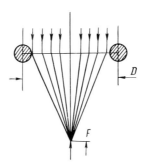

Fig. 7. Ion scattering from atomic rows in projection on a transverse plane.

In this approximation, the focal length F,

$$F = \frac{\rho}{\tan \theta} \approx \frac{\rho}{\theta} = \frac{E_\perp a^2 d}{2\pi C} , \qquad (7)$$

does not depend on the impact parameter (Shulga 1975). In this case, two surface rows act as a lens with a focal length F depending on $E_\perp = E_0 \sin^2\psi$ and on the semichannel size. If the focal length is equal to the semichannel depth ($F = h$), then all trajectories will be focused (in the transverse plane) on the row forming the bottom of the semichannel. On reflection from the bottom row at an angle $\theta > \psi$, the diverging ion flux will again pass through the focusing lens of the two rows in the top layer. In this case, a maximum of scattered intensity is observed. This effect was studied both theoretically and experimentally (Shulga 1975, 1976a, 1976b, 1977, 1978, 1983a,b, Mashkova et al. 1969, 1970a, 1970b, 1971, Balashova et al. 1979, Yamamura and Takeuchi 1979, 1980, M.W. Thompson and Pabst 1978, Pabst 1975, Evdokimov 1982).

The focusing properties of the semichannel were first revealed in computer calculations of ion scattering from single crystals (Yurasova et al. 1967, Parilis et al. 1969, Karpuzov et al. 1969b) and afterwards experimentally (Mashkova et al. 1969, 1970a, 1970b, 1971). The ion focusing effect was investigated theoretically by Shulga (1975, 1976a, 1976b, 1977, 1978), and Yamamura and Takeuchi (1979, 1980), and for light particles and for higher energies by Thompson and Pabst (1978) and Pabst (1975).

Shulga (1978) studied theoretically the effect of ion focusing in a semichannel which was simulated by five to six atoms. This called for a consideration of large scattering angles ($\theta = 75°$). The scattering intensity was shown to have two strong maxima corresponding to the energies of focusing at the first and second atom pairs.

An investigation of the focusing conditions for different interaction potentials was made by Shulga (1975). The dependence of $E_\perp$ on the semichannel parameters was obtained for three types of interaction: inverse square, Molière and Born–Mayer potentials.

The continuous potential of a row constructed basing on the Born–Mayer interaction has the form

$$U(r) = \frac{2Ar}{d} K_1(r/b) , \qquad (8)$$

where K_1 is MacDonald's function. In this case,

$$E_\perp = \frac{2\pi Ah(a - b)}{db} \exp\left(-\frac{a}{b}\right) . \qquad (9)$$

For the Molière function the continuous potential can be written as

$$U(r) = \frac{2Z_1 Z_2 e^2}{d} \sum_{i=1}^{3} \alpha_i K_0 \left(\frac{\beta_i r}{a_s} \right), \tag{10}$$

and

$$E_\perp = \frac{h 2\pi Z_1 Z_2 e^2}{d a_s} \sum_{i=1}^{3} \alpha_i \beta_i \exp\left(-\frac{\beta_i}{a_s} a \right), \tag{11}$$

where a_s is the screening parameter, eq. (15) of chapter 1.

Experimental studies of the focusing energy E_f (Mashkova et al. 1982) showed that it agrees with the values of E_f derived from calculations (Shulga 1982, 1983b) which were based on the Born–Mayer potential matched with the Firsov function, and from calculations (Shulga 1977) making use of the inverse square approximation of Lindhard's potential.

Since the calculated values of the focusing energy are sensitive to the parameters of the potential used, one can reconstruct the interaction potential based on the experimental value of E_f. This problem was solved theoretically (Shulga 1982) using the experimental data obtained by Mashkova et al. (1982). It thus became possible to determine the parameters in the Born–Mayer potential, as well as the screening parameter in the Thomas–Fermi potential with a higher precision.

Shulga (1978) carried out a computer simulation study of the dependence of focusing on thermal vibrations. This effect was shown to grow progressively weaker with increasing temperature.

The conditions of focusing in the scattering of high-energy ($E \sim 300\,\text{keV}$) light atoms were considered by M.W. Thompson and Pabst (1978). Figure 8 presents different types of focusing trajectories (in the transverse plane) for scattering from semichannels with a bottom row. Case a corresponds to the incidence direction parallel to the semichannel axis, cases b, c, and d to

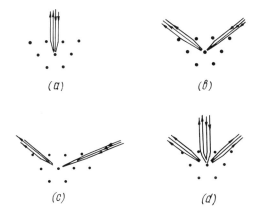

(a) (b)

(c) (d)

Fig. 8. Schematic representation of surface semichanneling in a transverse plane.

different entrance and exit angles. A simple focusing model is proposed based on the continuous string potential approximation. Computer simulation was used to study the conditions favoring the appearance of the intensity maxima in the spatial distribution of scattered ions, which correspond to type a, b, c, or d focusing. An analysis of the experiment of Marwick et al. (1972) made by M.W. Thompson and Pabst (1978) revealed the existence of maxima in these directions.

Yamamura and Takeuchi (1980) studied theoretically the dependence of ion focusing on the total reflection coefficient R as a function of azimuthal angle of the target ξ near the semichannel direction. It was found that, with the crystal rotated near the semichannel direction, the total reflection coefficient grows sharply at $\xi = \pm \xi_m$ if $E_\perp > E_f$, and at $\xi = 0$ if $E_\perp \leqslant E_f$, where

$$\xi_m = \left(\frac{2hE_\perp}{dE_f} \right)^{-1} \sqrt{\frac{1}{3} \left(\frac{E_\perp}{E_f} - 1 \right)^3}. \tag{12}$$

The reflection coefficient reached an absolute maximum at $\xi = 0$ and $E = E_f$.

Summing up, one can say that a simple analytical theory of the focusing of paraxial ion beams in surface semichannels has been developed which permits the calculation of the focal length and focusing energy for particular interatomic potentials. The results of the calculations are in a good agreement with computer simulation and experimental data, a comparison with experiment making it possible to determine the parameters of the potential and the degree of perfection of the surface structure.

3. Subsurface channeling and hyperchanneling

When the azimuthal grazing angle exceeds the critical angle of planar channeling, but is still smaller than that of axial channeling, ions move along channels in the surface layer. While the general direction of motion remains unchanged, the ion can now transfer from one channel to another.

Subsurface channeling was studied by computer simulation (M.W. Thompson and Pabst 1978, Sizmann and Varelas 1976). This effect was shown to manifest itself in anomalously high energy losses for a substantial part of the scattered ions and in a broader spectrum which exhibits a long tail in its low-energy part due to long trajectory lengths.

Sizmann and Varelas (1976) proposed a simple model to calculate the energy spectrum which gives a good agreement with the experiment of Marwick et al. (1972). If the grazing angle is less than the critical angle of surface channeling, the trajectory will pass above the second layer, above an atomic row or in a potential valley between two surface atomic rows. In the latter case, the critical angle can be determined from the condition $E_\perp =$

U_{min}, where U_{min} is the minimum height of the surface potential barrier along the scattering plane. Under certain conditions the particle will move only within one semichannel. This phenomenon was called hyperchanneling (Evdokimov et al. 1979, Karpuzov et al. 1978).

The hyperchanneling effect was studied both theoretically and experimentally (Evdokimov et al. 1979, 1981, Karpuzov et al. 1978, Evdokimov 1983). It manifests itself in increased intensity of scattered particles at very small grazing angles which do not correspond to the focusing conditions, and in the appearance of a distinct peak in the low-energy part of the spectrum.

This effect was investigated in the three-dimensional binary collision model (Karpuzov and Yurasova 1971b) and in the continuous string approximation (Evdokimov et al. 1981). The trajectory of a particle moving in a potential valley between atomic rows which make up the walls of the semichannel is determined by equipotential surfaces (fig. 9); these surfaces, however, depend only weakly on the presence or absence of a bottom row, so that the trajectory is primarily determined by interaction with the two surface rows.

Evdokimov et al. (1979) carried out a computer simulation and an experimental study of the dependence of the intensity I of $E_0 = 6\,\text{keV}\,\text{Ar}^+$ ions backscattered from the Ni (110) plane in the $\langle 001 \rangle$ direction on grazing angle ψ. Figure 10a compares the results of the calculation of $I(\psi)$ with those of the experiment. At $\psi < 2°$, the maximum possible intensity is reached, the entire surface acting as one reflecting plane. As the angle of incidence increases, the ions start to penetrate into the regions where the equipoten-

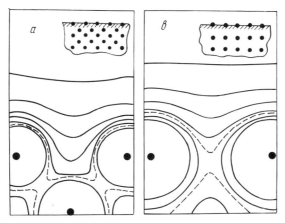

Fig. 9. Possible 'potential valley' configurations on the crystal surface: (a) the plane bottom of the valley lies above the atomic ridges of the subsurface atom layer; (b) the surface and subsurface layers lie in the same plane.

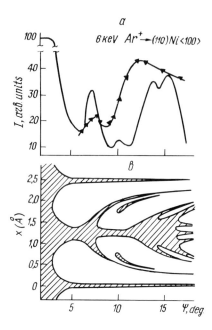

Fig. 10. (a) Dependence of the scattered ion intensity on grazing angle (full curve denotes results from calculation, triangles from experiment), and (b) angular dependence of the positions and widths of the strips on the target spot contributing to scattering (shaded areas).

tial surfaces are bent, the intensity in the reflection plane decreasing. In this case, the main contribution to scattering comes from the target points located along the axes of the rows and the bottom of the potential valleys between the rows, the contribution of the latter region being predominant throughout the angular range studied. As ψ continues to increase, deep potential valleys become accessible and the trajectories contract more effectively toward the scattering plane, the maximum in the hyperchanneling intensity being observed at $\psi = 7.5°$.

The minimum in intensity at $\psi \cong 10–11°$ is due to the ions beginning to penetrate the bottom of the potential valley. While surface hyperchanneling does not operate anymore, subsurface channeling between the first two atomic layers becomes possible. This effect has a noticeable intensity only within a limited range of angles.

The large intensity maximum observed both experimentally and in calculations at $\psi = 12–15°$ is due to the well-known semichanneling or wedge focusing, when particle trajectories are focussed onto the atomic rows in the second layer.

The shaded regions in fig. 10b represent the fraction of the target area contributing to scattering. The width of these regions is proportional to the scattering intensity.

The investigation of the dependence of scattered particle intensity on grazing angle (Evdokimov et al. 1979, Evdokimov 1985, Snowdon et al. 1989d, 1989c, 1989a, Oura et al. 1988, Shulga 1990b) made it possible to determine the scattering contributions from various mechanisms (ion focusing in semichannels, subsurface channeling, hyperchanneling) for various grazing angles.

4. Azimuthal anisotropy in scattering

The studies discussed above considered the specific features of scattering as a function of grazing angle for one chosen semichannel. It would be of interest to investigate the dependence of the scattering characteristics on the kind and dimensions of semichannels. This possibility was offered by the experimental study (Hou et al. 1978) of the energy spectra of $E_0 = 3$ keV Ne^+ ions backscattered at the same angle $\theta = 15°$ from the Ni (110) plane at a fixed incidence angle $\psi = 7.5°$ and with an azimuthally rotated target, which makes the scattered beam pass successively through different semichannels.

The energy spectra obtained by Hou et al. (1978) for different angles are similar (i.e., contain one maximum in the region of the QS and QD peaks which are not resolvable under these conditions) and differ only in peak height, thus revealing an azimuthal anisotropy in scattering intensity.

Nizhnaya et al. (1982a, 1982b) studied by computer simulation the scattering from semichannels oriented in different crystallographic directions corresponding to different azimuthal angles φ (figs. 1 and 11); these studies were carried out for the parameters corresponding to the conditions of the experiments of Hou et al. (1978). The calculations were performed with potential (19) of chapter 1 for $r_c = 1.7$ Å. The results obtained are as follows.

(1) In a given geometry, scattering at an angle $\theta = 15°$ (for all directions) adds up as a sequence of deflections to small angles. The inelastic energy losses calculated by Firsov's expression do not exceed in each interaction the ionization potential.

(2) The shape of the scattering trajectory depends on the kind of the semichannel, its width and interatomic separations in the rows forming the given semichannel. For wide [100] and [110] semichannels, the trajectory of a particle scattered from the bottom row passes below the crystal surface. In narrow semichannels, the trajectories of the ions aimed at the center of the semichannel pass above the surface, and the higher they pass, the more narrow the channel is. Interestingly, these trajectories differ very slightly

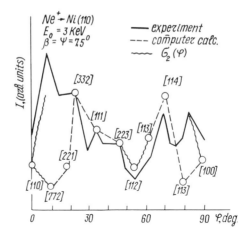

Fig. 11. Scattered particle intensity in the main peak versus azimuthal scattering angle φ; full curve: experiment, broken curve: computer simulation, thin curve: $\sigma_2(\varphi)$ plot.

whether there is a row on the bottom or not, since the projectile interacts in this case only with two surface rows, which correlates with the data of Evdokimov et al. (1979). Actually, in the case of a narrow semichannel the trajectory does not pass inside it, so that there is no interaction with the bottom row. The latter turns out to be blocked by surface rows.

(3) The energy spectrum of the particles scattered at a given angle $\theta = 15°$ has a double-peak structure corresponding to QS and QD scattering. The peaks lie close in energy, $10 < E_{QD} - E_{QS} < 50$ eV, in the region 2.94–2.96 keV, which is slightly in excess of the energy of single scattering from an isolated atom. The small value of $E_{QD} - E_{QS}$ did not permit Hou et al. (1978) to resolve the QS and QD peaks under their experimental conditions where one principal maximum is observed, its intensity being different in different crystallographic directions.

In the first approximation, neglecting the interaction of the projectile with several neighboring rows, the intensity of the particles scattered at a given angle φ can be represented in the form

$$I = n\sigma,$$ (13)

where n is the number of rows per unit length, σ is the scattering cross section for one row (which in the continuous approximation has the dimension of length), which means that scattering at an angle θ for a given φ is due to primarily ions interacting with the ridges of the rows oriented in the given direction. For grazing angles smaller than Lindhard's critical angle ψ_L (Lindhard 1965), the interaction with a row is described by the continuous potential model. In this case, the scattering cross section for the potential

$U(r) = C/r^2$ can be written as

$$\sigma = \frac{dp}{d\varphi_\perp} = \sqrt{\frac{C}{E_\perp}} \frac{\pi^2}{[\pi^2 - (\varphi_\perp + \eta_\perp)^2]^{3/2}} , \qquad (14)$$

where $\eta_\perp$, $\varphi_\perp$, and $E_\perp$ are the incidence and reflection angles and the energy in the transverse plane, respectively (fig. 1 of chapter 4). If the scattering plane passes through the row axis ($\eta_\perp = 0$, $\varphi_\perp = 0$), then $\sigma \sim \sqrt{C}$. Since the number of the rows n is proportional to the interatomic separation in a row d_i, whereas the constant C in the potential is inversely proportional to d_i, we obtain from eqs. (13) and (14) that $I(\varphi) \sim \sqrt{d_i}$. Although the experimentally obtained dependence $I(\varphi)$ (Hou et al. 1978) correlates with the quantity $\sqrt{d_i}$, this approach is valid only for the directions φ in which the interatomic separations in a row are not too large, and the condition $\psi < \psi_L$ is met. Besides, one not always can neglect the interaction of the projectile with several neighboring rows (the so-called zigzag collisions).

A numerical experiment (Nizhnaya et al. 1984, Nizhnaya and Umarov 1985) was performed to clarify the boundaries of applicability of this approach. The calculations showed that for wide [110], [100] and [111] semichannels the main contribution to scattering comes from a narrow strip along the surface row and a wider strip responsible for scattering from a semichannel and two surface rows, whereas, in narrow semichannels oriented along high-indexed directions, the entire width of the semichannel contributes to scattering, which implies that the approximation involving scattering from isolated rows cannot be used in these cases to evaluate $I(\varphi)$. A comparison of the numerical calculations of scattering in semichannels oriented in different crystallographic directions with the experimental data of Hou et al. (1978) (fig. 11) shows that approximation (13) describes adequately the behavior of $I(\varphi)$ in the region where fairly low-indexed directions are considered, i.e., such semichannels for which the interatomic separations in the rows are not too large ($d_i \lesssim 2.5a$, where a is the lattice constant in Ni) and, accordingly, the semichannel width is not too small.

The intensity maxima along [110] ($\varphi = 7°$) and [100] ($\varphi = 85°$) cannot be accounted for by scattering in the high-indexed semichannels oriented in these directions. These maxima can be interpreted by considering scattering in the [110] ($\varphi = 0°$) and [100] ($\varphi = 90°$) semichannels in the case where the plane of incidence makes an angle $\Delta\varphi \neq 0$ with the semichannel axis.

Nizhnaya and Umarov (1985) studied the dependence of scattering intensity on azimuthal angle $\Delta\varphi$ near the semichannel axis, $\Delta\varphi = |\varphi - \varphi_{\langle klm \rangle}|$. For specular scattering where the azimuthal angle of incidence is equal to that of escape, $\eta_\perp = \varphi_\perp$, straightforward analytical estimates of $I(\Delta\varphi)$ were obtained in the continuous potential approximation, $U = C/r^2$. In the string approximation, the cross section of scattering in the transverse plane can be written as

$$\sigma_1(\Delta\varphi) = \pi^2(C/E_0)^{1/2}(\psi^2 + \Delta\varphi^2)^{-1/2}[\pi^2 - 4\varphi_\perp^2]^{-3/2}, \tag{15}$$

where $\varphi_\perp = \tan^{-1}[\sin(\Delta\varphi)/\tan\psi]$.

The cross section of scattering from one atomic row $\sigma_1(\Delta\varphi)$ falls off slowly with increasing $\Delta\varphi$. The cross section of scattering in the transverse plane from two atomic rows,

$$\sigma_2(\Delta\varphi) = \tfrac{1}{2}\pi^2(C/E_0)^{1/2}(\psi^2 + \Delta\varphi^2)^{-1/2}[\pi^2 - (\varphi_\perp + \tfrac{1}{2}\pi)^2]^{-3/2}, \tag{16}$$

grows with increasing $\Delta\varphi$.

If the ion energy in the transverse plane, $E_\perp \simeq E_0(\psi^2 + \Delta\varphi^2)$, is lower than the energy of focusing in a semichannel, E_f, or if $\psi < \psi_f = (E/E_0)^{1/2}$, the projectile ion interacts primarily with two surface rows, the bottom row in the semichannel becomes blocked, and the scattered intensity behaves as $\sigma_2(\Delta\varphi)$, i.e., one should observe a growth of intensity with increasing azimuthal angle of the target relative to the semichannel axis direction. Using eq. (7), for the case of Ne^+–Ni scattering with $E_0 = 3$ keV, the focusing angle for the [110] semichannel is $\psi_f = 11°$, whereas for [100] it is $\psi_f = 15°$. Therefore, for $\psi = 7.5°$ the maxima close to these directions are accounted for by the behavior of $\sigma_2(\Delta\varphi)$. Nizhnaya and Umarov (1985) investigated the dependence of the azimuthal scattering anisotropy $I(\varphi)$ on grazing angle ψ. The total yield of scattered particles decreases with increasing ψ. For a given grazing angle ψ the maximum of intensity $I(\varphi)$ will correspond to the direction of the semichannel which will scatter under the focusing conditions. In the case in question [Ne–Ni (110), $E_0 = 3$ keV] for $\psi < 10°$ scattering in all directions occurs not under the conditions of focusing. The anisotropy in $I(\varphi)$ is due here to the difference in the trajectory types for the ions scattered on different semichannels. At $\psi = 10°$ one observes two sharp maxima along [110] and [100] where the conditions favor focusing. Thus, the calculations and their comparison with experiment permit establishing the areas of applicability of a scattering model used to explain the azimuthal anisotropy in the intensity of ion scattering from semichannels on the single-crystal surface, as well as the correlation between this effect and the type and dimensions of the semichannels.

C.-C. Chang et al. (1985) studied the azimuthal anisotropy in He^+, Ne^+, Ar^+ ion scattering from the Rh (111) plane. The dependence of the scattering intensity on initial ion energy in the range $E_0 = 0.5$–6.0 keV was measured experimentally for the [110] and [211] directions ($\varphi = 0°$, $\varphi = 30°$). The ion intensity ratio in the [110] and [211] directions, $K = I_{[110]}/I_{[211]}$, was shown to correlate with the projectile velocity v, the $K(v)$ dependence being approximately the same for all the ions considered (fig. 12a).

In order to study the effect of velocity on the scattering intensity, a molecular dynamics calculation of the $I_{[110]}/I_{[211]}$ versus v dependence was

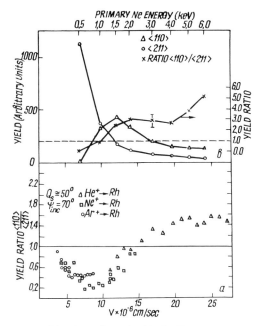

Fig. 12. $I_{\langle 110 \rangle}/I_{\langle 211 \rangle}$ scattered ion intensity ratio (C.-C. Chang et al. 1985): (a) experiment and (b) calculation.

carried out (fig. 12b). It was found that in the [211] direction (open semichannel) all particles are scattered primarily by the topmost layer. As the energy and, hence, the velocity of scattered particles increase, their yield decreases, just as does the scattering cross section. In the [110] direction (semichannel with a bottom row), a similar fall-off in intensity is observed to occur for $E_0 > 1.5$ keV. The structure of the [110] semichannel is such that the second and even third layers contribute substantially to scattering, therefore, the total yield in this direction in the energy range specified is higher than that along [211]. The decrease of the $I_{[110]}$ intensity with velocity for $E_0 < 1.5$ keV is explained by the chosen geometry of the experiment. For low velocities, the [110] direction becomes very smooth, and particles escape at a close-to-specular angle ($\theta = 40°$), the scattering intensity at $\theta = 50°$ being small. As E_0 increases, the angular distribution broadens revealing a growth of intensity up to $E_0 = 1.5$ keV, whereas for $E_0 \geqslant 1.5$ keV a mechanism of decreasing cross section with increasing E_0 becomes operative. In the [211] direction, the interatomic separations are greater than those along [110], and, therefore, the angular distribution is broader including scattering at close-to-specular angles θ even at low energies. C.-C. Chang et al. (1985) showed that simple considerations concerning the conditions of blocking and focusing are insufficient for the interpretation of the azimuthal anisotropy in scattering as a function of the projectile velocity, making it necessary to carry out a comprehensive study by computer simulation of all types of trajectories contributing to scattering.

Investigation of the azimuthal anisotropy in ion scattering from the surface of a single crystal permits one to reveal the dependence of the spatial and energy distributions of reflected ions on the type and dimensions of the semichannels oriented in different crystallographic directions. The existence of orientational effects in the dependence of the intensity of particles scattered from a single-crystal surface on the azimuthal angle of the target enables one to use this dependence for probing the surface structure (Derks et al. 1989).

5. Elastic and inelastic energy losses

An analysis of medium-energy (tens of keV) ion interaction with crystals should include both elastic and inelastic energy loss mechanisms. By the criterion of Seitz (1949), the elastic mechanism of energy losses is dominant in this energy range for ions with atomic numbers $Z \geqslant 10$. However, under the specific conditions of correlated glancing ion scattering from the single-crystal surface the reverse pattern becomes possible where the inelastic mechanism will predominate. The reasons for this are the large number of collisions involved and the fact that small impact parameters cannot be reached along the scattered ion trajectory.

Inelastic processes of ion interaction with crystals exhibit the so-called trajectory effects. Basically, this means that inelastic processes (excitation, ionization, transfer or exchange of electrons between the collision partners) and the associated inelastic energy losses depend on the actual trajectory of the scattered particle (Nizhnaya et al. 1979c). Summation of the inelastic energy losses along the trajectories of particles undergoing correlated small-angle scattering from a crystal surface is necessary to account for the so-called density effect when, e.g., the observed degree of ionization in closely packed directions exceeds the calculated value (Hou et al. 1978, Nizhnaya et al. 1982a).

Under the conditions of small-angle ion reflection from the surface of single and polycrystals there exists, besides the general trend of decreasing relative energy losses with decreasing scattering angle, also a region of grazing angles where their anomalous growth is observed (Evdokimov et al. 1969). It turned out that the relative magnitude of the anomalous energy losses depends on crystal orientation and increases with decreasing initial ion energy. Evdokimov et al. (1979) related this effect to the mechanism of surface hyperchanneling (SHC) which dominates for very small grazing angles. The particles undergoing SHC exhibit substantially elongated trajectories close to the surface and, accordingly, higher elastic and inelastic energy losses. It is assumed that the appearance in the spectra of reflected particles of a substantial fraction of ions with relatively large energy losses can be accounted for by the 'long-range' trajectories which become possible when conditions for certain specific kinds of surface channeling set in (Mashkova and Flerov 1983a). The nature of the anomalously high ion energy losses and their contribution to the elastic and inelastic energy losses still remain unclear and cannot be interpreted in terms of scattering from a continuous potential (Evdokimov 1987). It is difficult also to investigate the specific features of the trajectories of the ions suffering glancing scattering from a surface without invoking discrete models.

The present section gives the results of a study of elastic and inelastic energy losses and of the specific features of the ion trajectories appearing in scattering from discrete model potentials on an atomic row, a semichannel and a channel on the surface of a single crystal at small grazing and scattering angles, as well as of the contribution of the various scattering mechanisms to the experimentally observed anomalous energy losses (Dzhurakhalov et al. 1989a, 1990a).

Figure 13 shows energy distribution histograms of $E_0 = 15$ keV Ar$^+$ ions scattered specularly from the Cu (100) plane in the $\langle 110 \rangle$ direction into a

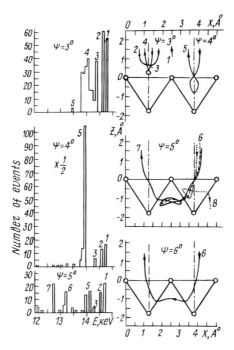

Fig. 13. Energy distribution histograms of $E_0 = 15$ keV Ar$^+$ ions scattered specularly from the Cu (100) plane in the $\langle 110 \rangle$ direction. Presented to the right of the spectra are the most characteristic ion trajectories projected on a plane perpendicular to the axis of the $\langle 110 \rangle$ row.

detector of angular aperture $\pm 0.5°$. Presented to the right of the spectra are the most characteristic ion trajectories projected on a plane perpendicular to the axis of the $\langle 110 \rangle$ row. The ions scattered along the atomic row ridges contribute to the peaks labelled 1. They lie in the highest-energy part of the spectrum and fall off in height with increasing ψ. The peaks located to the left and labelled 2–4 correspond to the ions undergoing SHC. They also decrease in amplitude with increasing ψ. The ions involved in SHC, rather than penetrating into the crystal, move in the valleys formed by the equipotential surfaces which characterize their interaction with the neighboring rows in the surface monolayer. The characteristic SHC trajectories 2–4 projected on a transverse plane are arranged in the order of increasing separation of their target points from the row axis. Trajectory 2 is characterized by a single interaction with atoms of the neighboring rows. Ions with trajectories of type 3 interact twice with one row. Trajectory 4 corresponds to ions which interact twice sequentially with atoms of two neighboring rows. In the latter case, the ions are seen to be focused in the $\langle 110 \rangle$ direction, the focus point lying about 0.5 Å above the surface plane.

At $\psi = 4°$, the shape of the spectrum undergoes a substantial change because of a sharp increase in the number of trajectories of the new type 5, and a decrease of the contribution coming from type 1–4 trajectories. While this type of trajectory appears already at $\psi = 3°$, it becomes most pronounced for $3.5° \leqslant \psi \leqslant 4.5°$, and shifts toward higher energies with increasing ψ. As seen from the projection of trajectory 5 on a transverse plane, ions begin to penetrate in this case the potential barrier formed by atomic rows in the top layer and to interact with atoms in the bottom row of a semichannel. Their transverse energy exceeds the limiting energy of surface semichanneling E_{lim} (Mashkova and Flerov 1983b). The figure shows that the ions with type 5 trajectories are focused by surface rows with the focus point lying slightly above the surface, and propagate afterwards in a diverging flux toward the walls of the semichannel. However, the transverse energy of such particles is still not high enough for them to penetrate the potential barrier of the semichannel walls, so that on reflection from the latter they will continue their motion in the semichannel along $\langle 110 \rangle$ and, on escaping out of it, will again be focused efficiently by the surface atomic rows into the plane of incidence, which manifests itself in a sharp growth of peak 5 at $\psi = 4°$. Type 5 trajectories differ in shape and character from SHC trajectories, the sharp growth in their number suggesting the existence of a peculiar refocusing effect which results in a pronounced narrowing of the spatial distribution of scattered particles. The angular range where this refocusing effect is observed is not large, so that at $\psi = 5°$ it is not seen at all; here the particles begin to overcome the potential barrier of the semichannel walls bringing about the appearance of trajectories of a new type, which, in its turn, affects substantially the shape of the spectrum.

In addition to peaks 1–5, new, lower energy peaks 6 and 7 are seen to appear in the spectrum at $\psi = 5°$. On penetrating the side walls of the semichannel, particles with type 6 and 7 trajectories move in the neighboring channel and then, on scattering, either return back into the original semichannel (trajectory 6 for $\psi = 5°$) or penetrate into the neighboring one (trajectories 7 for $\psi = 5°$, and 6 for $\psi = 6°$). The pattern of particle motion along type 7 trajectories is more complex than that along trajectories 6; indeed, in the latter case the part of the trajectory within the channel is shorter, the particles actually crossing it. Trajectories of types 6 and 7 were revealed in computer simulation studies (Yurasova et al. 1968). In their shape and general pattern, these trajectories cannot be classed among those typical of subsurface hyperchanneling (Evdokimov et al. 1979) which were also observed in our calculations. The fact is that the part of the subsurface hyperchanneling trajectories enclosed in a channel is a few times (depending on ψ) longer than that for type 6 or 7 trajectories, the corresponding region of aiming points being isolated spatially from those of type 6 and 7 trajectories. Moreover, as shown by calculations, the subsurface hyperchanneling trajectories are characterized by a substantial spread in energy, and, therefore, they form a low-energy tail in distributions rather than peaks as is the case with type 6 and 7 trajectories. There exist also trajectories of type 8 corresponding to the ions which, on overcoming the potential barrier of the semichannel walls, penetrate into deeper-lying layers and thus are not backscattered.

Figure 14a displays a computed angular dependence of the coefficient of specular reflection ($\theta = 2\psi$), K_N, of $E_0 = 15$ keV Ar^+ ions from Cu (100) plane in the $\langle 110 \rangle$ direction into a detector of angular aperture $\pm 1.5°$ (full curve), as well as the contributions to it coming from the various scattering mechanisms (curves 1–4). Panel b of fig. 14 shows the corresponding averaged elastic (broken curves) and inelastic (full curves) energy losses. Just as in the experiment of Evdokimov (1984), at very small grazing angles ($0 \leqslant \psi \leqslant 2°$) no effect of crystal structure is felt, the reflection occurring as from an absolutely plane surface. As ψ increases, features appear in the curves which are due to the contributions of the various scattering mechanisms that vary with ψ.

The contribution of the row (crosses) is seen to decrease monotonically with increasing grazing angle. The averaged inelastic energy losses ε corresponding to scattering from a row (full curve 1) do not practically depend on ψ and are equal to ~240 eV. It would seem that the inelastic energy losses should grow smoothly with decreasing ψ and a corresponding increase in the number of collisions. However, although the number of collisions does indeed increase from $n = 8$ at $\psi = 10°$ *to* $n = 19$ for $\psi = 3°$, this only compensates for the substantial decrease of inelastic losses in a single collision, and this is what results in the observed dependence $\varepsilon(\psi)$. The

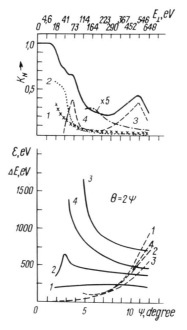

Fig. 14. (a) Angular dependence of the coefficient of specular reflection ($\theta = 2\psi$), K_N of $E_0 = 15$ keV Ar$^+$ ions from Cu (100) $\langle 110 \rangle$ (full curve) and the contributions to it coming from various scattering mechanisms (curves 1–4) and (b) corresponding averaged elastic (broken curves) and inelastic (full curves) energy losses.

elastic energy losses in scattering from a row (broken curve 1) for $2° \leqslant \psi \leqslant 8°$ are substantially lower than the inelastic ones; however, as ψ continues to increase they grow dramatically, so that in the region $10° \leqslant \psi \leqslant 12°$ they exceed already the inelastic losses by a factor 1.5–3.5. The contribution of SHC is the largest in the region $\psi = 2$–$3°$ (the shoulder on the broken curve in fig. 14a), but it falls off sharply with increasing ψ and does not exceed that for an atomic row starting from $\psi = 4°$. The averaged inelastic energy losses corresponding to SHC (full curve 2) first grow with increasing ψ, pass through a maximum at $\psi = 3°$, and then fall off. On the whole, in the grazing-angle region studied, they exceed by about a factor two those for scattering from an atomic row. The growth of inelastic energy losses in SHC can be attributed to the fact that the ion trajectories approach ever closer the surface with increasing ψ, the maximum losses corresponding to the situation where the trajectories come closest of all to the plane connecting the atomic rows in the top layer. The decrease in inelastic energy losses with a further increase in ψ is explained as due to the decrease in the number of particles undergoing SHC, because some of them penetrate the potential barrier of the surface layer (trajectory 5 in fig. 13). The elastic energy losses in SHC (broken curve 2 in fig. 14b) in the glancing scattering region are also substantially lower than the inelastic ones.

The contribution due to particles transferring from one semichannel into another in the course of scattering (type 6, 7 trajectories) to the specular reflection coefficient is relatively small and passes through a maximum at $\psi = 5\text{--}6°$ (chain curve). These trajectories, just as the ones related with subsurface hyperchanneling, involve the highest inelastic energy losses (full curve 3 in fig. 14b). Note that the inelastic energy losses corresponding to type 6 and 7 trajectories are much higher than the elastic ones (broken curve 3) throughout the interval of ψ studied.

The angular dependence of the contribution to K_N from the ions undergoing surface semichanneling (Yurasova et al. 1968, Parilis et al. 1969) (the broken curve in fig. 14a) exhibits a double-peaked structure due to the interaction with the second layer. The peak at $\psi_{cal} = 4°$ is related to the limiting energy E_{lim} of surface semichanneling, and the peak at $\psi_{cal} \approx 11°$, to the focusing energy E_f, i.e., to the energy at which the density of the particles striking the bottom row in a semichannel is the highest, and the ion focusing effect manifests itself in the scattering characteristics (Mashkova and Molchanov 1980). A similar double-humped dependence of the contribution of the second layer to K_N, but at $E_0 = 200$ keV, was revealed by Yamamura and Takeuchi (1983). Mashkova and Flerov (1983b) derived the energies E_{lim} and E_f experimentally from the spatial distributions of Ar^+ ions backscattered from Cu (100). Against the background of a monotonic growth of the spatial distribution Γ_φ with increasing grazing angle, two narrow transition regions were observed which exhibited a strong local decrease of Γ_φ. For $E_0 = 15$ keV and the $\langle 110 \rangle$ direction, these regions were observed at $\psi_{exp} \approx 6$ and $14°$, which corresponds to $E_{lim\,exp} = 160 \pm 30$ eV and $E_{f\,exp} = 850 \pm 90$ eV in the transverse plane. The computed values of ψ_{cal} are less than the experimental ones and correspond to $E_{lim\,cal} = 73 \pm 9$ eV and $E_{f\,cal} = 546 \pm 24$ eV. The focusing energy calculated by the expression proposed by Shulga (1983b, 1986b) with Molière's approximation replaced by the Biersack–Ziegler potential, turned out to be $E_{f\,cal} = 536$ eV. The values of ψ_{cal} were obtained without the inclusion of thermal vibrations of the lattice atoms. However, as shown by Yamamura and Takeuchi (1983), thermal vibrations do not affect the focusing angle while reducing substantially the particle reflection coefficient.

Mashkova et al. (1982) were the first to determine experimentally the semichannel focusing energy E_f from measurements of the ion scattering intensities under the conditions favorable and, conversely, unfavorable for focusing. The value $E_{f\,exp} = 590 \pm 140$ eV obtained in this way agrees well with our calculated $E_{f\,cal} = 546 \pm 24$ eV and with the values $E_f = 570$ and 640 eV yielded by the calculation of Shulga (1975) who used the Born–Mayer potential matched with the Firsov potential, and by the calculation

(Balashova et al. 1979) based on the inverse square approximation of the standard Lindhard potential, while disagreeing with the $E_{f\,exp}$ determined by Mashkova and Flerov (1983b).

It should be pointed out that a straightforward comparison of experimental and calculated values of E_{lim} and E_f meets with difficulties for two reasons. The first of them deals with a correct choice of the interaction potential to which the focusing energy is very sensitive. The second stems from the neglect in calculations of ion neutralization during scattering. Indeed, in calculations, one determines the total number of particles scattered under the conditions of focusing without taking into account their charge state, which provides information on the true focusing properties of a semichannel. At the same time, in an experiment one detects only the particles scattered in the ionic state, in other words, on the process of their focusing is superimposed the effect of neutralization which, as pointed out by Yamamura and Takeuchi (1983), can shift the value of the focusing angle towards larger angles.

The dependence of inelastic energy losses of the particles undergoing surface semichanneling on ψ (full curve 4) is similar to that for the particles of type 6 and 7 (curve 3), while lying below it. The elastic energy losses (broken curve 4) are here likewise substantially lower than the inelastic ones in the glancing scattering region and begin to dominate in the total losses only starting from $\psi > 10°$.

Thus, the elastic energy losses are considerably smaller than the inelastic ones in the region of glancing scattering. The fact that the inelastic losses exceed the elastic ones for small ψ in the medium-energy range is due to an increase in the number of collisions and the particle trajectory length in the surface region, as well as to the impossibility of reaching small impact parameters in the course of scattering. The predominance of the inelastic energy losses should reveal itself in the efficiency of the various inelastic processes accompanying the glancing ion scattering from a single-crystal surface.

The substantial increase of the path length of particles scattered from surface layers of a crystal at grazing angles suggests that they are slowed down, a process characterized by the energy expended by a particle per unit path length. The specific inelastic energy losses of particles in an amorphous target determined by the relation of Lindhard and Scharff (1961) for $E_0 = 15\,keV$ Ar^+ ions interacting with copper are $-(d\varepsilon/dx) \approx 25\,eV/\text{Å}$. To obtain the specific inelastic energy losses in the case of a single crystal the inelastic energy losses summed over the various trajectories for a given ψ were divided by the trajectory lengths and averaged over the particular scattering mechanisms. The values thus found were averaged for various grazing angles. The resultant specific energy losses turned out to be $-(d\varepsilon/dx) \approx 12\,eV/\text{Å}$, which is one half that for an amorphous target. According to

Kumakhov and Komarov (1979) the electronic stopping cross section for the case of channeling is approximately one half that for an amorphous target, depending on Z.

Figure 15a displays experimental (Evdokimov et al. 1969) (circles) and calculated (crosses and triangles) dependences of the relative energy losses, $(E_0 - E)/E_0$, on grazing angle for the case of specular scattering ($\theta = 2\psi$) for $E_0 = 15$ keV Ar^+ ions backscattered from Cu (100) in the $\langle 110 \rangle$ direction. The chain curve shows the losses calculated with the atomic row model. Figure 15b illustrates the elastic (curve 1) and inelastic (curve 2) contributions to the total relative energy losses. The experimental data were obtained by averaging over the energy distributions. The calculated curves were constructed by averaging the losses over the various scattering mechanisms in accordance with their relative contributions to the spectra cut off at $E = 13.5$ keV (triangles) and at $E = 10$ keV (crosses) (Dzhurakhalov et al. 1988c, 1990a). It should be kept in mind that the calculation was carried out for the total number of scattered particles, and the experiment for the ionic component only. The energy distribution of the ionic and neutral components are known to be similar (Chicherov 1972). The dependence obtained for $E = 13.5$ keV resembles the experimental one, however, the calculated values of the relative losses are smaller by about 30–40%. This curve

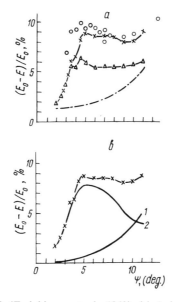

Fig. 15. (a) Experimental (Evdokimov et al. 1969) (circles) and calculated (crosses and triangles) dependences of the relative energy losses versus grazing angle for the case $\theta = 2\psi$ for $E_0 = 15$ keV Ar^+ ions scattered from Cu(100)$\langle 100 \rangle$ and (b) elastic (1) and inelastic (2) contributions to the total relative energy losses (b).

originates from ions with type 1–6 stable trajectories which produce the corresponding peaks in the spectrum.

The inclusion of type 7 trajectories responsible for the peak in the spectrum below 13 keV and of the subsurface hyperchanneling trajectories contributing to the low-energy tail ($E \leqslant 12$ keV) in the distribution brings the calculated curve (crosses) closer to experiment. Note that the calculation was performed for a perfect single-crystal surface neglecting the presence on it of adsorbed atoms and point defects of the type of vacancies and atomic steps. Dzurakhalov et al. (1988b, 1990a) showed that glancing scattering from a surface with atomic steps results in an increase in the number of channeling particles (because of the possibility for them to penetrate into the crystal under the step), and a variation of the steps in height and separation to a shift of the corresponding peaks in the spectrum and, hence, in a change of the relative energy losses. The inclusion of adatoms and atomic steps into the scattering model should also apparently help in understanding the nature of the experimentally observed anomalous energy losses in glancing scattering from polycrystal (Mashkova and Flerov 1983a).

As seen from fig. 15b, the main contribution to the anomalous energy

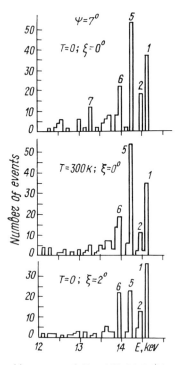

Fig. 16. Energy distribution histograms of $E_0 = 15$ keV Ar$^+$ ions, scattered from the Cu (100) plane in the $\langle 110 \rangle$ direction at $\xi = 0°$, $T = 0$ K; $\xi = 0°$, $T = 300$ K and $\xi = 2°$, $T = 0$ K.

losses comes from inelastic losses. The elastic losses grow monotonically with increasing ψ and θ and cannot account for the anomalous energy losses. As for the inelastic losses, they pass through a maximum in the region of ψ corresponding to the anomalous losses and provide a predominant contribution to the losses. The maximum inelastic energy losses are due to particles with trajectories 5–7, as well as to those undergoing subsurface hyperchanneling. The relative contribution of such trajectories falls off with increasing ψ starting from $\sim 11°$, quasisingle and quasidouble collisions in which the elastic energy losses are higher than the inelastic ones becoming predominant.

Note that the ions scattered along trajectories 6 produce peaks in the energy spectrum which disappear neither when thermal lattice vibrations are included, nor when the incident beam is slightly turned away from the $\langle 110 \rangle$ row direction (fig. 16). As seen from fig. 16, the inclusion of thermal vibrations (correlated, with the anisotropy in the surface atom displacements taken into account) results primarily in a certain decrease of peaks 2 and 6 and of the low-energy tail, while taking into consideration the beam misalignment with respect to the crystal (azimuthal angle of incidence $\xi = 2°$) reduces peak 5 corresponding to semichanneling. The inclusion of both factors reduces substantially the number of the particles undergoing subsurface hyperchanneling. Nevertheless, this does not reduce noticeably the anomalous energy losses. As is evident from fig. 16a (the point at $\psi = 7°$), the magnitude of the losses at $\xi = 2°$ does not differ from those at $\xi = 0°$ by more than 1%.

Thus, the increase of the losses originates from the growth of the inelastic energy losses, their being greater than the elastic contribution, and the specific features of the trajectories under glancing scattering from single-crystal surfaces.

CHAPTER 6

Atom Scattering from a Nonideal Surface

1. Surface of a solid

The surface layers of a solid possess specific mechanical, electrical, optical, and thermodynamic properties other than its bulk (Duke 1974, Kopetsky 1979, Gomoyunova 1982, Morrison 1977). From many estimates, this difference in properties extends to a depth of 40–50 Å, and even in this layer the elemental composition may not be the same as in the bulk (Belenky 1981, Partensky 1979, Bolshov et al. 1977, Adamson 1976, Kislovsky 1976, Houston et al. 1973, Kesmodel and Somorjai 1975, Davis and Noonan 1983, Stensgaard 1983). This is also true for the structure of the surface and its crystallographic characteristics. The position of an atom in a unit cell on the surface differs from that in the bulk. An atom may be displaced either parallel to the crystal surface (reconstruction) or normally to it (relaxation) (Morrison 1977, Kislovsky 1976, Davis and Noonan 1983, Stensgaard 1983, M.L. Xu and Tong 1985, Diehl et al. 1985). In metal crystals, relaxation may reach 10% of the interplanar spacing in the bulk.

Reconstruction results in the formation of various periodic structures on the surface with periods different from the lattice constant in the bulk. Surface reconstruction is more characteristic of semiconductor crystals and is less typical of metals.

The rms amplitude of thermal vibrations of the surface atoms (along the normal to the surface) is larger than that in the bulk by a factor 1.5–2 (Kopetsky 1979, McRae 1964, Mros and Mros 1981, Tabor and Wilson 1972, Babanskaya et al. 1983).

The surface of a real crystal on an atomic level is nonuniform and irregular. It always has terraces, i.e., plane parts of the surface separated by one or several interatomic spacings (Hren and Ranganathan 1968, Wagner 1979). Atomic steps and vacancies form on the surface (Kaminsky 1965).

Atoms residing in various sites on the crystal surface have different numbers of neighbors and, accordingly, different binding energies (Knake and Stransky 1959).

Determination of the concentration of single or group point defects (steps, vacancies and their clusters) on the surface is a very complex problem. Only the ion projector permits direct observation of point defects or of their small complexes (Hren and Ranganathan 1968). In some materials, the largest complexes can be studied with an electron microscope. It should be pointed out that ion bombardment degrades or distorts the surface (Mashkova and Molchanov 1980, Carter and Colligon 1968, McCraken 1975).

As shown by calculations (Bitensky and Parilis 1978a,b, Drozdova and Martynenko 1980), the concentration of surface point defects under dynamic equilibrium setting in under intense ion bombardment when the surface undergoes rapid layer-by-layer sputtering may exceed by far that in the bulk of the solid. This places a constraint on the possibilities of the corpuscular spectroscopy based on ion scattering, since ion bombardment may cause substantial electronic and structural surface changes (U.A. Arifov 1968, Mashkova and Molchanov 1980, Hren and Ranganathan 1968).

2. State of the surface after ion bombardment

The surface of a solid bombarded by an ion beam results in its etching and a change of topography (McCraken 1975, Predvoditelev and Opekunov 1977, Yurasova et al. 1965, Pleshivtsev 1968, Farnsworth and Hayek 1967). Sputtering produces in the surface and near-surface atomic layers radiation-induced damage of various kinds, namely, vacancies and their complexes and steps; note that the nonequilibrium concentration of such defects may be very high which manifests itself, e.g., in a broadening of the diffracted electron beam (Farnsworth and Hayek 1967, Hauffe 1971) and a substantial change in the pattern of the angular and energy distributions of scattered ions (Mashkova and Molchanov 1980, Petrov and Abroyan 1977, Carter and Colligon 1968). The presence of such nonuniformities should be taken into account in a theoretical consideration of the various surface phenomena (scattering, sputtering, etc.). The ion beam used to study the interaction of ions with a surface results also, as a rule, in the formation of defects.

Predvoditelev and Opekunov (1977) analyzed the studies of surface erosion under bombardment by 1–100 keV medium-mass ions. According to the cascade theory of radiation defect formation and sputtering (Carter and Colligon 1968, Pleshivtsev 1968, Sigmund 1972a,b,c), erosion is caused by the sputtered target atoms which acquire in the course of collision cascade development an energy high enough to overcome the surface binding forces.

The formation of surface relief under ion bombardment is initially associated exclusively with a statistical building of surface defects. The ions bombarding the surface produce on it cones, furrows, networks (Yurasova 1976, Navinsek 1972) ~1–10 μm in size. Hauffe (1971) describes theoretical models for the formation of such a surface relief which can be accounted for by a difference in sputtering coefficients due either to initial surface irregularities or radiation-induced damage created by ion bombardment. Nobes et al. (1969) and Du Commun et al. (1974) proposed a mode of formation of equilibrium surface topography under ion bombardment based on the dependence of the sputtering coefficient on the angle between the ion beam and the target. This model can be applied, however, only to describe changes in the surface contour much larger than the interatomic spacings, i.e., macroscopic changes, whereas the effects of major significance for our phenomenon are those occurring on the atomic scale.

Bitensky and Parilis (1978a,b) carried out a computer simulation study of the dependence of the point defect distribution in the surface layers of a single crystal on the irradiation dose. The surface was divided into elements of area F, so that one ion falling on such an element at a grazing angle ψ corresponded to the initial irradiation dose,

$$\varphi_0 = \frac{1}{F \sin \psi} \text{ ions/cm}^2 . \tag{1}$$

The trajectories of the projectile ion and the recoil atom, and the collision cascades were computed based on the binary collision model by the technique described by Lenchenko and Akilov (1972). The crystal temperature was taken into account in the usual way by introducing random Gaussian displacements of the atom positions from the site coordinates and was assumed to be 300 K. The potential of Firsov (1957) was used, the inelastic energy losses being calculated by Firsov's theory (Firsov 1959). On the completion of a cascade spontaneous annealing was simulated by assuming the interstitial atoms and vacancies at less than lattice constant separation to have recombined.

The calculation was carried out for the case of the Cu (100) face bombarded by $E_0 = 10 \text{ keV Ar}^+$ ions in the $\langle 110 \rangle$ direction. The calculation yielded the sputtering coefficient and the layer-by-layer vacancy distribution versus irradiation dose. It was found that the vacancy concentration in several near-surface layers reaches saturation after removal of only about two or three top layers, which required, depending on the actual angle ψ, a dose of 5×10^{14}–3×10^{15} ions cm^{-2}. A further increase of the dose does not change the equilibrium concentration while displacing the total layer structure as a result of sputtering at a rate

$$v = Sj \sin \psi \, nd \text{ Layers/s} . \tag{2}$$

where S is the sputtering coefficient, j the ion current density, n the number of atoms in a unit volume, and d the interplanar spacing. Figure 1 shows vacancy concentration distributions in the first three layers as a function of irradiation dose $\varphi = \varphi_0 N_i$ (N_i is the number of incident ions). The calculation was performed for a target area comprising 5×5 unit sells for $\psi = 10°$ and $\varphi_0 = 1.8 \times 10^{14}$ ions/cm^2. The equilibrium vacancy concentrations (%) and corresponding values of the sputtering coefficient S are as follows: for $\psi = 10°$, $\eta_j = 79$ for the first layer, 26 for the second, 10 for the third, 5 for the fourth, $S = 9.3$; for $\psi = 15°$, $\eta_j = 81$ for the first layer, 34 for the second,

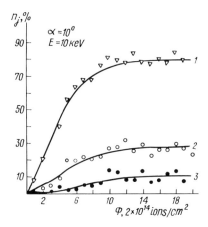

Fig. 1. Vacancy concentration distribution in the near-surface layers of single-crystalline copper versus irradiation dose (the numbers identify the layers).

14 for the third, 6 for the fourth, $S = 14.7$; for $\psi = 30°$, $\eta_j = 80$ for the first layer, 29 for the second, 11 for the third, 4 for the fourth, $S = 12.1$. We see that for the surface layer we may actually take the second, larger spacing layer (e.g., in the case of $\psi = 15°$ its density drops by 34%), above which individual atoms of the first layer stand out.

Drozdova and Martynenko (1980) studied also theoretically the surface microstructure under ion bombardment. They calculated the concentration of moving single atoms and atomic clusters on the irradiated surface of a solid. The following expression was derived for the maximum concentration of single atoms and clusters on the surface,

$$I_{\text{max}}^{(k)} = \left(\frac{k}{\beta_k + k} \right)^k \left(\frac{\beta_k}{\beta_k + k} \right)^{\beta_k} . \tag{3}$$

where k is the number of atoms in a cluster; β_k is the number of vacancies around a cluster. For a simple cubic lattice, $\beta_1 = 4$, $\beta_2 = 6$, and so on. For an fcc lattice, the maximum concentration of single surface atoms was found to be $I_{\text{max}}^{(1)} = 8\%$, which exceeds by far (by orders of magnitude) the equilibrium concentration in the absence of ion bombardment (Knake and Stransky 1959). However, the concentration of clusters on the surface falls off rapidly with increasing number of atoms in the clusters. Note that by the surface model of Knake and Stransky (1959) the equilibrium concentration of single atoms on the surface is

$$I_{\text{eq}}^{(1)} \simeq e^{0.6H/kT} , \tag{4}$$

where H is the sublimation energy, and T is the target surface temperature. By the same model, only single atoms ($k = 1$) may be considered as moving, whereas clusters with $k \geqslant 2$ are practically fixed. The activation energy of migration of individual atoms over the surface is $u = 0.1H$ and the energy of their bonding to the surface $W = 0.4H$. Inclusion of the mobility of single atoms reduces their concentration, but nevertheless, for $T < T^{**}$, it is higher than the equilibrium concentration on the irradiated surface. The characteristic temperature is

$$T^{**} = \frac{H}{\ln(50\nu\tau)} \; ; \quad \tau = (jSa^2)^{-1}, \tag{5}$$

where ν is the atomic oscillation frequency, and τ is the time taken to sputter one layer. The single atom concentration on the surface ($\approx 20\%$) obtained by Bitensky and Parilis (1978) is more than twice that calculated by Drozdova and Martynenko (1980). This difference can be attributed to the fact that the latter authors considered an atom to be single if it was surrounded by four vacancies. However, since a vacancy can be a neighbor to several atoms, the effective number of vacancies surrounding an atom may vary from $N_b = 1$ to $N_b = 4$. In this case, the single atom concentration will vary within $0.08 \leqslant n_1 \leqslant 0.5$, which is in accord with the value $n_1 = 0.2$ given by Drozdova and Martynenko (1980).

The results obtained suggest that, after the defect concentration has reached dynamic equilibrium, the characteristics of the secondary processes (scattering, ion and electron emission, various orientational effects) should not depend on the irradiation dose, a conclusion supported by experiments (Mashkova and Molchanov 1980). These authors described in detail a sputtering technique to clean targets involving bombardment by heavy inert-gas ions. As a result of sputtering, the surface of the target is continuously etched away, thus limiting the process of defect build up, so that a certain time after the beginning of irradiation an equilibrium sets in.

This can be illustrated by a study by Heiland and Taglauer (1973a) of the dependence of the energy distributions of $E_0 = 1$ keV Ar^+ ions backscattered from Ni (100) on the irradiation dose, and simultaneously of the bombardment-induced surface structure changes by LEED. It was found that the surface damage caused by ion bombardment results (starting from doses of $\sim 10^{15}$ ions/cm^2) in an increase of single scattering intensity and, accordingly, a decrease of multiple scattering, thus evidencing the formation of point defects and extended defect structures. Figure 2a shows dose dependences of the total single (I_B) and multiple (I_M) Ar^+ ion scattering intensities integrated over energy intervals from 0.2 to $0.25E_0$, and from 0.3 to $0.75E_0$, respectively. One readily sees that the single scattering intensity from more or less isolated atoms grows exponentially with irradiation dose, whereas the intensity of multiple scattering falls off gradually. Accordingly, the diffraction spots grow in width and become asymmetrical. Similar studies of the damage of a germanium surface, carried out by Jacobson and Wehner

(1965) and Bellina and Farnsworth (1972), revealed the exponential dependence to be true only for low irradiation doses. Figure 2b shows the model of the surface structure produced in ion bombardment which was proposed by Heiland and Taglauer (1973a). The structure is seen to contain vacancies and divacancies in the ⟨100⟩ direction. It should be pointed out that the

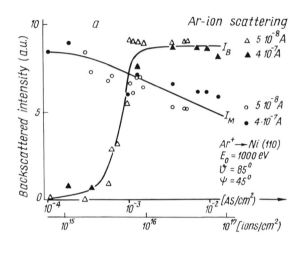

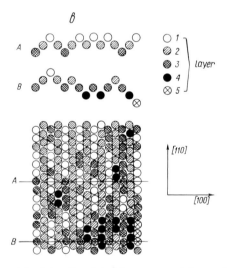

Fig. 2. (a) Dependence of the intensity of Ar⁺ ions scattered from a Ni crystal on irradiation dose (I_B denotes single scattering from isolated atoms, I_M multiple scattering from ordered areas of the surface); (b) model of surface structure produced by ion bombardment (with the layer numbers specified); and (c) irradiation dependence of the relative number of defects measured for He⁺, Ne⁺, Ar⁺ ions for different E_0 and target temperatures.

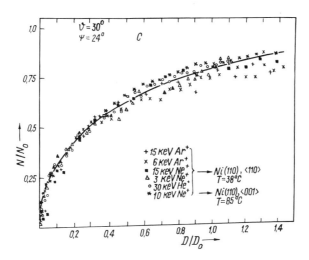

Fig. 2. (*Continued*).

structure of the surface is not damaged under ion bombardment to the full extent (i.e., it does not become amorphous). Even at the maximum doses, corresponding to ~100 ions per one surface atom, one still observed characteristic LEED patterns and the multiple scattering intensity increased insignificantly, thus evidencing the presence on the surface of rows consisting at least of a few atoms. Interestingly, the sum $I_B + I_M$, rather than remaining constant, increases with increasing dose and reaches a maximum at ~10^{16} ions/cm^2, which is accounted for by a change in the degree of neutralization of scattered ions.

Boers et al. (1980) determined the number of atomic steps forming on the Cu (100) face under bombardment by $E_0 = 10$ keV Ar$^+$ ions. They measured the energy distributions of the total scattered flux and of its ionic component as a function of grazing angle ψ for a constant scattering angle $\theta = 30°$. The number of steps was estimated from the intensity of single-scattering peaks in the energy spectra of all scattered particles at $\psi = 7$ and $23°$. Scattering at an angle $\theta = 30°$ at small and large grazing angles ($\psi < 10°$ and $\psi > 20°$) becomes possible because ion bombardment introduces irregularities into an ideal, infinite atomic row, making it finite and terminating in steps upward and downward. The heights of the single-scattering peaks due to atoms located at the edges of such steps (edge atoms) are proportional to the number of steps, scattering cross section, and coefficient of shadowing,

$$I = I_0 \frac{n}{\sin \psi} f(\psi)\sigma(\theta)\,\Delta\Omega\ , \tag{6}$$

where I is the scattered ion intensity, I_0 is the initial beam density, n is the scatterer density per unit area, $\sigma(\theta)$ is the differential scattering cross section, $\Delta\Omega$ is the solid angle subtended by the detector, and $f(\psi)$ is the shadowing coefficient.

The intensities of the single-scattering peaks for small and large ψ are compared with that of the QS peak for $\psi = 15°$, for which $f(\psi) = 1$ was assumed because of the surface damage affecting insignificantly the total spectrum. The ratio of the intensity at $\psi = 15°$ to those at other values of ψ yields an estimate of the number of 'visible' atoms on steps, $nf(\psi)$.

We present below estimated numbers of 'visible' atoms (i.e., those that contribute to QS scattering) on upward (small ψ) and downward (large ψ) steps (table 1). These values were obtained by integrating individual pulses in the single-scattering peak and discriminating the background, its uncertainty accounting for the considerable uncertainty in the evaluation of the number of the step atoms. The average number of the steps formed in ion bombardment was found to be $\simeq 2 \times 10^{14}$ cm^{-2}. An additional uncertainty may apparently come from single scattering on isolated atoms (surrounded by vacancies) of the first layer which concentration is fairly high (Van der Weg and Bierman 1969, Heiland and Taglauer 1973a, Jacobson and Wehner 1965, Bellina and Farnsworth 1972, Verhey et al. 1980).

The presence of a considerable number of vacancies in the surface layers of metals subjected to ion bombardment is evidenced also by the studies of Dahl and Sandager (1969). They investigated large angle ($\theta = 100°$) scattering of $E_0 = 100$ keV Ar$^+$ ions from the first two layers on the Au (100) face under close-to-normal beam incidence. The energy distributions measured at small escape angles ($\beta \leqslant 20°$) revealed a single-scattering peak with $E_1 = 61$ keV; for some crystal orientations the peak did not shift toward higher energies because of the effect of the neighboring atoms in the row. It was believed that double scattering including strong scattering from the first atom and weak scattering from the second would be observed at such small escape angles. The presence of a single-scattering peak at the escape angles

Table 1
The estimated numbers of accessible atoms on the steps and relative peak intensities for different glancing angles.

ψ (deg)	Relative intensity	Number of accessible atoms on steps, $nf(\psi)$, 10^{15} cm^{-2}
15	1	1.53
6	0.012	0.0184
4	0.003	0.0046
24	0.15	0.23
26	0.13	0.20
28	0.07	0.11
29	0.03	0.046

$\beta = 2$ and $4°$ where it should be absent implies that the surface structure is damaged and contains vacancies in atomic rows along the corresponding directions, their number being ~20% of monolayer coverage. This estimate is based on a summation over all charge states of the scattered particles and assumes the energy distribution for Ar atoms to be similar to that for Ar^+ ions.

As already pointed out, a damaged surface structure (steps, vacancies) can be studied by LEED and low-energy ion scattering. The second method appears promising since the formation of defects and the surface analysis occur simultaneously, i.e., the same ion beam is used for both purposes, and the annealing of damage even at close-to-room temperatures may be neglected (U.A. Arifov 1968, Mashkova and Molchanov 1985, Verhey et al. 1980).

Low-energy ion scattering was used by Verhey et al. (1980) to study the creation of surface damage on Ni (110) under ion bombardment as a function of ion mass and energy and target temperature. The steps were detected in the experimental geometry where one analyzed the ions back-scattered at small escape angles to the surface. In this case, all regularly arranged surface atoms, with the exception of those at step edges, are blocked. The measured energy distribution contains only one peak whose width remained constant during the bombardment and the height served as a measure of the number of damaged surface sites. Figure 2c presents a dependence of the relative number of damaged sites on dose measured for He^+, Ne^+, Ar^+ ions with different E_0 and for different target temperatures. All points are seen to fit within experimental error up to doses $D = D_0$ to one full curve approximated by the expression

$$N = N_0[1 - \exp(-D/D_0)]^{1/2}, \tag{7}$$

where N is the number of damaged sites, D is the dose, N_0 and D_0 are normalization parameters depending on the scattering conditions, namely, N_0 is the number of damaged sites at saturation, and D_0 is the characteristic dose. The data were obtained by isolating the background level at a low ion dose $D = 10^{12}$ ions/cm^2 and normalizing to the damage peak height at saturation N_0 and for a certain characteristic dose D_0.

The results are discussed in terms of a model including the migration and clustering of vacancies that form in the course of surface sputtering, as well as their subsequent annealing through diffusion of thermally activated adatoms. The number of steps was assumed to be proportional to the square root of the number of vacancies n produced in the topmost layer, i.e., $N \sim n^{1/2}$. For n, the following expression was derived,

$$n = n_0[1 - \exp(-D/D_0)], \tag{8}$$

where n_0 is the number of atoms per unit surface area, N_0 and D_0 being the main parameters describing the defect formation curve, they were subjected to a more detailed study. It was found that N_0 does not depend on projectile mass and energy, nor on target temperature up to 100°C. At the same time, the characteristic dose turned out to depend strongly on target temperature and ion mass, and only weakly on ion energy.

Thus, ion bombardment produces on the surface of crystals various defects, in particular, atomic steps and vacancies in surface rows. The reflected part of the incident ion beam can be employed for qualitative and even quantitative analysis of these defects. This becomes possible, as we will show below, due to a relation between the defect parameters and characteristics of the scattered ion beam.

3. Effect of the surface structure on the scattered ion distribution

As follows from the preceding consideration, ion bombardment produces on the crystal surface various point defects, primarily individual atoms, pairs of atoms, broken-off rows of a few atoms each, steps, vacancies, and their clusters in surface rows. The surface being nonideal, an analysis of experimental data carried out in terms of ideal models cannot provide an interpretation for a number of features in the angular and energy distributions of scattered ions. Among them is the existence of a two-peak structure in the energy spectrum down to the smallest grazing and scattering angles, i.e., in the shadow region, the strongly nonmonotonic dependence of double-scattering intensity on the grazing and scattering angles, and so on (Mashkova and Molchanov 1972a, 1975, 1980). Mashkova and Molchanov (1975) observed the scattering of 30 keV Ar^+ ions from the Cu (100) face at $\psi = 14°$ to occur down to $\beta = 3°$ ($\theta = 17°$), while for an ideal surface the scattered beam is limited by the escape angle $\beta = 11°$ ($\theta = 25°$). Experimental studies of the energy spectra of ions backscattered at a given angle θ as a function of ψ revealed a fine structure in the spectra at such values of ψ where one should have observed only one peak if the surface were ideal. In the scattering to $\theta = 25°$, they also found a structure in the range $2° \leqslant \psi \leqslant 22°$, while the models based on the concepts of an ideal surface limit this interval to $11 \leqslant \psi \leqslant 14°$. It should be pointed out that variation of the interaction potential and inclusion of thermal lattice vibrations (Parilis et al. 1975, Nizhnaya et al. 1979a, Taglauer and Heiland 1972) change the angular intervals within which double scattering is seen to occur while not providing a full explanation for the experimentally observed scattering in the shadow region.

S.H.A. Begemann and Boers (1972), who used the atomic row model proposed by Kivilis et al. (1970, 1967), were apparently the first to simulate on a computer the ion scattering from nonideal structures on the surface of a single crystal. They found experimentally that even at current densities $\sim 10^{-9}\,\text{A}/\text{cm}^2$ bombardment produces changes in the surface which may strongly affect, under certain conditions, the energy and angular distributions of scattered ions. Small-angle ($\theta = 30°$) scattering of $E_0 = 6\,\text{keV}\,\text{Ar}^+$ and Kr^+ ions from an atomically clean Cu(100) face was considered. The energy distribution of the scattered ions as a function of ψ were measured. In full agreement with Mashkova and Molchanov (1975, 1980), a large number of ions were detected outside the theoretically predicted scattering angle region. Moreover, the energy spectrum revealed the presence of several peaks at energies exceeding that of single scattering, which implies the existence of multiple-scattered ions at such grazing angles where no such ions should be according to the model of an infinite atomic row.

Figure 3.15 from Petrov and Abroyan (1977) shows the experimentally obtained and calculated (S.H.A. Begemann and Boers 1972) dependences of the energy of scattered Kr^+ ions on ψ obtained for various nonideal structures on Cu(001). As follows from this figure, scattering at an angle $\theta = 30°$ within the infinite row model is possible only in a narrow interval of grazing angles (approximatively specular scattering $\psi \cong 15°$ for the Kr^+ ions). The inclusion of vacancies in the row permits broadening of the oval in the $E(\psi)$ dependence; however, the experimentally observed structure of the distributions for small ($\psi < 10°$) and large ($\psi > 18°$) grazing angles can be accounted for only in calculations of scattering from upward and downward steps. A comparison of the energies measured for different peaks as a function of ψ with simple calculations for a broken-off row allowed identification of these peaks as originating from a surface with vacancies and steps. Calculations carried out for broken-off rows containing up to six regularly arranged atoms can explain at least seven experimentally observed multiple-scattering peaks. The presence of such a large number of peaks is explained by the low current density ($2 \times 10^{-9}\,\text{A}/\text{cm}^2$) at which only one of 10^5 surface atoms is struck by a projectile ion. The current density in the studies of Mashkova and Molchanov (1975) and Balashova et al. (1977c), where only one double-scattering peak was observed, usually was $2 \times 10^{-4}\,\text{A}/\text{cm}^2$, with every other surface atom struck by a projectile. Observation of only single- and double-scattering peaks under these conditions evidences the absence, in the course of ion bombardment, of surface structures containing more than one atom in the immediate vicinity. This conclusion is supported by studies of the bombardment-induced surface damage carried out by SIMS (McCraken 1975, Pleshivtsev 1968) and LEED (Farnsworth and Hayek 1967, Hauffe 1971).

We studied (Marchenko et al. 1973) the effect of a defected row on the spatial distributions of 30 keV Ar^+ ions scattered from the Cu(100) face in the $\langle 100 \rangle$ direction at the polar ψ and azimuthal ξ grazing angles equal to

13° and 4°, respectively. The defect was simulated by the removal of one of the atoms in the row, the vacancy being located either immediately before the target area or directly in it. We calculated the dependence of the escape angle β on the azimuthal scattering angle φ, the angles ψ and ξ being measured from the {110} axis. In the first cases, the scattering differed insignificantly from that on an ideal row; this difference consisted in the presence of a small number of ions penetrating the row (negative values of the β angle). In the second case, however, the penetration through the row predominated, with only a small part of the ions being backscattered.

Eckstein et al. (1974) used a computer simulation to study the scattering of 6 keV Ne^+ and Kr^+ ions at $\psi = 8°$ from defected rows of V and Cu atoms, respectively. They simulated vacancies (absence of one atom in a long row) and upward and downward steps. Figure 3a presents the energy dependence on scattering angle θ for Ne^+ ions on V rows with vacancies and steps, and fig. 3b similar dependences for Kr^+ ions on Cu. Also shown are similar dependences for fast recoil atoms. In the Ne^+–V case, the recoil energy is lower than the scattered ion energy; however, in the Kr^+–Cu case the $E_1/E_0(\theta)$ ovals for the scattered atoms and the recoils intersect several times, which complicates the interpretation of the measured spectra. A detailed study was carried out (S.H.A. Begemann and Boers 1972) for Kr^+ ions scattering from the Cu(100) face together with energy and mass analysis, however, no target ions were detected at $\theta = 30°$ and different grazing angles ψ. Eckstein et al. (1974) point out several reasons for that, in particular, the possibility of neutralization for the recoil ions may be higher than that for the scattered ones, however, one could suggested another reason as well. In their computer simulation, Eckstein et al. (1974) used the Molière potential for an ideal row, while the scattering angles were derived from the impact parameters based on the 'magical formula' of Lindhard et al. (1963). The impact parameter in each subsequent collision with atoms in the row was calculated by expression (8) of chapter 3, which is valid for double scattering. When considering sequential multiple scattering from atoms in a row one should take in eq. (8) of chapter 3 instead of θ_i a sum of the scattering angles acquired in all previous collisions, $\theta = \Sigma_{i=1}^{n-1} \theta_i$. Then the reflected beam will turn out to be limited to a much more narrow interval of scattering angles, and, as shown by the experiment of S.H.A. Begemann and Boers (1972), there will be no recoil atoms with an energy close to that of the scattered ions.

Parilis et al. (1975), and Umarov (1975) employed a two-atom scattering model to describe the experimentally observed energy and angular distributions of scattered ions. The scattering of ions from defected structures with vacancies was simulated by a sequential collision with two atoms located at a distance D, which is a multiple of the interatomic separation in an ideal crystal. When simulating scattering from steps, the position of one

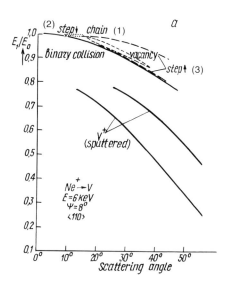

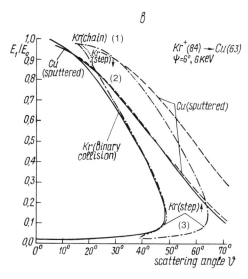

Fig. 3. E/E_0 versus scattering angle θ plots for (a) Ne$^+$–V and (b) Kr$^+$–Cu. (1) row with vacancies, (2) downward step, (3) upward step.

of the atoms was varied relative to the surface plane (xOy plane) by prescribing a coordinate $z \neq 0$. Figure 4a shows the dependences of the scattered-ion energy on scattering angle θ calculated within the ideal two-atom model with inclusion of vacancies and their linear clusters. Also shown, for comparison, are the results of a calculation for an ideal and a

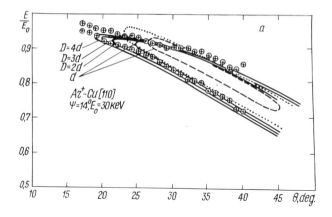

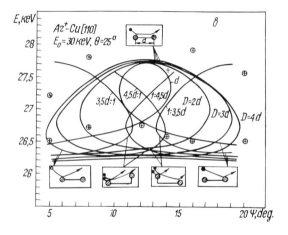

Fig. 4. Dependence (a) of the Ar$^+$ ion energy on the scattering angle for the row and two-atom models including vacancies and their linear clusters and (b) of the energy of the ions scattered at an angle $\theta = 25°$ on grazing angle ψ, with vacancies included. Crosses denote experiment.

heated row (Kivilis et al. 1970, Parilis et al. 1967). The calculations were carried out in the binary collision approximation for the Ar$^+$–Cu [110] pair for $E_0 = 30$ keV, $\psi = 14°$. When using the ideal row model, the reflected beam is limited on both the small and large scattering angle sides, whereas in the two-atom model the limitation is observed only on the side of small angles. Inclusion of thermal vibrations in the atomic row model increases the separation between peaks as well as broadens the interval of angles where scattering is possible. However, as seen from fig. 4a, scattering at angles $\theta < 23°$ based on the model of a heated row cannot likewise be explained. Calculation involving the two-atom model with vacancies permits one to interpret the experimentally observed scattering in the shadow region (Balashova et al. 1977c); note that as the number of the neighboring vacancies increases, the minimum scattering angle decreases, the double-scattering peak practically does not change its position, whereas the QS scattering peak shifts toward lower energies. This results from the fact that the effect of the second atom on QS scattering diminishes with increasing interatomic separation.

Figure 4b shows the dependences of the energies of the ions backscattered at $\theta = 25°$ on grazing angle ψ for various interatomic spacings D. The calculated $E(\psi)$ curves obtained for two-atom scattering represent loops, their branches corresponding to single and double collisions; note that as one approaches specular scattering ($\psi = \frac{1}{2}\theta$) from the sides of both large and small angles the single- and double-scattering peaks shift toward high energies. One readily sees that as the interatomic spacing increases, the interval of ψ angles where double scattering is possible increases too, the single-scattering peaks shifting toward lower energies. The same figure presents the calculations of $E(\psi)$ for upward and downward atomic steps which were simulated by two atoms with coordinates $x_1 = 0$, $z_1 = 0$ and $x_2 = 3.5-4.5d_{\langle 110 \rangle}$, $z_2 = \frac{1}{2}a$, respectively. The upper branches of the open ovals correspond to double scattering, and the lower ones to single colli-sions. In the region $10 \leq \psi \leq 15°$ near the specular scattering ($\psi = \frac{1}{2}\theta$), the peak positions for single and double scattering from a step differ from those for vacancies. Indeed, for a step $4.5d_{\langle 110 \rangle}$ long, close to $\psi = 12.5°$, the peaks are shifted toward one another. Particularly strongly shifted is the double-scattering peak. This displacement can be revealed experimentally since it exceeds considerably the analyzer resolution ($\approx 1\%$). Thus, there is a certain correlation between the type of the surface defect structure and the pattern of the ion-energy distributions.

Mashkova and Molchanov (1980) showed experimentally that the relative intensity of double scattering R is strongly nonmonotonic as a function of grazing angle ψ and scattering angle θ. In the beginning, R grows as θ decreases from 40 down to 25°, and ψ from 22 to 15°; in the region of specular scattering it becomes greater than unity and is practically in-dependent of the angles θ and ψ, while dropping sharply as these angles decrease still further. As seen from fig. 5, the calculations based on the ideal row and two-atom models cannot explain the experimentally observed nonmonotonic behavior and the presence of double scattering in the shadow region (for $\theta < 25°$, $\psi < 11°$ and $\psi > 14°$). As one turns over to models with defects (vacancies and steps) the magnitude of R decreases because of increasing interatomic separation, while the interval of angles θ and ψ where double scattering is possible increases. The $R(\theta)$ curves fall off monotonical-ly with increasing θ, while the $R(\psi)$ curves are shaped as concave arcs with the minimum corresponding to the specular scattering angle. The $R(\psi)$

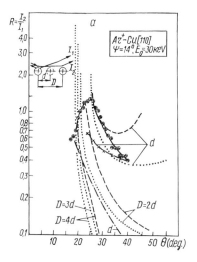
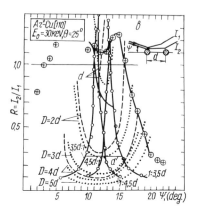

Fig. 5. Dependences of $R = I_2/I_1$ on (a) θ and (b) ψ. Circles are from experiment (Mashkova and Molchanov 1980), broken curves from atomic row calculation, dotted curves from two-atom calculation, lines with crosses from atom step calculation, chain curves from analytical estimates using eq. (30) from chapter 3, full curves from sum envelopes.

dependence for atomic steps differs from that for vacancies, namely, it consists of two branches going up as one approaches the specular scattering angle from the sides of both large and small grazing angles ψ. The left-hand branches ($3.5d:1$ and $4.5d:1$) correspond to the upward step, and the right-hand ones ($1:3.5d$ and $1:4.5d$) to the downward step. As follows from fig. 5, only calculations carried out for larger separation structures which simulate the presence of vacancies on the surface can explain the left-hand dropping branch of the experimental $R(\theta)$ and $R(\psi)$ curves; note that in the latter case one should also take into account the presence of steps. The heights of the single-scattering peaks from defected structures are proportional, in accordance with eq. (6), to their number, scattering cross section, the number of particles scattered in the ionic state, and the shadowing coefficient.

Nizhnaya et al. (1979a) simulated the effect of surface structure defects on the angular and energy distributions of scattered ions from a defected row with atoms located at distances $D = nd$ (where $n = 1, 2, 3, 4$), a multiple of the interatomic separation d, as well as from broken-off rows consisting of 2, 3, 4, etc. atoms. It was shown that, in contrast to the ideal or defected rows responsible for the two peak structure in the energy distribution, the number of peaks in the spectrum due to scattering from a broken-off row is $(2n - 1)$, and the number of loops in the dependence of energy on ψ is $(n - 1)$, where n is the number of atoms in a row. Figure 6a presents the

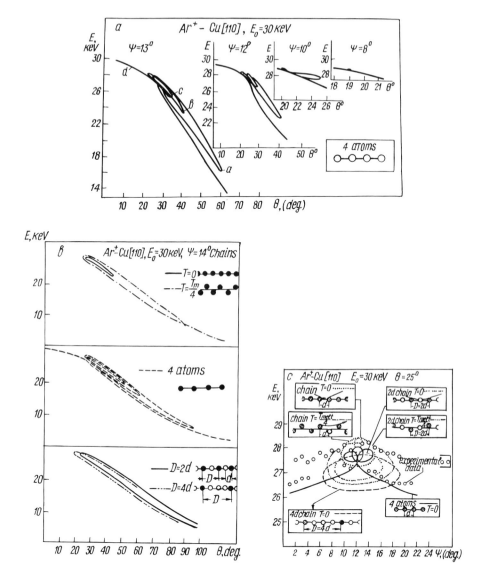

Fig. 6. $E(\theta)$ plots for (a) a four-atom row for different grazing angles, (b) for an ideal row for $T = 0$, and $T = \frac{1}{4}T_{\text{melt}}$, and a four-atom row with inclusion of vacancies, and (c) $E(\psi)$ plot for different surface models (circles denote experiment).

dependences of the energy of Ar^+ ions backscattered from a broken-off row of four Cu atoms in the $\langle 110 \rangle$ direction on scattering angle θ for various ψ and for $E_0 = 30\,keV$. Loop a corresponds to single and double scattering from the first pair of atoms, loop b from the second pair, loop c from the third pair, and loop d to single scattering from the last atom in the row. The different energy positions of the peaks are due to different blocking action of the preceding atoms. There is no blocking for the first pair of atoms, so that the peak positions in the energy spectrum agree with those calculated with the two-atom model; for atoms 2 and 3 the blocking narrows the $E(\theta)$ loop and shifts the peaks toward higher energies (b), the pattern of scattering from atoms 3 and 4 being similar to that from an infinite row. The number of the peaks decreases with decreasing ψ and E_0 while energy resolution of the peak deteriorates as the number of atoms n in a broken-off rows grows. To the scattering from defected rows corresponds the broadened $E(\theta)$ oval for both large and small scattering angles, which accounts for scattering in the shadow region for an ideal row (fig. 6b). The ideal row at $T = 0\,\mathrm{K}$ is seen to limit the $E(\theta)$ oval at angles $\theta_{\min} = 45°$, whereas for a row with a twice larger spacing the corresponding limiting angles are 22° and 90°. As the interatomic spacing in the row increases, the ovals shift toward lower energies. Figure 6c shows an experimental $E(\psi)$ dependence and the corresponding calculated curves for various surface defect structures. The calculation was performed for 30 keV Ar^+ ions back-scattered from the Cu (100) face in the $\langle 110 \rangle$ direction at an angle $\theta = 25°$. In the case of the ideal row the $E(\psi)$ dependence is seen to be shaped as a small oval close to specular scattering. As the spacing in the row increases, the ovals extend toward both large and small ψ. The model of a broken row consisting of four atoms yields for $T = 0\,K$ a curve with single-scattering branches and three loops.

Inclusion of thermal vibrations also results in a broadening of the $E(\psi)$ oval. Indeed, for an amplitude of thermal vibrations $\sigma = 0.05a_i$ for an ideal row the oval lies between the angles $\psi_{\min} = 10°$ and $\psi_{\max} = 15°$, while for a row with double spacing $(D = 2d)$ we have $\psi_{\min} = 7°$ and $\psi_{\max} = 19°$. The experimentally observed scattering in the shadow region is explained only as due to the participation in scattering of all the above-mentioned nonideal structures simultaneously. The pattern of the calculated $E(\psi)$ dependences and their comparison with experiment reveals that the participation of a nonideal structure in scattering varies with the variation in experiment geometry, which is apparently connected to the intensity of their formation and the degree of shadowing on the irradiated surface. The complex structure of the experimental energy spectra cannot be explained by using in calculations only one model, which suggests that various nonideal structures formed in ion bombardment participate in scattering. This complicates greatly the surface analysis since the ions scattered from different nonideal structures may get into the same energy interval while the height of a peak will depend not only on the scattering cross section but also on the relative abundance of the corresponding structure on the surface.

The selection of the interaction potential and of the type of thermal lattice vibrations affects considerably the character of the angular and energy distribution of scattered ions. Calculations performed for defected rows with inclusion of thermal vibrations and with different interaction potentials show that the increase of interatomic spacing and vibration amplitude, and the reduction of blocking associated with the application of harder potentials, result in an increase of the angular dimensions of the $E(\theta)$ ovals, which means that they act in the same direction; this complicates unambiguous identification of a surface structure on the one hand, and of the interaction

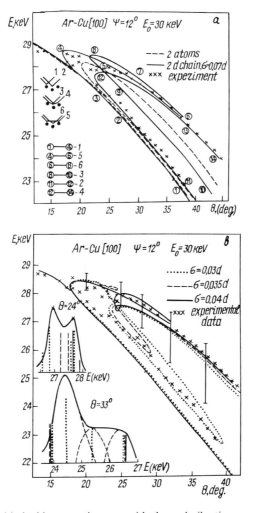

Fig. 7. $E(\theta)$ plots for (a) double-separation row with thermal vibration amplitude $\sigma = 0.07d$ Firsov potential), and (b) ideal row with $\sigma = 0.03d$, $0.035d$, $0.04d$ (model Vajda potential with $r_c = 1.7 \text{ Å}$).

potential on the other. This is illustrated in fig. 7, which compares the calculations of the $E(\theta)$ dependence for various nonideal structures and interaction potentials with experiment. As seen from fig. 7a, the two-atom model is not capable of accounting for the characteristic break in the $E(\theta)$ curve in the angular region $\theta = 22°–23°$, whereas calculation for a defected row making use of the Firsov potential with an rms vibration amplitude $\sigma = 0.07d$ agrees satisfactorily with experiment. However, using the model potential of Vajda, one can explain satisfactorily the experimental curve even with an ideal row with $\sigma = 0.03d–0.04d$ (fig. 7b). However, while in the first case the amplitude $\sigma = 0.07d$ at room temperature corresponds to a Debye temperature $\theta_D = 50$ K, in the second case $\sigma = 0.04d$ corresponds $\theta_D = 138$ K which is close to the experimental value $\theta_D = 133$ K (Verhey et al. 1975) and to the value $\theta_D = 192$ K obtained by Jackson (1974).

Thus, the preference given to a surface structure under the experimental conditions of Balashova et al. (1977c) is connected with the selection of the interaction potential and the rms amplitude of thermal lattice vibrations which depends at a given target temperature on θ_D.

Therefore, taking into account the nonideal structure of the surface of single crystals (the presence of vacancies, their clusters and steps) permits one to explain the various features in the experimentally observed angular and energy distributions of scattered ions. The theoretically established and experimentally confirmed relation between the multiple scattering of medium-energy ions and surface defect structure makes it possible to use it to advantage in the analysis of the structure of nonideal single-crystal surfaces, and in detection of point defects of the type of vacancies and their clusters, as well as of atomic steps by the medium-energy ion scattering technique.

CHAPTER 7

Scattering of Light Atoms

1. Energy spectra of scattered light atoms

At present, measurements of neutral and charged components of the scattered beam have established some regularities concerning the double differential yield for light atoms $R = d^2R/d\theta \, d\varepsilon$:

(1) Energy spectra of light backscattered atoms contain a cupola with a maximum at the relative escape energy $\varepsilon_m = E_m/E_0$, which decreases with an increase of the primary energy E_0 and the scattering angle θ (Zhabrev et al. 1974, Zhabrev 1979, Eckstein et al. 1976, 1978);

(2) the half width of the cupola-like distributions $\Delta\varepsilon_{1/2}$ changes non-monotonously with increasing E_0 and θ (Zhabrev et al. 1974, Zhabrev 1979);

(3) with increasing glancing angle α, the values θ, ε_m and $\Delta\varepsilon_{1/2}$ slightly increase (Zhabrev 1979);

(4) with increasing atomic number of the bombarding particles (Z_1) and target atoms (Z_2), ε_m is shifted towards the high-energy region (Eckstein and Verbeck 1978);

(5) in the spectrum, besides the cupola, near the single-scattering energy on the surface, $\varepsilon_1 = E_1/E_0$, an additional maximum has been observed, the so-called surface peak. This peak is isolated from the cupola only for low energies $E_0 < E_k$, where $E_k \approx 5$ keV (Eckstein et al. 1978).

The theoretical studies of energy spectra of light atoms scattered from a solid surface could be divided into three groups.

1.1. Monte Carlo computer simulation

Monte Carlo computer simulations of reflection have been intensively developed by Oen and Robinson (1976c), and J.E. Robinson et al. (1976). This method is applicable for any scattering angle θ, φ, energy E_0 and any ion–target combination. The type of differential scattering cross section in a separate collision event, the choice of particle stopping power and even the polycrystal models appeared to be different in most papers. It should be noted that in most papers concerning computer simulation the integral characteristics of atom backscattering, particle reflection and energy reflection coefficients have been calculated. The regularities in the behavior of the magnitude and the detailed comparison of theory with experiment were reported by Mashkova (1979). Difficulties arising in the calculation of double differential reflection coefficients using such method are usually a result of insufficient statistics, although, in principle, such calculations seem to be possible. So, e.g., in Robinson's group (J.E. Robinson et al. 1976), after each collision event the sign of the projective range has been determined. The negative range was related to the particle escaping the solid. For such particles, energy E and scattering angles θ and φ were calculated. The energy spectra of H, He, Ne, N, and Ar scattered on an Au polycrystal target with $E_0 = 5$ keV, $\theta = 90°$ were calculated using an analogous technique (Buck et al. 1982) that agreed well with the experiments. However, as far as we know, a systematical comparison between experimental data and computer calculations of light-atom energy spectra has never been made.

1.2. Multiple scattering model

Multiple-scattering theory for particles moving in a semifinite medium has been developed only for small glancing α and escape ψ angles. To calculate the function of the angle and energy distributions, the integro-differential equation is written as follows,

$$\frac{\partial f}{\partial t} = \boldsymbol{v}\, \frac{\partial f}{\partial \boldsymbol{r}} + \frac{\partial \boldsymbol{v}}{\partial t}\, \frac{\partial f}{\partial \boldsymbol{v}} = \left(\frac{\partial f}{\partial t}\right)_{\text{coll}}, \tag{1}$$

where $(\partial f/\partial t)_{\text{coll}}$ is the collision integral.

The first results allowing to calculate the energy spectra of reflected atoms were reported by Firsov et al. (1973, 1976). The authors have developed two approximate solution procedures of eq. (1) distinguished by $(\partial f/\partial t)_{\text{coll}}$.

(i) The first procedure is the diffusion approximation. In this approximation the integrand $(\partial f/\partial t)_{\text{coll}}$ is expanded over small values of the parameter θ (θ is the scattering angle in the laboratory system) up to terms θ^2. Then $(\partial f/\partial t)_{\text{coll}}$ could be expressed via an angular Laplace operator and eq. (1) becomes merely differential. Firsov (1970) pointed out that this approximation is valid for potentials $U = D/r^n$ at $n = 1$; at $n \geq 2$ its application seems to be questionable. For Coulomb scattering, $n = 1$ in the symmetrical case $\alpha = \psi$ and the result is given by

$$R = \frac{d^2 R}{d\Omega\, d\varepsilon}\, K_1 x^{-5/2} \exp(-y/g_0 l_0). \tag{2}$$

Here y is a function depending on the scattering angle and energy, $x = 1 - \sqrt{\varepsilon}$, $K_1 = 1 - \frac{9}{4}x + \frac{5}{4}x^2$, $l_0 = 2\sqrt{E_0/K}$ is the total ion effective range,

$$g_0 = \tfrac{1}{2}\pi\, \frac{N_0 Z_1 Z_2^4 e^2}{E_0^2}\, \ln(1 + 0.7 E_0/30\text{ eV}\, \sqrt{Z_1^{2/3} + Z_2^{2/3}}).$$

Formula (2) has been generalized by Kurnaev et al. (1985) for nonsymmetrical scattering,

$$R = \frac{1 - 2.25x + 1.25x^2}{x^{5/2}} \exp\left\{ \frac{-3(1 - 1.5x + 0.5x^2)\alpha^2 - 3(1 - x^2)\alpha\theta}{g_0 l_0 x} \right.$$

$$\left. + \frac{(1 - 2.25x + 1.25x^2)\theta^2}{g_0 l_0 x} \right\}. \tag{3}$$

Formulae (2) and (3) give cupola-like distributions with a broad maximum. Under conditions when application of the Coulomb potential seems to be justified (low atomic numbers Z_2 of the target atoms and E_0 of the order of tens of keV), eqs. (2) and (3) describe the experimental data well.

(ii) The second method of solution of eq. (1) was used when in the above expansion $v = m_1/M$ is considered to be a small parameter, rather than θ^2, i.e., valid at light-atom scattering on heavy targets, i.e., for $n = 2$. In this approximation, eq. (1) is written as

$$v\psi \frac{\partial f}{\partial z} + v\varphi \frac{\partial f}{\partial y} + v \frac{\partial f}{\partial x} = v g_1(l) \int_{-1}^{1} \int_{-\pi}^{\pi} \frac{[f(\psi', \varphi') - f(\psi, \varphi)] \, d\psi' \, d\varphi'}{[(\psi' - \psi)^2 + (\varphi' - \varphi)^2]^{3/2}}. \tag{4}$$

In eq. (4), both elastic and inelastic energy losses of particles were taken into account. The value of $g_1(l)$ is connected to the type of scattering cross section, for $n = 2$,

$$g_1(l) = \frac{0.4 N_e \pi Z_1 Z_2 a_0^2 l}{(Z_1^{2/3} + Z_2^{2/3}) E(l)}, \tag{5}$$

where l is the distance passed by a particle in the solid. At continuous stopping, according to Lindhard and Scharff (1961), $E(l) = E_0(1 - l/l_0)^2$. Firsov et al. (1976), by solving eq. (4), obtained

$$F_1(\theta, l) = \frac{1}{2\pi} \frac{\omega(l)}{(\omega^2 + \theta^2)^{3/2}} ;$$
$$\tag{6}$$
$$F_2(s, l) \simeq \frac{1}{2\pi} \frac{\bar{n}\omega(l)}{(\omega^2 \bar{n}^2 + s^2)^{3/2}} ,$$

where F_1 is the angular distribution of the particles inside the solid, integrated space-wise.

In eq. (6), $\omega(l) = 2\pi \int_0^l g_1(s) \, ds$ is the parameter coinciding with the effective scattering angle for particles which passed the path l in the solid, F_2 is the spatial distribution of particles integrated over the angles, s is the distance counted off over the normal to beam incidence direction, $\bar{n} = \int_0^l g(s)(1 - s) \, ds / \int_0^l g(s) \, ds$. Equations (5) and (6) could be compared with experimental data of ion bombardment from the back side.

The function $R(\varepsilon, \theta, \varphi)$ is included within the boundaries from the condition of eq. (4). Firsov et al. (1976) showed that R satisfied a rather complicated integral equation, analytical solution of which could be derived using additional simplified assumptions. This equation is written as

$$R = \frac{4\sqrt{3}g_1(0)\,\psi}{\pi[4\pi^2\omega^2(1-\sqrt{E})+\varphi^2+4(\psi^2-\psi\alpha+\alpha^2)]^2}\,. \tag{7}$$

The agreement of eq. (7) with experiment seems to be worse than of eqs. (2) and (3). Function (7) provides a distribution monotonously increasing with ε, whereas the experimental spectra have a maxima, as usual.

The theory of multiple scattering at small angles has been developed further by Remizovich et al. (1980a,b). These authors, as opposed to Firsov (1970), have supposed that the condition $\theta_{eff} \ll \alpha$ (θ_{eff} is the effective scattering angle in the solid) allowed to use the diffusion approximation regardless of the interaction potential. Besides, the requirement $\langle\theta_s^2\rangle l_0 \ll 1$ is needed ($\langle\theta_s^2\rangle$ is the mean square of the scattering angle per unit length), allowing to solve the problem in the small-angle approximation. Then, for the function $N(z, \psi, \varphi, E)$ being the particle flux density at depth z, the transfer equation is written as

$$\psi\frac{\partial N}{\partial z} = \frac{\langle\theta_s^2(E)\rangle}{4}\left(\frac{\partial^2 N}{\partial z^2}+\frac{\partial^2 N}{\partial\varphi^2}\right)+\frac{\partial}{\partial E}\left(\frac{\mathrm{d}E}{\mathrm{d}l}\,N\right). \tag{8}$$

The last term in eq. (8) describes the inelastic particle stopping in the continuous medium slowing-down approximation. All information about the single-scattering event is contained in the parameter $\langle\theta_s^2(E)\rangle$. The boundary condition

$$N(z=0, \psi, \varphi, E) = N_0\delta(\psi-\alpha)\delta(\psi)\delta(E-E_0)\,, \quad \psi<0\,,$$
$$= N_0\,\frac{\alpha}{\psi}\,R(\eta, \varphi, E)\,, \qquad\qquad \psi>0\,. \tag{9}$$

is added to eq. (8). Here, N_0 is the flux density in the incident beam, and R is a determinable distribution function for the reflected particles. Equation (8), with account of eq. (9), leads to the integral equation for the function R.

Remizovich et al. (1980a,b) have developed a new original technique, allowing to get an accurate analytical solution of this equation in two cases.

(i) The last term is neglected in eq. (8), this corresponds to elastic scattering (Remizovich et al. 1980a). Then all particles are escaping with energy $E = E_0$. R is given by

$$R(E, \alpha, \psi, \varphi) = \frac{\psi^{1/2}}{12\pi^2\alpha^{1/2}}\left[\frac{\omega_1^2}{1+\omega_1^2}+\omega_1^3\arctan\omega_1\right]\delta(E-E_0)\,, \tag{10}$$

where $\omega_1 = [3\psi\alpha/(\psi^2+\alpha^2+\varphi^2/4-\psi\alpha)]^{1/2}$. It is clear that eq. (10) is of limited use. The consideration of the slowing down seems to be necessary to calculate the real energy spectra.

(ii) A solution which takes this in account has been found by Remizovich et al. (1980b) using the following assumption as simplification,

$$\langle \theta_s^2(E) \rangle = \langle \theta_s^2(E_0) \rangle = \text{const} . \tag{11}$$

It is considered that eq. (11) could be realized, because at $\alpha \ll 1$ the observed spectra have a sharp maximum at $E \approx E_0$. In this approximation, R is given by

$$R = \frac{\sqrt{3}}{9\pi^2 l_0 \sqrt{\varepsilon}} \frac{\exp\{-[4(\psi^2 - \psi\alpha + \alpha^2) + \varphi^2]/4\sigma\alpha^2(1 - \sqrt{\varepsilon})\}}{(1 - \varepsilon)^{5/2} l_0^{3/2}}$$

$$\times \text{erf}\left[\left(\frac{3\psi}{\alpha\sigma_0(1 - \sqrt{\varepsilon})}\right)^{1/2}\right] . \tag{12}$$

Here $\sigma_0 = \langle \theta_s^2(E) \rangle / 4l_0\alpha^2$, and $\text{erf}(x) = 2/\sqrt{\pi} \int_0^x e^{-t} \, dt$ is the normalized error integral.

Remizovich et al. (1980b) compared eq. (12) with the experiment of Zhabrev et al. (1974). For $\langle \theta_s^2(E) \rangle$ dependence,

$$\langle \theta_s(E) \rangle = \frac{1.82 a_0^2 I N_0 Z_1 Z_2 (1 + \nu)}{(Z_1^{2/3} + Z_2^{2/3})E} \approx \frac{1}{E} , \tag{13}$$

corresponding to the potential $U = A/r^2$. Equation (12), with account of eq. (13), agrees well with experiment, in spite of the fact that within the limits of the calculated spectrum $\langle \theta_s^2(E) \rangle$ changes by a factor of two.

Remizovich and Shakhmamctyev (1983) have dropped the simplifying assumption, eq. (11), and found a simple approximation formula for $R(\theta, \varphi, \psi)$ at arbitrary dependence $\langle \theta_s^2(E) \rangle$. First, these authors, using arbitrary $\langle \theta_s^2(E) \rangle$, obtained the accurate solution of the problem of the particle distribution function for infinite medium. Then this solution was written for cases where eq. (11) could be satisfied. The obtained formula differs from eq. (12) for the reflection function from a semiinfinite medium by a factor $\text{erf}[(3\psi/\alpha\sigma_0(1 - \sqrt{\varepsilon})^{1/2}]$ only. Further, they naturally assumed that in general the semiinfinity of the medium could be also considered by multiplying the solution for infinite medium by a factor containing the function $\text{erf}(x)$.

As a result, they have obtained the following formula,

$$R = \frac{|\psi/\alpha|}{(4\pi\sigma_0)^{3/2}} \frac{1}{(A_0\Delta^2)^{1/2}} \exp\left[-\frac{\varphi^2}{4\alpha^2\sigma_0 A_0} - \frac{s^2 A_0 - 2s(1 + |\psi/\alpha|)A_1}{4\sigma_0\Delta}\right.$$

$$\left. + \frac{(1 + |\psi^2/\alpha^2|)A_2}{4\sigma_0\Delta}\right] \text{erf}\left[\left(\frac{3|\psi/\alpha|}{\sigma A_0(s)}\right)^{1/2}\right] . \tag{14}$$

Here,

$$s = \frac{1}{l_0} \int_E^{E_0} \left| \frac{dl}{dE} \right| dE ; \qquad A_n(s) = \int_0^s (s - s') \frac{n \langle \theta^2(E)/s' \rangle}{\langle \theta^2(E_0) \rangle} \, ds' ;$$

$$n = 0, 1, 2 ; \qquad \Delta^2 = A_0(s) A_2(s) - A_1^2(s) .$$

Equation (14) was compared with the experimental scattering spectra of $D^+ \to Si$, $E_0 = 30$ keV, $\alpha \leqslant 10°$, $\theta \leqslant 20°$. The parameter σ_0 has been fitted in such a way that the maxima of theoretical and experimental curves coincided. Good agreement has been achieved in the shape of the spectrum.

Tilinin (1983) made an attempt to be outside the scope at the small-angle approximation and to consider the case $\alpha = \frac{1}{2}\pi$, which seems to be practically important. The transfer equation could be solved in case of

$$l_0/l_t = \sigma^* \gg 1 , \tag{15}$$

where $l_t = (n_0 \sigma_t)^{-1}$ is the transport length, and $\sigma_t = \int_0^\pi (1 - \cos\theta)(d\sigma/d\Omega)\,d\Omega$ is the transport cross section. Condition (15) has been interpreted in such a way that most of the particles have time to undergo backscattering before considerable slowing down. Then from eq. (15) the main assumption for simplification used in this paper could be derived, according to which the total cross sections both for elastic and inelastic interactions are independent on ion energy,

$$\sigma_{el}(E) = \sigma_{el}(E_0) = \text{const} , \qquad \sigma_{inel}(E) = \sigma_{inel}(E_0) = \text{const} . \tag{16}$$

In the approximation of eq. (16), the solution of the transfer equation $R = R_0$ could be obtained in a explicit form, R_0 being expressed by the integrals of the elementary and special function product. Further, Tilinin (1983) has proposed to take into account the reflection probability for particles which have passed the path $l \leqslant l_t$, and $R(\varepsilon, \psi, \alpha)$ to be written in a closed form,

$$R(\varepsilon, \psi, \alpha) = R_1(\varepsilon, \psi, \alpha) + (1 - q)R_D(\varepsilon, \psi, \alpha) . \tag{17}$$

Here $R_1(\varepsilon, \psi, \alpha)$ is the reflection function, estimated within the model of single scattering with beam attenuation, R_1 describes the scattering in the high-energy region of the spectrum and corresponds to the surface peak observed in experiments by Buck et al. (1982) and Eckstein et al. (1978). The value q which describes the relative number of particles scattered over the angle $\theta < \alpha$ on the path $l \leqslant l_t(1 - \cos\alpha)$ depending on the glancing angle, seems to be a more suitable parameter than σ^*. The proposed theory holds to provide the correct results at $\alpha \geqslant 60°$. In our opinion this statement seems not always to be valid. So, at the comparison of the results by Tilinin (1983) with experiments on H^+ and He^+ scattering on heavy targets for

$E_0 \geq 10$ keV and $\alpha = 90°$ some difficulties would arise, since condition (16) would be essentially violated. Experimental data by Eckstein et al. (1976) and Eckstein et al. (1978) have shown that many particles are reflected at $\varepsilon = E/E_0 = 0.1$–$0.3$, i.e., a change in $\sigma_{el}(E)$ of nearly an order of magnitude.

1.3. Single scattering model

The single-scattering model has been applied long ago for light atoms with high energies. McCraken and Freeman (1969) considered scattering at angle θ along l_1 to occur as a result of a single collision event with a cross section

$$\mathrm{d}\sigma/\mathrm{d}\Omega = \sigma(E_1, \theta) = f(\theta)/E_1^n .$$

In addition, the motion of particles is considered to be along a straight line both before and after scattering (see fig. 1). If we confine ourselves to the consideration of only inelastic energy losses using the Lindhard–Scharff law (1961), then for $R(\varepsilon, \theta)$ at $\varphi = 0$ we could get

$$R(\varepsilon, \theta) \simeq N_0 \sigma(E_1, \theta)\, \mathrm{d}l_1/\mathrm{d}\varepsilon \; \frac{1}{\sqrt{\varepsilon}(\sqrt{\varepsilon} \sin \psi + \sin \alpha)^{2n}} . \tag{18}$$

Here $n = 2$ for $U = D_2/r^2$. The simplicity of eq. (18) and the lack of a general theory of multiple scattering at large angles provoked the attempts for its application in the middle-energy region.

Let us consider a typical energy spectrum (Eckstein et al. 1976): $H^+ \rightarrow Fe$, $E_0 = 15$ keV, $\alpha = 90°$, $\theta = 135°$, and compare the values of eq. (18) with those of experiment at $n - 1$. From fig. 2 it is seen that eq. (18) describes quite well the greater part of the energy spectrum $\varepsilon > \varepsilon_m \approx 0.1$. Moreover, using eq. (18) we could calculate both the particle and energy total reflection coefficients and obtain good agreement at $E_0 \geq 10$ keV (Eckstein and Biersack 1983). This means that a simple model (McCraken

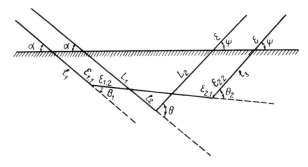

Fig. 1. Single- and double-scattering scheme.

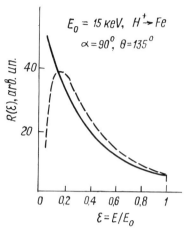

Fig. 2. Comparison of the energy spectra measured by Eckstein et al. (1976) (broken curve) with calculations using formula (18) (full curve).

and Freeman 1969) gives quantitative agreement with middle-energy experiment. This fact seems to be quite clear at first sight. Actually, the application of the model of single scattering seems to be possible at $E_0 > 100$ keV, when the inequalities $\omega(\varepsilon) \ll \theta$, $\rho(\varepsilon) \ll L(E)$ are satisfied practically for the whole spectrum, i.e., the scattering being quasisingle and the trajectories being almost straight lines. Using eq. (5) and with account of $\omega(\varepsilon) = 2\pi \int_\varepsilon^1 g_1(\varepsilon') l |\mathrm{d}l/\mathrm{d}\varepsilon'| \, \mathrm{d}\varepsilon'$, we obtain

$$\omega(\varepsilon) = 0.2\pi/\xi_1 (Z_1^{2/3} + Z_2^{2/3}) \sqrt{E_\mathrm{B}/E_0} (1 - \sqrt{\varepsilon}) . \tag{19}$$

In table 1, the values $\omega(\varepsilon)$ are listed under the condition: $Z_1 = 1$, $Z_2 = 26$

Table 1

Dependence of the effective multiple-scattering angle on relative energy.

ε	ω (°)
1	0
0.9	42
0.8	86
0.7	133
0.6	184
0.5	239
0.4	300
0.3	370
0.2	452
0.1	559

(Eckstein et al. 1976). The value $\xi_1 = 1.8$ [using the experimental data of Gott and Yavlinsky (1973)].

From table 1 it is seen that in the greater part of the spectrum $\omega(\varepsilon) > 1$; i.e., the particles which passed the path $L(\varepsilon)$ in the solid could be scattered at $\theta = 135°$ in multiple scattering at small angles with reasonable probability F_1. It should be noted that the function F_1 describes the probability for any particle that has passed in the solid a path $L(\varepsilon) = AB$ to be scattered at an angle θ, regardless of the orientation of the curve $L(\varepsilon)$ relatively to the solid surface. However, the backscattered particles, for which the points A and B lie on the surface, are of interest for us.

Apparently, for $\alpha = \frac{1}{2}\pi$ these particles, which turn to the surface in time (approximately at the first half of the path) mainly contribute to backscattering. Scattering at an angle $\theta_1 \geqslant \alpha$ is considered to be the condition of such a turn. Then one strong deflection in multiple scattering is of great importance. The probability of backscattering equals approximately the probability of a particle's turn to the surface and thus could be roughly estimated using eq. (18). The additional scattering angle $\theta - \theta_1$ appears to be small. Scattering at this angle is a result of the multiple scattering over the whole trajectory. Under the condition $\omega(\varepsilon) \gg |\boldsymbol{\theta} - \boldsymbol{\theta_1}|$, the half width of the distribution function for multiple scattering depends slightly on ε at $\varepsilon \to 0$, $F_1 \to \frac{1}{2}\pi$. Therefore, multiple scattering does not change $R(\varepsilon, \theta)$ from eq. (18) considerably in the greater part of the spectrum.

In conclusion, using the partial systematical quantitative agreement of eq. (18) with the energy experiments, several assumptions could be formulated:

(i) at large glancing angles α only one strong deflection at the angle $\theta > \alpha$ along the first half of the trajectory is a condition for backscattering;

(ii) the probabilities of every possible combination of θ_1 and $(\theta - \theta_1)$ are approximately equal;

(iii) the lengths of all trajectories with the given ε are $L \approx L_1 + L_2$ (see fig. 1).

2. Single and double scattering

To calculate $F(\varepsilon, \theta, \varphi, q)$, which itself is a complicated problem, there are several approximations. It is obvious that for an ion passing the condensed medium with keV energy the scattering is essentially multiple and must be calculated as such. At backscattering, the angles are large, $\theta \geqslant \frac{1}{2}\pi$. In the theory of multiply scattering (Firsov et al. 1976) the effective scattering angle $\omega(l) \approx \frac{1}{2}\pi$ is reached only at the end of a long path $l(\varepsilon)$.

Meanwhile, as direct computer simulation shows, the character of the most probable backscattering trajectories is such that the particle undergoes scattering at a large angle already elsewhere at the first stage of its path. The function $F(\varepsilon, \theta, \varphi, q)$ must describe the probability of exactly such trajectories.

The role of large-angle scattering in the formation of trajectories of a given length is to break it and turn towards the surface at a given point. In the point of violent scattering the main part of the total scattering angle $\theta \pm \delta$ is picked up and the residual small angle $\pm \delta$ is gained at multiple scattering along the trajectory with a probability of about 1. Thus, the probability of such event is defined, mainly, by the probability of violent scattering. Unlike the one-collision approximation, where only one trajectory satisfies the condition $(\varepsilon, \theta, \varphi) = \text{const}$, in the two-deflection approximation a family of double-scattering trajectories $\Phi(\varepsilon, \theta, \varphi, \theta_1, \theta_2, \varphi_1, \varphi_2)$ corresponds to it (see fig. 1).

With the consideration of elastic energy losses, only at the violent collisions the equation defining the family of functions $\Phi(\varepsilon, \theta, \varphi, \theta_1, \theta_2, \varphi_1, \varphi_2)$ is given by

$$\varepsilon = c^2(\theta_1)c^2(\theta_2), \tag{20}$$

where

$$c(\theta_i) = \frac{\nu \cos \theta_i + \sqrt{1 - \nu^2 \sin^2 \theta_i}}{1 + \nu}, \tag{21}$$

and $\nu = m_1/M$ is the ratio between the impinging ion mass and the target atom mass. The angles θ_1, θ_2 and the azimuthal angle of the primary scattering, φ_1, are connected by the equation

$$\cos \theta_2 = \cos \theta_1 \cos \theta + \sin \theta_1 \sin \theta \cos(\varphi - \varphi_1).$$

In this approximation there are two characteristic values in the high-energy part of the spectrum: the so-called single-scattering energy $\varepsilon_1 = c^2(\theta)$ and the maximum energy of elastic double scattering $\varepsilon_{\max} = c^4(\tfrac{1}{2}\theta)$, corresponding to the symmetrical scattering $\theta_1 = \theta_2 = \tfrac{1}{2}\theta$. Approximation (20) is applied to heavy ions. For light ones it is valid for $\varepsilon_1 < \varepsilon < \varepsilon_{\max}$. The family of functions $\Phi(\varepsilon, \theta, \varphi, \theta_1, \varphi_1)$ is defined by the following equation,

$$\cos(\varphi - \varphi_1) = [2 \sin \theta_1 \sin \theta]^{-1}$$

$$\times \left[\frac{\varepsilon^2(1 + \nu)^2}{\nu(\nu \cos \theta_1 + \sqrt{1 - \nu^2 \sin^2 \theta})} - \frac{1 - \nu}{1 + \nu} \right.$$

$$\left. \times \frac{\nu \cos \theta_1 + \sqrt{1 - \nu^2 \sin^2 \theta_1}}{\nu \varepsilon^2} - 2 \cos \theta_1 \cos \theta \right]. \tag{22}$$

In the plane $\varphi_1 = 0$, by neglecting the factors $\approx \nu^2$, the angular boundaries of the family Φ are defined by the angles

$$\theta_1^{\max} = \tfrac{1}{2}\theta - \arccos \frac{4\nu - 1 + \varepsilon}{4 \cos \tfrac{1}{2}\theta} ,$$

$$\theta_2^{\max} = \tfrac{1}{2}\theta - \arccos \frac{4\nu - 1 + \varepsilon}{4 \cos \tfrac{1}{2}\theta} . \tag{23}$$

With the consideration of only inelastic energy losses according to $dE/dl = K\sqrt{E}$ along the path $l = l_1 + l_2 + l_3$ the family $\Phi(\varepsilon, \theta, \varphi, \alpha)$ is defined by the equations,

$$\varepsilon = [1 - (l_1 + l_2 + l_3)/l_0]^2 ,$$

$$l_2 = \frac{l_1 \sin \alpha - (l - l_1) \sin \psi}{\sin \theta_1 \cos \alpha \cos \varphi_1 - \cos \theta_1 \sin \alpha - \sin \psi} , \tag{24}$$

$$l_3 = \frac{l_1 \sin \alpha + l_2 \sin \beta}{\sin \psi} ,$$

$$\sin \beta = \sin \alpha \cos \theta_1 - \sin \theta_1 \cos \alpha \cos \varphi_1 ,$$

where $l = 2\sqrt{E_0}/K$ is the total effective range of the ion in the solid. Its angular boundaries in the planar case are

$$\theta_1^{\max} = \alpha + \arcsin \frac{l_1 \sin \alpha}{l - l_1} ,$$

$$\theta_2^{\max} = \psi + \arcsin \frac{l_1 \sin \alpha}{l - l_3} . \tag{25}$$

This equation may be applied for light ions at large l.

In the general case, when both types of energy losses must be taken into consideration, the equation for $\Phi(\varepsilon, \theta, \varphi)$ is written as

$$\varepsilon = [c(\theta_1)c(\theta_2)(1 - l_1/l_0) - c(\theta_2)l_2/l_0 - l_3/l_0]^2 . \tag{26}$$

In the calculations, it is necessary to consider the energy and angular $\Delta\theta$ and $\Delta\varphi$ resolution of the analyzer $\Delta\Omega$ that somewhat expands the family $\Phi(\varepsilon, \theta, \varphi)$.

Formula (21) is obtained from eq. (26) at $l_0 \to \infty$ and eq. (24) at $\nu \to 0$. The function of beam attenuation must be written as

$$P = \exp\left[-\int_0^l dl \int_{\theta_1^{\max}}^{\pi} d\theta \int_{\varphi_1(\theta_1 l_1)}^{2\pi} N\sigma(E, \theta) \sin \theta \, d\varphi \right], \tag{27}$$

where the lower limit of integration is the boundary of the double-trajectory family $\varphi_1(\theta_1, l_1)$. In the general case the integration is performed over the whole region of alteration of variables, lying out of Φ. The approximations of single and double deflection are especially suitable for the description of charge state formation, since they allow to locate the point of violent ionizing collision, which plays an important role in the theory based on electron capture and loss at ion scattering inside the solid.

3. Energy distribution of scattered atoms

The method to calculate the function $F(\varepsilon, \theta, \varphi, q)$ consists in pointing out on each trajectory some violent collisions, which are considered as independent events. In the ith violent collision the elastic energy loss occurs at scattering over the angle θ_1, but between the violent collisions only the inelastic stopping occurs. The total scattering probability contains the product of the following type: $P_{i,1}\sigma(E_{i,1}, \theta_i)P_{i,2}$, where $\sigma(E_{i,1}, \theta_i)$ is the scattering cross section, and $P_{i,1}$ and $P_{i,2}$ are the functions of beam attenuation before and after the ith scattering. In the particular case of double scattering (see fig. 3)

$$F\,dq = N^2 P_{1,1}\sigma(E_{1,1}, \theta_1)P_{1,2}P_{2,1}\sigma(E_{2,1}, \theta_2)P_{2,2}$$

$$\times \sin\theta_1 \sin\theta_2\, d\theta_1\, d\theta_2\, d\varphi_1\, d\varphi_2\, dl_1\, dl_2, \tag{28}$$

where N is the concentration of the scattering centers, $\sigma(E_{1,1}, \theta_1)$ and $\sigma(E_{2,1}, \theta_2)$ are the cross sections of scattering over the angles θ_1 and θ_2 for the potential $U = A/R^2$, written as

$$\sigma(\theta_1) = \frac{B(Z_1, Z_2, \nu)}{E_0\varepsilon_{1,1}\sin^3\frac{1}{2}\beta_1}, \qquad \sigma(\theta_2) = \frac{B(Z_1, Z_2, \nu)}{E_0\varepsilon_{2,1}\sin^3\frac{1}{2}\beta_2}, \tag{29}$$

where $\beta_{1,2}$ are the scattering angles in the center-of-mass system, $\beta_{1,2} = \theta_{1,2} + \arcsin(y\sin\theta_{1,2})$, $B = 4.1e^2 a_{T-F} Z_1 Z_2(1 + \nu)/32\pi$; and Z_1 and Z_2 are

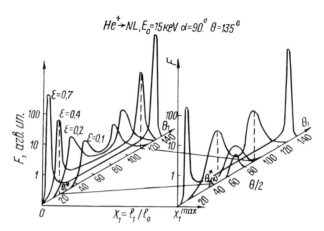

Fig. 3. The form of the scattering function $F(\varepsilon, \theta, \varphi, \theta_1, l_1)$.

the atomic numbers of the ion and target atom, respectively. Here $\varepsilon_{1,1} = (1 - l_1/l_0)^2$ and $\varepsilon_{2,1} = [c(\theta_1)(1 - l_1/l_0) - l_2/l_0]^2$ are the relative energies before the first and the second scattering, respectively.

The functions P, describing the beam attenuation on the paths l_1, l_2 and l_3, are given by

$$P_{1,1} = \exp[-d(\varepsilon_{1,1}^{-1/2} - 1)(\text{cosec } \tfrac{1}{2}\Delta_1 - 1)] ,$$

$$P_{1,2} = \exp\{-d[(\tfrac{1}{2}\varepsilon_{1,2} + \tfrac{1}{2}\varepsilon_{2,1})^{-1/2} - \varepsilon_{1,2}^{-1/2}][\text{cosec } \tfrac{1}{2}\Delta_1 - 1]\} ,$$

$$P_{2,1} = \exp\{-d[\varepsilon_{2,1}^{-1/2} - (\tfrac{1}{2}\varepsilon_{1,2} + \tfrac{1}{2}\varepsilon_{2,1})^{-1/2}][\text{cosec } \tfrac{1}{2}\Delta_2 - 1]\} ,$$

$$P_{2,2} = \exp[-d(\varepsilon^{-1/2} - \varepsilon_{2,2}^{-1/2})(\text{cosec } \tfrac{1}{2}\Delta_2 - 1)] .$$

(30)

Here $\varepsilon_{1,2}$ and $\varepsilon_{2,2}$ are the relative energies after the first and second scattering, $d = 0.277(Z_1^{2/3} + Z_2^{2/3})(\pi\xi_e)^{-1}(E_B/E_0)^{1/2}$, the Bohr energy $E_B = \tfrac{1}{2}mv_0^2$, $v_0 = 2.2 \times 10^8$ cm/s, $\xi_e = 1$–2 is the parameter of the Lindhard–Sharff–Shiott theory (Lindhard et al. 1963). The factors $P(\Delta_1)$, where $\Delta_1 = \chi\theta_1$ also allow to eliminate the divergence of the function $F(\theta_1, \theta_2)$ at $\theta_1, \theta_2 = 0$, which reflects the impossibility to follow a rectilinear section of a path in the condensed medium without scattering.

In fig. 3 the scattering function $F(\varepsilon, \theta, \varphi, \theta_1, l_1)$ obtained by the numerical calculation for the plane $\varphi_1 = 0$ at $0.1 < \varepsilon < 0.7$ for two values of the relative path $x_1 = l_1/l_0$ before the first violent scattering is illustrated. The calculation has been made for $\alpha = 90°$, $\psi = 45°$, $\nu = 0.067$, $(\text{Ne}^+ \rightarrow \text{Ni})$, $E_0 = 15$ keV. It is seen that at large ε the scattering is a quasisingle one. With increasing ε, i.e., with increasing of the path in a solid, the scattering becomes more symmetrical double and at $\varepsilon \rightarrow 0$ the angles $\theta_1 \rightarrow \theta_2 \rightarrow \tfrac{1}{2}\theta$.

Using eqs. (21)–(30), the energy spectrum of scattered particles $R(\varepsilon, \theta, \varphi)$ has been calculated,

$$R_{11}(\varepsilon, \theta, \varphi) = \mathrm{d}^2 R/\mathrm{d}\Omega \, \mathrm{d}\varepsilon$$

$$= \int\!\!\int\!\!\int_{\Phi(\varepsilon,\theta,\varphi)} F \sin^{-1}\theta |\mathrm{d}l_2/\mathrm{d}\varepsilon| \, |\mathrm{d}\varphi_2/\mathrm{d}\varphi| \, |\mathrm{d}\theta_2/\mathrm{d}\theta| \, \mathrm{d}l_1 \, \mathrm{d}\varphi_1 \, \mathrm{d}\theta_1 .$$

(31)

The integration is performed over the trajectory family $\Phi(\varepsilon, \theta, \varphi)$. The results of the computer numerical integration may be compared with the experimental energy distributions. They correctly describe the main observed regularities:

(1) a cupola-like shape of energy spectra with a maximum at $\varepsilon_m(E_0, 0)$;
(2) the reducing of $\varepsilon_m(E_0, \theta)$ with E_0 and θ increasing;
(3) the nonmonotonic changing of the half width of $\Delta\varepsilon_{1/2}$ spectra with E_0 and θ changing.

The obtained results agree on a whole with the calculation for the plane case ($\varphi_1 = \varphi_2 = \varphi = 0$), where only inelastic energy losses have been taken into consideration (Parilis and Verleger 1980). The main differences are the following:

(i) the consideration of the elastic energy losses has allowed to obtain the correct shape in high-energy tail, where the scattering is purely double and $R(\varepsilon, \theta)$ sharply falls off as in the high-energy part of heavy-atom spectra (fig. 4);

(ii) in this calculation ($\varphi_1 = \varphi_2 \neq 0$), a better agreement with the experiment on the half width is obtained due to the fact that the integration region $\Phi(\varepsilon, \theta)$ at $\varepsilon \to \varepsilon_{max}$ is reduced more rapidly than in the planar case. In particular, the approximative analytical formula for $R_{II}(\varepsilon, \theta)$ deduced without consideration of the elastic energy losses [eq. (31) from the paper by Parilis and Verleger (1980)] in this case requires the following form ($\varphi = 0$),

$$R_{II}(\varepsilon, \theta) = \frac{\exp[-d(\varepsilon^{-1/2} - 1)(\mathrm{cosec}\,\tfrac{1}{4}\chi\theta - 1)]\,(1 - \varepsilon^{1/2})^2}{\varepsilon^{1/2}(\varepsilon^{1/2}\sin\psi + \sin\alpha)^3(1 - \varepsilon)^{1/2}}. \tag{32}$$

Formula (32) describes better the experimental data on hydrogen scattering and its isotopes.

In table 2, the comparison of energy spectra characteristics, calculated using equations from a paper by Parilis and Verleger (1980) and eq. (32) at $\chi = \tfrac{3}{4}$, $\xi_e = 1.7$ with the experimental data from Zhabrev (1979), where the scattering $D^+ \to Cu$ has been considered with $E_0 = 5$–20 keV, $\alpha = 15°$, $\theta = 20°$–$110°$ has been shown. From table 2 it is seen that the calculation

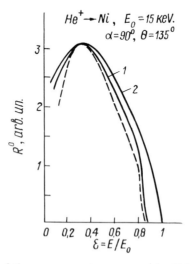

Fig. 4. The comparison of the energy spectra measured by Eckstein et al. (1978) (broken curve) with the calculations using formulae (31) (full curve 1) and (32) (full curve 2).

Table 2
The position of the maximum ε_m and half width $\Delta\varepsilon_{1/2}$ of the energy distribution versus the scattering angle.

θ (°)		$E_0 = 5$ keV			$E_0 = 20$ keV		
		Theory[a]	Experim.[b]	Calcul.[c]	Theory[a]	Experim.[b]	Calcul.[c]
20	ε_m	0.83	0.90	0.90	0.79	0.90	0.83
	$\Delta\varepsilon_{1/2}$	0.23	0.12	0.17	0.36	0.28	0.28
40	ε_m	0.76	0.80	0.81	0.50	0.72	0.62
	$\Delta\varepsilon_{1/2}$	0.38	0.25	0.29	0.62	0.43	0.43
60	ε_m	0.67	0.70	0.70	0.28	0.56	0.50
	$\Delta\varepsilon_{1/2}$	0.46	0.37	0.40	0.47	0.54	0.49
80	ε_m	0.58	0.61	0.60	0.19	0.40	0.36
	$\Delta\varepsilon_{1/2}$	0.52	0.48	0.46	0.32	0.56	0.49
100	ε_m	0.54	0.52	0.53	0.15	0.23	0.25
	$\Delta\varepsilon_{1/2}$	0.53	0.53	0.49	0.26	0.51	0.41

[a] Zhabrev (1979).
[b] Parilis and Verleger (1980).
[c] by eq. (31).

using eq. (31) gives a somewhat better agreement with experiment than using the equations from the paper by Parilis and Verleger (1979a).

Comparison of the calculation using eq. (31) with the experimental spectra from the work by Verbeek et al. (1980), where the scattering of $H^+ \rightarrow Au$ has been considered at $\alpha = 90°$, $\theta = 105°-155°$, is illustrated in fig. 5. The calculation has been made at $\xi_e = 1.4$, as proposed in work by

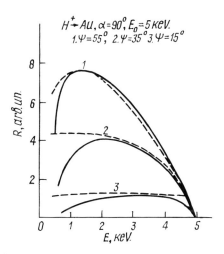

Fig. 5. Comparison of the energy spectra measured by Verbeek et al. (1980) (broken curve) with calculations using formula (31) (full curve).

Eckstein et al. (1980) and at $\chi = 1$. At large $\psi = \theta - \alpha$, the agreement with experiment is good. The agreement of the calculation with the majority of experimental data allows to use the developed method of calculation of the energy spectra of scattered ions and to use the function F for calculation of the ionization degree, which will be discussed in the next sections.

It is interesting to calculate also the approximation of one and two deflections, with consideration of the beam attenuation along the whole path, of the particle deflection coefficient,

$$R_N = 2\pi \int_\alpha^\pi \sin\theta \, d\theta \int_0^1 R(\theta, \varepsilon) \, d\varepsilon ,$$ (33)

and energy reflection coefficient,

$$R_E = 2\pi \int_\alpha^\pi \sin\theta \, d\theta \int_0^1 \varepsilon R(\theta, \varepsilon) \, d\varepsilon .$$ (34)

The results of the calculation R_N (see fig. 6a) and R_E (see fig. 6b) for $N^+ \rightarrow Au$ at $E_0 = 1$–20 keV, $\alpha = 90°$, $\xi_e = 1$, $\chi = 1$ and their comparison with

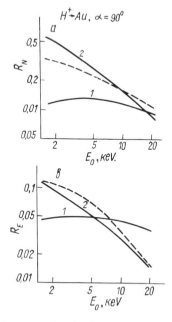

Fig. 6. The measured (broken curves) and calculated (full curves) (a) R_N and (b) R_E values. Curve 1 gives $R(\theta, \varepsilon)$ calculated by the single-scattering model; curve 2 $R(\theta, \varepsilon)$ by the double-scattering model.

the experimental data by Bohdansky et al. (1976) are shown in fig. 6. It is seen that the energy dependence of both coefficients is described slightly better in the approximation of two deflections.

4. Surface peak

The important peculiarity of the two energy spectra, recorded in a paper by Eckstein et al. (1978), of neutral He particles is a small surface peak in the narrow energy region $\varepsilon \rightarrow \varepsilon_1$ at $E_0 < 5$ keV. It is difficult to see it by direct calculation, since at $\varepsilon \approx \varepsilon_1$ the integrand in eq. (31) is sharply nonmonotonic (see fig. 3) and the correct calculation of the integral requires too much computer time. Parilis and Verleger (1980) have shown that the reason for the appearance of this peak is the possibility to obtain the same values of ε for the trajectories with different combinations of elastic and inelastic energy losses, which is possible only at small initial energies $E_0 < 5$ keV.

The numerical calculation of the function F has shown that in region $\varepsilon \approx \varepsilon_1$ the most probable trajectories of double scattering group near the plane $\varphi_1 = 0$ and contain a large-angle violent scattering $\theta = \theta_1$ and a small one $\theta_1(l)$, which increases proportionally to the path l, i.e., at $\varphi_1 = 0$ and $\theta_1^*(l)$ the function F has a sharp maximum, besides the effective value of the small angle $\theta_1^* \approx 1$ (see fig. 3). Such double scattering is called quasisingle. The multipliers which depend only on the large angle $\theta - \theta_1(l)$, change slowly and might be taken out of the integral (31). At small l, $P_{1,2} = P_{2,2} = 1$, we substitute eq. (31) as

$$R(\varepsilon, \theta) \simeq N\sigma[\theta - \theta_1(l)] \, \mathrm{d}l/\mathrm{d}\varepsilon \, I(\theta_1) . \tag{35}$$

Here the first three multipliers present the probability of violent scattering over the angle $\theta - \theta_1(l)$ and the integral $I(\theta_1)$ gives the probability of an additional scattering over a small angle $\theta_1(l)$. In the paper by Parilis and Verleger (1980) a formula for the double differential coefficient of double scattering, $R_{11}(\varepsilon, \theta)$, has been derived.

The equation $\mathrm{d}R_{11}(\varepsilon, \theta)/\mathrm{d}\varepsilon = 0$ has the only solution $\varepsilon = \varepsilon_m(E_0, \theta)$ which passes through a maximum. This conclusion might be applied to the angle θ_1 and then the path $\lambda(E_0, \theta_1) = l_0(1 - \sqrt{\varepsilon_m(E_0, \theta_1)}$ might be understood as the effective range for scattering over the angle θ_1. For small angles, $\theta_1 < 10°$, the length $\lambda \approx 10$–30 Å.

The analysis of the probability of scattering over the single trajectories $F(\theta_1, l_1)$ has shown that the most probable scattering trajectories in the region $\varepsilon \approx \varepsilon_1$ have the length $l \geqslant \lambda(\theta_1)$. This means that scattering over an angle θ_1 occurred with a probability near 1, i.e., $I(\theta_1) = 1$. On the other hand, the angle θ_1 could be considered as the most probable scattering angle on the path l. As was mentioned, θ_1 is proportional to the path,

$$\theta_1 = \frac{bl}{E_0}, \tag{36}$$

where the coefficients b, calculated within the model of double scattering (b_{II}) and by multiple-scattering theory (b_{m}) have close values,

$$b_{\text{II}} = \frac{4.1 N Z_1 Z_2 a_0^2 I_0}{\chi (Z_1^{2/3} + Z_2^{2/3})^{1/2}}, \qquad b_{\text{m}} = \frac{0.8 \pi^2 N Z_1 Z_2 a_0^2 I_0}{(Z_1^{2/3} + Z_2^{2/3})^{1/2}}. \tag{37}$$

Here Z_1 and Z_2 are the atomic numbers, a_0 is the radius of the Bohr orbit, $I_0 = 27.2$ eV.

The function $\varepsilon(l)$ for quasisingle scattering while neglecting the terms of the second order of magnitude ν, $\nu(l/l_0)$, $(l/l_0)^2$, is given by

$$\varepsilon(l) = 1 - 2l/l_0 - 4\nu \sin^2[\tfrac{1}{2}\theta - \tfrac{1}{2}\theta_1(l)]. \tag{38}$$

In this approximation $\varepsilon_1 = \varepsilon(0) = 1 - 4 \sin^2 \tfrac{1}{2}\theta$.

Let us show that the function $\varepsilon(l)$ has a maximum ε_{max} at $E_0 < E_{\text{cr}}$ (E_{cr} is a certain critical energy). Substituting eqs. (19) and (17) in the equation $d\varepsilon/dl = 0$, we obtain the maximum value ε_{max} corresponding to the path $l = l_{\text{m}}$,

$$l_{\text{m}} = E_0[\theta - \arcsin(k E_0^{1/2}/2\nu b)]/b. \tag{39}$$

It is clear that l_{m} must satisfy the condition $l_{\text{m}} > 0$. While solving this inequality for the energy E_0 we get the expression for the critical energy E_{cr},

$$E_0 \leqslant E_{\text{cr}} = (2\nu b \sin \theta / K)^2. \tag{40}$$

At the top of fig. 7, the dependence $l(\varepsilon)$ at $E_0 < E_{\text{cr}}$ is illustrated. It is seen that in the region $\varepsilon_1 < \varepsilon < \varepsilon_{\text{max}}$ the function $l(\varepsilon)$ is a double-valued one. The anomalous behavior of $\varepsilon(l)$, i.e., its increase with increasing l at $0 \leqslant l \leqslant l_{\text{m}}$ is caused by the fact that the increase of energy losses with increasing l is compensated by the reduction of the elastic energy losses at the account of the reduction of the angle of the violent scattering. Due to the double-valued $l(\varepsilon)$ in eq. (31), one should integrate over two branches of the function Φ, that causes the appearance of a peak at $E_0 < E_{\text{cr}}$.

At the bottom of fig. 7, the function $R(\varepsilon)$, given by formula (16) is illustrated. The surface peak occupies the whole region $\varepsilon_1 < \varepsilon < \varepsilon_{\text{max}}$. The jump of $R(\varepsilon)$ at $\varepsilon = \varepsilon_1$ is caused by the double value of the derivative $d\varepsilon/dl$.

At $\varepsilon \to \varepsilon_{\text{max}}$, $R(\varepsilon) \to \infty$ according to the dependence $R(\varepsilon) \simeq (\varepsilon_{\text{max}} - \varepsilon)^{-1/2}$. Thus this divergence is eliminated at any averaging, e.g., over energy or over the resolving power of the analyzer $\Delta\varepsilon_{\text{a}}$. The physical reason of the surface peak is quite clear; at $E_0 < E_{\text{cr}}$ in region $\varepsilon \to \varepsilon_1$ there is a great number of scattering trajectories per given energy range with different

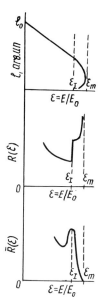

Fig. 7. The dependences $l(\varepsilon)$, $R(\varepsilon)$ and $\bar{R}(\varepsilon)$ at $E_0 < E_{\mathrm{cr}}$.

combinations of elastic and inelastic energy losses, corresponding to the same value ε. Thus, the surface peak is a rainbow effect.

In fig. 8 the case of $E_0 > E_{\mathrm{cr}}$ is illustrated, here $l(\varepsilon)$ changes in the usual way, it decreases monotonically, and there is no surface peak.

Let us estimate the value of E_{cr} for the conditions of the experiment by Eckstein et al. (1978): $Z_1 = 2$, $Z_2 = 28$, $\nu = 0.068$, $K = 8\pi a_0^2 N \xi_{\mathrm{e}} (Z_1^{1/2} + Z_2^{1/2})^{-3/2} E_{\mathrm{B}}^{-1/2}$, $\xi_{\mathrm{e}} = 1\text{--}2$ (a parameter of the LSS theory), $E = \frac{1}{2} m_1 v_0^2$, $v_0 = 2.2 \times 10^8$ cm/s.

So we get $E_{\mathrm{cr}} = 10.66/\xi_{\mathrm{e}}^2$ keV. The experimental value $E_{\mathrm{cr}} \approx 5$ keV is obtained at $\xi_{\mathrm{e}} = 1.46$, which in turn was obtained from the fitting of the position of the cupola calculated using formula (12) with the experimental one at $\varepsilon_{\mathrm{m}} = 0.6$. For a comparison of $R(\varepsilon)$ with experiment in the region $\varepsilon \approx \varepsilon_1$, we average $R(\varepsilon)$ over the resolving power of the energy analyzer,

$$\bar{R}(\varepsilon) = \int_{\varepsilon_1 - \Delta\varepsilon_{\mathrm{a}}/2}^{\varepsilon_1 + \Delta\varepsilon_{\mathrm{a}}/2} R(\varepsilon)\, \mathrm{d}\varepsilon / \Delta\varepsilon_{\mathrm{a}} . \tag{41}$$

At the bottom of figs. 7 and 8 the result of such averaging is shown at $\Delta\varepsilon_{\mathrm{a}} = 0.067$. In fig. 9, the whole spectrum of the scattering of He at $E_0 = 3$ keV is presented. The results of the calculation of $\bar{R}(\varepsilon)$ using formula (41) with $R(\varepsilon)$ calculated with the use of formula (35) and the numerical

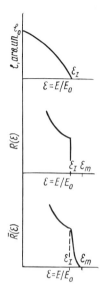

Fig. 8. The dependences $1(\varepsilon)$, $R(\varepsilon)$ and $\bar{R}(\varepsilon)$ at $E_0 > E_{cr}$.

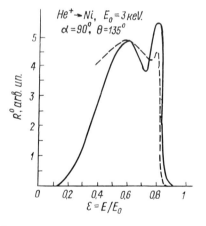

Fig. 9. Comparison of the energy spectra measured by Eckstein et al. (1978) (broken curve) with calculations using formula (41) (full curve).

calculation $R(\varepsilon)$ using eq. (31) at $\varepsilon < \varepsilon_1$ are fitted at $\varepsilon = \varepsilon_{min}$, in the point where $\bar{R}(\varepsilon)$ passes the minimum. It would be noted that at $\varepsilon < \varepsilon_1$ the curves joined together have close values of the derivative $dR/d\varepsilon$. The center of gravity of the surface peak is close to $\varepsilon = \varepsilon_1$ for all initial energies E_0 in the range $0 < E_0 < E_{cr}$.

Thus, why here the term 'surface peak' is justified, although, as it is seen from fig. 7 none of the particles, scattered on the surface, do contribute in the spectrum region $\varepsilon \approx \varepsilon_1$. With increasing E_0 the half width of these peaks decreases since

$$\lim_{E_0 \to E_{cr}} \varepsilon_{max}(E_0) = \varepsilon_1 \quad \text{at} \quad E_0 < E_{cr} .$$

The obtained results explain satisfactorily the properties of the surface peak in the energy spectra of scattered helium, but they appear not only for helium. The peaks of such type are universal peculiarity of the high-energy part of spectra also for other scattered atoms. They could be observed if both the elastic and the inelastic energy losses are comparable, i.e., at $E_0 < E_{cr}$. Formula (40) allows to estimate E_{cr} for any pair of colliding atoms. So for scattering on heavy targets over large angles ($\theta > 90°$) the value of the critical energy for hydrogen is $E_{cr} = 100$ eV, and for Ne and Ar $E_{cr} \geqslant 100$ keV. It would be of great interest to examine experimentally whether the surface peak appears at the scattering of H with energies less than 100 eV and whether it would disappear for heavy atoms at high energies (>100 keV).

In the energy spectra of heavy atoms, scattered by a solid surface, a narrow peak called the single-scattering peak has been observed a long time ago at $\varepsilon(\theta) = \varepsilon_1(\theta)$. A long time its shape has been explained on the basis of the conceptions of single and double scattering, besides, it has been shown that the region $\varepsilon > \varepsilon_1$ corresponds to the combination of angles of double scattering θ_1, $\theta_2 < \theta$, but in the region $\varepsilon < \varepsilon_1$ one of the angles (θ_1, θ_2) could be larger than θ (Parilis 1965). Since for heavy atoms the elastic energy losses exceed considerably the inelastic ones, and the penetration depth into the solid is small, the appearance of a single-scattering peak is explained well only by elastic energy losses.

It is known that for single-crystal targets some single-scattering peaks have been observed when the direction of bombardment coincides with the crystal axes. The reason of the appearance of the peaks is quite different: the bulk atoms are shaded by the atoms of the first layer and they do not have any contribution to the formation of the surface peak in the case of a polycrystalline solid.

If on a polycrystalline surface there are some microcrystals, the axes of which coincides with the direction of the impinging beam, then in the part $\varepsilon' < \varepsilon < \varepsilon_1$ of the spectrum the number of the scattered particles reduces sharply,

$$\varepsilon' = 1 - 2l/l_0 - 4\nu \sin^2 \tfrac{1}{2}\theta ,$$

where $l = x_0(\text{cosec } \alpha + \text{cosec } \psi)$ is the path of the particle, reflected from the lower boundary of the microcrystal with the depth x_0. Then in the spectrum a gap has been formed with width $\Delta\varepsilon = \varepsilon_1 - \varepsilon' = 2l/ l_0(\text{cosec } \alpha + \text{cosec } \psi)KE_0^{-1/2}$.

The gap width is defined by the statistical weight of the ordered microcrystals on the surface. The gap could be discovered experimentally under the condition $\Delta\varepsilon > \Delta\varepsilon_a$ that gives the upper boundary for the initial energy E_0 at which the isolated surface peak could be observed,

$$E_0 < \frac{\sin^2\alpha \, \sin^2\psi}{(\sin\alpha + \sin\psi)^2} \left[\frac{2\pi I_0 a_0^2 N Z_1 Z_2 \xi_e x_0}{\Delta\varepsilon_a \sqrt{E_B}(Z_1^{2/3} + Z_2^{2/3})^{3/2}} \right]^2 . \qquad (42)$$

Here, unlike eq. (40), with decrease of the scattering angle $\theta = \alpha + \psi$, the condition for the observation of the peak becomes better. Therefore, experimentally studying the surface peak at different θ one could determine which of the mechanisms, the rainbow effect or the partial atomic ordering in the near-surface region, have the main contribution to the observed peak.

CHAPTER 8

Charge State of Atoms Scattered by Solid Surfaces

1. Introduction

Knowledge of the regularities of the charge state formation is necessary for a proper interpretation of the experiments on ion scattering from the surface of single crystals and polycrystals. On the other hand, the ionization degree of the scattered particles itself is very sensitive to the composition, structure, state and electronic properties of the surface and may be used for its diagnostics.

A limited number of the available experiments allows us to establish some regularities in the dependence of the positive ionization degree of the scattered ions η^+ on their final velocity v, energy E, and exit angle ψ, as well as on the angle of incidence α and initial energy E_0 for both heavy (Ne, Na, Ar, K, Xe, etc.) and light (H, D, He) atoms in the energy range 1–100 keV.

The early theory of charge state formation has been connected with Auger or resonance neutralization and triple (Yavlinsky et al. 1966), tunnel (Trubnikov and Yavlinsky 1967) or simple (Kurnaev et al. 1985) ion recombinations which occur outside the solid because it was believed that there can be no bound state of electrons and ions (protons) inside the metal (Brandt 1975).

However, as will be shown below, the experimental data can be explained only using the concepts of electron capture and loss by ions moving inside a solid, no matter whether the electron is captured into a real atomic bound state or into a state associated with the moving ion. In any case, the possibility of electron capture and loss in the matter is of great importance. It is manifested by the dependence of η^+ on E_0 and the ion trajectory inside the solid, as well as by the nonmonotonic function $\eta^+(E)$ for the heavy ions and the existence of so-called surface peak in the energy distribution of light ions scattered by solid surface.

Another important feature of the mechanism of charge state formation, which has been put in the base of the model developed, was the description of electron capture and loss elementary processes as occurring in interactions of the ion with individual atoms of the solid. This model, first proposed by Van der Weg and Bierman (1969) and Van Veen and Hoack (1972), in contrast to the Hagstrum theory, opens a way to explain and predict some orientational effects in the charge state of atoms scattered from the single-crystal surface. It was widely used in computer simulations of the phenomenon, giving the possibility to find a connection between the crystal structure of the surface (both ideal and damaged) and the ionization degree of the scattered ions.

The ionization degree of the ion, which underwent a violent binary collision with an atom of the solid, in the first calculations usually used to be taken equal to 1, under a more detailed analysis turned out to be less, depending on the collision energy, scattering angle, initial ionization state of the projectile, atomic numbers of the colliding partners, etc., giving a very interesting possibility for surface composition analysis using the charge state of the scattered atoms.

In general, the charge state of the scattered atoms was found to be a very sensitive tool for surface analysis, sometimes much more precise than not only the integral scattered flux energy and angular distributions, but also other methods of surface analysis, e.g., Auger spectroscopy.

It has been shown that the ionization degree of the scattered atom carries even information on the position of an atom or molecule adsorbed on the solid surface and the molecule orientation relative to the crystal axes.

2. Charge state of ions scattered by metal surfaces

2.1. Introduction

In scattering of heavy atoms in the energy range 10–100 keV by metal surfaces, both neutrals and charged ions were observed (U.A. Arifov et al. 1961a). Panin (1962) has observed charged ions from +1 to +4. Datz and Snoek (1964) have noted that the ion charge is less in the part of the energy spectrum caused by multiple scattering. They have proposed a program of comparative studies of ionization on gases and on metal surfaces. This program was performed by Van der Weg and Bierman (1969), who have found that the mean charge of ions scattered on metals is about half as much as for those scattered on gases. They have also proposed a model for the charge formation process in which the multiple-charged ions are formed in violent collisions followed by step-by-step neutralization, either by Auger or resonance processes. Van der Weg and Bierman (1969) were the first who emphasized the orientational effects in ion charge formation. They have considered the neutralization of ions as sensitive to the atomic surface structure. Unfortunately, in the corresponding experiment a blocking effect is apparently present, rather than the orientational dependent neutralization. A similar effect has been found by Dahl and Sandager (1969).

The specially developed technique for recording the neutral component of a scattered beam made it possible to obtain some interesting results (Chicherov 1972, Lukyanov and Chicherov 1973, Buck et al. 1975). The two main regularities observed by these authors are:

(1) the ionization degree η^+ is approximately a linear function of the normal component of the scattered particle velocity $v_\perp$;

(2) the nonmonotonic function $\eta^+(E)$ in the energy spectrum results in a decrease of η^+ on either side of the main peak and an increase in the high-energy tail of the spectrum.

2.2. Basic principles

A model for the description of the ion charge distribution on scattering at a metal surface with $E_0 = 10$–100 keV [developing and improving the method of Van der Weg and Bierman (1969) and taking into account the connection between the ion charge states and scattering kinematics] has been proposed by Kishinevsky et al. (1976), Parilis (1980a), and Parilis and Verleger (1979a,b, 1980, 1983, 1985c,d). The main model principles are as follows:

(1) The violent ion–atom collision results in multiple electron excitation mainly in the first excited shell (Edwards and Rudd 1968), a great number of M-shell electrons and some L-shell electrons being promoted to excited levels. Using Slater's approximation (Slater 1930) it can be shown that some of these highly excited levels are within the filled part of the metal conduction band. Thus an excited atom within the metal, being practically neutral, carries a considerable energy. The ionization energies of some excited states of Ar are presented in table 1.

(2) Relaxation of such highly excited states occurs via Auger transitions in M- and L-shells. The one-vacancy filling rate via an L–MM transition in

Table 1
The ionization energies of Ar excited states.

Excited state	Ionization energy of one (4s4p) electron (eV)	Ionization energy of one 3d electron (eV)
$(3s3p)^7(4s4p)^1$	4.16	–
$(3s3p)^6(4s4p)^2$	4.54	–
$(3s3p)^5(4s4p)^3$	4.79	–
$(3s3p)^4(4s4p)^4$	4.62	–
$(3s3p)^3(4s4p)^5$	4.35	–
$(2s2p)^7(3s3p)^8(4s4p)^1$	4.80	–
$(2s2p)^6(3s3p)^8(4s4p)^2$	5.94	–
$(2s2p)^5(3s3p)^8(4s4p)^3$	7.05	–
$(3s3p)^3(3d)^5$	–	3.6
$(3s3p)^3(3d)^3(4s4p)^2$	4.54	8.2
$(3s3p)^4(3d)^4$	–	3.3
$(3s3p)^4(3d)^2(4s4p)^2$	4.54	5.9
$(3s3p)^5(3d)^3$	–	2.4
$(3s3p)^5(3d)^1(4s4p)^2$	4.54	3.4

an atom containing n L-vacancies is

$$W_L(n) = nW_L(1) . \tag{1}$$

Here $W_L(1) = 2.55 \times 10^{14} \, \text{s}^{-1}$ for Ar (McGuire 1971, 1972).
   Similarly, for M-vacancies,

$$W_M(m) = mW_M(1) .$$

   It is necessary to take into consideration the fact that the M-vacancy is filled both via direct atomic Auger transitions and via interatomic Auger transitions. There is no essential difference between these transitions within the metal but out of the metal they are separated because the resonance charge exchange into excited states precedes the atomic transitions. The effective distance of direct Auger neutralization is s_a and the resonance charge exchange effective distance is s_t. The direct Auger neutralization rate is (Hagstrum et al. 1965, U.A. Arifov et al. 1973b)

$$
\begin{aligned}
W_M(1) &= A_M , & \text{for } s \leqslant s_a , \\
       &= A_M \exp[-a_M(s - s_a)] , & \text{for } s > s_a ,
\end{aligned}
\tag{2}
$$

where A_M and a_M are the well-known Auger neutralization parameters: $A_M = 1-2 \times 10^{15} \, \text{s}^{-1}$, $a_M = 2 \, \text{Å}^{-1}$ and $s_a = 2-3 \, \text{Å}$. The same formula as eq. (2) is used for Auger transitions preceded by resonant neutralization with $s_t \approx 5 \, \text{Å}$ (Parilis 1969a, Kishinevsky 1974b). In both cases, $W_M(1) \approx 1-2 \times 10^{15} \, \text{s}^{-1}$. The difference between W_L and W_M is very important for the formation of scattered ion charge spectra.
   (3) Ions escaping a metal with vacancies in the inner shells have the N-shell filled to a degree defined by charge-exchange resonance processes.
   The processes with $W_N \approx 10^{16} \, \text{s}^{-1}$ occur in the region near s_t. The N-electron capture probability is assumed to be

$$P_N = 1 \quad \text{for excited levels in resonance with the conduction band,}$$

$$\quad\; = 0 \quad \text{for the opposite case.} \tag{3}$$

N-electron capture may result in the formation of single-excited or autoionizing states;
   (4) The relaxation of the single-excited state at large distances from the metal occurs via radiative transitions without changing the ion charge. Autoionization state decay increases the ion charge by 1 in each transition.
   (5) Thus, the final charge state of an atom escaping from the metal is defined by the number of retained L- and M-vacancies and by the number of electrons on excited levels at $s > s_t$. Since neutralization processes are irreversible, the number of retained vacancies is defined by the time of ion travel within the metal and in the regions $s < s_a$ and $s < s_t$. This time

depends on the depth x at which the violent collision has occurred and on both shape and location of the trajectory.

2.3. Charge distribution of scattered ions

On the basis of the considered model we have calculated the charge distribution of the scattered particles. First, we have calculated the portion of particles $p_{n,m}$ which have retained n vacancies on the L-shell and m vacancies on the M-shell. Assuming that in each Auger transition in the M-shell a single vacancy is filled and in the L-shell two new M-vacancies are formed, we obtain

$$\mathrm{d}p_{n,m}(s) = \{-[W_{\mathrm{L}}(n) + W_{\mathrm{M}}(m)]p_{n,m}$$
$$+ W_{\mathrm{M}}(m + 1)p_{n,m+1} + W_{\mathrm{L}}(n + 1)p_{n+1,m-2}\} \, \mathrm{d}t \, . \tag{4}$$

Having integrated, we obtain

$$p_{n,m}(s) = \sum_{l=n}^{n_0} \sum_{k=k_1}^{k_2} C_{n,m}^{l,k} \exp[-(l\tau_{\mathrm{L}} + k\tau_{\mathrm{M}})] \, . \tag{5}$$

Here,

$$k_1 = m - 2(l - n) \, , \quad \text{for} \quad m - 2(l - n) \geqslant 0 \, ,$$
$$= 0 \, , \quad\quad\quad\quad\quad \text{for} \quad m - 2(l - n) < 0 \, ,$$
$$k_2 = m_0 + 2(n_0 - l) \, , \quad \text{for} \quad m_0 + 2(n_0 - l) \leqslant 8 \, ,$$
$$= 8 \, , \quad\quad\quad\quad\quad\ \text{for} \quad m_0 + 2(n_0 - 1) \geqslant 8 \, ,$$

n_0 and m_0 are respectively the maximal numbers of L- and M-vacancies in the initial excited state formed after violent collision at depth $s = -x$,

$$\tau_{\mathrm{L}} = \int W_{\mathrm{L}}(1) \, \mathrm{d}t = \frac{1}{v_\perp} \int_0^{s_t} W_{\mathrm{L}}(1) \, \mathrm{d}s = \frac{v_{\mathrm{L}}}{v_\perp} \, ; \quad \tau_{\mathrm{M}} = \int W_{\mathrm{M}}(1) \, \mathrm{d}t \, ,$$

$C_{n,m}^{l,k}$ are coefficients defined by the recurrence relations,

$$C_{n_0,m_0}^{n_0,m_0} = p_{m_0,n_0}(-x) \, ,$$

$$C_{n,m}^{l,k} = \frac{(m + 1)C_{n,m+1}^{l,k} + \chi(n + 1)C_{n+1,m-2}^{l,k}}{(m - k) + \chi(n - 1)} \, , \tag{6}$$

if $l \neq n$, $k \neq m$ simultaneously,

$$C_{n,m}^{n,m} = p_{m,n}(-x) - \sum_{l \neq n} \sum_{k \neq m} C_{n,m}^{l,k} \, ; \quad \chi = \frac{A_{\mathrm{L}}}{A_{\mathrm{M}}} \, .$$

The solution of eq. (4) is simplified when the vacancies are only in one shell,

$$p_m(s) = C_m^m \exp(-m\tau_M) + \sum_{k=m+1}^{m_0} C_m^k \exp(-k\tau_M) . \qquad (7)$$

A numerical calculation was done for the experiments by Van der Weg and Bierman (1969), where the relative number of ions Ar^{2+}, Ar^{3+}, Ar^{4+} scattered from the (100) face of a Cu single crystal depending on scattering angle θ) has been measured ($E_0 = 60$ and 90 keV, $\alpha = 45°$). In this calculation it was assumed that a single-scattering event occurs in the first layer of an ideal single crystal, because $\alpha = 45°$ corresponds to beam incidence along an open direction in the single crystal.

The initial distribution $p_{n,m}(-x)$ may be obtained from the data of Van der Weg et al. (1969) on Ar and Cu atomic collisions. Comparison of these data with data of Fastrup et al. (1971) and Garcia and Fortner (1973) makes it possible to obtain the following estimation,

$$E_0 = 90 \text{ keV}: \quad \sum_m p_{3,m} = 0.55 ; \quad \sum_m p_{2,m} = 0.35 ; \quad \sum_m p_{1,m} = 0.1 .$$
$$\qquad (8)$$
$$E_0 = 60 \text{ keV}: \quad \sum_m p_{3,m} = 0.55 ; \quad \sum_m p_{2,m} = 0.35 ; \quad \sum_m p_{1,m} = 0.1 .$$

Initially, it was assumed that the M-shell is completely excited and that the M-vacancy distribution is Gaussian with a maximum at $m_0 = 4$. The obtained vacancy state distributions were transformed into final charge distributions in accordance with the considered scheme, i.e., in the region $s > s_t$ an (n, m)-excited atom comes on picking up $(n + m)$ or $(n + m - 1)$ electrons depending on the location of the excited level of the neutral atom relative to the conduction band (see table 1).

The autoionization states than decay via Auger transitions. Calculation with $s_t \approx 5$ Å, $W_L = 2.55 \times 10^{14}$ s^{-1}, $W_M = 10^{15}$ s^{-1} has shown that the contribution of states with $m \geqslant 2$ is about 5%.

Table 2 contains the values of the most probable charge at a given number

Table 2
The most probable charge at a given number of vacancies.

State (n, m)	Number of electrons captured in excited states	The most probable charge of a scattered ion
(3, 1)	4	+4
(3, 0)	3	+4
(2, 1)	3	+3
(2, 0)	2	+3
(1, 1)	1	+2
(1, 0)	1	+2

of vacancies (n, m) for $m < 2$. It is seen that within the accuracy up to terms with index $m \geq 2$, excited states with $n = 3$ leads to charge $+4$, with $n = 2$ to $+3$ and with $n = 1$ to $+2$. Simple formulas for the distribution p_m derived from eq. (7) and (8) are given by

$$E_0 = 90 \text{ keV}: \quad p_3 = 0.55 \exp(-3\tau_L) ;$$

$$p_2 = 2 \exp(-2\tau_L) - 1.65 \exp(-3\tau_L) ;$$

$$p_1 = 2.45 \exp(-\tau_L) - 4 \exp(-2\tau_L) + 1.65 \exp(3\tau_L) .$$

$$(9)$$

$$E_0 = 60 \text{ keV}: \quad p_3 = 0.35 \exp(-3\tau_L) ;$$

$$p_2 = 1.60 \exp(-2\tau_L) - 1.05 \exp(-3\tau_L) ;$$

$$p_1 = 2.25 \exp(-\tau_L) - 3.2 \exp(-2\tau_L) + 1.05 \exp(-3\tau_L) .$$

Taking into account that the double Auger process contribution in each L-vacancy relaxation is about 10%, we obtain charge distribution P_{+q}:

$$P_{+4} = 0.75 p_3 + 0.18 p_2 ;$$

$$P_{+3} = 0.82 p_2 + 0.10 p_1; \qquad (10)$$

$$P_{+2} = 0.90 p_1 .$$

The results of a comparison of the calculated distributions with experiment are shown on figs. 1 and 2. Here $v_L = 1.1 \times 10^7$ cm/s.

Analysis of the performed calculation shows that a multiple-charged ion yield with $q \geq 2$ is connected with a vacancy formation in the L-shell of Ar. Van der Weg and Bierman (1969) have connected such ions with M-vacancies. This is an essential difference which implies that the E_0 and θ thresholds in multiple-charged ion formation coincides with the L-vacancy formation threshold. When there are no long-living L-vacancies, any M-shell

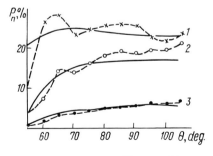

Fig. 1. Distributions of multicharged ions (1) Ar^{2+}, (2) Ar^{3+} and (3) Ar^{4+}. Full curves denote theoretical calculations; broken curves the experimental data by Van der Weg and Bierman (1969) ($E_0 = 60$ keV, $Ar^+ - Cu$ (100), $\alpha = 45°$).

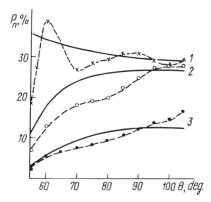

Fig. 2. The same as in fig. 1 for $E_0 = 90\,\text{keV}$.

excitation results mainly in the emergence of Ar^+ and Ar^0. Such conditions are fulfilled in the next section.

Ten years later, the correctness of our model, developed in 1975 (Kishinevsky et al. 1975, 1976), has been confined by the results of the experiments done by de Zwart et al. (1985). They investigated the scattering on Ne^{q+} ($q = 1-9$), Ar^{q+} ($q = 1-11$) and Kr^{q+} ($q = 6-11$) ions on polycrystalline W ($E_0 = 20\,\text{keV}$, the scattering angle $\theta = 30°$, the glancing angle $\alpha = 15°$). At $q \leqslant 8$, the ion component of the scattered particles contains only single-charged and double-charged ions. The number of the last ones was significantly, apparently by one order of magnitude, less. They were observed even at $q = 1$ (de Zwart 1987a,b). At $q \geqslant 9$, i.e., at the creation of inner-shell vacancies (in the K-shell of Ne, L-shell of Ar and M-shell of Kr), the number of double-charged ions has sharply increased and essentially less triple-charged scattered ions were observed.

The interpretation of these experiments proposed by de Zwart et al. (1985) is in a good agreement with our concept. Since the life-time of inner-shell vacancies is relatively long compared to the life-time of the outer-shell vacancies, a fraction of the inner-shell vacancies survives the collision event giving rise to double-charged ions and apparent of the triple-charged ones.

2.4. Ionization degree and ion trajectory

Experimental results by Chicherov (1972), Lukyanov and Chicherov (1973) and Buck et al. (1975) cannot be explained by the simple model of adiabatic neutralization of Shekhter (1937) or by the nonadiabatic modification model of Trubnikov and Yavlinsky (1967). For an explanation of these experiments it is necessary to take into account the connection of charge states with the shape of the ion trajectory and its location relative to the metal surface. For this purpose it is sufficient to take into account single and multiple collisions in a polycrystalline metal (Parilis 1965).

Let the incident particle with energy E_0 suffer a violent collision which results in scattering at an angle θ at a depth x (fig. 3). Then $N\sigma(\theta)\,\mathrm{d}l_0$ is the probability of scattering on the path $\mathrm{d}l_0$, $\sigma(\theta)$ is the effective differential cross section, N is the metal density, $\exp(-l_0/\lambda_0(E_0))$ is the probability to retain the energy E_0 at path l_0; $\exp(-l_1/\lambda_1(E))$ is the probability to retain the energy E on the path l_1, E is the post collision energy, $\lambda_0(E_0) \sim E_0$ and $\lambda_1(E) \sim E$ are the effective ranges for retention of E_0 and E within the accuracy of the experiment.

Let us write the probability $P_+(\theta, x)$ for argon to escape as Ar^+ when there is no vacancy in the L-shell. Since the first excited level $(\mathrm{Ar}^0)^*$ is higher than the top of filled part of the conduction band of Cu, Pt and Au, Ar^0 may be formed only via direct Auger neutralization at $s \leqslant s_a$. On the other hand, the levels $(\mathrm{Ar}^{n+})^*$ appear to be within the conduction band, so that Auger neutralization with charge exchange and resonance neutralization transfer all other states into Ar^+ at $s \leqslant s_t$.

Thus, taking into account eq. (7), we have

$$P_+(\theta, x) = \sum_{m=1}^{m_0} C_m^m \exp[-\tau(x)] + \sum_{m=1}^{m_0} \sum_{k=m+1}^{m_0} C_m^k \exp[-k\tau(x)]$$

$$= \sum_{m=1}^{m_0} D_m \exp[-m\tau(x)] \,, \tag{11}$$

and for the probability of Ar to escape from the metal as a neutral atom,

$$P_0(\theta, x) \approx 1 - P_+(\theta, x) \,. \tag{12}$$

Here

$$\tau(x) = \int_{-x/v_\perp}^{\infty} W_M(1)\,\mathrm{d}t = A_M x/v_\perp + v_0/v_\perp \,,$$

$$v_0 = \int_0^{\infty} W_M(1)\,\mathrm{d}s \,, \tag{13}$$

where $W_M(1)$ is the direct Auger neutralization rate defined by eq. (2).

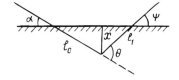

Fig. 3. The single-scattering event in a polycrystal.

From eqs. (11)–(13) follows the coefficient of scattering in the state of ion,

$$J_+(\theta) = \frac{N\sigma(\theta)}{\sin \alpha} \sum_m D_m \exp(-mv_0/v_\perp)$$

$$\times \int_0^\infty \exp(-x/\lambda_0 \sin \alpha - x/\lambda_1 \sin \psi - mA_M x/v_\perp)\,dx$$

$$= \frac{N\sigma(\theta)\lambda_0 b}{\sin \alpha} \sum_m \frac{D_m v_\perp}{v_\perp + mbv_1} \exp(-mv_0/v_\perp). \tag{14}$$

In eq. (14) $b = E \sin \alpha \sin \psi/(E \sin \psi + E_0 \sin \alpha)$, ψ is the escape angle and $v_1 = A_M \lambda_0$.

The total scattering coefficient is

$$J(\theta) = J_+(\theta) + J_0(\theta) = \frac{N\sigma(\theta)\lambda_0 b}{\sin \alpha}, \tag{15}$$

and the degree of ionization

$$\eta^+(v_\perp) = \frac{J_+}{J_+ + J_0} = \sum_m \frac{D_m v_\perp}{v_\perp + mbv_1} \exp(-mv_0/v_\perp). \tag{16}$$

In eq. (16), the value v_0 is connected with the neutralization out of the metal, v_1 is connected with the neutralization in the depth of the metal.

If the scattering occurs as a result of two collisions with metal atoms, then (see fig. 4) $N\sigma_1(\theta_1) \exp(-l_0/\lambda - l_1/\lambda_1)$ is the probability of scattering retaining the energy E_0 on path l_0 and retaining the energy E_1 on path l_1. $N\sigma_2(\theta)$

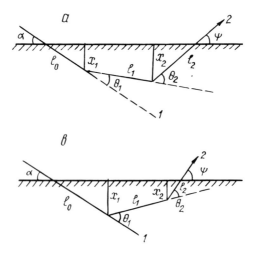

Fig. 4. The double-scattering event in a polycrystal: (a) $\beta > 0$; (b) $\beta < 0$.

$\exp(-l_2/\lambda_2)$ is the probability of scattering, retaining the energy E_2 on path l_2, θ_1 and θ_2 being the first and the second scattering angles, respectively.

In the independent-event approximation, J_+ for the trajectory with the given combination of scattering angles θ_1 and θ_2 may be expressed as in a paper by Parilis (1965),

$$J_+(\theta_1, \theta_2) = \frac{N^2 \sigma(\theta_1)\sigma(\theta_2)}{\sin \alpha}$$

$$\times \sum_m D_m \exp(-mv_0/v_\perp) \int_{d_{min}}^{\infty} dl_1 \int_{x_1}^{\infty} dx_1$$

$$\times \exp\left(-l_1/\lambda_1 - x_1/\lambda_0 \sin \alpha - l_2/\lambda_2 \sin \psi - \frac{mA_M x_2}{v_\perp}\right),$$

$$(17)$$

where d_{min} is the minimal distance between metal atoms.

The angle β is the angle between the first scattering direction and the metal surface,

$$\sin \beta = \cos \theta_1 \sin \alpha - \cos \alpha \sin \theta_1 \cos \varphi_1 , \tag{18}$$

where φ_1 is the azimuthal angle of the first scattering.

We consider two cases: $\beta > 0$ and $\beta < 0$ (see fig. 4a and 4b)

(a) in the case of $\beta > 0$ as $x_2 = x_1 - l_1 \sin \beta \geqslant 0$, then $x_1(l_1) = l_1 \sin \beta$ and we obtain

$$J_+(\theta_1, \theta_2) = N^2 \sigma(\theta_1)\sigma(\theta_2)\lambda_0^2 bb_1 \sin^{-1}\alpha \exp(-d_{min}/\{\lambda_0 b_1\})$$

$$\times \sum_m \frac{D_m v_\perp}{v_\perp + mbv_1} \exp(-mv_0/v_\perp) , \tag{19}$$

$$b_1 = E_1 \sin \alpha (E_0 \sin \alpha + E_1 \sin \beta)^{-1} .$$

The total collision coefficient is

$$J = N^2 \sigma(\theta_1)\sigma(\theta_2)\lambda_0^2 bb_1 \sin^{-1}\alpha \exp(-d_{min}/\{\lambda_0 b_1\}) , \tag{20}$$

and the ionization degree

$$\eta^+(v_\perp) = \sum_m \frac{D_m v_\perp}{v_\perp + mbv_1} \exp(-mv_0/v_\perp) . \tag{21}$$

(b) in the case of $\beta < 0$, the value $x_1(l_1) = 0$,

$$J = N^2 \sigma(\theta_1)\sigma(\theta_2)\lambda_0^2 bb_2 \sin^{-1}\alpha , \tag{22}$$

$$b_2 = \frac{EE_1 \sin \psi}{E_0(E \sin \psi + E_1 \sin \beta)} ; \qquad J^+ = \eta^+ J \tag{23}$$

and

$$\eta^+(v_\perp) = \sum_m D_m \frac{v_\perp^2 \exp[-m(v_0 + A_m d_{min} \sin \beta)v_\perp^{-1}]}{(v_\perp + mb_2 v_1 \sin \beta)(v_\perp + mbv_1)} . \tag{24}$$

In formulae (16), (21) and (24), both neutralization and stripping on the path l_1 after violent collision should be taken into consideration. Since, however, ion stripping cross sections are small, it is enough to consider only the stripping of neutrals. For the state with a single M-vacancy, the system of eq. (4) becomes

$$dp_{0,1}(s) = [-W_M(1)p_{0,1} + W_\ell(1 - p_{0,1})] \, dt ,$$

where $W_\ell = N\sigma_\ell v$ is the rate of electron loss. Thus, eq. (16) becomes

$$\eta^+(w, v_\perp) = \frac{w}{1 + w} + \frac{1}{1 + w} \frac{v_\perp}{v_\perp + bv_1'} \exp\left(-\frac{v_0'}{v_\perp}\right),$$

where

$$w = \frac{W_M(1)}{W_\ell} ; \quad w_1' = [W_M(1) + W_\ell]\lambda_1 \quad \text{and} \quad v_0' = [W_M(1) + W_\ell]s_a .$$

It is well-known that the single probability ionization cross section $\sigma_\ell \sim v$ and for $E_0 = 30$ keV $\sigma_\ell \approx 10^{-16}$ cm^2. Actually σ_ℓ is less since on the pass l_2 violent collisions are avoided.

Therefore, $w < 0.1$ and for $E_0 < 30$ keV the stripping contribution to the formulae (16), (21) and (24) can be neglected. However, $W_\ell \sim v^2$ and at large velocities it is necessary to use for states with many vacancies the function $\eta^+(w, v_\perp)$ instead of $\eta^+(v_\perp)$. The relations (16), (21) and (24) make it possible to conclude:

(1) The function $\eta^+(v_\perp)$ has an approximately linear shape if the contribution to neutralization from the path out of the metal is essentially less than the contribution from the path inside the metal, i.e., $v_0 < v_1$. The linear shape comes purely from integrating over the depth x. In general, the near-linear shape of $\eta^+(v_\perp)$ comes from the existence of several trajectories with the same α, θ, E and ψ, but different values of τ.

(2) The ionization degree for multiple scattering is always less than for single scattering at the same angle because of the smaller degree of excitation and trajectory orientation relative to the surface.

(3) On the left hand from the maximum II (region I, fig. 6), where $|\beta|$ increases, η^+ monotonically decreases and on the right hand from II η^+ has a minimum III, and then increases at $\beta \to 0$ (region IV).

A comparison of theory and the experiments is made in figs. 5 and 6.

Under the same conditions of the experiment by Lukyanov and Chicherov (1973) ($E_0 = 30$ keV, Ar $\to$ Pt, $\alpha = 10°$, $\theta = 68°$), we can make an estimation, using the formula (16).

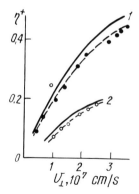

Fig. 5. The dependence of η^+ on $v_\perp$. Full curves denote theoretical calculations; broken curves the experimental data: (1) $Ar^+ \to Pt$, $\alpha = 10°$, $\theta = 68°$, $\psi = 0–58°$, $E_0 = 30$ keV (Lukyanov and Chicherov 1973); (2) $Ar^+ \to Au$, $\alpha = 45°$, $\theta = 90°$, $\psi = 45°$, $E_0 = 6–32$ keV (Buck et al. 1975).

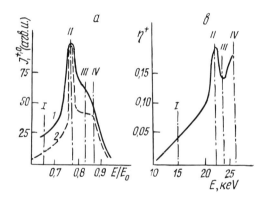

Fig. 6. Experimental energy spectra of charged and neutral particles. (a) $Ar \to Cu$, $E_0 = 16$ keV, $\alpha = 15°$, $\theta = 40°$; curve (1): Ar^0, curve (2) Ar^+ (Chicherov 1972). (b) $Ar^+ \to Au$, $E_0 = 32$ keV, $\alpha = 45°$, $\theta = 90°$ (Buck et al. 1975).

It was considered that the electrons in the M-shell of Ar are completely excited and the Gaussian distribution of the excited electrons was replaced by a 'mean' excitation $m_0 = 4$. This corresponds to saturation of the first inelastic characteristic energy loss (Fastrup et al. 1971). We describe the initial excitation in this manner in all cases considered in this section. It is clear that the basic consequences from eqs. (16), (21) and (24) are not connected with the choice of the initial excitation. Then $D_1 = 4$, $D_2 = -6$, $D_4 = -1$. The theory fits the experiment at $v_0 = 0.75 \times 10^7$ cm/s, $v_1 = 5 \times 10^8$ cm/s. Values v_0 and v_1 may be expressed as $v_0 = (s_a + 1/a) A_M = A_M/a^*$; $v_1 = A_M \lambda_0$. Then taking $A_M = 1.5 \times 10^{15}$ s^{-1}, $a^* = 2$ Å^{-1}, we obtain

$\lambda_0 \approx 33.3$ Å, $\bar{x} = \lambda_0 \sin \alpha = 5.6$ Å. For the calculation of the energy spectra of ions and neutrals under the same conditions of the experiment by Chicherov (1972) ($E_0 = 16$ keV, Ar → Cu, $\alpha = 15°$, $\theta = 40°$), we describe the observed energy spectra in terms of a double-collision model. Then the coefficient of scattering in a state of an ion with the final energy E is given by

$$J_+(E) = \oint J_+(\theta_1) \sin \theta_1 \frac{d\varphi_1}{dE} d\theta_1 . \tag{25}$$

The scattering coefficient in the state of a neutral atom is

$$J_0(E) = \oint J_0(\theta_1) \sin \theta_1 \frac{d\varphi_1}{dE} d\theta_1 , \tag{26}$$

and the total scattering coefficient

$$J(E) = J_+(E) + J_0(E) . \tag{27}$$

In eqs. (25) and (26) we integrate over the equal energies cones (Parilis 1965),

$$\cos \varphi_1 = \frac{1}{2 \sin \theta \sin \theta_1} \left[\frac{\varepsilon(1 + \mu)^2}{F(\theta_1)} + \frac{(1 - \mu)F(\theta_1)}{(1 + \mu)\varepsilon} - 2 \cos \theta \cos \theta_1 \right] , \tag{28}$$

where

$$\varepsilon = \sqrt{E/E_0} , \qquad \mu = m_{Cu}/m_{Ar} , \qquad F(\theta_1) = \cos \theta_1 + \sqrt{\mu^2 - \sin^2 \theta_1} .$$

Different ionization degrees corresponding to different spectrum regions are explained by different location of the cones, eq. (28), relative to the metal surface. The results of comparison of the calculated spectrum with the experimental one (fig. 6a) for $v_0 = 0.75 \times 10^7$ cm/s, $v_1 = 4 \times 10^8$ cm/s are given in table 3. The results of experiments by Lukyanov and Chicherov (1973) and Chicherov (1972) were confirmed by Buck et al. (1975) for the case of Ar$^+$ → Au, $\theta = 90°$, $\alpha = 45°$, $E_0 = 6$–32 keV. The line drawn through the tops of the main peaks in the experiment by Buck et al. (1975) has the same meaning as $\eta^+(v_\perp)$ in the experiment by Lukyanov and Chicherov (1973). The only difference is that in the experiment by Lukyanov and

Table 3
Results of comparison of the calculated with the experimental spectrum.

	$(J_+/J_0)_{exp}$	$(J_+/J_0)_{theor}$
I	0.33	0.328
II	1	1
III	0.67	0.684
IV	0.89	1.082

Chicherov (1973) $v_\perp$ is varied by changing ψ, and in case of experiment by Buck et al. (1975) by changing E_0. Analogously, we have made an estimate for the experiment by Buck et al. (1975) using eq. (16) (the lowest curve in fig. 5). Agreement is reached at $v_0 = 3.5 \times 10^7$ cm/s and $v_1 = 2 \times 10^8$ cm/s for $E_0 = 32$ keV. The spectrum shape (see fig. 6b) has been calculated for $E_0 = 32$ keV, with the same v_0 and v_1 (see table 4).

The agreement is relatively good in regions III and IV. In region I, however, the calculation gives an extremely high number of ions. This may be connected both with the decrease of excitation in double scattering and with the contribution of multiple scattering in this energy region. In multiple scattering the ionization degree is always less.

2.5. Charge states and surface structure

A comparison of values for v_0 and v_1 obtained to fit the data of the experiment by Lukyanov and Chicherov (1973) and Buck et al. (1975) (see figs. 6a and 6b) is of interest, see table 5. It is clear that this difference cannot be explained by different values of A.

The energy spectrum of scattered ions is known to be sensitive to the degree of surface destruction (S.H.A. Begemann and Boers 1972, Parilis et al. 1975). The relation established above between the ionization degree in different regions of the energy spectrum on one hand, and the scattering kinematics on the other demonstrates the dependence of charge state on surface structure (Kishinevsky et al. 1975).

Since the bombarding beam density in the experiment by Lukyanov and Chicherov (1973) is greater than in the experiment by Buck et al. (1975),

Table 4
Comparison between experimental and calculated spectrum shape.

	η^+_{exp}	η^+_{theor}
I	0.04	0.07
II	0.197	0.197
III	0.135	0.130
IV	0.175	0.208

Table 5
Comparison of values for v_0 and v_1 obtained from experiments by Lukyanov and Chicherov (1973) and Buck et al. (1975).

Experiment	v_0(cm/s)	v_1(cm/s)
Lukyanov and Chicherov (1973)	0.75×10^7	5×10^8
Buck et al. (1975)	3.5×10^7	2×10^8

the surface must have a larger number of atomic vacancies. Let us try to establish a connection between the neutralization process and the atomic surface roughness. For this it is convenient to introduce the variable atomic density in form $N(s) = Nf(s)$, where N is the density of the bulk and $f(s)$ is a smooth function. In this case, the terms $x/\lambda_i = xN\sigma_i$ in eqs. (16)–(24) should be replaced by

$$\int_{-x}^{0} N(s)\sigma_i \, ds = \frac{1}{\lambda_i} \int_{-x}^{0} f(s) \, ds \, ,$$

where λ_i are the effective ranges in undamaged matter. Let us consider that the Auger neutralization rate $W(s)$ is proportional to the atomic density, i.e.,

$$W(s) = A_M f(s) \, . \tag{29}$$

Under these conditions by introducing a variable,

$$z = \int_{-x}^{0} f(s) \, ds \, ,$$

it may be shown that $\eta^+(v_\perp)$ maintains its form, eq. (16), where $v_1 = A_M \lambda_1$ does not change and v_0 changes to

$$v_0 = A_M \int_{0}^{s_a} f(s) \, ds \, . \tag{30}$$

The integration in eq. (30) is carried out from $s = 0$, the boundary of the violent collision.

For example, we take $f(s) = 1 - \exp[-\gamma(s - s_a)]$, where γ is the parameter describing the degree of atomic surface roughness depending on the beam density, sputtering coefficient, etc. Thus,

$$v_0(\gamma) = A_M[s_a - 1/\gamma + 1/\gamma \exp(-\gamma s_a)] \, . \tag{31}$$

As there is no reason to consider the parameters A_M and s_a to be quite different for Pt and Au, differences in v_0 for the experiments by Lukyanov and Chicherov (1973) and Buck et al. (1975) could be explained by different values of γ.

Unfortunately, there are no available data for an accurate estimate of γ. We only know that the depth of the damaged layer is approximately the same as the depth of scattering. So all the scattering trajectories and the cones of double scattering are located in the damaged region. This fitting gives for the experiment by Lukyanov and Chicherov (1973) $\gamma = 0.19 \, \text{Å}^{-1}$ (about 37% of the ideal surface atomic density at $s = 0$) and for the experiment by Buck et al. (1975) $\gamma \geqslant 1 \text{Å}^{-1}$ (an approximately ideal surface). It should be noted that some trajectory calculations on the basis of the atomic row model have also shown that the trajectories with $\alpha = 10°$, $\theta = 68°$, $5° \leqslant \psi \leqslant 58°$ (Lukyanov and Chicherov 1973) could exist under the blocking conditions only for damaged rows (about 20% of ideal row density for $\psi = 8°$).

The difference in $v_1 = A_M \lambda_0$ may be explained in the following way. The parameter λ_0 is the effective range for an ion, after violent collision, to keep the energy E within the single-collision peak width ($\Delta E/E \approx 5$–10%). In the experiment by Buck et al. (1975), the main peak exactly corresponds to single scattering, but in the experiment by Lukyanov and Chicherov (1973) the peak is shifted to the right by approximately $\Delta E/E \approx 20\%$, that latter being evidence of multiple scattering. From this point of view, if the ratio between v_1 found by Lukyanov and Chicherov (1973) and v_1 found by Buck et al. (1975) equals 2.5, it appears to be reasonable.

It is necessary to draw attention to formula (29). It is linked closely with ideas about neutralization by single metal atoms (Van Veen and Hoack 1972, Parilis and Verleger 1973, Kishinevsky et al. 1974, 1975, de Wit et al. 1975, Kishinevsky and Parilis 1968b). It may also that the $W(s)$ shift at s_a in eq. (2) is connected with these ideas. It seems that this approach is the most convenient one for a single crystal or a deep damaged surface. In this approach, the rate of neutralization is

$$W(s) = \int W(R)N \, dV \quad \text{or} \quad W(s) = \sum_i W(R_i) \, .$$

Here, $W(R_i)$ is the rate of interatomic Auger neutralization for single atoms calculated by Kishinevsky and Parilis (1968b), where it was shown that for interatomic distance $R_i \to \infty$ the function $W(R_i) \simeq \exp(-\alpha R_i)$, and for $R_i \to 0$ it passes through a maximum at $R_i = R_a$.

Then, for a single crystal,

$$\tau = \int W \, dt = \sum_i \int_{R_{0i}}^{\infty} W(R_i) \, dt = \sum_i \tau_i \, ,$$

the summation being taken over all atoms along the ion trajectory, and where R_{0i} is the distance of the closest approach.

As an example we will take the scattering of Ar ions from atomic rows of a Cu single crystal, $E = 16$ keV, $\theta = 40°$, with

$$W(R_i) = A_M \, , \qquad\qquad\qquad \text{for } R_i \leqslant R_a \, ,$$
$$= A_M \exp[-\alpha(R_i - R_a)] \, , \quad \text{for } R_i > R_a \, ,$$

and with the values $A_M = 10^{15} \text{ s}^{-1}$, $\alpha = 2 \text{ Å}^{-1}$, $R_a = 2 \text{ Å}$.

(1) Let $\alpha = 20°$, the angle of incidence of Ar^+, on the [110] Cu row, $d = 2.56$ Å and for [100], $d = 3.61$ Å. Then

$$P_+[100] = 0.232 , \qquad P_+[110] = 0.056 ,$$

i.e., the ion charge distribution displays on orientational sensitivity (Kishinevsky et al. 1975).

(2) Let us consider under the same conditions the scattering on the [110] row of the following structures (a) aaa, (b) ava (a is an atom, v is an atomic vacancy). Then

(a) $P_+(\mathrm{aaa}) = 0.056 , \qquad E(\mathrm{aaa}) = 0.78 E_0 ,$

(b) $P_+(\mathrm{ava}) = 0.321 , \qquad E(\mathrm{ava}) = 0.75 E_0 .$

The shift $\Delta E = E_2 - E_1 \approx -500$ eV. It is possible that the shift of the main peak in the ion spectrum for a single crystal towards the left from the position of the neutral one (Chicherov 1972) can be explained in this way.

(3) Consider the influence of the blocking effect on the ion charge distribution for the ideal [110] row. For example, at $\alpha = 35°$, the angle $\theta = 40°$ is blocked since $\psi = 5°$. To make this trajectory possible, the minimum distance to the nearest-neighbor atom after violent collision must be

$$d_{\min} = \sqrt{\frac{\pi C}{E_0}\left(\frac{3}{2\psi}\right)^3} , \tag{32}$$

where C is a constant from the interatomic potential $U = C/R^2$). When using formulas (7), (30) and (31), it is necessary to exclude from eq. (30) the contribution in neutralization of all atoms located at distances $d < d_{\min}$. As a result, for $\psi \to 0$, $P_+(v_\perp)$ varies more slowly than shown in fig. 5, section 2.4.

2.6. Discussion

The calculation and considerations developed above are based on the very limited experimental data available. In fact, there are only about five experiments from which useful information for the development of a theory can be extracted. For the development of a more detailed picture some new experiments may be proposed:

(1) It would be interesting to find the threshold for multiple-charged ion yield caused by energy and angular thresholds of L-vacancy formation.

(2) As was shown in a paper by Fastrup et al. (1971), within the second period of the periodic table the L-vacancy is formed by a collision of two atoms in that atom whose atomic number is smaller. For instance, for the scattering of Ar ions from Al the number of multiple-charged ions Ar^{q+}

($q \geqslant 2$) must not be great over the whole energy range when L-vacancy formation is forbidden.

(3) For scattering from a complex surface the relative number of multiple-charged ions must be different in the peaks corresponding to scattering from each component. So for the collision $Ar^+ \rightarrow KCl$, according to the results by Fastrup et al. (1971), the number of Ar^{q+} ($q \geqslant 2$) in the peaks $Ar^+ \rightarrow K$ must three-fold exceed the number of Ar^{q+} in the peaks $Ar^+ \rightarrow Cl$.

(4) The experiment by Van Veen and Hoack (1972) shows that the number of Ar^+ ions in single- and double-scattering peaks is approximately the same. If in single scattering the probability of L-vacancy formation is much more than in double scattering (due to a lower threshold) the multiple-charged ion yield in these peaks will not be the same.

(5) Since the relative number of ions depends on the number of structural defects on the surface some measurements of the charge spectra as a function of incident beam density in high vacuum would be of great interest.

(6) It will be interesting to investigate experimentally the neutralization on a single crystal for trajectories (characterized by different τ values) passing over the rows of different linear density.

(7) In the case of surface contamination by residual gas atoms the corresponding τ values will be determined by the type of the adsorbed atoms. Therefore, an experimental investigation of the neutralization degree dependence on the character and degree of surface contamination would be of great interest.

3. The role of violent single collisions

A well-defined peak in the energy spectrum of ions scattered on a solid surface with energies between hundreds of eV and tens of keV enables us to study the single-scattering violent collisions with the target atoms. The scattering is followed by a change in the scattered ion charge, which has been studied for a long time.

Since Hagstrum (1954a), the charge of the scattered ion is thought to reach its value as a result of resonance or Auger neutralization at approaching and leaving the target surface. Hagstrum has developed his model for relatively low ion energies. The distance of closest approach to the target surface was considered to be energy independent. The change of ion velocity during the approach was neglected. The ionization rate was given by

$$R = A \exp(-as) , \tag{33}$$

where A and a are constants, and s is the distance from the surface.

Using such an approximation, the ionization coefficient equals

$$\eta^+ = \frac{n^+}{n^0 + n^+} = \exp\left[-v_c\left(\frac{1}{v_{1\perp}} + \frac{1}{v_{2\perp}}\right)\right],$$

(34)

where n^+ is the number of ions, n^0 is the number of neutral particles, $v_{1\perp}$ and $v_{2\perp}$ are the normal velocity component at approaching and leaving respectively and v_c is a constant.

The model describes the interaction of the scattered ion and the target as a whole, the target atomic structure being ignored.

However, the experimental data of the recent two decades testify in favor of a considerable contribution to the charge state formation by the processes occurring during the violent collision of the ion with a single target atom, when the electron shells of the colliding partners are overlapped.

Such processes could lead to the following effects:

(i) some intense oscillations of the relative charged-particle yield with the variation of the scattered ion energy due to quasi-resonance neutralization involving inner shells of the target atoms (Erickson and Smith 1975, Boardman et al. 1975, Eguiluz and Quinn 1975, Forstmann and Stenschke 1978);

(ii) an increase of the scattered ion ionization degree at ion energies of the order of several keV and higher, which could not be described by Hagstrum's formula (34) (Luitjens et al. 1980a,b).

MacDonald and co-workers (1981, 1983, 1984) and O'Connor et al. (1988) have performed a series of experiments concerning the determination of the violent collision contribution to the scattered-particle charge state formation.

Their experimental set-up permitted to observe particles at different escape angles at both fixed incidence and scattering angles, which was achieved by rotating the analyzer in the plane normal to the incident beam. A single-scattering peak was distinguished in the energy spectra of the scattered particles and the process accompanying both the incidence direction and the scattering by target were considered to be the same for all particles. Therefore, all discrepancies in the results should be attributed to the particle–target interaction during its escape at different escaping angles. For the consideration of the violent interaction contribution one can introduce into the Hagstrum formula a factor $k(v, \theta)$. For two different escaping angles one can get

$$\eta_1^+ = k(v, \theta)\exp\left[-v_c\left(\frac{1}{v_{1\perp}} + \frac{1}{v_{2\perp}'}\right)\right],$$

$$\eta_2^+ = k(v, \theta)\exp\left[-v_c\left(\frac{1}{v_{1\perp}} + \frac{1}{v_{2\perp}''}\right)\right],$$

(35)

that gives

$$\eta_1^+ / \eta_2^+ = \exp\left[-v_c\left(\frac{1}{v'_{2\perp}} - \frac{1}{v''_{2\perp}} \right) \right].$$ (36)

If one considers further that the neutralization at approaching does not differ from the neutralization at leaving in principle, the phenomena connected with the violent interaction could be separated and analyzed.

The above-mentioned authors have drawn the following conclusions:

(i) the contribution of the violent collisions to neutralization is rather substantial and can exceed both the contributions at approach and leave;

(ii) the parameter v_c is not constant at variation of velocity v within wide limits, the approximation $v_c \simeq v$ is more reasonable;

(iii) the contribution of the reionization during the violent collision connected with inelastic energy loss becomes significant at energies exceeding several keV.

The above-mentioned experimental results could be described within the Hagstrum model by applying it to binary collisions using the distance of the closest approach of the ion with the target atom instead of s in formula (33) (Godfrey and Woodruff 1981a, Woodruff 1982). In this case, the characteristic velocity v_c, being a function of s, becomes dependent on the collision velocity, i.e., on the primary ion energy.

This model needs consideration of the interaction with neighbor atoms for an explanation of the dependence of the final ionization degree on both incidence and escaping angles. The model calculations, which consider a cluster consisting of 25 atoms, have shown that with different values of parameters A and a in formula (33) one can get a qualitative agreement with the experimental data both for energy and angular dependencies. In this model the reionization has not been taken into account, which restricted its field of application to low energies.

The azimuthal dependence of the ion scattering on a single-crystal surface also indicates the necessity of consideration the interaction with the neighbor atoms. Engelmann et al. (1987) when studying the scattering of Na^+ and Ne^+ on the single-crystal Cu (110) surface at energies of 600–1000 eV have shown that the ionization degree of the scattered particles depended on the crystallographic direction and was maximal at coincidence of the incidence plane with the direction of closest approach. This indicates the multiple interaction along both the atomic chains and the surface semichannels. It has also been shown that the multiple collisions with one main collision (or two for the double-scattering peaks) and preceding or following scattering at small angles considerably contribute to the energy peaks corresponding to single and double scattering.

With the increase of scattered-ion velocity the influence of both electron excitation and reionization rises during the violent collision, which was obviously observed in the experiments on scattering of Ne^+ on Cu at energies of 5–10 keV (Luitjens et al. 1980a); the scattering of Ar^+ on Cu^+ at energies of 5–10 keV (Luitjens et al. 1980b) and scattering of Ar^+ on La, Yb and Au (Kumar et al. 1984).

The evidences of the reionization are the following:

(i) a sharp increase of the ion component in the scattered particles with increasing energy of primary ions;

(ii) a coincidence of η^+ under ion and neutral particles bombardment at energies sufficiently high for practically total reionization;

(iii) a lack of symmetry in the angular dependence of η^+ on the angle of incidence, which also testify in favor of the high reionization degree.

In the above investigations the dependence of the energy peak location as well as its intensity on the crystallographic direction has been observed. In addition, it was shown that the peak was shifted towards lower energy relative to the location of the single-collision peak, the shift being different for neutral and charged particles, which evidenced the existence of the inelastic energy losses as well as the interaction with the neighboring target atoms.

The reionization is very clearly displayed by the existence of the threshold for ions which underwent multiple collision (Van Leerdam et al. 1990). In the energy peak of the scattered ions there is a distinct left shoulder, abruptly canceled at the definite threshold energy. The left shoulder of the spectrum corresponds to multiple collisions at which the particle is almost always neutralized. Thus, for escaping the target in an ionized state the scattered particle should undergo a violent reionizing collision as the last one. The observed threshold energy is independent of the primary energy of the scattered particle and corresponds to the minimum energy of binary collision, at which reionization is possible.

A sophisticated theoretical analysis of the scattered-ion charge state formation at energy less than 5 keV by H.-W. Lee and George (1985) has shown that for relatively low energies there is no need to improve the Hagstrum formula with consideration of the violent binary collision. The experimental data on the dependence of the ionization degree of the scattered particles on energy, scattering and escaping angle are well explained by the model, which assumes the interaction with the target as a whole, but considers the characteristic velocity v_c in formula (34) to depend on the distance of the closest approach z_0 with the target surface, i.e., on velocity,

$$v_c = A/a \exp(-az_0) ,$$

$$z_0 = -1/b \ln[m_1(v_i^2 + v_f^2) \sin^2\psi/(4B)] , \qquad (37)$$

where A and a are free parameters, B and b are parameters connected with

Born–Mayer potential at the target surface, m_1 is the mass of the scattered particle, v_i is the primary velocity, v_f is the final velocity, and ψ is the angle of incidence.

Parilis and Verleger (1985a,b,c,d) have developed a theory of scattered-ion charge formation, based on the mechanism of neutralization, the ions which have either conserved the original charge or acquired it during a one-hundred-per-cent ionization in a violent collision. A version was stipulated where the ionization in the violent collision is less than 100% and depends on the collision energy.

Alimov et al. (1990) have calculated the charge state formation of scattered atoms using the following assumptions. The charge state formation is considered to be divided into three stages: approaching, the violent collision and leaving. For the energies of interest (up to 3 keV) the partial Auger neutralization of ions approaching the surface and leaving it takes place. At binary collision, the charge redistribution between the colliding particles occurs, followed by a partial reionization of the scattered particles due to the inelastic energy transfer.

The total probability of charge conservation by the ion during the three-stage process is given by

$$I = [I_a P^*(1) + (1 - I_a)P(1)]I_0 . \tag{38}$$

Here I_a and I_0 are the probabilities of charge conservation at approach and leave; $P^*(1)$ and $P(1)$ are the probabilities of escaping as single-charged ion after the violent collision for the approach ion and neutral atom, respectively.

The neutralization at approaching and leaving, determining the probabilities I_a and I_0, could both occur due to resonance or quasiresonance electron exchange and Auger neutralization. The quasiresonance charge transfer manifests itself as an oscillating dependence of the ionization coefficient on the collision velocity. In most cases, the neutralization is determined by the Auger processes.

As it is shown by Kishinevsky and Parilis (1968b) and Kishinevsky et al. (1983), the probability of the Auger transition drastically depends on the interatomic distance. For the considered energy-range dependence, $W(R)$, of the Auger transition probability on interatomic distance, this could be approximated by an exponent,

$$W(R) \simeq \exp(-\mu R) , \tag{39}$$

where μ is a parameter.

Then

$$v_c/v_0 = \int_{R_0}^{\infty} \frac{W(R)}{v(R)} \, \mathrm{d}R \simeq \int_{R_0}^{\infty} \frac{e^{-\mu R}\sqrt{1 - V(R)/E_0 - p^2/R^2}}{(1 - V(R)/E_0)v_0} \, \mathrm{d}R . \tag{40}$$

Here, $V(R)$ is the potential of atomic interaction, p is the impact parameter, R_0 is the closest distance of approach, v_0 and E_0 are the initial velocity and energy, respectively.

Having integrated, one can get for the head-on impact

$$v_c \simeq a K_1(a) . \tag{41}$$

where $K_1(a)$ is the MacDonald function, $a \simeq v_0^{-1}$ and in the velocity range of interest $K_1(a) \simeq v_0^2$. This leads to a linear dependence of v_c on v_0, which could explain the left (ascending) part of the diagram in fig. 7. It is necessary to consider the processes of both reionization and charge transfer at violent collision, i.e., the probabilities $P^*(1)$ and $P(1)$ for the explanation of the right part of the diagram.

At violent collision a quasimolecule is formed and a part of the collision energy is used for electron shell excitation, which can lead to the detachment of one electron. After the collision the charges (both these brought by the ion and those formed due to electron emission) are distributed between the colliding partners and determine the probability for a particle undergoing violent collision to be in the charged state.

To calculate the probabilities $P^*(1)$ and $P(1)$ a statistical model, being a modified model by Russek (1961), has been developed. The following assumptions have been taken as basis:

(i) the probability of a given final state after the quasimolecule decomposition (the degree of ionization and excitation of the colliding particles) is proportional to the number of ways to achieve this state via the redistribution of small portions of the excitation energy over the emitted electron and the collision partners;

(ii) all ways of energy distribution are equally probable;

(iii) the kinetic energy of the emitted electron is confined by available energy only, but the residual excitation energy could not exceed the next ionization potential.

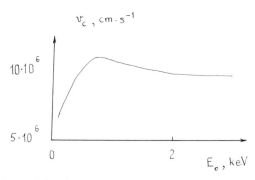

Fig. 7. The dependence of the phenomenological parameter v_c on initial energy of the scattered ion (Ar$^+$ → Pd).

For the probability of *n*-multiple ionization, it gives

$$P(n) = \frac{1}{(N'-1)!(N''-1)!(n-1)!N'!N''!n!}$$

$$\times \sum_{k=0}^{n-1} \binom{n-1}{k}(-1)^k \frac{(I'')^{N''+k}}{N''+k}$$

$$\times \sum_{l=0}^{n-1-k} \binom{n-1-k}{l}(-1)^l \frac{E_k^{n-1-k-1}(I')^{N'+l}}{N'+l} . \qquad (42)$$

Here, N' and N'' are the numbers of the outer-shell electrons of the colliding atoms $(N = N' + N'')$, n is the number of emitted electrons, I'_a and I'' are the ionization potentials, and $E_k = E_{inel} \pm \Sigma I_v$ is the energy of excitation.

The result of the calculation of the dependence of the ionization probability on electron excitation energy is shown in fig. 8. At low E_{inel} (low initial velocities of the scattered ions), one gets $0 < P^*(1) < 1$; $P(1) \approx 0$, the reionization does not take place and $v_c(v_0)$ is determined by Auger neutralization. With increase of v_0 the probabilities $P^*(1)$ and $P(1)$ increase, which is perceived phenomenologically as the decrease of $v_c(v_0)$.

Experiment has showed (fig. 9) that the energy peaks corresponding to scattered particles escaping the metal surface as neutral atoms or ions coincide. On the basis of the developed concepts this could be explained by the fact that the inelastic energy losses, used for electron excitation in the quasimolecule, depend only on the initial state of the colliding particles, but the charge state of the scattered particle is determined by the charge redistribution at the quasimolecule decay. Thus, the different charges of the scattered particles could correspond to just the same inelastic energy losses.

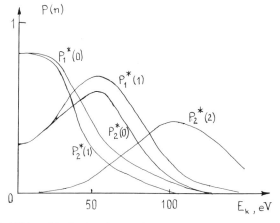

Fig. 8. The probabilities of ionization due to violent collision versus electron excitation energy ($Ar^+ \rightarrow Pd$).

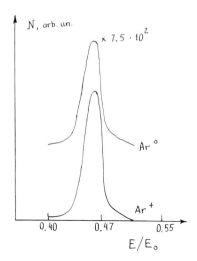

Fig. 9. Energy peaks of neutral and single-ionized scattered atomic particles (Ar$^+$ → Pt, $E_0 = 2$ keV, $\theta = 138°$).

4. Charge state of light atoms

4.1. Electron capture and loss in solid

We have introduced the electron capture and loss cross sections, or their mean rates, connected with the cross section via the formula $W = N\sigma v$ (Parilis and Verleger 1980). Electron capture and loss rates depending on the distance to the surface s are given by

$$\begin{aligned} W_c(s) &= A, & s < 0, \\ &= Af_1(s), & s > 0; \end{aligned} \tag{43}$$

$$\begin{aligned} W_\ell(s) &= B(E), & s < 0, \\ &= B(E)f_2(s), & s > 0. \end{aligned} \tag{44}$$

Here, A and $B(E)$ are the electron capture and loss rates inside the metal. If there is no bound electron level inside the metal, then the plane $s = 0$ is located at the distance from the metal where the level appears. The functions $f_1(s)$ and $f_2(s)$ describe the decreasing W_c and W_ℓ outside the metal at the account of both the decreasing of the electron wave functions in the metal and in atom overlapping, and at the account of atomic level shift. The function $f_1(s)$ and $f_2(s)$ are considered to decrease nonmonotonically.

The calculation of A inside the metal in the first order of perturbation theory is impossible, since in the considered velocity range, $v < 10^8$ cm/s, the perturbation is not a weak one. Let us consider A to be a parameter independent of ion energy E. For Auger capture $A \simeq 10^{15} \text{ s}^{-1}$, and for electron capture via tunneling $A = 10^{-16} \text{ s}^{-1}$. The electron loss processes are of quite another character. The electron being captured by an ion, cannot return back in the occupied part of the condition band according to the Pauli exclusion principle. To escape in the continuum it is necessary to transfer the energy $\Delta\varepsilon_n$. The rate of this slow nonadiabatic transition must depend on energy E, and at $E \to 0$ the amplitude $B(E) \to 0$.

Suppose the electron loss cross section to be $\sigma_\ell \simeq \sqrt{E}$ and, respectively, $B(E) \simeq E$. Here we try to evaluate σ_ℓ and $B(E) = N\sigma_\ell v$ with a formula of the Firsov type (Firsov 1959) for the ionization cross section in atomic collisions. Then the cross-section σ_ℓ may be written as follows,

$$\sigma_\ell = \int_0^{P_1} L(p)p \, dp . \tag{45}$$

Here, $L(p)$ is the ionization probabilty, normalized per time unit for the collision parameter p connected with the inelastic transfered energy $\Delta\varepsilon_n$, P_1 is the root of the equation $\Delta\varepsilon_n = I_1$, I_1 is the atomic ionization potential. In the modification proposed by Kishinevsky (1962), it reads

$$\Delta\varepsilon_n = \frac{bv(1 - 0.68U(r)/E)}{(1 + r/a_1)^3} . \tag{46}$$

Here, $r = \sqrt{p^2 + r_0^2}$ is the distance of closest approach, r_0 is the distance of closest approach at head-on collision, $U(r)$ is the potential energy of the interacting atoms, $a_1 = 1.32a_0(Z_1^{1/6} + Z_2^{1/6})Z_2^{-1/6}$. In eq. (46), the velocity v is expressed in units of 10^7 cm/s and $\Delta\varepsilon_n$ in eV.

If, according to the paper by Firsov (1959), $L(p)$ is chosen as

$$\begin{aligned} L(p) &= 1 , \quad j\Delta\varepsilon_n \geqslant I_1 , \\ &= 0 , \quad j\Delta\varepsilon_n < I_1 , \end{aligned} \tag{47}$$

then, in the approximation $p \gg r_0$ and $U(r)/E \ll 1$, from eq. (45) one would get

$$\sigma_n = \pi a_1^2 \left[\left(\frac{E}{E'} \right)^{1/6} - 1 \right]^2 . \tag{48}$$

The parameter j in eq. (47) describes the part of the inelastically transferred energy per scattered atom, from the paper by Filippenko (1962) we consider $j = Z_1/(Z_1 + Z_2)$. The value E in an extrapolated threshold for the ionization, calculated in the approximation $p \gg r_0$, $E' = \frac{1}{2}m(v')^2$, $v' = I_1/jb$.

Equations (47) and (48) simulate the situation, where the ionization probability rapidly reaches saturation with incrasing energy, i.e., it becomes equal to 1. In this case, it is convenient to accept

$$L(p) = 0 , \qquad j \Delta \varepsilon_n < I_1 ,$$

$$= \frac{j \Delta \varepsilon_n}{I_1} - 1 , \quad I_1 \leqslant j \Delta \varepsilon_n \leqslant 2 I_1 , \qquad (49)$$

$$= 1 , \qquad j \Delta \varepsilon_n > 2 I_1 .$$

Such form of $L(p)$ has been accepted by Nizhnaya et al. (1982a,b), which allowed to calculate more correctly the heavy-atom ionization at scattering over small angles ($\theta \leqslant 30°$).

Then the cross section is written as

$$\sigma_\ell = \pi a_1^2 [(E/E')^{1/6} - 1]^3 , \qquad \text{for } E' < E < E" ,$$

$$= \pi a_1^2 \, 0.15(E/E')^{1/3} - 0.175(E/E')^{1/6} , \quad \text{for } E' > E" . \qquad (50)$$

Here, $E"$ presents the solution of the equation $\Delta \varepsilon_n(E", p = 0) = 2 I_1$. Having chosen σ_ℓ as in eq. (48) or in eq. (50), one could estimate also $B(E)$.

4.2. Ionization probability outside the solid

The ionization probability Q^+ is the solution of the equation

$$v_\perp \, dQ^+/ds = -[W_c(s) + W_\ell(s)]Q^+ + W_\ell(s) , \qquad (51)$$

at the initial condition $Q^+(s = 0) = Q_1^+$, $v_\perp = v \sin \psi$.

For the level lying below the Fermi level, $W_c \gg W_\ell$. Let us take $W_c + W_\ell \approx W_c$ and substitute eqs. (43) and (44) in eq. (51), then

$$Q^+ = \exp\left[-v_\perp^{-1} \int_0^\infty A f_1(s) \, ds\right]$$

$$\times \left\{Q_1^+ + v_\perp^{-1} \int_0^\infty \exp\left[v_\perp^{-1} \int_0^s A f_1(x) \, dx\right] f_2(s) \, ds\right\} . \qquad (52)$$

In a paper by Parilis and Verleger (1980), Q^+ has been calculated in the simplest supposition $f_1(s) = f_2(s) = f(s)$. Let us see how Q^+ is changing if $f_1(s)$ and $f_2(s)$ are different. The function $f_1(s)$ for the Auger or resonance neutralization rate are usually written as in a paper by Grozdanov and Janev (1977),

$$f_1(s) \approx e^{-\alpha s} . \qquad (53)$$

Let us chose for $f_2(s)$ the analogous dependence

$$f_2(s) \approx e^{-\beta s} . \qquad (54)$$

The function $f_2(s)$ of such a type for the electron loss rate has been obtained in a paper by Schroeer et al. (1973), where the probability of atom ionization near the metal is taken in the adiabatic perturbation theory. Let us substitute eqs. (53) and (54) in eq. (52) and, having substituted the variable $y = \exp(-\alpha s)$, we should get

$$Q^+ = Q_1^+ \exp(-v_0/v_\perp) + (B/A)(v_0/v_\perp) \int_0^1 \exp(-yv_0/v_\perp)y^{n-1}\, dy. \quad (55)$$

Here, $v_0 = A/\alpha$, the parameter $n = \alpha/\beta$ describes the relative velocity of the capture and loss rate decreasing with the distance s,

$$\int_0^1 \exp(-yv_0/v)y^{n-1}\, dy = (v/v_0)\gamma(n, v/v_0), \quad (56)$$

$\gamma(n, v/v_0)$ is an incomplete gamma function.

(a) Let n be an integral number. Then, using the γ function (Ryzhik and Gradstein 1971), we get

$$Q^+ = \frac{B}{A}(n-1)!\left(\frac{v_\perp}{v_0}\right)^{n-1}$$

$$+ \left[Q_1^+ - \frac{B}{A}\left(\frac{v_\perp}{v_0}\right)^{n-1}(n-1)!\right]\sum_m^{n-1}\frac{1}{m!}\left(\frac{v_0}{v_\perp}\right)^m e^{-v_0/v_\perp}. \quad (57)$$

Then, for $n = 1$,

$$Q^+ = \frac{B}{A} + \left(Q_1^+ - \frac{B}{A}\right)e^{-v_0/v_\perp}, \quad (58)$$

which coincides with the solution obtained in a paper by Parilis and Verleger (1980),

$$Q^+ = \frac{B}{A}\left(\frac{v_\perp}{v}\right) + \left(Q_1^+ - \frac{B}{A} - \frac{B}{A}\frac{v_\perp}{v}\right)e^{-v_0/v_\perp}, \quad \text{for } n = 2, \quad (59)$$

$$Q^+ = 2\frac{B}{A}\left(\frac{v_\perp}{v_0}\right)^2 + \left[Q_1^+ - \frac{B}{A} - \frac{2B}{A}\frac{v_\perp}{v_0} - \frac{2B}{A}\left(\frac{v_\perp}{v_0}\right)^2\right]e^{-v_0/v_\perp},$$

$$\text{for } n = 3, \quad (60)$$

and so on.

(b) Let n be a nonintegral number. In this case, we failed to get a general simple expression like eq. (57). Nevertheless, we could show that in this case Q^+ was described by a similar formula. For example, we choose $n = \frac{3}{2}$,

$$Q^+ = Q_1^+ e^{-v_0/v_\perp}\frac{B}{A}\left(\frac{v}{v_0}\right)^{1/2}\gamma(\tfrac{3}{2}, v_0/v_\perp). \quad (61)$$

Let us use the following equations,

$$\gamma\left(\frac{3}{2}, \frac{v_0}{v_\perp}\right) = \frac{1}{2}\,\gamma\left(\frac{1}{2}, \frac{v_0}{v_\perp}\right) - \left(\frac{v_0}{v_\perp}\right)^{1/2} e^{-v_0/v_\perp}, \tag{62}$$

$$\gamma(\tfrac{1}{2}, v_0/v_\perp) = \pi^{1/2}\Phi\left[(v_0/v_\perp)^{1/2}\right], \tag{63}$$

$$\Phi\left[\left(\frac{v_0}{v_\perp}\right)^{1/2}\right] = 1 - \frac{1}{\pi}\exp\left(-\frac{v_0}{v_\perp}\right)\sum_{k=0}^{\infty}\frac{(1)^k\Gamma(k+1)}{(v_0/v_\perp)^{(k+1)/2}}\exp\left(-\frac{v_0}{v_\perp}\right). \tag{64}$$

So, at chosen capture and loss rates, Q^+ is the sum of two terms, the first of which has a dependence on the normal velocity of the type $\simeq(B/A)(v_\perp/v_0)^n$ and the second one of the type $\exp(-v_0/v_\perp)$. Thus, in the proposed model, the dependence $Q^+(v_\perp)$ is determined only by the parameter n.

The above consideration is also valid for the primary part of the trajectory before the impinging particles enter the metal. In this case, in eqs. (52)–(64) the probability Q_1^+ is replaced by Q_0^+, which is the primary ion fraction in the impinging beam (see fig. 10). Usually in experiments $Q_0^+ = 1$, sometimes (Luitjens et al. 1980e) $Q_0^+ = 0$, under the bombardment by two-atomic molecular ions $Q_0^+ = \frac{1}{2}$.

Besides, in the given equations, it is necessary to take $v_\perp = \sqrt{2E_0/m_1}\,\sin\psi$. The calculated value Q_4^+ has been used as the primary value of Q^+ at calculation of the ionization probability inside the solid.

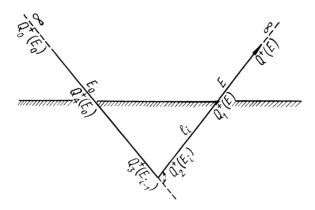

Fig. 10. Scheme of the formation of a charge state for the particles which underwent one violent scattering: $Q^+(E)$ is the ionization probability of the scattered beam, $Q_1^+(E)$ that at leaving the surface, $Q_2^+(E)$ – after violent scattering, $Q_3^+(E)$ – before violent scattering, $Q_4^+(E)$ – at incidence of the surface, and $Q_0^+(E)$ is the initial ionization degree of the bombarding beam, E_{i-1} and E_i are the ion energies before and after the violent scattering, respectively, and l_i is the path in the solid after violent scattering.

4.3. Ionization probability inside the solid

To calculate Q^+ and then η^+, it is necesary to know $Q_1^+(s = 0)$, which determines the charge state of the particles impinging the metal with energy E_0 and leaving the metal boundary with energy E.

If inside metal there is no effective bound state, then $Q_1^+ = 1$. In the opposite case, Q_1^+ is the solution of the equation

$$(dQ_1^+/dt) = -[A + B(E)]Q_1^+ + B(E),\tag{65}$$

at the initial condition $Q_1^+(t = 0; E = E_i) = Q_2^+(E_i)$.

The starting point for charge formation $(E = E_i)$ is connected with the last violent collision in solids, which resulted in the ionization of particles with the probability $L(p, E_{i-1})$, dependent on the energy before the violent collision, E_{i-1} (see fig. 10), E_i is the energy after this collision, $E_0 > E_{i-1} > E_i$.

If the beam containing $Q_3^+(E_{i-1})$ ions and $[1 - Q_3^+(E_{i-1})]$ neutrals impinges in a violent collision center, then

$$Q_2^+(E_i) = Q_3^+(E_{i-1}) + L(p, E_{i-1})[1 - Q_3^+(E_{i-1})].\tag{66}$$

The probability $Q_3^+(E_{i-1})$ is also defined by eq. (65) with the initial conditions $Q_3^+(E = E_0) = Q_4^+(E_0)$, if there are no other violent collisions on the trajectory. The analysis of scattering trajectories in solids has shown that on the trajectory there could be one, two or several violent ionizing collisions and sometimes there is no such collision. The role of the violent collisions depends on what formula for $L(p, E)$ [eq. (47) or eq. (49)] is used, i.e., whether the violent collision is fully or partial ionizing. In the first case, one should take into consideration only the last violent collision (see fig. 10) and the charge state before it has not much significance. In the second case, one should take into consideration the history of the ionization along the trajectory and the dependence on the initial charge and the angle of incidence has influence as well.

Below eq. (65) will be solved with $B(E)$, calculated for the first case, i.e., with $B(E)$ from eq. (48). The solution for the second case is more complicated.

In the simple case when there is only one violent collision, in which, however, the ionization probability does not reach 1, it is necessary to apply the second solution for the explanation of the experiment data by Eckstein et al. (1978). In a paper by Parilis and Verleger (1980), an equation like eq. (65) has been solved without the consideration of continuous changing σ_c and σ_ℓ at the account of the stopping process in the solid. The solution presented the sum of two terms, the first of which described the equilibrium charge independent of energy E_0, the second term is the surface peak, considerable only at $E \to E_0$. In a paper by Zhabrev (1979), an equation like eq. (65) has been solved numerically by the Runge–Kutta method. The author concluded, that the consideration of stopping made it impossible to establish the charge equilibrium and Q^+, for all values E, strongly depended on E_0. Let us show that at a natural assumption $A \gg B(E)$, eq. (65) is easily solved analytically also when taking in account the stopping. Moreover, this solution contains the same equilibrium term as the solution obtained in a paper by Parilis and Verleger (1980).

For a consideration of the stopping in solid after the violent collision we apply the Lindhard–Sharff formula,

$$dE/dl = K\sqrt{E} . \tag{67}$$

Then, $dt = -dx/w_0x$, where $w_0 = K\sqrt{2/m_1}$.

Thus, having chosen σ_ℓ in the form of eq. (48), we should calculate $B(E)$ and substitute eq. (66) into eq. (65), obtaining the following equation,

$$dQ_1^+/dx = h_1Q_1/x - h_2h^{-1/2}[(x/E')^{1/6} - 1]^2 . \tag{68}$$

Here, $h_1 = A/w_0$, $h_2 = \pi[Na_1^2(2/m_1)^{1/2}/w_0]$.

The solution of eq. (68) is given by

$$Q_1^+(E) = (E/E_1)^{h_1}\left\{Q_2^+ - h_iE_i^{h_i}\int_{E_i}^{E} x^{-(h_1+1/2)}\left[\left(\frac{x}{E'}\right)^{1/6} - 1\right]^2 dx\right\}. \tag{69}$$

The integral in eq. (69) is easily calculated. Let us take into account that the order of magnitude of $w_0 \simeq 10^{13}\,\text{s}^{-1}$, thus $h_1 \geqslant 10^2 \gg 1$ and $h_1 + \frac{1}{2} \approx h_1$. Then, $Q_1^+(E)$ requires the simple form

$$Q_1^+(E, E_i) = Q_{1,e} + (E/E_i)^{h_1}[Q_2^+ - Q_{1,e}^+(E_i)] . \tag{70}$$

The first term on the right-hand side of eq. (70) is for equilibrium charge, depending on the ion energy E,

$$Q_{1,e}^+ = (E/E_1)^{1/2}[(E/E')^{1/6} - 1]^2 = B(E)/A . \tag{71}$$

The parameter E_1,

$$E_1 = 0.001m_1Z_2^2A/N^2a_0^4(Z_1^{1/6} + Z_2^{1/6})^4 , \tag{72}$$

is defined by the electron capture rate A inside the metal.

As is seen from eq. (71), the energy dependence of the equilibrium charge state has a threshold, $Q_1^+(E)$ is defined only at $E > E'$. The appearance of the threshold resulted from the application of the Firsov-type formula for the ionization cross section. It is necessary to note that E' in eq. (48) and eq. (71) is not a true value of the threshold E as obtained in theories by Firsov (1959) and Kishinevsky (1962), generally $E < E_t$. Since the experimental data by Eckstein et al. (1978) testify for the existence of a threshold for the particle reflection as ions, it might be worth to evaluate E_t. The inelastically transferred energy at the collision of two atoms is given by eq. (46).

The threshold energy E_t is the solution of equation $\Delta \varepsilon_n^{max} = I_1$, $E_t = E$ is given in the approximation $U(r) \ll E$ and $r_0 \ll p$, which is not valid at small energies E. The inelastically transferred energy accepts the maximum value at $r = r_{max}$, which is defined from the following condition,

$$d(\Delta \varepsilon_n)/dr|_{r=r_{max}} = 0 . \tag{73}$$

The determination of r_{max} from eq. (73) comes to the solution of the following cubic equation,

$$3r_{max}^3 - 3.4 r_{max} r_0^2 - 1.36 a_1 r_0^2 = 0 , \tag{74}$$

if the considered energy range $U(r)$ is chosen as

$$U(r) = D/r^2 , \tag{75}$$

where $D = 6.4 Z_1 Z_2 (1 + v)/(Z_1^{1/2} + Z_2^{1/2})^{2/3}$ eV Å, $r_0 = (D/E)^{1/2}$.

At the condition $r_0 < a_1$, the solution of eq. (74) is given by

$$r_{max} = r_0 (0.22 a_1/r_0)^{1/3} (\{1 + [1 + (r_0/a_1)^2]^{1/2}\}^{1/3}$$
$$+ \{1 - [1 - (r_0/a_1)^2]^{1/2}\}^{1/3}) . \tag{76}$$

By substituting eq. (76) into eq. (46) at $\varepsilon_n = I_1$ and solving it to find E_0, one could define E_t.

Far from the threshold, $E \gg E_t > E'$, eq. (71) may be written as

$$Q_{1,e}^+ = E^{5/6}/E_1^{1/2}(E')^{1/2} \simeq E^{5/6} ,$$

which is close to the dependence $Q_1^+(E)$, as obtained in a paper by Parilis and Verleger (1980). The second term in eq. (70) as in a paper by Parilis and Verleger (1980) is for the nonequilibrium portion, dependent on E_0 (since $E_i \simeq E_0$) or path l_1 in the solid after violent collision on the given trajectory,

$$Q_1^+(E, E_i) = (E, E_i)^{h_1}[Q_2^+ - Q_{1,e}^+(E_i)]$$
$$= (1 - l_i/l_0)^{2h_1}[Q_2^+ - B(E_i)/a] . \tag{77}$$

Since h_1 is large, $h_1 \simeq 10^2 – 10^3$, the nonequilibrium term is essential only for those trajectories on which the violent scattering occurs at the very end of the path (E is close E_i), in particular, it is essential for all scattering trajectories on the surface. Otherwise, when $E \ll E_i$ and $l_i \gg 0$, $\Delta Q_1^+ \to 0$.

Function (77) is rapidly decreasing with decreasing E/E_i, i.e., with increasing of the path length l, on which ΔQ_1^+ is reduced to such extent that it becomes one order of magnitude smaller than the equilibrium charge value $Q_{1,e}^+$, which one could call the establishing length of the equilibrium charge as introduced in a paper by Parilis and Verleger (1980). The order of magnitude of λ is several Å.

Thus, via the processes of electron capture and loss, the charge equilibrium state is established both inside and outside the solid, also at continuous stopping.

The above calculated probability of positive ionization along the trajectories would be used for the calculation of the ionization degree of the reflected atoms by averaging of Q^+ along all scattering trajectories, the relative probability of which is given by the function $F(\varepsilon, \theta, \varphi, E_0; q)$. The calculation of the ionization degree is the aim of the next section.

4.4. Equilibrium charge function

As it is shown in section 4.3, the positive ionization probability contains the equilibrium portion and nonequilibrium term $Q^+ = Q_e^+ + \Delta Q^+$.

Correspondingly, we get for the ionization degree

$$\eta^+ = \int_\Phi FQ^+ \, dq \bigg/ \int_\Phi F \, dq = \eta_e^+ + \Delta\eta^+ . \tag{78}$$

Since Q_e^+ does not depend on trajectory, q, at all,

$$\eta^+ = B(E)(n-1)!(v_\perp/v_0)^{n-1}/A . \tag{79}$$

The first term on the right-hand side of eq. (78) is independent on E_0 and represents the equilibrium ionization degree η_e^+ and the second term $\delta\eta^+$ is the nonequilibrium portion. Let us compare eq. (79) with data of those experiments in which the equilibrium ionization degree was recorded independent on E_0. First, we shall consider the problem which has been discussed for a long time how the ionization degree η^+ depends on the escape angle of the scattered particle or of the normal velocity $v_\perp$, particular, at $v_\perp \to 0$. The present theories (Yavlinsky et al. 1966) give $\eta^+ = \exp(-v_0/v_\perp)$, while in the experiment such dependence has not been observed. Although the results of the measurements for the different combinations metal–atom are slightly different, always a dependence $\eta^+(v_\perp)$ considerably weaker than $\exp(-v_0/v_\perp)$ is recorded in experiments by Eckstein and Matschke (1976), Verbeek et al. (1980), Zhabrev (1979), Eckstein et al. (1978), Lukyanov and Chicherov (1973).

In eq. (79), this dependence is defined by the parameter n, which characterized the relative rate of slowing down the electron capture and loss outside the metal.

(a) In experiments by Eckstein and Matschke (1976) and Zhabrev (1979) on H^+ and D^+ scattering, the independence of η^+ on ψ has been observed. In a paper by Verbeek et al. (1980) for $He^+ \to Au$, η^+ is independent of ψ

in a broad range, $5° < \psi < 65°$. The data of papers by Eckstein and Matschke (1976), and Zhabrev (1979) give a good fit with eq. (79) at $n = 1$.

(b) Experiments on scattering of H^+ on Au have shown a weak dependence of η^+ on ψ. The authors have noted that the independence or a weak dependence of η^+ on ψ for $H^+ \rightarrow$ Au is defined by the way of preparation of Au terget. In a paper by Verbeek et al. (1980), the bad reproduction of experimental results has been noted. The equilibrium term $Q_e^+ = \frac{1}{2}\sqrt{\pi}(B(E)/A)(v_\perp/v_0)^{1/2}$ for $n = \frac{3}{2}$ is in quite good agreement with the data of an experiment by Verbeek et al. (1980).

(c) The experimental data by Lukyanov and Chicherov (1973), where the linear dependence of η^+ on $v_\perp$ has been observed at scattering of Ar on Pt have also been explained with eq. (79) at $n = 2$.

Let us consider in detail the comparison of η^+, calculated with eq. (79), with that part of the distribution of $\eta^+(E)$ for $He^+ \rightarrow$ Ni which Eckstein et al. (1978) called the equilibrium charge. It should be noted that the results in the paper by Eckstein et al. (1978) have been reproduced by Kishinevsky et al. (1976). Thus, we can use $n = 1$ in eq. (79). We shall calculate $B(E)$ using the electron loss cross section σ_ℓ by Firsov (1959), formula (48). Then for the equilibrium term we get

$$\eta_e^+ = (E/E_1)^{1/2}[(E/E_1)^{1/6} - 1]^2 . \qquad (80)$$

Comparison of eq. (80) with experiment is shown in fig. 11. It is seen, that

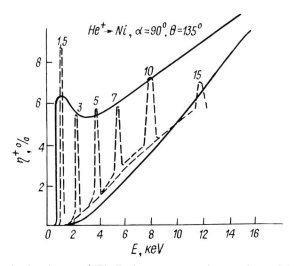

Fig. 11. The ionization degree $\eta^+(E)$. Broken curves are the experimental data by Eckstein et al. (1978), the bottom full curve is the equilibrium ionization degree obtained by eq. (80), the top full curve is the binding line of the surface peaks $\eta_{max}^+(E)$ from eq. (86). The figures are the values of the initial energy E_0.

the curve $\eta_c^+(E)$ has a threshold at $E = E_t(E)$ and then rises approximately linearly. The calculation of R_t using eqs. (46) and (76) gives for $E_t = 700$ eV, i.e., close to the observed value. The parameter E_1 has been found from fitting the theory with the experiment at $E = 10$ keV. From this comparison the electron capture rate parameter for the He^+ ion was found to be equal to $A = 1.76 \times 10^{15}$ s^{-1} which is characteristic for Auger neutralization. In analogy to the data by Verbeek et al. (1980), Kishinevsky et al. (1976) has calculated E_t for $\mathrm{He}^+ \rightarrow \mathrm{Au}$, $A \approx 2.84 \times 10^{15}$ s^{-1}.

Thus, eq. (79) allows to explain qualitatively the experimental data on the depencence of η^+ on final energy E and escape angle ψ of the scattered particles. For the parameter A, reasonable values have been obtained: $A = 1$–3×10^{15} s^{-1}; the parameter n, having changed in a narrow range $1 \leqslant n \leqslant 2$, allows to get all observed dependences $\eta_c^+(\psi)$. This means that at removing of an ion from the surface the electron capture rate decreases either with the same rate as the electron loss or somewhat slower depending on the surface state. It is quite natural, since the resonance and Auger captures occur up to the distances 5–10 Å (Kishinevsky et al. 1976), while electron losses require more close overlapping of electron shells (Kishinevsky and Parilis 1968a,b).

4.5. Surface peak in the ionization degree

Equation (79) does not describe fully the experimental data by Eckstein et al. (1978), where on the equilibrium charge state $\eta_c^+(E)$ the sharp surface peaks are imposed at each value $E \approx E_1$ (E_1 is the energy of single scattering). Besides, outside the peaks the curves $\eta_c^+(E, E_0)$ for different E_0 differ, so $\eta_c^+(E)$ might be considered to be their general asymptote (see fig. 11). Just in this part of the spectrum the more essential and nontrivial connection between the charge state of the reflected particles on one hand and the scattering trajectory on the other, which is the main subject of the present work, is exhibited.

Let us calculate the ionization degree by eq. (43), including the nonequilibrium part of the ionization degree $\Delta\eta^+$. Since the results of the calculation will be compared with the data from Eckstein et al. (1978), we confine ourselves to the case $n = 1$. Using Q_1^+ from eq. (70), one should get

$$\eta^+(E, E_0, \psi) = \eta_c^+(E, \psi) + \int_\Phi (E/E_i)^{h_1}[1 - \eta_c^+(E_i, \psi)]F\,dq$$

$$\times \exp(-v_0/v_\perp) \bigg/ \int_\Phi F\,dq\,. \tag{81}$$

Here, $E_i \geqslant E$ is the energy after the last violent fully ionizing collision on the given trajectory, the initial energy for the charge state formation inside the metal.

It is clear that η_e^+ mainly contributes in eq. (78) only when $v_\perp \ll v_0$, where $v_0 = A/\alpha \approx 10^8$ cm/s for resonance and $v_0 \approx 10^7$ cm/s for Auger neutralization. In the opposite case it is necessary to consider $\Delta\eta^+$.

It is necessary to take into consideration the nonequilibrium term in eq. (78) only when E_i is close to E, since $h_1 > 10^2$. The term $\Delta\eta^+$ depends on the position of the energy level of the scattered particle inside the solid and the type of the scattering function F. If inside the solid the particle exists only as ion, then $E_i = E$, i.e., the formation of charge state occurs only outside the solid and

$$\eta(E, \psi) = \eta_e^+(E) + [1 - \eta_e^+(E)] \exp(-v_0/v_\perp). \tag{82}$$

In this special case, $\Delta\eta^+$ also does not depend on E_0 and the type of the function F, the surface scattering is not separated from the bulk one, $\eta^+(E, \psi)$ is a smooth monotonically increasing function of E, and the surface peak is not observed.

If inside the solid a particle could exist as atom, then for the trajectories of surface scattering $E = E_i = E_1$, and inside the solid $E_i > E$, which leads to the appearance of the surface peak, the shape of which depends on the type of the function F. The term 'the surface peak' means in this case the extremely high ionization degree in the surface scattering, where the charge equilibrium state is not established yet. Let us compare the calculation $\eta^+(E, E_0)$ from eq. (81) with the experimental data by Eckstein et al. (1978), using for F expression (9) from chapter 1, which is given in the double-scattering theory. The earlier obtained value $A = 1.76 \times 10^{15}$ s^{-1} for this case allows to determine the parameters $h = 110$, $v_0 = 0.88 \times 10^8$ cm/s at $\alpha = 2$ Å^{-1}. The comparison of inelastic energy loss, calculated using eq. (46) with the ionization potential I_0 allows to introduce the criterion for the definition of the last violent collision, i.e., the energy E_i.

The consideration of the scattering trajectory shape, made in section 1, provides the probability to determine the position of the point of the last fully ionizing collision on the trajectory. In the double-scattering approximation,

$$
\begin{aligned}
E_i &= E_{2,2} = \varepsilon_{2,2} E_0, \quad \text{if } j\,\Delta\varepsilon_n(\theta_2) > I_0, \\
&= E_{1,2} = \varepsilon_{1,2} E_0, \quad \text{if } j\,\Delta\varepsilon_n(\theta_2) < I_0, \text{ but } j\,\Delta\varepsilon_n(\theta_1) > I_0, \\
&= E_0, \quad\quad\quad\quad \text{in other cases}.
\end{aligned}
\tag{83}
$$

The meaning of eq. (83) is that the last fully ionizing collision is identical with the second or the first bend on the double-scattering trajectory, depending where the inelastic energy exceeding I_0 is released. Having performed the numerical integration in eq. (81), we get the set of curves

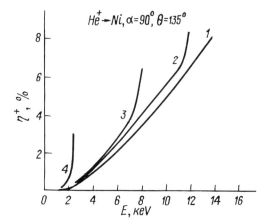

Fig. 12. The dependence $\eta^+(E)$ on the initial energy E_0. Curve 1 denotes the equilibrium ionization degree η_e^+ by formula (80); curve 2 $E_0 = 15$ keV; curve 3 $E_0 = 10$ keV, and curve 4 $E_0 = 3$ keV. The corresponding experimental curves are shown in fig. 11.

$\eta^+(E, E_0)$ (see fig. 12). Here, at each value of $E = E_0 C^2(\theta)$, the surface peaks have been observed. At $E < E_i$, the curves $\eta^+(E, E_0)$ sharply decrease and tend to the general asymptote $\eta_e^+(E)$, besides outside the peak region at given E the value η^+ is more or less equal to E_0, forming the family of curves $\eta^+(E, E_0)$ which is in agreement with the experiment by Eckstein et al. (1978). It should be noted that such regularity has been observed more pronounced at scattering of Ar on Au (Buck et al. 1975). The behavior of the curves $\eta^+(E, E_0)$ is easily explained by the fact that the exceeding of $\eta^+(E, E_0)$ over η_e^+ is determined by the term $\Delta\eta^+$, which at $h_1 = 110$, as it follows from eq. (81), is large, if the last violent collision took place when E_i is close to E, i.e., the path of the surface $l_i < \lambda$, where λ is the length for establishing the charge equilibrium. The analysis of double-scattering trajectories and the function $F(\theta_1, \theta_2, \varphi, q; E_0)$ from eq. (46) have shown that for given E_0 with removing E_1 toward low energies relative to the contribution of the trajectories, in which the second, fully ionizing collision appeared to be near the surface, is sharply reduced. At given E, the contribution of such trajectories is decreasing with the increase of the initial energy E_0. In both cases it takes place due to the increasing of the total trajectory length. As a result, the exceeding of η^+ over the equilibrium value falls down with the increase of both $(E_1 - E)$ and E_0. In the region of surface peaks, $E \approx E_1$. The measurements in papers by Eckstein et al. (1978) and Verbeek et al. (1980) have shown that η_{max}^+ is changing non-monotonically with increasing E_0 and unlike η_e^+ does depends on ψ.

As it was mentioned in the region $E \approx E_1$, the double-scattering trajectories, containing the large angle $\theta \pm \delta$ and a small angle δ prevail, besides the scattering probability is determined mainly by the probability of scattering over the large angle. Therefore, we can estimate it by simple analytical formulas of a single scattering. In the model of single scattering, where the inelastic energy losses are considered, each value E in the energy spectrum corresponds to a single-scattering trajectory. In particular, the energy E_1 corresponds to the scattering trajectory on the surface. This means that in eq. (81) the function $F(q)$ is of the δ-function type and for $Q_2^+ = 1$,

$$\eta_{max}^+(E) = \eta_0^+(E) + (E/E_1)^{h_1}[1 - \eta_e^+(E)] \exp(-v_0/v_\perp). \tag{84}$$

Formula (84) gives some very narrow and high surface peaks $\eta^+(E_1) = \eta_{max}^+$. For a comparison of eq. (84) with the experiment, we should average it over the resolving power of the energy analyzer $\Delta \varepsilon_a = \Delta E_a / E_0 = 0.067$ (Eckstein et al. 1978).

Thus, we get

$$\eta_{max}^+ = \eta_e^+ + \frac{2[1 - \eta_e^+(E_1)] \exp(-v_0/v_\perp)}{\Delta \varepsilon_a (h_1 + 1)} [1 - (1 - \Delta \varepsilon_a / 2\varepsilon_1)^{h_1 + 1}]$$

$$\times \exp(-v_0/v_\perp) \tag{85}$$

The mean value $\bar{\eta}_{max}^+(E_1, \psi, E_0)$ is close to the experimental one besides the widths of the surface peaks $\Delta E \simeq E_0$. However, unlike stated by Eckstein et al. (1978), $\bar{\eta}_{max}^+(E)$ increases nonmonotonically with E, which is caused not only by the finite resolved power of the apparatus. Apparently, the main reason is that not only trajectories of the reflection from the surface layers contributes to the peak region $E = E_1$, but also multiple-scattering trajectories from the depth, for which the ionization degree is considerably less.

The effect of scattering from the depth could be considered in the same approximation, having performed the integration over the depth at which the violent scattering over large angle occured accompanied by an additional scattering over a small angle, not exceeding angle δ, such that the double-scattering energy differs from $E_1(\theta)$ not more than ΔE_a. Such integration had been made in a paper by Kishinevsky et al. (1976), with the consideration of only elastic energy losses, which is justified in the surface peak region. The result is given by

$$\eta_{max}^+ = \eta_e^+(E_1) + v_\perp \exp(-v_0/v_\perp)/(v_\perp + b_1 v_1). \tag{86}$$

Here, $b_1 = E_1 \sin \alpha \sin \psi/(E_0 \sin \alpha + \sin \psi)$, α is the glancing angle, $v_1 = A\lambda_0$, and $\lambda_0 = (N\bar{\sigma})^{-1}$ is the effective range for the additional scattering over the angle larger than δ. The cross section of such scattering equals

$$\bar{\sigma} = 2\pi \int_\delta^\pi (d\sigma/d\Omega)\, d\Omega = 8\pi d_1(\operatorname{cosec} \tfrac{1}{2}\delta - 1)/E_0.$$

$$d\sigma/d\Omega = d_1/E_0 \sin^3 \tfrac{1}{2}\theta, \quad d_1 = 4.11 e^2 a_{T-F} Z_1 Z_2 (1 + v)/32\pi.$$

The angle δ is the angular width of the family of scattering trajectories, corresponding to the given final energy $E_1 \pm \frac{1}{2}\Delta E_a$, where $\Delta E_a = \Delta \varepsilon_a E_0$. One could estimate the angle $\delta = \arcsin(\Delta \varepsilon_a/4\nu \sin \theta)$ for $He^+ \to Ni$, $\nu = m_1/m_2 = 0.067$, $\theta = 135°$, we get $\delta \approx 20°$. At the given δ and with use of parameters obtained earlier, $A = 1.76 \times 10^{15}$ s^{-1} and $v_0 = 10^7$ cm/s, the results obtained from eq. (86) are in good agreement with the experiment both on absolute value and on type of dependence $\eta_{max}^+(E)$ (see fig. 11).

This is a possibility to compare eq. (86) also with the experimental data by Verbeek et al. (1980), where dependence of η_{max}^+ on the escape angle β for $He^+ \to Au$, $\alpha = 90°$, $\beta = 90° - \psi$; $5° \leqslant \psi \leqslant 65°$ for different E_0 has been measured. The experiment revealed the anomalous behavior of the curves $\eta_{max}^+(E_0, \beta)$; at $\beta < 60°$, with increasing E_0, the ionization degree η_{max}^+ decreases. In fig. 13, the results of the calculation are shown for previously determined values $A = 2.84 \times 10^{15}$ s^{-1}, $v_0 = A/\alpha = 1.42 \times 10^7$ cm/s and $\delta = 6°$.

Both the theoretical curves and experimental ones at small angles β give the larger ionization degree for the smaller escape energy. This, at first glance, surprising result comes naturally from eq. (86), where $v_1 \sim E_0$ and with increasing of E_0 the preexponential multiplier decreases. This effect is connected with the increase of the most probable scattering depth with increasing initial energy E_0. It should be noted that formally, at $E_0 \to \infty$, $\Delta \eta^+ \to 0$ and $\eta_{max}^+ \to \eta_e^+$, i.e., with increasing energy E the surface peak should disappear. Just this has been observed in the experiments by Eckstein et al. (1978) and Ball et al. (1972).

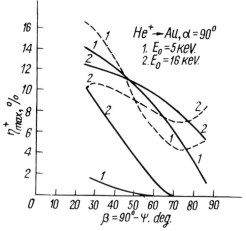

Fig. 13. The dependence of the surface peak height η_{max}^+ (E_0, β) on the escape angle $\beta = 90° - \psi$. Broken curves are the experimental data (Verbeek et al. 1980). The two top full curves are calculated using eq. (86). The two bottom full curves correspond to $\eta_{max}^+ = \exp(-v_0/v_\perp)$.

Since the calculation of the scattering has been made using a simplificated formula, one could expect only to get a qualitative agreement between eq. (86) and the data from Verbeek et al. (1980). As it is seen from fig. 13, the theoretical curves at $E_0 = 5$ and 16 keV differ to a lesser extent than the experimental ones. Nevertheless, the obtained results are considerably closer to the experiment than the commonly used function $\exp(-v_0/v_\perp)$, by which it is impossible to get a nonmomotonic dependence $\eta^+_{max}(E_0)$ at all. It is only due to the connection of charge state formation with scattering processes inside the solid, i.e., the effect of the existence of a bound state in a helium atom.

Formulae (81)–(86) have been obtained in the assumption that the violent collision is a fully ionizing one, i.e., $Q_2^+(E_i) = 1$. This assumption is valid only for scattering over large angles of light atoms having at $E_0 > 1$ keV sufficiently high velocities. At the calculation of the ionization degree by Nizhnaya et al. (1982a, b), another case has been considered, where all particles, scattered in a violent collision over angle θ became ions. A simple formula for η^+_{max} in this case was obtained only for surface scattering that took place at the bombardment of a single crystal along the open channel axis,

$$\eta^+_{max}(E_0) \approx \left\{\exp\left(\frac{-v_0}{v_{\perp,i}}\right) + L(E_0, \theta)\left[1 - \exp\left(-\frac{v_0}{v_{\perp,i}}\right)\right]\right\} \exp\left(\frac{-v_0}{v_\perp}\right). \quad (87)$$

Here, $L(E_0, \theta)$ is given by eq (49), $v_{\perp,i} = v \sin \alpha$ and $\exp(-v_0/v_{\perp,i})$ are the ion fraction in the beam, impinging on the surface atom. The case $Q_2^+(E_i) < 1$ is for scattering of slow heavy ions over comparatively small angles. Luitjens et al. (1980e) have measured $\eta^+_{max}(E_0)$ for $Ne^+ \rightarrow Ni$ (100), $\theta = 30°$, $\alpha = 15°$, $E_0 = 5$–10 keV. The experiment gives a more rapid increase of $\eta^+_{max}(E_0)$ than follows from formulae (85)–(86). In fig. 14, the result of a comparison of the calculation by eq. (87) with the experiment at $v_0 = 0.9 \times 10^7$ cm/s is presented.

The good agreement between the calculation and the data from Luitjens et al. (1980e) testifies the importance of the violent collision ionization at small E_0 and θ.

4.6. Ion spectra

The calculation of the double-differential coefficient of ion scattering,

$$R^+(\varepsilon, E_0, \theta, \psi) = \int_\Phi F(\varepsilon, \theta, E_0; q)Q^+(\varepsilon, E_0, \psi; q)\,dq , \quad (88)$$

proceeds the calculation of η^+, but the regularities of the function $R^+(\varepsilon, E_0, \theta)$ are of special interest if the aim is the calculation of the ion spectra. The matter is that at present in most experimental works only ion

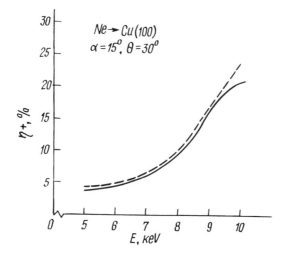

Fig. 14. The ionization degree of Ne η^+_{max} (E). The broken curve gives experimental data by Luitjens et al. (1980c). The full curve gives results from a calculation using eq. (87).

spectra, i.e., the coefficient $R^+(\varepsilon, E_0, \theta)$ have been measured, but the results are compared with theory where the scattering coefficient has been calculated, i.e., $R(\varepsilon, E_0, \theta)$. It is clear that such comparison is not always correct and in some cases it could lead to some erroneous conclusions. It is based on an a priori assumption of the independence of the charge state on the scattering process and the only case it might be used for $\bar{\beta}$ for alkali ions, for which $\eta^* \approx 1$. On the other hand, the calculated value of $R(\varepsilon, E_0, \theta)$ could be compared with the energy spectra of atoms, scattered as neutrals $R^0(\varepsilon, E_0, \theta)$ only for gas atoms, when $\eta^0 \approx 1$.

Let us consider as an exaxmple the ion spectrum calculated by eq. (88) for scattering of $He^+ \rightarrow Ni$, $\alpha = 90°$, $\theta = 135°$, $E_0 = 3\,keV$ and $E_0 = 15\,keV$ under the conditions of the experiment by Eckstein et al. (1978) (see fig. 15). From fig. 15 it is seen that R^+, calculated with the use of F from Parilis and Verleger (1985a,b) and Q^+ from Parilis and Verleger (1980) has a complicated form; besides the cupola-like maximum a narrow peak appears at $E \approx E_1$ and a gap between them, in agreement with the experiment.

Peaks of such type have been observed long ago in the experiments on ion spectra and have been explained only by predominant surface single scattering (Mashkova and Molchanov 1972a). The experimental data by Eckstein et al. (1978), and Verbeek et al. (1980) and the present calculation have shown that this is not true.

The main reason of the appearance of the surface peak in the ion spectra is the sharp increase of $Q^+(E)$ in the region $E \approx E_i$. It remains also valid in

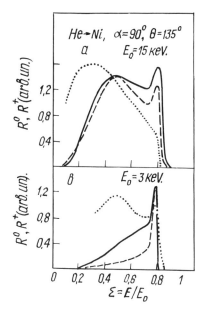

Fig. 15. The energy spectra of scattered particles. Experimental data (Eckstein et al. 1978): broken curves give $R^+(E)$, dotted curves give $R^0(E)$. Full lines give theoretical calculations by eq. (88). (a) $E_0 > E_{cr}$; (b) $E_0 < E_{cr}$.

the case where in $R(\varepsilon)$ there is a surface peak, which could be seen in neutral spectra (Eckstein et al. 1978). In the present work it has been shown that at $E_0 > E_{cr}$ in the region $E \approx E_1$ the value $R(\varepsilon)$ changes monotonically. Here the only reason for the appearance of the surface peak is the anomalous high ionization probability in the surface scattering, where the charge equilibrium state has no time to establish.

At $E_0 < E_{cr}$, in the region $E \approx E_1$, on the curve $R(\varepsilon)$ a small surface peak also appeared, caused by rainbow effect. In this case the main reason for the surface peak in the curves $R^+(\varepsilon)$ is the sharp increase of Q^+, which is more rapid than the increase of $R(\varepsilon)$ in this region.

4.7. Conclusion

Thus, the proposed simple model of charge formation of reflected particles in the rival processes of electron capture and loss has allowed to explain from a single point of view the numerous experimental data on the ionization degree of light atoms scattered on a solid surface. However, the present results, together with results of Kishinevsky et al. (1976), allow to deduce the peculiarities of $\eta^+(E, E_0, \psi)$ both for light and heavy particles.

Actually, the energy spectrum of the scattered gas atom could be divided into three regions:

(1) Region I of multiple scattering, $E_t < E < E_1$. Here the dependence $\eta^+(E)$ is usually close to linear, and the dependence on E_0 and ψ are weak, since at multiple scattering the equilibrium charge state is established.

(2) Region II, $E \approx E_1$. Here the dependence $\eta(E, \psi)$ is more complicated, sometimes it is close to the experimental one. The dependence η^+ on ψ and E_0 is strong and sharp increases in $\eta^+(E, E_0, \psi)$ in region II are explained by the nonequilibrium character of the charge state formation at single scattering and, correspondingly, by the strong dependence on the shape of the scattering trajectory.

(3) Region III, $E > E_1$, corresponds to plural scattering on the surface atoms. Here the dependence $\eta^+(E)$ is a nonmonotonic one (Buck et al. 1975, Chicherov 1972).

The universal function $\eta^+(E)$ for atoms, the ground energy level of which lies below the Fermi level, is schematically shown in fig. 16.

The difference between light and heavy gas atoms is in principle quantitative. So, for light particles (H, D^+, He), region I occupies the larger part of the energy spectrum, and region II and III are very narrow ones and could be resolved experimentally. For heavy atoms (Ne, Ar), these regions are large and well resolved. Often some contributions in the data compared have been measured in different spectrums.

Such different behavior of the function $\eta^+(E)$ in different parts of the energy spectrum testifies discretely in favor of the existence of a close

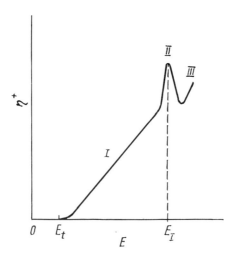

Fig. 16. The three regions in the dependence of the ionization degree of scattered atoms on energy E (scheme). E_t is the threshold energy, E_1 is the single-scattering energy.

connection between the charge state formation processes and the scattering regularities. The regions II and III, where the charge equilibrium state is not established, are the most sensitive to the scattering trajectory shape. Thus, here $\eta^+(E)$ for single crystals reveals some orientational effects and contains the information about the surface state structure (Hou et al. 1978, Nizhnaya et al. 1982a,b), which is valid for its diagnostics.

5. Orientational effects in the charge state of ions scattered by a single crystal

5.1. Introduction

In the present section, a theory of charge state formation at scattering on surface and near surface layers, based upon the concept of electron loss and capture by ions passing along some scattering trajectories approaching or moving away from the lattice atoms, has been developed.

The comparison of the calculations with experimentally observed energy distributions of scattered ions allows, in principle, the determination of surface structures which contribute in the scattering process. The method of determination the defect types and their distribution on real surface is based upon the high sensitivity of energy distribution to the atomic structure of the crystal surface.

It should be noted, that the application of this method is not quite correct, because in the experiment the ion current is recorded but the calculations give the total flux of scattered particles. These two types of spectrum are not similar, since the ionization degree changes nonmontonically along the energy scale. In addition, the theory by Kishinevsky et al. (1976) proves that different surface structures give very different values for η^+, the degree of ionization. Consequently, when any defect structure considerably contributes to the ion spectra this does not necessarily mean that there is much evidence of it on the surface. On the other hand, when any structure does not contribute to the ion spectra, one fails to make a conclusion about its absence on the surface.

This difficulty may be overcome by determining the probability of scattering per ion charge state for each type of structure. In this section, we present some estimations based on the method by Kishinevsky et al. (1976) of charge state calculation. These results were presented in papers by Nizhnaya et al. (1979b,c, 1982a,b). It was noted that the charge distribution for scattering on a real surface should be studied using the model of neutralization in binary encounters with metal atoms along the ion trajectory. In this model, the nonmonotonicity of $\eta^+(E)$ is connected with different locations of scattered particle trajectories relative to the solid surface. In this way different surface defect structures give different η^+.

This method is applied to scattering on surface semichannels. Specific features of this scattering, as well as the scattering of atomic rows, consist in the fact that the scattering is at glancing angle and the particle trajectory is formed in a number of correlated small-angle collisions.

For comparison with the experiments by Hou et al. (1978), the calculations of scattering and charge state have been made for Ne^+ ions and Ni single crystals taking into account their contamination with a small amount of N_2 molecules adsorbed on the surface.

5.2. Model of neutralization by single atoms

In velocity region $v < 10^8$ cm/s, where after a violent collision the electron loss may be neglected, the charge state of scattered particles is formed only by the electron capture where the rate $W(t)$ is the sum of the partial rates of independent capture processes from single atoms, $W(t) = \Sigma_k W[R_k(t)]$.

This summation is carried out over all atoms which contribute to neutralization. $W(R_k)$ may be considered analogous with the well-known atomic charge exchange or interatomic Auger transition rate (Kishinevsky and Parilis 1968b) when the electron vacancy in the scattered particle is deep enough and equals

$$W(R_k) = \frac{2\pi}{\hbar} \left| \int \gamma_M^i(1)\psi_M^i(2) \left| \frac{e^2}{r_{1,2}} \right| \psi_A^f(1)\psi_M^f(2) \, du_1 \, du_2 \right|^2 .$$

Here, $r_{1,2}(R_k)$ is the distance between the two electrons at the given interatomic separation. Subscripts i and f correspond to initial and final electron states and A and M correspond to scattered particle and metal surface atoms, respectively. The estimation of this matrix element made as by Kishinevsky and Parilis (1968b) gives the following. At distances less then the sum of atomic radii $R_a = R_A + R_M \approx 2$–$3$ Å, $W(R_k)$ changes slowly. Asymptotic behavior of $W(R_k)$ at $R \to \infty$ is determined by the greatly reducing function of the final bound state $\psi_A^f \approx \exp(-\alpha R_k)$, where $\alpha \approx \sqrt{I}$, and I is the energy of bound state. As a result, $W(R_k)$ is well approximated by the function

$$W(R_k) = B , \qquad\qquad \text{for } R_k < R_a ,$$
$$= B \exp[-\beta(R_k - R_a)] , \quad \text{for } R_k > R_a ,$$

where $\beta \approx 2\alpha$, and B has the meaning of the interatomic Auger transition rate. It was shown by Kishinevsky and Parilis (1968b) that the interatomic Auger transition rate is larger for small interatomic distances than in separate atoms due to the reducing of the system symmetry. This has also been found in the experiments by Afrosimov et al. (1976). Usually, $B = 1$–2×10^{15} s^{-1}. Let us consider the simple case of neutralization of an atom with a single vacancy, formed after a violent collision on the surface. The corresponding ionization degree equals

$$\eta^{+} = \eta_{0}^{+} \exp\left(-\int_{0}^{\infty} W(t)\, dt\right),$$ (89)

and then

$$\int_{0}^{\infty} W(t)\, dt = \int_{0}^{\infty} \sum_{k=1}^{N} W(R_{k})\, dt = \sum_{k=1}^{N} \frac{2}{v} \int_{R_{0k}}^{\infty} W(R)/\sqrt{R^{2} - R_{0,k}^{2}}\, R\, dR .$$ (90)

The integration in eq. (90) is carried out along the trajectory of the scattered ion, η_{0}^{+} is the ionization degree just after the violent collision, v is the ion velocity, and $R_{0,k}$ is the distance of closest approach with the kth atom. The starting point, $t = 0$, is the moment of the last violent collision. The labelling of a collision as a violent (ionizing) or weak (neutralizing) one is rather arbitrary. A collision with scattering angle θ_{1} is labelled as a violent one if the inelastic energy loss $\delta E(R_{0,k})$ exceeds the ionization potential I and $\eta_{0}^{+} \sim \Delta E$. For two neighboring violent collisions $\eta_{0}^{+} = \eta_{1}^{+} + (1 - \eta_{1}^{+})\eta_{2}^{+}$. Computer calculations show the existence of one or two violent collision events on each scattering trajectory. If the change of velocity after a violent collision could be neglected and the trajectories considered to be straight lines, then by substituting eq. (88) in eq. (90) we obtain

$$\int_{0}^{\infty} W(t)\, dt = \frac{2B}{v} \sum_{k=1}^{N} \left\{ \sqrt{R_{a}^{2} - R_{0,k}^{2}} + \int_{R_{0,k}}^{\infty} \frac{\exp[-\beta(R - R_{a})]}{\sqrt{R^{2} - R_{0,k}^{2}}}\, R\, dR \right\},$$

$$\text{for } R_{0,k} < R_{a},$$

$$= \frac{2B}{v} \sum_{k=1}^{N} \exp(-\beta R_{0,k})K_{1}(\beta R_{0,k}), \qquad \text{for } R_{0,k} > R_{a},$$ (91)

where K_{1} is a Mcdonald function.

5.3. Neutralization on surface atomic steps

In comparing the neutralization in a continuous medium (Kishinevsky et al. 1976) and eq. (90), the following conclusions may be drawn:

(1) the ionization degree for each trajectory is now determined not by $v_{\perp}$ (the normal component of the scattered particle velocity v), but by the set of $R_{0,k}$ and v;

(2) The values of η^{+} for scattering at a given angle from different structures may differ considerably due to different sets of $R_{0,k}$;

(3) For the same reason η^{+} may drastically differ for the trajectories of quasisingle and quasidouble scattering on the given structure.

We present the calculations on η^+ carried out within the framework of the program for computer calculations of scattering described by Nizhnaya et al. (1976). The potential was of the same type as in a paper by Vajda (1964). For example, we consider the following case of scattering: $Ar^+ \rightarrow Cu$, $E_0 = 6$ keV, (100), [110] under the same conditions as in the experiment by S.H.A. Begemann and Boers (1972). The starting point of neutralization was connected with the last ionizing collision, in which an inelastic energy loss, calculated using the Firsov theory (Firsov 1959), was greater than the argon ionization potential.

In fig. 17, the calculations of the energy spectra are displayed for scattering angle $\theta = 30°$ and incidence angle $\psi = 2$–$30°$. By the full curves, we denote the theoretical $E(\psi)$ function. Crosses present the experimental ion spectra (S.H.A. Begemann and Boers 1972), containing the quasisingle- (QS) and quasidouble- (QD) scattering peaks.

It is interesting that for the step-down there is only a QS peak, while the calculation, without taking into account the charge state, also shows the existence of the QD peak for the steps of both types. Lately, new experimental results by Luitjens et al. (1978) show the existence of a QD peak for the step down in neutrals which proves this calculation. The reason for this is that in neutralization calculations for the step up the ion peaks of QS and QD scattering are comparable, but for the step down the QD ion peak is less than the QS one by some orders of magnitude (see fig. 18). Such great

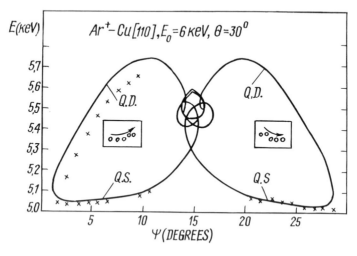

Fig. 17. The calculated (full curves) and experimental (crosses) (S.H.A. Begemann and Boers 1972) ion energy spectrum for Ar^+ scattered from the (100) face of a Cu single crystal in the plane [100] versus incidence angle φ for initial energy $E = 6$ keV and scattering angle $\theta = 30°$ in quasisingle (QS) and quasidouble (QD) collisions on step-up and step-down surface atomic structures.

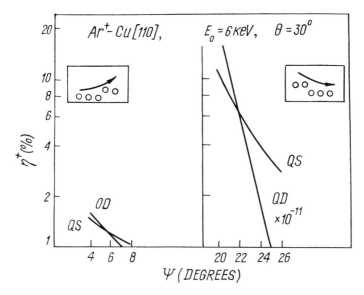

Fig. 18. The ionization degree η^+ for both QS and QD trajectories of scattering on a step up (left) and step down (right) calculated for $R_a = 2\,\text{Å}$, $\alpha = 2\,\text{Å}^{-1}$, $B = 1.5 \times 10^{15}\,\text{s}$.

difference is due to the fact that on the step down for the QD trajectory after the last violent collision the impact parameters are considerably less than in all other cases. This is different to the interpretation given by Luitjens et al. (1979a), where the absence of violent collision for this QD trajectory is presumed. In their simplified model the neutralization along the QD trajectory occurs both before and after the scattering event and the ion charge comes from the initial charge state from the incident ion.

5.4. Orientational effects

The important feature of charge formation is the orientational effects in the ionization degree at scattering on a single crystal.

The first attempt to obtain experimentally the orientational effects has been made by Van der Weg and Bierman (1969). They observed some oscillations of the scattered ion beam with the change of the azimuthal escaping angle φ. Unfortunately, it could not be simply interpreted, since the scattering occurs in the blocking directions.

The orientational effects were also observed in experiments by Yurasova et al. (1972) where, according to the paper by Kishinevsky et al. (1976), it was proved that the scattering along the close-packed direction results in the preferential neutralization.

Luitjens et al. (1979b) observed $\eta^+(\varphi)$ for scattering of $Ne^+ \to Cu$, (100),

$E_0 = 5$–10 keV, $\theta = 30°$, $\psi = 15°$. The results reveal a very good qualitative agreement with the prediction of the drastic difference in η^+ for the [100] and [110] directions made by Kishinevsky et al. (1976).

Recently, some interesting results were obtained by Matschke et al. (1977) for scattering of H_2^+ molecules on single-crystalline Au (110), $E_0 = 10$ keV, $\theta = 135°$, $\psi = 90°$.

They found that:

(i) The dependence $\eta^+(\varepsilon)$, when compared for poly- (p) and single (s) crystals, shows a difference both in absolute value and in shape. $\eta^+(\varepsilon)$ with increasing relative escape energy $\varepsilon = E/E_0$ changes linearly, $\eta_p^+(\varepsilon) = k_p \varepsilon$ and $\eta_s^+(\varepsilon) = k_s \varepsilon$, $k_s < k_p$ in the region $\varepsilon \ll 1$, but at $\varepsilon \to 1$ a sharp increase of η_s^+ was observed. Such peculiarity in energy spectra of the scattered particles was called a surface peak.

(ii) For scattering from a single crystal the dependence $\eta_s^+(\varphi)$ for the surface peak is nonmonotonic. The following parameters change with the variation of φ; the maximal $\eta^+ = A$ (the total height of the surface peak) and the height C_{max} measured from the equilibrium level (see fig. 19). Both values have a main minimum at $\varphi = 35°$ ([112]) and a main maximum at $\varphi = 55°$ ([111]) ($\varphi = 0$ corresponds to the direction [100]).

Let us try to interpret this result in our model. For the description of light-particle charge state formation for $E_s \gtrsim 10$ keV, which corresponds to $v \gtrsim 10^8$ cm/s, it is necessary to take into account the electron loss. We

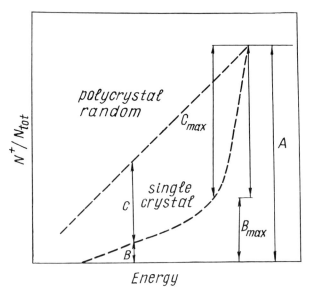

Fig. 19. Schematic diagram of η^+ versus escape energy from Matschke et al. (1977) for a definition of A, C_{max}, C and B.

introduce, after Kishinevsky et al. (1976), the following functions: W_c which is the electron capture rate and W_ℓ which is the electron loss rate in matter. Then η^+ will reduce to the form

$$\eta^+ = W_\ell/(W_\ell + W_c) + W_c/(W_c + W_\ell)\exp\left[-\int(W_c + W_\ell)\,dt\right]. \tag{92}$$

The first term on the right-hand side of eq. (92) describes the equilibrium charge distribution $\eta^+_{eq} = W_\ell/(W_\ell + W_c)$ and the second one,

$$\Delta\eta^+ = W_c/(W_c + W_\ell)\exp\left[-\int(W_c + W_\ell)\,dt\right],$$

describes the nonequilibrium portion, which corresponds to the value C_{max} by Matschke et al. (1977) (see fig. 19). Let us make some simplifications; since $\eta^+_{eq} = 0.003$, $W_c \gg W_\ell$. Then

$$\Delta\eta \approx \exp\left(-\int W_c\,dt\right) = \exp\left(-1/v\sum_k\int_{R_{0,k}}^{\infty}R_{0,k}W(R)R\,dR/\sqrt{R^2 - R^2_{0,k}}\right), \tag{93}$$

where $W(R)$ is electron capture rate from a single atom. As before, the summation is carried out over all atoms which contribute to the neutralization after a violent collision. For each trajectory we obtain $\Delta\eta^+_j$ and the final result is

$$\Delta\eta^+(\varepsilon) = \sum_j c_j\,\Delta\eta^+_j(\varepsilon), \tag{94}$$

where $\Delta\eta^+_j$ is the contribution of protons, scattered from the jth layer.

For a comparison of the calculations using eq. (93) with the experimental data of Matschke et al. (1977), it is necessary to establish the type of trajectories which contribute into the surface peak region $\Delta\varepsilon$. Since for protons the elastic energy losses may be neglected, we use the Lindhard–Sharff formula for the pass length of proton in metal; $\Delta l = \Delta\varepsilon\sqrt{E_0}/k\sqrt{\varepsilon}$.

Under the experimental conditions of Matschke et al. (1977) $\Delta l \approx 50$ Å, this corresponds to double passing through about five surface layers in a metal at single backscattering. It should be noted that this estimation was made for a polycrystal. In single crystals the energy losses are less, therefore, Δl is greater. This means that for each trajectory which passes through the deeper layers, $l \gg \Delta l$, the charge state reaches its equilibrium: $t = 1/v \gg (W_\ell + W_c)^{-1}$ in eq. (92).

At normal incidence on the (110) face only the first two layers are open. For them backscattering at $\theta = 135°$ is possible as a result of a single violent collision and the following collision by passing the neighboring atoms. We consider the first collisions as ionizing and the others as neutralizing. The trajectories of ions, scattered on the first two layers, may be considered to be straight lines and the distance of closest approach with each atom is easy to calculate (see fig. 20). The trajectories of ions, scattered from deeper layers, are inevitably broken lines. η^+ for such trajectories is rather close to the equilibrium value. These trajectories contribute to the left slope of the surface peak.

Figure 21 shows that the functions $\Delta\eta_1^+$ and $\Delta\eta_2^+$ for layers I and II change differently with φ. The oscillations similar to experimental ones were only found in the function

$$\Delta\eta^+ = \tfrac{1}{2}(\Delta\eta_1^+ + \Delta\eta_2^+),$$

with $W(R_k) = D/R_k^2$.

The amplitude of oscillations for $\delta\eta^+$ and C_{max} are comparable. The disagreement of oscillations in absolute values shows that the first two layers make a small but nevertheless most anisotropic contribution to the surface peak. On the other hand, the limited resolution of the set-up used gives no means of resolving the sharp and high peak corresponding to the two surface layers, but permits the measurement of its oscillations. The high absolute value of the ionization degree in the nonequilibrium surface peak agrees with the extrapolation of the data of experiment by Buck et al. (1975) for heavy ions up to the velocity $v \approx 10^8$ cm/s.

For a calculation of η_p^+/η_s^+ for the rest part of the energy spectrum (see fig. 19) connected with the trajectories of multiple scattering, where impact parameters may have different values, it is convenient to write $\eta^+(E)$ using the electron capture and loss cross sections σ_c and σ_ℓ,

$$\eta^+(E) = \frac{\sigma_\ell}{\sigma_\ell + \sigma_c} + \frac{\sigma_c}{\sigma_\ell + \sigma_c} \exp\left[\oint (\sigma_c + \sigma_\ell)N \, dl\right]. \tag{95}$$

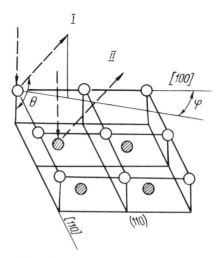

Fig. 20. The trajectories of single scattering from layers I and II.

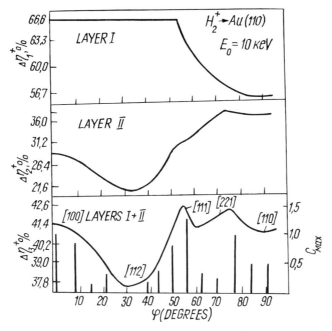

Fig. 21. The nonequilibrium portion of the ionization degree $\Delta\eta^+$ for layers I and II, calculated for $b_1 = 2.3\ \text{Å}, D = 0.3\ \text{cm}^2/\text{s}$ and their comparison with $C_{\max}$ from Matschke et al. (1977).

The formula, connecting the cross section σ with the corresponding transiton rate for a single atom, is

$$\sigma = 4\pi/v \int_{p_{\min}}^{\infty} p\,\mathrm{d}p \int_{p}^{\infty} W(R)R\,\mathrm{d}R/\sqrt{R^2 - p^2}\ . \tag{96}$$

$p_{\min}$ is a minimal impact parameter in a metal, depending on its structure (M.W. Thompson 1968). For the calculations of σ_c we take the above used $W(R)$,

$$\begin{aligned} W_c(R) &= D/R^2\ , & \text{for } R < b_1\ , \\ &= 0\ , & \text{for } R > b_1\ . \end{aligned} \tag{97}$$

Then

$$\sigma_c = 4\pi D b_1/(1 - \pi p_{\min}/2b_1)v\ , \tag{98}$$

if $p_{\min} < b_1$. Considering

$$\begin{aligned} W_\ell(R) &= B\ , & \text{for } R < b_2\ , \\ &= 0\ , & \text{for } R > b_2\ , \end{aligned}$$

$$\sigma_\ell = 4\pi B b_2^3/\{[1 - (p_{\min}/p_2)^2]^{3/2}\,3v\}\ . \tag{99}$$

In a polycrystal any impact parameter is regarded as possible ($p_{\min,p} = 0$).

In a single crystal, at chanelling $p_{min,s} \neq 0$ and then

$$\sigma_{c,p} = 4\pi D b_1 / v , \qquad \sigma_{c,s} = \sigma_{c,p}(1 - \pi p_{min}/2b_1) ,$$

$$\sigma_{\ell,p} = 4\pi B b_2^3 / 3v , \qquad \sigma_{\ell,s} = \sigma_{\ell,p}[1 - (p_{min}/b_2)^2]^{3/2} .$$

(100)

If we take into account that $\sigma_c \gg \sigma_\ell$, then for equilibrium

$$\eta_s^+ = \eta_p^+ [1 - (p_{min}/b_2)^2]^{3/2}/(1 - \pi p_{min}/2b_1) .$$

(101)

From eq. (101) it is easy to conclude that $\eta_s^+ < \eta_p^+$, when

$$p_{min} > \pi b_2^2 / 3 b_1 .$$

The ratio η_s^+/η_p^+ can be compared with the ratio $B/(B+C)$ from the experiment by Matschke et al. (1977) (see fig. 19). Since p_{min}, b_1 and b_2 only slightly depend on energy E, then this ratio is also energy independent in accordance with the experiment. The oscillations of C_{max} (see fig. 21) were fitted with $b_1 = 2.3$ Å. The minimal parameter p_{min} is known to be approximately equal to the Thomas–Fermi screening radius (M.W. Thompson 1968) $p_{min} = 0.47$ Å. From eq. (101), the experimental value for $B/(B+C) = 3.5$ gives for b_2 a reasonable value 0.6 Å.

5.5. Charge state of particles scattered by single crystal surfaces

The majority of the observed regularities in charge state formation at scattering by polycrystals have been explained by a model of competing electron loss and capture processes in the bulk and on the surface.

In case of a single crystal, it is convenient to introduce electron loss and capture rates $W_\ell(R)$ and $W_c(R)$ in interaction with a single atom. The differential equation describing the ionization degree after scattering of the ith atom η_i^+ is given by

$$\frac{d\eta_i^+}{dt} = -\eta_i^+ W_c(R) + [1 - \eta_i^+(R)]W_\ell(R) ,$$

(102)

and its solution is as follows:

$$\eta_i^+(R_{0,i}) = \exp\left[-\frac{2}{v_i} \int_{R_{0,i}}^{\infty} \frac{(W_c + W_\ell)R \, dR}{\sqrt{R^2 - R_{0,i}^2}}\right]$$

$$\times \left\{\eta_{i-1}^+ \frac{2}{v_i} \int_{R_{0,i}}^{\infty} W_\ell \exp\left[-\frac{2}{v_i} \int_{R_{0,i}}^{R} \frac{(W_c + W_\ell)R_1 \, dR_1}{\sqrt{R_1^2 - R_{0,i}^2}}\right] \frac{R \, dR}{\sqrt{R^2 - R_{0,i}^2}}\right\} .$$

(103)

Here, v_i is the velocity of a particle scattered on the ith atom, $R_{0,i}$ is the distance of closest approach to it, η_{i-1}^+ is the ionization degree after interaction with the $(i-1)$th atom along the trajectory.

The final ionization degree for a particle escaping the metal after scattering along the given trajectory is calculated by a consequent application of eq. (103). It is clear that η^+ for each trajectory depends not only on the final energy E and escaping angle φ as in a polycrystal, where η^+ is averaged over all impact parameters, but also on the given set of values $\{R_{0,i}\}$, which is different for different trajectories. Thus, η^+ will display some orientational effects. To simplify we divide all collisions along the trajectory of neutralizing and ionizing ones in the following way.

(i) The collision is considered to be a neutralizing one, if $E_1(R_{0,i}) \ll I$, $E_1(R_{0,i})$ is the inelastic energy loss, estimated by the Firsov formula (Firsov 1959), I is the ionization potential. For the neutralizing collisions $W_c \gg W_\ell$ is considered. Thus taking $W_\ell = 0$ in eq. (103), we obtain the ionization degree after a neutralizing collision in the following form,

$$\eta_i^+(R_{0,i}) = \eta_{i-1}^+ P_i(R_{0,i}),$$ (104)

where $P_i(R_{0,i})$,

$$P_i(R_{0,i}) = \exp\left[-\frac{2}{v_i} \int_{R_{0,i}}^{R} \frac{W_c R\,\mathrm{d}R}{\sqrt{R^2 - R_{0,i}^2}}\right] = \exp[-\varphi_c(R_{0,i})],$$ (105)

is the probability that the neutralization on the ith atom has not happened.

(ii) The collision is considered to be an ionizing one, if $E_1(R_{0,i}) > I$. For the ionizing collisions $W_\ell \gg W_c$ is considered. Then taking $W_c = 0$ in eq. (103) we obtain the ionization degree in the following way,

$$\eta_i^+(R_{0,i}) = Q_i + (1 - Q_i)\eta_{i-1}^+,$$ (106)

where $Q_i(R_{0,i})$,

$$Q_i(R_{0,i}) = 1 - \exp\left[-\frac{2}{v_i} \int_{R_{0,i}}^{R} \frac{W_\ell R\,\mathrm{d}R}{\sqrt{R^2 - R_{0,i}^2}}\right] = 1 - \exp[-\varphi_\ell(R_{0,i})],$$ (107)

is the probability of ionization in the ith collision.

When calculating Q_i, it should be taken into account that if the energy level of the atom lies below the Fermi level, the electron loss (ionization) would require the transfer from nuclear motion to electrons causing their transition above the Fermi level.

In the Firsov theory the ionization is considered to be semiclassical and connected with the inelastic energy transfer $E(p)$, depending on the collision parameter p. One can determine

$$E_1(p) = \frac{Z_1}{Z_1 + Z_2} E(p) \,,$$

the share of transferred energy per projectile atom.

It is necessary to establish the connection between $E(p)$ and $W_\ell(R)$. For the energy region of interest at scattering over small angles θ, the function $E(p) < 2I_1 < I_2$ and $Q(p)$ could be chosen as

$$Q(p) = 0 \qquad\qquad \text{at } p > p_1 \,,$$

$$\qquad = \frac{E_1(p) - I_1}{I_1} \qquad \text{at } p < p_1 \,, \qquad\qquad (108)$$

where p_1 is the solution of the equation

$$E_1(p) = I_1 \,. \qquad\qquad\qquad\qquad (109)$$

Using the modified Firsov formula,

$$E(p) = \frac{bv}{(1 + ap)^3} \,, \qquad\qquad\qquad\qquad (110)$$

we obtain from eq. (108)

$$p_1 = \left[\sqrt[3]{\frac{Z_1 C v_i}{(Z_1 + Z_2) I_1}} - 1 \right] \Big/ a \,, \qquad\qquad (111)$$

where

$$b = 0.3 Z_2 (\sqrt{Z_1} + \sqrt{Z_2})(Z_1^{1/6} + Z_2^{1/6}) \,,$$

$$\qquad\qquad\qquad\qquad\qquad\qquad\qquad\qquad (112)$$

$$a = \frac{0.757 \sqrt{Z_2}}{a_0 (Z_1^{1/6} + Z_2^{1/6})} \,.$$

Here Z_1 and Z_2 are the atomic numbers of the impinging particle and the target, respectively, I_1 and I_2 are the first and the second ionization potentials of the impinging particle, and a_0 is the Bohr radius. Then, substituting $Q(p)$ from eq. (108), we find $W_\ell(R)$ by solving the integral equation

$$\ln[1 - Q(p)] = -\frac{2}{v_i} \int_p^\infty \frac{W_\ell(R) R \, \mathrm{d}R}{\sqrt{R^2 - p^2}} \,, \qquad \tfrac{1}{2} v[1 - Q(p)] = F(p) \,. \quad (113)$$

$$\int_x^\infty \frac{F(p) p \, \mathrm{d}p}{\sqrt{p^2 - x^2}} = \tfrac{1}{2} \pi \int_x^\infty W_\ell(R) E \, \mathrm{d}R \,. \qquad\qquad (114)$$

Considering

$$\lim_{p \to \infty} [1 - Q(p)p] = 0 \,,$$

we get

$$W_\ell(R) = \frac{2v_i}{\pi} \int_R^{P_1} \frac{(dQ/dp)\,dp}{[1 - Q(p)]\sqrt{p^2 - R^2}} , \tag{115}$$

and

$$W_\ell(R) = \frac{6v_i a}{\pi} \int_R^{P_1} \frac{dp}{(1 + ap)\sqrt{p^2 - R^2}} . \tag{116}$$

In the region $ap < 1$,

$$W_\ell(R) = \frac{6v_i a}{\pi} \int_R^{P_1} \frac{(1 - ap)\,dp}{\sqrt{p^2 - R^2}}$$

$$= \frac{6v_i a}{\pi} \left\{ \ln \frac{p_1 + \sqrt{p^2 - R^2}}{R} - a\sqrt{p_1^2 - R^2} \right\} . \tag{117}$$

Let us compare the obtained electron loss rate $W_\ell(R)$ with electron capture rate $W_c(R)$. Figure 22 illustrates $W_\ell(R)$ and $W_c(R)$ for the typical case $B = 0.05 \times 10^{15}\,\mathrm{s}^{-1}$. It is seen that at $p > p_1$ only electron capture processes occur because $W_c \gg W_\ell$; and at $p < p_1$ the rate W_ℓ increases sharply and we could consider $W_c \ll W_\ell$. The shape of the functions W_c and W_ℓ actually enabled us to divide the collisions into ionizing and neutralizing

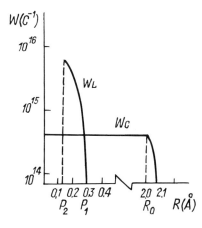

Fig. 22. Dependence of the electron capture $W_c(R)$ and loss $W_\ell(R)$ versus the distance R between the colliding atoms for scattering $Ne^+ \rightarrow Ni$, $E_0 = 3\,\mathrm{keV}$, p_1 and p_2 are the impact parameters, at which the inelastic energy transfer $E_1(p)$ equals the first and second ionization energy, respectively.

ones, considering that in the region $p < p_1$ only atom ionization occurs, and in the region $p_1 < p < R_a$ only neutralization occurs.

For the region $p_2 > p > p_1$, where p_2 is the impact parameter corresponding to the energy transfer equal to the second ionization potential, only single ionization occurs. The region $p < p_2$, where ionization of higher multiplicity is possible, has not been considered here. This region becomes considerable at scattering over large angles, when in the reflected beam multiple-charged ions are observed. To get orientational effects in the scattering of multiple-charged ions it is necessary to replace eq. (108) by the system of equations for functions of the multiple-ionization degree η^{n+}. In the equations the transitions from the multiplicity n^+ into the previous one $(n-1)^+$ and the following one $(n+1)^+$ are described by the corresponding electron capture and loss rates $W_e[n^+ \rightarrow (n-1)^+]$, $W_c[n^+ \rightarrow (n+1)^+]$ as was done by Kishinevsky et al. (1976).

Let on the given trajectory a particle interacts with N atoms, besides on the atoms with numbers $k, l \ldots s$ the ionization occurs. Then, the ionization degree at escaping from the solid is

$$\eta^+ = \eta_0^+ (1 - Q_k)(1 - Q_l) \cdots (1 - Q_s) \prod_{i=1}^{N}{}' P_i + Q_k (1 - Q_l) \cdots (1 - Q_s)$$

$$\times \prod_{i=k+1}^{N}{}' P_i + Q_l \cdots (1 - Q_s) \prod_{i=l+1}^{N}{}' P_i + Q_s \prod_{i=s+1}^{N}{}' P_i . \tag{118}$$

Here η_0^+ is the ionization degree of the incident beam, the prime means that from the product the terms corresponding to atoms with numbers $k, l, \ldots, s$ are excluded. The first term in eq. (118) describes the ion fraction conserved from the initial beam and each following term describes the ion fraction, formed in the $k, l, \ldots, s$ ionizing collision, respectively.

The following simple cases are possible:

(1) If $Q_k = Q_l = \cdots = Q_s = 1$, then in eq. (118) all terms but the last one are equal to zero and the ionization degree does not depend on the initial charge and on the charge state of the particle before the last violent collision, in which all neutral particles become ionized. The final charge state is determined by the neutralization on $(N - s)$ atoms. Using such an approach, the charge state has been calculated by Nizhnaya et al. (1979b).

(2) Some computer calculations have shown that, as a rule, in the energy range $E_0 = 10\text{--}50$ keV, on the trajectories of quasisingle and quasidouble scattering there is one or two violent collisions. Let the ionizing collisions occur on the kth and lth atoms, then eq. (118) will be given by

$$\eta^+ = \eta_0^+ (1 - Q_k)(1 - Q_l) \prod_{i=1}^{N}{}' P_i + Q_k (1 - Q_l) \prod_{i=k+1}^{N}{}' P_i + Q_l \prod_{i=l+1}^{N}{}' P_i .$$

$$\tag{119}$$

Equation (119) has been used by Nizhnaya et al. (1979b) for a calculation of the ionization degree after two violent collisions, where the kth and lth atoms are neighbors. In this case, the ionization degree after the second violent collision is

$$\eta^+ = \eta_k^+ + (1 - \eta_k^+)\eta_i^+ . \tag{120}$$

(3) In the low-energy region $E_0 = 5$ keV and for small scattering angles a violent collision may not occur anywhere on the trajectory. Then,

$$\eta^+ = \eta_0^+ \prod_{i=1}^{N} P_i , \tag{121}$$

and the ionization degree of the scattered particles is determined only by the initial degree of ionization. Nizhnaya et al. (1982a, b), using eq. (121), have carried out the calculation for glancing low-energy scattering of ions with ground energy level far below the metal conduction band. In this case $W_c(R)$ is the rate of the Auger neutralization process,

$$\begin{aligned} W_c &= B , &&\text{for } R < R_a , \\ &= B \exp[-\beta(R - R_a)] , &&\text{for } R > R_a , \end{aligned} \tag{122}$$

and

$$\begin{aligned} \varphi_c(R_{0,i}) &= \frac{2B}{v_i} \sqrt{R_a^2 - R_{0,i}^2} + \int_{R_a}^{\infty} \frac{\exp[-\beta(R - R_a)]R\,dR}{\sqrt{R_a^2 - R_{0,i}^2}} , &&\text{for } R_{0,i} < R_a , \\ &= \frac{2B}{v_i} \exp(\beta R_a)R_{0,i}K_1(\beta R_{0,i}) , &&\text{for } R_{0,i} > R_a . \end{aligned} \tag{123}$$

Here, v_i is the particle velocity after the ith collision, $R_{0,i}$ is the distance of closest approach in the ith collision on the given trajectory, K_1 is the Mcdonald function, R_a is approximately the sum of radii of the colliding atoms and B and β are the parameters describing Auger neutralization rates in ion–single atom encounter (Kishinevsky and Parilis 1968).

When eq. (105) is applied consequently along the trajectory using eq. (110), the sum $\Sigma_i \varphi_c(R_{0,i})$ arises in the exponent,

$$\eta^+ = \eta_0^+ \exp\left[-\sum_i \varphi_c(R_{0,i})\right] . \tag{124}$$

It is evident that $\Sigma_i \varphi_c(R_{0,i})$ and η^+ for each trajectory depend on values B and β, but some general conclusions can be drawn just from eqs. (110)–(124):

(a) For different trajectories of scattering at a given angle the degree of ionization η^+ has different values, since for given B and β it is determined by the set of the values $R_{0,i}$. Thus, η^+ must display an orientational effect.

(b) It is clear, that $\Sigma_i \varphi_{ci}$ is larger and η^+ is, respectively, less for those trajectories, which are connected with less values $R_{0,i}$. Therefore, the ionization degree at scattering from the surface rows is the less the closer the trajectory approaches the rows.

(c) Since $\Sigma_i \varphi_{ci}$ depends on the total number of atoms participating in neutralization, then η^+ for the given row becomes considerably less if the row is located on the semichannel bottom. In this case, the atoms of two rows of the first layer, forming the semichannel, additionally contribute to neutralization, and η^+ is less when the semichannel is narrower.

(d) Nevertheless, there is no direct connection between the ionization degree and the semichannel width, because wide semichannels are formed by dense rows providing a greater number of neutralizing collisions.

More detailed information on such effects may be obtained by direct calculation. This was done for the experimental data of Hou et al. (1978), where the Ne$^+$ scattering has been considered at the initial energy $E_0 = 3$ keV, for scattering angle $\theta = 15°$ at glancing incidence ($\varphi = 7°$) on the Ni single-crystal surface (110). The azimuthal angle φ varied within the limits $0° \leqslant \varphi \leqslant 90°$. This enabled us to consider different surface semichannels.

5.6. Ion scattering in semichannels

The mechanism of charge formation is closely connected with the shape of the scattering trajectory, which in its turn is determined by the type of scattering potential. In a paper by Nizhnaya et al. (1979a), it has been shown that for scattering of heavy ions on an atomic row the potential $U(R)$,

$$U(R) = \frac{A_1}{r^2} + \frac{A_2}{r} + A_3 , \tag{125}$$

for a certain set of parameters A_1, A_2 and A_3 fits the experiments well. The radius r_0, at which $U(r_0) = 0$, is the cut-off parameter of the potential and the value $r_0 = 1.7$ Å for Ar$^+ \rightarrow$ Cu is close to the value obtained by Yurasova et al. (1972). All calculations of scattering trajectories in semichannels here have been carried out by the method by Nizhnaya et al. (1979a), using the potential from eq. (125) with $r_0 = 1.7$ Å. The trajectories of scattering along the ridges of surface and bottom atomic rows, forming the semichannels corresponding to minima and maxima of the experimental curve (Hou et al. 1978) were considered. The contribution in scattering of all atoms with impact parameter $p < r_0$ has been taken into account.

The calculation has shown the following:

(1) At given geometry the scattering at angle $\theta = 15°$ (for all directions) is

a result of some deflection at small angles. Inelastic energy losses, calculated with the Firsov formula (Firsov 1959) in each interaction do not exceed the ionization potential of Ne.

(2) The shape of the trajectory depends on the type of the semichannel, on its width and on the distance between the atoms in the rows, forming the given semichannel.

In fig. 1 of chapter 5 two types of semichannels are displayed: (a) an open one and (b) one with a bottom row. A is the distance between the row atoms, d is the semichannel width.

For the wide semichannels [100] and [110] the trajectory of particles scattered from the bottom row passes close to it, below the crystal surface. In the narrow semichannels ($d < 0.3$ Å) the trajectories with the impact point in the center of the semichannel passes beyond the surface. The narrower the semichannel, the higher the trajectory. It is interesting that the trajectories in the narrow semichannels with a bottom row and without it differ slightly because in this case the incident ion interacts only with two surface rows in agreement with Evdokimov (1976). Actually, the trajectory does not lower the narrow semichannel and the interaction with the bottom row does not occur ($p > r_0$). The bottom row appeared to the blocked by the surface ones.

(3) The energy spectra of the scattered particles (ions + neutrals) at a given angle φ and $\theta = 15°$ has a double-peaked structure corresponding to quasisingle (QS) and quasidouble (QD) scattering. The peaks are close ($10 \text{ eV} < E_{QD} - E_{QS} < 50 \text{ eV}$) and are located in the region 2.94–2.96 keV, somewhat beyond the energy of single scattering on an isolated atom, this testifies for the scattering on the surface semichannels. In the experiment by Hou et al. (1978), the small value of $E_{QD} - E_{QS}$ does not allow to resolve the peaks of QS and QD scattering. When calculating the charge state we have taken into account the trajectories of such type.

5.7. Anisotropy of ionization degree at scattering in semichannels

The calculation of η^+ has been carried out using eqs. (119)–(124), taking into consideration the fact that all collisions along the glancing trajectory of surface scattering are neutralizing ones. The value η^+ for semichannels was obtained as $\eta^+ = \frac{1}{2}(\eta_I^+ + \eta_{II}^+)$, where η_I^+ is the value for the surface row and η_{II}^+ is the value for the bottom one. Both η_I^+ and η_{II}^+ were calculated using formula $\eta_i^+ = C_1 \eta_{QS}^+ + C_2 \eta_{QD}^+$, where C_1 and C_2 are the statistical weights of QS and QD scattering, η_{QS}^+ and η_{QD}^+ being the corresponding ionization degrees.

Equations (122)–(124) contain three parameters: B, R_a and β. The value η^+ for a polycrystal measured in the experiment by Hou et al. (1978) gives the opportunity to calculate some of these parameters.

For the ionization degree at scattering on a polycrystal, the following formula has been deduced by Kishinevsky et al. (1976),

$$\eta^+ = \frac{v_\perp}{v_\perp + Cv_1} \exp(-v_0/v_\perp) . \tag{126}$$

Kishinevsky et al. (1976) have considered that after a violent single scattering the particles become completely ionized. If on a trajectory there is no violent ionizing collision, an analogous calculation gives

$$\eta^+ = \frac{v_{\perp,i}v_{\perp,f}}{v_{\perp,i}v_{\perp,f} + Cv_1(v_{\perp,i} + v_{\perp,f})} \exp(-v_0/v_{\perp,i} - v_0/v_{\perp,f}) . \tag{127}$$

Here

$$v_{\perp,i} = v \sin \psi , \qquad v_{\perp,f} = v \sin \alpha , \tag{128}$$

$$C = \frac{E_f \sin \psi \sin \alpha}{E_f \sin \alpha + E_i \sin \psi} , \tag{129}$$

where E_i, E_f, v_i, v_f are the initial and final energies and velocities, respectively, v_1 and v_0 are the parameters describing the neutralization inside and outside metal, respectively,

$$v_1 = A_c \lambda_0 , \tag{130}$$

$$v_0 = \int_0^\infty W_c(S) \, dS . \tag{131}$$

Here A_c is the mean neutralization rate in the bulk and $W_c(S)$ is the same rate outside the metal depending on the distance from the surface S.

The parameters v_1 and v_0 can be calculated using B and R_0.

(1) Calculation of A_c is done from the equation

$$A_c = N\sigma_c v , \tag{132}$$

where σ_c is electron capture cross section,

$$\sigma_c = 2\pi \int_0^\infty q(p)p \, dp , \tag{133}$$

and $q(p)$ is the probability of Auger neutralization in a collision with a single atom (Kishinevsky and Parilis 1968a,b), p is the collision parameter,

$$q(p) = 1 - \eta^+ = 1 - \exp\left[-\frac{2}{v} \int_{R_0}^\infty \frac{W_c R \, dR}{\sqrt{R^2 - R_0^2}}\right] . \tag{134}$$

If the trajectory does not contain large-angle scattering, i.e., if $R_0 \approx p$, the exponent is small and

$$2q(p) \approx \frac{2}{v} \int_p^\infty \frac{W_c R \, dR}{\sqrt{R^2 - p^2}} \ . \tag{135}$$

Substituting eq. (135) in eq. (133) and changing the order of integration, we obtain the following formula

$$\sigma_c = \frac{4\pi}{v} \int_0^\infty W_c(R) R^2 \, dR \ . \tag{136}$$

From eqs. (136) and (132), it follows that

$$A_c = 4\pi N \int_0^\infty W_c(R) R^2 \, dR \ . \tag{137}$$

Formula (137) may be deduced in another way, considering A_c to be the sum of neutralization rates from the single atoms,

$$A_c = \sum_i W_c(R_i) = N \int W_c(R) \, du = N \int_0^{\pi/2} d\varphi \int_0^\pi \sin\theta \, d\theta \int_0^\infty W_c(R) R^2 \, dR$$

$$= 4\pi N \int_0^\infty W_c(R) R^2 \, dR \ . \tag{138}$$

If we chose for simplicity $\beta - 0$, then

$$\begin{aligned} W_c(R) &= B , \quad \text{for } R < R_a , \\ &= 0 , \quad \text{for } R > R_a . \end{aligned} \tag{139}$$

and

$$\sigma_c = \frac{4}{3} \frac{\pi B R_a^3}{v} ; \qquad A_c = \tfrac{4}{3}\pi N B R_a^3 . \tag{140}$$

(2) According to Kishinevsky et al. (1976), λ_0 is the effective range of a particle in the metal,

$$\lambda = (N\sigma_0)^{-1} , \tag{141}$$

where σ_0 is the elastic scattering cross section for the beam attenuation. A more detailed theory of beam attenuation has been presented by Parilis and Verleger (1980). From this theory

$$\sigma_0 = 2\pi \int_\Delta^\pi \frac{d\sigma}{\delta\Omega} \, d\Omega , \tag{142}$$

where Δ is the maximum angle of scattering without beam attenuation. Choosing

$$\frac{d\sigma}{d\Omega} = \frac{C}{E_1 \sin^3(\frac{1}{2}\theta)} , \qquad (143)$$

$$C = \frac{0.113a_0^2 Z_1 Z_2(1 + \nu)}{\pi(Z_1^{2/3} + Z_2^{2/3})^{1/2}} , \qquad (144)$$

we obtain

$$\sigma_0 = \frac{8\pi C}{E_i} \left(\frac{1}{\sin(\frac{1}{2}\Delta)} - 1 \right) \qquad (145)$$

with $\Delta = \chi\theta$; $\chi = \frac{1}{2}$ as by Parilis and Verleger (1980), ν is the mass ratio, Z_1 and Z_2 are the atomic numbers of the projectiles and the target atom, respectively.

(3) Calculation of $W_c(S)$ and v_0 is carried out as follows. Formulae (137) and (138) give A_c for the neutralization in bulk. Outside the metal, in the near-surface region, A_c will decrease with increasing S. We define W_c in this region by carrying out the integration in eq. (111) over the semiinfinite medium. Then

$$W_c(S) = 2\pi N \int_0^{\pi/2} \sin\theta \, d\theta \int_{S/\cos\theta}^{\infty} W_c(R)R^2 \, dR . \qquad (146)$$

In approximation (141),

$$W_c(S) = \frac{1}{3}\pi NB(R_a - S)^2(2R_a + S) . \qquad (147)$$

The value W_c from eq. (147) coincides with the value A_c from eq. (140) inside the metal, starting from $S \le -R_a$,

$$v_0 = \int_0^{\infty} W_c(S) \, dS = 0.25\pi NBR_a^4 . \qquad (148)$$

By using the experimental value for polycrystals $\eta^+ = 0.032$ (Hou et al. 1978), one could obtain from eq. (127) only a certain curve $B(R_a)$. The number of points lying on this curve determine the combination of B and R_a, fitting the experiment on the polycrystal. Comparison of the calculated dependence $\eta^+(\varphi)$ with the single crystal experiment enables one to find the only values B and R_a. The best agreement with experiment gives $B = 0.25 \times 10^{15} \text{ s}^{-1}$, $R_a = 2 \times 10^{-8}$ cm, obtained with $A_c = 0.754 \times 10^{15} \text{ s}^{-1}$, $\sigma_c = 4.93 \times 10^{-16} \text{ cm}^2$, $\lambda_0 = 5.2 \times 10^{-8}$ cm, $\sigma_0 = 3.32 \times 10^{-16} \text{ cm}^2$, $v_1 = A_c\lambda_0 = 3.93 \times 10^7$ cm/s, and $v_0 = 0.28 \times 10^7$ cm/s. $\eta_s^+ = \exp(-v_0/v_{\perp,i} - v_0/v_{\perp,f}) = 0.08$ is the ionization degree for surface neutralization.

Kishinevsky and Parilis (1968a,b) have shown that the Auger transition rate decreases with increasing Auger-electron energy. Kishinevsky et al. (1976) have obtained the value $B = 1.5 \times 10^{15}\,\text{s}^{-1}$ for argon. Since the ionization potential and, therefore, the Auger-electron energy for neon is greater, the value $B = 2.5 \times 10^{15}\,\text{s}^{-1}$ should be considered as a reasonable one.

In fig. 23, the experimental data and calculated values of η^+ for different semichannels are given. As could be seen from the figure the ionization degree is changing nonmonotonously with φ with the highest maximum at $\varphi = 55°$ corresponding to the widest open semichannel [112]. The next-highest maximum corresponds to an open semichannel [332]. The deepest minimum is for the semichannel [111] with a bottom row, in which the greatest number of metal atoms effectively participate in the neutralization process. In this channel the bottom row is not completely blocked by the surface rows and the trajectory of scattered particles is passing lower in the semichannel [114]. The last is three times narrower and its bottom row is completely blocked. Thus, $\eta^+_{[114]} > \eta^+_{[111]}$. The semichannels [110] and [100] are the widest among the ones with a bottom row and although the first one is wider, nevertheless, $\eta^+_{[110]} < \eta^+_{[111]}$. In such wide semichannels the incident ion moves closely to the bottom row and the ionization degree from the second layer $\eta^+_{II} = 0$. The scattering from the first layer is like the scattering from some isolated rows, but in the more dense packed direction [110]

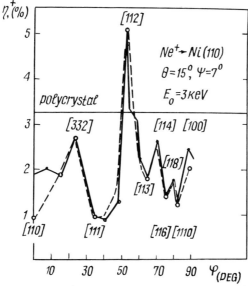

Fig. 23. The ionization degree η^+ versus the azimuthal angle φ for Ne$^+ \rightarrow$ Ni (110), $E_0 = 3$ keV. Full curve: experimental data by Hou et al. (1978), broken curve: calculated values.

($\varphi = 0°$) and [100] ($\varphi = 90°$). In these directions, the experimental value of η^+ exceeds the calculated one.

A possible explanation of this effect is that the time of flight between two consecutive collisions along these close-packed directions, 10^{-15} s, is comparable with the characteristic time of Auger relaxation of the atom excited in these collisions.

5.8. Charge state of ions scattered on absorbed molecules

In the ion energy spectra measured in the experiment by Hou et al. (1978), besides the peak some additional low-energy maxima have been observed. The authors connect them with light absorbed atoms on the surface. It is interesting that these maxima can be slightly distinguished in ion spectra and especially distinctly distinguished in the diagrams $\eta^+(E)$ for different angles φ. It should be noted that this additional maximum at $\varphi = 52°$ has a pronounced double-packed structure, typical for double scattering. It is natural to assume that the most widespread light impurity in the experiment by Hou et al. (1978) is nitrogen, penetrating from the stripping cell. Actually, calculation has shown that the position of the additional maximum fits well the energy of single scattering of Ne^+ on an atom N. The double-peaked structure for $\varphi = 52°$ corresponds to scattering of Ne^+ from a molecule N_2, oriented probably along the semichannels [112], $\varphi = 55°$, the nearest to the direction $\varphi = 52°$. Some calculations of η^+ for scattering of Ne^+ both on N and N_2 has shown that the molecules N_2 are apparently located just inside the semichannel, because if they were located above the surface rows, it would give for the low-energy peak, in contradiction with experiment, a greater ionization degree than in the main peak, and if they were implanted into one of the rows, the scattering from the molecule would be blocked by row atoms and could not be detected.

Some scattering calculations have been carried out for the molecule N_2 located in three different ways: above the semichannel axis [112] (positions 1, 2 and 3 in fig. 24). In each position, the height of the molecule over the surface, h, has been varied from 0 to $0.4a$ with a path $0.05a$ (a is the lattice constant).

(1) In position 1, the distance between the atoms in a molecule N_2 is 2.15 Å. The incident ion slightly glances the nitrogen atoms and the contribution in scattering comes mainly from the surface rows forming the main maximum in the energy spectrum. The double-peaked structure at $\varphi = 52°$ was obtained only for the height $h = 0.4a$, but the calculated η^+ is 5–7 times larger than the experimental value.

(2) In position 2, the distance between the atoms in the molecule is 4.3 Å. The double-peaked structure at $\varphi = 52°$ was obtained already for the height $h = 0.15a$ and η^+ coincides perfectly well with experimental values:

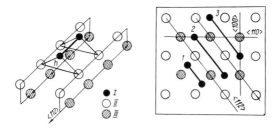

Fig. 24. Scheme of location of an adsorbed molecule N_2 on the (110) face of a single crystal Ni, (1), (2) and (3) the possible position of the nitrogen atoms among the atoms of Ni.

$\eta_{QS}^{+} = 0.014$ and $\eta_{QD}^{+} = 0.0136$. $E_{QS} = 2.82$ keV, $E_{QD} = 2.84$ keV. In position 2, the atoms N are located above the bottom [110] row and above the surface [100] ($\varphi = 0°$ and $\varphi = 90°$) which has shown that the nitrogen peak location in the energy spectrum fits the experiment and slightly depends on the type of row above which the impurity atom is located, but η^{+} appears to be extremely sensitive to this location. For instance, in the direction [110] the calculated value of η^{+} is very large (0.16) because the neutralization comes from the bottom row but not from neighboring surface rows (the channel [11] is the widest one). Hence, neither position 1 nor position 2 fits the experiment by Hou et al. (1978) for all φ.

(3) In position 3, the nitrogen atom is located above the surface [110] row and above the bottom [100] row. In this position, good agreement with experiment for $\varphi = 0°$, 52° and 90° was achieved In the direction [110] the neutralization comes from the surface row atoms, and in the direction [100] it comes from neighboring surface rows.

The best agreement with the experiment was achieved, when the molecule N_2 has been located along the axis of semichannel [112] at a height 1.5a from the surface position 3.

Therefore,

(1) The model of neutralization from single atoms satisfactorily describes the orientational effects in the charge state of scattered particles. This has been shown for large-angle scattering of Ar^{+} and H^{+} by Nizhnaya et al. (1979b), and here for Ne^{+} for the glancing angle scattering on surface semichannels.

(2) The function of ionization degree η^{+}, as has been mentioned by Kishinevsky et al. (1976), is sensitive to the surface structure and chemical composition. The calculation of η^{+} testifies for the recorded existence of light impurities on the surface in small quantities which are not detectable by Auger spectroscopy (Hou et al. 1978). A proper calculation allows us even to determine the orientation of the absorbed molecule N_2 relative to the surface rows and its height over the surface.

6. Charge state of alkali atoms scattered by solid surfaces

6.1. Ionization degree of alkali atoms

In the previous subsection, it has been shown that for scattered noble-gas ions the ionization degree η^+ depends on several parameters: the primary energy, the orientation of both the target and detector with respect to the bombardment direction, the shape and the character of the scattering trajectory. The ionization degree of noble-gas atoms scattered on a crystal surface with $E_0 \sim 1$ keV is small, being on the order of several percents.

In contrast with this complicated behavior, alkali atoms are scattered preferably in the ionized state as has been noted by U.A. Arifov et al. (1962a), and Terzic et al. (1979a,b).

The difference in behavior of η^+ for alkali ions and noble-gas ions could be apparently explained by the fact that the ionization energy of alkali atoms is much less than for noble-gas atoms, being of the same order or less than the work function of the most targets. This means that the charge formation of alkali ions occurs in the resonance charge transfer (Los and Geerlings 1990).

Alkali atoms scattered mainly as ions are widely used for primary beam. The use of alkali ions allows us to make use of the beam current in small doses that significantly decreases the surface damage. Low primary energy ions ($E_0 \sim 100$ eV) are used to decrease the contribution to scattering made by the deep crystal layers and to get information about the topmost layers. Usually, within the method of ion-scattering spectroscopy, the interpretation of both angular and energy distribution of the scattered alkali ions is based on the cross sections of scattering only, the ionization degree η^+ being 100%. However, it appeared to be valid only for K^+ ions, but for Na^+ and Li^+ the ionization degree was recorded to be much less (U.A. Arifov et al. 1976).

Algra et al. (1982a,b,c) have obtained the typical values of ion fraction: 63% for lithium, 76% for sodium, and 99% for potassium at primary energies of 5 keV. In addition, η^+ for Li^+ and Na^+ decreases with decreasing energy.

6.2. Model of neutralization

One can consider the charge state of the scattered alkali ions following the approach developed by Algra et al. (1982a,b,c), Rasser and Remy (1980), and Overbosch et al. (1980). Since the electron velocity in a metal is much larger than the velocity of the scattered ion, the charge transfer could be considered in a motionless atom approximation.

At a given distance of the ion from the surface, z, the equilibrium level occupation $n(z)$ is determined by the convolution of the probability that the valent electron has an energy E and the corresponding level in the metal is free,

$$n(z) = \int\limits_{-E_F-\varphi}^{0} f(E)\rho(E, z)\, dE ,$$ (149)

where E_F is the Fermi energy, φ is the metal work function, $f(E)$ is the density of states in the empty in the empty part of the conduction band in the metal, $\rho(E, z)$ is the density of states of the valence level of alkali ions, i.e., the probability that the electron being on the valence level has an energy from E to $E + dE$, where E is the electron energy in relation to the vacuum level (fig. 25). Near the metal surface the valence level of an alkali ion is broadened and rises up. According to the uncertainty relation, its energy width is given by

$$\Gamma = \hbar W ,$$ (150)

where $\hbar$ is Planck constant, and W is the electron transition rate at a distance z from metal,

$$W = W_0 \exp(-az) ,$$ (151)

where $1/a$ is a distance of the order of the Bohr radius, W_0 is the amplitude of resonance charge transfer rate, and W_0 amounts to $10^{18}\,\text{s}^{-1}$.

The level shift, ΔE, could be calculated according to the Coulomb law at the account of image forces,

$$\Delta E = 14.4/(4z + \delta) ,$$ (152)

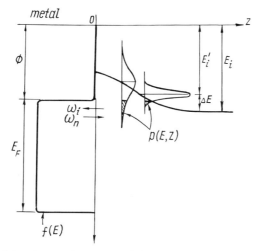

Fig. 25. Scheme of the electron distribution $f(E)$ in the metal and the density of states $\rho(E, z)$ of the valence level of an alkali atom in the vicinity of the surface. The shaded areas of $\rho(E, z)$ indicate the overlap with $f(E)$, and depends on the separation z.

where $\delta = 0.1-1$ Å. The parameter δ regulates the value ΔE at z close to zero. Then, the position of the valent level is $E'_i = E_i - \Delta E$. According to the Lorenz distribution, the density of the broadened level states is given by

$$\rho(E, z) = \frac{1}{\pi} \frac{\frac{1}{2}\Gamma}{(E - E'_i)^2 + (\frac{1}{2}\Gamma)^2} . \tag{153}$$

The distribution of conduction electrons in the metal is chosen both as the Fermi distribution (at zero temperature),

$$f(E) = 1 , \quad \text{at } E_F \leq E \leq \varphi ,$$
$$= 0 , \quad \text{at } E_F > E > \varphi , \tag{154}$$

and the Gibbs distribution for high target temperatures,

$$f(E) = \{1 + 0.5 \exp[(E + \varphi)/(kT)]\}^{-1} , \tag{155}$$

where k is Boltzmann constant, and T is the target temperature.

The abundance of neutrals in the scattered flux could be determined by integration of the equilibrium occupation of the valence level, $n(z)$, multiplied by a weight function, $F(z)$. The degree of neutralization at infinity, η^0,

$$\eta^0 = \int_0^\infty n(z)F(z) \, dz . \tag{156}$$

The analysis of $F(z)$ by Overbosch et al. (1980) has shown that $F(z)$ has a sharp shape and could be substituted by a δ-function,

$$F(z) = \delta(z - z^*) . \tag{157}$$

Then, from eq. (156) one can get

$$\eta^0 = n(z^*) . \tag{158}$$

Such a model was called the freezing-point model. In this model, the final charge state is considered to be determined in the point z^*. The distance z^* is determined by the condition that the ion state stops changing for a time expended by the ion during passing the distance where the effective exchange takes place. Then

$$W = a\upsilon_\perp , \tag{159}$$

where $\upsilon_\perp$ is the normal ion-velocity component.

From eqs. (151) and (159), one can get

$$z^*(\upsilon_\perp) = -1/a \ln(a\upsilon_\perp/W).$$

For $T = 0$, integral (149) is taken analytically,

$$\eta^0 = 1/\pi\{\arctan [(2/\Gamma)(-\varphi - E'_i)] - \arctan[(2/\Gamma)(-\varphi - E_f - E'_i)]\} .$$

When $T > 0$, the neutralization degree η^0 could be calculated numerically using formula (158).

6.3. Experimental

Algra et al. (1982a,b,c) have obtained the dependencies of the ionization degrees of K^+, Na^+ and Li^+ scattered from Cu (100) with the TOF spectrometer. The results obtained are well described by the freezing-point model. The calculations of the occupations probability for Li, Na and K valent levels as a function of distance z gave the results shown in fig. 26. The three curves represent three characteristic cases where the ionization potential is less than the work function of the solid: $E_i < \varphi$ (lower curve, potassium); $E_i > \varphi$ (upper curve, lithium); and E slightly larger than φ (middle curve, sodium).

At $E_i > \varphi$, for large z the functions $\rho(E, z)$ and $f(E)$ completely overlap, hence $n(z) = 1$. This means that the atom is in the neutral state. For decreasing z, the increase of ΔE and Γ results in a reduction of the overlap and $n(z)$ decreases.

At $E < \varphi$ (for potassium), for large z the function $\rho(E, z)$ and $f(E)$ do not overlap, hence $n(z) = 0$. This means that the atom is in the ionized state. For smaller z, the overlap increases due to the fact that the width Γ increases faster than the shift ΔE.

At $E \approx \varphi$ (for Na), the behavior of $n(z)$ appears to be intermediate between the first two cases.

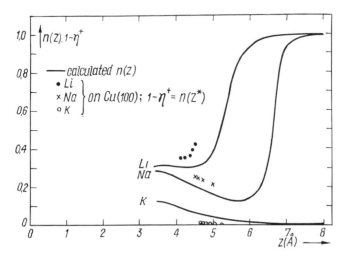

Fig. 26. The calculated and measured occupation probability $n(z)$ of the valence level of Li^+, Na^+ and K^+ ions as a function of the separation z between the atom and the substrate.

The experimental values of $\eta^0(z) = 1 - \eta^+$ are plotted as dots and coincide with the calculated ones.

The energy spectra of ions Na^+ scattered on Cu (100) surface with $E_0 = 5$ keV at the angle $\theta = 30°$ have been measured for the total flux of the scattered atoms and for neutrals only (fig. 27a). The ion fraction η^+ calculated from these energy spectra in the considered energy range have a

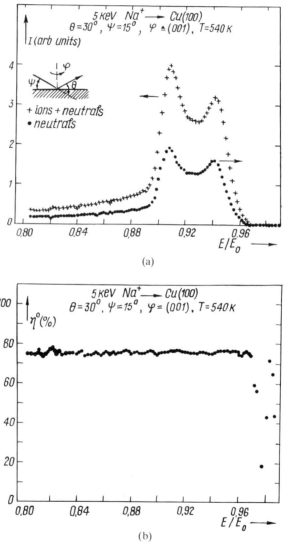

(a)

(b)

Fig. 27. (a) The energy spectra of all scattered particles ($+$) and of neutrals only ($\bullet$) obtained with 5 keV sodium ions. (b) The ion fraction η^+ of sodium ($E_0 = 5$ keV) as a function of the relative energy E/E_0 calculated from the energy spectra of (a).

constant value of $75.9 \pm 0.5\%$ (fig. 27b). In the low-energy part of the spectrum small fluctuations are observed due to statistical variations in the scattering signal. This is also the case for the high-energies at the right side of the surface peaks.

Algra et al. (1982b) have measured the mean values of η^+ in the described way for lithium, sodium and potassium. For all primary energies E_0, angles of incidence ψ, and azimuthal orientations φ used in the experiments, η^+ has been found to be constant within the energy range of the spectra $(0.7 < E/E_0 < 1)$.

The mean η^+ values appear to depend weakly on E_0 and ψ, but not on φ. The values of η^+ for sodium at $\psi = 15°$ and φ corresponding to [001] and [011] are the same for both crystal orientations and decrease from $78.2 \pm 0.4\%$ for 2 keV to $74.5 \pm 0.4\%$ for 10 keV.

The value of η^+ for potassium is very high, about 99%, independent of ψ and decreases slightly with E_0 from $99.7 \pm 0.1\%$ for 1.5 keV to $98.4 \pm 0.3\%$ for 10 keV. Finally, η^+ for lithium is the same for all three azimuthal orientations and increases from $75.3 \pm 2\%$ for 3 keV to $64.6 \pm 0.8\%$ for 10 keV.

Aliev et al. (1991), when studying the ionization degree of alkali ions, have used the method proposed previously by U.A. Arifov et al. (1976). This method comprises two targets M_1 and M_2 placed at an angle to each other, the first target being bombarded by alkali metal ions, the ions and neutrals reflected from it reach the second one, and being scattered on it reach a collector.

By applying a voltage delaying the positive ions merely neutrals reaches the target M_2. The scattering current from the second target induced no other than by neutrals (i_0) and, separately, the current induced by the total flux of $i_0 + i_+$ reflected by the first target at angle $\theta = 90°$ was measured.

Algra et al. (1982a,b,c) have shown that the energy spectra of the scattered alkali ions and neutrals at given angle θ were similar. This means that the scattering from the second target under ion and neutral bombardment actually would depend only on their intensities. The ratio of the currents is the ionization degree of the particles scattered from the first target at a given angle of scattering, θ. Then the ionization degree $\eta^+ = 1 - \eta^0$ could be obtained by measurement of the currents and on the collector.

The measurement of η^+ was made for the scattering of Na^+ on an Mo polycrystalline first target. The ion detection was performed on the second target representing a tungsten strip.

6.4. Discussion

The results of the experiment have been discussed within the framework of the model of neutralization described above. The calculations of η^+ made using this model showed that it appeared to be extremely sensitive to the work function φ, i.e., to the location of the valent level relative to the upper boundary of the conduction band. From fig. 28, it is seen that with the change of φ not only the value of the ionization degree changes but also the behavior of the curve $\eta^+(E)$. So, at $\varphi \leqslant 4.5$ eV, the ionization of Na$^+$ scattered on Mo increases with the increase of E_0. In this case, the valence level of the ion lies lower than the upper boundary of the conduction band and the faster the particle is moving away from the metal surface the less time it exists in the effective neutralization region. Within this model, with increasing $v_\perp$, z^* decreases. At small z, the level is more broadened and lifted and most of it lies opposite the unfilled region, i.e., the neutralization is less and η^+ is larger. At an increase of the work function, the valent level lies above the upper boundary of the conduction band, moreover, it lies higher the less z is (the lift of the level is due to the image forces), therefore, an additional energy is needed for metal electrons to be transferred on it. Thus, with the increase of the energy, the neutralization increases but η^+ decreases, which has been observed at $\varphi = 4.6$ eV, when for Na$^+$ the function $\eta^+(E_0)$ is similar to that for K$^+$. Since the experiment was carried out without additional control of the surface work function, the comparison of the experimental data on scattering of Na$^+$ on Mo with the calculations performed for different values of φ has been made. A good agreement was achieved for $\varphi = 4.5$ eV. This value diverges with the tabulated data for the polycrystal ($\varphi = 4.33$ eV) and such difference could be caused by peculiar experimental conditions (e.g., oxygen or carbon contamination).

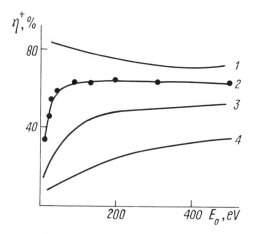

Fig. 28. The ion fraction η^+ of sodium scattered by a polycrystal Mo surface over the angle $\theta = 90°$ versus the primary energy E_0. The points are experimental data by Aliev et al. (1991). Full curves are the calculation results for different values of the work function: (1) $\varphi = 4.6$ eV, (2) $\varphi = 4.5$ eV, (3) $\varphi = 4.4$ eV, and (4) $\varphi = 4.2$ eV.

To check how precisely the scattering at the given angle θ occurred, i.e., just how strong the value of η^+ changes with the change of θ within the detector slit, $\Delta\theta = \pm 10°$, the proper calculation was made which showed that at $\varphi = 4.5$ eV $H^+(E_0)$ within this angular interval slightly depends on θ for K^+, Na^+ and Li^+ (see fig. 29). The calculation of $\eta^+(E)$ made for scattering at the given angle describes the experiment rather well. However, at $\varphi = 4.33$ eV, the divergence is rather strong, particularly for Na^+ and Li^+, hence to compare η^+ with experiment it is necessary to calculate it using the following formula,

$$\eta^+(E_0) = \int_{\theta-\Delta\theta}^{\theta+\Delta\theta} \sigma(\theta)\eta^+(\theta)\,d\theta\ .$$

The temperature dependence $\eta^+(E_0)$ is shown in fig. 30. For Na^+ it was measured for energies $E_0 = 20$–200 eV in the temperature interval 500–2000 K. At $E_0 = 20$ eV, it was practically independent on the target temperature within this interval, but at $E_0 = 200$ eV it slightly decreases (3–4%) with the increase of temperature from 500 up to 2000 K.

The calculation has shown that $\eta^+(E_0)$ increases with the temperature increase at $E_0 < 20$ eV and decreases at $E_0 > 20$ eV. For $E_0 = 20$ eV, it appeared to be almost unchangeable, which agrees with the experimental data. For $E_0 \approx 200$ eV, η^+ decreases with temperature increase more intensively than in the experiment. The temperature dependence of $\eta^+(E)$ for

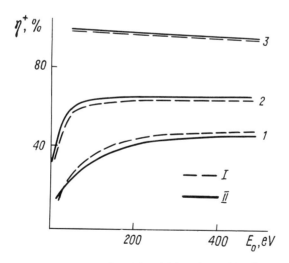

Fig. 29. The calculated ion fraction η^+ of (1) Li^+,(2) Na^+ and (3) K^+ ions scattered on an Mo polycrystal surface over an angle $\theta = 80°$ (broken curves) and $\theta = 100°$ (full curves) versus the primary energy E_0.

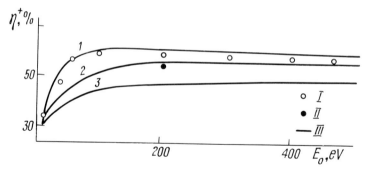

Fig. 30. Dependencies $\eta^+(E)$ for Na^+ ions scattered by an Mo polycrystal surface. Open and full points are experimental data at $T = 500$ and $2000\,K$, respectively, full curves are the calculation results at $\varphi = 4.5\,eV$, (1) $T = 0\,K$, (2) $500\,K$ and (3) $2000\,K$.

Na^+ depends strongly on the work function of the target. So, for $\varphi = 4.33\,eV$, it appeared to be weaker than for $\varphi = 4.5\,eV$.

Thus, the behavior of the experimental temperature dependence agrees better with the calculation for $\varphi = 4.33\,eV$, and the absolute value η^+ one could get only for $\varphi = 4.5\,eV$. Further measurements of $\eta^+(T)$ for different values of E and $\eta^+(E_0)$ for different scattering angles θ would provide significant information about the alkali-ion neutralization processes.

The strong sensitivity of $\eta^+(E)$ to the value of φ could be used for determining the work function of the target at ion bombardment.

CHAPTER 9

Scattering of Swift Molecules by Solid Surfaces

1. Introduction

While the interaction of gas molecules with solid surfaces has been investigated rather well (Toennies 1974, Barker and Auerbach 1984), systematic studies of the phenomena connected with scattering of swift molecular ions by solid surfaces have been carried out only recently. This phenomenon appears to be interesting because:

(i) it contains new effects compared to the effects observed under atomic-ion bombardment;

(ii) it provides the possibility to study the peculiarities of atomic-ion scattering using molecular-ion bombardment;

(iii) the scattered molecules carry the information on solid surface structure; and

(iv) some applications are possible in the field of production of neutral or negatively charged atomic or molecular beams.

In recent experiments, the scattering of swift molecular ions on solid surfaces without dissociation has been discovered. The peculiarity of this puzzling event is that the molecules were scattered at such large angles that the elastic energy losses exceed the dissociation energy by more than one order of magnitude.

While Panin (1961) predicted the possibility of nondissociative scattering, and Molchanov and Soszka (1965) recorded this effect for nitrogen molecular ions, only Eckstein et al. (1975) have measured experimentally the dependence of the yield of the scattered molecules D_2^+ and D_3^+ on the scattering angle which allowed a simple estimation of this effect to be provided (Bitensky and Parilis 1978a).

Heiland et al. (1979) have investigated the dependence of the nondissociative scattering of H_2^+ by nickel and tungsten single crystals on energy and scattering angle. These experiments allowed a quantitative theory of nondissociative scattering of swift molecules to be worked out (Bitensky and Parilis 1980, 1981a, 1984). The experimental results on scattering of heteromolecular diatomic 3 keV CO^+ molecules from a clean polycrystal surface were presented by Jo et al. (1985). In a paper by Heiland et al. (1979), the orientational effects in the molecule scattering yield were recorded for the first time. This was investigated in a more detailed way by Balashova et al. (1982a) for N_2^+ on a Cu surface. In the last paper, the so far highest energy (30 keV) of the scattered molecules was achieved. If the molecules do dissociate, the energy spectrum of the scattered ions would be broader than for the bombardment by atomic ions with the same velocity. In the scattered beam, some ions with energy exceeding the initial energy per atom have been observed. Panin (1962) noted this effect and it has been investigated in more detail by Feijen (1975), Heiland et al. (1979) and Evstigneev et al. (1982). The spectrum broadening has been explained in a paper by Feijen (1975) as Coulomb repulsion of the atoms ionized during scattering or as the transition of the molecule into an excited state (Heiland et al. 1979). However, in fact this can be explained in a rather simple way as mutual shuttle-type collisions of the atoms during scattering (Bitensky and Parilis 1980, 1981a, Bitensky et al. 1985, 1987, 1988, 1992a).

It should be noted that the dissociationless scattering of hydrogen molecules at glancing incidence has been observed even at very high energies: $E = 2.1$ MeV (Mannami et al. 1986) and $E = 500-700$ keV (Kimura and Mannami 1984).

Dodonov et al. (1988) have reviewed the large part of the experimental investigations of molecules scattering on a solid surface in which apparently molecule electronic excitation did not essentially contribute to the dissociation.

In section 2, the mechanism of scattering of swift molecules without dissociation based on the consequent correlated collisions of molecule atoms with surface atoms is considered. Within the framework of this mechanism the formula is written for the relative yield of molecules scattered without dissociation as a result of single and double collisions with the surface atoms.

It is shown that under certain conditions the orientational effects in the yield of molecules scattered by single-crystal surfaces should be stronger than for atomic-ion scattering. The molecular-ion scattering gives the possibility of evaluating the degree of ionization of the scattered atoms. The motion of the recoil atoms is taken into consideration. The scattered molecule yield is calculated as a result of double scattering over a wide range of scattering angles, at glancing scattering angles in the string approximation and in the atomic chain approximation with consideration of the thermal vibrations on the basis of computer simulation. The influence of the focusing effect on molecule scattering in surface semichannels is shown. It is shown that the phenomenon is an essentially surface effect and the bulk does not contribute to it.

The molecules scattered by solid surfaces without dissociation should be excited vibrationally and rotationally. The distribution of the scattered molecules over vibrational and rotational states depending on the orientation of the scattering plane relatively to the single crystal directions is calculated in section 3.

In section 4 the evolution of the energy distribution of the scattered atoms following impact dissociation of the fast molecules on a solid surface for nonplane scattering with arbitrary atom masses in the molecule and in the solid is described. All these calculations are based on consideration of elastic energy losses only.

However, there are experimental data obtained by Heiland et al. (1987) in which it has been shown that at grazing incidence and for not too high energies the molecules dissociate via electronic transitions.

In section 5, the models of molecule scattering including the electronic processes and some experiments which give opportunity to define the influence of electronic excitation on the dissociation mechanism are analysed.

2. Scattering of swift molecules within the elastic collision model

The relative probability of molecule scattering without dissociation is determined by the solid angle in which the molecular axes are orientated when the change of momentum of each atom is considerable, but the energy of the relative motion does not exceed the dissociation energy.

This model is valid when the kinetic energy normal to the surface sufficiently exceeds the molecular dissociation energy. The scheme of molecule scattering as a result of a single collision with the surface atom is shown in fig. 1.

2.1. Comparison between molecular and atomic scattering

The ratio R_{21} of the number of undissociated molecules to the number of atoms scattered at the same angle after dissociation is

$$R_{21} = \Delta\Omega_1/2\pi , \tag{1}$$

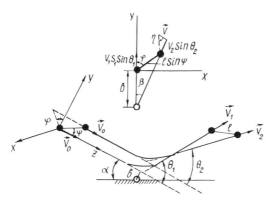

Fig. 1. The scheme of scattering of a molecule.

where $\Delta\Omega_1$ is the small solid angle, in which the molecule axes have to be orientated at dissociationless scattering.

The corresponding impact area for the second atom of the molecule $\Delta S_1 = l^2\,\Delta\Omega_1$ (l is the interatomic distance in the molecule).

As a result of independent scattering of the atoms the velocity of the second atom would be directed into a certain solid angle $\Delta\Omega_2$ around the velocity of the first atom and

$$\Delta S_1 \approx R_1(\theta,\varphi)\,\Delta\Omega_2\,, \tag{2}$$

where $R_1(\theta,\varphi)$ is the Jacobian of the transformation from impact parameter coordinates $\{b_x, b_y\}$, to the polar θ and azimuthal angle φ of the scattering. This defines the cross section for scattering of the atom at angle θ. The solid angle $\Delta\Omega_2$ is defined from the condition of scattering without dissociation,

$$\tfrac{1}{4}m_1(v_1 - v_2)^2 \leqslant \varepsilon\,, \tag{3}$$

where m_1 is mass of the molecule atom, $v_{1,2}$ are the velocities of the two atoms after scattering, and ε is the dissociation energy. Since the scattering angles are small, one can consider that the relative velocity $v = v_1 - v_2$ is normal to the molecular center of mass velocity V_0 and then $\Delta\Omega_2 = 2\pi\varepsilon/E_0$ (E_0 is the initial energy per atom). The largest possible angle between the velocities of the atoms of the molecule scattered without dissociation equals

$$\theta_0 = \left(\frac{\Delta\Omega_2}{\pi}\right)^{1/2} = \left(\frac{2\varepsilon}{E_0}\right)^{1/2}. \tag{4}$$

Taking into account eqs. (2) and (3), we get from eq. (1)

$$R_{21}(\theta,\varphi) = \frac{\varepsilon}{l^2 E_0}\,R_1(\theta,\varphi)\,. \tag{5}$$

Thus, the relative number of molecules scattered without dissociation is proportional to the atomic scattering intensity.

The dissociation of the molecules is supposed to influence weakly the scattered particle characteristics. Thus, for the molecular scattering intensity R_2 we get

$$R_2 = R_{21}R_1 = \frac{\varepsilon}{l^2 E_0}\,R_1^2\,. \tag{6}$$

The relative number of scattered molecular ions $I_{21} = I_2/I_1$ with $I_1 = \eta_1^+ R_1$, $I_2 = \eta_2^+ R_2$ and $I_{21} = (\eta_2^+/\eta_1^+)R_{21}$, where η_2^+ and η_1^+ are the degrees of ionization of the molecules and atoms, respectively. Supposing $\eta_1^+ \approx \eta_2^+$ we get $I_{21} = R_{21}$, and thus the comparison between the experimental value I_{21} and calculated R_{21} becomes reasonable. Taking into account eq. (5) we should find that $I_{21} \sim R_1$ and so we get an opportunity for both theoretical and experimental investigation of these quite different coefficients.

Balashova et al. (1982b) measured both the intensities I_2 of the scattered molecular N_2^+ and I_1 of atomic N^+ ions versus the scattering angle under molecular ion bombardment of a Cu polycrystal surface with 30 keV N_2^+ ions for different glancing angles α. In fig. 2, the ratio $I_{21} = I_2/I_1$, taken from data by Balashova et al. (1982b), is shown. One can draw the following conclusions:

(i) the intensity of scattering R_1 depends weakly on α;

(ii) for large scattering angles $R_1 \simeq \theta^{-3}$;

(iii) the decrease of scattered ion intensity at small escape angles $\zeta = \theta - \alpha$ is mainly connected with the decrease of the ionization degree.

Taking into account eq. (6) one could get the following expression for the degree of ionization,

$$\eta_2^+ = \frac{\varepsilon I_1^2}{l^2 E_0 I_2} .\tag{7}$$

From eq. (7) follows the interesting possibility of evaluating the degree of ionization η_1^+ by measuring the intensity of the scattered molecular ions I_2 and atomic ions I_1 without recording the neutrals scattered from the surface.

The dependence of the degree of ionization η_1^+ on the escape angle

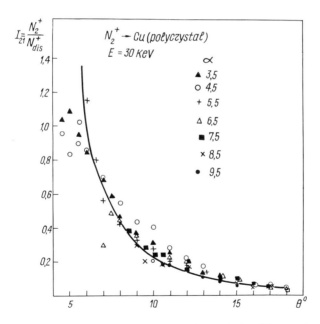

Fig. 2. Relative yield of molecular ions versus scattering angle for different glancing angles α. The full curve presents the law θ^{-3}, the data are taken from the experiment by Balashova et al. (1982b).

defined by eq. (7) is shown in fig. 3. It is seen that at small ζ the degree of ionization does not depend on the glancing angle. It means that at glancing escape ($\zeta \leqslant 5°$), the charge equilibrium of the scattered particles is established. It should be noted that the value of η_1^+ calculated by eq. (7) characterizes the degree of ionization of the scattered atoms and also contains information about the dissociation. The dependence of $\eta_1^+ \simeq I_1^2/I_2$ and $\eta_{dis}^+ \simeq I_{dis}^2/I_2$ on the glancing angle α is illustrated in fig. 4. Here, I_1 is

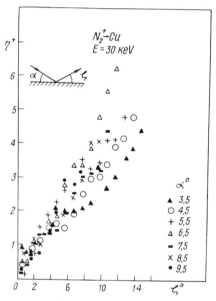

Fig. 3. The degree of ionization η^+, calculated from experimental data by Balashova et al. (1982b) using formula (7) versus the escape angle ζ for different glancing angles α.

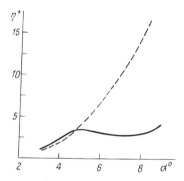

Fig. 4. The degree of ionization η^+ versus glancing angle α calculated from the experimental data by Balashova et al. (1983) using formula (7). The full curve represents I_1^2/I_2; the broken curve I_{dis}^2/I_2.

the intensity of N^+ ions scattered under N^+ bombardment and I_{dis} is the intensity of N^+ ions scattered under N_2^+ bombardment on the (100) face for a Cu single crystal in the direction $\langle 100 \rangle$ (Balashova et al. 1983). It is seen that η_{dis}^+ and η_1^+ differ, which indicates an interaction of atoms with the molecule during scattering and the influence of atomic interaction on electron capture by the molecule on its dissociation.

By neglecting the elastic energy losses and using a method described elsewhere (Bitensky and Parilis 1980, 1981a), a formula for the intensity of nondissociative scattering was deduced,

$$R_2(\theta_x, \theta_y) = \frac{1}{2\pi l^2} \int_0^\varepsilon dE_r \int R_1(\theta_1) R_1(\theta_2) \delta(\theta - \tfrac{1}{2}(\theta_1 + \theta_2))$$

$$\times \delta[E_r - \tfrac{1}{2} E_0 (\theta_1 - \theta_2)^2] \, d\theta_1 \, d\theta_2 \,. \qquad (8)$$

Here, $\theta_{x,y}$ is the component of the scattering angle $\theta = v/V_0$ in the plane normal to the incident beam, $\delta(x)$ is a delta function. Performing the integration with the δ function and expanding the integrand in a power series of $(E_r/2E_0)^{1/2}$ up to terms of second order, we get

$$R_2(\theta_x, \theta_y) = \frac{\varepsilon}{l^2 E_0} \left\{ R_1^2 + \left[R_1 \left(\frac{\partial R_1^2}{\partial \theta_x^2} + \frac{\partial R_1^2}{\partial \theta_y^2} \right) - \left(\frac{\partial R_1}{\partial \theta_x} \right)^2 - \left(\frac{\partial R_1}{\partial \theta_x} \right)^2 \right] \frac{\varepsilon}{8E_0} \right\}.$$
$$(9)$$

For the angles $\theta_{x,y}$ corresponding to the maximum R_1, the derivatives $\partial R_1 / \partial \theta_x = \partial R_1 / \partial \theta_y = 0$. Expressing the second derivatives by the peak width $g_{x,y}$ as $\partial^2 R_1 / \partial \theta_{x,y}^2 \approx 4 R_1 / g_{x,y}^2$, we get for the maximum of R_2 the expression, preceding eq. (6),

$$R_2 = \frac{\varepsilon}{l^2 E_0} [1 - \tfrac{1}{4} (\theta_0 / g_1)^2] R_1^2 \,, \qquad (10)$$

where $g_1^{-2} = g_x^{-2} + g_y^{-2}$, and the angle θ_0 is defined by eq. (4).

From eq. (10), it follows that at $\theta_0 \ll g_1$ the intensity R_2 is proportional to R_1^2 [as in eq. (6)] and, therefore, the orientational effects in the yield of scattered molecules should be more pronounced than for atomic scattering. These effects were observed by Balashova et al. (1983) and are displayed in fig. 5.

If θ_0 is comparable with g_1, then the dependence R_2 on R_1 would appear to be more complicated, since g_1 is known to have an opposite orientational effect (Mashkova and Molchanov 1980). This would smooth out R_1 oscillations. Consideration of the energy spread of the scattered particles leads to an expression analogous to eq. (10) where the ratio ε/g_E is a small parameter (g_E is the width of the energy distribution for atomic scattering).

From eq. (6), it follows that the peak width g_2 of the angular or energy

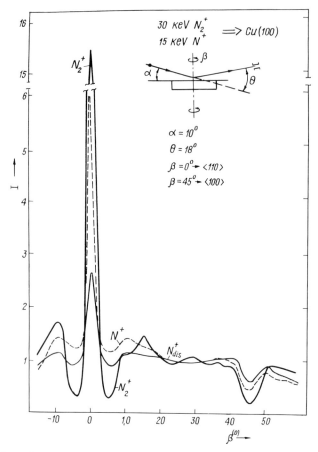

Fig. 5. The dependence of the intensities of the molecular ions N_2^+, dissociated ions N_{dis}^+ and atomic ions N^+ on the azimuthal target rotation angle β calculated from experimental data by Balashova et al. (1983).

distributions of the scattered molecules should be narrower than the width g_1 of the distributions of the scattered atoms. For rather narrow peaks, the widths g_2 and g_1 are connected by the following relation,

$$g_2 = g_1 \sqrt{2}. \tag{11}$$

The results of the experiments by Balashova et al. (1982a) on scattering of 30 keV N_2^+ molecular ions by a Cu single crystal have shown that the orientational effects in the molecular yield are in fact more pronounced than in the atomic yield recorded under bombardment with N^+ ions of the same velocity. The molecular ion yield in accordance with eq. (7) is proportional to the square of atomic ion yield,

$$I_2 = \frac{\varepsilon}{l^2 E_0 \eta_1^+} I_1^2. \tag{12}$$

However, the anisotropy of I_2 is also defined by the anisotropy of the degree of ionization η_1^+, which is known to be very considerable (Hou et al. 1978), and is explained by the differences in the consequent electron capture and loss during ion scattering in different crystallografic directions (Nizhnaya et al. 1982a).

This should be taken into account when finding the correlation between I_2 and I_1. On the other hand, eq. (7) may be used for an estimation of the η_1^+ anisotropy. The calculations of η_1^+ from the data by Balashova et al. (1982a) using eq. (7) gives the degree of ionization versus the azimuthal angle of the target rotation as shown in fig. 6.

It is interesting that the function $\eta_{dis}^+(\beta)$ obtained by substitution of I_{dis} in eq. (7) has a somewhat different shape. This is due to the influence of the atomic interaction in scattering and dissociation processes, as in fig. 4.

In another experiment by Heiland et al. (1979), the scattering of 400 eV H_2^+ ions from a tungsten single crystal at azimuthal target rotation has been investigated. In this case, as follows from eq. (4), the angle θ_0 does not allow the last factor in eq. (10) to be neglected. Thus, the anisotropy $I_2(\beta)$ should be less pronounced than in $I_1(\beta)$. Indeed this has been observed by Heiland et al. (1979). Since the widths of the energy distributions g_1 and g_2 for H^+, H_2^+ exceeded considerably the dissociation energy, they should be connected by eq. (11).

In table 1, the ratios of the experimental widths g_1 and g_2 of the energy

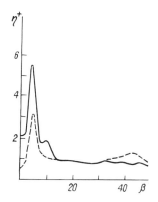

Fig. 6. The degree of ionization η^+ versus the angle of azimuthal target rotation β calculated from experimental data by Balashova et al. (1983) using formula (7). The full curve represents I_1^2/I_2; the broken curve I_{dis}^2/I_2.

Table 1
Ratios of the experimental widths, g_1/g_2, of the energy and
angular distributions of the scattered particles.

Distribution	E (keV)	Scattering system	β (deg)	g_1/g_2
Energy distribution[a]	0.6	$H_2^+ \rightarrow Ni$	–	1.45
		$H_2^+ \rightarrow W$	–	1.35
Azimuthal angular distribution[b]	30	$N_2^+ \rightarrow Cu$	20	1.37
			0	1.0

[a] Heiland et al. (1979).   [b] Balashova et al. (1982a).

(Heiland et al. 1979) and the angular (Balashova et al. 1982a) distributions
of the scattered particles are given.

From table 1 it is seen that the ratio of distribution widths g_1 and g_2 fits
well the value $\sqrt{2}$ obtained from formula (11). When the direction of
bombardment coincides with the low-index plane (110) ($\beta = 0$) the peak
widths for ion bombardment $g_1 = 4°$ is comparable with the angular accept-
ance of the analyzer (3°) and $\theta_0 = 2°$. In this case, the ratio g_1/g_2 is deduced
from eq. (10) rather than from eq. (6) and the exact value for it is not $\sqrt{2}$,
but depends on the shape of the distribution.

In fig. 7 the experimental distribution of the scattered N_2^+ and N^+ ions
(Balashova et al. 1982a) versus the azimuthal angle of scattering is compared with the distribution calculated with eq. (6) for two azimuthal angles
of target rotation: (a) $\beta = 20°$ and (b) $\beta = 4.5°$. The good agreement with
experiment for the peak width ratio and the angular distribution calculated
by eq. (12) is in accord with the weak dependence of the ionization degree
η_1^+ on the azimuthal escape angle φ.

If a molecule consists of atoms with different masses m_1 and m_2, then the

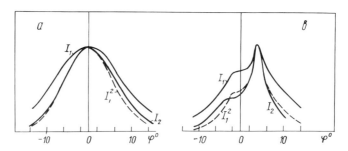

Fig. 7. Azimuthal angular distribution of nitrogen ions scattered from a Cu (100) surface for
different orientations of the axis $\langle 110 \rangle$ relative to the plane of incidence. (a) $\beta = 20°$; (b)
$\beta = 4.5°$. The experimental date are from Balashova et al. (1983): I_1 for N^+, $E = 15$ keV, I_2 for
N_2^+, $E = 30$ keV.

condition of nondissociative scattering, eq. (3), is written as

$$E_r = \frac{m_1 m_2}{2(m_1 + m_2)} (\boldsymbol{v}_1 - \boldsymbol{v}_2) \leqslant \varepsilon .$$

By analogy with eq. (8), we obtain for the yield of scattered molecules the following formula,

$$R_m(\theta) = \frac{(1 + \nu)^2 \varepsilon}{2 \nu l^2} R_{a1}(\theta) R_{a2}(\theta) ,$$

where $\nu = m_1/m_2$, E is the molecular energy, and R_{a1} and R_{a2} are the yields of scattered atoms with masses m_1 and m_2, respectively.

As follows from the later formula, if the atom mass difference is large ($\nu \gg 1$), then the yield of the scattered heteronuclear molecules exceed essentially the yield of homonuclear ones.

2.2. Single scattering

The expression for the relative yield of the scattered molecules after a single collision with a target atom, neglecting the latter motion, may be obtained from eq. (5) taking into account that $R_1(\theta) = |db/d\theta| b/\theta$. Then

$$R_{21}(\theta) = \frac{\varepsilon b}{l^2 E_0 \theta |d\theta/db|} . \tag{13}$$

This equation is deduced according to Bitensky and Parilis (1981a). Choosing the initial velocity direction as a polar axis, the molecular axis orientation is determined by a polar angle ψ and an azimuthal angle φ (fig. 1). The impact parameter for the second atom is given by

$$b_2 = [b^2 + (l\psi)^2 + 2bl \cos \varphi]^{1/2} \approx b + l\psi \cos \varphi , \tag{14}$$

where b is the impact parameter for the first atom. In eq. (14), we took into account that $\psi \ll 1$. The angle between the scattering plane for the first and second atom equals $\beta = (l\psi/b) \sin \varphi$. Substituting $\boldsymbol{v}_{1,2} = V_0 \theta_{1,2}$ in eq. (3), we get

$$\tfrac{1}{2} E_0 [(\theta_1 - \theta_2)^2 + \theta_1 \theta_2 \beta^2] \leqslant \varepsilon . \tag{15}$$

Expanding $\Delta\theta = \theta_1 - \theta_2$ in a power series on $\Delta b = l\psi \cos \varphi$ up to the term of first order of magnitude, we obtain from eq. (15)

$$\left(\frac{d\theta}{db} l\psi\right)^2 \cos^2\varphi + (\theta l\psi/b)^2 \sin^2\varphi \leqslant \theta_0^2 . \tag{16}$$

Since $\psi \ll 1$, then the solid angle $\Delta\Omega_1$ in eq. (1) equals the area of the region defined by inequality (16). As is seen from eq. (16), this region is an ellipse with semiaxes $A = \theta_0/(l \, d\theta/db)$ and $B = \theta_0 b/(l\theta)$, therefore, we get immediately the following expression,

$$R_{21} = \frac{1}{2\pi} \int_{\Delta\Omega_1} \psi \, d\psi \, d\varphi = \frac{1}{2} AB = \frac{\varepsilon b}{l^2 E_0 \theta |d\theta/db|} , \tag{17}$$

which is the same as in eq. (13).

Using the power potential $U(r) = U_0 r^{-n}$ in the momentum approximation, we get

$$\theta = cb^{-n} , \tag{18}$$

where $c = U_1/E_0$, $U_1 = \sqrt{\pi} \Gamma[\frac{1}{2}(n+1)]/\Gamma(\frac{1}{2}n)U_0$, $\Gamma(x)$ is a gamma function. Calculating the derivative $d\theta/db$ we get, from eq. (17),

$$R_{21}(\theta) = \frac{\varepsilon U_1^{2/n}}{nl^2 E_0^{(n+2)/n} \theta^{2(n+1)/n}} . \tag{19}$$

For $n = 2$, the relative yield $R_{21} \simeq \theta^{-3}$ agrees well with the dependence I_{21} on the scattering angle θ in fig. 2.

2.3. Movement of recoil atoms

If the ratio $\mu = m_1/M$ (M is the mass of the target atom) is not too small, the target recoil velocity $u = \mu v_0 \theta_1$ should be taken into account. Here, θ_1 is the scattering angle for the first atom. In the reference frame connected with the moving target atom, the second atom has the velocity $v_0' = v_0 - u$, which forms an angle α_1 with the velocity v_0, $\alpha_1 = u/v_0 = \mu\theta_1$. The target atom displacement during the time of flight of the first atom leads only to a change in polar axis orientation with invariable solid angle $\Delta\Omega_1$ [see eq. (1)].

In this case, the impact parameter of the second atom collision and angle β are determined by the formulae

$$b_2 = b_1 + l(\mu\theta - \cos\varphi) ,$$
$$\beta = [(\psi l \sin\varphi/b)^2 + (\mu\theta)^2]^{1/2} . \tag{20}$$

Taking into account eq. (20), we obtain from eq. (15)

$$\left(\frac{d\theta}{db}\right)^2 l^2 (\mu\theta - \psi\cos\varphi)^2 + \theta^2\left(\frac{l^2}{b^2}\psi^2\sin^2\varphi + \mu^2\theta^2\right) \leq \theta_0^2 . \tag{21}$$

The left-hand part of eq. (21) is positive and has the minimum value $\mu^2\theta^4$ at $\varphi = 0$, $\psi = \mu\theta$. Thus, the maximum angle of nondissociative scattering is given by

$$\theta_m = \left(\frac{2\varepsilon}{\mu^2 E_0}\right)^{1/4} . \tag{22}$$

Usually, the molecular ion scattering dissociation is considered by the 'spectator' model (Gerasimenko and Oksyuk 1965, Green 1970) in which the scattering occurs on one of the atoms since the other does not participate in the scattering process. The maximum angle of scattering θ_m [see eq. (22)] exceeds considerable the limiting scattering angle θ_0 [see eq. (4)] in the 'spectator' model. The maximum molecular energy loss at nondissociative scattering $\Delta E_m = 4\mu E_0 \theta_m^2 = 2(8\varepsilon E_0)^{1/2}$ exceeds considerable the dissociation energy.

By calculating the solid angle $\Delta\Omega_1$, taking into account condition (21), we get from eq. (1) the expression for the relative yield of the scattered molecules,

$$R'_{21}(\theta) = R_{21}(\theta)[1 - (\theta/\theta_m)^4],\qquad(23)$$

where $R_{21}(\theta)$ is defined by eq. (13). From eq. (23), it follows that at $\theta/\theta_m \ll 1$ the motion of the recoil atom may be neglected, e.g., for scattering of H_2^+, D_2^+ by Au, W, Ni targets, as was done by Bitensky and Parilis (1981a).

It should be noted that eqs. (22) and (23) may be deduced from the formulae obtained by Bitensky and Parilis (1982) for the nondissociative scattering on a diatomic molecule, by transition into the frame of reference connected with the ion.

In fig. 8, the calculated yield of scattering of $N_2^+ \to Cu$, $E = 30\,keV$, without (curve 1) and with (curve 2) consideration of the recoil atom motion using formulae (19) and (23), respectively, is presented. In the calculation,

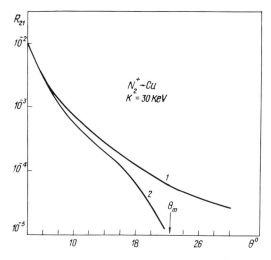

Fig. 8. The relative yield R_{21} versus angle θ for single scattering (1) without and (2) with consideration of the motion of recoils.

the inverse-square approximation of the Firsov potential has been used ($n = 2$). The existence of the maximum scattering angle, which in this case equals $\theta_m = 22.6°$, is seen in fig. 8.

Another peculiarity of nondissociative scattering follows from eq. (23): at scattering angles $\theta > \theta_m$ the nondissociative scattering becomes possible only as a result of both plural and quasisingle collisions, smoothing out to a certain extent the orientational effects.

This result was confirmed by Sass and Rabalais (1988), where the single-scattering peak was observed at nondissociative scattering of N_2^+ by Au ($\mu = 0.07$) and was not observed in case of $N_2^+ \rightarrow C$ ($\mu = 1.17$).

2.4. Double scattering

At glancing incidence bombardment, it is necessary to consider the multiple collisions of the molecule atoms with the surface. In the paper by Bitensky and Parilis (1981a), the condition of nondissociative scattering as a result of double collisions with the target atoms is given by

$$(\theta_1 - \theta_2)^2 + \left(\theta + \theta_{12} \frac{\alpha d}{b_{12}}\right)^2 \frac{l^2}{b_{11}^2} \psi^2 \sin^2\varphi \leqslant \theta_0^2 ,$$

where $\theta_{ij}(i, j = 1, 2)$ is the scattering angle of the ith atom of the molecule on the jth target atom, b_{ij} are the corresponding impact parameters, θ_i are the scattering angles, θ is the scattering angle of the molecule, d is the distance between the surface atoms, α is the glancing angle, and $b_{12} = b_{11} + d(\theta_{11} - \alpha)$. The scattering scheme is illustrated in fig. 9.

The condition for nondissociative scattering, eq. (3), requires $\Delta\theta = \theta_1 - \theta_2$ to be small. Thus, $\Delta\theta$ may be expanded in a power series of the impact parameter difference for two molecule atoms $\Delta b = l\psi \cos \varphi$. In the first approximation $\Delta\theta = (d\theta/db) \Delta b$. By using the power potential for symmetrical scattering ($\alpha = \frac{1}{2}\theta$), we get the following expression,

$$d\theta/db_{11} = -4z(1 - z)/d , \tag{24}$$

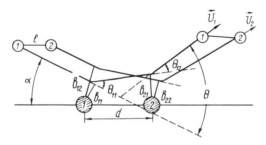

Fig. 9. Scheme of double scattering.

where $z = (\theta/\theta_k)^{(n+1)/n}$, and θ_k is the value of the scattering angle at which the derivative $d\theta/db_{11} = 0$.

It equals

$$\theta_k = \frac{4}{nd} \left(\frac{nU_1 d}{2E_0} \right)^{1/(n+1)} . \tag{25}$$

Considering eqs. (23) and (24) in the first approximation we obtain the expression for the relative yield,

$$R_{21}(\theta) = \frac{n\varepsilon d^2}{16 l^2 E_0 z^2 |1 - z|(1 + z/n)} . \tag{26}$$

It should be noted that the application of a power potential has allowed the dependence of R_{21} on θ to be obtained in a closed form. The dependence R_{21} using eq. (26) is shown in fig. 10 (curve 1).

The expression for $R_{21}(\theta)$ for the screened Coulomb potential was obtained by Bitensky and Parilis (1981a) for scattering of molecules H^+ and D^+ on Ni and Au surfaces. The experimental data by Eckstein et al. (1975) and Heiland et al. (1979) gave the possibility to check the formulae obtained by Bitensky and Parilis (1981a) for the relative yield of nondissociative scattering for diatomic and triatomic hydrogen molecules in quasisingle, double and multiple collisions with a polycrystalline surface. In fig. 11, the

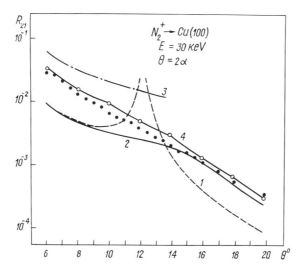

Fig. 10. Comparison of different models with the experimental data from Balashova et al. (1983) (points). Theory: (1) double scattering according to eq. (26); (2) eq. (31); (3) string approximation [eq. (38)]; (4) atomic chain.

Scattering of swift molecules

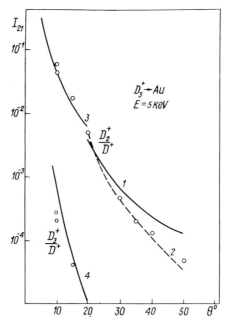

Fig. 11. Relative yield versus scattering angle θ. Theory: (1) single, (2) double, (3) and (4) multiple scattering; (points) experiment by Eckstein et al. (1975).

comparison of the calculations with the data by Eckstein et al. (1975) is displayed for the screened Coulomb potential.

Analogously to single scattering, we obtain from eq. (24) for the relative yield the following formula,

$$R_{21} = \frac{\varepsilon b_{11}}{l^2 E_0 |d\theta/db_{11}|(\theta + \theta_{12}\alpha d/b_{12})} , \qquad (27)$$

where $\theta = \theta_{11} + \theta_{12}$,

$$\theta_{11,12} = \frac{Z_1 Z_2 e^2}{E_0 a} K_1\left(\frac{b_{11,12}}{a}\right) ,$$

$$|d\theta/db_{11}| = \frac{Z_1 Z_2 e^2}{E_0 a^2} \left\{ f(b_{11}) + f(b_{12})\left[1 - \frac{Z_1 Z_2 e^2 d}{E_0 a^2} f(b_{11})\right]\right\} , \qquad (28)$$

where $f(x) = (a/x)K_1(x/a) + K_0(x/a)$, $K_n(x)$ is the McDonald function, and a is the screening parameter.

As is seen from fig. 11, the model of double scattering for angles $\theta \geqslant 20°$ agrees well with experimental data. It should be noted that in this region $d\theta/db \neq 0$ because for a chosen value $a = 0.25$ Å, the derivative $d\theta/db = 0$ at $\theta = 16°$.

The calculation of R_{21} using eq. (27) has been made for H_2^+ scattering on a (111) surface of a Ni single crystal in the energy range $E = 0.2$–$2\,keV$. The dependence $R_{21}(\theta)$ for different energies at a given glancing angle $\alpha = 15°$ is shown in fig. 12.

In the calculations, the values $\varepsilon = 2.7\,eV$, $l = 1.04\,\text{Å}$, $a = 0.45\,\text{Å}$ were used. The increase of the screening parameter as compared to $a = 0.25\,\text{Å}$ for $D_2^+ \to Au$ corresponds to a transfer toward a metal with less atomic number. Equation (27) fits the experiment by Heiland et al. (1979) for scattering angles $\theta \leqslant 40°$ and outside the specular reflection $\theta = 2\alpha$ when $|d\theta/db|$ is small. In fig. 12, it is seen that the relative yield has a maxima at $\theta = 2\alpha$ and decreases with increasing of both the scattering angle and the energy of bombardment.

The infinity in R_{21} should not occur at all, because the critical glancing angle α_k, at which it may occur, is larger than the glancing angle $\alpha = 15°$, used in the experiment by Heiland et al. (1979). Therefore, for all considered energies, $E = 0.2$–$2\,keV$, the dependence $\theta(b_{11})$ is a monotonic one and the curves $R_{21}(\theta)$, calculated in the first approximation, have only at $\theta = 2\alpha$ a finite maximum, which is eliminated in the next approximation. Figure 12 displays the restrictions of probability of nondissociative scattering for large scattering angles and small bombarding energies. For such values of θ and E_0, the momentum approximation becomes too rough and needs an exact calculation of the collision integral. Secondly, it is necessary to take into account the interaction of molecule atoms during the scattering. Thirdly, for scattering angles close to θ_k [see eq. (25)], the expanding of $\Delta\theta = \theta_1 - \theta_2$ over Δb up to terms of first order of magnitude becomes

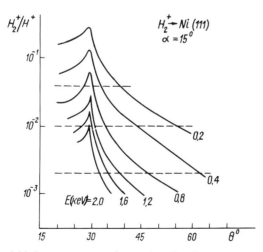

Fig. 12. Relative yield R_{21} versus scattering angle θ for different energies of a molecule H_2^+ scattered from the (111) surface of an Ni single crystal using eq. (27).

insufficient. The last refinement can be treated analytically (Bitensky and Parilis 1984).

As it is known from the double-scattering theory by Martynenko (1973) and Parilis et al. (1975), the dependence $\theta(b_{11})$ becomes nonmonotonic with the increase of the glancing angle α at $\alpha > \frac{1}{2}\theta_k$. At $\theta = \theta_k$, the derivative $d\theta/db_{11} = 0$ and R_{21} becomes infinite. This means that in the vicinity of the point $\theta = \theta_k$ the first approximation becomes insufficient. It has been noted by Bitensky and Parilis (1981a) that for getting rid of this infinity it is necessary to consider higher-order terms in the expanding $\Delta\theta$ over Δb.

In the momentum approximation and using the power potential, we can calculate the impact parameter b_{11} as a function of θ for symmetric scattering. In this case,

$$\theta_{11} = \theta_{12} = c/b_{11}^n = \alpha ,$$

and

$$\theta = \theta_{11} + \theta_{12} = 2c/b_{11}^n .$$

Therefore, we can expand $\Delta\theta$ over Δb in the vicinity of the point $b_0 = (\alpha/c)^{1/n}$. Since $\theta = \theta_k$ is the inflection point, then in the expansion one should include the terms up to the third order. By calculating these for the potential with $n = 2$, we get the condition for nondissociative scattering as

$$(\gamma_1 x + \gamma_2 x^2 + \gamma_3 x^3)^2 + \gamma_4^2 y^2 \leqslant (\theta_0/\theta)^2 , \tag{29}$$

where $x = \psi \cos \varphi$, $y = \psi \sin \varphi$ and the coefficients γ_i are defined by the following formulae,

$$\gamma_1 = -2(1-z)h;$$

$$\gamma_2 = 3(1-z)(1-2z)h^2 ;$$

$$\gamma_3 = -(1-2z)(8z^2 - 17z + 4)h^3 ; \tag{30}$$

$$\gamma_4 = (1 + \tfrac{1}{2}z)h$$

and $h = 2lz/\theta d$.

The solid angle $\Delta\Omega_1$ [see eq. (1)] equals the area of the region defined by inequality (29). Then the relative yield R_{21} is defined by the following expression:

$$R_{21} = \frac{1}{\pi\gamma_4} \int_{x_1}^{x_2} \sqrt{(\theta_0/\theta)^2 - (\gamma_1 x + \gamma_2 x^2 + \gamma_3 x^3)^2} \, dx , \tag{31}$$

where $x_{1,2}$ are the roots of the integrand. In the first approximation, inequality (29) represents an ellipse and from eq. (31) we get eq. (26). At $z = 1$, the coefficients $\gamma_1 = \gamma_2 = 0$, the integral in eq. (31) has a finite value and is given by

$$R_{21}(\theta_k) = \frac{1.82\theta_0^{4/3}}{\pi\gamma_4(1)|\gamma_3^{1/3}(1)|\theta_k^{4/3}} = 0.06\theta_0^{4/3}\theta_k^{2/3}d^2/l^2 \ . \tag{32}$$

An estimation of the integral in eq. (31) has shown that R_{21} may be written approximately as

$$R_{21} \approx C\frac{|x_1| + x_2}{\pi\gamma_4} \ , \tag{33}$$

where C is constant ($0.79 \leqslant C \leqslant 0.93$).

While calculating $R_{21}(\theta)$ using eq. (31) for the scattering angles $\theta > \theta_k$, one should take into account the possibility of appearance of the additional roots of the integrand.

Figure 10 illustrates the dependence of R_{21} on the scattering angle θ, calculated by eqs. (26)–(31) for symmetric reflection of N_2^+ with energy $E = 30$ keV on the Cu single-crystal surface (100) (the scattering plane coincides with the direction $\langle 100 \rangle$). As it is seen in the approximation of eq. (31) the dependence of R_{21} remains monotonic even at $\theta \geqslant \theta_k$ ($\theta_k = 12.3°$), which agrees satisfactorily with the experiment by Balashova et al. (1983). At small glancing angles, the yield of nondissociative scattering appears to be lower than the experimental data. This is explained by the fact that the double-scattering approximation becomes insufficient and it is necessary to consider plural collisions.

2.5. Single scattering of molecules by two neighboring surface atoms

In general, the reflection of molecules seems to be possible when molecule atoms are scattered in pairs by two surface atoms, each molecule atom being scattered due to a quasisingle collision. In this case, one cannot consider the recoil motion and condition (3) in designations of eq. (29) leads to the inequality

$$\left[\frac{d\theta}{db}\frac{l}{2}x^2\right]^2 + \left[\frac{\theta l}{b}y\right]^2 \geqslant \theta_0^2 \ . \tag{34}$$

By calculating the relative yield analogously to eq. (31) and neglecting the mutual interaction of molecule atoms, we get

$$R_{21}''(\theta) = 2.2[R_{21}(\theta)]^{3/4} \ , \tag{35}$$

where R_{21} is determined by eq. (13). Since $R_{21} \ll 1$, then from eq. (35) it follows that R_{21}'' considerably exceeds R_{21} for a molecule scattering by one atom. During the time interval between the scattering of the first and the second molecule atoms the distance between the molecules decreases to $l' \approx \theta_1 l \ll l$. This should lead to the strong interaction of molecule atoms and diminish the probability of molecular nondissociative scattering. The contribution of the considered collision could be essential for nondissociative molecule–molecule scattering when the interatomic distances are equal or close to each other.

2.6. Scattering of molecules on an unbroken string

Let us consider the scattering of molecules by a single-crystal surface at small glancing angles within the Lindhard string approximation (Lindhard 1965). In this approximation particles with the transverse energy $E_\perp = E_0 \alpha^2$ are scattered by the string potential $U(r) = Ar^{-2}$, and neglecting the mutual interaction of the atoms in a molecule one may get the condition of nondissociative scattering [see eq. (3)] as

$$\sin \tfrac{1}{2}\gamma \leqslant (\varepsilon/2E_\perp)^{1/2} ,$$

where γ is the angle between the velocities of the two atoms after scattering

$$\gamma = \tfrac{1}{2}\pi\rho(E_\perp/A)^{1/2} , \quad \rho = l \sin\psi \sin\varphi .$$

We get

$$|\sin\psi \sin\varphi| \leqslant D , \tag{36}$$

where $D = (2\varepsilon A)^{1/2}/\pi l E_\perp$. While calculating the relative yield R_{21}, it is necessary to consider that, at $0 \leqslant \varphi \leqslant \arcsin D$, the polar angle $\psi \leqslant \tfrac{1}{2}\pi$, since, unlike the scattering on atoms, the scattering on the string does not depend on ψ. The calculation of the integral from eq. (1) using eq. (36) leads to the following expression,

$$R_{21}(\theta) = D(\theta) = \frac{4\sqrt{2\varepsilon A}}{\pi l E_0 \theta^2} . \tag{37}$$

The influence of the mutual interaction of the atoms on the dissociation of the molecule may be estimated by considering the change of the distance between the atoms during scattering,

$$\Delta l = l[1 - (\cos^2\psi + \sin^2\psi \sin^2\varphi)^{1/2}] ,$$

which results in the vibrational excitation of the molecule.

Approximating the interaction in the molecule by a parabolic potential, we replace the dissociation energy ε in eq. (37) by the expression $\varepsilon - \tfrac{1}{2}k\,\Delta l^2$, where k is the constant of quasielastic force, and we get

$$R_{21}(\theta) \approx 0.4D(\theta)/g^{1/4} . \tag{38}$$

Here $g = kl^2/2\varepsilon$.

In the previous section, in the calculation of the single and double scattering of the molecule on the target atoms the consideration of mutual interaction of atoms in the molecule was neglected because it resulted in a correction or order ψ^4 while $\psi \ll 1$.

From fig. 10, it is seen that eq. (38) gives an overestimated value of the relative yield of the nondissociative scattering. The problem is that the Lindhard string approximation is applicable only for very small glancing angles, e.g., for channeling. For scattering on a solid surface, the glancing angles are rather large and it is necessary to consider scattering on a discrete atomic chain (Kivilis et al. 1967).

2.7. Nondissociative scattering on a discrete atomic chain

For investigating the nondissociative scattering at small glancing angles, taking into account the interaction between the atoms in the molecule, a calculation has been made on the basis of a numerical solution of the differential equations of molecular movement during its collision with the discrete surface atomic chain (Bitensky and Parilis 1981a, 1984). For small scattering angles, the motion of recoil atoms may be neglected (as shown in section 2) and the equations of motion are given by

$$m_i \frac{d^2 \mathbf{r}_i}{dt^2} = -\sum_{j=1}^{K} \nabla U(|\mathbf{r}_i - \mathbf{r}_{0j}|) - \nabla U_{12}(|\mathbf{r}_1 - \mathbf{r}_2|), \quad i = 1, 2, \quad (39)$$

where K is the number of atoms in the chain, $U(|\mathbf{r}_i - \mathbf{r}_{0j}|)$ is the interaction potential of the ith molecule atom with the jth target atom as an inverse-square approximation of the Firsov potential, $U_{12}(|\mathbf{r}_1 - \mathbf{r}_2|)$ is the Morse potential for the interaction of the atoms in the molecule,

$$U_{12}(r) = \varepsilon[e^{-2(r-l)/\delta} - 2e^{-(r-l)/\delta}]. \quad (40)$$

For the molecule N_2^+, the parameter $\delta = 0.35$ Å.

The equation of motion [see eq. (39)] has been solved numerically for different orientations of the molecular axis. The total molecular energy in the center of mass system, E_t, was calculated using the values obtained for coordinates and velocities of the atoms. It was assumed that the molecule would not dissociate if at large distance from the target, $E_t \leq 0$.

The calculation was made taking into account the thermal vibrations of the target atoms which have been simulated by means of correlated displacements with a correlation length equal to the distance d between the atoms in the chain as in the paper by Nizhnaya et al. (1979a,b,c,d).

The result of the calculation of $R_{21}(\theta)$ for scattering on the discrete chain ($K = 5$) is illustrated in fig. 10. It is seen that at small glancing angles the relative yield for the discrete chain (curve 4) passes above curve 2, corresponding to double scattering, but is lower than for the string (curve 3), besides, it agrees well with the experimental value. With increasing glancing angle, the scattering on the atomic chain does not differ from the double

scattering on two neighboring atoms and the yields for the discrete chain
(curve 4) and two atoms (curve 2) coincide.

The agreement of both curves with the experimental data at large
scattering angles, and also of the curve (4) for the discrete chain in a broad
range of scattering angles, should be considered as rather satisfactory. It
should be noted that all calculations have been made without any fitting
parameters.

The results of the numerical calculation of eq. (39) using the screened
Coulomb potential is shown in fig. 13. As it is seen from fig. 13, the curves
for large scattering angle θ are sufficiently far below those calculated from
eq. (27) (see fig. 12) and the maxima are smoothed. The dependences of the
scattering angle on molecular energy E as a result of numerical calculations
for three values of R_{21} (these values are shown on fig. 13 by the broken
lines) are illustrated in fig. 14 being in agreement with the experiment by
Heiland et al. (1979). The analogous calculation was made for $D_2^+ \to Au$
and gave a dependence $R_{21}(\theta)$ close to the analytical one from eq. (27).
This is explained by the fact that the maxima R_{21} do not appear in the range
of angles near $\theta_k = 16°$ and the bombarding energy is sufficiently high,
$E = 3.2$ keV.

Besides the computer simulation (Bitensky and Parilis 1981a, 1984), there
are some papers in which this model has been developed and as a result
some details of molecular scattering became more clear. Jakas and Harrison
(1985) corroborated the conclusion concerning the important role of both
recoil motion and mutual interaction of molecule atoms. The computer
simulation revealed the orientational effects in molecule scattering by a
single-crystal surface. The trajectory focusing is found to affect essentially
the molecule scattering near the semichannel direction (Shulga 1985,

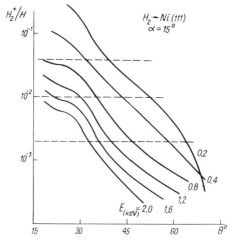

Fig. 13. The same as fig. 12 but as a result of a numerical calculation (see section 2.7).

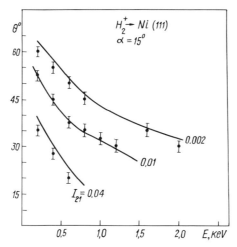

Fig. 14. The scattering angle θ versus energy E for three values of the relative yield. Full curves represent the numerical calculation; and the points the experiment by Heiland et al. (1979).

1986a,b, Evdokimov 1983). Van den Hoek et al. (1988) have performed some trajectory calculations of O_2 scattering from $Hg\,(111)$ at energies 100–300 eV. The calculations show a larger dissociation probability for scattering of molecules than has been observed experimentally for negative ion formation in collisions $O_2^+ \rightarrow Ag$. The authors suggest that the experimentally observed O^- formation is possible due to the impulse interaction. It was also observed that in molecular dissociation the value of transverse energy $E_\perp = E_0 \sin^2\alpha$ is important. The calculations overestimate the degree of dissociation observed in the experiment on negative ion formation without explicit consideration of the charge transfer process.

2.8. Multiple scattering

Under glancing incidence, the molecular scattering occurs essentially due to multiple collisions. It was proposed by Bitensky and Parilis (1981a) that in this case the molecule dissociates not as a result of individual collisions with a surface atom but rather due to straggling of the molecule atom velocities.

By considering the molecule atoms to be scattered independently and the velocity distribution described by the function $f(\boldsymbol{v})$, one obtains the expression for the relative yield,

$$R_{21} = \int_0^\varepsilon dE \int f(\boldsymbol{v}_1)\, f(\boldsymbol{v}_2)\, \delta[E - \tfrac{1}{4}m(\boldsymbol{v}_1 - \boldsymbol{v}_2)^2]\, d\boldsymbol{v}_1\, d\boldsymbol{v}_2 \,, \tag{41}$$

where $\delta(x)$ is a delta function. If we suppose that the velocity distribution is Gaussian, $f(\boldsymbol{v}) = \pi^{-3/2}\sigma^{-3} \exp[-(\boldsymbol{v} - \boldsymbol{v}_0)^2/\sigma^2]$, the relative yield is given by

$$R_{21} = \frac{2}{\sqrt{\pi}} \int_0^x e^{-t}\sqrt{t}\,dt = \frac{2}{\sqrt{\pi}}\,\gamma(\tfrac{3}{2}, x)\,, \tag{42}$$

where $\gamma(\tfrac{3}{2}, x)$ is an incomplete gamma function, $x = 2\varepsilon/m\sigma^2$.

Since the straggling of the scattered particles decreases with the scattering angle decrease, therefore, in the first approximation we could suppose that the dispersion is proportional to the scattering angle $\sigma(\theta) = kv_0\theta$, where k is a fitting parameter. From eq. (42) we obtain

$$R_{21} = \frac{4}{3\sqrt{\pi}}\left(\frac{\varepsilon}{k^2 E_0}\right)^{3/2} \theta^{-3}\,, \quad \text{for } x \ll 1\,, \tag{43a}$$

$$= 1 - \frac{2}{\sqrt{\pi}}\,e^{-\varepsilon/(k^2 E_0\theta^2)}\,, \quad \text{for } x \gg 1\,, \tag{43b}$$

In case of three-atomic molecular scattering, we consider it to be scattered without dissociation if the relative energy of the third atom and the diatomic molecule does not exceed the dissociation energy. By analogy with eq. (41), the relative yield R_{31} becomes

$$R_{31} = \frac{4}{\sqrt{\pi}} \int_0^x e^{-t}\sqrt{t}\,\gamma(\tfrac{3}{2}, x - t)\,dt = 1 - e^{-x}(\tfrac{1}{2}x^3 + x + 1)\,, \tag{44}$$

for small x the eq. (44) leads to

$$R_{31} \approx \frac{1}{6}\left(\frac{\varepsilon}{k^2 E_0\theta^2}\right)^3\,. \tag{45}$$

As is seen from fig. 11, formulae (43a) and (45) describe the experimental data from Eckstein et al. (1975) for $\theta \leqslant 15°$ rather well for the same value of the fitting parameter $k = 0.6$ (curves 3 and 4). Formulae (42) and (44) were deduced supposing the nondissociative scattering to be a pure surface phenomenon. Is it possible for a molecule to survive scattering in the bulk? The following calculation shows that it is not. Indeed, for scattering in the bulk, the straggling both of velocities and escape points need to be taken into account. The theory by Remizovich et al. (1980a) provides this possibility and gives the following expression for R_1,

$$R_1 = \frac{24\zeta}{\pi^{5/2}[\langle\theta_s^2\rangle]^2(x - x_0)^4}\,\mathrm{erf}\!\left(2\sqrt{\frac{3\zeta\alpha}{\langle\theta_s^2\rangle(x - x_0)}}\right)$$

$$\times \exp\left\{-\frac{4}{[\langle\theta_s^2\rangle]^2(x - x_0)}\left[\zeta^2 - \zeta\alpha + \alpha^2 + \varphi^2\right.\right.$$

$$\left.\left. -\frac{3\varphi(y - y_0)}{x - x_0} + \frac{3(y - y_0)^2}{(x - x_0)^2}\right]\right\}\,, \tag{46}$$

where $\{x_0, y_0\}$ and $\{x, y\}$ are the coordinates of the impact point and the escape point, respectively, $\langle \theta_s^2 \rangle$ is the mean square scattering angle per unit length, and erf(x) is the error function.

For the yield of nondissociative scattering as in eq. (8), one gets

$$R_2(\zeta, \varphi, x, y) = \frac{2\pi^2 l^2 \varepsilon}{E_0} R_1^2(\zeta, \varphi, x, y). \tag{47}$$

By integrating R_2 over x and y in the dimensionless units $\psi = \zeta/\alpha$, $\chi = \phi/\alpha$, one obtains

$$R_2(\psi, \chi) = \frac{283\zeta\sqrt{3}l^2\varepsilon[\langle\theta_s^2\rangle]^2}{\pi 2^{18} E_0 \alpha^7} \frac{\psi^2}{[\psi^2 - \psi + 1 + (\frac{1}{2}\chi)^2]^{11/2}}. \tag{48}$$

By integrating R_2 over ψ and χ, one obtains

$$R_2 = \frac{163\sqrt{6}l^2\varepsilon[\langle\theta_s^2\rangle]^2}{\sqrt{\pi}2^{14}E_0\alpha^7}. \tag{49}$$

Using the result of the integration of $R_1(x, y)$ over x, y (Remizovich et al. 1980a), one obtains for symmetrical scattering $R_{21} \simeq \theta^{-7}$ in drastic disagreement with the experiment by Eckstein et al. (1975). This means that the nondissociative scattering is a surface phenomenon.

2.9. Scattering in surface semichannels

The scattering in surface semichannels is well-known for atomic reflection (Parilis et al. 1969). Some of its peculiarities are connected with focusing effect (Mashkova and Molchanov 1974, Suurmeijer and Boers 1973).

Let the first atom be aimed at the middle of a semichannel of width $2a$ and depth h, but the impact point on the second atom is shifted on $\rho = 1 \sin \psi \sin \varphi$.

The impact parameter for a collision with the bottom string (M.W. Thompson and Pabst 1978) is given by

$$\rho_1 = \frac{2\rho^3}{a^2} + \frac{3\rho^5}{a^4}. \tag{50}$$

For simplicity, we have assumed that the transverse energy of the atom equals the focusing energy (Shulga 1975) $E_f = 2\pi hA/a^3$. The angle between the velocities of the scattered atoms after the reflection from the bottom string and the second focusing by the surface strings is determined by

$$\gamma = \frac{12\pi\rho^5}{a^4} \left(\frac{E_\perp}{A}\right)^{1/2}. \tag{51}$$

Taking into account eq. (51), we obtain the condition of nondissociative scattering in the form of eq. (36) where $D(\theta)$ is replaced by $D_f = (a^4D/$

$12l^4)^{1/5}$. The relative yield of the semichannel, $R_{21} = D_f$, is larger than the yield corresponding to the scattering by the single string [eq. (37)]. This effect can be used to explain the drastic increase of the relative yield observed by Balashova et al. (1982a) for the scattering of N_2^+ ions on a Cu single-crystal surface (100), when the plane of incidence coincided with the direction $\langle 110 \rangle$. In this case $a = 1.28$ Å, $h = 1.8$ Å, $E_f = 0.22$ keV, $D = 3.5 \times 10^{-2}$, $D/D_f \approx 10$.

A computer simulation within the Lindhard atomic-string model confirmed the strong directional effects in molecule scattering by single-crystal surfaces (Shulga 1986a).

2.10. Conclusion

Thus, the phenomenon of nondissociative scattering of swift molecules discovered and studied in recent years can be explained using some simple relations connecting the momenta of the atoms scattered in single, double and plural collisions on the surface.

A comparative study of the energy and angular distributions of the reflected molecular ions and both the ionized atomic fragments of the dissociated molecules and the ion scattered at ion bombardment with the same velocity gives some new opportunities for investigating the interaction of atomic particles with the surface. For instance, it can serve for studying the degree of ionization of the reflected atoms without measuring the neutral flux.

The very fine and proper surface effect of nondissociative scattering is a new and promising tool for solid surface diagnostics.

The scattering of swift molecules by the surface could be a unique source of molecules in the highest rotational and vibrational exited states.

3. Highest rotational and vibrational excitation of swift diatomic molecules scattered by solid surfaces

3.1. Introduction

In the previous section, a mechanism of swift-molecule nondissociative scattering based on some consequent correlated collisions of molecule atoms with the surface atoms has been considered.

The molecule is scattered without dissociation if the relative kinetic energy of its atoms does not exceed the dissociation energy. The velocities of the atoms are not exactly equal, therefore, the scattered molecule should be excited both vibrationally and rotationally (Bitensky and Parilis 1985).

So far, the rotational state distributions have been measured only with low-energy molecules (Kleyn et al. 1981, 1982, Barker et al. 1983). In these experiments, the rotationally excited states for NO scattered from Ag (11ī) have been studied over the incident kinetic energy range of 0.1 to 1.7 eV. It was found that the distribution over rotational states for high rotational quantum numbers differs essentially from the Boltzmann one.

In this section, the distribution of the scattered molecules over the vibrational and rotational states is considered.

3.2. Single scattering

For simplicity, we consider at first scattering of molecules by a single atom, although the nondissociative scattering in a single collision does not describe the phenomenon completely (Bitensky and Parilis 1980, 1981a, 1984).

The orientation of the molecular axis relative to the direction of bombardment is defined by the polar ψ and the azimuthal φ angles ($\varphi = 0$ is the plane of incidence); the impact parameter of the first atom is b (fig. 1).

According to the energy and momentum conservation laws, the energy of the ith atom of the molecule is

$$E_i = E_0(\sqrt{1 - \mu^2 \sin^2\theta_i} + \mu \cos \theta_i)^2/(1 + \mu)^2 ,$$

$\mu = m_1/M < 1$, M is the mass of target atom and E_0 is the initial energy of molecule atom. The condition of nondissociative scattering, eq. (3), requires both the scattering angles for the first and for the second atoms, θ_1 and θ_2, their difference $\theta_1 - \theta_2$ and the angle ψ to be small. Thus, for elastic energy losses we get $E_t \approx \mu E_0 \theta_i^2$. The velocities of the scattered atoms are $v_i \approx v_0(1 - \frac{1}{2}\mu\theta_i^2)$, where v_0 is the initial velocity.

These velocities have the following vector components: $\boldsymbol{v}_1$ ($v_1 \sin \theta_1$, 0, $v_1 \cos \theta_1$) and $\boldsymbol{v}_2$ ($v_2 \sin \theta_2 \cos \beta$, $v_2 \sin \theta_2 \sin \beta$, $v_2 \cos \theta_2$), where β is the angle between their projections in the plane XOY (see fig. 1). The angle $\beta \approx (l\psi/b) \sin \psi$ is also small and the relative velocity $\boldsymbol{v} = \boldsymbol{v}_1 - \boldsymbol{v}_2$ has the following components,

$$\boldsymbol{v} \{v_0(\theta_1 - \theta_2), -v_0\theta(l\psi/b) \sin \varphi, v_0(1 + \mu)\theta(\theta_1 - \theta_2)\} ,$$

where l is the interatomic distance in the molecule and $\theta = \frac{1}{2}(\theta_1 + \theta_2)$ is the angle over which the molecular center of mass is scattered. The relative velocity component perpendicular to the molecule axis, $l\{l \sin \psi \cos \varphi, l \sin \psi \sin \varphi, l \cos \psi\}$, causes the molecular rotation and the component parallel to the molecular axis causes its vibration. By resolving v over these two components, one gets the following formulae for the rotational energy,

$$E_j = \frac{1}{2}E_0\left[(\theta_1 - \theta_2)^2 + \left(\frac{\theta l\psi}{b} \sin \psi\right)^2\right], \tag{52}$$

and for vibrational energy

$$E_v = \tfrac{1}{2}E_0\left[(\theta_1 - \theta_2)\psi\cos\psi\,\frac{\theta l\psi^2}{b}\sin^2\varphi + (1+\mu)\theta(\theta_1 - \theta_2)\right]^2 . \qquad (53)$$

From eqs. (52) and (53), one could see that the vibrational energy E_v has a second order of smallness compared with the rotational energy E_j, i.e., the molecule dissociates via rotation [case III of the predissociation process in terms of paper by Herzberg (1950)].

It follows from eqs. (52) and (16) that the rotational energy of a molecule, E_j depends on its orientation relatively the bombardment direction. The distribution of the molecules over E_j could be deduced as

$$f(E_j) = \frac{1}{\Omega}\int_\Omega \delta\left\{E_j - \left[\left(\frac{d\theta}{db}\,l\psi\cos\varphi\right)^2\left(\frac{\theta}{b}\,l\psi\sin\varphi\right)^2\right]\tfrac{1}{2}E_0\right\}\psi\,d\psi\,d\varphi ,$$

$$(54)$$

where the integration is performed over the solid angle defined by eq. (16), and $\delta(z)$ is a delta function. Performing the integration in the generalized polar coordinates with the δ function, we get a distribution of molecules over E_j that is far from the Boltzmann equilibrium distribution; $f(E_j)$ is independent both of E_j and of the initial energy E_0, and equals

$$f(E_j) = 1/\varepsilon . \qquad (55)$$

Since for large quantum rotational numbers j the energy $E_j \simeq j^2$, then, as follows from eq. (55), the distribution over the rotational states is linear: $N_j \simeq j$. The distribution of the scattered molecules H_2^+ over the rotational states in comparison with the Boltzmann distribution,

$$N_j \simeq (2j + 1)\exp[-Bj(j + 1)/kT] ,$$

where B is the rotational constant, is shown in fig. 15. The distributions are

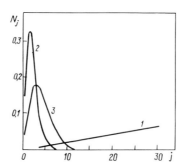

Fig. 15. Distributions of molecules H_2^+ over the rotational states. (1) Surface scattering; (2) Boltzmann distribution: $T = 300$ K; (3) $T = 1000$ K.

normalized to the same area, which is equal to 1. It is seen that the nondissociative scattering gives the inverse population of the rotational states. The maximum j is determined by the limit of the rotational pre-dissociation.

As follows from the paper by Herzberg (1950), a stable state is still possible even for the rotational energies exceeding the dissociation limit. Besides, the higher the vibrational excitation is, the lower the value j at which the centrifugal potential barrier could be overpassed is. Therefore, when calculating the relative yield of the nondissociative scattering, it is necessary to consider the dissociation energy ε depending on both the initial distribution over the vibrational states and the scattering angle.

The rotation plane passes through the molecular axis and the direction of the relative velocity (see fig. 1). Thus, the rotation of the scattered mole-cules is polarized in such a way that the rotational momentum is perpen-dicular to the velocity of the center-of-mass motion direction. However, the distribution over the directions of rotational momenta is a uniform one. Indeed, the angle between the rotational plane and the scattering plane is,

$$\eta = \arcsin(v_0 \sin \theta_2 \sin \beta / v) \ .$$

Considering that

$$v \approx v_0 [(d\theta / db \ l\psi \cos \varphi)^2 + (\theta l\psi \sin \varphi / b)^2]^{1/2} \ ,$$

$$\sin \theta_2 \approx \theta_2 \approx \theta \ \text{and} \ \sin \beta \approx \beta \approx (l\psi / b) \sin \varphi, \text{we get} \tag{56}$$

$$\eta = \arctan(A \tan \varphi) \ ,$$

where $A = \theta / (b|d\theta/db|)$. Having used a power potential for the interaction of the molecule atom with the atom of the solid, $U(r) = U_0 r^{-n}$, we get in the momentum approximation $\theta \approx b^{-n}$ and $|d\theta/db| \approx nb^{(n+1)}$, and then $A = n^{-1}$. Since, as follows from eq. (56), the angle η is independent of the polar angle ψ, then the distribution of the molecules over the rotation momentum direction $\Phi(\eta)$ is given by

$$\Phi(\eta) = F(\varphi) \, d\varphi / d\eta \ , \tag{57}$$

where $F(\varphi) \approx \frac{1}{2}\psi_m(\varphi)$ is the distribution over the azimuthal angle φ of the direction of the molecule axis and ψ_m is maximum value of ψ, deduced from eq. (16).

The normalized distribution is determined by

$$F(\varphi) = \frac{A}{2\pi(\cos^2\varphi + A^2 \sin^2\varphi)} \ . \tag{58}$$

Considering eqs. (56) and (58), we get from eq. (57) that $\Phi(\eta) = 1/(2\pi)$, i.e., the distribution of the scattered molecules over the direction of rotation indeed is uniform. However, the vibrational excitation depends on the direction of the rotational axis. Indeed, the vibrational energy $E_v = \hbar\omega(v + \frac{1}{2})$, and, therefore, the vibrational quantum number v according to eqs. (53) and (56), depends on the angle η. The same holds for E_v^{max} and v_{max} being the maximal values of E_v and v correspondingly calculated from eq. (53) for the maximal angle of nondissociative scattering ψ_m from eq. (16). In fig. 16, the polar diagram of E_v^{max} versus η for $N_2^+ \rightarrow Cu$, $\hbar\omega = 0.27$ eV, and $E = 30$ keV [the conditions of the experiment by Balashova et al. (1982b)] for different scattering angles θ, is presented. The points of intersection of the ovals and the circles mark the angles η^{max} correspondingly to v^{max} for each θ. As is seen from fig. 16, the molecule with the rotation axis perpendicular to the scattering plane ($\eta = 0$) has the largest vibrational energy E_v, but when the rotation axis lies in the scattering plane and is perpendicular to the bombardment direction ($\eta = \frac{1}{2}\pi$), the amount of energy is not enough to excite even the lowest vibrational state.

Thus, at nondissociative surface scattering, a specific type of polarization consisting in the dependence of the vibrational excitation on the rotation axis orientation could be displayed.

3.3. Double scattering

At glancing bombardment, it is necessary to consider the multiple scattering. For a single-crystal surface, both the polar and the azimuthal orientational effects appear.

Using eqs. (24)–(29) deduced in section 2.4 for double scattering, one can show that the rotational energy E_j and the vibrational energy E_v are determined by

$$E_j = [(\gamma_1 x + \gamma_2 x^2 + \gamma_3 x^3)^2 + \gamma_4^2 y^2]\tfrac{1}{2}E_0\theta^2 , \qquad (59)$$

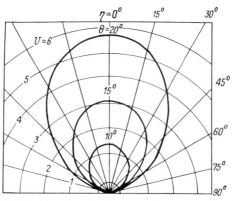

Fig. 16. Polarization of vibrational excitation at surface scattering. Full curve represents $E_v^{max}(\eta)$ for different θ.

$$E_v = \{(\gamma_1 x + \gamma_2 x^2 + \gamma_3 x^3)[x + (1 + \tfrac{1}{2}\mu)\theta] + \gamma_4 y^2\}\tfrac{1}{2}E_0\theta^2 , \tag{60}$$

where $x = \psi \cos \varphi$, $y = \psi \sin \varphi$, θ and the coefficients γ_i are defined by eqs. (4) and (30), respectively.

Having induced a dimensionless variable $t = E_j/\varepsilon$, one can show that the distribution over the rotational energy is given by

$$f(t) = \frac{1}{q(1)} \frac{\partial q}{\partial t} , \tag{61}$$

where

$$q(t) = \int_{x_1}^{x_2} \sqrt{t - (\theta/\theta_0)^2 (\beta_1 x + \beta_2 x^2 + \beta_3 x^3)^2} \, dx . \tag{62}$$

The limits $x_{1,2}$ in integral (62) are roots of the integrand and $\beta_i = \gamma_i/h^i$ ($i = 1, 2, 3$).

The integration results in a distribution

$$\begin{aligned}
f(E_j) &= 1/\varepsilon , & \text{for } z \ll 1 \quad (\theta \ll \theta_k) , \\
&= (2/3\varepsilon)(\varepsilon/E_j)^{1/3} , & \text{for } z = 1 \quad (\theta = \theta_k) .
\end{aligned} \tag{63}$$

The numerical integration of eq. (62) at $z > 1$, when it is necessary to take into account the multiplicity of the integrand, has shown that the function $f(E_j)$ may be approximated by $f(E_j) \approx 1/\varepsilon$.

From eq. (63), it is seen that the distribution of the scattered molecules over the rotational energy essentially depends on the scattered angle θ; hence it must influence the intensity of the rotational excitation at different incidence and escape angles (the polar orientational effect).

Since γ_i depend on the distance between the surface atoms, which is different in different crystallografic directions, the distribution of molecules over the rotational states depends also on the azimuthal angle of target rotation around the surface normal for a fixed angle of incidence (the azimuthal orientational effect). These orientational effects are displayed in fig. 17 for $N_2^+ \rightarrow Cu$, $E_0 = 30$ keV.

Since the energy of the last stable level decreases with an increase of the vibrational number (Herzberg 1950) and as follows from eq. (53) and fig. 16, the energy E_v increases with increasing scattering angle, the break of the rotational structure in the emission spectra should also depend on the scattering angle (the anisotropic spectral effect).

It has been found (Heiland and Taglauer 1972) that the yield of the molecules N_2^+ scattered by an impurity-covered Ni single-crystal surface increases with increasing nitrogen coverage and it is insensitive to the presence of sulphur on the surface. From the papers by Bitensky and Parilis (1980, 1981a, 1984, 1985), it follows that at escape and near the surface, the molecule is still aligned with the escape direction and only further on its

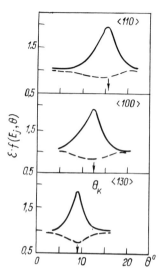

Fig. 17. Polar and azimuthal effects in the distribution of scattered molecules $N_2^+ \rightarrow Cu$ over the rotational energy for two values of E_j: (full curves) $E_j = 0.05\ \varepsilon$ and (broken curves) $E_j = \varepsilon$.

path becomes arbitrarily orientated due to rotation. Such an alignment of the scattered molecules just after the scattering should influence their interaction with the adsorbed atoms and molecules. This, apparently, causes the quenching of both the rotational and the vibrational excitation of the molecules, which increases the nondissociative scattering yield by the surface covered with the proper atoms.

3.4. Conclusion

Thus, the excitation of swift molecules scattered by a solid surface is mainly rotational, while a vibrational excitation is negligible.

At scattering by a solid surface, an inverse population of the highest rotational states, unattainable at thermal excitation, could be achieved.

Some polarization and orientational effects in both rotational and vibrational excitation of molecules scattered by a single-crystal surface could be observed, using the modern methods of laser-induced fluorescence (Kleyn et al. 1981).

It should be noted that heteronuclear molecules with their distant infrared emission spectra are very convenient to study the high rotational states excitation at scattering by a solid surface.

The study of the surface scattering excited highest rotational states could give some information about the potential at large interatomic distances in the molecule (Bernstein 1966, Way and Stwalley 1973).

4. Energy distribution and ionization degree of scattered atoms under bombardment of molecules

4.1. Introduction

The experiments by Heiland et al. (1979), Feijen (1975), Evstigneev et al. (1982), Dodonov et al. (1988), and Willerding et al. (1984a,b) show that the energy distributions of the ion produced by molecule dissociation are broader than the distributions of the ions scattered at bombardment by atomic ions having the same velocities.

Figure 18 illustrates the energy spectra obtained by Heiland et al. (1979) for H_2^+ and H^+ bombardment of W (100) with energy $E_i = 600$ eV. The H^+ parts of the spectra are also shown when incident H_2^+ ions are accelerated by 1200 eV. As is seen, the full width at half maximum of the H^+ peak resulting from H_2^+ bombardment at double incident energy is the largest if compared with those of H_2^+ and H^+ spectra resulting from incident H^+ at the same energy. One can see also that some fraction of scattered H^+ ions resulting from H_2^+ bombardment has an energy E which is larger than the initial energy of the molecule atom E_0.

Heiland et al. (1979) attributed this broadening to an electronic transition produced in the molecule on scattering from the surface, leading to a subsequent mutual repulsion of the molecule nuclei, and hence to a change in their kinetic energy.

The energy of a proton or hydrogen atom originating from a dissociation process in the laboratory system is given by

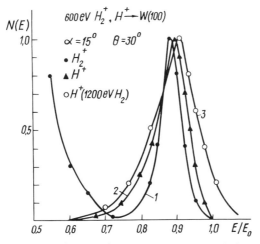

Fig. 18. Energy spectra for H^+ and H_2^+ scattered from W (100) (Heiland et al. 1979): (1) molecules ions H_2^+ ($E = 600$ eV); (2) H^+ resulting from incident H^+ ($E = 600$ eV); (3) H^+ resulting from incident H_2^+ ($E = 1200$ eV, $E_0 = 600$ eV).

$$E = E_0 + \tfrac{1}{2}E_{rep} \pm (2E_0 E_{rep})^{1/2} \cos \psi , \qquad (64)$$

where ψ is the angle between the beam direction and the molecular axis, and E_{rep} is the repulsive energy between the particles, resulting when dissociation occurs.

It is possible that such a mechanism works at low bombarding energies or small glancing angles, for in this case the maximum energy that can be transferred from one atom to another is considerably lower than their energy after scattering. This mechanism will be considered in the next section.

Experiment reveals a spectrum broadening comparable to the initial energy. For example, the measurements by Evstigneev et al. (1982) of the energy spectra of nitrogen, oxygen, and carbon ions scattered under molecule bombardment showed some ions with an energy of approximately 15 keV, whereas the initial energy of the atom in the molecule was $E_0 = 10$ keV. It is clear that such an energy could not be obtained under simple repulsion of atoms from antibonding states.

Bitensky and Parilis (1980, 1981a) have advanced the hypothesis that this broadening is the result of shoving of the scattered atom by another atom from the molecule and redistribution of their kinetic energies, and obtained some estimates, while Bitensky et al. (1985) showed that it is possible to transfer to the atom, in this manner, an appreciable fraction of the initial kinetic energy of the molecule, independently of the change of electronic structure.

From a paper by Bitensky et al. (1988) we present here a general theory of the collisional broadening mechanism. We describe the evolution of the energy distribution of the scattered atoms following impact dissociation of fast molecules on a surface of a solid in general form, in the case of spatial scattering with an arbitrary ratio of the atom masses in the molecule and in the solid. The influence of additional collisions of the partners on the degree of ionization of the scattered atoms is analyzed.

4.2. Formation of the energy spectrum of scattered atoms

In the general case, the energy and angular distribution is determined by the double differential cross section,

$$\sigma(E_0, E, \theta_1, \theta) = \iint\limits_{\Phi(E_0, E, \theta)} f(E_0, E, \theta_1, \theta)\sigma_2(\theta_2, E_1) \, d\theta_1 \, d\theta_2 , \qquad (65)$$

where θ is total scattering angle; $f(E_0, E, \theta_1, \theta)$ is the probability of scattering of an atom by the solid at an angle θ_1; θ_2 and $\sigma_2(\theta_2, E_1)$ are the angle and cross section for an atom scattered by collision with its partner; and E_0 and E are the initial and final energies of the particle. The integration is over all possible scattering trajectories with given final values of E and θ allowed by the kinematic relations. It is also required that close mutual approach by the molecule atoms be possible after one of them is scattered in the solid. This calls for close values of the emission times and points. The function $f(E_0, E, \theta_1, \theta)$ and the trajectory family $\Phi(E_0, E, \theta)$ are different for the molecules consisting of heavy or light atoms. In the first case, it is possible to use the approximation of single or double scattering by atoms of one or two surface layers of a polycrystal (Mukhamedov et al. 1979). The probability $f(E_0, E, \theta_1, \theta)$ of scattering at an angle θ is determined by the cross section $\sigma_1(\theta_1)$ for scattering of the molecule atom by an atom of the solid, or by the cross section for double scattering, and $\Phi(E_0, E, \theta)$ is one trajectory of scattering at an angle θ or a family of a two-fold scattering trajectories.

In the case of light atoms, $f(E_0, E, \theta_1, \theta)$ describes the probability of multiple scattering, and Φ is the aggregate of multiple-scattering trajectories corresponding to a given E_0, E, and θ. The length of these trajectories reach thousands of angstroms.

When the surface is bombarded by a molecule consisting of both heavy and light atoms, such as ArH, the scattering of each atom should be described in its own approximation. In this case the *shuttle acceleration* of the light atom is most pronounced and could contribute even to cluster induced fusion (Bitensky and Parilis 1990).

4.3. Scattering of heavy atoms

Scattering of heavy atoms by a surface is well described in a first-order approximation by a single collision with the surface atom. The problem of broadening of the atom spectrum in collision of a fast molecule with an individual atom was solved by Bitensky et al. (1987). It was shown there that mutual collisions of the molecule atoms cause a considerable broadening of the energy distribution of the atom scattered at an angle θ. The distribution extends to the high-energy part to values of E that exceed not only the single-scattering energy $E_1(\theta)$, but also the initial energy E_0 of the molecule per atom. To find the relations between the kinematical parameters, we use the kinematic diagram of scattering (see fig. 19). Let m_1, m_2 and M be the masses of molecule atoms and target atom, respectively; $\mathbf{OA} = \boldsymbol{v}_0$ the initial velocity of the molecule; the atom with mass m_1 is scattered by the target atom. In the center-of-mass system K' of both the molecule atom and the target atom, the velocity of the molecule atom is equal to

$$\mathbf{OA} = \boldsymbol{v}_0' = \frac{1}{1 + \mu} \boldsymbol{v}_0 ,$$

where $\mu = m_1/M$. After collision with the target atom, the molecule atom is scattered over angle θ_1 and requires velocity $\mathbf{O'F} = \boldsymbol{v}_1'$, the velocity values are equal, $|\boldsymbol{v}_1'| = |\boldsymbol{v}_0'|$. The region of available values $\boldsymbol{v}_1$ is a sphere with radius $|\boldsymbol{v}_0'|$ and center in point O'.

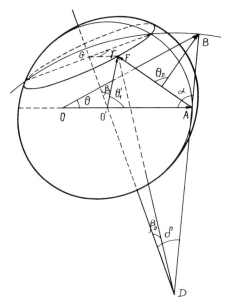

Fig. 19. Kinematical diagram of collision of a molecule with an atom.

After the first collision the molecule atom collides with the second molecule atom moving as before with velocity v_0' and then is scattered over an angle θ_2 in the molecule center-of-mass system. This leads to the turn of relative velocity $v_r = v_0' - v_1'$ by an angle θ_2. The vector $\mathbf{a} = \mathbf{AB}$ is the velocity of the molecule atom after these two collisions in the system K', then $\mathbf{OB} = \mathbf{OA} + \mathbf{AB}$ or $v = v_0 + \mathbf{a}$ is the resulting velocity of the molecule atom in the direction of the detector determined by the scattering angle θ. It should be noted that a multitude of values v_1' corresponds to one and the same value v. This multitude of points represents a circle in the velocity space formed by transaction of the sphere of radius v_1' with the sphere of radius $DB = AB/|1 - \nu|$, where $\nu = m_1/m_2$. The center D of this sphere lies on continuation of AB from the point A or B if ν is less or larger than 1, respectively. The diagram in fig. 19 corresponds to the case $\nu < 1$. The condition of transaction of these spheres limits the available values of resulting velocity $v = (v, \theta)$. The limit of the region of available values v is determined by

$$v^2 = a^2 + v_0^2 - 2av_0 \cos \alpha ,$$

$$\cos \theta = (v_0 - a \cos \alpha)/v , \tag{66}$$

where $a = uv_0(1 + \nu \cos \alpha)$, $u = 2/[(1 + \mu)(1 + \nu)]$. The angle α changes from 0 to π. The boundary determined by $a(\alpha)$ is the Pascal snail which at $\nu = 1$ transforms into a cardioide and at $\nu \to 0$ becomes a circle. These particular cases were considered by Bitensky et al. (1985). Excluding $\cos \alpha$ in eq. (66), we obtain the equation relating the scattering angle to the particle energy,

$$\cos \theta = [1 - ux(1 + \nu x)]/s^{-1/2} , \tag{67}$$

where $s = E/E_0$, and

$$x = \frac{u(1 - \nu u) \pm [u^2 + \nu u(\nu u - 2)(s - 1)]^{1/2}}{\nu u(\nu u - 2)} . \tag{68}$$

Both signs should be used in eq. (68) for $\nu > \nu^*$ and the minus sign for $\nu \leqslant \nu^*$. Then

$$\nu^* = \{[(3 + \mu)^2 + 8\mu(1 + \mu)]^{1/2} - (3 + \mu)\}/4\mu .$$

For $\mu \to 0$, we have $\nu^* = \frac{1}{3}$. For particular cases,

$$\begin{aligned} \cos \theta &= s^{1/2} - s^{-1/2} \pm (2/s - 1)^{1/2} , &&\text{for } \mu \to 0, \ \nu = 1 , \\ \cos \theta &= (s - u^2 + 1)/(2s^{1/2}) , &&\text{for } \nu \to 0 . \end{aligned} \tag{69}$$

A plot of $E(\theta)/E_0$ for $\mu \ll 1$ and different ν is shown in fig. 20. If $\nu \geqslant \nu^*$, the maximum of s is

$$s_{\mathrm{m}} = 1 + \mu/[\nu(2 - \nu u)] , \tag{70}$$

and is reached at $\theta = \theta_{\mathrm{m}}$,

$$\theta_{\mathrm{m}} = \arccos\left\{\left[1 + \frac{u(1 - \nu u)}{\nu(2 - \nu u)^2}\right]/s^{1/2}\right\} . \tag{71}$$

If $\nu < \nu^*$, then $\theta_{\mathrm{m}} = 0$, and $s_{\mathrm{m}} = [u(1 - \nu) + 1]^2$. For $\nu \to 0$ and $\mu \to 0$ at $\theta_{\mathrm{m}} = 0$, in particular, we have $s_{\mathrm{m}} = 9$, while for $\mu \to 0$ and $\nu = 1$, which corresponds to a homonuclear molecule incident on a very heavy target, a value s_{m} is reached at $\theta = 45°$, i.e., one of the atoms of the molecule receives its total kinetic energy.

We consider the angle between the planes $O'FA$ and BFA as the azimuthal angle of the second scattering. The multitude of values v'_1 corresponding to one and the same value v' is determined by the angle γ between the planes $O'DF$ and $O'DB$. It could be shown that the relation between $\{\theta_1, \theta_2, \varphi_2\}$ and $\{v, \theta, \gamma\}$ is given by

$$\cos \theta'_1 = \sin \beta_1 \sin(\alpha \mp \delta) \cos \gamma - \cos \beta_1 \cos(\alpha \mp \delta) ,$$

$$\sin \tfrac{1}{2}\theta_2 = (1 + \nu) \sin < FAB , \tag{72}$$

$$\cos \varphi_2 = \frac{\cos \alpha - \sin \tfrac{1}{2}\theta'_1 \cos < FAB}{\cos \tfrac{1}{2}\theta'_1 \sin < FAB} ,$$

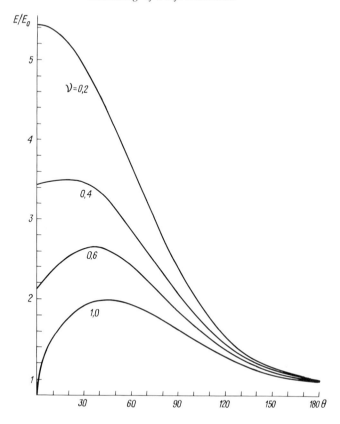

Fig. 20. The relative energy E/E_0 versus the scattering angle θ for different mass ratios v of the molecule atoms.

where

$$a^2 = v_0^2 + v^2 - 2v_0 v \cos \theta \, ,$$

$$\cos \alpha = (1/a)(v_0 - v \cos \theta) \, ,$$

$$\sin \sphericalangle FAB = \frac{r \sin \sphericalangle FDB}{2v_0' \sin \frac{1}{2}\theta_1'} \, ,$$

$$\cos \sphericalangle FDB = \cos \beta_2 \cos \delta + \sin \beta_2 \sin \delta \sin \gamma \, ,$$

$$\sin \delta = \frac{v_0'}{c} \sin \alpha \, ,$$

$$\cos \beta_1 = \pm \frac{r^2 - v_0'^2 - c^2}{2v_0' c} \, ,$$

$$c^2 = v_0'^2 + \frac{v^2 a^2}{(1-\mu)^2} + \frac{2v_0' a v}{1-v} \cos \alpha ,$$

$$\sin \beta_2 = \frac{v_0'}{r} \sin \beta_1 .$$

In eqs. (72), the upper and lower signs correspond to $v < 1$ and $v > 1$, respectively. For $v = 1$, formulae (72) are simplified,

$$\cos \theta_1' = \sin \beta \sin \alpha \cos \gamma - \cos \beta \cos \alpha ,$$

$$\cos \varphi_2 = \frac{\cos \alpha - a/2v_0'}{\cos \frac{1}{2}\theta_1' \sin \frac{1}{2}\theta_2} ,$$

(73)

where

$$\cos \beta = a/v_0' - \cos \alpha ,$$

$$\cos \tfrac{1}{2}\theta_2 = \frac{a}{2v_0' \sin \frac{1}{2}\theta_1'} .$$

Using eqs. (72), we calculate the double differentiated (over angle and energy) cross section according to the paper by Bitensky et al. (1987).

If the flux of molecules with density n impinges on the target atom, then the flux of atoms scattered at solid angle $d\Omega_1'$ near the direction $\{\theta_1', \varphi_1'\}$ (θ_1' and φ_1' are the polar and azimuthal scattering angles, respectively) is given by

$$dN_1 = n\sigma_1(\theta_1') \, d\Omega_1' .$$

(74)

As the molecule orientation is isotropic, the flux density of scattered atoms impinging on the second molecule atom is $dN_1/4\pi l^2$. After this double scattering the atom flux in the solid angle $d\Omega_2$ near direction $\{\theta_2, \varphi_2\}$ is given by

$$dN_2 = \frac{dN_1}{4\pi l^2} \sigma_2(\theta_2) \, d\Omega_2 = \frac{n}{4\pi l^2} \sigma_1(\theta_1') \, \sigma_2(\theta_2) \, d\Omega_1 \, d\Omega_2$$

$$= n \frac{\sigma_1(\theta_1')\sigma_2(\theta_2)}{4l^2} \frac{\sin \theta_1' \sin \theta_2}{\sin \theta} \frac{\partial(\theta_1', \varphi_1', \theta_2, \varphi_2)}{\partial(E, \theta, \varphi, \gamma)} \, dE \, d\Omega \, d\gamma .$$

(75)

Using eq. (75), we obtain for the cross section the following formula,

$$\sigma(\theta, E) = \frac{1}{4\pi l^2 \sin \theta} \int_0^{2\pi} \sigma_1(\theta_1') \, \sigma_2(\theta_2) \sin \theta_1' \sin \theta_2 \frac{\partial(\theta_1', \varphi_1', \theta_2, \varphi_2)}{\partial(E, \theta, \varphi, \gamma)} \, d\gamma .$$

(76)

Since $d\varphi_1'/d\varphi = 0$ and other derivatives on φ equal 1, the Jacobian in integral (76) can be simplified,

$$\frac{\partial(\theta_1', \varphi_1', \theta_2, \varphi_2)}{\partial(E, \theta, \varphi, \gamma)} = \frac{\partial(\theta_1', \theta_2, \varphi_2)}{\partial(E, \theta, \gamma)} .$$ (77)

The cross sections σ_1 and σ_2 are calculated using the inverse-square approximation of the Firsov potential $U(r) = A_{1,2}/r^2$,

$$\sigma_i \sin \theta_i = \frac{A_i \pi^2 (\pi - \theta_i)}{E_i (2\pi\theta_i - \theta_i^2)^2} , \quad i = 1, 2 ,$$

$$E_1 = \frac{E_0}{1 + \mu} , \qquad E_2 = \frac{4E_0}{(1 + \nu)(1 + \mu)^2} \sin^2 \tfrac{1}{2}\theta_1' .$$ (78)

The power potential is a good approximation for the relative energies which are considerably higher than the dissociation energy when the attractive part of molecule potential could be neglected. The Jacobian, eq. (77), for $\nu = 1$ is given by

$$\frac{\partial(\theta_1', \theta_2, \varphi_2)}{\partial(E, \theta, \gamma)} = \frac{2 \sin^2\alpha \sin \beta \sin \gamma}{am_1 v_0' \sin^2 \tfrac{1}{2}\theta_2 \sin \varphi_2 \sin^2\theta_1'}$$ (79)

The integral in eq. (76) could be estimated analytically in the vicinity of the energy spectrum maximum. The maximum in the energy spectrum of particles scattered by an angle θ corresponds to an energy $E = E_1(\theta)$ required by an atom after single collision with a target atom. The minimal angle θ_2 of the molecule atom scattered by its partner corresponds to the maximal impact parameter equal to l. For this energy region, the integrand in eq. (76) strongly decreases with increase of γ. Expanding the integrand over γ in the vicinity $\gamma = 0$ and using eq. (79), we get

$$\sigma(\theta, E) = \frac{\sqrt{2}\sigma_1(\theta_{10}') \sin \theta_{10}' \, \sigma_2(\theta_{20}) \sin \theta_{20}}{\pi l^2 m a v_0' \sin \theta \cos \tfrac{1}{2}\theta_{10}'} .$$ (80)

Since the energy E is close to E_0 in the high-energy part of the spectrum, it follows that $\theta_2 \ll 1$ and, taking for simplicity $A_1 = A_2 = U_2$, one obtains from eq. (80) the approximative equation

$$\sigma(E, \theta) = \sigma_{\max} \left(\frac{\pi U_2}{2l^2} \right)^2 \frac{\sigma_1(\theta_1) E_1 E_0 \sin^2\theta}{\sigma_1(\theta) E_c^2 (E - E_1)^2} ,$$ (81)

where $\sigma_{\max}$ is the scattering cross section corresponding to the maximum impact parameter (equal to l) of the second collision,

$$E_c = \tfrac{1}{2}[E_0 + E_1 - 2(E_0 E_1)^{1/2} \cos \theta_1] ,$$

$$\theta_1 \approx \theta + (E - E_1)/[2(E_0 E_1)^{1/2} \tan \tfrac{1}{2}\theta] .$$

The minimal scattering angle θ_{2min}, corresponding to the impact parameter l, equals $\theta_{2min} = \pi U_2/(2E_c l^2)$. Using this value θ_{2min}, we obtain the minimal energy loss of a molecule atom in the second collision,

$$\Delta E_{min} = \frac{\pi(E_1 E_0)^{1/2} U_2 \sin\theta}{2E_c l^2} \, .$$

For small scattering angles θ, we get

$$\Delta E_{min} = \frac{\pi U_2}{l^2 \theta} \, . \tag{82}$$

From eq. (82), one can calculate that while the relative width $g_1 = E/E_0$ of the energy spectrum of scattered atomic particles does not depend on E_0 (Feijen 1975), the relative broadening $g_2 = \Delta E_{min}/E_0 \simeq E_0^{-1}$ increases with a decrease of E_0 according to the experiment by Feijen (1975).

It is known that the energy spectra of scattered particles resulting from atomic-ion bombardment displays the orientational effects (Dodonov et al. 1984). Since the spectrum broadening is a result of additional scattering of one molecule atom by another, the azimuthal effects in the spectrum width of the scattered ions resulting from molecule bombardment should be smoothed, which agrees with the experiment by Dodonov et al. (1984).

Experiments by Evstigneev et al. (1982) permit a comparison of the calculated spectrum broadening with the data obtained by bombarding various metals of nitrogen molecules. Figure 21 shows the dependence of σ/σ_{max} on the relative value E/E_1 for different E_0 and for a scattering angle $\theta = 40°$. The figure shows that the deduced eq. (81) describes satisfactorily both the form of the high-energy part of the spectrum and the dependence of the cross section on the initial energy.

The mass ratio for the considered combinations of bombarding ions and targets is $\mu \ll 1$. In this case, $E_1 \approx E_0$, therefore, the dependence of the high-energy part of the spectrum for different energies is practically independent of the target material. It should be noted that the agreement was obtained without any fitting parameter whatever, at one and the same constant potential, $U_2 = 33$ eV.

Evstigneev et al. (1982) measured also the ion energy spectrum for bombardment of Au by N_3^+ molecular ions. The presence of one more atom in the molecule should cause a large broadening of the spectrum because of the additional scattering by the third atom of the molecule. This broadening was in fact observed, but its dependence on E_1, the same as for N_2^+, and its relatively small value indicate that it is due only to a decrease of the maximum impact parameter in the molecule, which amounts to $\frac{1}{2}\sqrt{3}l$ for a triangular configuration of atoms. In fact, the experimental ratio $\sigma_{N_3^+}/\sigma_{N_2^+} = 1.9$ is close to value $(2/\sqrt{3})^4 \approx 1.8$ obtained from eq. (81). Copper

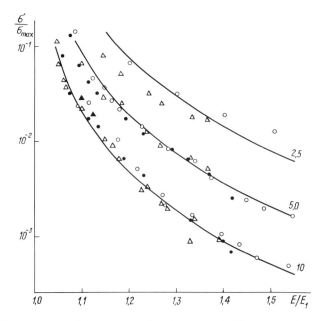

Fig. 21. High-energy part of the spectra at bombardment of different metal targets by N_2^+ ions. Experimental data are from Evstigneev et al. (1982): (○) Cu, (●) Rh, (△) Au, (■) Ti. Full curves represent calculations using eq. (81) for different primary energies in keV.

was bombarded by ArH^+ molecules (Vlasov et al. 1982) to simulate the scattering of slow protons by a solid surface. Our result shows that in scattering of heteronuclear molecules with atoms of greatly differing mass, the spectrum of the scattered light particles should be broadened much stronger than for homonuclear molecules. For example, under the conditions of the experiment by Vlasov et al. (1982), where $\theta = 40°$, the maximum energy as it follows from eq. (69) is $E_{max} \approx 7E_1$. It is also possible to consider the broadening of the argon peak for bombardment by ArH^+ molecules compared with bombardment by atomic Ar^+ ions. To attribute this lengthening of the tail of the energy spectrum to decay from the repulsive state (Heiland et al. 1979) at an initial ArH^+ energy 20 keV in the experiment by Vlasov et al. (1982), the energy required for such a repulsion would be about 100 eV.

It is known that the energy spectrum of heavy atoms scattered by a polycrystal surface does not reduce to a single-scattering peak, but is broadened towards higher energies, or even contains in this region an additional maximum attributed to double collisions with the atoms of solid. The double-scattering model, in which the solid is represented by a diatomic molecule (Mukhamedov and Parilis 1981) is also of independent interest for the physics of atom-molecule collisions. This calculation used to describe the broadening of the energy spectrum of atoms in a collision of a molecule with a molecule (Bitensky et al. 1987). The result of the later calculation can be used to describe the broadening of the energy spectrum of the scattered heavy atoms when a surface is bombarded by molecules, with allowance in the spectrum for double collisions of one of the atoms of the molecule with the atoms of the solid.

As is shown by Bitensky et al. (1987), in double scattering it is necessary to take into account the lag of the second atom of the molecule relative to the atom that undergoes double scattering in the solid. The requirement that the instants of time be close limits the first scattering angle and hence the orientation of the dumbbell molecule that simulates the solid. As a result, when eq. (65) is integrated, the region Φ is narrowed down, and not the entire double-scattering spectrum contributes to the broadened spectrum. With allowance for the double scattering, the cross section [see eq. (65)] is described by Bitensky et al. (1987, 1992a).

It should be noted that the influence of additional mutual scattering of the molecule atom on both energy and angular spectra formation has been shown in computer simulations of the reflection I_2 with $E = 9$ eV from a solid surface (Elber and Gerber 1983).

4.4. Scattering of light atoms

Light atoms are scattered by a solid surface at glancing angles via multiple collisions. This is accompanied by a spread of the points of the emergence of the particles from the solid. The probability of collision of molecule atoms on emerging from a solid is $p \approx l^2/S$, where S is the area of the region of particles emergence. An expression for the scattering coefficient, with allowance for the coordinate of the atom emergence points, was obtained by Remizovich et al. (1980a). It yields $S \approx \alpha_0^5/\langle \theta_s^2 \rangle^2$, where α_0 is the glancing angle and $\langle \theta_s^2 \rangle$ is the mean squared scattering angle per unit path. Since $S \ll l^2$ in this case, the probability of particle collision after emergence is $p \ll 1$. Thus, additional scattering of the molecule atoms by one another on emergence, such as considered in the preceding section, has low probability. At the same time, the spectra of the scattered particles upon bombardment by atomic and molecular particles differ substantially (Schneider et al. 1982). In this paper, the largest broadening of the spectrum was observed at the lowest energies, corresponding to long paths of the scattered particles inside the material. Clearly, the difference is due to the interaction of the molecule atoms with one another as they pass through the material, and not when they are scattered at the surface. A rigorous analysis of such a process entails a solution of the kinetic equations and is a complicated problem. The influence of the collisions of molecule atoms with one another on their energy distribution after emerging from the solid was considered by us under the following approximations. The angular divergence of the molecule atom trajectories is such as that the path L^*, along which the distance between them exceeds l and mutual collisions become unlikely, is equal to $L^* = [lL(E)/\theta]^{1/2}$, where $L(E)$ is the atom path corresponding to a final energy E in the solid. In this case $L(E) = L_0[1 - (E/E_0)^{1/2}]$, where L_0 is the total path of an atom with initial energy E_0, and $L^* \ll L(E)$.

The substantial influence of the interaction of the molecule atoms in the initial section of the trajectory was observed by Escovitz et al. (1979). These authors investigated the energy spectra of protons passing through a carbon film bombarded by H_2^+ molecules of an energy of 12.5 keV/atom. Broadening of the energy distribution was observed already in passage through a film 15 Å thick, and the largest measured energy of the transmitted protons reached 13.5 keV. The calculations by Escovitz et al. (1979) show that, as a result of multiple screening of the Coulomb scattering of protons at dissociation of the H_2^+ molecule, the deviation angle is 5–8° already on a path of the order of 10–20 Å, and the transverse energy in the center-of-mass system reaches 60–80 eV. The mutual scattering of ions takes place thus on the initial path of the trajectory. This leads, in contrast to mutual collisions of heavy atoms on a surface (considered in the preceding section), to a change in the trajectories of the multiple scattering of the ion in the substance, and this change determines in the case of light ions their final energy.

The action of the mutual scattering reduces in the first approximation to a change in the glancing angle α_0. We have then in eq. (65) $f = P(\theta_1)R(a, E)$, where $P(\theta_1)$ is the probability of scattering at an angle θ_1, $R(\alpha, E)$ is the backscattering coefficient of Remizovich et al. (1980b), and $\alpha = \theta - \alpha_0$. The scattering cross section is expressed in the form

$$\sigma(\theta, E) = \iint P(\theta_1) R\left(\frac{\alpha}{\alpha_1 + \alpha_0}, \frac{E}{E_2}\right) \frac{\partial(\theta_1, \theta_2)}{\partial(\alpha_1, E_2)} \sigma_2(\theta_2) \, d\alpha_1 \, dE_2 . \qquad (83)$$

Here $P(\theta_1) = \delta_1/(\theta_1^2 + \delta_1^2)$, which is an approximation of the probability of the deflection of the atom by an angle θ_1 over a path L^* as a result of multiple screened Coulomb scattering (Escovitz et al. 1979). The parameter $\delta_1 = U_1 \pi n L^*/E_0$ depends on the initial (E_0) and final (E) energies of the scattered atom via L^* (n is the number of atoms per unit volume). The function R of Remizovich et al. (1980b) is of the form

$$R(\psi, s) = \frac{3^{1/2}\psi}{2\pi^{3/2}s^{1/2}} \frac{\exp\{-(\psi^2 + \psi + 1)/\kappa s_1\}}{\kappa s_1^2} \mathrm{erf}\left[\left(\frac{3\psi}{\kappa s_1}\right)^{1/2}\right] , \qquad (84)$$

where $\psi = \alpha/\alpha_0$, $s = E/E_0$, $s_1 = 1 - \varepsilon^{1/2}$, κ is a dimensionless parameter and $\mathrm{erf}(x)$ is the probability integral. The integration region in eq. (84) is determined with the aid of a kinematic diagram similar to that given in fig.

19. The integration variable α_1 is the angle of scattering of the molecule atom by a partner in the laboratory system, and the region of integration with respect to energy in the combination of two joined segments, $[E, E_0(1 - 2\sin\alpha_1)]$ and $[E_0(1 + 2\sin\alpha_1), E_{2max}]$. The lower integration limit E_{2max} is given by

$$E_{2max} = E_0(\cos\alpha_1 + \sin\alpha_1 \cotan\beta)^2, \tag{85}$$

where β is the root of the equation $\alpha_1 = \sin\beta \tan\beta \tan[\frac{1}{2}(\beta + \alpha_1)]$. Since the function $R(\psi, \varepsilon)$ decreases rapidly with the increase of the scattering angle, small angles contribute to the integral in eq. (83). At $\alpha_1 \ll 1$, the energy E_{2max} can be approximated by the following expression,

$$E_{2max} = E_0[1 + (2\alpha_1^2)^{1/3}]. \tag{86}$$

In the integration over the angle it must be recognized that the angle of scattering of the molecule atoms by one another is $\theta_2 \geqslant \theta_{min}$, where $\theta_{min} = 2U_2/E_0 l^2\theta^2$ is the minimum scattering angle, determined by the finite distance l between the molecule atoms. As $\theta_{min} \to 0$, the contribution of the small angles increases without limit, the cross section $\sigma_2(\theta_2)$ becomes δ-function-like, and the integration in eq. (83) yields a nonbroadened function $R(\alpha, E)$ corresponding to scattering of atomic particles. The cross section $\sigma(\theta, E)$ in eq. (83) was calculated for bombardment of a polycrystalline Na surface by D_2^+ ions with the initial energy $E_0 = 6$ keV/atom at a glancing angle $\alpha_0 = 5°$ and a scattering angle $\theta = 10°$ [as by Schneider et al. (1982)].

The parameter κ in eq. (84) was chosen to make the function $R(\alpha, E)$ describe correctly the energy distribution of the scattered particles in bombardment by atomic ions. The calculation result is shown in fig. 22. It

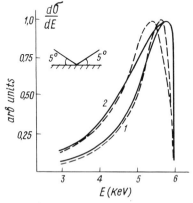

Fig. 22. Energy distribution of D^+ ions scattered on a polycrystalline Na surface bombarded by (1) atomic D^+ ions and (2) molecular D_2^+ ions of energy 6 keV/atom. Full curves represent the experiment by Schneider et al. (1982), broken curves calculations (1) by Remizovich et al. (1980a,b,c); and (2) by eq. (83).

can be seen that the low-energy part of the spectrum is satisfactorily described by eq. (83). The reason for the deviation from experiment in the high-energy part is that the ions scattered with an energy close to the initial one have a short path in the material and the change in the energy of such particles can no longer be regarded as a result of a trajectory change and expressed in terms of the change in the effective glancing angle. In addition, the function $R(\alpha, E)$ (Remizovich et al. 1980b) itself differs in the high-energy part from experimental dependence.

In the region $E > E_0$, the broadening of the spectrum can take place, just as for heavy particles, only as a result of a direct redistribution of the energy among the colliding partners. An estimation of the broadening of the spectrum in this region can be obtained from the equations of sections 4.2 and 4.3. It turns out to be small, but experimentally observable at $E > E_0$ (see fig. 22).

4.5. Ionization degree of scattered particles

The interaction of the molecule atoms with one another when scattered on the surface of the solid can influence various processes. In particular, the ionization degree of the scattered particles can be different, depending on whether the solid is bombarded by molecule or atomic ions. This should be particularly manifested if a collision between the molecule atoms occurs on the final section of the trajectory, e.g., at heavy-molecule scattering on a solid surface. The difference in the population of the excited states of scattered atoms under the atomic and molecular ion bombardment was observed by Christensen et al. (1986).

The evolution of the charge state of a scattered particle in processes in which electrons are captured and lost was considered by Parilis and Verleger (1985d,c) and Nizhnaya et al. (1982b). Electron loss occurs in a hard collision of an incident particle with a surface atom, when the inelastically transferred energy $\Delta\varepsilon_{in}$ exceeds the ionization energy J.

The following equation was obtained by Parilis and Verleger (1985d,c) for scattering of an ion by surface atoms,

$$\eta_a^+(E) = \left\{ \exp\left(-\frac{v_c}{v_{\perp i}}\right) + F(E_0, \theta)\left[1 - \exp\left(-\frac{v_c}{v_{\perp i}}\right)\right] \right\} \exp\left(-\frac{v_c}{v_{\perp f}}\right),$$

$$(87)$$

where $v_{\perp i} = v_0 \sin \alpha$ and $v_{\perp f} = v_1 \sin(\theta - \alpha)$ are the components, normal to the surface, of the incident ion velocity v_0 and scattered ion velocity v_1, respectively; v_c is a parameter characterizing the electron capture, $\exp(-v_c/v_{\perp i})$ is the fraction of ions in the beam incident on the surface atom, and $F(E_0, \theta)$ is the ionization probability of an atom with energy E_0 scattered by a surface atom at an angle θ. The probability $F(E_0, \theta)$ was calculated from the equation (Nizhnaya et al. 1982a)

$$F(E_0, \theta) = \Delta\varepsilon_{in}/J - 1 .$$

As is shown in section 4.2, in surface bombardment by molecules, deflection of a particle by an angle θ is the result of scattering of a molecule atom by a surface atom at an angle θ_1 and additional collision with the partner in the molecule. Since $\theta_1 > \theta$, the collision of the molecule atom with the surface atom is harder than in atomic ion scattering by the same angle θ. As a result, the probability of electron loss is increased. This should increase the degree of ionization η_M^+ of the scattered particles in bombardment by atomic ions, at equal initial energy E_0 per atom at equal scattering angle θ.

The degree of ionization η_M^+ is calculated by summing over different molecule atom trajectories leading to the scattering by an angle θ. As a result we get

$$\eta_M^+ = \left\{ \int_0^\pi \exp[-(v_c/v_{\perp f}) \exp(-az(\gamma))] \sigma(E, \theta, \gamma) \right.$$

$$\left. \times [B + (1 - B)F_2(E_0, E, \gamma)] \, d\gamma \right\} [\sigma(E, \theta)]^{-1} \tag{88}$$

where B,

$$B = \left\{ \frac{1}{2} \exp\left(-\frac{v_c}{v_{\perp i}}\right) + \left[1 - \frac{1}{2} \exp\left(-\frac{v_c}{v_{\perp i}}\right) \right] F_1(E_0, E, \gamma) \right\}$$

$$\times \exp\left\{ -\frac{v_c}{v_\perp(\gamma)} [\exp(-az(\gamma)) - 1] \right\}, \tag{89}$$

is the possibility of the particle remaining charged prior to collision with its partner; $v_{\perp i}$, $v_\perp$, $v_{\perp f}$ are the components normal to the surface of the atom, respectively, in incidence on the surface, after scattering by a surface atom, and after scattering by its molecular partner; $z(\gamma)$ is the distance from the surface to the point of collisions between the molecule atoms; $a \approx 2 \text{ Å}^{-1}$, F_1 and F_2 are the ionization probabilities of the first and second collision. The scattering cross section $\sigma(E, \theta, \gamma)$ was calculated by Bitensky et al. (1987).

Figure 23 shows the dependence of the degree of ionization $\eta_M^+(E)$ on the energy of an atomic ion N^+ scattered by the surface of a copper single crystal, calculated from eq. (88) for different energies of bombardment by N_2^+ molecule ions. As is seen from fig. 23, molecular effects are manifested in the degree of ionization $\eta_M^+(E)$. Their gist is that $\eta_M^+(E)$, unlike $\eta_a^+(E)$, depends on the initial energy E_0. A dependence of $\eta_a^+(E)$ on E_0 was observed also by Eckstein et al. (1978). It was shown by Parilis and Verleger (1985c), however, to be due to the nonequilibrium character of electron capture and loss in scattering by the surface, and to be manifested by the

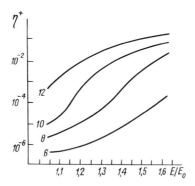

Fig. 23. Degree of ionization η_{M}^{+} of particles scattered by a single-crystal Cu surface bombarded by N_{2}^{+} molecule ions versus the energy of the scattered particles for different primary energies in keV.

so-called surface peaks in $\eta_{a}^{+}(E)$ when E is approximately equal to the single-scattering energy E_{1}. On the contrary, the dependence of $\eta_{M}^{+}(E)$ on E_{0} manifests itself in the region $E > E_{0}$ and is due to additional collisions.

5. Molecule dissociation via electron transition at grazing scattering

5.1. Introduction

In the previous section, the mechanism of such scattering based on the consequent correlated elastic collisions of molecule atoms with surface atoms has been developed. This approximation is valid when the normal kinetic energy $E_{\perp} = E_{0}\alpha^{2}$ considerably exceeds the dissociation energy ε (E_{0} being the initial energy per molecule atom, and α is glancing angle).

The broader energy distribution of scattered atomic particles resulting from molecule dissociation than that from primary ion scattering was explained by additional mutual collisions of the molecule atoms. The findings of these models agree satisfactorily with the experimental data when $E_{\perp} \gg \varepsilon$.

If the normal energy $E_{\perp} \leqslant \varepsilon$, then in the elastic collision model only an essentially small fraction of molecules could dissociate due to multiple scattering (see section 2.8) while the experiments by Willerding et al. (1984a,b, 1986), Willerding (1984), Schubert et al. (1986), Heiland et al. (1987), and Akazawa et al. (1988) have shown that a considerable fraction of the molecule does dissociate. Willerding et al. (1984a) reported results for beams of H_{2}^{+} and N_{2}^{+} grazing incident angles $\alpha < 5°$ and energies $E < 1\,keV$ on an Ni (111) surface. At energies of a few 100 eV, about equal yields of H_{2}

and dissociated H were observed from incident H_2^+. For N_2^+, up to 90% N_2 and 10% N can be obtained. In addition, the nondissociated fractions of N_2 and O_2 are different in experiments by Schubert et al. (1986). As the masses of atoms N and O are close then these results mean that some inelastic electronic processes should affect essentially the molecule dissociation.

The dependence of molecules surviving on the work function, as observed by Schubert et al. (1986), is another proof of the essential contribution of the electronic dissociation mechanism. When covering the Ni surface with about 1 monolayer of Cs, i.e., changing the work function from 5.35 eV to approximately 2 eV, the authors found an appreciable increase of the N_2 molecular peak. The experiments with O_2^+ as primary ion give completely different results. From the clean and cesiated metal surface they did not observe scattered molecules, there were also no surviving ions. These results have been confirmed in experiments of Schubert et al. (1989). In these experiments the scattering of 480 eV CO^+ and CO_2^+ at grazing incidence from Ni (111) +K and clean Ni (111) has been studied. The dissociation of CO^+ was weakly changed by the presence of K, whereas in the CO_2^+ case dissociation was strongly increasing with K coverage.

The electronic dissociation mechanism has been proposed by Heiland et al. (1979). In this model, the resonant neutralization and Auger neutralization processes as the most important processes have been considered. The model predicts, in addition to dissociation, the formation of neutral atoms in the ground electronic state and the formation of electronically excited molecules. This model we consider in more detail in the next section.

Another approach to the molecule surface interaction is given by Holloway and Gadzuk (1985) by combining the charge transfer theory with the classical trajectory theory. They have considered the transfer of an electron from a solid to a neutral molecule (harpooning transition). This transfer takes place before the molecule hits the repulsive wall of the surface and cause the molecule dissociation.

Lucas (1979) has considered the interaction of a molecular ion with the surface plasmons of the solid. From this model it follows that molecular ion survival should depend on the plasmon properties of the solid. Heiland et al. (1987) have compared the obtained experimental results with the findings following from the Lucas model. They have observed nearly the same H_2^+ molecule yield at 500 eV H_2^+ scattering on Al and Ni surfaces, whereas, according to the Lucas model, the H_2^+ yield from these two metals should differ by a factor of four.

5.2. Dissociation due to molecule neutralization into nonbound states

An explanation for the striking difference between the N_2 and H_2 yields was given by Willerding et al. (1984a) using an extension of the charge-exchange models between solid and atoms to the case of molecules. In this paper, the model proposed by Heiland et al. (1979) was developed.

The ionization energies of the molecular states in question as a function of the internuclear distance, and a simplified band structure of Ni are shown in fig. 24. With H_2 as a final product, mainly the ground state $X\,{}^1\Sigma_g^+$ and nonbound triplet state $B\,{}^3\Sigma_u^+$ are important for charge capture. Willerding et al. (1984a) have considered that the electron capture in ${}^3\Sigma_u$ is the main cause of dissociation, and Auger capture into ${}^1\Sigma_g$ the primary mechanism for neutral H_2 formation.

On the other hand, N_2 allows no electron capture into antibound states under the assumption of conservation of the configuration. Only if during the particle solid interaction one of four $1\Pi_u$ electrons is excited, the dissociative states ${}^5\Sigma_g^+$ and ${}^7\Sigma_u^+$ are accessible. With the assumption that N_2^+

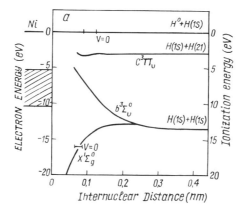

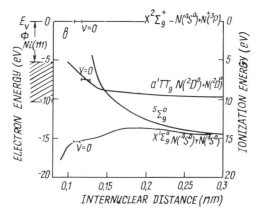

Fig. 24. Simplified band structure in Ni and ionization energies versus internuclear distance for the most important states of (a) H_2 and (b) N_2 (Willerding et al. 1984a).

is in the $X\,^2\Sigma_g^+$ state, the binding state accessible by single electron capture are the ground states $X\,^1\Sigma_g^+$, $B\,^3\Pi_g$, $a\,^1\Pi_g$.

Therefore, N_2 may dissociate only from sufficiently vibrationally and rotationally excited $B\,^3\Pi_g$ and $a\,^1\Pi_g$ states ($v = 12$ and 6, respectively).

The difference between the surviving of N_2 and H_2 was explained by Schubert et al. (1986) on the base of symmetry conservation rules. The consideration of conservation of the symmetries of the initial and final stages gives that transition from the initial state $X\,^2\Sigma_g^+$ of N_2^+ to the final state $X\,^1\Sigma_g^+$ of N_2 is possible whereas $X\,^2\Pi_g$ (O_2^+) to $X\,^3\Pi_g^-$ (O_2) by electron capture is essentially forbidden. Dissociation attachment into numerous Π states which leads to $O + O$ are, however, possible.

The same mechanism has been used by Schubert et al. (1986, 1989) for explaining different dependences of different molecules surviving on work function. Shi et al. (1989) have explained their experimental results on neutralization and ionization processes in the scattering of D_2^+ and D_3^+ on a Mg surface in the framework of this model.

Willerding et al. (1986) have derived the relative energy distribution of N atoms from measured atomic spectra. The relative energy distribution did not depend on the initial molecule energy E_0 in the range below 1 keV. This allowed to conclude that at low transverse energies the processes leading to dissociation were electronic ones. However, at higher energies the expanding of distribution tails have shown that the dissociation was strongly influenced by vibrational excitation.

Using the mechanism of molecule dissociation via electron capture into antibonding states Imke et al. (1986a,b) have provided a parameter-free model of electronic interaction of a molecular diatomic ion with a helium-like metal surface. For low incident velocities ($E_\perp < 10$ eV) the probabilities of the singlet $X\,^2\Sigma_g^+$ and triplet $B\,^3\Sigma_u^+$ states of H_2 were calculated. Capture occurs primarily to the singlet state for molecular axis orientations near parallel to the surface, and to the triplet state for orientations normal to the surface.

Thus, the model of Heiland et al. (1979) explains at least qualitatively the peculiarities of molecule scattering at low transverse energies. However, in this model the Auger relaxation of the excited antibound state has not been considered. Meanwhile, the rate of Auger relaxation of the electronically excited molecule including the exchange of electrons with spin-flip near the metal surface is of the order of $10^{15}\,\text{s}^{-1}$ (Hagstrum 1954a,b) and the number of molecules $R_2 \simeq \exp(-v_c/v_\perp)$ that remain in the excited states at grazing bombardment should be very small [for hydrogen, $v_c \approx 3 \times 10^7$ cm/s and $v_\perp \approx 4 \times 10^6$ cm/s (Eckstein 1981), $v_\perp$ is the normal velocity]. It appeared to be in contradiction with experiments by Willerding et al. (1984a), Willerding et al. (1984b), Willerding (1984), Willerding et al. (1986), Schubert et al. (1986), Heiland et al. (1987), and Akazawa et al. (1988). In addition, from the experiment by Willerding (1984), it follows that the relative number of scattered molecules decreases with $v_\perp$ decrease according to the power law rather than the exponential one. These discrepancies with experiment could be avoided by taking into account both the Auger relaxation and the repeated collision-induced electron spin-flip exchange and singlet–singlet excitation into repulsive states, e.g., $B\,'\Sigma_u^+$ of H_2.

In the next section, the swift-molecule dissociation degree at scattering on the solid surface due to inelastic processes with account of both the relaxation of the excited state and the repeated excitation has been calculated according to Bitensky et al. (1990).

5.3. *Account of repeated excitation*

Let P_{el} be a probability of the nondissociative scattering of a diatomic molecule due to elastic collisions with the surface atoms and Q^* be a probability that the molecule is scattered in the electronically repulsive state. Then the ratio of the number of scattered atoms R_1 to the number of nondissociated molecules R_2 is given by

$$R_{12} = R_1/R_2 = \frac{2[1 - P_{el}(1 - Q^*)]}{P_{el}(1 - Q^*)} \,. \tag{90}$$

The probability P_{el} was calculated by Bitensky and Parilis (1980, 1981a) and section 2.8 [eq. (43b)],

$$P_{el} = (2/\sqrt{\pi})\gamma(\tfrac{3}{2}, x) \approx 1 - (2/\sqrt{\pi})\exp(-x) \,, \tag{91}$$

where $\gamma(\tfrac{3}{2}, x)$ is the incomplete gamma function, $x = 2\varepsilon/(kE_\perp a)$, and k is a constant. As is seen from eqs. (90) and (91), at grazing scattering $(x \gg 1)$ $P_{el} \approx 1$ and the ratio R_{12} should be determined mainly by the electronic inelastic processes, $R_{12} \approx 2Q^*$.

For calculation of Q^* it is necessary to take into account not only neutralization of the molecular ion both into repulsive and ground states (Imke et al. 1986a,b), but also the relaxation and repeated excitation from the ground state into repulsive antibound states due to electronic inelastic processes. The competition of excitation and relaxation is similar to the competition of electron capture and loss at the charge state formation of the scattered atoms (Parilis and Verleger 1985c) and is described by the same mathematics.

The relaxation ω_r and excitation ω_e rates depending on the distance from the surface s are as follows,

$$\omega_r(s) = Af_1(s) \,, \tag{92}$$

$$\omega_e(s) = B(E)f_2(s) \,, \tag{93}$$

where A and $B(E)$ are the relaxation and excitation rates at $s = 0$, respectively. For Auger relaxation $A \approx 10^{15} \text{ s}^{-1}$, for resonance neutralization into the ground state $A \approx 10^{16} \text{ s}^{-1}$ (Hagstrum 1954a,b).

The excitation of molecules from the ground state requires the transfer of a certain amount of energy from nuclear motions to electrons. The rate of this nonadiabatic excitation should increase with the molecule energy E. Let us assume for simplicity that the molecule has only one electrically excited repulsive state. The probability that the molecule after scattering remains in this state is determined by solution of the following equation,

$$v_{\perp} \, dQ^*/ds = -\omega_{\text{r}}Q^* + (1 - Q^*)\omega_{\text{e}} \,. \tag{94}$$

Further, we assume that $Q^* = Q_0^*$ is the initial condition for eq. (94). The dependence of $\omega_{\text{r,e}}(s)$ on the distance s was chosen in the form $f_1(s) = \exp(-as)$ and $f_2(s) = \exp(-bs)$. Then we get the solution of eq. (94) (Parilis and Verleger 1985c),

$$Q^* = Q_0^* \exp(-v_{\text{c}}/v_{\perp}) + B(E)/A(v_{\perp}/v_{\text{c}})^{\nu-1}\gamma(\nu, v_{\text{c}}/v_{\perp}) \,, \tag{95}$$

where $v_{\text{c}} = A/a$, $\nu = a/b$ and $\gamma(\nu, v_{\text{c}}/v_{\perp})$ is the incomplete gamma function. From eq. (95), it is seen that if $B(E) = 0$, then Q^* depends on $v_{\perp}$ exponentially and, therefore, only account of the repeated excitation leads to the power dependence Q^* on $v_{\perp}$. Neglecting the first term in the right-hand side of eq. (95) and using at $v_{\text{c}}/v_{\perp} \gg 1$, the relation $\gamma(\nu, v_{\text{c}}/v_{\perp}) = \Gamma(\nu)$, where $\Gamma(\nu)$ is a gamma function, we get

$$Q^* = B(E)/A\Gamma(\nu)(v_{\perp}/v_{\text{c}})^{\nu-1} \,. \tag{96}$$

From eqs. (95) and (90) it is seen that the ratio R_{12} decreases with decreasing $v_{\perp}$ according to the power law. The dependence of Q^* on energy E is determined both by the parameter ν and the function $B(E)$, while the dependence of Q^* on the escaping angle α_{f} is determined only by ν. As $v_{\perp} \approx v_0 \alpha_{\text{f}}$ (v_0 is the molecule center of mass velocity) then $Q^* \sim \alpha_{\text{f}}^{\nu-1}$. In fig 25, the dependence of R_{12} on the scattering angle θ for H_2^+ taken from the paper by Willerding (1984) is shown (for specular reflection $\alpha_{\text{f}} = \frac{1}{2}\theta$). It is seen that $R_{12}(\theta)$, indeed, could be described by a power dependence with $\nu = 2.5$ and $\nu = 3$ for N_2^+ and H_2^+, respectively. The value $\nu = 1$–2 has been obtained by Parilis and Verleger (1985c) for the ionization degree of light atoms scattered from a solid surface. To find out the dependence of R_{12} on energy E, we could write the excitation rate as $B(E) = nv_0\sigma_{\text{e}}$, where n is the number of atoms of the solid per unit volume, and σ_{e} is the excitation cross section. Assuming that $\sigma_{\text{e}} \approx v_0$, as, e.g., in the Firsov theory, we get $B(E) = cE$ and from eqs. (90) and (96) we obtain $R_{12} \simeq E^{(\nu+1)/2}\theta^{\nu-1}$. In fig. 26, the dependence of R_{12} for $N_2^+ \rightarrow Ni$ both (a) on θ at $E_0 = 275 \text{ eV/atom}$ and (b) E_0 at $\theta = 10°$ are plotted for $\nu = 2.5$. In fig. 27, for $N_2^+ \rightarrow Ni$ the dependence of R_{12} both (a) on θ at $E_0 = 330 \text{ eV/atom}$ and (b) on E_0 at $\theta = 10°$ are plotted for $\nu = 3$. As is seen, the experimental data of Willerding

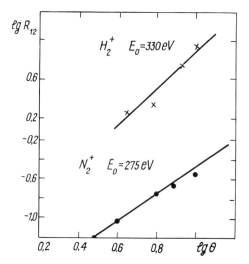

Fig. 25. The ratio R_{12} versus the scattering angle θ for H_2^+ ($E_0 = 330$ eV/atom) and N_2^+ ($E_0 = 275$ eV/atom) (Willerding 1984).

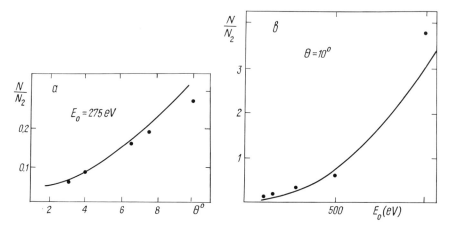

Fig. 26. The ratio R_{12} for $N_2^+ \rightarrow$ Ni (a) versus the scattering angle θ at $E_0 = 275$ eV/atom and (b) versus primary energy E_0 at $\theta = 10°$. Points represent the experiment by Willerding (1984); curves the theory.

(1984) were described satisfactorily by eq. (90) with eq. (96). We used the following values of the parameters: $v_c = 10^7$ cm/s, $A = 0.4 \times 10^{15}$ s^{-1}, for $N_2^+ \rightarrow$ Ni $c = 3 \times 10^{12}$ (eV s)$^{-1}$ and for $H_2^+ \rightarrow$ Ni $c = 3.5 \times 10^{12}$ (eV s)$^{-1}$. It should be noted that in these calculations there are only two fitting parameters: ν and the combination $c/(A v_c^{\nu-1})$.

   The molecules scattered on a solid surface could be excited not only into a

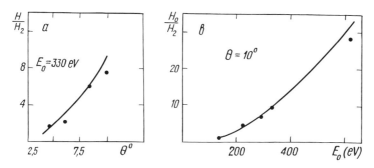

Fig. 27. The ratio R_{12} for $H_2^+ \to Ni$ (a) versus scattering angle θ at $E_0 = 330$ eV/atom and (b) versus E_0 at $\theta = 10°$. Points represent the experiment by Willerding (1984); curves the theory.

repulsive state but also into some bound states. In a paper by Shamir et al. (1982), the processes of neutralization of N_2^+ with $E < 50$ eV on different metal surfaces were considered and it was shown that the probability of neutralization in different bound states depended on the structure of electron state density. The processes of relaxation and excitation lead to the population of bound excited states depending on the electron structure of metals. It should result in a change of the molecule scattering as well as in $E_\perp > \varepsilon$. For example, if we take into account only the elastic collisions and consider that the molecule dissociates from the ground state, then the relative yield of the scattered molecular ions N_2^+ at bombardment of Ag should be by a factor of 1.5 larger than at bombardment of Cu (Bitensky and Parilis 1981a, 1984) while as it follows from the experiment by Dodonov et al. (1988), R_{12} for $N_2^+ \to Ag$ and $N_2^+ \to Cu$ ($E = 30$ keV) are approximately equal. Apparently, this could be explained by the fact that the inelastic cross section for $N_2^+ \to Ag$ is larger than for $N_2^+ \to Cu$, therefore, the probability of molecule excitation into a bound state for $N_2^+ \to Ag$ increases due to $B(E)$ increase. This results in a decrease of the effective dissociation energy and hence R_{12} decreases. The same arguments can be used for an understanding of the nearly equal R_{12} behavior for $N_2^+ \to Au$ and $N_2^+ \to C$ ($E = 1.5-4.5$ keV, $\theta = 22°$) observed by Sass and Rabalais (1988).

We would like to note that the relaxation and repeated excitation of the excited bound states should lead to the dependence of the radiation intensity on v according to the power law as it follows from eq. (86).

As is shown, the excitation of the molecule into a repulsive state and, therefore, the electronic dissociation are the result of a competition between collision-induced excitation and Auger relaxation along the molecule scattering trajectory. In case of a single crystal the surface scattering along the atomic row or in the surface semichannel is a result of a set of correlated collisions, in which the competition balance is ruled by the consequence of impact parameters. It is well-known from both the experimental and theoretical investigations that in this case some orientational effects in the processes depending on the impact parameter, angular and energy distributions (Dodonov et al. 1988), charge state formation (Hou et al. 1978, Nizhnaya et al. 1979c) and nondissociative molecule scattering (Heiland et al. (1979), Dodonov et al. 1988), are displayed.

The electronic excitational dissociation should contain some analogous azimuthal orientational effects. When rotating the target around the surface normal, keeping both the glancing and escaping angles constant, the degree of dissociation should oscillate near a certain mean value. In addition, a fine density effect for scattering along the close-packed crystallographic direction should be expected. It consists in a drastic, up to double, increase of the excitation probability due to the accumulation of the electronic excitation when the time of flight between two consequent collisions is less than the Auger relaxation time.

Another interesting effect could be observed in molecule scattering on a single-crystal surface (Bitensky et al. 1992d). The light molecules could dissociate at scattering along the surface atomic rows or surface semichannels also by the resonance excitation into a repulsive state, a version of the Okorokov effect (Okorokov 1965). This excitation consists in a resonance between the transition frequency $\Delta E/(2\pi\hbar)$ and frequency v_0/d of the periodical perturbation field which affects the molecule moving along a row with the interatomic distance d. Thus, resonance occurs at $v_0 = \Delta Ed/(2\pi\hbar)$. For H_2 at transition from the ground state $X_1 \Sigma_g^+$ into the continuum $B' \Sigma_u^+$ we get the energy difference $\Delta E = 15$ eV. With $d = 3$ Å we have obtained that the resonance is reached at a molecule energy of $E_0 = 5$ keV/atom. The orientational peculiarity and resonance velocity behavior of this effect should make its experimental recording easy.

CHAPTER 10

Sputtering of Clusters and Biomolecules under Ion Bombardment

1. Introduction

It is well-known that at sputtering of a solid under ion bombardment a considerable number of particles are emitted as multiatomic clusters, both neutral and charged (Staudenmaier 1972, Herzog et al. 1973, Gerhard and Oechsner 1975, Wittmaack 1979a,b, Katakuze et al. 1985, Colton et al. 1986, W. Begemann et al. 1986, Dzhemilev et al. 1987, Gnasser and Hofer 1989).

The investigation of cluster emission from solids under ion bombardment is important for a better understanding of the processes of interaction of accelerated particles with solids. There is also a practical interest in studying the characteristics of complex ions for producing thin films and epitaxial layers. The clusters carry valuable information on adsorption processes, surface oxidation, catalysis, molecule stability, physical and chemical composition as well as the structure of a solid.

In section 2, we consider the sputtering of small clusters, mainly dimers, within the framework of the statistical model (Staudenmaier 1972, Gerhard 1975, Können et al. 1975). The influence of the different sputtered atom moments on dimer formation probability has been shown (Bitensky and Parilis 1978b).

The direct dimer knocking-out is considered in detail as a process of swift-ion scattering by a molecule without its dissociation (Bitensky and Parilis 1982). The *push-and-stick* (Bitensky et al. 1992b) and the *plane-shaving* model of ejection of high-energy dimers has been proposed. However, the statistical model of dimer formation cannot explain the emission of large clusters observed experimentally. The yield of large clusters must be determined not by independent individual collisions but by some collective effects of momentum transfer to a group of atoms ejected intact.

In section 3, a mechanism for large cluster emission during sputtering of solids under ion bombardment via formation and emergence of shock waves originating from some high-density energy spikes is proposed.

In section 4, the same model is developed to describe the biomolecule desorption under heavy-ion bombardment. The dependences of the yield on angle of incidence, energy and energy loss, charge state of incident ions, film thickness as well as angular, energy and spatial distributions of biomolecules have been calculated and compared with the experimental data. The destruction of macromolecules located near heavy-ion track under δ electrons has been discussed.

2. Small-cluster sputtering

When considering the sputtering of clusters containing several atoms, Staudenmaier (1972) has assumed that a cluster is formed of sputtered atoms, belonging to the same collision cascade if the time interval between ejection is sufficiently small and if the kinetic energy of their relative motion does not exceed the dissociation energy. On the basis of this model, the dimer energy distribution has been obtained by Können et al. (1975), and the relative yields of dimers and trimers were calculated by Gerhard (1975) in assumption that the points of sputtered atoms emission were distributed uniformly over a certain area. Computer simulation by Harrison and Delaplain (1976) has corroborated this model.

The possibility of direct knocking out of dimers by cascade atoms has been discussed by Joyes (1971) in the framework of the model in which a low-velocity atom moves parallel to the diatomic molecule axis. However, in the papers by Können et al. (1975) and Gerhard (1975) the collision cascade resulting in sputtering has not been considered actually and the straggling of the atom ejection moments have not been taken into account. According to Joyes (1971), it is impossible to estimate the probability of direct dimer knocking out.

2.1. Account of surface ejection points distribution

We take into account these distributions following the paper by Bitensky and Parilis (1978b). As distinct from the assumption made by Gerhard (1975) on uniform distribution of atom ejection coordinates, according to Matzumura and Furukawa (1973) it is described by a Gauss function,

$$f(\rho) = S/(\pi \overline{R^2}) \exp(-\rho^2/\overline{R^2}) , \tag{1}$$

where S is the sputtering yield, $(\overline{R^2})^{1/2}$ is the dispersion, $\rho^2 = x^2 + y^2$, and x and y are the atom ejection coordinates on the solid surface. For simplicity, we consider the case of normal bombardment.

The probability that two sputtered atoms are ejected from points separated by a distance r is determined by

$$P_1(r) = \int f(\boldsymbol{\rho})f(\boldsymbol{\rho}')\delta(\boldsymbol{\rho} - \boldsymbol{\rho}' - r)\, \mathrm{d}\boldsymbol{\rho}\, \mathrm{d}\boldsymbol{\rho}' = S^2/(\pi \overline{R^2}) \exp(-r^2/\overline{R^2}) . \tag{2}$$

If one of the particles is put in the center of a circle, then the probability that the other particle will be inside a circle with radius r_0 (r_0 is chosen in such a way that for $r > r_0$ one can consider the particles as noninteracting) equals to

$$P_2(r_0) = 2\pi \int_0^{r_0} P_1(r)\,dr = S^2[1 - \exp(-r_0^2/2\overline{R^2})]\,. \tag{3}$$

Taking into account that $r_0^2 \ll \overline{R^2}$, we get the formula for the dimer relative yield,

$$Y = P_2/S \approx Sr_0^2/(2\overline{R^2})\,. \tag{4}$$

From eq. (4), one can get the formula deduced by Gerhard (1975) if the area through which the sputtered atoms are ejected is determined by $F = \pi\overline{R^2}$.

For inclined incidence, the dispersion over the x and y axes are character-ized by different values: $\overline{R_x^2}$ and $\overline{R_y^2}$, respectively. It is easy to show that in this case we obtain eq. (4) in which $\overline{R^2} = (\overline{R_x^2 R_y^2})^{1/2}$.

Now we estimate the influence of difference in the atom ejection moments on dimer-formation probability. Using a computer simulation, Bitensky and Parilis (1978b) have shown that the distribution of ejection moments is determined by

$$\psi(t) = (1/T)\exp(-t/T)\,, \tag{5}$$

where T is the mean cascade developing time. Similarly to the calculation, eq. (2), we get the following formula for the probability that the ejection moments for the two atoms are separated by a time interval τ:

$$W(\tau) = (1/T)\exp(-\tau/T)\,. \tag{6}$$

These two atoms form a dimer if the momentum of the atom to be ejected the first satisfies the condition $p \leqslant p_0 = Mr_0/\tau$. Taking into account this restriction, the dimer-formation probability is given by

$$P_2' = \int_0^\infty W(\tau)\,d\tau \int \varphi(p)\,dp\,, \tag{7}$$

where $\varphi(p)$ is the distribution over the sputtered atom momentum. The inner integral is calculated over the upper hemisphere with radius p_0. For the particles ejected along the surface normal we can calculate P_2' by substituting $W(\tau)$ from eq. (6) and $\varphi(r) = 8ME_b\, p/[\pi(p^2 + 2ME_b)^3]$ in the eq. (7) (Können et al. 1975).

Then

$$P_2' = 8ME_b/T \int_0^\infty e^{-\tau/T}\,d\tau \int_0^{p_0} p^3/(p^2 + 2ME_b)^3\,dp$$

$$= 1/T \int_0^\infty e^{-\tau/T}/[1 + (\tau/t_b)^2]^2\,d\tau\,. \tag{8}$$

Here, M is the mass of the sputtered atom, E_b is the surface binding energy, $t_b = r_0 \sqrt{M/2E_b}$ is the time interval during which an atom with energy E_b passes the distance r_0. Thus, account of the difference in the atom ejection moments leads to an additional dependence of dimer formation probability on the E_b through eq. (8). Integral (8) cannot be calculated analytically. At $T \ll t_b$, when the difference in the atom ejection moments could be neglected, $P_2' \approx 1$. At $T \gg t_b$, the probability $P_2' \approx t_b/T$. Then the relative yield $Y \approx (\frac{1}{2} S) Y_0 t_b/(R^2 T)$, therefore, the atom ejection moments could essentially reduce the dimer formation probability.

2.2. Direct dimer knocking out

The knocking out of a dimer from the solid surface by a cascade atom has been studied by Bitensky and Parilis (1978b). The probability of knocking out of high-energy dimers was calculated in the case where the cascade atom moves under small angle to the axis connecting the two surface atoms.

To describe such dimer knocking out in detail we have to consider first the process of nondissociative scattering of swift ions by a diatomic molecule, which is possible at such molecule axis orientation when the molecule does not dissociate but acquires a kinetic energy exceeding the dissociation energy (Bitensky and Parilis 1982).

We consider small-angle scattering, where the momentum approximation is valid but the corresponding impact parameters are much less than the interatomic distance in the molecule. Thus, the ion–molecule interaction potential could be presented as the sum of interaction potentials of the ion with each molecule atom.

As it was shown by Gerasimenko and Oksyuk (1965), the quantum correction to the dissociation probability for the majority of molecules could be neglected.

The nondissociative collision for a homonuclear molecule takes place if the relative kinetic energy of the atoms does not exceed the binding energy. For small scattering angles, the condition is

$$\tfrac{1}{2} \mu E_0 (\theta_1 \boldsymbol{n}_1 - \theta_2 \boldsymbol{n}_2)^2 \leq \varepsilon , \tag{9}$$

where $\mu = m_1/m_2$, m_1 and m_2 are the ion and molecule atom masses, respectively, E_0 is the projectile energy, ε is the dissociation energy, $\boldsymbol{n}_{1,2}$ are unit vectors of recoil directions, and $\theta_{1,2}$ are the scattering angles in collision with the first and second atom, respectively. Since the recoil energy $E_2 \ll E_0$, both the displacement of the first recoil atom during the time of flight between the two collisions and the elastic energy loss in the first collision could be neglected.

Since at nondissociative scattering the difference between the impact parameters b_1 and b_2 in a collision of the ion with molecule atoms is small, the polar angle of the orientation of the molecule axis relative to the projectile direction $\psi \ll 1$. By expanding $\cos(\boldsymbol{n}_1, \boldsymbol{n}_2)$ and $b = b_2 - b_1$ in power series of ψ and θ (the scattering angle $\theta = |\theta_1 \boldsymbol{n}_1 + \theta_2 \boldsymbol{n}_2| = 2\theta_1$) up to terms ψ^2 and θ^4, we get condition (9) in the following form,

$$l^2(d\theta/db_1)^2(\tfrac{1}{2}\theta - \cos\varphi)^2 + \tfrac{1}{4}\theta^2[\psi(l/b_1)^2\sin^2\varphi + \tfrac{1}{4}\theta^2] \leq 2\varepsilon/\mu E_0, \quad (10)$$

where l is the interatomic distance in the molecule, φ is the azimuthal angle of the molecular axis relative to the plane passing through the ion velocity $\boldsymbol{v}_0$ and $\boldsymbol{b}_1$. As follows from eq. (10), the maximum angle of scattering without dissociation (at $\psi = 0$ and $\cos\varphi = \tfrac{1}{2}$) equals

$$\theta_m = 2(2\varepsilon/\mu E_0)^{1/4}, \quad (11)$$

and exceeds the limiting angle θ' in the spectator model (Gerasimenko and Oksyuk 1965, Green 1970). The value $\theta' = 2(2\varepsilon/\mu E_0)^{1/2}$ one could obtain from eq. (9) at $\theta_2 = 0$. Nondissociative scattering is possible for molecules with axes orientated in a solid angle Ω_1 for which the values of angles ψ and φ satisfies inequality (10). Therefore, the probability of the process is given by

$$P(\theta) = 2\Omega_1/4\pi \approx 1/2\pi \int_{\Omega_1} \psi \, d\psi \, d\varphi. \quad (12)$$

For the potential $U = U_0 r^{-n}$, the scattering angle has been determined by

$$\theta_{1,2} = cb_{1,2}^{-n}, \quad (13)$$

where $c = U_1/E_0$, $U_1 = \sqrt{\pi} U_0 \Gamma(n + \tfrac{1}{2})/\Gamma(\tfrac{1}{2}n)$, and $\Gamma(x)$ is the gamma function. Substituting b_1 from eq. (13) into eq. (11) and integrating eq. (12) we get

$$P(\theta) = 1/(8n)\theta_l^{2/n}\theta^{2(n-1)/n}[(\theta_m/\theta)^4 - 1], \quad (14)$$

where $\theta_l = cl^{-n}$. The cross section averaged by the solid angle Ω_1 is

$$\sigma(\theta, \Phi') = 1/(2\pi) \int_{\Omega_1} b_1/\theta| \partial(b_1, \Phi)/\partial(\theta, \Phi')| \, d\Omega, \quad (15)$$

where Φ and Φ' are the azimuthal angles, $|\partial(b_1, \Phi)/\partial(\theta, \Phi')|$ is the Jacobian of transformation from variables $\{b_1, \Phi\}$ to (θ, Φ'). It could be shown that in the above approximation the Jacobian approximately equals $\partial b_1/\partial\theta$. As follows from the double-scattering model (Martynenko 1973, Mukhamedov et al. 1979), at some angles θ_1 and ψ the derivative $\partial\theta/\partial b_1 = 0$ and the cross section becomes infinite. It is easy to show that the angle θ_l is expressed by

$$\theta_i = 4/n(\tfrac{1}{2}n\theta_i)^{1/(n+1)} , \tag{16}$$

and the derivative $\partial\theta/\partial b_1$ is given by

$$\partial\theta/\partial b_1 = 2\theta_i/b_1[1 - (\theta/\theta_i)^{(n+1)/n}] . \tag{17}$$

Note that from eq. (16) it follows $E_0\theta_i^{n+1} = \text{const.}$

Taking into account eq. (17), we get from eq. (15) the cross section of the nondissociative scattering as

$$\sigma(\theta) = \sigma_1(\theta)2^{2/n}P(\theta)/[1 - (\theta/\theta_i)^{(n+1)/n}] , \tag{18}$$

where $\sigma_1(\theta)$ is the cross section of the single scattering. Although the probabilty of nondissociative scattering $P(\theta) \ll 1$, the singularity of cross section at $\theta = \theta_i$ gives a good opportunity to observe this process.

Since at $n \leqslant 2$, with increasing E_0, the angle θ_i decreases more rapidly than the angle θ_m does, the peculiarity of the cross section should appear at energy E_0^*, determined by $\theta_i = \theta_m$,

$$E_0^* = \{[\mu/(2\varepsilon)]^{1/4}2/(nl)(\tfrac{1}{2}nlU_1)^{1/(n+1)}\}^{4(n+1)/(3-n)} . \tag{19}$$

Using the relation $\varphi_2 = \tfrac{1}{2}\pi - \tfrac{1}{4}(1 + \mu)\theta$, it is simple to obtain from eq. (18) the cross section as a function of the molecule center-of-mass recoil angle φ_2.

The dependence of the cross section on molecule recoil energy E_2 could be obtained taking into account that the molecule center-of-mass energy $E_2 = \tfrac{1}{2}\mu E_0\theta^2$. This dependence is

$$\sigma(E_2) = \sigma_1(E_2)2^{2/n-3}/n\theta_1^{2/n}(2E_2/\mu E_0)^{(n-1)/n}$$
$$\times [(E_m/E_2)^2 - 1]/[1 - (E_2/E_i)]^{(n+1)/2n} , \tag{20}$$

where $E_{m,i} = \tfrac{1}{2}\mu E_0\theta_{m,i}$ exceeds essentially the dissociation energy. Note that $E_m = (8\mu\varepsilon E_0)^{1/2}$ does not depend on the chosen potential and E_i does not depend on E_0 for the Coulomb potential at $n = 1$. The dependence $\sigma(E_2)/\sigma_1(E_2)$ on E_2 is shown in fig. 1 for the scattering of a nitrogen ion by a nitrogen molecule. We have used in the calculation $l = 1.12$ Å, $\varepsilon = 8.7$ eV, $n = 2$. At these values and for $E_0 = 4$ keV, we get $E_0^* = 2.7$ keV, $\theta_m = 30°$, $E_m = 530$ eV, $E_i = 500$ eV.

It should be noted that the molecule can dissociate due to the electronic excitation into continuum at low-keV energies when the ion velocity is much less than the orbital one, the electronic excitation can be neglected. The appearance of the peculiarities in the cross section of small-angle scattering by two Coulomb centers was discussed by Demkov (1981). However, the focal points on the impact parameter plane correspond to scattering when

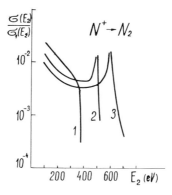

Fig. 1. The dependence of ratio $\sigma(E_2)/\sigma_1(E_2)$ on the molecule recoil energy E_2 for different primary ion energies E_0: (1) $E_0 = 2\,\text{keV}$, (2) $E_0 = 4\,\text{keV}$, (3) $E_0 = 8\,\text{keV}$.

the molecule axis is perpendicular to the ion velocity. In our case, the fast recoil just appears at the consecutive collisions with the molecules atoms, when the molecule axis is aligned parallel to the primary ion velocity and nondissociative scattering at large angles is possible.

The above-deduced expressions could be applied also to the nondissociative scattering of swift molecule ions by replacing θ by $2(m_2/m_1)\theta$ and E_0 by $(m_1/m_2)E_0$. The scattering of swift molecules without dissociation was considered in chapter 9.

2.3. Ejection of high-energy dimers at grazing incidence

When ions are bombarding a nonideal single-crystal surface along the plane of incidence coinciding with a low-index direction, the movement of an ion depicted schematically in fig. 2, is possible. After undergoing reflection from the surface, the ion finds itself beneath an atomic step, and as a result of successive correlated collisions knocks two adjacent atoms out of the crystal without disrupting the bonds between them (Bitensky and Parilis 1981b). A consequence of this is that a flux of sputtered particles should contain dimers with an energy substantially exceeding the average energy of the sputtered clusters. In order for two atoms to be ejected as a dimer, their relative energy E_r must not exceed the dissociation energy ε,

$$E_r = \tfrac{1}{4}m_2(\boldsymbol{v}_1 - \boldsymbol{v}_2)^2 \leqslant \varepsilon . \tag{21}$$

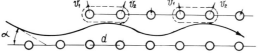

Fig. 2. Scheme of the ejection of high-energy dimers.

Here m_2 is the atomic mass of the solid, and $v_{1,2}$ are the recoil velocities of the two atoms. For small scattering angles they are $v_{1,2} = \mu v_0 \theta_{1,2}$, where $\mu = m_1/m_2$, v_0 is the ion velocity, m_1 is the ionic mass, and $\theta_{1,2}$ are the scattering angles in the laboratory system. In the paper by Bitensky and Parilis (1978b) it was shown that an ejected dimer has a maximal energy when the turning point of the trajectory lies midway between two atoms. The relative energy then takes the form

$$E_r = \tfrac{1}{2} \mu E_0 \theta_m^4 , \tag{22}$$

where θ_m is the scattering angle at the midpoint determined by eq (11).

In this case, the center-of-mass energy of the dimer is given by

$$E_c = 2\mu E_0 \theta_m^2 . \tag{23}$$

Using eqs. (22) and (23), we obtain from condition (21) the expression for the maximal energy of a dimer,

$$E_c^{\max} = (8\mu E_0 \varepsilon)^{1/2} , \tag{24}$$

which is the same as in previous subsection.

In order to obtain the conditions for fast dimers to be ejected, we consider the motion of an ion as was done by Grishin and Skripka (1975). If an ion with energy E_0 impinges a chain of atoms at an angle α, then the impact parameters b_n for the successive collisions are related by

$$b_n - b_{n+1} = d\left(\alpha - \sum_{i=1}^{n} \theta_i\right) , \tag{25}$$

where d is the interatomic distance in the chain, and θ_i is the scattering angle in the ith collision. Using the inverse square approximation of the Firsov potential, $U(r) = U_0/r^{-2}$, and the momentum approximation, we obtain $\theta_n = \pi U_0/2E_0 b_n^2$. Introducing the function $f = \theta^{-1/2}$, we get from eq. (25) the equation

$$-\frac{df}{dn} = A\left(\alpha - \int_1^n \frac{dn}{f^2}\right) , \tag{26}$$

where $A = d(2E_0/\pi U_0)^{1/2}$. Upon solving eq. (26) with the initial condition $f_1 = Ab_1/d$ (b_1 is the impact parameter in a collision with an atom located at the edge of surface step), we obtain the expression

$$1/f_m = 1/f_1 + (\tfrac{1}{2} A\alpha^2) , \tag{27}$$

for the value of f at a turning point.

Since $\theta_m = f_m^{-2}$, we find from eq. (22), using eq. (21), that ejection of dimers is possible at grazing angles $\alpha < \alpha_2$, where

$$\alpha_2 = \left\{ \frac{2}{A} \left[\left(\frac{2\varepsilon}{\mu E_0} \right)^{1/3} - \frac{d}{Ab_1} \right] \right\}^{1/2} . \tag{28}$$

For bombardment energies of the order of a few keV, the second term inside the square root in eq. (28) can be neglected, and then

$$\alpha_2 \approx (2\pi U_0/d^2)^{1/4} (2\varepsilon/\mu)^{1/16} E_0^{-5/16} . \tag{29}$$

Using eqs. (23) and (27), we obtain that the dependence of the dimer energy on the grazing angles is given by

$$E_c = \mu d^4 E_0^3 \alpha^8 / 2\pi^2 U_0^2 \tag{30}$$

and the value E_c^{max} in eq. (24) is attained at $\alpha = \alpha_2$. With decreasing glancing angle, the ion trajectory passes far from the surface atoms and the dimer knocking out is possible when

$$E_c > 2E_b - \varepsilon . \tag{31}$$

Using eq. (30), we get that the ion can knock out the dimer at $\alpha > \alpha_1$, where

$$\alpha_1 = [2\pi^2 U_0^2 (2E_b - \varepsilon)/(d^4 \mu E_0)]^{1/8} . \tag{32}$$

If a silicon single crystal is bombarded by Ar^+ ions of energy $E_0 = 4\,keV$, then taking the values $\varepsilon = 2\,eV$, $d = 2.5\,Å$, $d/b_1 \approx 1$, we obtain $\alpha_1 = 5°$, $E_c^{max} \approx 300\,eV$, $\alpha_2 = 16°$. We note that in the experiment by Wittmaack (1979a,b) the most probable energy of Si_2^+ dimers for ion bombardment normal to the surface did not exceed $1\,eV$, and the half width of the spectrum was $\approx 5\,eV$.

Thus, ejection of fast dimers should give rise to a peak in the high-energy part of the cluster energy distribution, or at least to a significant broadening of the peak.

Analogously to calculations carried out in section 3 of chapter 9, it could be shown that the excitation of knocked out dimers is mainly rotational. The rotation energy approximately equals to relative energy of motion of the molecule atoms. Therefore, from eq. (22) with account of eq. (27) we obtain that the molecule rotational energy is given by

$$E_{rot} = \tfrac{1}{2} \mu E_0 (1/f_1 + \tfrac{1}{2} A\alpha^2)^8 \approx \mu E_0^5 d^8 / (2^5 \pi^4 U_0^4) \alpha^{16} . \tag{33}$$

As follows from eq. (33), the rotational energy strongly depends on the glancing angle α and achieves the maximal value $E_{rot,max} = \varepsilon$ at $\alpha = \alpha_2$. Thus, the rotational excitation of knocked out dimers differs significantly from the rotational excitation via others mechanisms (De Jonge et al. 1988, Hoogerbrugge and Kistemaker 1987).

The computer simulation was carried out for a detailed description of dimer knocking out. The elementary cell was determined on the face of a single crystal near the surface step. The impact points were chosen uniformally for the ion impinging under glancing angle α. The ion trajectories were calculated on the basis of the binary collision model. The potential of the ion interaction with crystal atoms was used in the form of eq. (19) of chapter 1. The thermal vibration were simulated as usual, using the normal distributed displacements with a temperature being equal to 300 K.

It was considered that the dimer knocking out takes place when conditions (21) and (31) were fulfilled. The energy, ejecting angle of dimers, as well as the dimer yield Y_2 per one primary ion were calculated.

The calculations were carried out for the face (100) of a Cu single crystal, the plane of incidence coincided with the $\langle 110 \rangle$ direction. For this case, the step height is the largest one and equals to lattice constant.

The results of the calculation show that the yield of knocked out dimers depends nonmonotonically on the glancing angle. As is seen from fig. 3. The dimer knocking out is possible when $\alpha_1 < \alpha < \alpha_2$, where $\alpha_{1,2}$ are described by eqs. (28) and (32), respectively. The yield increases with the increase of primary ion mass.

Figure 4 shows the calculated distribution of dimers ejected at grazing angle $\alpha = 10°$. Thermal fluctuations were neglected. The histograms reveal the presence of a peak in the high-energy part of the energy distribution, and it is clear from fig. 4 that most of the high-energy dimers are ejected at near-normal directions.

The results of the calculation reveal that fast dimers are ejected by ions with impact points lying close to a ridge in an atomic chain, and a single ion can knock out several dimers (see fig. 2). As the grazing angle increases, the peak corresponding to fast dimers moves toward higher energies and becomes more pronounced, although the height decreases. There is a

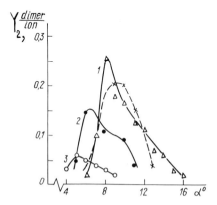

Fig. 3. The dependence of dimer yield on glancing angle for Ar^+ and Xe^+ primary ions and different bombardment energy. The full curves are for Ar^+: (1) $E_0 = 5$ keV, (2) $E_0 = 10$ keV, (3) $E_0 = 20$ keV. The broken curve is for Xe^+: $E_0 = 10$ keV.

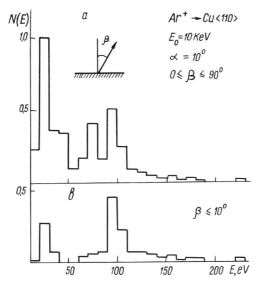

Fig. 4. Energy distribution of (a) all ejected dimers and of (b) dimers ejected at angles <10°
relative to the normal of the crystal surface.

definite grazing angle at which ejection of high-energy dimers disappears.
The calculated results confirm the analytically obtained dependence both of
the maximal grazing angle on bombardment energy and of the dimer energy
on grazing angle [eqs. (29) and (30)].

The simulation without taking into account the thermal vibration is similar
to the analytic consideration of ion scattering by an atomic chain. In this
case, at $\alpha_1 < \alpha < \alpha_2$, the dimers are knocked out by all ions moving along
the chain ridge. The high-energy peak (see fig. 4) corresponds just to the
dimers knocked out by these ions. The dimer knocking out with account of
thermal vibrations is analogous to the case of nondissociative ion scattering
by diatomic molecule considered in the section 2.2. In this case, the
peculiarity in the energy spectrum appears at a primary ion energy $E_0 \geqslant E_0^*$
[eq. (19)]. For $Ar^+ \rightarrow Cu_2$, the energy $E_0^* \approx 180$ keV, therefore, the high-
energy peak is not observed although the spectrum is broadened signifi-
cantly (see fig. 5).

Although the yield of dimers knocked out at grazing bombardment as
follows from fig. 3 is nearly the same as the yield of dimers sputtered at
normal bombardment, their energies differs significantly. For comparison,
the energy distribution of Cu_2^+ obtained experimentally by Dzhemilev
(1980) is shown in fig. 5. It is seen that the knocked-out dimers could be
separated not only by peak shift but also by spectrum width. In addition,
from fig. 5, it follows that the maximal dimer energy increases with increase
in primary ion mass according to eq. (24).

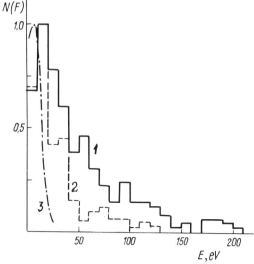

Fig. 5. Energy distribution of ejected dimers with account of lattice thermal vibrations. Calculation ($E_0 = 10$ keV, $\alpha = 10°$) for different primary ions: (1) Xe^+, (2) Ar^+, (3) data for Xe^+ from the experiment by Dzhemilev et al. (1987) ($E_0 = 8$ keV, $\alpha = 40°$).

The angle distribution of knocked-out dimers differs from the cosinusoidal one as is seen fig. 6. The calculations were carried out for single-crystalline Cu bombarded under $\alpha = 10°$ by 10 keV Xe^+ ions.

An orientation effect is observed when fast dimers are ejected. It consists in the dependence of dimer yield on azimuthal rotation angle of the target

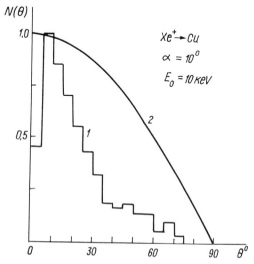

Fig. 6. The dependence of dimer yield on ejection angle relative to the surface normal ($Xe^+ \rightarrow Cu$, $E_0 = 10$ keV, $\alpha = 10°$): (1) calculation, (2) cosinusoidal distribution.

(ejection of dimers should reach a maximum when the plane of incidence coincides with directions of reflection with low indices) and in a dependence of dimer energy on the distance between atoms in the chain, as is implied by eq. (30).

Indeed, from fig. 7 it follows that at $\beta = 6°$, where β is the deflection angle of the incidence plane from the $\langle 110 \rangle$ direction, the yield decreases over one order of magnitude.

These peculiarities of the knocked-out dimers make their experimental observation possible over the background of dimers sputtered via other mechanisms, e.g., by *push-and-stick* (Bitensky et al. 1992b).

According to the *plane-shaving* mechanism which we have considered, the yield of clusters containing more than two atoms should also increase when the surface of a single crystal is bombarded at grazing angles.

Since ejection of fast dimers is best achieved when a nonideal surface of a single crystal is bombarded by ions, the detection of such dimers may prove to be useful in surface diagnostics.

3. Large-cluster sputtering

3.1. Introduction

As was mentioned in the previous section, the calculations of cluster yield both within the statistical model and the model of direct knocking out give satisfactory agreement with experimental results for two- and three-atomic

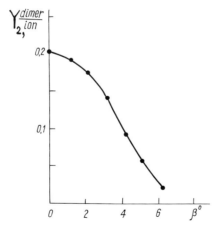

Fig. 7. The dependence of dimer yield on azimuthal angle of Cu single-crystal rotation at 10 keV Xe⁺ bombardment under $\alpha = 10°$. The angle $\beta = 0°$ corresponds to the $\langle 110 \rangle$ direction on the face (100).

clusters only, the calculated large-cluster yields getting smaller by some orders of magnitude than is experimentally found.

It has been suggested by Bitensky and Parilis (1978b) that direct knocking out, as well as the statistical recombination are not the only possible mechanisms for the formation of large clusters ($k > 3$), which are formed not so much in a single collision as via some collective effects of momentum transfer and energy deposition by a swift particle penetrating the solid.

The analysis of the experimental results (Dzhemilev and Kurbanov 1979, 1984, Hofer 1980) has shown that the clusters are formed with the greatest probability under heavy-ion bombardment. Just in this case, a sharp increase of the sputtering yield caused by its nonlinear dependence on elastic energy loss has been observed (Bay et al. 1976, Johar and Thompson 1979), the sputtering yield being considerably higher than the Sigmund theory (Sigmund 1969) predicts. Therefore, it is naturally to suppose that the large-cluster yield is connected with the nonlinear part of the total sputtering yield.

The nonlinear effects are considered in several papers. It was assumed that for heavy-ion bombardment overlapping of elastic collision cascades takes place since the ion mean free path becomes comparable with the interatomic distance. The formation of such regions of high-density energy deposition strongly influences the character both of radiation damage in solids and sputtering (D.A. Thompson 1981). When considering such processes a nonlinear theory should be applied, because, in the case of Brinkman displacement spikes (Brinkman 1954), the application of linear kinetic equations (Sigmund 1969) becomes incorrect.

A nonlinear increase of the sputtering yield with the number of atoms in the bombarding molecules has also been recorded (Andersen and Bay 1973, D.A. Thompson and Johar 1979). Some craters on a gold surface under bombardment with bismuth molecular ions have been observed with the mean and maximum number of sputtered atoms being about 4×10^3 and 3×10^4, respectively (Merkle and Jager 1981).

The overlap of the collision cascades and the formation of a high-density energy deposition region could be a source of a shock wave (Guinan 1974). Goldansky et al. (1975) has shown that the energy dissipation from an energy spike is more effective through a hydrodynamical mechanism than through a thermodynamical one.

The arrival of the shock wave on the solid surface and its reflection is known to cause surface tension. When this tension exceeds the critical value, cracking of the surface and separation of macroscopic pieces of the solid have been observed (Zel'dovich and Raizer 1966). The formation of shock waves in solids under ion bombardment has been used to explain crater formation (Ronchi 1973) and the sputtering of solids (Hayashiuchi et al. 1977, Carter 1979, 1983, Kitazoe and Yamamura 1980, Kitazoe et al.

1981). Bitensky and Parilis (1987) have proposed a shock-wave mechanism for both large-cluster emission and the desorption of biomolecules under ion bombardment. The principal idea of this mechanism is that the correlation of the velocities on the shock-wave front enables a superthermal momentum to be transferred to large amounts of matter without destruction.

3.2. Shock-wave mechanism of cluster emission

The proposed mechanism of cluster emission during sputtering of solids is based on the following assumptions:

(i) Cluster emission occurs as a result of a nonlinear effect – the formation of some elastic collision spikes and both formation and emergence of the shock waves on the surface. In the correlated movement at the shock wave front, a group of atoms could acquire considerable kinetic energy without breaking their bonds.

(ii) Under ion bombardment some high-density elastic collision spikes occur as a result of a dense cascade generated by one or more energetic recoils near the surface. This occurs with high probability, if the free path for elastic collisions is comparable with the mean interatomic distance in the solid. In contrast to this, it occurs with a smaller probability depending on the probability of two or more violent collisions near the solid surface.

(iii) Under molecular bombardment an elastic collision spike is formed as a result of the overlap of collision cascades, formed by each atom.

The necessary condition for the shock-wave generation is the achievement of a critical mean kinetic energy of atoms ε_c, which is a free parameter of the theory.

From the hydrodynamical effects (Kitazoe and Yamamura 1980, Kitazoe et al. 1981) the internal energy carried by a shock wave equals

$$W = C(h-1)^2(h+5)/[4h(4-h)] , \tag{34}$$

and the pressure P is

$$P = CN(h-1)(h+5)/[2(4-h)] , \tag{35}$$

where $C = 1/(KN)$, K is the compressibility of a solid, and $h = N_1/N$, N_1 and N are the atomic density of the compressed matter behind the shock-wave front and in the normal target state, respectively. The initial value $h = h_c$ is determined from

$$\varepsilon_c = Ch_c(h_c+5)/[4(4-h_c)] . \tag{36}$$

For the further description of the shock wave an assumption should be made about its symmetry, which under middle-energy ion bombardment and due to cascade development is naturally spherical (Kitazoe and Yamamura 1980, Kitazoe et al. 1981). Under high-energy heavy-ion or fission fragment bombardment, the shock wave has a cylindrical symmetry (Hayashiuchi et al. 1977).

In the case of spherical symmetry the shock-wave energy dissipates according to

$$W(R) = \frac{W_c(R_c/R)^2}{1 + 1.5(2W_c/C)^{1/2} \log(R_c/R)},$$ (37)

where $B = \frac{16}{3}(2W_c/C)^{1/2}$, W_c is the initial value of W for $h = h_c$, R is the distance from the center, and R_c is its initial value determined by

$$R_c = [3(dE/dx)_n/(2\pi N\varepsilon_c)]^{1/2},$$ (38)

where $(dE/dx)_n$ is the nuclear stopping power.

3.3. Large-cluster yield

The shock wave emerging from the surface breaks off a chunk of matter with its boundary determined by

$$P(R_s) \cos \theta = P_{cr}.$$ (39)

The integrity of the cluster emitted from the shock-wave epicenter of the surface is determined by the condition for its nondisruption,

$$2P(R_k) \sin \beta = P_{cr},$$ (40)

where θ and β are the angles between the surface normal and vectors $\boldsymbol{R}_s$ and $\boldsymbol{R}_k$, respectively (see fig. 8), P_{cr} is the critical pressure at which the scabbing of the solid occurs. P_{cr} is a parameter of the given material and by order of

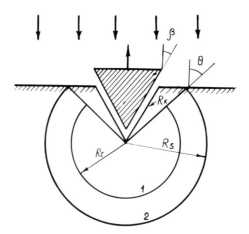

Fig. 8. Ejection of a large cluster caused by a shock wave emerging towards a solid surface; (1) primary compressed region, (2) region of shock-wave propagation.

magnitude is about κ/d, where κ is the specific surface energy and d is the lattice constant. If the initial spike is a sphere with radius R_c, from eq. (36) the number of atoms in the cluster emitted via shock waves from the epicenter on the surface is given by

$$k = \tfrac{1}{3}\pi N[(dE/dx)_n/(2\pi N\varepsilon_c)]^{3/2}\tan^2\omega ,$$ (41)

where $\omega = \min(\theta, \beta)$ should be calculated using eqs. (31)–(34), (39), and (40). For a rough approximation, $P(R) \approx P_c R_c/R$ and from eqs. (39) and (40) it follows that

$$\omega = \arccos^{1/2}\eta , \quad \text{for } \tfrac{4}{5} < \eta < 1 ,$$

$$= \tfrac{1}{2}\arcsin\eta , \quad \text{for } \eta \leqslant \tfrac{4}{5} ,$$ (42)

where $\eta = P_{cr}/P_c$.

The sputtered matter escapes from a cone volume with the vertex angle 2ω as a whole and is thus a cluster. Equation (41) with $\omega = \theta$ gives the total number of atoms S_n sputtered by a shock wave.

It should be noted that the total energy acquired for the emission of a cluster containing k atoms is $E_k < kE_b$ (E_b is the sublimation energy), since escape as a cluster requires a smaller number of bonds to be broken. Actually, the energy of formation of two free faces $2\kappa s$, is required (s is the lateral cone surface). So, for the formation of a cone-shaped cluster, the energy

$$E_k = 2\pi\kappa[3k/(\pi N\cos\omega\sin^{1/2}\omega)]^{2/3}$$ (43)

is necessary. Considering $k = 100$, $E_b = 4$ eV, $\kappa = 0.13$ eV/Å^2, $\omega = 20°$, $N = 0.06\,\text{Å}^{-3}$, we get $E_k/(kE_b) \approx 0.4$. Equation (39) is, therefore, the condition for disrupting the bounding to the cone surface.

It should be noted that if $\eta > \tfrac{4}{5}$, only one large cluster is emitted from the area near the shock-wave epicenter on the surface. If $\eta \ll \tfrac{4}{5}$, apart from this, some other clusters of approximately the same size could be emitted from the surrounding area.

In this case the condition for the integrity, eq. (40), of clusters emitted from the peripheral area has the form $\Delta\rho|\text{grad } P_t(\rho)| = P_{cr}$, where $\Delta\rho$ is the cluster radial dimension on the surface. The radial pressure $P_t(\rho) \approx P_c R_c\rho/(R_c^2 + \rho^2)$ has a maximum and the gradient of the pressure is 0 at $\rho = R_c$, where the cluster dimension of cause is not infinite, but at maximum. If we take the azimuthal dimension of a cluster as $\rho\,\Delta\varphi$, then the total number of clusters of the dimension increases with ρ when $\rho > R_c$. Thus the emission of clusters under a shock wave comes mainly from the peripheral area.

The ratio η shows the intensity of the shock wave and, therefore, the importance of the cluster emission contribution to the total sputtering flux.

In the proposed mechanism the energy spike under ion bombardment is formed as a result of energy transfer to one or several atoms located at a small depth. Let the mean free path of an ion with energy E_0 in an amorphous medium be $\lambda(E_0)$, then the probability of its collisions with an atom down to the depth d_1 is

$$P_1 = 1 - \exp(-d_1/\lambda) . \tag{44}$$

The probability for the transfer of the energy E_1 to this atom is given by

$$w(E_0, E_1)\, dE_1 = \sigma(E_1)\, dE_1/\sigma_t(E_0, E_d) , \tag{45}$$

where $\sigma(E_1)$ is the differential cross section of the scattering with energy loss E_1, $\sigma_t(E_0, E_d)$ is the total cross section of the energy transfer greater than the displacement energy E_d, $\lambda = (N\sigma_t)^{-1}$.

If the mean free path of the recoils is small, the probability for spike creation is determined by the probability for energetic recoil generation with energy E_1 in one collision (Winterbon 1980),

$$Q(E_1)\, dE_1 = P_1(E_0)w(E_0, E_1)\, dE_1 . \tag{46a}$$

For the other case, the appearance of two or more energetic recoils near the surface is necessary. After the first collision the ion has energy $E_0' = E_0 - E_1$ and the probability of its collision with a second atom, located at distance d_2, with the transfer of energy E_2 is defined by formulae analogous to eqs. (44) and (45), respectively.

The total probability is the product of the probabilities of events,

$$Q(E_1, E_2)\, dE_1\, dE_2 = P_1(E_0)w(E_0, E_1)P_2(E_0')w(E_0', E_2)\, dE_1\, dE_2 , \tag{46b}$$

and the elastic energy loss in eq. (41) is $(dE/dx)_n = (dE/dx)_{E_1} + (dE/dx)_{E_2}$.

The distribution over the number of atoms in the cluster, k, in these two cases is defined by the functions

$$f(x) = \int_{E_d}^{\gamma E_0} Q(E_1)\delta[k - S(E_1)]\, dE_1 , \tag{47a}$$

or

$$f(k) = \int_{E_d}^{\gamma E_0} dE_1 \int_{E_d}^{\gamma(E_0 - E_1)} Q(E_1, E_2)\delta[k - S(E_1, E_2)]\, dE_2 , \tag{47b}$$

where the functions $S(E_1)$ and $S(E_1, E_2)$ are the right-hand parts of eq. (41) and $\delta(x)$ is the Dirac δ function, $\gamma = 4m_1 m_2/(m_1 + m_2)^2$, and m_1 and m_2 are the ion and the target atom masses, respectively. For the power potential $U(r) = U_0 r^{-n}$, the cross-section $\sigma(E)\, dE = C_n E_0^{-1/n} E^{-(1+n)/n}\, dE$ and the elastic energy losses $(dE/dx)_n$,

$$(dE/dx)_n = NnC_n/(n-1)\gamma^{(n-1)/n}E^{(n-2)/n} , \qquad (48)$$

are given by Winterbon et al. (1970). If the exponent n is different for different energy regions, then analytical calculations of integrals (47) seems to be impossible. However, considering that, at $n = 3$, eq. (48) satisfactorily describes the energy dependence of both the cross section and the elastic energy loss up to 10 keV, we get from the eq. (47) that the function $f(k)$ at large k has an asymptotical form,

$$f_k \simeq k^{-c} , \qquad (49)$$

where $c = \frac{5}{3}$ and $\frac{7}{3}$ for eqs. (45a) and (45b), respectively. Although in eq. (49) the values of the exponent are approximate, the importance of the result obtained is that the model provides a power-law decrease of the cluster yield in accordance with the experimental data by Herzog et al. (1973) and Staudenmaier (1972) (where the exponent equals 2), whereas in the model of the statistical cluster formation (Gerhard 1975) the cluster yield decreases as an exponental law $f(k) \approx g^k$ ($g \ll 1$).

The power-law behavior of cluster-mass distributions was obtained by Urbassek (1988, 1989) and Dunlap et al. (1983). Urbassek (1988) had supposed that the formation of a high-density energy region nearby the surface was the necessary condition for cluster emission. If the state of the substance in this region corresponds to the critical point of the liquid–gas phase transition, then particularly large clusters would be emitted. Dunlap et al. (1983) had described the mass distribution of sputtered clusters using a percolation model, where each bond of the energized target lattice may be broken with a certain probability. King et al. (1987) have considered the sputtering of large molecules and clusters induced by a relaxation process in which the deposited energy was transferred to the internal vibration modes of the perturbed region. Within this model, clusters are desorbed from the very surface, therefore, the experimentally observed crater formation cannot be explained.

Comparison of absolute values of the cluster yield with experiments is difficult because so far only charged-cluster yields have been measured. In the experiment by Gerhard and Oechsner (1975) the yield of small neutral clusters with $k = 2, 3$ has been measured. As for Giber and Hofer (1980), the data have been obtained indirectly, which does not allow the neutral cluster yield to be determined unambiguously.

3.4. Cluster emission from nonideal surfaces

The cluster yield must depend on the surface topography. Such structures as cones and pyramids are known to be formed on the surface under certain

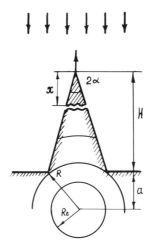

Fig. 9. A cone top breaking off as a result of energy cumulation at the shock-wave front propagating towards the cone vertex.

conditions of ion bombardment (Gvozdover et al. 1976, Whitton and Carter 1980).

Let us assume, that at depth a under a cone with vertex angle 2α and height H a high-density cascade is formed and a shock wave emerges (see fig. 9). As it propagates towards the vertex the cumulation effect increases the wave-energy density. If it reaches a certain magnitude, the cone top breaks off and and flies away. This occurs within the cross section, located away from the cone vertex a distance x, equal to

$$x = H(u/2\kappa)^{1/2} , \tag{50}$$

where $u = W(R)N$ and $R = (a^2 + H^2 \tan^2\alpha)^{1/2}$. The number of atoms in the removed cone top is given by

$$k = (1/3\pi)Nx^3 \tan^2\alpha = k_1(u/2\kappa)^{3/2} , \tag{51}$$

where $k_1 = (1/3\pi)NH^3 \tan^2\alpha$ is the total number of atoms in the cone.

From eq. (51) follows that with H increasing, the number of particles in the cluster initially also increases up to a certain value k_{max} at $H = R_c/\tan \alpha$, and then according to eq. (34) decreases logarithmically. Taking $\varepsilon_c = 16$ eV and $dE/dx = 300$ eV/Å, from the data by Kitazoe et al. (1981) we get $W_c \approx 2$ eV, $R_c \approx 12$ Å. Assuming that $a \approx 0$, $\alpha = 20°$ (Witcomb 1975) $\kappa = 0.13$ eV/Å^2, $N = 0.06$ Å^{-3}, we get $k \approx 80$. According to the proposed mechanism, the cluster yield should correlate with the formation of cones on the surface exposed to ion bombardment.

The cluster originating from a conic region, adjoining the epicenter on the surface, should escape the surface in a direction close to the normal. The other clusters escape the surface in other directions but their angular distribution must differ from the cosinusoidal law.

The cluster yield must sharply increase for molecule ion bombardment, in which nonlinear effects are enhanced (D.A. Thompson 1981). In an experiment comparing cluster yields for molecule ions with atomic yields at the same velocity a considerable enhancement could be expected.

Since for ion bombardment of a single crystal the probability of violent ion collisions decreases if the bombardment direction coincides with a low-index direction, then in the cluster yield some orientational effects would be expected.

The shape of the escaped cluster, e.g., conic, is a nonequilibrium one. Besides, the cluster is emitted being strongly vibrationally excited. After the cluster escapes the surface, its rearrangement into a more compact equilibrium shape is possible. The extra energy ΔE is released being equal to the surface-energy decrease connected with the cluster shape change. For an approximate evaluation one would consider an equilibrium cluster shape as a spherical one. Then

$$\Delta E = \pi\kappa[3k/(\pi N)]^{2/3}[(1 + \sin \alpha)/(\tan^{1/3}\alpha \cos \alpha) - 2^{2/3}]. \tag{52}$$

Taking $\alpha = 20°$ (Witcomb 1975), and $k = 100$, we get $\Delta E \approx 20$ eV. This energy seems to be sufficient for break off of about five atoms or break up of a cluster into two or more subclusters of different size.

Since the time rearrangement could not be shorter than 10^{-12} s, there is enough time for distributing this energy to the electron subsystem excitation and electron emission. As a result, the degree of ionization for the clusters must be much higher than for atoms, since they acquire their change at a considerable distance from the surface where the electron capture from the metal becomes scarcely probable.

Thus, the proposed mechanism allows some regularities in large-cluster emission and their high ionization degree to be explained, some peculiar feature of the yield in connection with the crystal structure and the surface topography to be revealed, and the possible high cluster yield with molecular bombardment to be discussed. Unfortunately, the available experimental data on the emission of large clusters is still poor.

4. Desorption of biomolecules under heavy-ion bombardment

4.1. Experimental data and theoretical models

Recently, the desorption of bioorganic molecules under heavy-ion bombardment has attracted great interest. This phenomenon is of great importance for understanding the interaction of heavy ions with biological substances and nondestructive mass-spectrometry of thermally labile organic molecules. Since the pioneering experiment on high-energy heavy-ion desorption of bioorganic molecules was reported by Torgerson et al. (1974), detailed studies of its dependencies on primary ion velocity and energy (Håkansson and Sundqvist 1982, Nees et al. 1984, Becker et al. 1986), the angle of incidence (Nieschler et al. 1984a, Håkansson et al. 1982) and charge state (Håkansson et al. 1981, Nieschler et al. 1984b, Wien et al. 1987) has been carried out. It has been shown by Salehpour et al. (1986) that the molecules were desorbed mainly as neutrals. The review concerning this phenomenon was given by Wien (1989).

Despite some theoretical papers (Krueger 1979, Hedin et al. 1985, Johnson 1988), the mechanism of this phenomenon is not yet completely clear. It is difficult to understand how the fast heavy ions or fission fragments which deposit energy of the order of 10^2–10^3 eV/Å cause the deposition of nonvolatile and thermally labile molecules with a mass of 10^4 a.m.u. [the largest recorded molecule of porcine phospholipase A_2 has a molecular weight of 13980 a.m.u. (Sundqvist et al. 1984)].

Krueger (1979) has proposed a model in which the molecular ion was polar bound to the surface and the fast heavy ion caused a high-frequency polarization of the electron plasma that resulted in the desorption of the molecular ion. However, as Håkansson and Sundqvist (1982) have noted, it was difficult to explain how nonpolar molecules like ergosterol could be desorbed in this model. It is known that keV-energy ions cause also induced desorption of organic molecules (Kamensky et al. 1982, Standing et al. 1982), although the inelastic energy loss in this case could be neglected.

Håkansson and Sundqvist (1982) and Håkansson et al. (1982) have analyzed their experiments on the base of the previously developed thermal spike model (Ollerhead et al. 1981, Besenbacher et al. 1981, Seiberling et al. 1980) and the ion explosion model (W.L. Brown et al. 1980) and have shown, that these models could not explain the whole set of experimental data.

In an experiment by Håkansson et al. (1982), it has been observed that the desorption yield dependence on inelastic energy loss was not universal for all studied samples from valine (MW 117) to insulin (MW 5733) as it was expected from the above-mentioned models, but it was determined rather by the size of the molecule, than by its mass. In the models proposed by Ollerhead et al. (1981), Besenbacher et al. (1981), Seiberling et al. (1980) and W.L. Brown et al. (1980), the molecule size was not taken into account at all.

In the most developed theory, that of Hedin et al. (1985), the size of the desorbed molecules has been taken into account. In this theory the molecule desorption is due to the breaking of bonds by the shower of the secondary electrons generated by a passing heavy ion. The biomolecule desorption yield is determined by the probability of acquiring some 'hits' from the secondary electrons, which was calculated using the ion-track model developed by Katz (1978). The theory by Hedin et al. (1985) explained the dependencies on the energy loss and the molecular size, observed in the same laboratory (Håkansson et al. 1982) and the dependence on the ion velocity measured in the experiments by Alberts et al. (1982) and Duck et al. (1980). It should be noted, however, that for a biomolecule the secondary electron shower is as well a destructive agent as a desorbing one. Both these effects should depend on the primary-ion parameters in a similar way.

Desorption of some large molecules is also observed under keV-energy ion bombardment, where the secondary electron flux is low. Yet, the source of energy for the desorbed large molecule, all atoms of which should acquire simultaneously the same superthermal energy of the aligned motion, is not clear.

Recently, P. Williams and Sundqvist (1987) have proposed the popcorn model in which a biomolecule is assumed to expand after becoming vibrationally excited by low-energy secondary electrons from the heavy-ion track. This rapid expansion causes the biomolecule pushing away against a right substrate or its subsurface neighbors and, thus, desorption takes place.

S.-L. Lee and Lucchese (1988) have undertaken a simulation of the post-excitation vibrational motion of peptide units inside a biomolecule using as model an one-dimensional chain containing 40 coupled anharmonic oscillators. It has been found that the expanding biomolecule is desorbed like an exploding popcorn.

However, it is still not clear how to avoid the destructive effect of the intensive secondary-electron shower necessary for molecule expanding due to synchronous excitation of a large number of vibrational modes. It seems that this model describes the desorption of rather large molecules. It could be shown as follows.

The increase of molecule radius $\Delta \alpha = \alpha a \, \Delta T$, where α is the coefficient of linear expansion, a is the molecule radius, and ΔT is the temperature increase independent of a. The center-of-mass velocity $v_c = 2\Delta a / t_c = 4\alpha a \, \Delta T \nu$, where ν is the frequency of fundamental mode. In the popcorn model (P. Williams and Sundqvist 1987), $\nu = (1/a)(E/\gamma)^{1/2}$, where γ is the density and E is the Young modulus of the material. The kinetic energy of the molecule $W_k = \frac{1}{2}Mv_c^2 = (\frac{32}{3})\pi a^3 \alpha^2 \, \Delta T^2 E$. The condition for a molecule to be pushed away from the substrate is $W_k \geq \pi a^2 U$, where U is the binding energy per surface unit. Then we obtain that the desorption is possible only if $a \geq a_{min} = 3U/(32\alpha^2 \, \Delta T^2 E)$. Taking from P. Williams and Sundqvist (1987) the values $U = 1.6 \times 10^{-2}$ eV/Å^2, $\alpha = 5 \times 10^{-5}$ K^{-1}, $\Delta T = 10^3$ K, $E = 10^{10}$ N/m^2, we get $a_{min} \approx 10$ Å. This minimal radius corresponds to a molecule with mass $M \approx 10^4$ a.m.u. Meanwhile, the available data do not corroborate any sharp difference in the yield regularities for large and small biomolecules. It is interesting to note that an analogous mechanism has been proposed by Vorobjova et al. (1986).

Bitensky and Parilis (1987) have proposed a shock-wave mechanism for the desorption, being in fact the sputtering of bioorganic molecules at emergence at the surface of a shock wave originated from high-density energy spikes in the vicinity of a fast heavy-ion track. The principal idea of this mechanism is that correlation of the velocities of the shock-wave front enables a superthermal momentum to be transferred to large molecules without destruction. It has been shown that with reasonable values of parameters describing the process of the shock-wave formation and propagation, both the theoretical desorption yield dependence on energy loss, the angle of incidence, primary-ion charge and the desorbed-molecule energy distribution were in satisfactory agreement with the experimental data. This mechanism was developed by Bitensky et al. (1989a).

Säve et al. (1987a,b,c) have studied the desorption yield dependence on sample film thickness and have shown unambiguously that the biomolecules were emitted from some craters of 100 Å depth. The same result, but with another value of depth (150–200 Å), was obtained by Bolbach et al. (1988).

It should be noted that Johnson (1989) has proposed a pressure pulse model nearly identical to the shock-wave model by Bitensky and Parilis (1987).

4.2. Shock-wave formation

The shock-wave mechanism has been proposed for the description of nonlinear effects in sputtering of solids under keV-energy heavy-ion bombardment (Hayashiuchi et al. 1977, Carter 1979, 1983, Kitazoe and Yamamura 1980, Karashima 1982). It has been brought in question by Winterbon (1980) and was discussed by Sigmund (1987). In our opinion, the probability of shock-wave generation under keV ion bombardment is small because the probability of the transfer of almost the total ion energy to the several neighboring subsurface atoms is rather small. The emission of large clusters under ion bombardment is a result of this rare 'catastrophic' event (Bitensky and Parilis 1987). Only under heavy-ion bombardment with energies of the order of 100 keV, where the elastic collision mean free path is comparable with the interatomic distance, then shock-wave effects are displayed in sputtering (Hayashiuchi et al. 1977, Carter 1979, 1983, Kitazoe and Yamamura 1980, Winterbon 1980) and radiation damage (Zhukov and Boldin 1987, Yanushkevich 1979).

As for the effect of MeV heavy ions on biological substance, there are some experimental data (Büker 1977, Akoev et al. 1985) that the damage takes place on a distance from the ion path exceeding the maximum δ-electron range.

It is known that in condensed matter the part of the energy deposited in fast heavy-ion track could dissipate via acoustic- (Lyamshev 1987) or shock- (Goldansky et al. 1975) wave formation. The exact mechanism of shock-wave formation is not yet completely clear. Apparently, Askaryan (1957) was the first one who supposed the shock wave to occur due to a fast expansion resulting from both ionized atom repulsion and some microexplosions of intensively heated microscopic domains of high-density ionization in the fast heavy-ion track.

Under high-energy heavy-ion bombardment, the formation of an energy spike leading to the generation of a shock wave may be described in the following way: as a fast heavy ion passes through an insulator, the accumulated energy is initially the potential energy of Coulomb repulsion of the ionized atoms. In a nonmetallic solid the time for its neutralization is rather long and after 10^{-12} s this potential energy is transferred via a Coulomb explosion into kinetic energy spent in track creation, formation of high-density collision cascades and shock-wave creation (Hayashiuchi et al. 1977).

Another mechanism leading to the energy transfer from the electronic system to the ions has been proposed by Katin et al. (1989). These authors have considered the dynamics of the electron and lattice temperature at relaxation of the excited region via a low-temperature ionization wave. Inelastic sputtering of solids by fast ions is reviewed by Baranov et al. (1988).

It has been obtained by Kovalyov and Brill (1984) that at fast heavy ions passing through water, the shock wave originates when

$$(c/v)^2(dE/dx) \geqslant 3.6 \times 10^4 \text{ eV/nm} ,$$

where v and c are the primary ion and light velocity, respectively, and dE/dx is the electronic energy loss. It is known that about half of the total energy loss is deposited inside the track core.

In condensed matter a considerable pressure on the shock-wave front is achieved at small excess of the atomic density of the compressed matter, N_1, over the normal density, N. For example, at $h = N_1/N = 1.04$, the pressure $p = 10^5$ bar (Hayashiuchi et al. 1977). The mass flow velocity v_s is connected with the front velocity D by the following relation (Zel'dovich and Raizer 1966),

$$v_s = D(h - 1)/h . \tag{53}$$

Since $\delta = (h - 1) \ll 1$, $v_s \ll D$. Therefore, we can neglect the delay time and consider the shock-wave propagation in the geometrical acoustic approximation.

As has been assumed by Hayashiuchi et al. (1977) and Kitazoe and Yamamura (1980), the shock wave originates when the mean energy per atom ε_c in a domain with radius R_c exceeds the binding energy. In this case,

$$R_c = [3(dE/dx)/(4\pi N\varepsilon_c)]^{1/2} .\tag{54}$$

It has been shown by Bitensky and Parilis (1987) that the initial pressure inside this domain is connected with ε_c by the following relation,

$$P_c = \delta_c/K = 2[2\varepsilon_c/(KN) - 1]/(3K) ,\tag{55}$$

where K is the compressibility, and δ_c is the value of δ inside the domain. Subdividing the track into domains with radius R_c, which are the source of spherical shock waves, and considering that the maximum pressure on the shock-wave front decreases with increasing distance l according to the power law $P = P_c(R_c/l)^s$ (Stanyukovich 1975), we get for the normal P_n and the tangential P_t pressure component at the point with coordinates $\{x, y, z\}$ the following equations (Bitensky et al. 1989a),

$$P_n = P_c R_c^{s-1} \int_0^\infty (t \cos \psi - z)/l^{s+1} \, dt ,\tag{56a}$$

$$P_t = P_c R_c^{s-1} \int_0^\infty [(t \sin \psi - y)^2 + x^2]^{1/2}/l^{s+1} \, dt ,\tag{56b}$$

where $l = [(t \cos \psi - z)^2 + (t \sin \psi - y)^2 + x^2]^{1/2}$, and ψ is the angle of incidence of the fast heavy ion. The plane xy coincides with the sample surface, the ion trajectory lies in the yz plane (see fig. 10).

The region from which the ejection of matter occurs is determined by the condition

$$P_n(x, y, z) \geq P_{cr} ,\tag{57}$$

where P_{cr} is the scabbing critical pressure.

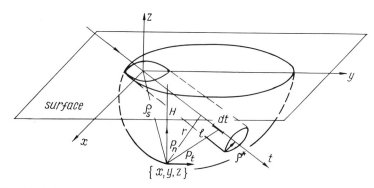

Fig. 10. Scheme for shock-wave generation by a fast ion passing through solid.

At normal incidence ($\psi = 0°$), condition (57) together with eq. (56a) determines a hemisphere with radius

$$\rho_s = R_c[(P_c/P_{cr})/(s-1)]^{1/(s-1)} . \tag{58}$$

The number of particles ejected from the formed crater is

$$Y_0 = (\tfrac{2}{3})\pi N R_c^3[(P_c/P_{cr})/(s-1)]^{3/(s-1)} . \tag{59}$$

The separate biomolecules are emitted as the hydrogen bonds between such molecules are much weaker than the interatomic bonds inside the molecules.

It should be noted that as follows from eqs. (38), (58) and (59) if $P_c = $ const, then at any s the crater radius $\rho_s \simeq (dE/dx)^{1/2}$ and $Y_0 \simeq (dE/dx)^{3/2}$. It is not out of the question that with the possible dependence of the initial energy ε_c on dE/dx the dependence of Y on dE/dx could be somewhat different depending on s.

4.3. Biomolecule destruction by shock wave and in δ-electron flux

The biomolecule can be destructed by shock wave if the parameters on its front (gradient of pressure, temperature) exceed the critical values.

In the simplest case, the condition of destruction for a molecule with radius a is

$$|\text{grad } P(x, y, z)|2a \geqslant P_d , \tag{60}$$

where P_d is the destructive tension. At normal incidence, from eqs. (56a, b) for $s = 2$ we obtain

$$P_n = P_c R_c/(r^2 + z^2)^{1/2} , \tag{61a}$$

$$P_t = (P_c R_c/r)[1 + z/(r^2 + z^2)]^{1/2} , \tag{61b}$$

where r is the distance from the track axis. Condition (59) together with eqs. (61a) and (61b) determines the radius ρ_d of the molecule destruction region. In particular, on the surface, $z = 0$, the radius ρ_d is given by

$$\rho_d = 2(P_c R_c a/P_d)^{1/2} . \tag{62}$$

From eqs. (54) and (62) it follows that $\rho_d \simeq (dE/dx)^{1/4}$. Since it is necessary for molecule destruction to break a covalent bond, but for its desorption only a weaker hydrogen one, then $P_d \gg P_{cr}$. Besides, δ-electrons generated by a fast heavy ion passing through matter should cause biomolecule destruction.

The theoretical description of radiation effects in a macromolecular substance induced by fission fragments is based on the theory of radial distribution of deposited energy developed by Katz and Kobetich (1968). In this theory a certain critical dose required for a considerable change of the general features of substance is introduced as a parameter.

We suppose that the molecule distribution occurs as a result of Coulomb repulsion of two ionized neighboring atoms, which appear due to either direct ionization or inner-shell vacancy formation with a subsequent Auger cascade (Parilis 1969a). These Varley-type mechanisms were proposed for under-threshold defect formation in alkaline halloid crystals (Klinger et al. 1985). The main factor of this mechanism is the competition between the electron and ion relaxation time. According to Yunusov et al. (1976), the probability that neutralization does not occur before elementary damage happened is given by

$$P_\tau = \exp(-\tau_+/\tau) , \tag{63}$$

where $\tau_+ = 10^{-13}$ s is the time for ions to acquire a kinetic energy sufficient to overcome the chemical binding barrier, and τ is the mean neutralization time. For biomolecules as insulator substances it should be expected that $\tau_+/\tau \ll 1$, hence $P_\tau \approx 1$. According to Miterev et al. (1974), the density deposited by δ-electrons at a distance r from the track axis equals

$$\mathscr{E}(r) = (dE/dx)/[8\pi r^2 \ln(L/r_{ad})] = p_1(r)JN , \tag{64}$$

where J is the mean ionization energy. If the distance between two ionized atoms is less than l_{cr}, then elementary damage could happen. In eq. (64), L is the maximum range of δ-electrons, r_{ad} is the Bohr adiabatic radius, and p_1 is the probability for an atom to be ionized. Therefore,

$$p_1(r) = \mathscr{E}(r)/(JN) . \tag{65}$$

The critical distance l_{cr} is determined by the value of the ion Coulomb repulsion force sufficient for chemical bond breaking off,

$$e^2/(\kappa l_{cr}^2) \geqslant Q/\Delta , \tag{66}$$

where e is the electron charge, Q is the height of the chemical binding barrier, κ is the dielectric constant corresponding to a time of 10^{-13} s, and Δ is the barrier width. The probability for a randomly chosen atom to be ionized and another ion to be found within a distance less than l_{cr} from it equals (Bitensky et al. 1989b)

$$p_2(r) = p_1[1 - (1-p_1)^m] \approx mp_1^2 , \tag{67}$$

where $m = \frac{4}{3}\pi N l_{cr}^3$ is the number of atoms inside the sphere of radius l_{cr}. The probability for a molecule with the radius a located at distance r from the ion track to survive is given by

$$\begin{aligned} p &= \exp[-\beta Z p_2(r)] , &&\text{for } r > 2a , \\ &= 0 , &&\text{for } r \leqslant 2a , \end{aligned} \tag{68}$$

where Z is the number of atoms in the molecule, and the factor β is connected with the repulsion geometry and determined by the probability for two ionized atoms to be covalent-bond neighbors. The probability p is about unity at distance $r \geqslant p_i$, where

$$p_i = (\beta Zm)^{1/4} \{(dE/dx)/[8\pi JN \ln(L/r_{ad})]\}^{1/2} . \tag{69}$$

4.4. Biomolecule desorption yield

The yield of desorbed molecules per incident ion is

$$Y = N_m \int_V \eta(r) p(r)\, dV , \tag{70}$$

where $p(r)$ is the survival probability, see eq. (68), N_m is the number of molecules per unit volume, $\eta(r)$ is the molecule ejection probability ($\eta^{+,-}$ in a charged state, η^0 as a neutral one). Integral (70) is calculated over a volume determined by conditions (57) and (60). The total number of desorbed molecules is determined by eq. (70), where η is replaced by $\eta^+ + \eta^- + \eta^0 = 1$. Integral (70) cannot be calculated in the general form. Assuming approximately $p = 1$ for $r > \rho_i$ and $p = 0$ for $r \leqslant \rho_i$, as well as $\eta = \text{const}$, we get for normal incidence

$$Y = \tfrac{2}{3}\pi\eta N_m \rho_s^3 [(1 + \rho^*/\rho_s)(1 - \rho^*/\rho_s - 2a/\rho_s)]^{3/2} , \tag{71}$$

where $\rho^* = \max\{\rho_i, \rho_d\}$, and ρ_s, ρ_d and ρ_i are determined by eqs. (58), (62) and (69), respectively (see fig. 10). It should be noted that the expression in brackets is the geometrical factor that has been introduced by Bitensky and Parilis (1987).

For a calculation of Y, the values of the following parameters should be known: the energy ε_c and compressibility K, to determine R_c and P_c by eqs. (54) and (55), respectively; the pressure P_{cr} and P_d, to determine ρ_s and ρ_d by eqs. (58) and (62). As is seen from eq. (69) for a calculation of ρ_i, a rather large number of parameters should be known. However, as it follows from eq. (71) for a calculation of Y, only three fitting parameters η, ρ_s and ρ^* are necessary. Assuming the elastic features of all films to be approximately the same, ρ_s can be fitted for a certain value of dE/dx and then from eqs. (54) and (58) with $\rho_s \simeq (dE/dx)^{1/2}$ it could be calculated for other values of dE/dx. Thus, the number of fitting parameters is reduced to two: η and ρ^*. Assuming $\rho^* = \rho_i$, we get that ρ_i/ρ_s is independent on dE/dx but depends on the number of atoms in the molecule, Z, according to eq. (69). In all calculations it was chosen that $\rho_s = 78.5$ Å for $dE/dx = 99.1$ MeV/(mg cm^2). This value of ρ_s agrees well with the radius of damaged region recorded by Salehpour et al. (1984) and with the crater size measured by Säve et al. (1987a,b,c). The molecule radius was calculated as by Hedin et al. (1985) with $a = 0.5 N_m^{-1/3}$.

4.5. Comparison with experiment

4.5.1. Dependence on the angle of incidence

It is interesting to compare the dependence of Y on dE/dx and v as they follow from eq. (71), with recent experimental data, but first of all we calculate dependence of Y on the angle of incidence ψ. Bitensky and Parilis (1987) have obtained some approximate expressions for the yield dependence on ψ for small ψ or $\psi \leqslant \frac{1}{2}\pi$. Equations (56a), (56b) and (46) give the possibility to trace this dependence for the whole range of ψ variation and in such a way as to determine the parameter s defining the shock-wave attenuation. Integral (56a) at $s = 2$ results in the following equation, the same as in the paper by Bitensky and Parilis (1987),

$$P_{\mathrm{n}} = (P_{\mathrm{c}}R_{\mathrm{c}}/\rho_{\mathrm{s}})\cos\psi + (P_{\mathrm{c}}R_{\mathrm{c}}/r)\cos\alpha\sin\psi\{1 - [1 - (r/\rho_{\mathrm{s}})^{2}]^{1/2}\}, \quad (72)$$

where r and α are the polar coordinates in the plane normal to the track axis (see fig. 10). From condition (57), taking into account eq. (72), one obtains the crater size and the desorption yield. Calculation of eqs. (56a) and (70) has shown that the dependence $Y(\psi)$ is different for different s. From fig. 11, it is seen that the theoretical dependence $Y(\psi)$ for $s = 2$ fits better the experiment by Håkansson et al. (1982). For simplicity, this value was taken in all calculations. However, as has been mentioned above for different substances and even for different film deposition techniques some different values s are valid.

4.5.2. Desorption yield dependence on energy loss and primary ion velocity

In figs. 12 and 13, the yield Y of valine clusters versus dE/dx and v are displayed. As is seen the experimental dependences are described by eq.

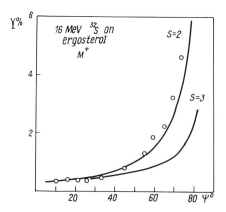

Fig. 11. The yield of ergosterol M^{+} versus ψ for ^{32}S ions with $E = 16$ MeV. Circles represent experimental data by Håkansson et al. (1982), full curves theory.

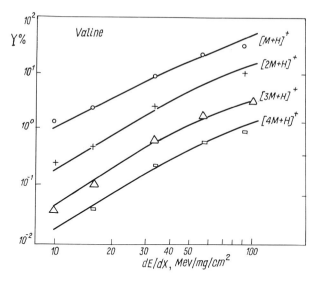

Fig. 12. The yield of valine and its clusters versus dE/dx for different ions (Hedin et al. 1985). Full curves represent calculations using eq. (71).

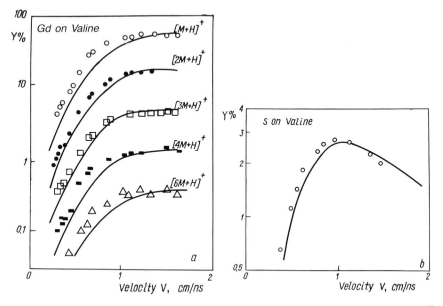

Fig. 13. The yield of valine and its clusters versus v for (a) Gd (Becker et al. 1986) and (b) ^{32}S (Alberts et al. 1982). Full curves represent calculations using eq. (71).

(71). The energy loss for the given velocity is calculated using the tables from Northcliffe and Schilling (1970).

Although the fitting parameter is ρ_i, it is more convenient to use the the ratio ρ_i/ρ_s which together with the second parameter η^+ is listed in table 1 (n is the number of valine molecules in the cluster). It should be noted that for the identical clusters the same values ρ_i/ρ_s fit the experimental data by Becker et al. (1986), Hedin et al. (1985) and Alberts et al. (1982). The probability η^+ can differ in these experiments because it depends to a great extent on experimental conditions.

The ratio ρ_i/ρ_s from table 1 satisfies the relation $\rho_i/\rho_s \simeq n^{1/4}$ according to eq. (69). It should be noted that in the paper by Bitensky and Parilis (1987) $\rho_i/\rho_s = $ const and the fitting parameter was a. The probability η^+ for valine agrees by the order of magnitude with the experiment by Salehpour et al. (1986), where it has been found that the charged fraction of the desorbed molecules was 10^{-4} of the total yield.

However, a satisfactory agreement of the yield dependence on dE/dx with the experimental data was also obtained using other available models: the ion track model (Hedin et al. 1985) and the thermal evaporation model (Lucchese 1987, Beining et al. 1988). It means that the dependence of Y on dE/dx is not crucial for the choice of the appropriate model. Further it would be shown that the dependence of Y on film thickness as well as the shape of the angle and spatial distribution of the emitted molecules are in favor of the shock-wave model (Bitensky and Parilis 1987, Bitensky et al. 1989a,c).

4.5.3. Dependence on film thickness

It is very important for the development of a desorption mechanism for a biomolecule to know whether the molecules are ejected just from the surface or also the depth of sample. In experiments by Säve et al. (1987a,b,c, Bolbach et al. 1988) it has been found that the yield increases with the increase in film thickness and saturation is reached at $100-200 \text{ Å}$.

Table 1
The values of cluster radius a, ratio ρ_i/ρ_s and ionization degree η^+ for valine clusters with different number of molecules, n.

n	a (Å)	ρ_i/ρ_s	$\eta^+ (10^{-4})$		
			fig. 12	fig. 13a	fig. 13b
1	2.7	0.17	0.8	0.7	0.27
2	3.4	0.20	0.46	0.47	–
3	3.9	0.22	0.08	0.22	–
4	4.3	0.24	0.12	0.072	–
6	4.9	0.26	–	0.035	–

This value is associated with the crater depth. As follows from the experiment by Säve et al. (1987b), the crater depth scales as $(dE/dx)^{1/2}$ according to the dependence $\rho_s(dE/dx)$ deduced from eqs. (54) and (58).

We consider that the experimental data by Hedin et al. (1987a,b) gives evidence in favor of biomolecule ejection from the depth and the crater formation. In this experiment a comparison has been made between the desorption from some multilayer films prepared with the electrospray technique and monolayers adsorbed on a nitrocellulose backing. The dependence of Y on dE/dx was steeper for these films both for LHRH ($M = 1182$ a.m.u.) and insulin ($M = 5733$ a.m.u.) samples. When describing the experimental data (Hedin et al. 1985) on the base of the track model, Hedin et al. (1987a) had to assume that the number of bonds to be broken for molecule desorption from the thick film is considerably larger than the monolayer covering a dielectric. For instance, in order to desorb an intact insulin molecule this quantity equaled six for a monolayer and eighteen for a film.

In our opinion, the difference between these two cases is caused by quite another reason. Indeed the desorption from the thick film occurs from the whole volume with formation of a crater in the biomolecule substance and is determined by eq. (71), but in case of a monolayer sample on a dielectric substrate the main part of the crater is located in the backing substance and the emission of biomolecules comes only from a thin surface layer. From the rest of the crater the particles of the dielectric backing are emitted. It would be interesting to check this fact experimentally. The total sputtering yield should depend on dE/dx in the same way as the molecule yield from the film. In case of a monolayer using eq. (70) we obtain the following formula

$$Y \approx 2\pi\eta^+ N_m a \rho_s^2 (1 + \rho_i/\rho_s)(1 - \rho_i/\rho_s - 2a/\rho_s). \qquad (73)$$

If the values ρ_s for the biomolecule substance and the dielectric backing are the same, then, as follows from eqs. (61) and (63), the ratio $g = Y_m^{1/2}/Y_f^{1/3}$ (Y_m is the yield for a monolayer on a dielectric, Y_f for a thick biomolecule film) does not depend on dE/dx. If the values ρ_s differs then according to eq. (71) and (73) the equality $g = const$ is the more exact the less the molecule size is. The desorption yield for LHRH and insulin together with g are given in table 2.

As is seen from table 2, the ratio g for LHRH with the accuracy within 10% indeed does not depend on dE/dx. For insulin, the ratio g increases with dE/dx decrease. This means that ρ_s for the dielectric backing is larger than for the biomolecule substance due to P_{cr} increase. Taking ρ_s for the monolayer on a dielectric backing as ρ_s for the thick film multiplied by a factor 1.2, we get a satisfactory agreement with the experiment by Hedin et al. (1987a). According to the law $\rho_s \simeq (dE/dx)^{1/2}$, the value $\rho_s = 64\,\text{Å}$ corresponds to $dE/dx = 66.6\,\text{MeV}/(\text{mg cm}^2)$ for the thick film and $\rho_s = 77\,\text{Å}$

Table 2

The values of molecule desorption yield Y_f, Y_m and the ratio $g = Y_m^{1/2}/Y_f^{1/3}$ for different primary ions.

Ions	dE/dx (MeV/mg/cm^2)	LHRH			Insulin		
		Y_f (%)	Y_m (%)	g	Y_f (%)	Y_m (%)	g
I	66.7	11.0	9.3	0.64	2.0	0.59	0.78
Br	50.9	6.2	6.9	0.66	–	–	–
Ni	42.5	–	–	–	1.4	0.26	0.86
S	25.4	3.3	3.0	0.54	0.54	0.4	1.0
O	11.6	0.63	1.1	0.57	–	–	–
C	7.8	0.065	0.46	0.78	–	–	–

for the monolayer on a dielectric backing is valid. The radius ρ_i for both cases was taken to be the same.

In fig. 14, it is seen that the eqs. (71) and (73) describe satisfactory the experimental dependence $Y(dE/dx)$. The following values of parameters have been used. For LHRH: $a = 5.8$ Å, $\rho_i = 0.27\rho_s$, $\eta^+ = 5.7 \times 10^{-4}$ and 9×10^{-4} for the film and monolayer, respectively. For insulin: $a = 9.7$ Å, $\rho_i = 0.3\rho_s$, $\eta^+ = 2.3 \times 10^{-4}$ and 6.5×10^{-4} for the film and monolayer, respectively. It should be noted that the ratio ρ_i/ρ_s increases with increasing molecule size but not as sharp as it follows from eq. (59).

Bitensky and Parilis (1987) have pointed out that the shock waves were not generated in the conducting materials. Therefore, when calculating the pressure for a film deposited on such backing in eqs. (56a, b) and the yield

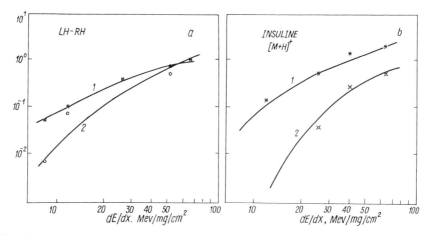

Fig. 14. The yield of Y versus dE/dx for nitrocellulose (1) adsorbed and (2) electrosprayed (a) LHRH and (b) insulin samples, experiment by Hedin et al. (1987a), full curves represent calculations by eqs. (71) and (73).

in eq. (70), the integration should be fulfilled only over the film thickness, i.e., from 0 to d. In this case, the Y versus d threshold behavior considerably differs in shape for the films deposited on polysterol and conducting materials (Säve et al. 1987a).

Integrating eq. (56a) with eq. (57) one gets that for $d \ll \rho_s$ the crater depth $H_c \approx d - 2a$ and its radius $r_1 \approx (d^2 \rho_1)^{1/3}$. Then the desorption yield is

$$Y = \pi \eta^+ N_m \rho_1^{2/3} d^{7/3} (1 - 2a/d) , \tag{74}$$

where $\rho_1 = \rho_s[(1 + \rho_i/\rho_s)(1 - \rho_i/\rho_s - 2a/\rho_s)]^{1/2}$. In addition, the molecules adsorbed in close proximity to the substrate interface being bonded weaker to the substrate than those adsorbed in the bulk due to the reflected wave. This allowed to explain on bases of the shock-wave mechanism the existence of the maximum in the dependence $Y(d)$ observed by Säve et al. (1987c).

The other effect that decreases the desorption of molecules deposited on a conducting backing is the penetration of electrons which source the partial neutralization of the accumulated positive charge. It weakens the intensity of shock wave by diminishing ε_c, P_c and, therefore, the yield Y. This effect decreases with increasing d when $d > \lambda_e$, where $\lambda_e = v_e \tau_e$ is the average penetration depth of conducting electrons into the film during the shock-wave formation time $\tau_e \approx 10^{-13}$ s, $v_e = 10^6$ cm/s being the velocity of the injected electrons. This causes an additional shift of the desorption threshold by $\lambda_e \approx 5$–10 Å.

When the film is deposited on a dielectric backing, the shock wave is also generated in the backing and integral (56a) should be calculated from 0 to infinity, but the biomolecules are desorbed only from a spherical layer (for normal incidence) with depth d and radius ρ_s. Using eqs. (70) and (57), we obtain for $d > \rho_1$ eq. (71) and for $d \leqslant \rho_1$

$$Y = \pi \eta^+ N_m d (\rho_1^2 - \tfrac{1}{3} d^2) , \tag{75}$$

where ρ_1 is the same as in eq. (74). The comparison of the calculations with experimental data is displayed in fig. 15. The shift of the sorption threshold for the film deposited on Au and Si is comparable with λ_e estimation.

The dependence of Y on dE/dx in the case of conducting backing is determined by the factor $[\rho_1(dE/dx)]^{2/3}$ from eq. (74) and at normalization this dependence disappears, which agrees with the experiment by Säve et al. (1987a). As can be seen from eq. (64), the threshold shift decreases and, therefore, the yield should increase with decreasing a. This agrees with the paper by Sundqvist et al. (1986) in which the exceeding of the yield for a fragment ($M = 130$ a.m.u.) over the intact molecule LHRH ($M = 1183$ a.m.u.) has been observed.

It should be noted that crater formation could not be explained within the framework of thermal spike models.

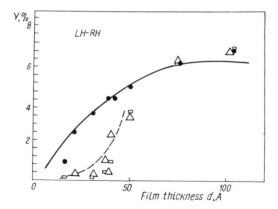

Fig. 15. The yield of LHRH versus film thickness on ($\triangle$) silicon backing, ($\square$) gold backing and ($\bullet$) polystyrene backing (Säve et al. 1987a). Broken curves represent calculations by eq. (74) and full curves by eq. (75).

4.5.4. Dependence on ion charge

Experiments investigating the dependence of Y on ion charge provide important information about the process of biomolecule desorption. As follows from eq. (71), for $a \ll \rho_s$ the yield $Y \simeq (dE/dx)^{3/2}$. It is known that $dE/dx \simeq q_0^2$, where q_0 is the initial ion charge, so $Y \simeq q_0^3$.

In the experiment by Becker et al. (1986) it has been been observed that the desorption yield from the back side to the film exceeds that from the front side. It could also explain by ion charge formation both during passing through matter and crossing the surface. Contrary to Betz–Grodzins model (Betz and Grodzins 1970) the experiments by Francke (1982), Zaikov et al. (1986, 1988) suggest that the ion charge inside the solid is larger than at emergence from the surface. Thus the shock wave is generated within the track of ions moving with a charge q_i larger than the equilibrium charge q_{eq} with which it enters the film after passing through a foil. If one even neglects the ion charge diminution caused by energy loss in the film, nevertheless the charge q_i near the back surface is larger than near the front one, where q_i is changed from q_{eq} much slower than it drops back to q_{eq} at the exit. Therefore, the yield from the back side should enhance. Zaikov et al. (1988) have obtained on the basis of an experiment by Zaikov et al. (1986), an empirical formula for the charge change $\Delta q = q_i - q_{eq} = Z_1 F(X)$ versus reduced velocity $X = v/(v_0 Z_1^{0.45})$ (where v_0 is Bohr velocity and Z_1 is the atomic number of primary ion). The function $F(X)$ is given by

$$F(X) = B \exp[-(X - 0.6)^2/1.5], \tag{76}$$

where B is a constant. We have used this expression to explain the enhancement of valine $(M-H)^-$ desorption yield from the back side depending on Gd ion velocity (Becker et al. 1986). From fig. 16 it is seen that the dependence of Y_{back} on velocity v, $Y_{back}(v) = Y_{front}(v)(1 + \Delta q/q_{eq})^3$ with $B = 0.02$, follows the experiment by Becker et al. (1986) satisfactorily. Here, $\Delta q/q_{eq} \approx 6\text{--}7\%$ corresponds to $Y_{back}/Y_{front} \approx 20\%$. As it was shown by Parilis (1973) and Baklitzky and Parilis (1976), the difference in ion charge also causes a larger electron emission from the back side, e.g., under Kr^{35+} ion bombardment with energy 1.6 MeV/n it could reach 200--300 electrons/ion according to the experiment by Della-Negra et al. (1988a). Both molecule desorption and electron emission are caused by charge change but this does not mean that the desorption is induced by electron flux.

The experiments by Håkansson et al. (1981), Nieschler et al. (1984a,b) and Wien et al. (1987) have shown that the dependence $Y(q_0)$ is more complicated. The desorption yield increases with q_0 slower than q_0 and the rate of increase is different for different q_0. For simplicity, we consider normal incidence. The dependence of ion charge $q(z)$ on ion path z is given by

$$q(z) = q_{eq} + (q_0 - q_{eq}) \exp(-z/\lambda_q) , \tag{77}$$

where λ_q is the charge equilibrium length. Since $R_c \simeq (dE/dz)^{1/2} \simeq q(z)$, then using eq. (56a) we get

$$P_n = P_c R_c^{eq} \int_0^\infty [1 + (q_0/q_{eq} - 1) \exp(-z/\lambda_q)]$$

$$\times (z - H)/[(z - H)^2 + r^2]^{3/2} dz , \tag{78}$$

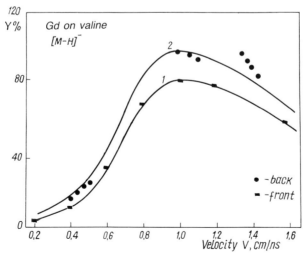

Fig. 16. The front and back yields of (M–H)$^-$ valine ions versus v for Gd ions (Becker et al. 1986). Full curves: (1) front, experiment; (2) back, theory, (points) back, experiment.

where H is the distance from a given point to the surface (see fig. 10). Integral (78) could be estimated as follows. If the ion charge insignificantly differs from the equilibrium one then for $r/\lambda_q \ll 1$ and $H = 0$ from eq. (77) we obtain

$$P_n = (P_c R_c^{eq}/r)[q_0/q_{eq} - (q_0/q_{eq} - 1)(r/\lambda_q) \ln(1.78\lambda_q/r)] . \tag{79}$$

The crater radius is determined from eq. (57) using eqs. (58) and (79),

$$\rho = \rho_0[1 - (1 - q_{eq}/q_0)(\rho_s/\lambda_q) \ln(1.78\lambda_q/\rho_s)] , \tag{80}$$

where ρ_0 and ρ_s are the domain radii corresponding to q_0 and q_{eq}, respectively; $\rho_0 = \rho_s(q_0/q_{eq})$. It could be shown that the crater depth is determined by an expression analogous to eq. (80). So, assuming that the region from which the molecules are ejected is a hemisphere with radius ρ, we get for Y expression (71) in which ρ_s should be replaced by ρ from eq. (80).

In fig. 17 it is shown that the theoretical curves describe rather satisfactory the experimental data by Wien et al. (1987) for such an approximate equation as eq. (80). The following values of parameters have been used for fitting: the radius ρ_s from section 4.5.2 for a multilayer film, for coronene $\rho_i/\rho_s = 0.22$, $\eta^+ = 5.9 \times 10^{-5}$, $\lambda_q = 180$ Å; for phenylalanine $\rho_i/\rho_s = 0.17$, $\eta^+ = 2.8 \times 10^{-5}$, $\lambda_q = 150$ Å.

The formulas obtained by Bitensky and Parilis (1987) and Bitensky et al. (1989a) can be used after some modifications for describing the experimental data from Della-Negra et al. (1990) on biomolecule desorption from Langmuir–Blodgett films.

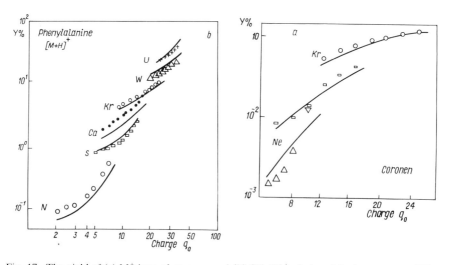

Fig. 17. The yield of (a) M^+ ions of coronene and (b) $(M-H)^+$ of phenylalanine versus q_0 (Wien et al. 1987). Full curves represent calculations by eqs. (71) and (79).

As it follows from Bitensky et al. (1989c) at grazing incidence of primary ions with equilibrium charge the crater boundary on the surface represents an ellipse. One of its semiaxes ρ_{10} depends on the angles of incidence, ψ, and the other one (being perpendicular to the beam direction) ρ_{20} does not,

$$\rho_{10} = \rho_s/\cos\psi , \qquad \rho_{20} = \rho_s , \tag{81}$$

where ρ_s is determined by eq. (58).

If the primary ion charge q_0 differs from equilibrium value q_{eq}, then analogously to the derivation of eq. (80) (Della-Negra et al. 1990) the semiaxes of ellipse are given by

$$\rho_{1,2} = \rho_{10,20}(q_0/q_{eq})[1 - (1 - q_{eq}/q_0)/x_{1,2} \ln(x_{1,2} + \sqrt{x_{1,2}^2 + 1})] , \tag{82}$$

where $x_{1,2} = \lambda_q/\rho_{10,20}$, and λ_q is the equilibrium distance. For $\psi = 0$, the semiaxes ρ_1 and ρ_2 are equal and achieve the value ρ from eq. (80) for $x_{1,2} \gg 1$.

In the case of two monolayers of molecules, M_1, on the top of $6M_2$, the yield of the parent ions M_1 is given by a formula which is equivalent to eq. (73),

$$Y = 2\pi\eta^- N_m L \rho_1 \rho_2 (1 + \rho_i/\rho_s)[(1 - \rho_i/\rho_s - b_1/\rho_1)(1 - \rho_i/\rho_s - b_2/\rho_2)]^{1/2} , \tag{83}$$

where L is the molecule length. The values $b_{1,2}$ determine the distance from the boundary on which the center of a molecule should be located for nondestructive emission.

For a spherical molecule, $b_1 = b_2 = 2a$ [eq. (73)]. If the structure of the molecule is linear, then for large angle of incidence, ψ, this distance $b_2 \approx L \tan\psi$ and for $\psi = 0$ it is constant: $b_{1,2} = a_0$. Therefore, approximately it could be written as

$$b_1 = a_0 + L \tan\psi ; \qquad b_2 = a_0 . \tag{84}$$

Here, a_0 is a fitting parameter depending on the crater shape at $\psi = 0$. Figure 18 shows that the results of a calculation which describes satisfactorily the experimental data (for $\psi \leqslant 70°$) of the yield dependence on both the initial charge and the angle of incidence. For $\psi = 78°$, eq. (83) overestimates the molecular ion yield because for large ψ the parameter $x_1 = \lambda_q/\rho_{10}$ becomes less than 1 and approximations made to evaluate eq. (82) are not valid. The primary-ion equilibrium charge state of ^{127}I at 127 MeV has been taken as $q_{eq} = 24$. For the other parameters the following values have been taken: $\rho_s = 100$ Å, $\rho_i/\rho_s = 0.17$, $a_0 = 24$ Å, $\lambda_q = 200$ Å. A value η^- has been adjusted to the experimental data for $\psi = 20°$ and $q_0 = 24$. The value $\rho_s = 100$ Å agrees with the saturation depth of the yield dependence of film thickness (Säve et al. 1987a) but is less than 1 in the experiment by Bolbach

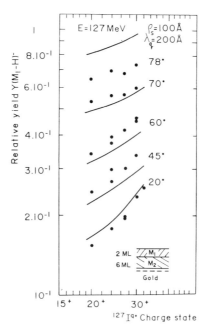

Fig. 18. Comparison of the calculated variation of yields with ψ and q_0 [full curves represent eq. (83)]. The target is made out of 2 ML of M_1 on 6 ML of M_2 on a gold substrate. Experimental data are from Della-Negra et al. (1990).

et al. (1987) in which the emission of the intact molecules from a depth larger than 150 Å was observed.

As is seen in fig. 19, the dependence of the yield of the parent ions $(M_1-H)^-$, the dimers $(2M_1-H)^-$ and even the trimers $[(M_1-H)+Cd]^-$ can be described by eq. (83) but with a larger value of $\lambda_q = 300$ Å. The observed steeper yield dependence of the dimers and trimers [compared with the curve of $(M_1-H)^-$] on the primary-ion charge state is reproduced if a bigger size of the molecules is used in the calculations. It was chosen $a_0 = 50$ Å and $L = 25$ and 50 Å, respectively, for a dimer and a trimer. That means that the dimer molecule has the same length as M_1 whereas the trimer is twice as long. The different fitting values of λ used for comparison with the experimental data (see figs. 18 and 19) can denote that eq. (77) is only a rough approximation for the dependence of ion charge on distance (Maynard and Deutsch 1989).

4.6. *Differential characteristics of sputtered biomolecules*

The differential charateristics of sputtered molecules can provide fruitful information about the sputtering process. Bitensky and Parilis (1987)

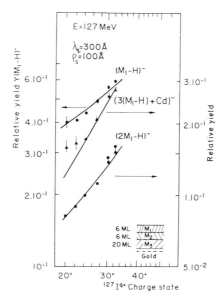

Fig. 19. Yields of the $(M_1-H)^-$ parent ions, the dimer and trimer as a function of the charge state of ^{127}I at 127 MeV ($\psi = 45°$) [full curves represent eq. (83)]. The target consists of superposed layers $6M_1/6M_2/20M_3$ on Au. Experimental data are from Della-Negra et al. (1990).

showed that the sputtering molecules should have a superthermal energy determined by shock-wave energy transfer, and their angular distribution would differ from the cosinusoidal one.

Indeed, in the experiments by Ens et al. (1989) on radial energy distribution depending on the angle of incidence this has been observed, testifying in favor of the shock-wave mechanism.

4.6.1. Angular and energy distribution of sputtered molecules

Detailed calculation of angular and energy distributions has been carried out by Bitensky et al. (1989c). For normal incidence, the normal, P_n, and tangential pressure, P_t, are determined by eqs. (61a), and (61b) and can be rewritten in the form

$$P_n = P_c R_c / R , \qquad P_t = (P_c R_c / r)(1 + Z/R) , \tag{85}$$

where $R = (r^2 + Z^2)^{1/2}$ (see fig. 20).

Using eq. (85), we get, for incident bombardment,

$$P_n(\psi, \alpha) = P_n \cos \psi + P_t \cos \alpha \sin \psi ,$$
$$P_t(\psi, \alpha) = P_n \sin \psi + P_t \cos \alpha \cos \psi , \tag{86}$$

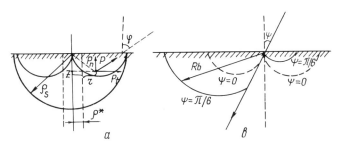

Fig. 20. The shape of a crater at different angles of incidence: (a) $\psi = 0°$; (b) $\psi = \frac{1}{6}\pi$.

where α is the azimuthal angle in the plane normal to the track. If θ is the polar angle, we get from eqs. (85) and (86), for $\alpha = 0$ (sputtering forward in the plane of incidence),

$$P_n(\psi, \alpha) = (P_c R_c / R) \sin(\psi + \tfrac{1}{2}\theta) / \sin \tfrac{1}{2}\theta \ ,$$

$$P_t(\psi, \alpha) = (P_c R_c / R) \cos(\psi + \tfrac{1}{2}\theta) / \sin \tfrac{1}{2}\theta \ . \tag{87}$$

For backward sputtering ($\alpha = \pi$), it is sufficient to replace ψ by $-\psi$.

Designating the escape angle as $\varphi = \arctan(P_t/P_n)$, we get from eq. (87) $\varphi = \tfrac{1}{4}\pi - \tfrac{1}{2}\theta - \tfrac{1}{2}\psi$. For particles originating from the surface, $\theta = \tfrac{1}{2}\pi - \psi$ and $\varphi = \tfrac{1}{4}\pi - \tfrac{1}{2}\psi$. This result was obtained by Johnson et al. (1989).

To calculate the angular and energy distributions of sputtered molecules, the region from which the particles escape should be determined. Bitensky et al. (1989a) have considered the emission from a crater determined only by the condition $P_n = P_{cr}$, where P_{cr} is the critical pressure of material scabbing. However, as is seen from eq. (85), $P_t \geq P_n$ and a part of these molecules could not escape, because their path should cross the region of undamaged matter where the total pressure $P < P_{cr}$. We suppose large molecules to escape without destruction only when their velocity is aimed into the region on the surface determined by the condition $P \geq P_{cr}$, i.e., along some regions with boundaries in the incidence plane,

$$R_b = \rho_s \sin(\tfrac{1}{2}\theta + \psi) \tan(\tfrac{1}{4}\pi + \tfrac{1}{2}\psi) / \cos \tfrac{1}{2}\theta \ , \tag{88}$$

where $\rho_s = P_c R_c / P_{cr}$.

The depth of the direct escape region increases with increasing ψ. Therefore, the saturation of the yield with film thickness should be achieved the later, the larger is ψ. Experimental tests of this fact seem to be interesting.

The distribution over the escape angle normalized per volume $V_0 = \tfrac{2}{3}\pi \rho_s^3$ with account of refraction on the surface is given by

$$f(\varphi) = V_0^{-1} \int\limits_{\theta^*}^{(\pi/2)-\psi} \sin\theta\, d\theta \int\limits_{\rho^*}^{R_b} R^2 \delta[\varphi - \arctan A(\theta, R)] , \qquad (89)$$

where $\delta(x)$ is a delta function, $A = P_t/(P_n^2 - P_{cr}^2)^{1/2}$, and the lower limits θ^* and ρ^* are connected with the molecule fragmentation region [eq. (71)]. With use of eq. (87), we obtain from eq. (89) the approximate formula,

$$f(\varphi) = \sin[2(\varphi + \psi)]\{\tan(\tfrac{1}{4}\pi + \tfrac{1}{2}\psi)\cos\varphi/[\sin(\varphi + \psi)]^3\} , \qquad (90)$$

for $\varphi \geq \tfrac{1}{4}\pi - \tfrac{1}{2}\psi$, and $f(\varphi) = 0$ for $\varphi < \tfrac{1}{4}\pi - \tfrac{1}{2}\psi$. The distribution $f(\varphi)$ is shown in fig. 21. It is drastic and depending considerably on angle of incidence ψ.

Ens et al. (1989) have measured by the deflective potential method the distribution over the molecule radial (tangential) energy E_t. If the time of the shock-wave pulse duration is small, then the particle velocity v is proportional to the pressure P. Therefore, in our theory, this distribution is given by

$$f(E_t) = V_0^{-1} \int\limits_{\theta^*}^{(\pi/2)-\psi} \sin\theta\, d\theta \int\limits_{\rho^*}^{R_b} R^2 \delta\{E_t - E_{cr}$$

$$\times [\rho_s \cos(\tfrac{1}{2}\theta + \psi)/(R\cos(\tfrac{1}{2}\theta))]^2\}\, dR , \qquad (91)$$

where E_{cr} is the critical energy to break away the molecule.

Integration of eq. (91) leads to a rather cumbersome expression so it is worth to use an approximate function,

$$f(\varepsilon_t) = [\varepsilon^{1/2} - \tan(\tfrac{1}{4}\pi + \tfrac{1}{2}\psi)][1 - \varepsilon_t(\rho^*/\rho_s)^2]^{1/2}\cos^2\psi/\varepsilon^3 , \qquad (92)$$

which has a threshold at $\varepsilon_{min} = \tan^2(\tfrac{1}{4}\pi - \tfrac{1}{2}\psi)$, a maximum at $\varepsilon_t \approx 1.5\varepsilon_{min}$ and also depend on ψ (see fig. 22), $\varepsilon_t = E_t/E_{cr}$.

The difference between the distributions for forward and backward

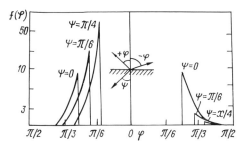

Fig. 21. The distributions of biomolecules over the escape angle φ for different angles of incidence ψ.

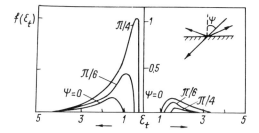

Fig. 22. The distributions of biomolecules over ε_t for different angles of incidence ψ.

sputtering increases with increasing ψ. It should be noted that $f(\varepsilon_t)$ appears to be considerably narrower than the experimental one (Ens et al. 1989). The source of both a narrow and a sharp-edged $f(\varphi)$ is the same: the neglection of scattering of the biomolecule along their path l toward the surface. If we take this into account, e.g., by replacing the δ function by a Gaussian-type function, we could get an asymmetric broadening of both distributions. As follows from Bitensky and Parilis (1987), the molecule escapes the area of nodestructive knocking out if their centers are located inside it at the distance $r \geq a$ from the boundary (a is the molecule radius). Then $E_{cr} \simeq (\rho_s - a)^{-2}$ increases with increasing molecule size, which results in the shift $f(E_t)$ towards high energy in accordance with experiment by Ens et al. (1989). The strong dependence of the most probable energy E_t on ψ could explain the broadening of $f(E_t)$ at sputtering under fission fragments recorded by Ens et al. (1989). It should be noted that results of the computer simulation made by Fenyö et al. (1989) based on the 'popcorn' model satisfactorily agree with experimental data on the radial energy distribution of sputtering molecules. It is explained by the fact that both within the shock-wave model and the popcorn model the molecules eject due to momentum transfer from the region perturbed by primary ions.

A formula analogous to eq. (91) allows to get the axial (normal) energy distribution $f(E_n)$. This distribution weakly depends on the angle of incidence and for $\psi = 0$ is given by

$$f(\varepsilon_n) = \varepsilon_n [1 - \varepsilon_n(\rho^*/\rho_s)^2]^{1/2}/[(\varepsilon_n + 1)^{5/2}(\varepsilon_n + 2)], \qquad (93)$$

where $\varepsilon_n = E_n/E_{cr}$.

The maximum of distribution (93) is achieved at $\varepsilon_n = 0.5$. The most probable axial energy of insulin molecules measured experimentally is $E_n = 3.4$ eV (Widdiyasekera et al. 1988) or 5.6 eV (Jacobs and Macfarlane 1988). Choosing $E_n = 4.5$ eV, we obtain a reasonable value for $E_{cr} = 9$ eV. In this case, the most probable radial energy $E_t = 2.4$ eV at $\psi = \frac{1}{4}\pi$ is smaller than the experimental value $E_t = 5$ eV (Ens et al. 1989). This shows that the broadening due to the scattering is, indeed, important.

4.6.2. *Spatial distribution of sputtered biomolecules*

In the previous subsection, the emission of nondestructed molecules was considered to be possible only in the free way direction toward the surface. This condition gives a radial energy distribution which is sharper than found in experiments (Ens et al. 1989), although it agrees with them qualitatively. Other molecules do not contribute significantly in the formation of both angular and energy distributions but smoothen them to a certain degree.

When molecules escape the crater they can undergo some destructive collisions resulting in angular and energy redistribution. Therefore, the spectra smoothing should be more pronounced for small and more stable molecules. Actually, it was observed when comparing the experimental spectra of valine (116 a.m.u.) and insulin (5733 a.m.u.) (Moshammer et al. 1989). The small molecules and the fragments emerge in directions far from the shock-wave direction.

In the present subsection the spatial distribution of biomolecules emitted from the very surface has been calculated within the framework of the shock-wave mechanism. In this case, the molecules do not interact with the molecule emerging from the deeper layers during emission. Therefore, the transformation of energy spectra should be minimal. This will allow us to compare the results of calculation with some special experiments. The method of unimolecular layer deposition on insulator substrate (Hedin et al. 1987a) and the Langmuir–Blodgett procedure of film preparation give the possibility to study the angular and energy distribution of biomolecules emitted from the very surface layer.

Let a fast heavy ion impinges a biomolecule surface layer at the angle of incidence ψ. We shall use a coordinate system with the axis OZ″ directed along the same direction as the ion velocity and the axis OX″ lying on the

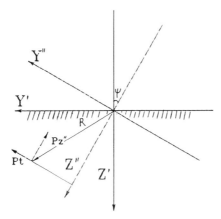

Fig. 23. Coordinate system.

surface (see fig. 23). According to the shock-wave mechanism of bio-molecule sputtering the source of the shock waves is the high-density energy spike around the ion track. The normal and radial components of pressure are determined by formulae from eqs. (61a) and (61b).

Let us introduce a new coordinate system (x', y', z'), with the z'-axis directing along the surface normal. In this system we get

$$P_{z'} = P_{z''} \cos \psi + P_t \cos \beta \sin \psi ,$$

$$P_{y'} = P_{z''} \sin \psi + P_t \cos \beta \cos \psi , \qquad (94)$$

$$P_{x'} = P_t \sin \beta ,$$

where β is the azimuthal angle in the (x'', y'', z'') system. Expressing coordinates $\{x'', y'', z''\}$ by $\{x', y', z'\}$ and supposing $z' = 0$ on the surface, we obtain the formulae for pressure components in the new coordinate system,

$$P_{z'} = \frac{P_c R_c}{\rho} \frac{\cos \psi}{1 - \cos \alpha \sin \psi} ,$$

$$P_{y'} = \frac{P_c R_c}{\rho} \frac{\sin \psi}{1 - \cos \alpha \sin \psi} , \qquad (95)$$

$$P_{x'} = \frac{P_c R_c}{\rho} \frac{\sin \alpha}{1 - \cos \alpha \sin \psi} ,$$

where ρ and α are the cylindrical coordinates in the plane $z' = 0$. The crater boundary is determined by the condition $P_z = P_{cr}$ (P_{cr} is the critical pressure for breaking away the molecule). For normal incidence ($\psi = 0$), the crater boundary is a circle. Designating the radius of this circle by ρ_s, we get from eq. (95) $\rho_s = P_c R_c / P_{cr}$. Introducing the dimensionless variables $y = y'/\rho_s$, $x = x'/\rho_s$, we derive from eq. (95) the equation of the boundary,

$$\rho_b = \frac{\cos \psi}{1 - \cos \alpha \sin \psi} \qquad (96)$$

or using Cartesian coordinates

$$x^2 + (y - \tan \psi)^2 \cos^2 \psi = 1 . \qquad (97)$$

Equation (97) describes the ellipse with semiaxes $b_t = 1$; $b_\ell = 1/\cos \psi$. It should be noted that the semiaxis b_t being normal to the plane of incidence is independent of ψ and the semiaxes b_ℓ lying in the plane of incidence increases with ψ increasing. The center of the ellipse is shifted at the distance $c = \tan \psi$ from the origin of the coordinates. The regions of molecule emergence from the surface are displayed in fig. 24a for different ψ. The spatial distribution is described by the function $f(v'_x, v'_y)$ being the molecule distribution over the velocity in the plane parallel to the sample

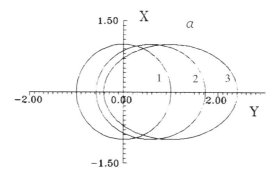

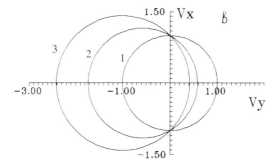

Fig. 24. Crater boundaries (a) on the surface, (b) boundaries $v(\varphi)$ in the velocity plane for the different ψ: (1) $\psi = 0°$; (2) $\psi = 30°$; (3) $\psi = 45°$.

surface. Since $P_{x',y'} = \eta D v'_{x',y'}$ where η is the sample density, and D is the shock-wave velocity, and introducing the polar coordinates (v, φ), we can obtain

$$f(v, \varphi)v \, dv \, d\varphi = \frac{1}{\pi b_t b_\ell} = \rho \, d\rho \, d\alpha \,, \tag{98}$$

using the reduced variables $v = v'/v_{cr}$, $v_{cr} = P_{cr}/\eta D$.

Taking into account eq. (97), we get

$$f(v, \varphi) = \frac{\cos \psi}{\pi} \frac{\rho}{v} \frac{\partial(\rho, \alpha)}{\partial(v, \varphi)} \,. \tag{99}$$

Here $\partial(\rho, \alpha)/\partial(v, \varphi)$ is a Jacobian, variables v and φ are given by

$$\tan \varphi = \frac{\sin \psi}{\cos \alpha - \sin \psi} \,,$$

$$v = \frac{1}{\rho} \frac{\sqrt{\sin^2 \psi - 2 \sin \psi \cos \psi + 1}}{1 - \cos \alpha \sin \psi} \,. \tag{100}$$

The function $f(v, \varphi)$ describes the transformation of the internal region of the ellipse, eq. (97), into the external part of the plane $\{v, \varphi\}$ limited by the curve $v_{\lim}(\varphi)$. The equation could be found for this curve from eqs. (96) and (100),

$$v_{\lim} = \frac{\sqrt{1 - \sin^2\psi \sin^2\varphi} - \cos\varphi \sin\psi}{\cos\psi} . \tag{101}$$

The calculation of the Jackobian, eq. (99), using eqs. (100) and (101) leads to the following expression of the distribution for $v > v_{\lim}$,

$$f(v, \varphi) = \frac{\cos\psi}{\pi v^4 (1 - \sin^2\psi \sin^2\varphi)} \left(1 - \frac{\cos\varphi \sin\varphi}{1 - \sin^2\psi \sin^2\varphi}\right), \tag{102}$$

and $f(v, \varphi) = 0$ for $v \leqslant v_{\lim}$. When passing to Cartesian coordinates, we derive the equation of boundary in the velocity plane,

$$(v_y + \tan\psi)^2 + v_x^2 = 1/\cos^2\psi , \tag{103}$$

i.e., represents itself as a circle with a center shifted at $\tan\psi$ along v_y and with the radius equal to $1/\cos\psi$. These boundaries are shown in fig. 24b for different ψ.

The spatial distribution of the molecules sputtered from the surface calculated with the use of eq. (102) are exhibited in fig. 25. As it is seen, the spatial distribution differ significantly from the cosine law. This feature seems to be essential for the developed shock-wave mechanism.

The calculated spatial distribution agrees qualitatively with the experimental one (Moshammer et al. 1989).

Experimental study of spatial distribution of molecules sputtered just from the surface and its comparison with theory is of great interest.

4.7. Charge state of sputtered molecules

The investigations of Hedin et al. (1987b) have shown that despite some analogous features in the dependences of the yields Y^+ and Y^- on angle of incidence, film thickness, velocity and heavy-ion charge, their dependence on energy loss is different. Although it is the only experimental results, it revealed that the charge formation process should influence the sputtered molecule yield. Johnson et al. (1989) have shown on the base of the pulse pressure propagation model that at large energy loss the neutral molecule yield $Y \simeq (\mathrm{d}E/\mathrm{d}x)^3$ because the crater radius $\rho_s \simeq \mathrm{d}E/\mathrm{d}x$. However, the distinctive behavior of $Y^{+,-}$ has not been discussed.

We intend to discuss here a modified shock-wave theory in which the difference in charged and neutral yield dependence on $\mathrm{d}E/\mathrm{d}x$ could be explained. In papers by Bitensky and Parilis (1987) and Bitensky et al. (1989a) the shock wave emerged if in the primary domain with radius R_c the

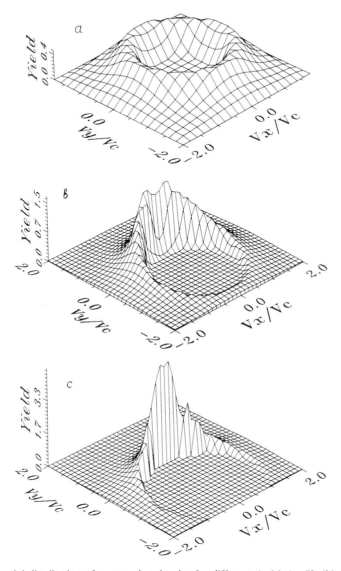

Fig. 25. Spatial distribution of sputtered molecules for different ψ: (a) $\psi = 0°$; (b) $\psi = 30°$; (c) $\psi = 45°$.

mean energy per particle exceeded a certain critical value ε_c. Then $\rho_s \simeq (dE/dx)^{1/2}$ and $Y \simeq (dE/dx)^{3/2}$. If we assume that the shock wave originates when the energy flux through the primary domain surface exceeds a critical value f_c, then

$$f_c = \Delta E/(\pi R_c^2) \approx (dE/dx)/(2\pi R_c) \tag{104}$$

and $R_c \simeq dE/dx$. Thus neutral molecule yield $Y \simeq (dE/dx)^3$, as in the experiment by Hedin et al. (1987b).

One could assume that the probabilities of the formation and survival of both positive and negative ions depend strongly but in different ways on the distance from the fast heavy-ion track and on the molecule ion path in the matter. There are some reasons for such an assumption:

(i) Formation of the quasimolecular ions like $(M-H)^+$ and $(M-H)^-$ occurs via the breaking, under nonequilibrium jerk, of the hydrogen bonds which bind the neighboring molecules. Then the proton recombines with a 'stranger' molecule (see fig. 26). In such a case the place of the most intensive charged molecule pair formation is located in the coaxial region determined by the condition: $P_h < |\mathrm{grad}\, P(x, y, z)2a| < P_d$, where P_d is the molecule destruction and P_h is the hydrogen bond breaking tension.

(ii) Since there is an excess of positive charge in the track vicinity the probability of formation and survival of positively charged molecules is large in this region.

(iii) If we use the phenomenological expression $Y^{+,-} = \eta^{+,-}Y$, then since $Y \simeq (dE/dx)^3$ we obtain that the ionization degree $\eta^{+,-}$ decreases with energy loss increase $\eta^{+,-} \simeq (dE/dx)^{-k}$, where $1 < k < 2$. According to Hedin et al. (1987b), it should be considered as an experimental fact. A possible explanation of this relation is the decrease of the charged-state survival probability with molecular ion velocity increase due to neutralization along the ion path l toward the surface. If we consider $\eta^{+,-} \simeq \exp(-t/\tau_{+,-})$, where $\tau_{+,-}$ is effective neutralization time, $t = 1/v$ is the time of motion along the trajectory of direct escape, $l \simeq \rho_s \simeq dE/dx$ (the energy distribution of sputtered molecules is known to be independent of dE/dx), then $\eta^{+,-}$ decreases with increasing dE/dx. If $\tau_- > \tau_+$, then Y^- increases faster than Y^+ does. Neutralization of negatively charged ion occurs in differential way, which leads to the effective survival length λ_- and λ_+.

According to Bitensky et al. (1989c) one could choose $\eta_{+,-}(r)p(r) = b \exp[-(r-\rho_d)/\lambda_{+,-}]$, where b is a constant, and ρ_d is the radius of the molecule destruction region. The yield of charged molecules is given by eq. (70). Calculations of integral (83) leads to the expression

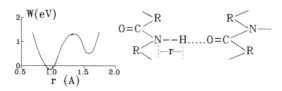

Fig. 26. The dependence of the potential of the hydrogen atom with a molecule on the distance from the nearest atom.

$$Y^{+,-} = 2\pi N_{\mathrm{m}} \rho_{\mathrm{s}}^3 \lambda'_{+,-} b \int\limits^{(1-\rho')/\lambda_{+,-}} (\lambda'_{+,-} u + \rho')$$

$$\times [1 - (\lambda'_{+,-} u + \rho')^2]^{1/2} e^{-u} \, \mathrm{d}u \,, \tag{105}$$

with the dimensionless $\lambda'_{+,-} = \lambda_{+,-}/\rho_{\mathrm{s}}$ and $\rho' = \rho_{\mathrm{d}}/\rho_{\mathrm{s}}$. For $\lambda'_{+,-} \ll 1$ (which apparently corresponds to positively charged molecule yields),

$$Y^+ \approx 2\pi N_{\mathrm{m}} \rho_{\mathrm{s}} \lambda_+ b \rho_{\mathrm{d}} [1 - (\rho_{\mathrm{d}}/\rho_{\mathrm{s}})^2]^{1/2} \,. \tag{106}$$

Since $\rho_{\mathrm{d}} \approx R_{\mathrm{c}}^{1/2}$ [eq. (62)] and $R_{\mathrm{c}} \simeq (\mathrm{d}E/\mathrm{d}x)$ then $\rho_{\mathrm{d}} \simeq (\mathrm{d}E/\mathrm{d}x)^{1/2}$ as follows from eq. (85) $Y^+ \simeq (\mathrm{d}E/\mathrm{d}x)^{3/2}$ in accordance with Beining et al. (1988). For $\lambda' \geqslant 1$, as follows from eq. (84), $Y^- \simeq (\mathrm{d}E/\mathrm{d}x)^3$. The intermediate values λ'_- give $Y^- \simeq (\mathrm{d}E/\mathrm{d}x)^n$ as by Hedin et al. (1987b).

Certainly, the alternative dependence of both ρ_{s} and the yield on energy loss needs an improvement on the yield dependences on different parameters as well as angular and energy distributions. However, for such alternation some new experimental data on both neutral and charged molecule yields are necessary.

CHAPTER 11

Auger Processes in Kinetic Electron Emission

1. Introduction

The electron emission under ion bombardment became an object of deep and thorough study in many laboratories all over the world because of the connection with rapid progresses in plasma research and in studies of the physics of interaction of charged and neutral particles with the lattice of solids. Interesting prospects are opened by electron emission under ion bombardment in the field of emission microscopy for the study of the structure of solids under arbitrary temperature conditions. Knowledge of the kinetic emission coefficient is required also in connection with the use of electron multipliers as sensitive ion detectors in mass spectrometers.

Certain features of this phenomenon have been elucidated, which enabled the development of a more or less adequate theory. Although nothing has been discovered in this field which could have revolutionized our concepts concerning the fundamental laws of matter, nevertheless, many interesting and sometimes unexpected facts have been found, and a number of original ideas have been formulated.

The present chapter is an attempt to review these ideas. The available experimental data are treated only briefly since they are discussed thoroughly in excellent reviews by U.A. Arifov (1961, 1968), Kaminsky (1965), Medved and Strausser (1965), Abroyan et al. (1967), Benazeth (1982), Krebs (1983), Hasselkamp (1988), Brusilovsky (1985, 1990a,b), Hofer (1990) and others.

2. Kinetic electron emission

The electron emission resulting from the collision of a fast ion with the surface of a solid and discovered by Villard (1899) in a discharge tube where it produced cathode rays, was afterwards observed by Rutherford (1905) as a phenomenon accompanying the passage of α particles through matter.

Information of the electron emission from solids under the bombardment by ions and atoms accumulating during 85 years of research was obtained by experimental techniques of different degrees of perfection. As revealed already in early studies, the emission is influenced to a very considerable extent by the degree of cleanness of the surface under investigation.

Adsorbed films of even very low density distort strongly the pattern of the phenomenon.

The problem of obtaining an atomically clean surface is closely associated with that producing a high vacuum, therefore, early works carried out at 10^{-4}–10^{-5} Torr are at present only of historical interest.

Data on the emission of the same particles obtained at different times are contradictory. As was reasonably pointed out by Massey and Burhop (1952), a set of data obtained by each worker is ordinarily self-consistent, however, these data may disagree strongly with those of other researchers. The brief summary of experimental data presented below is based on the most reliable results obtained in several laboratories.

2.1. Experimental data

The kinetic electron emission is described by the so-called kinetic emission coefficient or yield γ. This quantity representing the ratio of the number of ejected electrons to that of the atomic particles incident on the surface is studied as a function of their mass, energy, charge, atomic number and the angle of incidence, as well as the kind, temperature and condition of the surface. The energy distribution of the ejected electrons is also studied.

2.1.1. Dependence of γ on ion velocity

This is the major point since the energy dependence of the electron emission permits to draw conclusions concerning the mechanism of electron excitation. In particular, already at an early stage of research, the study of the dependence of emission on the energy (or velocity) of the ions resulted in the distinguishing between two principal kinds of electron emission, i.e., the potential emission and the kinetic emission. These two kinds of emission are based on two totally different mechanisms of electron excitation. They are assumed to be additive and independent of one another. We shall be interested here in the kinetic electron emission only.

Figure 1 presents a double-log plot of the dependence of the kinetic emission on ion velocity v_0 within a wide velocity range constructed using the data of different authors (Hagstrum 1953a,b, 1954b, Waters 1958, Higatsberger et al. 1954, Little 1956, Large 1963, U.A. Arifov 1961, 1968, U.A. Arifov and Rakhimov 1960, U.A. Arifov et al. 1962b,c,d, 1963a,b, 1964, Hill et al. 1939, Akishin and Vasilev 1959, Aarset et al. 1954, Petrov 1960a,b, 1962, Bosch and Kuskevics 1964, 1965, Brünne 1957, Scharkert 1966, Murdock and Miller 1955, Fogel et al. 1960a,b, Allen 1939, Rostagni 1934, Berry 1958, Burrekoven 1956, Bourne 1952, Philbert 1953, Sommeria 1954). The investigations by Baragiola et al. (1979), Alonso et al. (1980) and Ferron et al. (1980) should also be noted.

One may note here several characteristic regions of variation of $\gamma(v_0)$:

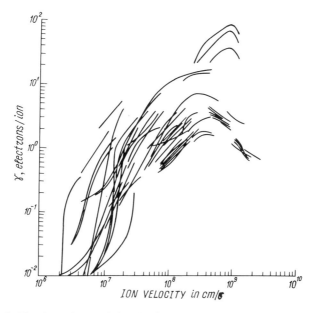

Fig. 1. The dependence of the kinetic emission coefficient on ion velocity.

The threshold of kinetic emission. The existence of a threshold for the kinetic emission from a clean metal may be considered as proved. Although this threshold is not very clearly pronounced, one can reliably indicate the region of the values of v_0 at which it is already practically zero. The actual position of the threshold on the experimental curves depends, beside other things, on the accuracy of measurements. In each experiment, the threshold is essentially that value of v_0 at which γ is just in excess of the lowest measurable magnitude.

From the most accurate data, $v_{min} = (0.7 \pm 0.3) \times 10^7$ cm/s. For the lightest ions, v_{min} is higher: for He$^+$ it equals $\approx 2 \times 10^7$ cm/s, for H$^+$ and D$^+$ it equals $\approx 3 \times 10^7$ cm/s (Cawthron 1971).

It should be stressed that the position of the threshold depends strongly on the actual cleanness of the surface. In fact, the discovery of the region where $\gamma = 0$ has become possible only after the production of atomically clean metal surfaces.

The quadratic region. Just beyond the region lying close to the threshold there is a region where $\gamma \simeq kv_0^2 -$ const. This region on the $\gamma(E_0)$ curves corresponds to a linear increase in emission and, therefore, it is sometimes referred to as the 'linear region'. The quadratic region extends approximately to $v_0 = (1.5-2) \times 10^7$ cm/s.

The linear region. This begins immediately beyond the quadratic one and extends up to $v_0 \approx (2.5-3) \times 10^8$ cm/s. On the graphs $\gamma(E_0)$, one observes here a slowed-down increase (of the kind $\sqrt{E_0}$) with a gradual transition to a maximum.

The maximum on the $\gamma(v_0)$ curve. It is observed at $v_0 \approx (10^8-10^9)$ cm/s which corresponds to a region of a few tens, hundreds of keV or even a few MeV (for different ions).

A fall-off. Beyond the maximum at high ion velocities one observes a fall-off down to quite low values of γ.

The dependence of kinetic electron emission on ion velocity was found to be the same even for the emission under large molecules and clusters (Thum and Hofer 1979).

2.1.2. Dependence of kinetic emission on ion charge

This dependence was investigated in specially designed experiments (Dunaev and Flaks 1953, Schram et al. 1966, Telkovsky 1956). In contrast to Dunaev and Flaks (1953) who observed sudden changes in emission in the case of (3–4)-fold charged ions (see fig. 2), Schram et al. (1966) have

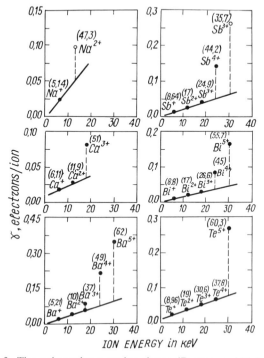

Fig. 2. The γ dependence on ion charge (Dunaev and Flaks 1953).

revealed an independence of emission on ion charge of Ne, Ar, Kr and Xe up to the charge states 8–9 within the ion velocity range $(1–5) \times 10^7$ cm/s (see fig. 3). Telkovsky (1956) has come to the same conclusion.

Recent investigations show that the straight line in fig. 3 is rather an envelope of a family of curves (see chapter 12). In the same chapter, the possible weak dependence of the kinetic emission on high charge for metals and the strong dependence for insulators and semiconductors was obtained.

The dependence (or independence) of the kinetic emission on the ion charge is closely associated with the problem of additivity of the potential and kinetic emission on the ion velocity. Indeed, the sudden changes observed by Dunaev and Flaks (1953) are revealed also with those multicharged ions which the highest ionization potential (shown in brackets in fig. 2) is equal to about 50 eV or even higher than that. The authors interpret these jumps as due to a contribution from the potential emission which should be quite high for high ionization potentials (Parilis 1967). The same effect of additivity is indicated by the experiments of U.A. Arifov et al. (1962c,d, 1963a,b, 1964), who compared the kinetic emission under bombardment of ions and neutral atoms, e.g., of Ar^+ and Ar^0. The general pattern of the dependence of electron emission on the ion charge is illustrated in fig. 4.

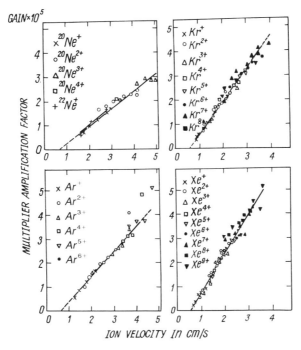

Fig. 3. The γ independence on ion charge (Schram et al. 1966).

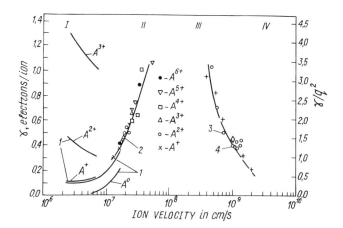

Fig. 4. The ratio of γ on ion charge q for different ion velocities. ($\bigcirc$) α particles; ($+$) protons; ($\square$, $\bullet$) fission fragments. Theory: Region I is the theory of Hagstrum (1954a); II the theory of Parilis and Kishinevsky (1961); III the region in which no theory is developed; IV the theory of Sternglass (1957). Experiment: Curves 1 data obtained by U.A. Arifov (1961, 1968); 2 data obtained by Schram et al. (1966); 3 data obtained by Hill et al. (1939); 4 data obtained by Anno (1962, 1963).

Figure 4 shows the dependence of γ on the charge q within various ranges of ion velocity v_0. As shown by Hagstrum (1954a), at low velocities the potential emission increases sharply with q and decreases with velocity the faster, the higher the ion charge is.

A comparison is made in fig. 4 between the data of U.A. Arifov et al. (1963a,b) and those of Schram et al. (1966) for emission under bombardment by argon ions. Since in the experiment by Schram et al. (1966) one used the electron multiplier which had not been calibrated by γ, we have matched for the purpose of comparison the values of γ at $v_0 = 10^7$ cm/s for Ar^+. Thus, the data by Schram et al. (1966) and U.A. Arifov (1968) supplement one another. Unfortunately, the experiments by Telkovsky (1956) have not been extended to low velocities.

As for the high-velocity region (lying to the right from the maximum in fig. 4), here apparently the emission rises proportionally to q^2. Figure 4 presents a graph of γ/q^2 for high-velocity protons (Hill et al. 1939), α particles and fission fragments (Anno 1962, 1963). As is well-known, the charge of the fission fragments is about 20–22, which may account for the extremely high electron emission observed under bombardment by fission fragments and reaching, according to Anno (1962, 1963) and Jarger and Anno (1966), 300–500 electrons per fragment. Experiments have been designed so that this figure included 40–50 electrons emitted at the moment of fission. This is the only possible explanation since the velocity of the fission fragments (10^9 cm/s) lies in the region where the emission falls off as $1/v_0$ and is very small under proton bombardment (see fig. 4).

Unfortunately, there is little information on the charge dependence of the emission in the intermediate velocity range 10^8–10^9 cm/s. However, it is in this range that the transition from the independence on q to the $\sim q^2$ relationship should occur. At present, it remains to be a gap in our knowledge of the kinetic electron emission.

Thus, we have apparently four regions differing in the dependence of the electron emission on ion charge:

(1) Velocity range 10^6–5×10^6 cm/s. The potential emission increases sharply with q. The theory of Hagstrum (1954a) provides a basis for the consideration of this increase and fall-off of the emission with velocity, which should be apparently obtained quite rigorously by taking directly into account the relative probabilities for the Auger neutralization of ions with different charge states of levels of different depth (Parilis 1967, Sternberg 1957).

(2) Velocity range 5×10^6–10^8 cm/s. A threshold and a keV region of kinetic emission, independence of the emission on charge. A possible contribution from the part of potential emission which still has not vanished owing to the increase of velocity. The independence of emission on charge is contained naturally in the theory of Parilis and Kishinevsky (1961) of two kinds of emission caused many arguments (U.A. Arifov 1961, Medved et al. 1963).

(3) Velocity range 10^8–5×10^8 cm/s. The theory due to Sternglass (1957) is applicable here. It yields $\gamma \sim q^2/v_0$. Whether the borderlines between these regions are sharp is unclear.

2.1.3. Dependence of kinetic emission on the kind of ion

The dependence of the kinetic emission on the kind of ion is intricate since it contains the dependence on both the mass M_2 and the atomic number Z_2 of the ion. In particular, the difference in ionic mass makes the relation which could be inferred from the comparison of the curves $\gamma(E_0)$ untypical since ions of different mass but equal energy passes different velocities. To exclude this effect, one should compare γ for different ions at the same velocity.

Therefore, one can deduce the dependence of the emission on the kind of ion only from experiments performed in the same apparatus, under the same conditions, but with different ions. Figure 5 shows the data on the dependence of the kinetic emission on the atomic number of the incident ion. Plotted along the ordinates axis is the quantity $\gamma(Z_1, Z_2)/\gamma(Z_1, Z_1)$ at $v_0 = $ const, and along the abscissae axis Z_2/Z_1. Here Z_1 is the atomic number of the target atoms, and Z_2 that of the ions.

The broken curve obtained from the theory (Parilis and Kishinevsky 1961,

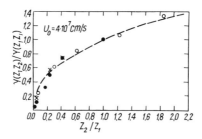

Fig. 5. The ratio of kinetic emission coefficients versus atomic number ratio of the colliding particles. The broken curve represents the theoretical calculation (Kishinevsky and Parilis 1962), (X) data obtained by U.A. Arifov (1961), (○) data obtained by Telkovsky (1956), and (●) data obtained by Schram et al. (1966).

Kishinevsky and Parilis 1962) fits well the experimental data by U.A. Arifov and Rakhimov (1960), U.A. Arifov et al. (1962b), Telkovsky (1956) and Schram et al. (1966) in a wide region except for $Z_2/Z_1 \leqslant 0.1$.

2.1.4. Isotopic effect
There is only one way of isolating in an explicit form the dependence on the ion mass which consists in studies on emission under bombardment of isotopes of an element. The isotopic effect is defined by the quantity

$$\beta = \left(\frac{\Delta\gamma}{\gamma} \right)_{v_0 = \text{const}} \bigg/ \frac{\Delta M_2}{M_2} \ .$$

Here $(\Delta\gamma)_{v_0 = \text{const}}$ is the difference of the values of γ for two isotopes at the same velocity v_0, ΔM_2 is the difference in mass of two isotopes of bombarding ions.

Ploch (1951) attempted to measure the isotopic effect; however, the surface contamination precluded its observation. A positive isotopic effect has been observed by Brünne (1957). The isotopic effect was studied also by Higatsberger et al. (1954), Palmer (1959), Campbell and Whittem (1961) and Fehn (1974). Debevec et al. (1966) have studied emission under bombardment of isotopes of Cr, Sr, Sm, Dy and W (see fig. 6).

The negative isotopic effect revealed by them is probably only apparent since measurements were carried out at a constant ion energy ($E_0 = 5$ keV) and hence this effect is due to a difference in the velocities of ions of different mass.

The sign of the isotopic effect,

$$\beta = \frac{M_2}{\gamma} \left(\frac{\partial\gamma}{\partial M_2} \right)_{v_0 = \text{const}} = \frac{M_2}{\gamma} \left(\frac{\partial\gamma}{\partial M_2} \right)_{E_0 = \text{const}} + \frac{v_0}{2\gamma} \left(\frac{\partial\gamma}{\partial v_0} \right)_{M_2 = \text{const}},$$

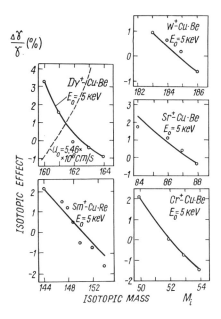

Fig. 6. The isotopic effect existing in the kinetic electron emission (Debevec et al. 1966). The broken curve represents the theoretical calculation for the following conditions: $(\Delta\gamma/\gamma)v_0 = $ const, where $v_0 = 5.46 \times 10^6$ cm/s.

depends on the magnitude of the second term on the right-hand side. The first term [measured by Debevec et al. (1966)] is always negative, and the second one is always positive (to the left of the maximum in fig. 1). Depending on the relative magnitude of the absolute values of the two terms, β may be either positive or negative. In the region near the threshold the derivative $\partial\gamma/\partial v_0$ is large and $\beta > 0$, but as one approaches the maximum, it becomes smaller, and the isotopic effect may become negative. There are no experimental data on the isotopic effect near the maximum and to the right from it.

As will be shown later, the theory of Parilis and Kishinevsky (1961) yields a satisfactory value for the isotopic effect. Unfortunately, Debevec et al. (1966) did not measure the dependence $\gamma(v_0)$, and, therefore, one cannot calculate β using only experimental data. If one takes the theoretical value for $\partial\gamma/\partial v_0$ near the threshold [this is the region where the experiments of Debevec et al. (1966) were carried out], one obtains a large isotopic effect.

A theoretical curve $(\Delta\gamma/\gamma)_{v_0 = \text{const}}$ for Dy is shown for comparison in fig. 6 (broken curve). The experiments of Krebs (1968) were designed to reveal the error introduced by the isotopic effect into the readings of the mass-spectrometer electron multiplier. Of better use for this purpose would be data on $(\partial \gamma / \partial M_2)_{v_0 = \text{const}}$ since particles escaping the mass spectrometer have the same velocities and are separated by the difference in the value e/M_2.

2.1.5. Energy spectrum of the emitted electrons

Unfortunately, no one has ever made a systematic experimental study of the energy distribution of secondary electrons and of its dependence on various parameters characterizing the process of kinetic electron emission. Only unrelated measurements for some ions and some metals have been performed. The spectra of kinetic emission were studied by Brünne (1957), Waters (1958), Telkovsky (1956), Pradal and Simon (1958), Slodzian (1958), Linford (1935), U.A. Arifov et al. (1963a, 1966a,b), Magnuson and Carlston (1963, 1965). Special attention was paid to the Auger fine structure of the spectra carrying important information on the composition and state of the surface (Wittmaack 1979a,b, Soszka 1981, Benazeth 1982, E.W. Thomas 1984, Dorozhkin 1985, Dorozhkin and Petrov 1983).

As follows from the data in fig. 7 for the case $Ar^+ \rightarrow Mo$ (U.A. Arifov et al. 1963a,b) the energy spectrum of the kinetic emission has a maximum at $E_k \lesssim 10$ eV, the position of which (as well as the shape of spectrum) almost does not depend on the energy of the incident ions within a very broad range. The maximum energy of electrons increases somewhat with increasing ion energy, however, this increase is, of course, not comparable with that of E_0.

The energy distribution of electrons in the kinetic emission bears some resemblance to the Maxwell curve. On these grounds the electrons are sometimes ascribed an equivalent 'temperature' (30 000 to 80 000 K) which, however, only weakly depends on the ion energy.

Some interesting results have been obtained by Snoek et al. (1966), who observed a maximum in the high-energy part of the spectrum of electrons emitted from gold under the bombardment of 60 keV Ar^+ ions (see fig. 8). The maximum lies near $E_k = 200$ eV and resembles the maximum in the

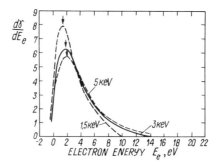

Fig. 7. Energy spectrum typical for the kinetic emission (U.A. Arifov et al. 1966b).

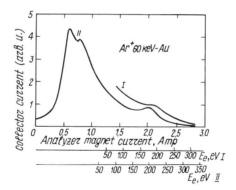

Fig. 8. The data on the electron spectrum measurement (Snoek et al. 1966). (I) the target potential is 0 V; (II) the target potential is +18 V.

characteristic energy loss curve observed by Afrosimov et al. (1964a,b,c) and Everhart and Kessel (1966). If it turns out to be really so, this will open new possibilities in the study of characteristic losses on solid targets. It would be very desirable to use the coincidence technique in the study of ion bombardment phenomena where it has not yet been applied.

On the other hand, an investigation of the Doppler shift of the maximum at 200 eV could provide an answer to the question of whether these electrons are emitted by the metal or by the incident Ar atom.

Abroyan (1961), who studied the kinetic emission from dielectrics, has revealed a regular dependence on the width of the forbidden band (see fig. 9). It should be noted that the kinetic emission from insulators, as well as

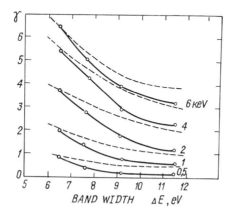

Fig. 9. The ratio of the kinetic emission from insulators to the conduction band width. The full curves represent the experiment by Abroyan (1961); the broken ones the theory by Parilis and Kishinevsky (1961).

from metals coated with insulating films, is several times that from a clean
metal (Batanov 1960, 1961a,b, 1962).

2.1.6. Dependence of γ on the angle of incidence

This dependence considered in some early papers in which the relation
$\gamma \sim \sec \psi$ was obtained, where ψ is the angle of incidence of the beam on the
surface. The data of Evdokimov et al. (1967b) for a polycrystalline surfaces
follow closely this law (see fig. 10), except for grazing incidence.

Very interesting data were obtained by Carlston et al. (1965) and
Mukhamadiev and Rakhimov (1966). First, the emission was observed to
increase with increasing packing density at the surface (see fig. 11). Second,

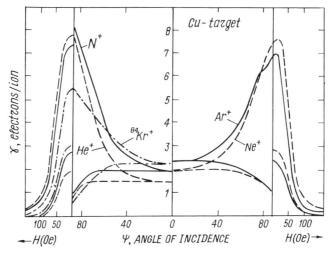

Fig. 10. The values of $\gamma(\psi)$ for a polycrystal surface (Evdokimov et al. 1967b). For N⁺, Ar⁺,
Ne⁺, He⁺ $E = 30$ keV, for Kr⁺ 25 keV. Lower curves $\gamma(\psi) \cos \psi$ illustrate the deviation from the
$\sec \psi$-dependence at glancing incidence.

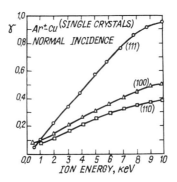

Fig. 11. The kinetic emission from different single crystal faces (Carlston et al. 1965).

and this is the most interesting fact, as one varied continuously the angle of beam incidence on the face of a single crystal there were revealed deviations from the sec ψ law, these deviations being periodical (Mashkova et al. 1963a,b, Mashkova and Molchanov 1963, 1964a, 1965, Kvlividze et al. 1964, Fert et al. 1962, Fagot et al. 1964, 1966, Fagot and Fert 1964, Zscheile 1964, Abroyan et al. (1963) (see fig. 12). This behavior was correctly interpreted as due to the action of atoms in the top atomic layers on atoms in deep layers (Rol et al. 1960, Fluit et al. 1963, Sonthern et al. 1963).

The next step was the investigation of the effect of target temperature on the structure effects observed by Mashkova and Molchanov (1964b) and Evdokimov et al. (1966) (see fig. 13). It turned out, as should have been expected, that with the increase of temperature the structure effects become smeared.

The study of the temperature dependence in the anisotropy of kinetic emission revealed an interesting possibility to follow the formation and annealing of lattice defects by the electron emission curves. It was found that below the defect annealing temperature a single crystal subject to ion bombardment yields a pattern (see fig. 14) characteristic for a polycrystal, and only above this temperature one sees the anisotropy again (Evdokimov et al. 1967c).

These are in short the main experimental data on which the theory of the kinetic emission of electrons from poly- or single crystals should be based.

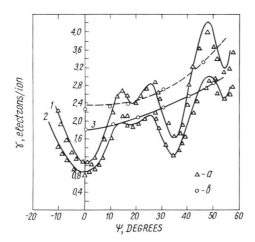

Fig. 12. The influence of the Cu single-crystal structure on the dependence of emission on the angle of incidence for Ar$^+$ ions. The calculations were made with the use of the Odintsov formula (Odintsov 1963, 1964): (1) $E_0 = 30$ keV; (2) $E_0 = 20$ keV; (3) polycrystal, 20 keV. The experimental points are (a) a single crystal; (b) a polycrystal surface; (broken curve) data by Telkovsky (1956) for polycrystals.

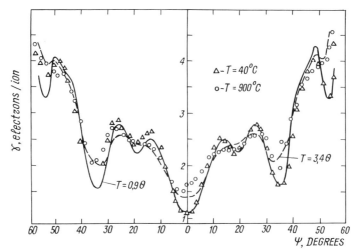

Fig. 13. The influence of the thermal vibrations on the structure effects in kinetic emission. Experiment by Mashkova and Molchanov (1964b) and Evdokimov et al. (1966) and theory (Agranovich and Odintsov 1965) (broken curve). θ is the Debye temperature, $\theta = 345$ K for Cu.

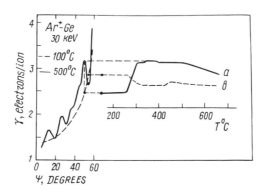

Fig. 14. The kinetic emission and lattice defects (Evdokimov et al. 1967b). Below the defect annealing temperature ($\sim 300°C$): (broken curve) typical for a polycrystal $\gamma(\psi) \sim \sec \psi$. Above 300°C: (full curve) typical for a single crystal.

Naturally, the theory developed in parallel with the accumulation of experimental data; a new, modified picture of the mechanism for the kinetic electron emission being formed at each stage of its development.

2.2. Theory

The first attempt to find a method to theoretically calculate this phenomenon is made by Kapitza (1923), who suggested that the ion (an α particle in early experiments) impinging on the target surface transfers a part of its energy to a very small volume of the metal causing very brief but intense heating (*local heating*). The resulting temperature calculated using a classical equation for heat conductivity turns out to be sufficiently high for thermionic emission to appear. Thus, the principal point of this bright hypothesis consisted in reducing the phenomenon of kinetic emission to the thermionic emission from a locally heated area. In this way, one could apply the Richardson formula for the calculation of emission.

It should be pointed out that Kapitza (1923) was fully aware of the inapplicability for this case of the Richardson equation derived from the equilibrium Maxwellian distribution. However, he did not intend to develop a rigorous theory and wanted merely to obtain a correct order-of-magnitude estimate for the emission.

Assuming the kinetic energy of the ion to be transferred to the electrons in this microscopic zone, Kapitza succeeded in obtaining a correct estimate for the emission and for the *temperature* of secondary electrons. He was of the opinion that the extremely strong effect of surface conditions on emission makes more detailed calculations unreasonable.

Morgulis (1934) has taken into account the cooling of the heated microscopic zone by heat conduction and radiation, making use of the temperature distribution in time and space from an instantaneous heat source. Now knowing this corresponding work of Kapitza (1923), he calculated also the electron emission from a locally heated area and obtained independently similar results.

A decisive role in the development of the thermal theory of kinetic electron emission was played by Sommermeyer (1936), who pointed out a very poor heat exchange between the metal lattice and the conduction electrons in the nonsteady state case. Since the ion energy is transferred directly to the metal atoms, the electrons can obtain this energy only via these atoms. Because of a large difference in mass between atoms and electrons the energy transferred to electrons is so small that the electrons in the zone concerned simply cannot become heated during the time of existence of this locally heated zone. This consideration struck a decisive blow on the thermal theory.

Morgulis (1934) soon joined Sommermeyer (1936) in renouncing the thermal theory.

Gurtovoj (1940) drew attention to the fact that the principal argument of the thermal theory, i.e., the Maxwellian energy spectrum of emitted electrons, is likewise invalid, since, first, the Maxwellian shape does not necessarily result from thermal motion (e.g., a similar distribution will have secondary electrons produced by fast electron bombardment), and second, a consistent theory of local heating taking into consideration fast removal of heat should not yield a Maxwellian velocity distribution of emitted electrons, and thus such a distribution is an argument against, rather than for the theory.

Because of the inconsistency of the thermal theory, Morgulis (1939) returned to the opinion expressed by Kapitza (1923) and Sommermeyer (1936) that the correct point of view on the electron emission under ion bombardment consists in considering the ionization produced by ions in collision with atoms of the metal.

Unfortunately the *impact-ionization* hypothesis was not developed at that time into theory, possibly because of the absence of a theory for the atomic ionization by ion impact.

Frenkel (1941) proposed a new *mechanical* theory of the kinetic emission based on a paper by Migdal (1939), who suggested to consider the ionization of an atom by neutron impact as a consequence of *shacking-off* its electron shell resulting from a sudden change in the motion of nucleus.

Frenkel (1941) found that the intensive electron emission due to such mechanism should occur only at ion energies of 0.1–1 MeV. The probability for ionization as a result of shaking-off in a central collision will be

$$\gamma = 0.28 \, \frac{mE_0}{M_2 V_i} \, ,$$

where m is the electron mass, M_2 the ion mass, and V_i the ionization potential. At $E_0 = 10$ keV, $\gamma = 3 \times 10^{-3}$. Even if one does not take into account the probability for a central collision to occur and the yield of electrons from the metal, this is still about two orders of magnitude smaller than the observed value. Thus, the proposed mechanism cannot account for the emission in the energy region where the motion of electrons in colliding atoms is adiabatic.

Izmajlov (1958a) pointed out that the real pattern of collision between two atoms differs strongly from the one suggested by Frenkel since, first, electron shells of the colliding systems come into contact, and only after this nuclei begin to slow down.

Avakyantz (1951a,b) mathematically developed the idea of Frenkel by considering the shaking-off as an internal conversion of bremsstrahlung quasiquanta emitted by the nucleus in the electron shells of colliding atomic systems. Avakyantz found the probability of ionization in collision of a proton with an atom to be

$$\gamma \approx \frac{Z_1^2}{30} \left(\frac{e^2}{\hbar c}\right)^2 \frac{E_0}{M_2 c^2} \left(\frac{mc^2}{V_i}\right) \sec^{-1} \frac{E_0}{V_i} \, .$$

In his criticism of Avakyantz's paper, Izmajlov (1958b) repeated the argument brought against the *shake-off* theory. He also pointed out the inapplicability of a hydrogen-like function for the description of the motion of electrons in cooling systems.

Von Roos (1957) made an attempt to calculate the kinetic emission coefficient using the quantum-mechanical method of perturbed steady states; however, the method was applied not rigorously enough and, therefore, it is difficult to draw a conclusion on the validity of this theory.

Sternglass (1957) developed the *impact-ionization* theory for the description of the electron emission from metals under fast ion bombardment (energy of the order of a few MeV). In his calculation of the ionization produced by an ion penetrating into the metal, the author applied the theory developed by Bohr and Bethe for the passage of the fast particles through the matter.

Sternglass (1957) obtained the following expression for γ,

$$\gamma = \frac{4AP\pi e^4 Z^2}{\bar{V}\alpha'\sigma_g} \left(\frac{Z_1}{I_0 E_{eq}}\right)^{1/2} \sim \frac{Z^2}{v_0} ,$$

where P is the probability for the electron to escape from the metal, the constant A is determined by the distribution of the secondary electrons in initial velocity (it is about $\frac{1}{2}$ for a symmetrical distribution), $\bar{V}$ is the average energy expended to produce an ion in a solid, Z is the effective charge of the incident particle, E_{eq} is the energy of the electron having the same velocity as the ion, I_0 is the Rydberg energy, Z_1 is the atomic number of the metal, σ_g is the cross-sectional area of the metal atom defined by the covalent radius, and α' is a constant characterizing the cross section for the scattering of secondary electrons.

The same method was used by Ghosh and Khare (1962) to calculate the electron emission from aluminum under bombardment of ions and atoms of hydrogen. A further development of the Sternglass theory (Sternglass 1957) is due to Jahrreiss (1964).

The theory of Sternglass (1957) explains in a correct way the presence of a maximum on the emission versus ion energy curve and the subsequent fall-off at very high energies. The escape of electrons from the metal is considered as a consequence of their diffusion to the surface. This model was improved by Schou (1980a,b).

However, the principal emission of electrons is observed in the 1 to 1000 keV range where the theory of Sternglass (1957) is not valid. In this range, the motion of electrons in colliding atoms is adiabatic, and specific methods are required for its description.

Izmajlov (1958b, 1959a,b, 1961) proposed a *radiation* theory of kinetic emission under bombardment of ions and atoms, in which the ejection of electrons is considered as a consequence of the absorption by free electrons in the metal of quanta and quasiquanta radiated as the ion steps as a result of its central collision with an atom from the surface. The latest version of the theory of Izmajlov (1961) takes into account also recoil of the target particles.

The difficult point of this theory is that it cannot explain the experimentally observed independence of emission on ion charge. The method used is not free from objections. Only central collisions of the ion with the metal atom are considered, since only such collisions can bring an ion to a sharp stop, and only they are effective according to the theory. The relative probability of such collisions is not estimated and it is assumed that each ion colliding with the surface experiences an instantaneous slowing-down. However, the fraction of a central collision in a thin surface layer providing secondary electrons is small.

The real character of ion motion even at a central collision is far from an instantaneous slowing-down. The intensity of the bremsstrahlung radiation field is quite sensitive to the magnitude of acceleration and, therefore, one cannot consider the replacement of the finite acceleration by an infinite one as justified. This precludes the extension of the bremsstrahlung radiation theory for the secondary electron emission, due to Fröhlich (1955), valid for fast electrons to slow ions.

The mechanism of Frenkel–Avakyantz–Izmajlov is apparently responsible for some fraction of the emission due to central collisions at a very high ion energy; however, one cannot probably account for all the emission in the main energy range in this way.

Thus, a large number of various ideas proposed at different times for the explanation of the phenomenon of kinetic emission not only did not result in the creation of a complete theory, but even have not brought about the development of a common view point on the mechanism of electron excitation, since the above-mentioned theories suggested models incompatible both with one another and with experiment.

The problem consisted, first of all, in finding a sufficiently efficient mechanism for the excitation and emission of electrons of a metal under bombardment by ions with 1–100 keV energy, and then in the explanation, on the base of such mechanism, of all experimental data characterizing this phenomenon. The first question which arises here is that of the source of electron emission (Parilis 1962).

Three possibilities should be considered here:

(1) Free electrons in the metal. These are the electrons forming a cloud on the metal surface and the conduction electrons inside the metal.

(2) Bound electrons. These are the electrons of the valence and deeper bands, as well as electrons bound on the surface.

(3) Electrons of the bombarding ions (atoms).

The above theories answer this question in different ways. The thermal and radiation theories consider free electrons as a source of the kinetic emission, the mechanical theory considers the electrons of the incident particle, and the impact-ionization hypothesis considers the bound electrons as the sources.

As will be shown below, none of these extreme viewpoints is correct. The process of electron emission consists of several steps, and electrons of a definite kind play a predominant role in each of them.

2.2.1. Mechanism of transfer of the ion kinetic energy to electrons

The energy which can be directly transferred to a free electron in collision with an ion is very small. Indeed, in an elastic collision of an ion of mass M_2 and energy E_0 with a free electron the latter will acquire the energy $E_k = (m/M_2)E_0$. Even for a light ion as Ne^+ this energy exceeds the work function of tungsten (4.5 eV) only at $E_0 \approx 150$ keV.

Thus only bound electrons remain.

The question which of them, those bound in the bulk or on the surface, play the predominant role was discussed in literature. Ploch (1951) and Gurtovoj (1940) were of the opinion that only surface electrons are excited. However, this is not so. Already the existence of the isotopic effect indicates that this is a bulk phenomenon. Indeed, the isotopic effect can be only explained as due to the ions of different mass loosing their energy and slowing down during their penetration into a solid in different ways. A considerable magnitude of the effect shows that a large fraction of electrons are excited in the inner layers of the lattice. The nomination of free and bound electrons are rather arbitrary, because during the short time of this collision duration (10^{-15} s) the so-called free electrons could be taken as bound.

As is well-known from the theory of atomic collisions, the mechanisms of electron excitation can be different depending on the relative magnitudes of the velocity of bound electrons v_m.

At $v_0 \gg v_m$, the potential acting on a bound electron changes suddenly, resulting in the excitation of ionization of the atom.

At $v_0 \lesssim v_m$ (it is this region which determines the kinetic emission), the motion of electrons follows adiabatically that of the nuclei, and the probability for the electron excitation calculated by means of the adiabatic theory decrease exponentially with increasing parameter $a \Delta E/\hbar v_0$. In this region the Born approximation is not valid, whereas special quantum-mechanical methods, such as, e.g., the method of excited steady states, find only limited application in particular cases for single- or two-electron systems. Meanwhile, the character of collisions between many-electron atoms is such that deformation, mutual interpenetration and perturbation are experienced practically by all electron shells surrounding the nuclei (at $E_0 \approx 1–100$ keV, the distance of closest approach $R_0 \approx 10^{-9}–10^{-10}$ cm).

The energy is transferred to electrons not by the impact of an ion as a whole, but through the interaction of its electrons as the electron shells pass through one another, the electrons of both colliding atoms participating in the energy-transfer process.

Under these conditions, the statistical consideration of the excitation of a gas of bound electrons surrounding a 'quasimolecule' formed by colliding atoms (Firsov 1959, Russek and Thomas 1958) turns out to be effective. On one hand, the statistical approach as here forced, since this is a many-electron process and the energy levels of excited states are located very close to one another. On the other hand, the statistical model of Thomas–Fermi permits to obtain in a comparatively simple way universal results independent of the individual features in the structure of electron shells of colliding atoms.

A similar consideration can be applied to the problem of kinetic emission, since electrons in the inner shells of heavy-metal atoms are weakly cooperated and form an ion core close in properties to the isolated atom. The modern way of describing the ionization of heavy-atom collision includes both the Firsov model (Firsov 1959) and the Fano–Lichten quasimolecular model (Fano and Lichten 1965), in which all electrons should be regarded as taking part in the energy transfer.

An intensive electron exchange occurs during a collision of an ion with the atom of a solid. Since electrons move almost adiabatically, this exchange is accompanied by a transfer of the mean momentum of translational motion acquired by the electron in another atom, $m(\dot{\boldsymbol{R}}_a - \dot{\boldsymbol{R}}_b)$, through the surface S separating the regions of action for the potentials of the two atoms. When charges differ not more than by a factor of four, the surface may be replaced by a plane equally distant from the nuclei and normal to the line drawn from one to another. Now certain effective dissipative forces will act between the atoms,

$$\boldsymbol{F} = \pm m(\dot{\boldsymbol{R}}_a - \dot{\boldsymbol{R}}_b) \int_s \tfrac{1}{4} nv \, \mathrm{d}S \,, \tag{1}$$

which are similar to friction forces. Here, $\dot{\boldsymbol{R}}_a$ and $\dot{\boldsymbol{R}}_b$ are the velocities of the atomic nuclei; n,

$$n = \frac{2^{3/2}(me\varphi)^{3/2}}{3\pi^2\hbar^3} \,, \tag{2}$$

the electron concentration in the Thomas–Fermi model; v,

$$v = \frac{3(3\pi^2)^{1/3}\hbar n^{1/3}}{4m} \,, \tag{3}$$

the mean value of the absolute electron velocity; φ,

$$\varphi = \frac{(Z_1 + Z_2)e}{r} \, \chi\!\left(1.13[Z_1 + Z_2]^{1/3}\frac{r}{a_0}\right), \tag{4}$$

the potential on the surface S expressed in terms of the Thomas–Fermi function, Z_1 and Z_2 are the nuclear charges, r is the distance from a point on the surface to one of the nuclei, and $a_0 = \hbar^2/me^2$.

The work expended by this force to move the atoms by $d\boldsymbol{R} = d\boldsymbol{R}_a - d\boldsymbol{R}_b$ will be the electron excitation energy,

$$\mathscr{E} = m \int \left(\int_S \tfrac{1}{4} nu \, dS \right) \dot{\boldsymbol{R}} \, d\dot{\boldsymbol{R}} \ . \tag{5}$$

Or, taking into account eqs. (2) and (3),

$$\mathscr{E} = \frac{m^2 e^2}{4\pi^2 \hbar^3} \int \left(\int_S \varphi^2 \, dS \right) \dot{\boldsymbol{R}} \, d\dot{\boldsymbol{R}} \ . \tag{6}$$

If we restrict ourselves to the case of rectilinear uniform motion of nuclei of colliding atoms, i.e., to collisions with large impact parameters resulting in low-angle scattering, then we shall have the following simple formula for the energy transfer obtained by Firsov (1959),

$$\mathscr{E} = \frac{0.35(Z_1 + Z_2)^{5/3} \hbar v_0/a_0}{[1 + 0.16(Z_1 + Z_2)^{1/3} R_0/a_0]^5} \ . \tag{7}$$

For further calculations we shall have to find the energy transfer for any impact parameter, and one cannot apply here the simplification of uniform rectilinear motion of the nuclei. A classical consideration of the motion of atoms (it is valid in this energy range for heavy particles) yields in this case the following expression for the energy transfer,

$$\mathscr{E} = \frac{\hbar v_0}{\pi a_0^2} (Z_1 + Z_2)^2 \int_{R_0} \frac{[1 - V(R)/E] \, dR}{\sqrt{1 - V(R)/E - p^2/R^2}} \int \chi^2(\rho)/\rho \, d\rho \ , \tag{8}$$

where E is the relative motion of the atoms, ρ the impact parameter, and $V(R)$ the repulsion potential acting at short distance between atoms.

In the limiting case of low-angle scattering, eq. (8) reduces to eq. (7).

Kishinevsky (1962) has generalized the Firsov formula to include the arbitrary impact parameters and atomic numbers of colliding atoms and from eq. (8) has obtained the formula

$$\mathscr{E} = \frac{0.3 \times 10^{-7} v_0 Z_1 Z_2 (\sqrt{Z_1} + \sqrt{Z_2})(Z_1^{1/6} + Z_2^{1/6})}{[1 + 0.67\sqrt{Z_1} R_0/a(Z_1^{1/6} + Z_2^{1/6})]^3} \left[1 - \frac{0.68 V(R_0)}{E} \right] . \tag{9}$$

Here, Z_1 is the largest and Z_2 is the smallest of the two atomic numbers, $V(R)$ is the interaction potential, R_0 is the distance of the closest approach, E is the energy in the center-of-mass of system, $a = 0.47 \times 10^{-8}$ cm, the velocity v_0 is measured in cm/s, and E in eV.

2.2.2. Excitation of electrons in the metal

One may ask how the calculated excitation energy, transferred to electrons at the expense of the kinetic energy of the ion, is realized.

The first attempt was to connect this energy with the heating of a small volume of electron gas to which it is transferred, to ascribe to the electrons a 'temperature' determined by the mean energy transfer per electron and, using the Fermi distribution for electrons in the metal, to calculate the emission of electrons in the same way as the thermionic emission is calculated.

However, the corresponding calculations have shown that the energy spectrum of ejected electrons obtained in this way changes continually with changing ion velocity (and hence with the energy transfer $\mathscr{E}$). With the increase of v_0 the maximum energy of electrons in this spectrum increases considerably, and the maximum of the spectrum shifts strongly to higher energies.

Meanwhile, as pointed out earlier, the peculiar feature of the energy spectrum of the kinetic emission is the weak dependence of its shape on the ion energy within a very broad range.

Thirty years ago, Parilis and Kishinevsky (1961) conceived that the described mechanism of excitation could be effective only for the bound electrons and could not be applied to free metal electrons, i.e., the excitation consists in the bound electron transition into the conduction band and could be considered as a kind of 'ionization' of metal cores. The emission is caused both by initially excited electrons and Auger transition to the vacancies formed in a core.

Baragiola et al. (1979) have criticized this theory. They asserted this theory to be incorrect since the inelastic energy losses were calculated with account of all electrons but distributed only among the core electrons. Then direct excitation of the projectile particle electrons and free electrons of the metal is forbidden. Therefore, the agreement with the experiment, from their point of view, was obtained by using incorrect values of the depth of the occupied band δ.

From our point of view, these suggestions should be considered as rather nonconstructive. First, the main idea of the mechanism of kinetic ion–electron emission is the release of electrons in binary collisions via inelastic energy loss calculated using the model proposed by Firsov (1959) for atomic collisions without consideration of free electrons. For short collision times $(10^{-15}-10^{-16}\ \text{s})$, there is no sense to divide the electrons in two types.

Second, the inelastic energy losses take place during close approach of two atoms, i.e., in a quasimolecule. The autoionization of a quasimolecule is indeed connected both with the Auger transitions and other quasimolecular mechanisms of ionization. This fact was corroborated by many investigations on inelastic atomic collisions, after the development of the Parilis–Kishinev-sky theory (Ogurtsov 1972, Lichten 1980, Afrosimov et al. 1983).

The role of the Auger processes in the emission of slow electrons was discussed further by Kishinevsky and Parilis (1968a,b) and Kishinevsky (1970). Therefore, in the Parilis–Kishinevsky theory (Parilis and Kishinevsky 1961) the affirmation of vacancy formation in the filled band of the metal as well as about Auger transitions onto these vacancies should be substituted by affirmation of the quasimolecular autoionization involving both the outer and inner electrons of the colliding particles.

All calculations performed by Parilis and Kishinevsky (1961) correspond to such, more contemporary quasimolecular interpretation, and the depth of the filled band, δ, used according to Harrower (1956) apparently was close to mean vacancy depth in a quasimolecule, which stipulated the agreement between the calculations and the experimental data.

In other words, the calculations by Parilis and Kishinevsky (1961) are still valid, but it is necessary to refine their interpretation with taking account of the excitation and electron emission mainly in the outer shells of a quasimolecule.

One can estimate the upper value of the cross section for single ionization of the quasimolecule by the above mechanism. We obtain

$$\sigma_1 = \pi p_1^2 , \tag{10}$$

where p_1 is the impact parameter at which $\mathscr{E}(p_1) = \delta_1 - \varphi$, δ_1 being the energy depth of the bound electron, φ is the work function of electrons in a metal.

Large impact parameter collisions are accompanied by an energy transfer which is less than $\delta_1 - \varphi$ and which is, therefore, insufficient for the excitation of a bound electron. This evaluation is similar to that of the cross section for single ionization of an atom in an indirect comparison with experiment by Firsov (1959).

However, the cross section calculated in this way turns out to be insufficient to account for the observed kinetic emission and its principal features, in particular at high energies.

To obtain a correct behavior of the emission curve, it turns out to be expedient to evaluate the cross section for the ejection of electrons into the conduction band, taking into account the possibility of ejection from the filled band of more than one electron in a single collision.

It is natural to use in the evaluation of the cross section for such excitation the formula

$$\sigma = 2\pi \int_0^{p_1} \mathscr{E}(p)/Jp \, \mathrm{d}p . \tag{11}$$

Here, J is the average ionization potential for the outer shells in an atom. This quantity, known from the theory of the passage of particles through matter, is about 20–30 eV for metals.

At small impact parameters and ion energies of a few keV, the ratio $\mathcal{E}(p)/J$ is considerably in excess of unity. Assuming $\frac{1}{4} < Z_1/Z_2 < 4$, one can obtain from eqs. (9) and (11) the following expression for the cross section,

$$\sigma(v_0) = \frac{1.39 a_0 \hbar}{J} \left(\frac{Z_1 + Z_2}{\sqrt{Z_1} + \sqrt{Z_2}} \right)^2 S(v_0) . \tag{12}$$

An additional calculation, taking into account a decrease of electron density in the quasimolecule near light ions (Kishinevsky and Parilis 1962), allowed to extend the theory to the case of light ions. In this case, the factor $f_1(Z_1, Z_2)$,

$$f_1(Z_1, Z_2) = \left(\frac{Z_1 + Z_2}{\sqrt{Z_1} + \sqrt{Z_2}} \right)^2 , \tag{13}$$

from eq. (12) is substituted by the factor

$$f_2(Z_1, Z_2) = \tfrac{1}{16} (\sqrt{Z_1} + \sqrt{Z_2})(\sqrt[6]{Z_1} + \sqrt[6]{Z_2})^3 , \tag{14}$$

which at $Z_2 = Z_1$ coincides with f_1, at $Z_2 \leqslant Z_1$ is close to it, and at $Z_2/Z_1 \ll 1$ strongly differs from it.

One distinguishes two principal regions of variation of $S(v_0)$. They differ in the dependence of the distance of closest approach of the nuclei R_0 on velocity v_0. At high velocities R_0 is close to p, whereas at low velocities it differs strongly from it.

To these regions correspond two regions for $S(v_0)$: the low-velocity region $v_0 < 3 \times 10^7$ cm/s where the relationship $\sigma(v_0)$ is close to quadratic, and the high-velocity region where it approaches a straight line. The curve $\sigma(v_0)$ has a clearly pronounced threshold. By definition, the threshold velocity corresponds to $p_1 = 0$. In a collision with a smaller velocity the energy transfer does not reach $\delta_1 - \varphi$ even at central collision.

One has to take the integral

$$S(v_0) = \frac{1.28}{a_0^2} v_0 (\sqrt{Z_1} + \sqrt{Z_2})^2 \int\limits_0^{P_1} p \, dp \int\limits_{R_0}^{\infty} \frac{[1 - V(R)/E] \, dR}{\sqrt{1 - V(R)/E - p^2/R^2}}$$

$$\times \int\limits_{R/2}^{\infty} \frac{\chi^2 [1.13(Z_1 + Z_2)^{1/3}(\rho/a_0)]}{\rho} \, d\rho \tag{15}$$

numerically. Despite a considerable difference in the mass of ions and atoms of metals, the curves $\sigma(v_0)$ lie fairly close to one another, so that, within the accuracy of the theory, it can be approximated by the function

$$S(v_0) = 5.25v_0 \arctan(0.6)(v_0 - v_{min}) \times 10^{-7}, \tag{16}$$

where v_{min} is the threshold velocity. It is different for different pairs and constitutes, e.g., for Ar^+, Kr^+ on Mo, W its value $(0.6–0.7) \times 10^{-7}$ cm/s (Parilis and Kishinevsky 1961), which is in good agreement with experiment. This formula fits the experiment also for the other threshold values (Cawthron 1971).

One used in these calculations the screened repulsion potential acting between atoms (Firsov 1958),

$$V(R) = \frac{Z_1 Z_2 e^2}{R} \chi\left([\sqrt{Z_1} + \sqrt{Z_2}]^{2/3} \frac{R}{a_0}\right). \tag{17}$$

2.2.3. Kinetic electron emission yield

If the following collision of an ion with an atom of the lattice takes place inside the meal at depth x, then secondary electrons will experience a series of collisions before escaping from the surface. It is usually assumed that the flux of secondary electrons decreases by the exponential law $e^{-x/\lambda}$ with a mean free path λ. Then the yield of the kinetic emission, i.e., the number of emitted electrons per incident ion will be

$$\gamma = \int_0^{x_n} \sigma(v) wN \exp(-x/\lambda) \, dx, \tag{18}$$

where w is the probability for an excited electron to escape from the metal, N is the number of atoms in 1 cm^3 of the metal, and x_n is the depth at which an ion still retains the power to ionize.

One should take into consideration the decrease of the velocity v with depth. The average loss of ion velocity resulting from collisions with atoms of the metal at a path length dx is

$$d\bar{v} = - \frac{M_1 M_2}{(M_1 + M_2)} vN\sigma_d \, dx. \tag{19}$$

σ_d is the diffusion scattering cross section which, at the assumed potential (Firsov 1958) in the energy range considered, reduces to

$$\sigma_d = \frac{1.24\pi a_0 e^2 Z_1 Z_2 (M_1 + M_2)}{v^2(\sqrt{Z_1} + \sqrt{Z_2})^{2/3} M_1 M_2}, \tag{20}$$

whence follows a simple law relating the decrease of velocity to the depth of penetration,

$$v_0^2 - v^2 = kx, \tag{21}$$

where v_0 is the initial velocity of the ion, and the quantity k,

$$k = \frac{2.48\,\pi N a_0 e^2 Z_1 Z_2}{(\sqrt{Z_1} + \sqrt{Z_2})^{2/3}(M_1 + M_2)} ,\qquad (22)$$

has the meaning of deceleration. By definition, x_n is the depth at which the velocity drops to v_{min}.

Expression (21) permits to go over to the integration in velocity,

$$\gamma = \frac{2Nw}{k} \int_{v_{min}}^{v_0} v\sigma(v)\,\exp[(v^2 - v_0^2)/(kx)]\,dv$$

$$= Nw\lambda[\sigma(v_0) - \Delta\sigma(v_0)] ,\qquad (23)$$

where $\Delta\sigma(v_0)$,

$$\Delta\sigma(v_0) = \exp[-v_0^2/(k\lambda)] \int_{v_{min}}^{v_0} \exp[v^2/(k\lambda)]\,\frac{d\sigma(v)}{dv}\,dv ,\qquad (24)$$

accounts for the decrease of the ionization cross section as a consequence of the ion slowing down in the metal.

According to eqs. (12) and (16),

$$\frac{d\sigma(v)}{dv} \sim \frac{dS(v)}{dv} \sim \frac{d}{dv}\,[v\,\arctan(0.6)(v_0 - v_{min}) \times 10^{-7}] .\qquad (25)$$

The last derivative is well-approximated by the function

$$\tfrac{1}{2}\pi - v_{min}/v[\tfrac{1}{2}\pi - 0.6 \times 10^{-7}\,v_{min}] .\qquad (26)$$

It follows that

$$\Delta\sigma(v_0) = \exp(-v_0/\{k\lambda\})\{\tfrac{1}{2}\pi\sqrt{k}\lambda[\Phi(v_0/\sqrt{k}\lambda) - \Phi(v_{min}/\sqrt{k}\lambda)]$$

$$- \tfrac{1}{2}v_{min}(\tfrac{1}{2}\pi - 0.6 \times 10^{-7}v_{min})$$

$$\times [Ei(v_0^2/\{k\lambda\}) - Ei(v_{min}/\{k\lambda\})]\} .\qquad (27)$$

where

$$\Phi(x) = \int_0^x e^{t^2}\,dt ; \qquad Ei(x) = \int_{-\infty}^x e^{t^2}\,dt/t .$$

The effect of the term $\Delta\sigma(v_0)$ is most significant at low velocities when the path length within which the incident particle can ionize the atoms of the metal does not exceed the depth from which electrons can still escape from the metal $x_n < \lambda$. At high velocities $x_n \gg \lambda$ and hence the ion velocity within the layer providing secondary electrons remains almost constant. Here, $\Delta\sigma(v_0) \ll \sigma(v_0)$.

Introducing the effective cross section for ionization,

$$\sigma^*(v_0) = \sigma(v_0) - \Delta\sigma(v_0) \,, \tag{28}$$

one can reduce the formula for the kinetic emission yield to a simple form

$$\gamma = N\sigma^*(v_0)\lambda w \,. \tag{29}$$

At the end of the seventies, Baragiola et al. (1979) have derived a semiempirical formula for the coefficient of kinetic emission of electrons,

$$\gamma = N\lambda w S_i(v)/J \approx 0.1 S_i(v) \, (\text{Å}/\text{eV}) \,, \tag{30}$$

where S_i is the total inelastic stopping power. The last relation is valid far from the threshold with an accuracy of $\pm 40\%$. From the Parilis–Kishinevsky theory (Parilis and Kishinevsky 1961) follows the proportionality of γ to elastic energy losses per unit path, as well as the approximate value of the proportionality factor, since the magnitude S_i is connected with the cross section σ [eq. (11)] by the relation

$$S_i = NJ\gamma \,.$$

It should be noted that relation (29) appears to be valid up to the threshold.

2.2.4. Escape of electrons from the metal

For comparison, the energy spectra of electrons knocked-out during binary collisions could be used, but as a rule, in the experiments with gas or solid targets, different kinds of atoms are used. Unfortunately, the experiments on atomic collisions in vapors with W, Ta, Mo and other metals investigating the electron energy spectra are unknown.

The first attempt to measure the energy distribution of the ionization electrons were made by Moe and Petsh (1958). A number of investigations in this field were performed in the seventies–eighties (Ogurtsov 1972, Afrosimov et al. 1983). Using the regularities which could be obtained on the basis of these data one can estimate w.

However, it should be noted that one can confine oneself to the consideration of w as a theory parameter. Its value is about 0.1–0.2. Indeed, the product $C = \lambda w/J$ appears in γ. Since the value λ is known to be approximate one, then instead of estimating w and matching of the value J, one could use one fitting parameter C when comparing with experiment. The value of this parameter appeared to be in reasonable agreement with the expected values of λ, w, J.

The energy spectra of electrons at atomic collisions are shown in fig. 15 (Moe and Petsh 1958, Berry 1961), and the cross section of electron emission (Rudd et al. 1966) is shown in fig. 16. These data and many analogous ones give evidence that at binary atomic collisions the energies of

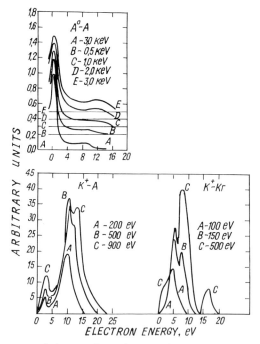

Fig. 15. Energy spectrum of electrons formed at atomic collisions (Moe and Petsh 1958, Berry 1961).

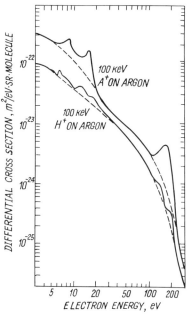

Fig. 16. Differential cross sections for electron emission at atomic collisions (Rudd et al. 1966).

most emitted electrons do not exceed 20–25 eV. When using a metal target they lead to the formation of energy spectra of ion–electron emission, in which comparatively slow electrons with energies of some eV predominate.

When considering the low-energy spectrum region of kinetic ion–electron emission ($E < 10$ eV), attention should be paid to the resemblance of this spectrum with the energy spectra of electrons at both secondary electron emission and photoelectron emission. There are only a few slow electrons even in the spectra of potential ion–electron emission for He^+, Ne^+ ions with high ionization potentials. The cascade mechanism of electron energy distribution formation from metals at all types of emission provides the explanation of these facts, i.e., the spectrum shape is caused not by initial excitation but by cascade development (Hasselkamp 1988).

As discussed above, the mechanism of emitted electron formation is the Auger processes occurring in inner shells of the quasimolecule target atom, projectile atom and other quasimolecular mechanisms of autoionization. It should be noted that in the case of a metal target the probability of Auger processes should increase in comparison with the analogous value in a quasimolecule, since together with the excited quasimolecular electron another metal electron and could be involved in their Auger relaxation with the filling of the vacant states of a quasimolecule. In turn, the Auger transition rates in a quasimolecule are significantly higher than in atoms, which was verified by investigations by Parilis et al. (1989b).

The early estimations of the probabilities of Auger recombination made by Parilis (1968) are of interest, too. This process is a two-electron one. Both electrons that are excited are in the initial state (quasimolecular and from conductive band). In the finite state, one electron fills the vacancy, the other being above the Fermi level.

The perturbation resulting in the transition consists in the screened Coulomb interaction of electrons,

$$V(r_1, r_2) = e^2/r_{1,2} \exp(-r_{1,2}/\lambda_e) , \tag{31}$$

where λ_e is the screening radius in the metal, and $r_{1,2} = |r_1 - r_2|$. According to the perturbation theory, the probability of transition (per unit time), in the first approximation, is

$$w = 2\pi/\hbar N_f(\varepsilon_k)|V_{if}(r_{1,2})|^2 , \tag{32}$$

where N_f is the density of final states and ε_k,

$$\varepsilon_k = \varepsilon' + \varepsilon'' + \delta - \varepsilon_0 , \tag{33}$$

is the energy of an excited electron, ε' and ε'' are the initial electron energies, ε_0 is the energy of the vacuum level, all reckoned from the bottom of the conduction band, and δ is the depth of the hole reckoned from vacuum level.

The matrix element of the Auger transition is

$$V_{if}(r_{1,2}) = \int \psi_i^*(r_1)\psi_i^*(r_2)V(r_{1,2})\psi_f(r_1)\psi_k(r_2)\,d\tau_1\,d\tau_2\ ,$$

where $\psi_i(r)$,

$$\psi_i(r) = v_i(r)\exp i(k, r)\ ,$$

are the Bloch wave functions of the initial state, ψ_f is the wave function of the final state of electron in the ionized atomic core, $\psi_k(r_2)$ is the wave function at free electron for $\varepsilon_k > \varepsilon_0$, which decays exponentially outside the metal at $\varepsilon_f < \varepsilon_k < \varepsilon_0$.

The density of the final states

$$N(\varepsilon_k) = \frac{4\pi(\frac{4}{3}\pi R^3)(2m)^{3/2}}{\hbar^3}\ \varepsilon_k^{1/2}\ ,\tag{34}$$

since $\varepsilon_k \gg \varepsilon_f$. Here, R is the radius of the region of electron interaction in the field of the hole. It may be taken equal to the screening radius,

$$R = \lambda_e\ .\tag{35}$$

A precise calculation of the probability w is beyond the scope of our problem which consists in an order-of-magnitude evaluation for comparison with the probability of the competing process of recombination with photon emission.

For this purpose, the matrix element V_{if} may be presented in the form

$$|V_{if}| = N_c e^2/R(\tfrac{4}{3}\pi R^3)\ ,\tag{36}$$

where N_c is the density of initial states, i.e., the density of electrons in the conduction band of the metal.

Now, from eqs. (32), (34), (35) and (36), we obtain

$$W = \frac{64\sqrt{2}\pi^3}{3}\ \frac{m^{3/2}e^4\varepsilon_k^{1/2}\lambda_e^7 N_c^2}{\hbar^4}\ .\tag{37}$$

In contrast to other recombination processes, the probability for Auger recombination is proportional to the square of the electron concentration in the conduction band N_c^2.

For tungsten, $N_c = 3 \times 10^{23}$ cm^{-3} for $\varepsilon_k = 10$ eV, $\lambda_c = 1$ Å, $W \approx 10^{15}$ s^{-1}. Hence the lifetime of a hole with respect to Auger recombination $\tau_a \approx 10^{-15}$ s.

The principle of detailed equilibrium offers another possibility for an evaluation of the probability.

The process opposite to the electron–hole recombination with transfer of energy of ~10–15 eV to another electron is that where the atomic core is ionized by the impact of an electron with the same energy.

The probabilities for the direct and reverse processes should be equal. Therefore,

$$W = \sigma_i v N_c \,, \tag{38}$$

where σ_i is the cross section for ionization (hole formation) by electron impact, and v is the electron velocity.

For $\varepsilon_k = 12 \text{ eV}$ $(v = 2 \times 10^8 \text{ cm/s})$, $\sigma_i \approx 10^{-17} \text{ cm}^2$ at $N_c = 3 \times 10^{23} \text{ cm}^{-3}$, $W = 6 \times 10^{14} \text{ s}^{-1}$ or $\tau_a \approx 1.7 \times 10^{-15} \text{ s}$ which coincides with the preceding estimate.

These estimates agree with the data of Pincherle (1955), Sosnovsky (1957) and Bess (1957), who showed that even in semiconductors the nonradiative impact recombination of electrons and holes plays a considerable role although there the concentration of electrons in the conduction band is much smaller than in metals.

As is well-known, the probability for radiative capture of an electron by a hole under the same conditions is many orders of magnitude smaller. The lifetime with respect to recombination with photon emission is $\tau_0 \approx 10^{-8} \text{ s}$.

Now, in a metal, the photon would be immediately absorbed by the photoeffect, which would have the same result as Auger recombination. Hence radiative recombination cannot compete with Auger recombination.

It is interesting that the time τ_a is comparable with the duration of atomic collision resulting in hole formation, so that emission occurs either during the collision or soon after it.

As a matter of fact, Auger recombination of an electron from the conduction band with a hole in the filled band represents only the final stage of a series of Auger transitions accompanying the process of energy partition among electrons in an atom excited in a collision with an ion.

At present, this viewpoint has obtained confirmation. Following a general idea on the role of the Auger effect in atomic collisions proposed by Weizel and Beeck (1932), the first attempt at a consistent development of this model has been made by Fano and Lichten (1965). It turned out that using simple assumptions on a two-stage ionization mechanism including the Auger effect one can obtain an explanation for some of its features, i.e., the formation of multicharged ions, low-energy characteristic energy losses, and the electron spectrum (Kishinevsky and Parilis 1966). Thus, from the viewpoint of the theory, all kinetic emission electrons can be considered as Auger electrons.

Nevertheless, we have divided the electron emission probability into two terms: w_a, the probability of emission by Auger recombination in the conduction band, and w_i, the probability for direct electron ejection into vacuum in an atomic collision. This was done (Parilis 1963) by recalculating the spectra of the type of Moe and Petsh (1958) and Berry (1961), taking into account the dependence of their mean energy on Z_1, the cascade multiplication of electrons in the metal and the possibility for overcoming the potential barrier. As a result, one obtained the probability

$$w = w_a + w_i . \tag{39}$$

Figure 17 shows the probabilities for direct escape of electrons from the metal (w_i) and for escape by the Auger effect (w_a) as a function of ionization energy (or the energy of formation of the electron–hole pair). It is seen that for the values of $\delta_1 - \varphi$ which are of interest to us, they are close. This means that Auger electrons from the conduction band and electrons ejected from the filled band yield approximately equal contributions to the emission. The total probability w, which should be inserted into eq. (30), is

$$w = w_i(\delta_1 - \varphi) + w_a(\delta_1) . \tag{40}$$

This is a general outline of our viewpoint concerning the mechanism of kinetic electron emission from polycrystalline metals under bombardment of ions with the energy 1–100 keV.

2.3. Comparison with experiment

One could explain within a framework a number of peculiarities of this phenomenon.

(1) A correct relationship was obtained for the dependence of the emission on the ion velocity including not only the location of the threshold but also detailed behavior of $\gamma(v_0)$. It is contained in the cross section $\sigma^*(v_0)$. One distinguishes here three regions of variation of $\sigma^*(v_0)$: near the threshold, due to the effect of $\Delta\sigma(v_0)$, the emission grows slowly (as a

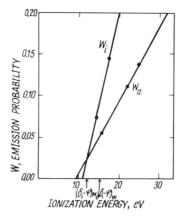

Fig. 17. The probabilities for both direct emission of electrons, w_i, and Auger recombination induced emission, w_a (Parilis 1963). The arrows indicate the values of $\delta_1 - \varphi$ for Mo and W.

fourth-order polynomial in v_0); then the region $(1-3) \times 10^7$ cm/s,

$$\sigma^*(v_0) \sim v_0^2 - (\tfrac{3}{2} v_{\min})^2 , \tag{41}$$

corresponding to a linear increase with energy; and finally, at high energies, where $\Delta\sigma(v_0)$ is negligible, the behavior of $\sigma^*(v_0)$ coincides with

$$\sigma(v_0) \sim v_0 \arctan[6 \times 10^{-8}(v_0 - v_{\min})] , \tag{42}$$

and approaches asymptotically a straight line $C(v_0 - v_1)$ whose continuation intersects the velocity axis at $v_1 = 1.05 \times 10^7$ cm/s, at the same point for different ions. It is exactly such behavior of $\gamma(v_0)$ which is observed in the experiments by U.A. Arifov (1968), Abroyan et al. (1967) and Schram et al. (1966). The falling-off part of $\gamma(v_0)$ at high velocities is described by the theory of Sternglass (1957).

(2) The independence of the kinetic emission on ionic charge follows naturally from the theory. The ion charge simply does not enter the equations.

(3) The dependence of γ on the kind of ion (to be precise, on its atomic number Z_2) is contained in $\sigma^*(v_0)$ in the factors f_1 and f_2 [see eqs. (13) and (14)]. As is shown in fig. 5, the theoretical curve fits well to the experimental data (U.A. Arifov 1968, Schram et al. 1966) at not very small values of Z_2/Z_1, but at $Z_2/Z_1 \ll 1$ (for H and He ions) the agreement between the theory and experiment is only quantitative. Alonso et al. (1980) have shown that better agreement between calculations and experiments could be obtained (particularly using lightest ions) if instead of inelastic energy losses (Firsov 1959), one used semiempirical formulas derived by Lindhard and Scharff (1961) or Yarlagadda et al. (1978). Using more accurate formulas for inelastic energy losses (or even the values of these losses at binary atomic collisions in gases measured experimentally) is not in contrast with the Parilis–Kishinevsky theory (Parilis and Kishinevsky 1961), but collaborates its application for the description of the emission under the lightest ion bombardment.

(4) The isotopic effect can be obtained from eq. (30),

$$\left(\frac{d\gamma}{dM_2} \right)_{v_0=\text{const}} = \frac{d\gamma}{dk} \frac{dk}{dM_2} . \tag{43}$$

From eqs. (12) and (28), we get

$$\beta = \left(\frac{\Delta\gamma}{\gamma} \right)_{v_0=\text{const}} \Big/ \frac{\Delta M_2}{M_2} = - \frac{M_2}{\sigma^*} \frac{d}{dk} \Delta\sigma \frac{dk}{dM_2} . \tag{44}$$

The magnitude of β calculated with this formula agrees well with the experimental data. For instance, Brünne (1957) obtained for ^{39}K and ^{41}K $\beta = 0.4 \pm 0.2$, and for ^{85}Rb and ^{87}Rb $\beta = 0.5 \pm 0.2$. The theory yields for these isotopes, respectively, 0.36 and 0.45.

(5) The soft spectrum of kinetic emission can be satisfactorily explained on the basis of the Auger mechanism. Thus, the above-mentioned theory can be apparently considered as satisfactorily describing the kinetic emission from polycrystalline metals. However, it was not intended to interpret the peculiarities of emission from ordered structures.

2.4. Emission from single crystals

This theory was applied to single crystals by Harrison Jr et al. (1965), who used formulas (9) and (11) to take into consideration the density of packing on each face of a single crystal and the change in the impact parameter probability distribution associated with a dense packing (see fig. 18). The first point permits to explaining the experimental data by Carlston et al. (1965) (see fig. 11). The second point does introduce significant change into the theory since the most intensive electron emission corresponds to small impact parameters whose probability does not depend on the packing density.

Harrison Jr and co-workers (1965) attempted to improve the theory using the model of Russek and Thomas (1958) to estimate the number of electrons released at a given value of the inelastic energy transfer; however, he could not reach a reasonable agreement with experiment on the base of their assumption on the independence of ionization energy on the number of electrons emitted. Kishinevsky and Parilis (1966) have also analyzed the model of Russek and Thomas (1958) and showed that the unnatural

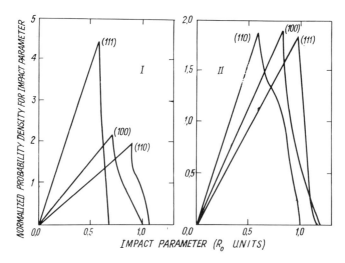

Fig. 18. The probabilities for impact parameters on different single crystal planes (Harrison Jr et al. 1965): (I) face-centered cubic lattice; (II) body-centered cubic lattice.

assumption on the uniformity of ionization energy can be eliminated on the base of the Auger mechanism.

Harrison Jr and co-workers (1965) point out a convincing agreement of the above mechanism of the kinetic electron emission with experimental data, which is far from being accidental. Furthermore, Harrison used the theory of kinetic electron emission to evaluate the repulsion potential acting between the ion and the lattice atoms. This point arouses argument since such an in no way straightforward method of calculating the potential can neither be considered precise nor simple.

A further development of the theory was associated with using the model of 'transparent' atomic layers (Odintsov 1963, 1964).

Mashkova et al. (1963a,b) proposed a phenomenological formula,

$$\gamma = \sum_i \beta_i(\psi) \frac{S_i(\psi)}{A_0 \cos \psi} \bar{E}_i(\psi) = \frac{\sum_i \beta_i S_i(\psi) \bar{E}_i(\psi)}{A_0 \cos \psi} . \tag{45}$$

where S_i is the open area of spheres in the ith layer, $\bar{E}_i$ is the mean energy transfer in an elastic collision, and β_i is the number of emitted electrons per 1 eV.

The Odintsov's formula is only semiempirical since it contains coefficients β_i taken from experiment. The peculiar feature of such a phenomenological formula is that it is not associated with a definite microscopic model and contains the anisotropy of emission as a whole. Its disadvantage consists in using the cross section of hard spheres which is the same for both the electron emission and calculation of 'transparency'.

Naturally, these cross sections are completely different. The first is the cross section for ionization of the metal atom $\sigma(v_0)$, whereas the second is that for ion scattering. Besides, the Odintsov's formula is not connected with any mechanism of excitation and emission of electrons. The emission of electrons is considered to be proportional to the elastic energy transfer, or the nuclear stopping energy, using the terminology of Lindhard et al. (1963), whereas it is proportional to the inelastic energy transfer (or electron stopping energy). These quantities do not coincide, far from being proportional, the first one is proportional to the initial energy, and the second to the ion velocity, eq. (7). Nevertheless, the calculations by Odintsov have played an important role acting as an incentive to new experiments which permitted for the first time to 'see' the atomic transparency of the crystal through electron emission. Qualitatively, the Odintsov formula agrees with experiment (see fig. 12).

Next, one had to find the relation between the transparency anisotropy and the kinetic electron emission theory.

Such connection with the theory was established by Martynenko (1966a,b). Starting from formulas (11)–(14), he obtained the following expression for the kinetic emission coefficient,

$$\gamma = \sum_i w \exp(-x/\lambda) \, p_i \, , \tag{46}$$

where w is the probability of electron emission by the Auger effect, and p_i is the probability of ionization of the atom in the ith layer.

In calculating the probability p_i, Martynenko used the screened Coulomb potential of Firsov (1958) to determine the shadow left by the metal atom in the ion beam and took into account the screening of the subsequent layers getting into this shadow. The radius of the shadow R_s at distance d is

$$R_s = 1.5 \left(\frac{\pi d R_0^2}{1 - \mu} \right)^{1/3} , \tag{47}$$

where $\mu = M_2/M_1$, and R_0 is the distance of the closest approach.

Finally,

$$\gamma = \frac{7.3w}{J} \frac{\hbar a_0}{\delta \cos \psi} \left(\frac{Z_1 + Z_2}{\sqrt{Z_1} + \sqrt{Z_2}} \right)^2 v_0 \arctan[6 \times 10^{-8}(v_0 - v_{min})]$$

$$\times \sum_i \exp(-d/\lambda) \, F(\delta_i, R_s) . \tag{48}$$

The function F is determined by the equality

$$
\begin{aligned}
F &= 0 , & \text{at } \delta < R_s - 0.9p_1 , \\
&= (\delta - R_s)/(1.6p_1) , & \text{at } R_s - 0.9p_1 < \delta < R_s + 0.7p_1 , \\
&= 1 , & \text{at } \delta > R_s + 0.7p_1 .
\end{aligned}
\tag{49}
$$

Here, δ is the projection of the distance between the screening and screened atoms onto the plane perpendicular to the direction of ion motion, and p_1 is the impact parameter for a collision resulting in ionization.

The calculations by Martynenko permitted to establish a more rigorous foundation for the transparency model on the basis of existing concepts concerning the mechanism of kinetic electron emission.

Unfortunately, these calculations are not extended to grazing angles of ion incidence. As shown by calculations of the trajectory of ion motion along a chain of surface atoms (Kivilis et al. 1967), the concept of the shadow should be replaced, in this case, by a more general concept of scattering on the chain, in which one observes a specific relation between elastic and inelastic losses.

Drentje (1967), processing his experimental data, has calculated, apparently independently of Martynenko, the shadow cast by a surface on the second layer. In doing this, he used the Firsov potential (Firsov 1958). Starting also from the theory of Parilis and Kishinevsky (1961), he obtained the following function,

$$\gamma(\psi) = \frac{w(\varepsilon)}{J \cos \psi} (1 + D)\Bigg[\sum_{j_1} \Delta E_1 \exp(-d_1/\lambda) + \sum_{j_2} \Delta E_2(\psi) \exp(-d_2/\lambda)$$

$$+ \sum_{j_3} \Delta E_3(\psi) \exp(-d_3/\lambda)\Bigg], \qquad (50)$$

where w is the probability of emission of an Auger electron, J is the average ionization potential, λ is the electron mean free path length, d_i is the depth of the ith layer, and ΔE_i is the average energy transferred to electrons.

Figure 19 shows a comparison between the experimental and theoretical curves of Drentje. The next problem consisted in describing the effect of thermal vibrations of the lattice on the shape of the curves $\gamma(\psi)$. This problem has been considered by both Odintsov and Martynenko, each within the framework of his own theory.

Agranovich and Odintsov (1965), starting from the Einstein approximation, without taking into account the correlation in thermal vibrations of the lattice atoms at temperatures below the Debye temperature and taking into account correlation at high temperatures, calculated the mean value of the square of projection displacement of the vector connecting the centers of the atoms on the plane perpendicular to the direction of particle incidence. Inserting this quantity into the Odintsov formula, they obtained curves presented in fig. 13 which agree well with experimental data of Mashkova and Molchanov (1964b) and Evdokimov et al. (1966). Martynenko (1966a,b) also considered thermal vibrations of the lattice atoms and calculated their rms displacement. Inserting this into the formula for γ, he obtained the curves of fig. 20. As seen from the figure, the curves $\gamma(\psi)$ smooth out with increasing temperature.

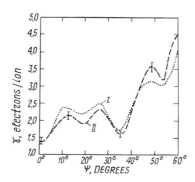

Fig. 19. (I) Theoretical curves and (II) experimental ones for $\gamma(\psi)$ using the data obtained by Drentje (1967). The energy of Ar^+ is 50 keV. The bombarding plane is Cu (100), rotation around (110).

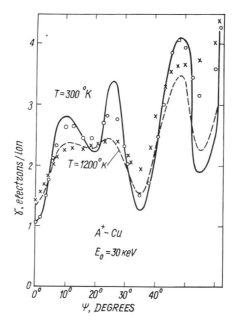

Fig. 20. Comparison of the Martynenko theory (Martynenko 1966b) with the experiments by Mashkova and Molchanov (1963, 1965) for the (100) plane of a Cu single crystal. The energy of Ar$^+$ is 30 keV; (full curve) $T = 300$ K, (broken curve) 1200 K.

2.5. Role of recoils

Relations (13) and (14) contain the dependence on atomic number of a target Z_2 (see fig. 21, curve 1). Dorozhkin et al. (1974) have attempted to check this dependence experimentally (see fig. 21, circles) and found the drastic disagreement with theory, namely: the dependences $\gamma(Z_2)$ differ qualitatively. However, the authors ignored the fact that under bombardment of different targets by the same ions it is necessary also to take into account the dependence of other parameters on Z_2: k/N [eq. (21)], w, J and in the first place λN, and all disagreement with the theory should be explained by recoils, highly overstating their contribution.

According to Bronstein and Freiman (1969), one can get for slow electrons

$$\lambda N = 1.9 \times 10^{-7} N_A Z_2^{-0.6} . \tag{51}$$

Here, $N_A = \rho/NA$ is the weight of an atomic unit (in grammes), A is the atomic weight, ρ is the material density, and λ is in cm.

For simplicity, the ratio w/J could be considered independent of Z_2.

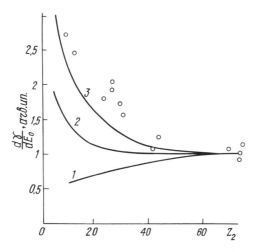

Fig. 21. Dependence of $d\gamma/dE_0$ on atomic number target Z_2 for Hg^+ with energy $E = 15$ keV. (○) Experiment by Dorozhkin et al. (1974) for targets: Re, W, Ta, Cu, Ni, Fe, Ge, Si and C. Theory: (1) calculations using formula (14), (2) without, and (3) with account of recoils.

Eventually a dependence more close to the experimental one could be obtained (see fig. 21, curve 2). The account of the recoils (curve 3) permitted to explain the remaining disagreement.

According to Vinokurov et al. (1976), the contribution to the emission at the account of recoils could be estimated as

$$\gamma_{rec}(v_0) = \int \gamma_1(v_{01}, x) N \frac{d\sigma}{d\theta} \, d\theta \, dx \,, \tag{52}$$

where $\gamma_1(v_{01}, x)$ is the coefficient of the emission from a recoil formed at the depth x with velocity $v_{01}(\theta)$, the scattering angle in the center-of-mass system $\theta = 2 \arcsin[(m_1 + m_2)v_{01}/2m_1v]$, and v is the projectile ion velocity at the depth x. When calculating the effective cross section of the scattering $d\sigma(\theta)$, the Firsov potential (Firsov 1958) was used, well-approximated by the following expression in the considered energy range,

$$V(r) \approx A/r^2 \,, \quad A = \frac{3.05 \times 10^{-16} Z_1 Z_2}{(\sqrt{Z_1} + \sqrt{Z_2})^{2/3}} \text{ eV cm}^2 \,, \tag{53}$$

then for $d\sigma(\theta)$ the following formula seems to be valid (Landau and Lifshitz 1960),

$$d\sigma(\theta) = \frac{4\pi^3 A(m_1 + m_2)}{v^2 m_1 m_2} \frac{\pi - \theta}{\theta^2(2\pi - \theta)^2} \, d\theta \,. \tag{54}$$

In this case,

$$\gamma_1(v_{01}, x) = \frac{0.174Nh a_0 wf(Z_1, Z_2)}{Jk_1} \exp\left(-\frac{v_0^2 - v^2}{k\lambda}\right)\zeta(v_{01}),$$

$$\zeta(v_{01}) = \int_{v_t}^{v_{01}} v_1 S(v_1) \exp\left(-\frac{v_{01}^2 - v^2}{k_1\lambda}\right)\cos\frac{\pi - \theta}{2}\,dv_1,$$

(55)

where k_1 is the deceleration of a recoil when it moves to the target.
By substituting eqs. (54) and (55) into eq. (52), one can get

$$\gamma_{rec}(v_0) = \frac{qh\pi}{a_0 e^4}\frac{(m_1 + m_2)^2}{m_1 Z_2^{2/3}}\int_{v_1}^{v_0}\exp\left(-\frac{v_0^2 - v^2}{k\lambda}\right)\frac{\eta(v)}{v}\,dv,$$

(56)

$$\eta(v) = \int_{v_t}^{v_m}\frac{\pi - \theta}{\theta^2(2\pi - \theta)^2}\frac{d\theta}{dv_{01}}\zeta(v_{01})\,dv_{01};\quad v_1 = \frac{m_1 + m_2}{2m_1}v_t,$$

$$q = 1.83 \times 10^{-14}\,\text{eV/cm}^2,$$

where $v_m = 2m_1 v/(m_1 + m_2)$ is the maximum velocity acquired by the recoil (at a head-on collision).
The rough formula was estimated to be valid for all angles θ with high accuracy,

$$\frac{\pi - \theta}{\theta^2(2\pi - \theta)^2}\frac{d\theta}{dv_{01}} \approx \frac{m_1}{4\pi(m_1 + m_2)}\frac{v}{v_{01}^2},$$

(57)

thus the expression for $\eta(v)$ is significantly simplified,

$$\eta(v) = \frac{v_m}{8\pi}\int_{v_t}^{v_m}\frac{\zeta(v_{01})}{v_{01}^2}\,dv_{01}.$$

(58)

The following calculation has shown that $\zeta(v_{01})$ nearly does not depend on v and $Z_2(m_2)$ in the case of interest when $m_1 > m_2$. It allows us to use the estimated dependence $\eta(v_m)$, shown in fig. 22, when calculating γ_{rec} for all targets (at $m_1 > m_2$) in the considered velocity range. In this case, $|\Delta\gamma_{rec}/\gamma_{rec}| \leq 0.06$.

The graph of the dependence $d\gamma(Z_2)/dE_0$ with the same energy of the Hg ions $E_0 = 15$ keV, taking into account the recoils, is presented by curve 3 in fig. 21. When calculating this dependence, it was considered that at the ionization by Hg ions (Dorozhkin 1967, Dorozhkin and Petrov 1965) $v_t \approx 0.4 \times 10^7$ cm/s, which appeared to be less than for other ions.

It should be noted that since $\sigma \sim \arctan[0.6 \times 10^{-7}(v_1 - v_t)]$ and $v < v_m$ at $m_1 > m_2$, then the ionization cross section of the fastest recoils, mainly

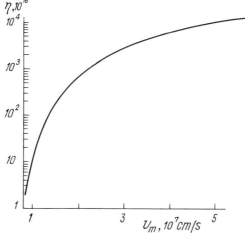

Fig. 22. The function $\eta(v_m)$.

contributed into γ_{rec}, gains saturation more readily than the cross section of the ionization directly by projectile ions. Thus, the relative contribution of recoils into emission decreases with the increase of v_0. At $m_1 < m_2$, $v > v_m$ and, therefore, γ_{rec}/γ increases, remaining small because of v_m is small.

The dependence $\gamma(Z_1, Z_2)$ obtained on the basis of statistical model of inelastic energy losses presents a monotonic function Z_1, Z_2. When taking into account the electron distribution over shells, Baklitzky et al. (1973) have shown that γ oscillates with the change of Z in accordance with the oscillation of outer shell size. The spread of experimental points relative to the curve obtained being insignificant can correspond to this oscillation.

As the above-mentioned calculations have shown, the contribution of the recoils to γ becomes significant at $m_1 \gg m_2$. The role of the recoils in the formation of the high-energy region of a spectrum of the emitted electrons is more significant inasmuch as the excitation with subsequent Auger electron emission appears to be more effective at the collisions of the same atoms (Viaris de Lesegno et al. 1974, Hennequin and Viaris de Lesegno 1974).

Recently, some new theories of kinetic electron emission have been proposed. The cascade theory by Holmen et al. (1979, 1981), Svensson and Holmen (1981, 1982) and Schou (1980a,b) dealt with calculations of both angular and energy distributions of the emitted electrons via solving sets of Boltzmann equations as Von Roos (1957) used to do earlier. This approach was used to estimate the contribution of recoils (Svensson et al. 1981).

2.6. Conclusion

The kinetic electron emission under ion bombardment, as follows from the foregoing, was studied extensively during many years and by many researchers.Nevertheless, there are some unexplored areas both in the map of known facts and in our understanding of these facts. Some of these areas can be pointed out now, some will come to light later. Here are some of them.

(1) There is a striking disproportion in the study of emission from metals and nonmetals. Dielectrics and semiconductors have not been studied enough in this respect. Meanwhile, metals fall short to nonmetals in the diversity of the electron spectra and lattice types. One more paradox consists in that the kinetic emission from nonmetals exceeds in magnitude by far that from metals.

(2) The range of incident ion velocities is studied not uniformly. Less than 10% of all investigations cover velocities in excess of 3×10^8 cm/s. Meanwhile, it is in this range that the maximum and falling-off part are located. Besides, the experimental level of these 10% is below that of the other 90%. This is apparently due to difficulties encountered in the work with accelerators or with products of nuclear reactions. By the way, nothing is known concerning the electron emission under neutron bombardment. Meanwhile, one may expect here interesting results since in the case of neutrons only recoil atoms can cause emission.

(3) A poor knowledge of the kinetic emission near the maximum is all the more disappointing in that it is somewhere here that the transition from the independence of emission on the ion charge to the $\sim q^2$ relationship occurs. The emission curves for a set of multicharged ions in this transition region should be quite impressive. A similar family of curves obtained in the region of transition from intermediate to low velocities would marvellously supplement the picture. Unfortunately, nobody has yet done it.

(4) We enter here the field of potential emission and touch on the delicate question of the additivity of these two kinds of emission. Unfortunately, arguments which took place here have not clarified the situation. The problem is closely connected with the question of up to what velocities there can occur Auger neutralization of ions on the surface. Apparently, it decreases quickly with increasing velocity.

(5) Although the diffusion of electrons to the surface and their escape from the solid is not inherent in the emission under ion bombardment, this problem is likewise poorly studied. The energy spectrum of the kinetic emission contains enough information on this mechanism, but is leveled off by a cascade of electron collisions to the shape common for any type of emission.

(6) The idea on the plasma mechanism of the emission has not yet been developed. However, if this mechanism exists, it is not specific for the kinetic emission under ion bombardment. We should apparently be interested rather in the difference of the kinetic emission from other types of electron emission than in their common features. Of principal interest here are the peculiarities connected with the mechanism of electron excitation.

(7) The investigation on the kinetic emission is developed in parallel with that of the ionization at atomic collisions. As a rule, the features which have already been studied in atomic collisions were revealed also in the electron emission. For instance, the theory of ionization by ion impact proved to be of use for the kinetic emission theory. On the other hand, some ideas have been earlier applied to the study of electron emission. As an example one can take the role of Auger effect which has first been pointed out in the case of kinetic emission, and only after this in the field of atomic collisions. The Auger neutralization of slow ions has been very well studied for metals, whereas in atomic collisions it was studied much later as transfer ionization. Following this analogy, one can point out a number of methods widely in use in atomic collisions research which are still not applied in ion bombardment studies. For instance, it is difficult to overestimate the importance of the coincidence technique for the study of angular and energy distributions and of the charge state of atoms and ions scattered or ejected from a solid by ion bombardment. This technique was used to advantage in the investigation of characteristic energy losses in atomic collisions which are still not studied in ion bombardment experiments.

(8) Great attention was paid in recent years to the structure effects of the emission from single crystals; however, they are studied by measuring the kinetic emission coefficient rather than angular and energy distributions. In contrast to the scattering or sputtering where both incident and escaping particles are governed by classic laws of motion of heavy particles, in electron emission only primary particles follow classic trajectories, whereas the electron emission depends on inelastic energy losses of heavy ions. This feature of the kinetic emission can make advantageous its further study with different structures and at different ion velocities.

(9) A completely new region is opened in the study of defect formation by the method of kinetic emission. The first experiments mentioned above are very interesting. The physics of radiation damage at present lacks a simulation of the mechanism of point defect formation. The electron emission can become a fine tool in these studies.

3. Shell effects

3.1. Electronic energy losses

3.1.1. Generalization of the Firsov formula
The Thomas–Fermi model is known not to take into consideration the shell structure in atoms. Taking into account the shell structure would be possible if the electron flux through the surface S is calculated by using the atomic model which reflects this structure. Cheshire et al. (1968) did so by substituting for n the electron density which was estimated using Slater

functions. In this approximation, eq. (1) takes the form

$$\mathscr{E}_{\text{inel}} = \tfrac{1}{2} m \int_R \dot{R} \, dR \int_S \left(\sum_i v_i |\psi_i|^2 + \sum_j v_j |\psi_j|^2 \right) dS \, . \tag{59}$$

Summation involves all electrons of the colliding atoms. Subscript i refers to electrons of the first atom and j to electrons of the second atom, v_i,

$$v_i = \frac{1}{m} \, \xi_i \left(1 + 2 \, \frac{l_i(l_i + 1) - n_i(n_i - 1)}{n_i(2n_i - 1)} \right)^{1/2} ,$$

is the mean-square velocity in orbit, and ψ_i,

$$\psi_i = c r^{n_i - 1} \, e^{\xi_i r} \, Y_{lm}(\theta, \varphi) \, ,$$

is the Slater wave function,

$$\xi_i = \frac{Z - s}{n_i a_0} \, ,$$

Z is the atomic number, s is the screening parameter, a_0 is the Bohr radius, r is the distance from the nucleus, l_i is the orbital quantum number, and n_i is an empirical value coinciding for the first three shells with the radial quantum number.

Cheshire et al. (1968) have estimated the integrals numerically in different cases of channeling for a comparison with the experiment by Eisen (1968).

Baklitzky and Parilis (1972), Baklitzky et al. (1973) and Komarov and Kumakhov (1973) simultaneously and independently obtained the analytical expression for $\mathscr{E}_{\text{inel}}$ by reduction of the integrals from eq. (59) to the modified Bessel functions.

The Z-dependence studies were reported further by Komarov and Kumakhov and co-workers in a series of papers. In particular, Komarov and Temkin (1976) have used the linear averaging of the velocity over Rutan–Hartree–Fock wave functions in the momentum representation mentioned.

The above authors have developed the inelastic loss theory in the intermediate velocity range, i.e., in the region of the curve bend of inelastic loss dependence on velocity. In this region the effect of two mechanisms has been considered. Good agreement was obtained with experiment in a wide velocity range, moreover, in the limits of low and high energy the modified theory transforms into the Firsov theory and Bethe theory, correspondingly.

Baklitzky and Parilis (1982) used another approach to find a solution for this problem. Being within the scope of the main ideas of the Firsov theory, one could obtain a formula describing the curve bend of the electronic energy loss or velocity, taking into consideration the time necessary to exchange electrons to lose the energy in the shell of the collision partner and supposing that electron ejection occurs during the autoionizing state decay via Auger processes.

Let us normalize the function ψ_i,

$$c^2 \int_0^\infty r^{2n_i-1} e^{\xi_i r} r^2 \, dr \int_{\varphi,\theta} Y_{lm}(\theta, \varphi) \, d\theta \, d\varphi = 1 . \tag{60}$$

By averaging over angles φ and θ, and confining to integer values of n, one gets,

$$c^2 = \frac{(2\xi_i)^{n_i+1}}{(2n_i)!} \frac{1}{4\pi} . \tag{61}$$

For energy $\mathscr{E}_{\text{inel}}$ we obtain

$$\mathscr{E}_{\text{inel}} = \frac{m}{8\pi} \int_{R_0}^\infty R \, dR \int_S dS \left[\sum_i v_i \frac{(2\xi_i)^{n_i+1}}{(2n_i)!} r^{2n_i-2} e^{-2\xi_i r} + \sum_j \cdots \right] . \tag{62}$$

Here, Σ_j is the summation over electrons on the second colliding atom, analogous to the summation over the electrons of the first one.

The surface S has been placed in the point of the maximum electron density on the straight line, connecting the nuclei of the colliding atoms.

As before, the surface S has been considered to be a plane, which divides the distance between atoms R at a ratio $\alpha/(1-\alpha)$ (Kishinevsky 1962), where α,

$$\alpha = \left[1 + \left(\frac{Z_1}{Z_2} \right)^{1/6} \right]^{-1} , \tag{63}$$

does not depend on R. The integration over R results in

$$\begin{aligned}
\mathscr{E}_{\text{inel}} &= \frac{m}{4} \int_{R_0}^\infty \dot{R} \, dR \int_{\alpha R} dR \left[\sum_i v_i \frac{(2\xi_i)^{n_i+1}}{(2n_i)!} r^{2n_i-2} e^{-2\xi_i r} + \sum_j \cdots \right] \\
&= \frac{m}{4} \int_{R_0}^\infty \dot{R} \, dR \left[\sum_i v_i \frac{2\xi_i}{2n_i} e^{-2\xi_i r} \sum_{k=0}^{2n_i-1} \frac{1}{k!} (2\alpha\xi_i R)^k + \sum_j \cdots \right] .
\end{aligned} \tag{64}$$

The velocity $\dot{R}$ changes during elastic collision,

$$\dot{R} = \frac{1 - V/E}{1 - V/E - \rho^2/R^2} u . \tag{65}$$

The Firsov screened potential has been used (Firsov 1957),

$$V(R) = \frac{(Z_1 + Z_2)e}{R} \chi[1.13(Z_1 + Z_2)^{1/3} R/a_0] . \tag{66}$$

Here, ρ is the impact parameter, u is the primary relative velocity, E is the collision energy, and χ is the Thomas–Fermi screening function.

Within the investigated region of atomic spacing

$$0.7 \frac{a_{T-F}}{(\sqrt{Z_1}+\sqrt{Z_2})^{2/3}} < R < 7 \frac{a_{T-F}}{(\sqrt{Z_1}+\sqrt{Z_2})^{2/3}} ,$$

where $a_{T-F}=0.885a_0$ is the Thomas–Fermi parameter, the Firsov potential is described with 10% accuracy by the expression

$$V(R) = \frac{0.45a_{T-F}e^2Z_1Z_2}{(\sqrt{Z_1}+\sqrt{Z_2})^{2/3}R^2} = \frac{A(Z_1,Z_2)}{R^2} , \tag{67}$$

where e is electron charge. By substituting eq. (65) in eq. (64) with account of eq. (67), the expression takes the form

$$\mathscr{E}_{inel} = \frac{mu}{4} \left\{ \sum_i \frac{v_i}{2n_i} \left[\sum_{k=0}^{2n_i-1} \frac{2\xi_i}{k!} \int_{R_0}^{\infty} dR \frac{R}{(R^2-R_0^2)^{1/2}} (2\alpha\xi_i R)^k e^{-2\xi_i r} \right. \right.$$

$$\left. \left. - \sum_{k=0}^{2n_i-1} \frac{2\xi_i}{k!} \int_{R_0}^{\infty} \frac{dR}{R} \frac{R_{min}}{(R^2-R_0^2)^{1/2}} (2\alpha\xi_i R)^k e^{-2\xi_i r} \right] + \sum_j \cdots \right\}. \tag{68}$$

Here, R_{min} is the distance of closest approach in a head-on collision.

By $(k+1)$-fold differentiation of the second term over the parameter $2\alpha\xi_i$, eq. (68) is reduced to the set of modified Bessel functions. After introduction of the dimensionless parameter $x_i = 2\alpha R_0 \xi_i$, the result is given by

$$\mathscr{E}_{inel} = \frac{mu}{4} \left\{ \sum_i \frac{v_i}{2n_i} [\varphi_1(n_i, x_i) - (2\alpha\xi_i)^2 R_{min}^2 \varphi_2(n_i, x_i)] \right.$$

$$\left. + \sum_j \frac{v_j}{2(1-\alpha)} \{\varphi_j(n_j, x_j) - [2(1-\alpha)\xi_j]^2 R_{min}^2 \varphi_2(n_j, x_j)]\} \right\}. \tag{69}$$

The introduced function equals

$$\varphi_1(n_i, x_i) = 1/n_i \sum_{k=0}^{2n_i-1} (-1)^k \frac{x_i^{k+1}}{k!} \frac{d^{k+1}}{dx_i^{k+1}} K_0(x_i) ,$$

$$\varphi_2(n_i, x_i) = 1/n_i \sum_{k=0}^{2n_i-1} (-1)^k \frac{x_i^{k-1}}{k!} \frac{d^{k-1}}{dx_i^{k-1}} K_0(x_i) , \tag{70}$$

where $K_0(x_i)$ is the modified Bessel function. The dependence of function $\varphi_1(n_i, x_i)$ on x_i is shown in fig. 23.

The functions $\varphi_1(n_i, x_i)$ and $\varphi_2(n_i, x_i)$ depend on the atomic numbers of colliding atoms, Z_1 and Z_2, since

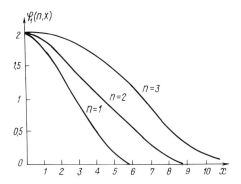

Fig. 23. Dependence of the dimensionless auxiliary function $\varphi_1(n, x)$ on the dimensionless parameter x for $n = 1, 2, 3$.

$$x_i = 2\alpha R_0 \xi_i , \tag{71}$$

where both α and ξ_i depend on Z_1 and Z_2.

The dependence of α on Z_1 and Z_2 is weak, R changes smoothly with the change of Z_1 and Z_2 charge. The shell effect reveals in the dependence of ξ_i. With increasing number of electrons as the shell is filled (i.e., within the limits of one n_i value) ξ_i increases linearly with increasing Z, which results in a sharp decrease of both the function φ_1 and the energy transmitted to each electron. Such decrease is caused by a decrease due to shell reduction with increasing Z of the electron flux through the surface S between the atoms. The increasing of the number of electrons in the outer shell with Z increase appears to be a competitive process.

At considerable relative alternations of this number, i.e., at the beginning of the periodic system chart, this process prevails over the decrease of electron density caused by shell reduction. As a result, the increase of inelastic energy losses with increase of atomic number has been observed. To the end of a period the relative alternation of electron number in the outer shell becomes small, and the shell reduction causes mainly the decrease of inelastic losses. The above-mentioned regularities explain the nonmonotonic dependence of inelastic energy losses on atomic number observed in experiment.

The term which contains functions $\varphi_2(n_i, x_i)$ appears as a result of slowing down of the relative atom motion at mutual approach. It is of importance only at both small relative velocities and impact parameters.

In order to obtain the mean inelastically transmitted energy, it is necessary to integrate eq. (69) over all the impact parameters within the elementary cell of the crystal,

$$\mathscr{E}_{inel} = \frac{\pi mu}{4} \left\{ \frac{1}{2\alpha_i} \sum_i v_i [(2\alpha\xi_i)^{-2} f_1(n_i, a_i) - R_{min}^2 f_2(n_i, a_i)] \right.$$

$$\left. + \frac{1}{2(1-\alpha)} \sum_j v_j [[2(1-\alpha)\xi_j]^{-2} - R_{min}^2 f_2(n_j, a_j)] \right\} .$$

(72)

In this formula

$$f_1(n_i, a_i) = \int_{a_i}^{\infty} \varphi_1(n_i, a_i) x_i \, dx_i ,$$

(73)

$$f_2(n_i, a_i) = \int_{a_i}^{\infty} \varphi_2(n_i, a_i) x_i \, dx_i ,$$

The dimensionless parameters a_i, a_j are determined by the minimum distances of closest approach for all impact parameters,

$$a_{i,j} = x_{i,j\,min} = 2\alpha R_{0\,min} \, \xi_{i,j} .$$

(74)

At binary collision, a_i is determined by the distance of closest approach in a head-on collision, in the case of channeling in the crystal by the distance from the channel wall to its center.

Shell effects appeared to be intensive at slight shell overlapping, i.e., when the distance of closest approach is not far less than the outer shell diameter. This is due to the outer shell filling and reduction determining the nonmonotonic dependence of the inelastic energy losses on atomic number. The energy transmitted in the deeper shells suppressed this effect.

Due to this fact the shell effects become particularly noticeable at low collision velocities, when the mutual penetration of atoms is not considerable, or in the crystal for well channeled particles for which the distance of closest approach is large, because of low transverse energy.

Figure 24 illustrates the graph of the stopping power of Si single crystal for ions of the second and third periods of the periodic system chart, bombarding the target with velocity $u = 2 \times 10^7$ cm/s. The stopping power decreases and Z oscillations enhance from the first to the second and sixth layers as the ions gradually enter the channel.

Ion flux redistribution as a result of scattering by the atoms of the single-crystal lattice at entering the crystal has been calculated using the shadowing model developed by Martynenko (1966a).

In accordance with this model, a surface atom forms a shadow behind it in the ion flux. The radius of the shadow in the next layer of a lattice interatomic spacing d equals

$$R_T = 1.5[\pi d R_{min}^2 / (\mu + 1)]^{1/3} ,$$

(75)

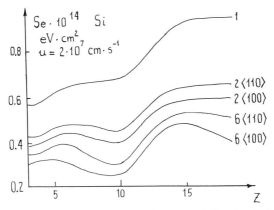

Fig. 24. Stopping power of the first, second and sixth layer of Si under the bombardment along channels $\langle 110 \rangle$ and $\langle 100 \rangle$ by single charged ions of the second and third periods of the periodic system chart with velocity $u = 2 \times 10^7$ cm/s.

where μ is the incident ion and lattice atom mass ratio. When entering the channel, the shadow radius increases, rising the lower limit of integration in eq. (73).

The shell effects reveal also at the collisions in gases, occurring with large impact parameters, i.e., for small scattering angles. Figure 25 illustrates the dependence of electronic energy losses on the atomic number for ions of the second and third period with velocity $u = 2 \times 10^7$ cm/s in neon for a scattering angle θ from 0.25° to 5° in the laboratory coordinate system. With θ increasing the oscillation process becomes weak and at $\theta = 5°$ it disappears.

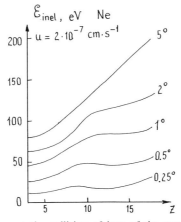

Fig. 25. Inelastic energy loss at the collision of ions of the second and third periods of the periodic system with Ne at different angles of scattering.

In addition to the above-described method, other methods have been developed for consideration of the shell structure when calculating electronic energy losses. Yarkulov (1970) has also used the modified Firsov formula for calculation the slow ion energy losses in solids. Kesselmann (1971) has inserted the Z dependence into the potential, which also affects the electron energy loss. Bhalla and Bradford (1968) considered Z dependence on the basis of the Lindhard–Sharff formula. Z oscillations were considered to be related to the Z dependence of the cross section for incident ion–electrons scattering by target atoms (Pathak 1974, Briggs and Pathak 1974).

3.1.2. Auger transitions

The Firsov formula in its original form as well as modified from eq. (59) implies a dependence of the electron energy loss on velocity which slightly differs from a linear one. Such dependence was observed for velocities up to $(2-5) \times 10^8$ cm/s (Fleischmann et al. 1972). At further velocity increase, the increase of electron energy losses slows down, and then passes through a maximum, and for velocities larger than 10^9 cm/s the dependence of electron energy losses on velocity becomes inertly proportional. In this velocity range, the Bethe formula deduced from the Bohr approximation is valid,

$$-\frac{dE}{dx} = \frac{4\pi e^2 Z_1^2 Z_2 N}{mu^2} \ln \frac{2mu^2}{I} , \tag{76}$$

where N is the number of target atoms per volume unit, and I is the ionization potential.

An attempt to apply the Firsov model in the intermediate velocity range is of interest. The original version of the Firsov theory has been developed for relatively small collision velocities ($u < v_0$, where v_0 is the Bohr velocity). In this connection, the collision duration was considered to be long enough so that the energy transmitted by small portions and shared among the electrons would have time to be concentrated on the electrons that gets energy sufficient for its escape (if the total transmitted energy was sufficiently high). The number of the excited electrons is small relatively to their total number.

The process of electron emission has two stages: in the first stage the part of atomic kinetic energy is transmitted to electrons, which leads to their excitation and formation of the autoionizing state with multiple excitation in outer shells; at the second stage this excitation energy is redistributed due to electron interaction, in addition, a fraction of electrons is ejected, i.e., the decay of the autoionizing state takes place.

Experimental data (Afrosimov et al. 1976) have indicated that most of the electrons ejected from outer shells belong to the continuum part of the electron energy spectrum and are ejected during the existence of the quasimolecule.

Possibly, this is due to the fact that the decay time of the autoionizing state, i.e., the time of Auger processes in the quasimolecule at small intermolecular distance, rises by an order of magnitude and could reach 10^{-15} cm/s, which is comparable with the quasimolecule life time at velocity 10^8 cm/s (Kishinevsky and Parilis 1968b).

Using the Firsov theory, the excitation energy could be estimated. This approximation seems to be valid until the collision time application of the Firsov theory to higher velocity requires both giving up the adiabatic consideration and taking into account the collision time to electron ejection ratio.

An attempt has been made to apply the classical theory in the transition region by Brice (1972). The time dependence has been introduced on the basis of general considerations. The quantum-mechanical calculation of electron flux passing through the separating plane gives

$$\Phi = -\frac{i\hbar}{2m} \left[\lambda_t^* \frac{\partial \lambda_t}{\partial Z} - \lambda_t \frac{\partial \lambda_t^*}{\partial Z} - (2ik_0|\lambda_t)^2 \right] dx\, dy \,, \tag{77}$$

where the partial electrons wave function equals

$$\lambda_t(r') = (2\pi)^{-3/2} \int_{-\infty}^{\infty} dk_x \int_{-\infty}^{\infty} dk_y \int_{-\infty}^{\infty} dk_z\, e^{ik\cdot r}\Phi(k) \,. \tag{78}$$

Here, $\Phi(k)$ is the Fourier integral for the electron wave function of the given orbital.

The hydrogen-like orbitals were used and a free parameter Z_{eff} has been introduced. The transition from the ascending branch to the descending one in the velocity dependence has been reached using three free parameters that allowed to give a fit to the experimental data with an accuracy of 1%.

If one takes into account the decay rate of the autoionizing state final formed in the quasimolecule at the collision, considering the electrons to be ejected as a result of binary Auger processes, it could be shown that the increase of velocity slows down and then passes through the maximum. In addition, the excited electrons were assumed not to obey the Firsov mechanism. While both the assumptions do not affect the Firsov formula at low velocity, they lead to a sharp decrease of the transmitted energy at velocities exceeding 10^8 cm/s (Baklitzky and Parilis 1976).

The balance equation at the collision takes the form

$$d\nu = (\nu_2 - \nu_1)\, dt \,, \tag{79}$$

where ν is the number of excited electrons in a quasimolecule, ν_1 is the number of electrons ejected from the quasimolecule via Auger processes, and ν_2 is the number of electrons excited according to the 'electron function' mechanism per unit.

In accordance with the assumption of binary electron transition, we get

$$v_1 = \omega v^2 ,\tag{80}$$

where ω is the Auger transition probability.

With consideration of the second assumption we get the integral equations for v_2,

$$v_2(t) = \mathscr{E}/\bar{I} .\tag{81}$$

Here, $\mathscr{E}$ is the energy transmitted according to the Firsov mechanism per time unit, N is the number of electrons involved into energy transfer (entering in overlapping shells), and $\bar{I}$ is the average ionization potential.

The solution of eq. (81) is given by

$$v_2(t) = \mathscr{E}/\bar{I}\, e^{-\mathscr{E}t/(N\bar{I})} .\tag{82}$$

For the balance equation, we get

$$dv/dt = -v^2\omega + \mathscr{E}/\bar{I}\, e^{-\mathscr{E}t/(N\bar{I})} .\tag{83}$$

Equation (83) is the Riccati equation. The Riccati equations are not solved in general terms, yet eq. (83) could be transformed into Bessel equations by substituting $du/dt = vu$,

$$\frac{d^2u}{dt^2} - u\,\frac{\mathscr{E}\omega}{\bar{I}}\, e^{-\mathscr{E}t/(N\bar{I})} = 0 .\tag{84}$$

The roots of this equation are cylindrical functions of purely imaginary variables,

$$u = z_0\left[i\,\frac{2N(\bar{I}\omega)^{1/2}}{\sqrt{\mathscr{E}}}\, e^{-\mathscr{E}t/(N\bar{I})} \right] .\tag{85}$$

The inverse transformation to variable v provides the following complete solution of balance equation with account of the initial conditions,

$$v(t) = \frac{[K_1(z)I_1(z_0) - I_1(z)K_1(z_0)]2Nz}{[K_0(z)I_1(z_0) + I_0(z)K_1(z_0)]z_0^2} .\tag{86}$$

Here,

$$z = z_0\, e^{-\mathscr{E}t/(N\bar{I})} , \qquad z_0 = \frac{2N(\bar{I}\omega)^{1/2}}{\sqrt{\mathscr{E}}} ,$$

$K(z)$ and $I(z)$ are the modified Bessel functions.

The solution of eq. (86) expresses the filling of the excited levels of a quasimolecule in the course of collision. For determination of the total number of ejected electrons during the collision process the integral

$$n = \int_0^T \nu_1 \, dt = \int_0^T \nu^2 \omega \, dt \qquad (87)$$

should be calculated.

After integration we get

$$n = \int_0^T \nu_1 \, dt = N(1 - e^{-\mathscr{E}t/(N\bar{I})}) - \nu(t) , \qquad (88)$$

where $T = 2r/u$ is the collision duration, u is the collision velocity, and r is the radius of the outer orbit. The excitation energy was considered to depend linearly on velocity (Firsov 1959).

The dependence of the number of ejected electrons according to eq. (88) on the velocity of incident atom is shown in fig. 26. It is seen that at low velocities there is a linear dependence on velocity, but a sharp decrease is observed at high velocities. With the designations $k = \mathscr{E}T/\bar{I}$ and $\tau = \omega T$, the product of these magnitudes

$$c = k\tau = (\mathscr{E}T/\bar{I})\omega T \qquad (89)$$

would be a Massey-type parameter, which contains the ratio of the collision time to Auger effect time, in first approximation being independent of collision velocity. Actually, the value k is a total dimensionless version of the electron excitation energy, according to the 'electron friction' mechanism in the course of the collision process being proportional to the collision time, i.e., inversely proportional to u.

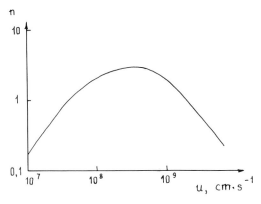

Fig. 26. Dependence of the number of ejected electrons on the velocity of the impinging atom at the collision of two Ar atoms.

Using this new designation, eq. (88) takes the form

$$n = N(1 - e^{k/n}) - \nu(k),$$ (90)

besides

$$z = (2N/k)\sqrt{c}\, e^{(-k/2N)}; \qquad z_0 = (2N/k)\sqrt{c}.$$

Being in such dimensionless form, eq. (88) contains only one parameter, c, which characterizes the colliding parts and could be presented as a family of curves with parameter c (see fig. 27). The location of the maximum and its absolute value are determined by both the efficiency of electron excitation and the number of parameters involved in the process. The following dependence on c (at $c \gg 1$) could be written for the location of the maximum,

$$(k/N)_{max} \sim \sqrt{c}.$$ (91)

One can check the fact that eq. (90) really would transform into the Firsov formula (i.e., using new designations $n = k$), if we consider the process to be adiabatic ($W \gg 1/T$) and the number of ejected electrons to be small ($k \ll N$). The first requirement leads to the inequality $c \gg 1$, from the second one it follows that $z_0 \gg 1$.

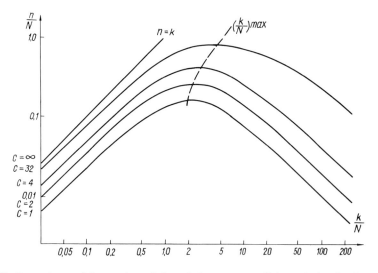

Fig. 27. Dependence of the number of ejected electrons on collision velocity (in dimensionless values).

For large z, we get

$$I_1(z) \approx \frac{e^z}{\sqrt{2\pi z}}, \qquad I_0(z) \approx \frac{e^{-z}}{\sqrt{2\pi z}}, \qquad K_1(z) \approx \frac{e^z}{\sqrt{2z}}\sqrt{\pi},$$

$$K_0(z) \approx \frac{e^{-z}}{\sqrt{2z}}\sqrt{\pi}. \tag{92}$$

From this equation it follows

$$\frac{\nu(z)}{N} = \frac{e^{2(Z_0-Z)}-1}{e^{2(Z_0-Z)}+1}\frac{2z}{z_0}. \tag{93}$$

By substituting z and z_0 in eq. (93), we get

$$\frac{\nu}{N} \approx \frac{e^{2\sqrt{c}}-1}{e^{2\sqrt{c}}+1}\frac{1-k/(2N)}{(N/k)\sqrt{c}}. \tag{94}$$

For eq. (90), this gives

$$\frac{n}{N} \approx \frac{k}{N}\left(1 - \frac{e^{2\sqrt{c}}-1}{e^{2\sqrt{c}}+1}\frac{1-k(2N)}{\sqrt{c}}\right), \tag{95}$$

with the consideration of $k \ll N$ and $c \gg 1$, the equality is given by

$$n = k. \tag{96}$$

Thus, it has been demonstrated that relation (90) is different from the original Firsov theory only by the two mentioned requirements and maintains all other physical assumptions taken as fundamentals of the Firsov theory.

For comparison with the experimental data it is convenient to use the summary graph of the experimental ionization cross section at atomic collisions (Fleischmann et al. 1972). This graph was obtained as a result of calculation of experimental data to the reduced cross section determined from

$$\sigma_{i,i+1} = n_{\text{eff}}\left(\frac{I_1}{13.6}\right)^{-\alpha}(Z_1^{2/3} + Z_2^{2/3})\sigma_{\text{TM}}(\varepsilon). \tag{97}$$

Here $\sigma_{i,i+1}$ is the cross section of single ionization of the ion with charge i, and α and n_{eff} are the empirical parameters. The dimensionless energy ε is determined by

$$\varepsilon = \frac{E}{MR^2(I_1/13.6)^2}, \tag{98}$$

where E is the energy of incident ion, M is the ion mass, and R is the characteristic collision length.

From the graph it is seen that curves for different colliding particles fit with certain spread the empirical curve

$$\sigma_{TM}(\varepsilon) = 3.2 \times 10^{14} \left[\frac{\varepsilon}{(E^{2/3} + 30^{2/3})^3} \right]^{1/2} , \tag{99}$$

which would be used for comparison with the theory as the summary experimental curve.

For comparison with the cross section σ_{TM}, it is necessary to rewrite the function $n(k)$ using coordinates $f(\varepsilon)$. The dependence of the cross section on k is given by

$$\sigma(k) = n/k\sigma_F(u) , \tag{100}$$

where σ_F is the Firsov cross section (Firsov 1959),

$$\sigma_F(u) = \sigma_0[(u/u_0)^{1/5} - 1]^2 , \tag{101}$$

where

$$u_0 = 23.3 \times 10^6 I/(Z_1 + Z_2)^{5/3} ,$$

$$\sigma_0 = 32.7 \times 10^{-16}/(Z_1 + Z_2)^{2/3} .$$

There is a relation between k and ε,

$$\sqrt{\varepsilon} = \hbar W k/(\bar{I}c) . \tag{102}$$

The summary graph with plotted recalculated curve $n(k)$ for $c = 1$ is shown in fig. 28.

The location of the maximum is given by

$$k(\max) \approx N[c/(1 - c_1)]^{1/2} , \tag{103}$$

where

$$c_1 = \frac{e^{2\sqrt{c}} - 1}{e^{2\sqrt{c}} + 1} \frac{1}{c} .$$

At $c \approx 1$, we get

$$k(\max) = \omega/(Ic)\sqrt{\mathscr{E}} \approx 2Nc . \tag{104}$$

An asymptotic expansion for small ε and small k has shown that within the theory at $k/n \ll 1$ $\sigma(k) \simeq k$ and according to the summary graph at $\sigma \ll 30$ we get $\sigma_{TM} \simeq k^2$.

From fig. 28 it is seen that the Firsov theory describes well the interaction for noble gases, in particular when the atomic number of target atoms and incident ones coincide. It should be noted that the theoretic curve corresponds to the total ionization cross section of both colliding particles, but

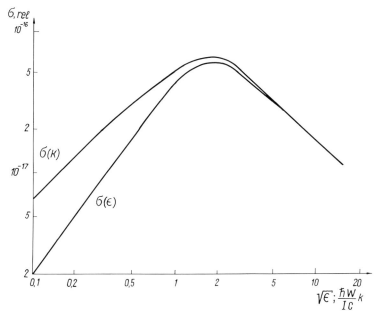

Fig. 28. Comparison of dimensionless ionization cross section with the dimensionless experimental one versus the dimensionless collision energy.

the experimental ones correspond to the ionization cross section of the incident ion only.

The question arises about the error originating from neglecting the dependence of collision time on the impact parameter, and considering ω to be constant during the collision process.

For consideration of both dependence of the collision time on the impact parameter and the probability of Auger processes on the distance between atoms in the course of collision, W needs to be averaged over the collision time and impact parameters, at which the overlapping of electron shell of the colliding atoms occurs, taking into consideration the T dependence on p.

The averaged value of W over time is given by

$$\overline{W(n)} = 2\pi \int_{0}^{p(R_b)} p\, dp \int_{0}^{T(p)} W(T)/T(p)\, dT/[\pi p^2 (R_{atom})]\,. \tag{105}$$

Here $p(R_b)$ is the impact parameter, at which the distance of the closest approach becomes equal to the radius of the atom boundary $(R_0 = R_b)$ with the given velocity u,

$$p(R_b) = (R_b^2 - R_{min}^2)^{1/2}\,, \tag{106}$$

where the potential $V(R) \simeq 1/R^2$ has been used, T is the collision time with the given velocity u and impact parameter p,

$$T = 2 \int_{R_0}^{R_b} \mathrm{d}R/\dot{R} . \tag{107}$$

For $\overline{W(u)}$, we get

$$\overline{W(u)} = 2\pi \int_0^{p(R_b)} p \, \mathrm{d}p \int_{R_0}^{R_b} W(R) \, \mathrm{d}R \Big/ \Big(\pi p^2(R_b) \dot{R} \int_{R_0}^{R_b} \mathrm{d}R/\dot{R} \Big) . \tag{108}$$

Here, $\dot{R}$ is the collision velocity,

$$\dot{R} = \frac{1 - V(R)/E}{[1 - V(R)/E - \rho^2/R^2]^{1/2}} \, u . \tag{109}$$

By changing the order of integration in eq. (108), we get

$$\overline{W(u)} = 2\pi \int_{R_{min}}^{R_b} W(R) \, \mathrm{d}R \int_0^{p(R_b)} p \, \mathrm{d}p \Big/ \Big[\pi p^2(R_b) \dot{R} \int_{R_0}^{R_b} \mathrm{d}R/\dot{R} \Big] . \tag{110}$$

Here,

$$p(R) = (R^2 - R_{min}^2)^{1/2} . \tag{111}$$

For the collision time we get

$$T(u, p) = 2/u(R_b^2 - R_{min}^2 - p^2)^{1/2} - p \arccos[p/(R_b^2 - R_{min}^2)]^{1/2} . \tag{112}$$

This expression is approximated by

$$T(u, p) = \pi/(2u)[(R_b^2 - R_{min}^2)^{1/2} - p] . \tag{113}$$

With such an approximation, the integral over p in eq. (110) is written as

$$\int_0^{p(R_b)} p \, \mathrm{d}p \Big/ \Big(\dot{R} \int_{R_0}^{R_b} \mathrm{d}R/\dot{R} \Big) = 4R/[\pi(R^2 - R_{min}^2)]$$

$$\times \int_0^{p(R_b)} \frac{(R_b^2 - R_{min}^2 - p^2)^{1/2}}{[(R_b^2 - R_{min}^2)^{1/2} - p]} \, p \, \mathrm{d}p = R . \tag{114}$$

For the Auger ionization probability, we get

$$\overline{W(u)} = 2/(R_b^2 - R_{min}^2) \int_{R_{min}}^{R_b} W(R)\, dR .$$ (115)

Then the Auger ionization probability averaged over time has been obtained to be equal to the mean value of $W(R)$ within the ring from R_{min} to R_b. Such simple dependence appears to be a consequence of the chosen potential,

$$V(R) \simeq 1/R^2 .$$

The mean time could be obtained by averaging eq. (112) over the impact parameters,

$$T(u, p) = 2\pi \int_0^{p(R_b)} Tp\, dp/[\pi p^2(R_b)] = \tfrac{4}{9}(R_b^2 - R_{min}^2)^{1/2}/u .$$ (116)

Thus, for the parameter c we get

$$c = \overline{W}\overline{T}k = \tfrac{8}{9}k/[u(R_b^2 - R_{min}^2)^{1/2}] \int_{R_{min}}^{R_b} W(R)\, dR .$$ (117)

As $k \simeq u$ and W depends on velocity via $R_{min}(V)$, the dependence of W on u becomes essential only at low velocities when $R_{min} \approx R_b$.

Figure 29 exhibits the change of $\overline{W}$, $W(R)$ and uT versus u or R_{min} for the colliding pair He and Be, for which the calculation of $W(R)$ has been made by Kishinevsky and Parilis (1968b). In addition, the relative velocity at which R_{min} has been reached was plotted along the X-axis. It is seen that at velocity higher than 10^7 cm/s the parameter c could be considered to be constant.

3.1.3. Interatomic electron flux

The bombardment of solid surfaces with neutral atoms is connected with the difficulty of their acceleration. As a rule, ions are used, including multiple-charged ones. For the calculation of inelastic energy losses of such ions both at the interaction with a solid target and at binary collisions, e.g., in gases, a number of refinements in the Firsov formula are needed (Baklitzky and Parilis 1986).

The original Firsov formula neglected the charge state of the colliding particles, because it was derived to explain the processes occurring at neutral atom collisions. Equation (59), in which the summation is made over each electron, allows, in principle, to take ionization into account; however, the value of α, determining the location of the plane which separated two

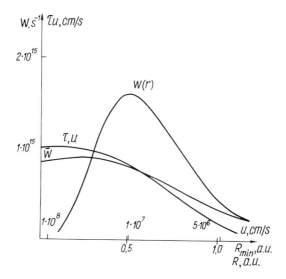

Fig. 29. The dependence of the Auger ionization probability W on interatomic distance R and the dependence of mean probability of Auger ionization $\bar{W}$ and reduced velocity $v\tau$ on the distance of closest approach R_{min} and the corresponding collision velocity at head-on impact for He–Be atoms.

neutral atoms was used. Therefore, for the calculation of losses at ion collisions a more detailed studying of both shape and location of the separating surface influences the energy loss.

The expression used for the electron velocity v appeared to be not quite satisfactory. The use of the mean square velocity increases the electron flux through the separating surface and conceals the shell effects. Accordingly, Komarov and Kumakhov (1973), Komarov and Temkin (1976) have both calculated the expression for velocity using bound energy and velocity averaging over the wave functions in momentum representation.

We assume to calculate the electron velocity using the Firsov formula, taking advantage of the shell model for the electron density, e.g., introducing the Slater functions as was done by Cheshire et al. (1968); Komarov and Kumakhov (1973) and Baklitzky et al. (1973),

$$v = 3^{4/3}\pi^{2/3}\hbar n^{1/3}/(4m) = 2.68n^{1/3} . \tag{118}$$

Then, for the electron flux through the surface S in the course of the process we get

$$F = \int_S \tfrac{1}{4}nv \, dS = 0.67 \int_S \left(\sum_i n_i + \sum_j n_j \right)^{4/3} dS , \tag{119}$$

where the densities are given by

$$n = \sum_i \frac{1}{4\pi} \frac{(2\xi_i)^{2n_i+2}}{(2n_i)!} r^{2n_i-2} e^{-2\xi_i r} + \sum_j \frac{1}{4\pi} \frac{(2\xi_i)^{2n_i+2}}{(2n_i)!} r^{2n_j-2} e^{-2\xi_j r},$$

(120)

where index i refers to the electrons of the first atom, index j to those of the second one, r is the distance from the nucleus, and $\xi_{i,j}$ and $n_{i,j}$ are calculated according to the Slater rules (Gombas 1949).

It should be noted that eq. (120) for the electron density has been derived under the assumption as used by Kishinevsky (1962), Komarov and Kumakhov (1973) and Baklitzky et al. (1973) that during quasimolecule formation in the course of collision the electron densities were superposed, which should be considered as certain approximation.

For comparison, in fig. 30 are plotted the curves for n and F, calculated according to the Thomas–Fermi model, which was taken as a basis in the Firsov theory, and using eqs. (119) and (120) for collisions of $N \rightarrow N$ versus the interatomic distance R. The surface S crosses over the internuclear axis point $R_1 = \alpha R$ (riding off from the nucleus Z_1). On the ground of symmetry, the surface S in the case under consideration is a plane separating the interatomic distance in two, i.e., $\alpha = \frac{1}{2}$. This surface Firsov considered to be a plane located in the middle of the atoms and having the property that the

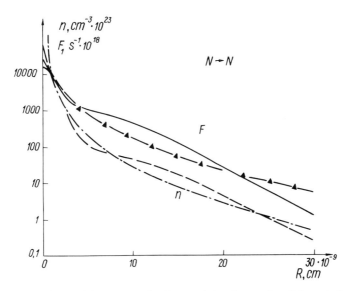

Fig. 30. The dependence of the electron density n and the electron flux F through the surface S on the interatomic distance R: for n, (—·—) according to the Thomas–Fermi model, (-----) using eq. (119); for F, (—▲—) according to the Thomas–Fermi model, (———) using eq. (120).

electron crossing it alters by a sudden charge which is part of its momentum which is connected with the relative nucleus movement. Both the location and shape of this surface determine what fraction of electrons of each atom participate in inelastic energy exchange. Thus, according to the Firsov theory the electron energy transmitted during collisions is determined by mutual velocity of the colliding atoms, and the number of electrons involved in the energy exchange is determined on the ground of the geometry. In the stated modified theory both assumptions retain.

If according to Parilis and Kishinevsky (1961) one takes into account the deflection of the colliding particle trajectory and for the electron flux substitutes eq. (119), then integral (59) could be transformed into

$$\mathscr{E}_{inel} = 1.34mu \int_{R_0}^{\infty} \frac{dR(R^2 - R^2_{min})}{R(R^2 - R^2_0)^{1/2}} \int_S \left(\sum_i n_i + \sum_j n_j \right)^{4/3} dS . \tag{121}$$

Kishinevsky (1962) proposed to place the surface in the point of minimum potential (or electron density) on a straight line which connects the nuclei. In this case, the surface also was considered to be a plane. In these calculations, made within the framework of the Thomas–Fermi model, the location of point R_1 slightly depends on R and could be given by

$$\alpha = [(Z_1/Z_2)^{1/6} + 1]^{-1} . \tag{122}$$

Let us consider the α dependence on the interatomic distance R. As is seen from numerical calculations for the shell model α exhibits a significant dependence on R. By way of example, in fig. 31 the $\alpha(R)$ dependence for the cases $Li^+ \to N$ and $Li \to N$ is shown, whereby α was calculated under the following conditions,

$$\alpha = R_1/R , \tag{123}$$

$$\frac{d}{dR_1} \left\{ \sum_i n_i(R_1) + \sum_j n_j(R - R_1) \right\} = 0 . \tag{124}$$

For comparison, in this figure the value α is shown as estimated by eq. (122). In the graph the influence of the shells is well-defined.

It should be noted that α deflection towards high values as well as towards low values always leads to an overestimation of the electron density and thereby to an overestimation of the flux through the surface S, which increases the inelastic energy losses. In this connection, the choice of the exact location of the plane S is of great importance since at integration over R the error compensation in the calculation of energy losses does not take place.

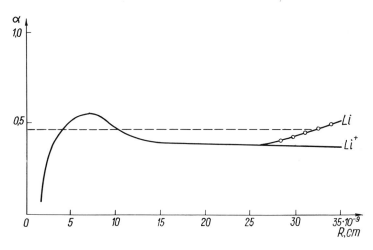

Fig. 31. The dependence of the α parameter of the surface location S on the interatomic distance R: (———) for $Li^+ \rightarrow N$, (—·—) for $Li \rightarrow N$, (–––––) according to Kishinevsky (1962) for both pairs.

An especially significant distinction of the curve $\alpha(R)$ from $\alpha = \mathrm{const}$ exists at small R. In the above-mentioned example, in this region, a approaches zero. This means that the minimum point of electron density between two centers is absent. Such behavior is a result of the superposition of the electron density of colliding atoms at shell overlapping. Actually, a fraction of electrons being between atoms is driven out, which results in a minimum at small interatomic distances. The consideration of this phenomenon within the framework of this theory seems not to be possible. Thus, at small R (of the order of 10^{-9} cm and less), the surface location between the colliding atoms is not determined.

The surface S passing through the minimum point on the axis connecting the nuclei of the colliding atoms (when such point does exist), in general, does not present itself as a plane. It seems more natural to determine this surface using lines of maximum gradient of electron density. In fig. 32, the cross sections of this surface by a plane containing both nuclei (Li–N) at different interatomic distances R is exhibited. It is seen that the surface changes the curvature sign as fast as the colliding particles approach and it could significantly differ from a plane. Nevertheless, the flux through the above surface differs slightly from one, the location of which has been chosen properly.

The distorted surface could be substituted by a plane and the value α could be considered to be a free parameter, then evidently the least deviation from the true value of flux $F_{\min}$ would correspond to the minimum

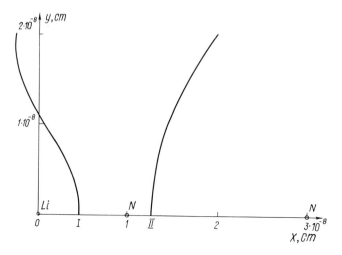

Fig. 32. The cross section of surface S versus the interatomic distance for Li–N: (I) $R = 10^{-8}$ cm; (II) $R = 3 \times 10^{-8}$ cm.

flux through the plane of variation of α dependence on R. α_{opt} could be chosen in such a way that F/F_{min} differs slightly.

The relative small difference between the integral through a plane and the one through a surface at optimal α seems not to be surprising. Actually, at the integration through the plane we get

$$F = \int_S \tfrac{1}{4} n v \, dS = 2\pi \int_{\alpha R}^{\infty} \tfrac{1}{4} n v r \, dr = 1.34 \int_{\alpha R}^{\infty} \left(\sum_i n_i + \sum_j n_j \right)^{4/3} r \, dr . \quad (125)$$

The value $\left(\sum n_i + \sum n_j \right)^{4/3}$ could be roughly approximated by the exponent for estimation (see fig. 30). Then,

$$F \simeq \int_{\alpha R}^{\infty} e^{-\mu r} r \, dr . \quad (126)$$

Integrand is small at large and small values r, besides, at $r \gg R$, the electron density distribution is of spherical symmetry, in this connection in this region the integral through the distorted surface should coincide with the integral through the plane. The integrand maximum is achieved at $r = 1/\mu$, and the integral within the limits $\tfrac{1}{2}\mu - \tfrac{2}{3}\mu$ contributes to the total integral of order $\tfrac{1}{2}F$. In order to obtain the shortest distortion of the integral through the plane from the one through the distorted surface it is necessary to draw the plane in such a way that at $r = 1/\mu$ it would cross over the surface S; moreover, the region adjacent to the surface S and the plane intersection line mainly contributes to the integral. Such plane location is close to the location corresponding to the minimum of flux.

It seems to be reasonable to find optimal value α not by flux, but by the end result, i.e., directly by inelastically transmitted energy $\mathscr{E}_{inel}$ by varying the expression for inelastically transmitted energy over α.

At integration of eq. (125), the integral through the surface, which is considered to be a plane, has been transformed in the following way,

$$
\int_S \left(\sum_i n_i + \sum_j n_j \right)^{4/3} dS = 2\pi \int_{\alpha R}^{\infty} \frac{\left[\sum_i n_i(r) \right]^{4/3} \left(\sum_i n_i + \sum_j n_j \right)^{4/3}}{\left[\left(\sum_i n_i \right)^{4/3} + \left(\sum_j n_j \right)^{4/3} \right]} \, r \, dr
$$

$$
+ 2\pi \int_{\alpha' R}^{\infty} \frac{\left[\sum_i n_i(r') \right]^{4/3} \left(\sum_i n_i + \sum_j n_j \right)^{4/3}}{\left[\left(\sum_i n_i \right)^{4/3} + \left(\sum_j n_j \right)^{4/3} \right]} \, r' \, dr'
$$

$$
= 2\pi \int_{\alpha R}^{\infty} k n_1^{4/3} r \, dr + 2\pi \int_{\alpha' R}^{\infty} k n_2^{4/3} r \, dr . \tag{127}
$$

This factor

$$
k = \left(\sum_i n_i + \sum_j n_j \right)^{4/3} \Bigg/ \left[\left(\sum_i n_i \right)^{4/3} + \left(\sum_j n_j \right)^{4/3} \right]
$$

varies within the limits $(0.79-1)^{-1}$ being the function of interatomic distance R. However, at $\Sigma_i n_i = \Sigma_j n_j$, when the distinction of the considered multiplier from 1 achieves maximum, the dependence on R disappeared. Then double integral (121) permits the change in order of integration, afterwards the integral is taken within the analytic functions, and the error appears to be significantly less than 20%,

$$
\mathscr{E}_{inel} = 2.68 m u \pi
$$

$$
\times \left\{ \int_{\alpha R}^{\infty} k n_1^{4/3} r \, dr [(\{r/\alpha\}^2 - R_0^2)^{1/2} - R_{min}^2/R_0 \arccos(\alpha R_0/r)] \right.
$$

$$
\left. + \int_{\alpha' R}^{\infty} k n_2^{4/3} r \, dr [(\{r/\alpha'\}^2 - R_0^2)^{1/2} - R_{min}^2/R_0 \arccos(\alpha' R_0/r)] \right\} .
$$

$$
\tag{128}
$$

Here, $\alpha' = 1 - \alpha$.

It is convenient to transform the integral to the limits $(0; 1)$,

$$\mathscr{E}_{inel} = 2.68 mu\pi R_0^2 \int_0^1 \{\alpha^2 kn_1^{4/3}(\alpha R_0/x) + \alpha'^2 kn_2^{4/3}(\alpha' R_0/x)\}$$

$$\times [R_0(1/x^2 - 1)^{1/2} - R_{min}^2/R_0 \arccos x] \, dx/x^3 .$$

$$(129)$$

There is no analytic solution of integral (129), and numerical integration is hampered by the rather slow convergence of the Taylor series of the integrand. The adaptable program based on the Newton–Cotes formula of the eighth order (Forsythe et al. 1977) has been used for the calculation. The electron flux density was calculated using the Slater wave functions, but eq. (129) permits the use of any wave functions.

For the calculation with much overlap of the shells, one could deduce an analytical expression for $\alpha(Z_1, Z_2)$, averaged over all impact parameters p.

Actually, at averaging over p we get,

$$\int_0^\infty \mathscr{E}_{inel}(p)p \, dp = 2.68 mu\pi$$

$$\times \left\{ \int_{\alpha R_{min}}^\infty kn_1^{4/3}r \, dr[(\{\tfrac{1}{3}\}\{r/\alpha\}^2 - R_{min}^2)^{3/2} \right.$$

$$- R_{min}^2(\{r/\alpha\}^2 - R_{min}^2)^{1/2} + R_{min}^3 \arccos(\alpha/rR_{min})]$$

$$+ \int_{\alpha' R_{min}}^\infty kn_2^{4/3}r \, dr[(\{\tfrac{1}{3}\}\{r/\alpha'\}^2 - R_{min}^2)^{3/2}$$

$$\left. - R_{min}^2(\{r/\alpha'\}^2 - R_{min}^2)^{1/2} + R_{min}^3 \arccos(\alpha'/rR_{min})]\right\} .$$

$$(130)$$

At deep overlapping of the shells one could consider $R_{min} = 0$. Then

$$\int_0^\infty \mathscr{E}_{inel}(p)p \, dp = 2.68 mu\pi \int_0^\infty [kn_1^{4/3}\alpha^3 + kn_2^{4/3}\alpha'^3]r^3 \, dr .$$

$$(131)$$

We ask $d/d\alpha \int_0^\infty \mathscr{E}_{inel}(p)p \, dp = 0$, i.e., we would find a minimum through α. We get

$$\alpha/(1 - \alpha) = \int_0^\infty kn_1^{4/3}r^4 \, dr \bigg/ \int_0^\infty kn_2^{4/3}r^4 \, dr .$$

$$(132)$$

The electron cloud density n could be roughly approximated by the exponent $e^{-\mu r}$. Considering that

$$\int_0^\infty n(r) r^2 \, \mathrm{d}r = eZ \, ,$$

we get

$$\alpha = 1 / [(Z_1/Z_2)^{5/12} + 1] \, . \tag{133}$$

Such an expression for α could be used for high velocities, in particular when both integral or averaged values are of interest. Due to the averaging performing over all shells, the shell structure in eq. (133) has not been exhibited in an explicit form.

The atomic shell structure is known to lead to a nonmonotonic dependence of inelastic losses on atomic number of colliding particles. The theory can explain such Z dependence as a consequence of the change of number of electrons, participating in the interaction, which is related on the one hand with outer shell filling and on the other hand with reduction of atomic volume at increase of atomic number within the shell.

In experiments (Eisen 1968) on the direction of the channel [110] of an Si single crystal, the bombardment of the ions with different atomic numbers was performed. The stopping power was determined for well-channeled ions, i.e., for ions moving along the channel center. The comparison of the shell theory with Eisen's experiment was made previously (Cheshire et al. 1968, Komarov and Kumakhov 1973, Baklitzky et al. 1973). While the location of minima and maxima in the theory coincided with the experiment, obtaining of experimentally observed energy losses with high oscillation amplitude failed.

The proposed theory due to both the refined expression for electron flux and precise consideration of the plane location provides results being more similar to experimental data, however, it failed to lead to a full agreement with the experimental results (see fig. 33). Account of the charge state of a particle passing the target along the channel results in a certain improvement. Indeed, the curve in fig. 33 exhibits the inelastic losses of the single-charged ion inside the channel. Ionized state conservation at passing of an atom with low ionization potential, like Li and Na, through the target is beyond doubt. The presence of a minimum on Na, but not on Ne, which is related to the absence of the Na outer electron, could serve as direct evidence. There is a quite different situation for atoms located at the end of a period in the periodic system chart. It would be necessary to take into account the actual charge state inside a crystal for these atoms, unfortunately, till now it seems to be impossible. One could say for general considerations that for these atoms the neutral state appears to be more probable,

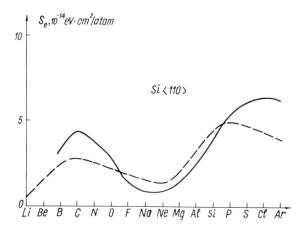

Fig. 33. The dependence of the stopping power of an Si single crystal of impinging ions under bombardment along channel $\langle 110 \rangle$: (———) experiment by Eisen (1968), (– – – – –) theory.

that leads both to the increase of inelastic losses and to better agreement with experiment. Apparently, the account of the electron redistribution at quasimolecule formation in the course of collision events can improve the agreement since the polarization leads to the largest change of the electron density in the atom outer section between the colliding nucleus, which should result in deepening of the minima on the theoretical curve in fig. 33.

The picture of a particle moving strictly along the channel axis seems to be idealized to a certain extent. In fact, the location of atoms within the crystal cell does not correspond to the exact axial symmetry, in this connection the impact parameter of a particle would vary within certain limits relative to the target atoms. The thermal vibrations of lattice atoms, neglected in fig. 33, lead to the same effect. Accordingly, it is desirable to verify the influence of the impact parameter on the shape of the inelastic energy loss curve versus the atomic number.

In fig. 34, the curves $\mathscr{E}_{inel}(Z)$ for the impact parameters from 0 to 2×10^{-8} cm are shown, which seems to be significant, because all curve shapes have been changed. In addition, in fig. 34, the theoretical curve for head-on impact, obtained using eq. (133) has been plotted for comparison.

For high velocities (10^8 cm/s) and head-on impact considering $R_0 = R_{min} \approx 0$ from eq. (128) we get

$$\mathscr{E}_{inel} = 2.68 mu\pi \left\{ \frac{1}{\alpha^2} \int_0^\infty kn_1^{4/3} r^2 \, dr + \frac{1}{\alpha^2} \int_0^\infty kn_2^{4/3} r^2 \, dr \right\}. \tag{134}$$

As previously, let us approximate the density n by $e^{-\mu r}$ considering

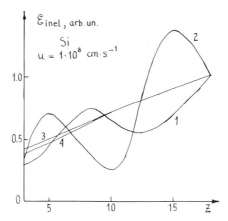

Fig. 34. Inelastic energy losses at collisions of atoms of the second and third periods of the periodic system chart with an Si atom with velocity $v = 10^8$ cm/s at different impact parameters: (1) $p = 2$ Å, (2) $p = 1$ Å, (3) $p = 0$, (4) $p = 0$ using eq. (75).

$$4\pi \int_0^\infty n(r)r^2 \, dr \simeq 4\pi \int_0^\infty e^{-\mu r} r^2 \, dr = (2!)/\mu^3 = eZ \,, \tag{135}$$

we get

$$\mathscr{E}_{inel} = 2.68 mu\pi ke(\tfrac{3}{4})^3 (Z_1/\alpha^2 + Z_2/\alpha'^2) \,. \tag{136}$$

As it is seen from fig. 33, a short deflection from the curve calculated using eq. (129) is observed only for light atoms for which the approximation $R_{min} \approx 0$ becomes rather rough.

In order to exclude the indeterminacy, connected with the processes occurring at the penetration of a bombarding particle into deep layers of a target, it is desirable to apply experimental data of interaction with a surface. Hogberg (1971) has reported experimental results on a thin polycrystalline film with variable thickness. The energy losses of the ions which passed through the film and escaped in normal direction to the surfaces were registered. The author has assumed that due to the drop of probability of multiple collisions inside the film, the emergence in the direction of bombardment, as the film thickness decreases, would be possible only at symmetric location of target atoms relative to the impinging particle. In addition, the authors considered the energy losses to be truly inelastic ones if the film thickness approaches zero.

We have calculated the energy losses for the parameters corresponding to the experiment by Hogberg (1971). The calculation results were compared with the experiment (see fig. 35). The inelastic energy losses were calculated using eq. (129), in addition, the elastic energy losses in the impulse

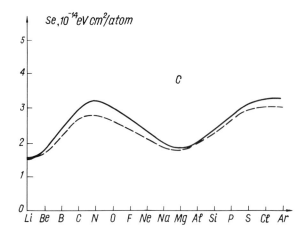

Fig. 35. The dependence of the stopping power of a super-thin amorphous carbon film on atomic number of impinging ion: full curve denotes experiment by Hogberg (1971), broken curve theory.

approximation,

$$E_{el} = (m_1/m_2)(\tfrac{1}{2}\pi)^2(3.05Z_1Z_2)^2/(Z_1^{1/2} + Z_2^{1/2})^{4/3}(Ep^4) , \qquad (137)$$

have been considered, for the heaviest atoms they were found to contribute up to 20% of the total losses.

It is worth noticing that the agreement between theory and experiment appeared to be better than the one in fig. 33. Apparently, the improved agreement between theory and experiment is caused by the fact that in the experiment in question the interaction of the impinging ions with the target occurs directly on its surface.

It should be noted that at the impact parameters typical for particles passing the crystal through the channel center or between two atoms located symmetrically in amorphous or polycrystalline targets, the estimated values of the inelastically transmitted energy obtained by eq. (129) differs highly from the ones from previous theories.

In fig. 36, the curves for the dependence of inelastically transmitted energy on the impact parameter for the original Firsov theory, the Kishinevsky formula (Kishinevsky 1962) and for eq. (129) are given. The difference in the curves at small parameters could be explained by the fact that in the original Firsov theory the trajectory distortion was not taken into account, which is the largest at small impact parameters and low velocities. The surpassing of the curves calculated by eq. (129) over the other curves within the narrow range 7×10^{-9} cm $< p < 14 \times 10^{-9}$ cm is caused by overlapping of L shells of the colliding atoms; the surpassing of these two curves over the curve calculated by eq. (129) at $p > 16 \times 10^{-9}$ could be explained by

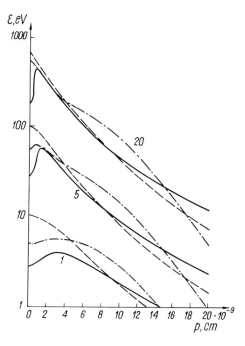

Fig. 36. The dependence of inelastic energy losses at collisions of two nitrogen atoms on the impact parameter for different collision velocities (in 10^7 cm/s): (– – – – –) according to Firsov (1959), (———) according to Kishinevsky (1962), (—·—) using eq. (71).

overvaluation of the electron density at large distances from the nucleus when using the Thomas–Fermi model.

3.2. Electron emission

3.2.1. Z_1 dependence

The coefficient of the kinetic ion–electron emission is calculated using the following formula (Parilis and Kishinevsky 1961),

$$\gamma = w/d \int_0^\infty e^{-x/\lambda} n(x) \, dx , \tag{138}$$

where w is the probability of electron escape from the target, d is the interatomic distance in a lattice, λ is the depth of electron escape, and $n(x)$ is the number of electrons knocked out at depth x. The integration to infinity is justified by the fast decrease of the integrand ($\lambda \ll D$, where D is the target thickness).

The considered dependence on the atomic number of the projectile is not observed in w at small doses but it is involved in the value $n(x)$. For $n(x)$, one can write

$$n(x) = \sum \mathscr{E}_{\text{inel}}(x) / I_{\text{mean}} . \tag{139}$$

Here, $\sum \mathscr{E}_{\text{inel}}(x)$ is the total inelastic energy released during the bombarding-particle–target-atom interaction at depth λ, and I_{mean} is the mean ionization potential.

It is evident that both numerator and denominator in eq. (139) depend on the atomic number of particles participating in the collisions.

The calculation of the coefficient of the kinetic electron emission for an Si single crystal, bombarded by atoms of the second and third periods of the periodic system chart in the channel direction $\langle 110 \rangle$ with velocity 10^8 cm/s has been made.

The mean energy inelastically transferred was calculated by the following formula,

$$\bar{\mathscr{E}}_{\text{inel}} = 2\pi \int_0^\infty \mathscr{E}_{\text{inel}}(p) p \, dp / d^2 . \tag{140}$$

This integral has been considered before (eq. 130), the interatomic distance d was calculated for the plane parallel to the surface being located on the depth x.

The particle entrance into the channel has been considered using the Martynenko shadow model, and the shadow radius has been calculated using eq. (75). Due to the shadowing of the crystal atoms lying below, the averaging of the inelastically transferred energy using eq. (139) at the lower integration limit has to be taken equal to $p_{\text{min}} = \sqrt{R_T^2 - R_{\text{min}}^2}$.

For the mean energy inelastically transferred at the interaction of the bombarding particle with each of the lattice atoms located within the lattice cell in the considered plane, one can get

$$\bar{\mathscr{E}}_{\text{inel}} = \int_{(R_T^2 - R_{\text{min}}^2)^{1/2}}^\infty \mathscr{E}_{\text{inel}}(p) p \, dp / d^2$$

$$= 2.68 mu\pi / d^2 \left\{ \int_{\alpha R_T}^\infty kn_1^{4/3}(r) r \, dr \left[(\{\tfrac{1}{3}\} \{r/\alpha\}^2 - R_T^2)^{3/2} \right. \right.$$

$$- R_T^2 (\{r/\alpha\}^2 - R_T^2)^{1/2} + R_{\text{min}}^2 R_T \arccos(\alpha/rR_T)]$$

$$+ \int_{\alpha' R_T}^\infty kn_2^{4/3}(r) r \, dr \left[(\{\tfrac{1}{3}\} \{r/\alpha'\}^2 - R_T^2)^{3/2} - R_T^2 (\{r/\alpha'\}^2 - R_T^2)^{1/2} \right.$$

$$\left. \left. + R_{\text{min}}^2 R_T \arccos(\alpha'/rR_T)] \right\} . \tag{141}$$

The results of the calculations are shown in fig. 37. In addition, in this figure the coefficient of the kinetic electron emission from the back side of a thin Si single-crystal film versus atomic number has been displayed for two cases. In the first case (curve 2), the film was considered to be so thin that the decrease of the kinetic energy of the bombarding particles moving along the channel center ($D = 5 \times 10^{-6}$ cm) could be neglected. At the same time, at such target thickness the ions which did not get into the channel center were hampered due to intensive both elastic and inelastic interactions with the target atoms and did not reach the back side of the film. This allows us to calculate the inelastically transferred energy using formula (129). The channel radius being equal to 2×10^{-8} cm is considered as the distance of closest approach.

In the second case (curve 3), the target was considerably thicker ($D = 10^{-3}$ cm). At such target thickness even well-channeled ions were near their stopping point, in this case their residual energy and thereby their inelastic energy losses significantly depend on the total energy losses along the previous part of the trajectory. Owing to the fact that the energy loss of the practically well-channeled ions is an exceptionally inelastic one, and these losses depend on the atomic number of the bombarding particle (curve 2) the smoothing of maxima and minima takes place.

3.2.2. Emission from thin films

Some homogeneous metallic and nonmetallic films with thickness down to 40 Å could be shot through ions with relatively low energies ($u \geqslant 5 \times 10^6$ cm/s). The ions passing through the film cause on this back side the same processes as on the front side including the emission of electrons.

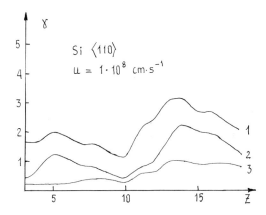

Fig. 37. The dependence of the coefficient of the kinetic ion–electron emission from the front and the back side of a single-crystal Si film on the atomic number of the bombarding particles from films with different thickness: (1) front side, (2) back side, $D = 5 \times 10^{-6}$ cm, (3) back side, $D = 10^{-3}$ cm.

Thus, the determination of the coefficient of the kinetic electron emission from the back side of a thin film could be divided in the two following stages (Baklitzky et al. 1975):

(i) the determination of the number of electrons $n(x)$, knocked out at depth x during the collision with target atoms due to the total ion energy losses from the surface to the collision point; and

(ii) the determination of the fraction of the electrons able to reach the surface, with energy exceeding the work function.

At passing through the crystal the ion beam is gradually disordered and escaping ions have both a spread in energy and escape angle, which cause some second-order corrections. Indeed, the dependence of the coefficient of the secondary electron emission γ on the angle is given by the following relation (here, φ is the angle between the beam direction and the normal to the target surface),

$$\gamma = \gamma_0 \cos^{-1} \varphi \,. \tag{142}$$

At small correction $\Delta\varphi$ for the angle φ, one can get

$$\gamma = \gamma_0 \cos^{-1} \varphi = \gamma_0 \{1 + \tfrac{1}{2}(\varphi + \Delta\varphi)^2 - \cdots\}$$
$$= \gamma_0 (1 + \tfrac{1}{2}\varphi^2 + \varphi\,\Delta\varphi + \cdots) \,. \tag{143}$$

As follows from eqs. (138) and (139), the coefficient could be considered to be proportional to the elastic energy losses (taking into account that λ is small) at small energy spread. In this turn, within the limits of the increased loss section with the velocities $u < 2 \times 10^8$ cm/s, the inelastic energy losses depend nearly linearly on the collision velocity, which results in first-order correction compensation at the energy spread of the bombarding particles. Thus, with velocity $u < 2 \times 10^8$ cm/s when calculating the coefficient γ from the back side of a film, one can use a mean value of the bombarding particle velocity and neglect the angular spread.

To determine $n(x)$, eq. (139), the inelastic energy losses on the back side of a film are needed to be calculated. To this end the total energy losses at passing through the film are necessary to be estimated. For this purpose, the equation,

$$\mathrm{d}(E_{el} + \mathscr{E}_{inel})/\mathrm{d}x = f(x) \,, \tag{144}$$

should be integrated over the film thickness.

One can consider that

$$\frac{\mathrm{d}\mathscr{E}_{inel}}{\mathrm{d}x} = \frac{\mathscr{E}_{inel}}{d} \,, \quad \text{and} \quad \frac{\mathrm{d}E_{el}}{\mathrm{d}x} = \frac{E_{el}}{d} \,, \tag{145}$$

where d is the interatomic distance. The solution of eq. (144) could be gotten by using eq. (142) for inelastic losses, and the following expression

(Firsov 1958),

$$E = 1.4\pi a e^2 \frac{Z_1 Z_2}{\psi} \frac{m_1}{m_1 + m_2} \frac{1}{d^2} \frac{\ln(1 + 2Bu^2)}{2Bu^2} ,$$

(146)

$$B = \frac{0.245\mu a}{1.4 Z_1 Z_2 \psi e^2} , \quad \psi = (\sqrt{Z_1} + \sqrt{Z_2})^{2/3} ,$$

for the elastic losses, where a is the Thomas–Fermi radius, e is the electron charge, and $\mu = m_1 m_2/(m_1 + m_2)$. In the general case, eq. (144) is integrated only numerically.

The calculations made for a great number of the colliding pairs as well as the tabulated data (Northcliffe and Schilling 1970) have shown that within a wide velocity range the sum of both elastic and inelastic losses slightly depends on ion velocity, therefore, the bombarding particle energy decreases linearly as the particle penetrates into the target. In its turn, $n(x)$ could be considered approximately linear over u,

$$n(x) = c_1(u - u_{thr}) ,$$

(147)

where u_{thr} is the threshold velocity at which the inelastically transferred energy is too low for inducing ionization.

The coefficient of kinetic electron emission from the front side of a film, i.e., from the side on which the beam of the bombarding particles is directed, according to eqs. (138) and (147), is written as

$$\gamma_{fr} = w/d \int_0^\infty e^{x/\lambda} n(x) \, dx ,$$

(148)

the coefficient from the back side of a film being written as

$$\gamma_{back} = w/d \int_0^\infty e^{(-D+x)/\lambda} n(x) \, dx .$$

(149)

By intending that the total energy loss is constant, $dE/dx = c$, eq. (149) can be written as

$$\gamma_{back} = w/d \frac{mc_1}{d} \int_0^\infty \exp(-m/\{2c\}[u^2 - u_1^2]/\lambda)(u - u_{thr})u \, du ,$$

(150)

where u_1 is the velocity with which the ions escape the film after passing through it.

If one introduces some additional designations,

$$k = u_1^2 m/(2c) , \quad a = c_1/\alpha , \quad \alpha^2 = m/(2c) ,$$

(151)

then the integration result would be written in the following compact form,

$$\gamma_{back} = w/da\sqrt{k}\{1 + \tfrac{1}{2}\sqrt{\pi}\exp(k/\lambda)\sqrt{\lambda/k}[1 - \Phi(\sqrt{k/\lambda})] - u_{thr}/(\alpha\sqrt{k})\}.$$

(152)

The introduced parameters have the following physical meaning: k is the path which the ions might pass inside a target with primary velocity being equal to the one on the back side of the film, i.e., the ion energy on the back side of a film expressed in path units; α is the coefficient which recalculates the square of ion velocity per unit path, and $\Phi(\sqrt{k/\lambda})$ is the probability integral.

When taking into account that λ is small, it is reasonable to expand eq. (152) in a series over λ/k,

$$\gamma_{back} = w\lambda a/d(\sqrt{k} - au_{thr}/\alpha) = w\lambda a/d(\sqrt{u_0^2/\alpha^2 - D} - u_{thr}/\alpha). \quad (153)$$

For emission from the front side of a film in the same approximation one can get

$$\gamma_{fr} = w\lambda a/d[(u_0 - u_{thr})/\alpha]. \quad (154)$$

From eqs. (153) and (154), the expression for the ratio of emission from the back side of a film to emission from the front side, one can derive

$$\frac{\gamma_{back}}{\gamma_{fr}} = \frac{\sqrt{u_0^2/\alpha^2 - D} - u_{thr}/\alpha}{(u_0 - u_{thr})/\alpha}. \quad (155)$$

The derived formulae were used in the investigation of the emission from an Si polycrystalline film. The results obtained are shown in figs. 37 and 38. The value λ according to Bronstein and Freiman (1969) was taken equal to ten monoatomic layers, and the probability of electron escape was taken equal to $w = 0.2$.

The graphs have shown that a sharp decrease in the emission from the back side of a film occurred only at significant decrease of the velocity of the escaping ions. The ratio $\gamma_{back}/\gamma_{fr}$ increases at increasing velocity and approaches 1 at high velocities.

The experiments by Lichtenstein and Tankov (1975) have revealed that the general behavior of the experimental curves could be explained by eq. (155). However, an important difference was found: in the experiment the ratio $\gamma_{back}/\gamma_{fr}$ exceeded 1 and approached 2 at high velocities. Thus, the experiment has shown that at high velocities more electrons escaped from the back side of a film than from the front, regardless of the energy decrease.

For the coefficient of the kinetic electron emission both from the front and back sides, eqs. (148) and (149) are valid, being different in the probability of electron emergence from the solid.

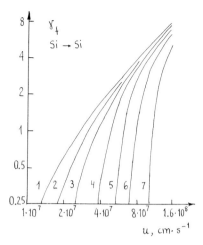

Fig. 38. The dependence of the kinetic electron emission from the back side of Si film on the velocity of impinging Si ion for films with different thickness: (1) γ_{fr}, (2) 100 Å, (3) 200 Å, (4) 400 Å, (5) 800 Å, (6) 1600 Å, (7) 3300 Å.

If $f(u)$ defines the electron velocity distribution in a center-of-mass system where the distribution is an isotropic one, then the probability of an electron emergence for the front and back sides of a film are given by

$$w_{\pm} = 2\pi \int_0^{\pi/2} \sin\theta \, d\theta \int_{v_{\pm}}^{\infty} v^2 f(v) \, dv \Big/ 4\pi \int_0^{\infty} v^2 f(v) \, dv , \qquad (156)$$

where

$$v_{\pm} = \left(\frac{\sqrt{2\varphi}}{m_e} \pm \frac{m_1}{m_1 + m_2} u \right) \Big/ \cos\theta ,$$

where the index $(+)$ is used for the front side of the film, and the index $(-)$ for the back one.

Here u is the relative velocity of the colliding particles, v is the velocity in the center-of-mass system, θ is the angle between v and the normal to the surface, m_e is the mass of electron, and φ is the work function.

Thus, the ratio of the total number of electrons escaping the target to the number of all electrons formed during the collision in the velocity space would be equal to the ratio of the number of electrons in half-space, which is cut-off by a plane located at a distance $(\sqrt{2\varphi}/m_e) \pm [m_1/(m_1 + m_2)]u$ from the center of electron distribution to the number of electrons in the whole velocity space, as it is written in eq. (156).

The function $f(v)$ is approximated by the following expression,

$$f(v) = c \exp(-\mu v) , \qquad (157)$$

in this case, parameters c and μ are determined from the experiments by Colombie et al. (1973), Louchet et al. (1972) and Evdokimov and Molchanov (1969).

One can get the ratio of emission coefficients from the back side and the front side,

$$\gamma_{back}/\gamma_{fr} = w_- \int_0^\infty e^{(-D-x)/\lambda} \, u(x) \, dx \bigg/ w_+ \int_0^\infty e^{-x/\lambda} \, u(x) \, dx \, . \tag{158}$$

The ratio of the integrals entered in eq. (158) in the approximation of small λ has been calculated in eq. (155). It appeared to be equal to the ratio of the number of electrons knocked out from the back and front sides of a film, respectively. As is seen from eq. (158), it is necessary to multiply this ratio by the corresponding ratio of probabilities of electron emergence, for which one can get

$$w_-/w_+ = \int_0^{\pi/2} \sin\theta \, d\theta \int_{v_-}^\infty v^2 \, e^{-\mu v} \, dv \bigg/ \int_0^{\pi/2} \sin\theta \, d\theta \int_{v_+}^\infty v^2 \, e^{-\mu v} \, dv \, , \tag{159}$$

in this case the parameter c is obtained from further calculations.

The integrals in eq. (159) are expressed by analytical functions, and for the ratio of the coefficients of the kinetic electron emission one can get

$$\frac{\gamma_{back}}{\gamma_{fr}} = \frac{n(D)}{n(0)} \frac{1 + \frac{1}{2}\mu v_-}{1 + \frac{1}{2}\mu v_+} \exp[-(v_- + v_+)] \, . \tag{160}$$

The parameter μ could be deduced from the spectra of electrons emitted from the front side of a film. At experimental measurement of the electron spectra not the radial velocity component in center-of-mass system is directly measured, but only the normal velocity component of emitted electrons is measured.

Thus, the following distribution is experimentally registered,

$$N(\omega) = c_1 \int_{v_+ + \omega}^\infty v \, dv \, e^{-\mu v} = c_1/\mu^2[1 + \mu(v_+ + \omega)] \exp[-\mu(v_+ + \omega)] \, , \tag{161}$$

where ω is the normal velocity component of electrons in the laboratory system outside a target, and c_1 is a constant.

The ratio of the number of electrons for two different values of the normal velocity component equals

$$\frac{N(\omega_1)}{N(\omega_2)} = \frac{1 + \mu(v_+ + \omega_1)}{1 + \mu(v_+ + \omega_2)} \exp[-\mu(\omega_1 - \omega_2)] \, . \tag{162}$$

Determination of μ from this relation is possible, e.g., by using the recursion formula,

$$\mu_{\nu+1} = \frac{1}{\omega_1 - \omega_2} \ln \frac{[1 + \mu_\nu(v_+ + \omega_1)]N(\omega_2)}{[1 + \mu_\nu(v_+ + \omega_2)]N(\omega_1)},$$ (163)

where the value obtained for large ω ($\omega \gg v_+$; $\mu\omega \gg 1$) could be used as a first approximation,

$$\mu_1 = \frac{1}{\omega_1 - \omega_2} \ln \frac{\omega_1 N(\omega_2)}{\omega_2 N(\omega_1)}.$$ (164)

To compare with the experiment by Lichtenstein et al. (1978), the ratio $\gamma_{back}/\gamma_{fr}$ for $H^+ \to C$ has been calculated at velocities from 10^7 to 5×10^8 cm/s (see fig. 39). In reality, this ratio becomes more than one at high velocities. It is seen that the dependence on velocity for both experimental and theoretical curves is similar, but the dependencies on the thickness of the shot film are different.

To determine the dependence of the ratio $\gamma_{back}/\gamma_{fr}$ on the mass of an impinging ion these ratios have been calculated for He^+, N^+ and O^+ for a velocity of 6×10^7 cm/s. By comparing with the experimental data one can obtain the values of $\gamma_{back}/\gamma_{fr}$ as shown in table 1.

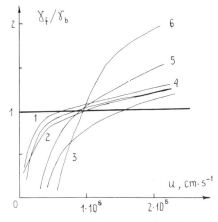

Fig. 39. The dependence $\gamma_{back}/\gamma_{fr}$ on ion velocity for $H^+ \to C$; theory (1) for film thickness 50 Å, (2) 100 Å, (3) 500 Å; experiment: (4) 40 Å, (5) 70 Å, (6) 500 Å.

Table 1
Values of $\gamma_{back}/\gamma_{fr}$.

	Ion		
	He^+	N^+	O^+
Experiment	1.22	1.45	1.97
Theory	1.25	1.89	1.97

Maximal disagreement between theory and experiment equals 30%. This is likely caused by the fact that for all colliding pairs the same value of the parameter μ, corresponding to He^+, was used.

For general reasons at the increase of atomic number of the colliding particles the contribution of high-energy electrons should increase, which should result in both the decrease of the parameter μ and the improvement of the agreement between theory and experiment.

It should be noted that the dependence of the ratio $\gamma_{back}/\gamma_{fr}$ on the ion mass is not a result of the dependence of the coefficient γ on the mass known from the experiments and theory (Parilis and Kishinevsky 1961). This dependence appears in eq. (160) via the values $n(D)$ and $n(0)$. As it is seen from this formula, in a first approximation the dependence does not influence the ratio of the coefficients. The observed increase of the ratio $\gamma_{back}/\gamma_{fr}$ with the increase of the mass of the impinging ion appeared to be a result of the anisotropy of the velocity of the emitted electrons under additional momentum acquired in the center-of-mass of the colliding particles.

The influence of the direction of the center-of-mass momentum on the coefficient of the kinetic electron emission is manifested in the γ dependence on the angle of incidence ψ of the impinging particles. At glancing incidence, the ion velocity component which is directed along the normal inside the front side of the film decreases. This leads to the decrease of value of v_+ in eq. (159),

$$v_+ = \left(\frac{\sqrt{2\varphi}}{m_e} + \frac{m_1}{m_1 + m_2} u \cos \psi \right) \Big/ \cos \theta . \tag{165}$$

In this case, the dependence $\gamma(\psi)$ on the angle of incidence becomes sharper than $1/\cos \psi$ (see fig. 40).

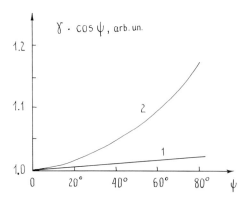

Fig. 40. The dependence of $\gamma_{fr} \cos \psi$ on the angle ψ for different velocities: (1) 10^7 cm/s, (2) 10^8 cm/s.

The results mentioned have shown that a simple kinematic effect consisting in anisotropy of the secondary electron distribution in the laboratory system connected with the isotropy of distribution of electrons, emitted during the collisions in the center-of-mass system, allows to obtain the excess $\gamma_{back}/\gamma_{fr}$ over one and its dependence on mass, which was not previously accounted for by theory.

The model described proved to be correct at comparison with the experimental spectra of the emitted electrons.

The dependence of the registrated coefficient γ on the retarding potential applied to the detector could be obtained experimentally. According to ion velocity, such dependence is described by eq. (161) for the emission from the front side of a film, and it is described by the analytical formula, where v_+ was substituted by v_- for the emission from the back side of the film. Thus, from theory it follows that the experimental dependence for the back side of a film should be the same as for the one for the front side if the first of the mentioned curves is shifted along the longitudinal axis, along which the retarding potential is plotted by $m\{[m_1/(m_1 + m_2)]u\}^2$.

The estimation of the curve shift has been made for the experiment by Meckbach et al. (1975) with the following data: $N^+ \rightarrow C$, ion energy $E = 257$ keV. The calculated shift is 1.04. At low value of the retarding potential, the theoretical value of the shift is close to the observed one.

The dependence of the ratio $\gamma_{back}/\gamma_{fr}$ on the film thickness at high velocities remained unexplained. None of the mechanisms considered could explain the increase of kinetic electron emission from the back side of a film with the increase of the film thickness. In this connection, the properties of film structure were involved to explain this effect (Baklitzky and Parilis 1982).

Inelastic energy losses depend very sharply on the impact parameter. Such peculiarity provides a high sensitivity of the inelastic energy losses to the disarrangement of the single-crystal structure. Ions with velocity 2×10^8 cm/s directed along the channel are able to shoot through a film with thickness of some thousand angstroms.

From the intensity of the emission induced by a channeled ion from the back side of the single-crystal film, one can get directly the information about the change in the crystal structure of the near-surface layer. In such a way one can, e.g., observe the epitaxial film growth by shooting through the film, e.g., with ions of the target substance.

The coefficient of the kinetic electron emission is given by eq. (138). Due to multiple ionization of the particle at large values of inelastic energy losses, the number of electrons knocked out is not proportional to the energy transferred inelastically, but rises much more slowly.

In table 2, the dependence of electron emission on lattice structure disarrangement is presented. The structure disarrangement has been simu-

Table 2
The dependence of the kinetic electron emission on lattice structure disarrangement.

Distance of the displaced atom from channel center (Å)	Coefficient of ion–electron emission versus ion velocity (cm/s)		
	5×10^7	10^8	2×10^8
0	8.81	15.31	25.22
0.2	8.65	13.47	20.27
0.5	5.22	7.99	12.32
1.0	2.88	4.49	7.28
2.0	1.43	2.14	3.82

lated by the displacement of one of the target atoms toward the channel center.

It is seen from table 2 that the displacement of even one of the eight cell atoms leads to a significant increase of the coefficient (by a factor of eight). It allows to observe the structure of the near-surface film layer using ion bombardment. Thus, by using the ions of the target substance as bombarding particles, one can observe the violence of the proper crystal structure during epitaxial film growth.

It is also seen that in the absence of crystal structure disarrangement (which corresponds to a distance of the atom from the channel axis equal to 2 Å) the values of the coefficient are small. The calculations using eq. (139) with account of eq. (129) have revealed that at the decrease of velocity down to 1.8×10^7 cm/s the electron emission terminates, i.e., for an ideal crystal structure this velocity is the threshold one. In this connection at proper fit of the velocity one can set-up such measuring conditions at which the appearance of the single-crystal structure disarrangement in the near-surface layer would be signaled by the emergence of the kinetic electron emission that is missing for the proper structure.

3.2.3. Charge dependence of kinetic electron emission

When considering the charge dependence of the coefficient of the kinetic electron emission it is reasonable to divide the velocity range of the bombarding particles, at which the kinetic emission can be observed, into three regions:

(i) 5×10^6–2×10^8 cm/s, where the Firsov mechanism is valid,

(ii) 10^9 cm/s and higher, where the Bethe–Bloch formula is used,

(iii) the intermediate region from 2×10^8 to 10^9 cm/s.

According to the Firsov model, the inelastically transferred energy is independent of charge up to velocities of 2×10^8 cm/s. The density of the electron cloud is indirectly determined by charge. The modified Firsov theory allows us to calculate the above dependence of the inelastic energy loss on charge using eqs. (129) and (131), and thereby the analogous dependence for the coefficient γ using eqs. (138) and (139).

At velocities allowing the use of the *electron friction* mechanism, the charge dependence is weak and leads to the decrease of the inelastically transferred energy with the increase of the charge of the colliding particles and, therefore, the coefficient of the kinetic ion–electron emission.

For velocities higher than 10^9 cm/s, the coefficient is calculated using the Sternglass formula (Sternglass 1957), which contains the squared charge. Parilis (1973) has made calculations of interest for multiple-charged ions with high velocities.

The charge dependence in the intermediate velocity range is of special interest since in this region the Firsov mechanism of ionization transfers into the Bete–Block mechanism, which displays the small square dependence on the charge.

When swift atomic particles pass through a solid, electron shell stripping occurs, proceeding until equilibrium with the neutralization processes comes. In this case, the equilibrium charge is reached, depending on the velocity of the impinging ion and its atomic number, but not on the ion mass. The charge reached could be determined using the following formula (Nikolaev and Dmitriev 1968),

$$\frac{\bar{q}}{Z} = \left[1 + \left\{ \frac{u}{u'Z^{0.45}} \right\}^{-1/k} \right]^{-k}, \tag{166}$$

where $k = 0.6$ is the universal constant, $u' = 3.6 \times 10^8$ cm/s, Z is the atomic number of the impinging ion, and $\bar{q}$ is an equilibrium charge.

When the equilibrium charge inside the target is reached, more than half of all electrons can be detached at velocities higher than 10^9 cm/s. This leads to a significant increase of the coefficient γ in comparison with the one for single-charged ions.

The results of the calculation of the coefficient γ for $Ar^+ \rightarrow Cu$ are shown in fig. 41. Curve 1 displays the dependence of the coefficient γ for a single-charged ion, i.e., for an ion impinging the front side of a target. Curve 2 corresponds to the electron emission from the back side of a thin polycrystalline Cu film, where the ion charge corresponds to the charge reached according to eq. (166).

Because one fails to deduce from first principles the charge dependence of the inelastic energy losses in the intermediate region, the joining of the two regions by an approximative formula transferring into the Sternglass formula at high velocities and into the formula which corresponds to the Firsov model at low velocities is needed,

$$\gamma = \frac{a(u - u_{thr})}{1 + b(u - u_{thr})^2}. \tag{167}$$

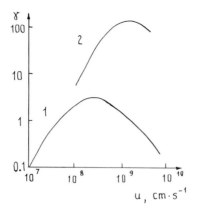

Fig. 41. The dependence of the coefficient of kinetic electron emission on the velocity of Ar^+ impinging on a thin Cu film: (1) for the front side of the film, (2) for the back side.

From the comparison with the above-mentioned theories the coefficients are written as

$$a = 7.35\lambda/(dI_{mean})\alpha_0\hbar\psi(Z) , \tag{168}$$

$$b = \frac{I_{mean}\alpha\sigma\sqrt{I_0M/(2Z_2)}}{\tau A\pi e\bar{q}^2} , \tag{169}$$

where α_0 is the Bohr constant, $\psi(Z) = (\sqrt{Z_1} + \sqrt{Z_2})^{2/3}$, α is the cross section of secondary electron scattering, σ is the covalent cross section of an atom of the solid, Z_2 is the atomic number of the target, τ is the coefficient of transparency of the secondary electrons through the surface, and I_0 is the Rydberg energy.

4. Orientational effects in ion-induced Auger electron emission

4.1. Introduction

In the early experiments by Louchet et al. (1972) and Viel et al. (1971), a fine structure in the spectra of the kinetic ion–electron emission consisting of a number of peaks with a width from 0.5 to 1 eV has been recorded. In the experiments by Hennequin and Viaris de Lesegno (1974), Viaris de Lesegno et al. (1974) and Viel et al. (1976), using more perfect methods, the main results of the earlier experiments have been confirmed, and the high-energy peaks have been explained by the Auger transitions after inner shell ionization in ion–atom collisions.

According to the theory by Parilis and Kishinevsky (1961), both the high-

and low-energy parts of the spectra are caused by Auger effect, respectively, in the inner and outer shells. It is necessary to note that all available experimental data refer to polycrystalline targets. The single-crystal ordered structure affects many processes under ion bombardment of the solids. For instance, the coefficient of the kinetic ion–electron emission depends on the ion beam orientation relatively to the crystal axis (Mashkova et al. 1963a).

This phenomenon has been explained by the screening of the lower atomic layers by the upper ones and a calculation based on the theory by Parilis and Kishinevsky (1961) has been given by Martynenko (1966a). The anisotropy of the Auger electron emission from a single crystal under ion and electron bombardment caused by the electron escape anisotropy has been described by C.-C. Chang (1971), Von Rock (1970), Mischler et al. (1979), Aberdam et al. (1978) and Gomoyunova et al. (1978a,b).

Our aim is to investigate the possibility of the orientational effects in the Auger electron emission from single crystals under ion bombardment due to different electron excitations for various orientations of the beam.

4.2. General theory

The Auger electron emission coefficient (the number of Auger electrons per incident ion) for the Auger transition $X \to YZ$ can be presented as

$$\gamma_{X \to YZ}(\varphi) = \sum_i \frac{Nd}{\cos \varphi} \, \sigma_i(\varphi) \omega_a P(E) \exp\left[-\frac{l_i}{\lambda(E)}\right], \tag{170}$$

where φ is the angle of incidence (fig. 42), N the atomic density, $\sigma_i(\varphi)$ the cross section of the vacancy formation in the X shell of an atom in the ith crystal layer, ω_a the probability of an Auger transition, E the Auger electron

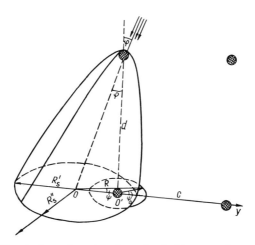

Fig. 42. The shadow of an atom from the first layer on the second layer.

energy, $P(E)$ the probability of the escape depth of an Auger electron in a metal, l_i the ith layer depth, and d the interlayer distance.

The cross section can be determined as

$$\sigma_i(\varphi) = \int_0^{2\pi} d\psi_0 \int_0^{\infty} P_x(\rho, E_0, Z_1, Z_2) I_i[R(\rho, \psi_0, \varphi)] \rho \, d\rho \,, \tag{171}$$

where $P_x(\rho, E_0, Z_1, Z_2)$ is the X shell ionization probability as a function of the impact parameter ρ, the incident ion energy E_0, the ion atomic number Z_1, and the metal atomic number Z_2; ψ_0 and ψ are the azimuthal angles. $I_i[R(\rho, \psi_0, \varphi)]$ is the dimensionless function of the bombarding flux density redistribution on the ith layer, and R is the distance from the epicentre (fig. 42).

In the coordinates (ρ, ψ_0)

$$R^2 = \rho^2 + d^2 \tan^2\varphi + 2\rho d \tan\varphi \cos\psi_0 \,,$$

and in the coordinates (R, ψ)

$$\rho^2 = R^2 + d^2 \tan^2\varphi - 2Rd \tan\varphi \cos\psi \,.$$

For the first layer ($i = 1$), the function $I_1[R] = 1$. As a result of the ion scattering on the upper layers, the initial flux density is redistributed and the atoms of the lower layers are completely or partially shaded. For the potential $V = V_0 r^{-2}$ which is applied for heavy-ion–atom collisions in the energy range from 1 to 100 keV, the radius of the shade at distance d behind the atom is equal to

$$R_s = 1.5[\pi d R_0^2/(1 + \mu)]^{1/3} \,,$$

where R_0,

$$R_0 = \left[\frac{3.05 Z_1 Z_2 \times 10^{-16}\ \mathrm{eV\ cm}^2}{E_0 (Z_1^{1/2} + Z_2^{1/2})^{2/3}} \right]^{1/2} \,,$$

is the distance of closest approach (Martynenko 1964). $\mu = M_1/M_2$, M_1 is the ion mass, and M_2 the metal atom mass. For light ions, $\mu \ll 1$ and the scattering can be taken as Rutherford scattering with the potential $V = V_0 r^{-1}$ and $R_s = 2(R_0 d)^{1/2}$, where $R_0 = (Z_1 Z_2 e^2/E_0)$ cm.

The values of the distance of closest approach R_0 and the shadow radius R_s for different E_0 and μ are given in table 3. The shape of the shadow as function of the angle φ in the plane of incidence for the $V = V_0 r^{-1}$ potential is determined by the equations

$$y_s(\varphi) = 2R_0 \frac{\sin\varphi}{\cos^2\varphi} \pm \left(\frac{4R_0 d}{\cos^3\varphi} \right)^{1/2} \,,$$

$$x_s(\varphi) = \pm 2 \left(\frac{R_0 d}{\cos\varphi} \right)^{1/2} \,. \tag{172}$$

Table 3
Distances of closest approach and shadow radii.

E_0 (keV)	$H^+ \to Al$ ($\mu \approx 0.08$)		$Ar \to Al$ ($\mu \approx 1.48$)	
	R_0 (10^{-11} cm)	R_s (10^{-9} cm)	R_0 (10^{-10} cm)	R_s (10^{-9} cm)
30	6.24	3.17	7.76	4.69
60	3.12	2.24	5.48	3.74
80	2.33	1.92	4.75	3.39
90	2.08	1.83	4.48	3.25
100	1.87	1.74	4.25	2.09

The shape of this shadow is an oval with major semiaxes

$$R_s'(\varphi) = \pm 2\left(\frac{R_0 d}{\cos^3 \varphi}\right)^{1/2} + 2R_0 \frac{\sin \varphi}{\cos^2 \varphi} ,$$

and minor semiaxes

$$R_s''(\varphi) = \pm 2\left(\frac{R_0 d}{\cos \varphi}\right)^{1/2} .$$

At the second layer ($i = 2$), the function $I_2(R)$ has for the $V = V_0 r^{-2}$ potential the form (Martynenko 1964)

$$\begin{aligned} I_2(R) &= 0 , & R < R_s , \\ &= [(1 + \tfrac{1}{2}R_s^3/R^3)(1 - R_s^3/R^3)]^{-1} , & R > R_s , \end{aligned} \tag{173}$$

and for the $V = V_0 r^{-1}$ potential (Lindhard 1969)

$$\begin{aligned} I_2(R) &= 0 , & R < R_s , \\ &= \tfrac{1}{2}[(1 - R_s^2/R^2)^{1/2} + (1 - R_s^2/R^2)^{-1/2}] , & R > R_s , \end{aligned} \tag{174}$$

i.e., it contains a shadow $I_2(R) = 0$ for $R < R_s$, a bright ring $I_2(R) \to \infty$ for $R = R_s$, and the initial flux density $I_2(R) = 1$ for $R \ll R_s$. Omitting backscattering, the normalization equation for $I_2(R)$ is

$$\pi \int_0^\infty (I_2 - 1) R \, dR = 0 . \tag{175}$$

The function $I_3(R)$ has a similar form which contains the shadows from both the first and the second layers.

The dependence of the K and L shell ionization probability $P_x(\rho)$ on the impact parameter was studied by Briggs and Taulbjerg (1975), Kochboch (1976) and Weber and Bell (1976). It is convenient to use the empirical linear function

$$P_x(\rho) = P(0)(1 - \rho/\rho_m) , \tag{176}$$

where ρ_m is the maximal impact parameter of the ionization collision.

The probability that electrons with energy E are passing over the surface potential barrier is (Parilis 1969a)

$$P(E) = 0, \qquad\qquad\qquad\qquad E < E_F,$$
$$= \tfrac{1}{2}[1 - (E_F/E)^{1/2}], \quad E > E_F, \qquad (177)$$

where E_F is the Fermi energy, and $P(E) \approx \tfrac{1}{2}$ when $E \gg E_F$.

The recombination of an X vacancy is accompanied either by the emission of an Auger electron via $X \to YZ$ transition, or by the radiation of a quantum of the X series. The Auger electron yield ω_a is connected with the fluorescence yield ω_f by the equation $\omega_a + \omega_f = 1$. The most appropriate form of the function $\omega_f(Z_2)$ is determined by

$$\omega_f/(1 - \omega_f)^{1/4} = A + BZ_2 + CZ_2^3, \qquad (178)$$

where $A_K = -6.4 \times 10^{-2}$, $B_K = 3.4 \times 10^{-2}$ and $C_K = 1.03 \times 10^{-6}$ for the K shell and $A_L = 0.132$, $B_L = 0.0092$ and $C_L = 0$ for the L shell.

4.3. Orientational effects

The Auger electron emission coefficient calculated using formulas (170)–(178) should display some orientational effects consisting in the non-monotonous change of $\gamma_{X \to YZ}(\varphi)$,

$$\gamma_{X \to YZ}(\varphi) = \sum_i [\gamma_{X \to YZ}(\varphi)]_i,$$

with incident angle. This is caused by the continuous change of the shading of the lower layers by the upper layers (fig. 42).

As is seen in figs. 43 and 44, the minima of $\gamma_{X \to YZ}(\varphi)$ correspond to such incident angles that the direction of the bombarding beam coincides with some crystal axis. By these orientations, the close ion–atom collisions are avoided and the $X \to YZ$ Auger yield $\gamma_{X \to YZ}(\varphi)$ reaches its minima.

The behavior of the function $\gamma_{X \to YZ}(\varphi)$ is determined by the relation between R_s and ρ_m. Two cases are possible.

(i) $R_s > \rho_m$. For this case the cross section determined by the general form in eq. (171) has the form

$$\sigma(\varphi) = 0, \qquad\qquad\qquad\qquad\qquad\qquad\qquad \delta \leq R_s \rho_m,$$
$$= \tfrac{1}{2}\sigma_{max}[(\delta - R_s + \rho_m)/\rho_m]^2, \qquad\qquad R_s - \rho_m \leq \delta \leq \rho_m,$$
$$= \delta_{max}(\delta - R_s + \rho_m)/2\rho_m, \qquad\qquad R_s \leq \delta \leq R_s + \rho_m, \quad (179)$$
$$= P(0)\left\{ \tfrac{1}{3}\pi\rho_m^2 + \tfrac{1}{4}\pi R^4\left[\frac{1}{4(\delta^2 - \rho_m^2)} - \frac{1}{8\rho_m\delta} \ln\left(\frac{\delta + \rho_m}{\delta - \rho_m}\right) \right] \right\}, \quad \delta \geq R_s + \rho_m,$$

where $\delta = d \tan \varphi$ (fig. 42) and

$$\sigma_{\max}(\varphi) = P(0)\left\{ \tfrac{1}{3}\pi\rho_m^2 + \tfrac{1}{4}\pi R_s^4 \left[\frac{1}{4(R_s^2 + 2R_s\rho_m)} \right. \right.$$
$$\left. \left. - \frac{1}{8(R_s + \rho_m)\rho_m} \ln\left(1 + \frac{2\rho_m}{R_s}\right) \right] \right\}.$$

The atoms of the second and following layers are completely shaded at $\varphi = 0$. Beginning with the angle $\varphi_1 = \arctan\{[R_s(\varphi) - \rho_m]/d\}$, the Auger

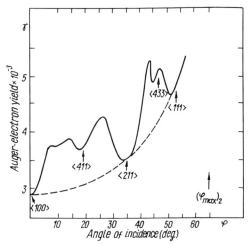

Fig. 43. $L_{2,3} \rightarrow VV$ Auger electron yield under 30 keV H^+ bombardment of the Al (100) surface at rotating around the $\langle 110 \rangle$ axis. Broken curve: yield from the unshaded surface layers. The arrows show the axes corresponding to minima.

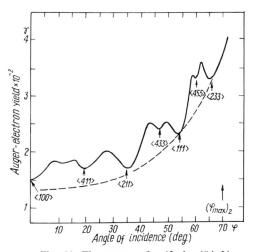

Fig. 44. The same as fig. 43, for 60 keV.

electron yield $\gamma_{X \to YZ}(\varphi)$ from the second layer drastically increases, reaching a maximum at $\varphi_2 = \arctan\{[R_s(\varphi + \rho_m)/d\}$, when atoms are entering the bright ring, and then decreases, reaching the average value for $I_2(R) = 1$.

If $R_s \gg \rho_m$, then $\Delta\varphi(\varphi_2, \varphi_1)$ is small, but φ_1 may be large.

If $R_s \approx \rho_m$, then $\Delta\varphi$ may have large values, but φ_1 is small. Therefore, the angles φ_1 and φ_2 characterize the shape of $\gamma_{X \to YZ}(\varphi)$, and similar angles are appropriate for deeper layers and the next minima.

(ii) $R_s < \rho_m$. In this case the atoms of the second layer are not shaded even at $\varphi = 0$. Only the atoms of the following layer are completely shaded. The partial shading diminishes the degree of anisotropy. On the other hand, in this case the minima of $\gamma_{X \to YZ}(\varphi)$ become smoother, which gives an opportunity for a more accurate determination of the atom location using the function $\gamma_{X \to YZ}(\varphi)$.

The orientational effect caused by the shading of the lower atom by the upper ones can also be observed under the azimuthal rotation of a single crystal with a fixed angle of incidence corresponding to one of the minima of $\gamma_{X \to YZ}(\varphi)$, e.g., to the direction $\langle 211 \rangle$. In this case, the increasing of the Auger electron yield caused by $\cos \varphi$ is avoided. Starting from $\theta = 0$ for $\langle 100 \rangle$ up to the azimuthal angle,

$$\theta_1 = \arctan\{[\sqrt{2}R_s(\varphi) - \rho_m]/d\} ,$$

the Auger electron yield is 0, then it drastically increases due to the entering of the second-layer atom into the bright ring, and then reaches a maximum at

$$\theta_2 = \arctan\{[\sqrt{2}R_s(\varphi) + \rho_m]/d\} .$$

At

$$\theta_3 \approx \arctan\{[\rho_m + 3\sqrt{2}R_s(\varphi)]/d\} ,$$

the Auger electron yield has its average value $\gamma_{X \to YZ}(I_2(R) = 1)$. If $R_s < \rho_m$, the behavior of the function $\gamma_{X \to YZ}(\varphi)$ is analogous to that described already in case (ii).

The energy dependence of the orientational effects comes from the fact that both R_s and ρ_m are functions of the ion incidence energy E_0. With increasing E_0, R_s decreases as $R_s \approx E_0^{-1/2}$, while ρ_m decreases as $\rho_m \approx E_0^{1/2}$ (for light ions). So the transition from case (i) $R_s > \rho_m$ to case (ii) $R_s < \rho_m$ will take place. After ρ_m passing its plateau, both R_s and ρ_m decrease, but ρ_m decreases faster, and at some energy E_0 the back transition from case (ii) $R_s < \rho_m$ to case (i) $R_s > \rho_m$ will take place. The curves $\gamma_{X \to YZ}(\varphi)$ calculated for $E_0 = 30$ and 60 keV are displayed in figs. 43 and 44, respectively. It is seen that the minima and maxima of $\gamma_{X \to YZ}(\varphi)$ become narrower and sharper with increasing E_0.

Of special interest is the excess of the minima $\langle 100 \rangle$ and $\langle 211 \rangle$ above the broken curves displaying the Auger yield from the upper unshaded layers. It can be seen that for $E_0 = 30$ keV, the bottoms of the minima coincide exactly with the broken curve, while for $E_0 = 60$ keV they are above it. This fact is connected with the following. For $E_0 = 30$ keV, in the directions $\langle 100 \rangle$ and $\langle 211 \rangle$ the second layer is shaded completely because both $R'_s(\varphi = 0) = 3.17 \times 10^{-9}$ cm and $R'_s(\varphi = 35°) = 4.27 \times 10^{-9}$ cm exceed $\rho_m(E_0 = 30 \text{ keV}) = 2.78 \times 10^{-9}$ cm, while for $E_0 = 60$ keV, in the same directions, the second layer is not shaded completely because both $R'_s(\varphi = 0) = 2.24 \times 10^{-9}$ cm and $R'_s(\varphi = 35°) = 3.03 \times 10^{-9}$ cm are less than $\rho_m(E_0 = 60 \text{ keV}) = 3.5 \times 10^{-9}$ cm.

Both for $E_0 = 30$ and 60 keV the bottoms of the minima $\langle 111 \rangle$ coincide with the broken curves, which means that the second layer is completely shaded because for $E_0 = 30$ keV and $\varphi = 54°30'$, $R_s = 7.17 \times 10^{-9}$ cm $> \rho_m$ and for $E_0 = 60$ keV and $\varphi = 54°30'$, $R_s = 5.05 \times 10^{-9}$ cm $> \rho_m$.

The minima $\langle 411 \rangle$ and $\langle 433 \rangle$ come from the third layer, $i = 3$. With increasing E_0, the fine structure of these minima becomes more perfect (compare figs. 43 and 44), due to the fact that R'_s is a decreasing function of E_0, and with increasing E_0 the overlapping of maxima and minima of the deeper layers diminish.

The oscillating structure of the function $\gamma_{X \to YZ}(\varphi)$ disappears at glancing incidence when the deeper layers are completely shaded and the Auger electrons are ejected from the unshaded surface layer only (broken curve). The maximum angle of incidence φ_{max} from which this blocking of deeper layers starts is determined by

$$[y_s(\varphi)]_i + 2\rho_m = C , \tag{180}$$

where C is the distance between the atoms in the layers and $[y_s(\varphi)]_i$ is the major axis of the shadow in the ith layer, see eq. (172). For the second layer, $i = 2$,

$$(\varphi_{max})_2 = \arccos\left\{\left[\frac{16 Z_1 Z_2 e^2}{E_0} \frac{d}{(C - 2\rho_m)^2}\right]^{1/3}\right\} , \tag{181}$$

and for the third layer, $i = 3$,

$$(\varphi_{max})_3 = \arccos\left\{\left[\frac{4(1 + \sqrt{2})^2 Z_1 Z_2 e^2}{E_0} \frac{d}{(C - 2\rho_m)^2}\right]\right\} , \tag{182}$$

$[\varphi_{max}(E_0)]_i$ is an increasing function of E_0 and a decreasing function of i.

For $E_0 = 30$ keV, $H^+ \to Al(100)$, $\langle 110 \rangle$, $L_{2,3} \to VV$,

$$(\varphi_{max})_2 \approx 65° , \quad (\varphi_{max})_3 \approx 62° .$$

For $E_0 = 60$ keV, $H^+ \to Al(100)$, $\langle 110 \rangle$, $L_{2,3} \to VV$,

$$(\varphi_{max})_2 \approx 71° , \quad (\varphi_{max})_3 \approx 67°30'.$$

It is seen that $(\varphi_{max})_2 > (\varphi_{max})_3$, therefore, the curve $\gamma_{X \rightarrow YZ}(\varphi)$ coincides with the broken curve starting from $(\varphi_{max})_2$.

4.4. Influence of thermal vibrations

Let it consider the influence of the thermal vibrations on the function $I_2(R)$. The vibrations change from R to $R + \Delta$, where Δ is determined by a Gaussian function,

$$\Phi(\Delta) = (2\pi\overline{\Delta^2})^{-1/2} \exp(-\Delta^2/2\overline{\Delta^2}) .$$

The average $I_2(R)$ is

$$\overline{I_2(R)} = \frac{1}{2} \int_{-\infty}^{+\infty} \left\{ \left[1 - \frac{R_s^2}{(R+\Delta)^2} \right]^{1/2} + \left[1 - \frac{R_s^2}{(R+\Delta)^2} \right]^{-1/2} \right\} \Phi(\Delta) \, d\Delta .$$

This integral can be calculated using the method of the steepest descent. As a result, we have

$$\overline{I_2(R)} = \tfrac{1}{2}[R_s/2(R_s - R)]^{1/2} \exp[-(R_s - R)^2/2\overline{\Delta^2}] , \quad R < R_s ,$$

$$= \tfrac{1}{2}(R_s^2/2\overline{\Delta^2})^{1/2} , \qquad\qquad\qquad R = R_s , \qquad (183)$$

$$= 1 + \tfrac{1}{8}(R_s/R)^4 \text{ [from eq. (4)]} , \qquad R \gg R_s ,$$

and $\overline{\Delta^2} = \overline{\Delta_x^2} + \overline{\Delta_y^2} + \overline{\Delta_z^2} = \tfrac{2}{3}\overline{l^2}$, where $\overline{l^2}$ is the mean-square deviation of the interatomic distance. According to Martynenko (1966a),

$$\overline{l^2} = \frac{3 \times 10^{-2} \text{ Å}}{T_M} T , \qquad T > T_D ,$$

$$= \frac{3 \times 10^{-2} \text{ Å}}{T_M} \tfrac{1}{4}T_D , \quad T < T_D , \qquad (184)$$

where T is the temperature, T_D the Debye temperature, and T_M the melting temperature.

Now, we consider the influence of the thermal vibrations on the function $\sigma(\varphi)$ determined by eq. (179),

$$\sigma(\varphi, \overline{\Delta^2}, \delta, R_s) = \overline{\sigma(\varphi, \delta + \Delta, R_s)} , \qquad\qquad (185a)$$

$$\sigma(\varphi, \overline{\Delta^2}, \delta, R_s) = \int_{-\infty}^{+\infty} \sigma(\varphi, \delta + \Delta, R_s)\Phi(\Delta) \, d\Delta . \qquad (185b)$$

Substituting eq. (179) into eqs. (185a) and (185b), we have

$$\overline{\sigma(T, \varphi)} = \sigma_1(\overline{\Delta^2}, \delta)$$

$$\approx [(\tfrac{2}{3}\pi\overline{\Delta^2})^{1/2} + 2\rho_m]^{-1}\sigma_{max}(\overline{\Delta^2})[\delta + (\tfrac{2}{3}\pi\overline{\Delta^2})^{1/2} - R_s + \rho_m],$$

$$0 \leqslant \delta \leqslant R_s + \rho_m,$$

$$= \sigma_2(\overline{\Delta^2}, \delta)$$

$$= \tfrac{1}{3}\pi\rho_m^2 P(0) + \tfrac{1}{4}\pi R_s^4 P(0) \times \left[\frac{1}{4(\delta^2 - \rho_m^2) - 4\overline{\Delta^2}} - \frac{R_s + 0.9\rho_m}{2\rho_m}\right.$$

$$\times \frac{1}{[\delta + (\delta^2 - 8\overline{\Delta^2})]^{1/2}} \left. \frac{1}{\{[\delta + (\delta^2 - 8\overline{\Delta^2})^{1/2}]^2 - 8\overline{\Delta^2}\}^{1/2}}\right],$$

$$\delta \geqslant R_s + \rho_m, \tag{185c}$$

where

$$\sigma_{max}(\overline{\Delta^2}, \delta) = \sigma_2(\overline{\Delta^2}, \delta = R_s + \rho_m).$$

As can be seen from eq. (185c), the function $\sigma(\varphi, \overline{\Delta^2}, \delta_s, R)$ becomes smoother when $\overline{\Delta^2}$ increases. Indeed, for $(\tfrac{2}{3}\pi\overline{\Delta^2})^{1/2} > R_s - \rho_m$, the function $\sigma_1(\overline{\Delta^2}, \delta) > 0$, which means that even the most shaded atoms become open due to the thermal vibrations. On the other hand, $\sigma_{max}(\overline{\Delta^2}, \delta) = \sigma_{max}(\delta = R_s + \rho_m)$, which means that due to thermal vibrations even the most open atoms are partially shaded. The curves $\gamma_{L_{2,3} \to VV}(\varphi)$ for $H^+ \to Al(100)$, $\langle 110 \rangle$, $E_0 = 30$ keV, calculated for $T = 300$ and 700 K, are displayed on fig. 45.

4.5. Contribution of deep layers

For the cubic crystal structure, the minima appear at the angles

$$\varphi_{iK} = \arctan[(K - 1)\tfrac{1}{2}\sqrt{2}(i - 1)], \tag{186}$$

where $i = 2, 3, \ldots$ is the layer index, and $K = 1, 2, 3, \ldots$ the digital number of the minimum starting with $K = 1$ for $\varphi = 0$.

The maxima appear at the angles

$$[\varphi_2]_{ij} = \arctan \frac{R_s'(\varphi) + \rho_m + (j - 1)d\tfrac{1}{2}\sqrt{2}}{d(i - 1)}, \quad j = K - 1, \tag{187}$$

$$[\varphi_2]_{ij} = \arctan \frac{(j - 1)d\tfrac{1}{2}\sqrt{2} - R_s'(\varphi) - \rho_m}{d(i - 1)}, \quad j = K + 1. \tag{188}$$

The anisotropy of the Auger electron emission from the upper layers is decreased by the contribution of the deeper layers. This decrease becomes

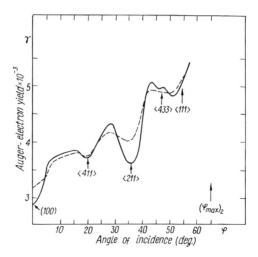

Fig. 45. The dependence of the $L_{2,3} \to VV$ Auger electron yield under 30 keV H^+ bombardment of the Al(100) surface at rotation around the $\langle 110 \rangle$ axis on temperature: (full curve) $T = 300$ K, (broken curve) $T = 700$ K.

most significant when the minima from the ith layer overlap the maxima from the second layer, or vice versa, i.e., when $\varphi_{iK} = [\varphi_2]_{2j}$. From eqs. (186) and (187), we have the number of overlapping layers,

$$i = \frac{(K-1)d\frac{1}{2}\sqrt{2}}{(j-1)d\frac{1}{2}\sqrt{2} + R'_s(\varphi) + \rho_m} + 1. \tag{189}$$

However, for the most important cases the ith layer depth $(i-1)d$ exceeds the effective escape depth $\lambda(E)$ and the contribution from the ith layer is negligible. For instance, for $H^+ \to Al$, $d_{100} = 4.04$ Å, $E_0 = 30$ keV: $\varphi \leqslant 35°$, $i = 6$ and $\varphi > 35°$, $i = 3$. In both cases the depth $(i-1)d$, which is 24.24 Å and 8.08 Å, respectively, exceeds the effective escape depth, which for $L_{2,3} \to VV$, $E = 63.5$ eV, equals $\lambda = 4$ Å.

4.6. *Role of recoil atoms*

The fine structure in the high-energy part of the spectrum of secondary electrons emitted from a solid surface under ion bombardment recorded in experiments by Mishin and Petrov (1975), Viel et al. (1976), Dorozhkin et al. (1978), Vrakking and Kroes (1979) and Powell (1978) has been proven to originate from Auger transitions which embrace the inner shells of the colliding particles.

The Auger spectra, formed under the bombardment of Al targets by Ar^+ ions of low and medium energy, have been intensively studied in papers by

Benazeth et al. (1980), Wittmaack (1980), Legg et al. (1980) and Viaris de Lesegno and Hennequin (1981). The Al Auger spectrum consists of two main peaks: a sharp peak at Auger electron energy $E = 63.5$ eV and a broad one at $E = 67$ eV. The interpretation of the observed Auger spectra and the contribution of symmetric and asymmetric collisions to the total inner shell ionization cross section of target atoms are the points of discussion. For instance, Viel et al. (1976) and Benazeth et al. (1980) have considered that the above-mentioned Auger peaks are due respectively to $L_{2,3} \rightarrow MM$ and $L_{2,3} \rightarrow VV$ Auger transitions in the bulk, while Legg et al. (1980) has considered that both peaks are due to Auger transitions in neutral sputtered Al atoms with the $1s^2 2s^2 2p^5 3s^2 3p^2$ structure.

Kitov and Parilis (1979, 1981a,b, 1982a,b, 1984a,b) found that the anisotropy of the Auger electron emission due to the different shadowing of the lower crystal layers by the upper ones at an Auger electron yield $\gamma_{X \rightarrow YZ}(\varphi)$ displayed an orientational effect, which is expressed in the interchange of some maxima and minima appearing at an ion beam incidence along the crystal axes. The deepest minima were observed for the ion beam passing along the low-index crystal axes. The calculations by Kitov and Parilis (1979, 1981a,b, 1982a,b) seem to be proven experimentally by Hou et al. (1986), Schuster and Varelas (1985), and MacDonald et al. (1985). According to this last paper, the yield of Al $L_{2,3} \rightarrow MM$ Auger electrons versus the angle of incidence contains minima for the ion beam passing along the axes $\langle 100 \rangle$, $\langle 344 \rangle$, $\langle 112 \rangle$ and $\langle 111 \rangle$ which fits the theoretical prediction. However, the experimental curve $\gamma_{L_{2,3} \rightarrow MM}(\varphi)$ increases considerably more rapidly than $\cos^{-1}\varphi$ at $\varphi > 35°$.

The aim of this chapter is to explain this sharp increase and also to understand which collisions – symmetrical or asymmetrical ones – promote the creation of a vacancy in the $L_{2,3}$ shell of the target atom and where Auger recombination occurs, inside or outside the solid (fig. 46).

The vacancy formation in the $L_{2,3}$ shell of the target atoms may occur

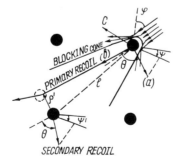

Fig. 46. Shadow by surface atoms in the incident beam: (a), (b) and (c) are the different directions of the inner-shell excited recoil movement.

both in asymmetrical collisions with bombarding ions and in symmetrical collisions with recoil atoms. The vacancy-carrying atom can undergo Auger transition both inside or outside the solid, being knocked out of recoil. The relative contribution of symmetrical and asymmetrical collisions in the relative Auger $L_{2,3} \rightarrow MM$ yield depends on ion energy and on the ratio of masses of colliding particles.

We consider that if Auger recombination of the vacancy occurs inside the solid, either in a resting or in a moving atom, it gives rise to a band-like $L_{2,3} \rightarrow VV$ Auger spectrum. If the Auger recombination occurs in a recoil atom outside the crystal, then an atomic-like $L_{2,3} \rightarrow MM$ Auger spectrum would be exhibited. First, we consider the asymmetrical collision induced $L_{2,3} \rightarrow VV$ Auger electron yield.

It is given by

$$\gamma_{L_{2,3} \rightarrow VV}(\varphi) = \frac{Nd}{\cos \varphi} \omega_a P(E) \sum_i \int_0^{2\pi} d\psi \int_0^{\theta_m} d\theta \int_{(i-1)d}^x dx$$

$$\times \exp\left[-\frac{x(\theta, \varphi, \psi)}{\lambda}\right] \sigma_{as}(\varphi, \theta) W_i(\theta, \varphi, \psi, x) , \qquad (190)$$

where N is the atomic density, d the distance between the crystal layers, ω_a the probability of filling the vacancy via an Auger process, $\sigma_{as}(\theta, \varphi)$ the cross section of the vacancy formation in the ith layer by the asymmetrical collisions, $P(E) = 0.5$ the probability of Auger electron escape from a single crystal, and ψ the azimuthal angle.

The cross section is given by

$$\sigma_{as}(\varphi, \theta) = \frac{d\sigma(\theta)}{d\theta} P_{L_{2,3}}(\rho(\theta, \varphi)) , \qquad (191)$$

where $d\sigma(\theta)/d\theta$ is the recoil differential cross section and $P_{L_{2,3}}$ the probability of $L_{2,3}$ shell ionization as a function of the recoil angle θ. $W(\theta, \varphi, \psi, x)$ is the probability that the Auger transition would occur in the crystal at depth x, λ is the effective escape depth of the Auger electrons and ρ is the impact parameter.

The summation is carried out over all crystal layers. The recoil differential cross section for the Firsov potential $V = V_0 r^{-2}$ in the momentum approximation is given by

$$\frac{d\sigma(\theta)}{d\theta} = \left(\frac{\pi R_0}{\pi - 2\theta}\right)^2 = \frac{3.05 Z_1 Z_2 \times 10^{-16}(1 + \mu)\pi^2}{(\sqrt{Z_1} + \sqrt{Z_2})^{2/3} E_0(\pi - 2\theta)^2} \text{ eV cm}^2 , \qquad (192)$$

where $\mu = m_1/m_2$, and Z_1, m_1 and Z_2, m_2 are the atomic number and mass of the projectile and target atom, respectively. E_0 is the energy of the projectile.

We define the function $P_{L_{2,3}}(\rho(\theta))$ as

$$P_{L_{2,3}}(\rho(\theta)) = 0 , \qquad \rho > \rho_m ,$$
$$= P_{L_{2,3}} = \text{const} , \qquad \rho \leqslant \rho_m . \tag{193}$$

The function $W(\theta, \varphi, \psi, x)$ gives the probability per unit depth that the Auger transition occurs at a depth x at time t,

$$W(t(\theta, \varphi, \psi, x)) = \frac{1}{\tau} \exp\left(-\frac{t}{\tau}\right) \frac{dt}{dx} , \tag{194}$$

where τ is the lifetime of the vacancy in the $L_{2,3}$ shell.

In fig. 46, the three directions of ionized recoil atoms in the crystal bulk are shown. For cases (a) and (b), the depth $x = (i-1)d + x_1$, where i is the number of the crystal layer in which the ionizing collision occurred and x_1 the projection of the ionized atom path on the surface normal. By proceeding in eq. (194) from variable t to x, we obtain

$$W_i(\theta, \varphi, \psi, x) = \frac{\exp\{-[x - (i-1)d]/S_0 f(\theta, \varphi, \psi)\}}{S_0 f(\theta, \varphi, \psi)} , \tag{195}$$

where

$$S_0 = \left(\frac{2E_{max}}{m_2}\right)^{1/2} \tau , \qquad E_{max} = \frac{4m_1 m_2}{(m_1 + m_2)^2} E_0 .$$

For case (c), the function $W_i(\theta, \varphi, \psi, x)$ equals

$$W_i(\theta, \varphi, \psi, x) = -\frac{\exp\{-[(i-1)d - x]/S_0 f(\theta, \varphi, \psi)\}}{S_0 f(\theta, \varphi, \psi)} , \tag{196}$$

where

$$f(\theta, \varphi, \psi) = \cos\theta(\cos\varphi \cos\theta - \sin\varphi \sin\theta \cos\psi) .$$

We define the probability $P_{L_{2,3}}$ from eq. (193) as

$$P_{L_{2,3}} \approx \bar{P}_{L_{2,3}} = \sigma_{ion}(\varphi)/\pi\rho^2(\theta_m) , \tag{197}$$

Since $P_{L_{2,3}}(\rho(\theta))$ depends only on ρ_m [see eq. (193)], eq. (190) – taking into account eqs. (194)–(197) – may be expressed in the following way,

$$\gamma_{L_{2,3} \to VV}(\varphi) = \frac{N d\omega_a P(E) \sigma_{ion}(\varphi)(\pi R_0)^2}{\cos\varphi \, \pi\rho^2(\theta_m)}$$

$$\times \sum_i \int_0^{2\pi} d\psi \int_0^{\theta_m} d\theta \int_{(i-1)d}^{x_m} \frac{W_i(\theta, \varphi, \psi, x) \exp(-x/\lambda)}{(\pi - 2\theta)^2} \, dx , \tag{198}$$

where $\sigma_{ion}(\varphi)$ is molecular-orbital cross section of the target atom ionization (Kitov and Parilis 1982b), $\pi\rho^2(\theta_m)$ the total recoil cross section, θ_m the maximal recoil angle, and

$$x_m = \infty, \quad \beta < \tfrac{1}{2}\pi, \quad \text{cases (a) and (b)},$$
$$= 0, \quad \beta > \tfrac{1}{2}\pi, \quad \text{case (c)},$$

with $\beta = \theta + \varphi$.

In eq. (198), the integration over x is elementary. As a result we obtain

$$[\gamma_{L_{2,3}\to VV}(\varphi)]_{as} = \frac{Nd\omega_a P(E)\sigma_{ion}(\varphi)(\pi R_0^2)}{\cos\varphi \; \pi\rho^2(\theta)_m} \sum_i \exp\left[-\frac{(i-1)d}{\lambda}\right]\Phi_i(\varphi),$$
(199)

where R_0 is distance of closest approach of an ion with the target atom.

The function $\Phi_i(\varphi)$ equals

$$\Phi_i(\varphi) = \int_0^{2\pi} d\psi \int_0^{\theta(\beta=\pi/2)} d\theta\left\{\left[1 + \frac{S_0}{\lambda} f(\theta, \varphi, \psi)\right](\pi - 2\theta)^2\right\}^{-1},$$

for $\beta < \pi/2$, cases (a) and (b); (200a)

$$\Phi_i(\varphi) = \int_0^{2\pi} d\psi \int_{\theta(\beta=\pi/2)}^{\theta_m} \left\{\exp\left[\frac{(i-1)d}{S_0 f(\theta, \varphi, \psi)}\left(\frac{S_0}{\lambda} f(\theta, \varphi, \psi) - 1\right)\right] - 1\right\}$$

$$\times \left\{\left[\frac{S_0}{\lambda} f(\theta, \varphi, \psi) - 1\right](\pi - 2\theta)^2\right\}^{-1}, \quad \text{for } \beta > \pi/2, \text{ case (c)}.$$
(200b)

The integration over θ and ψ needs a numerical calculation. The function $[\gamma_{L_{2,3}\to VV}(\varphi)]_{as}$ is shown in fig. 47.

The fast recoil atom with a vacancy in the $L_{2,3}$ shell can escape from the surface without Auger electron emission. In this case, the Auger recombination will occur outside the crystal and an atom-like $L_{2,3}\to MM$ Auger spectrum would be exhibited. If the fast recoil atom with a vacancy in the $L_{2,3}$ shell is formed in the ith layer, then the Auger electron yield equals

$$[\gamma_{L_{2,3}\to MM}(\varphi)]_{as} = \frac{Nd\omega_a \sigma_{ion}(\varphi)(\pi R_0)^2}{\cos\varphi \; \pi\rho^2(\theta_m)}$$

$$\times \sum_i \int_{\theta(\beta=\pi/2)}^{\theta_m} d\theta \int_0^{2\pi} d\psi \int_0^{-\infty} \frac{W_i(\theta, \varphi, \psi, x)}{(\pi - 2\theta)^2} \, dx, \quad (201)$$

and after integration over x we obtain

$$[\gamma_{L_{2,3}\to MM}(\varphi)]_{as} = \frac{Nd\omega_a \sigma_{ion}(\varphi)(\pi R_0)^2}{\cos\varphi \; \pi\rho^2(\theta_m)} \sum_i \Phi_i(\varphi), \quad (202)$$

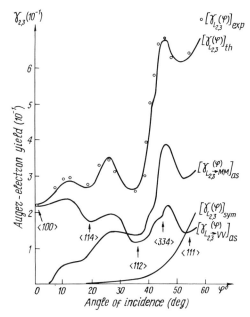

Fig. 47. The $L_{2,3}$ Auger electron yield under 35 keV Ar^+ bombardment of the Al (100) surface at rotating around the $\langle 110 \rangle$ axis: (○) $[\gamma_{L_{2,3}}(\varphi)]_{exp}$, the experimental yield (Hou et al. 1986); (full curves) $[\gamma_{L_{2,3} \to MM}(\varphi)]_{as}$ the atom-like spectrum yield, $[\gamma_{L_{2,3} \to VV}(\varphi)]_{as}$ the band-like spectrum yield, $[\gamma_{L_{2,3}}(\varphi)]_{sym}$ the $L_{2,3} \to VV$ and $L_{2,3} \to MM$ Auger yield, $[\gamma_{L_{2,3}}(\varphi)]_{th}$ the total calculated yield.

where $\Phi_i(\varphi)$,

$$\Phi_i(\varphi) = \int_0^{2\pi} d\psi \int_{\theta(\beta = \pi/2)}^{\theta_m} \frac{d\theta}{(\pi - 2\theta)^2} \exp\left(\frac{-(i-1)d}{S_0 f(\theta, \varphi, \psi)}\right), \tag{203}$$

is calculated numerically.

The ratio of $L_{2,3} \to VV$ to $L_{2,3} \to MM$ transitions in the total Auger electron yield calculated using formulae (198) and (201)–(203) is about 2. This figure fits well the experimental evaluation made by Wittmaak (1980). Unfortunately, both the $[\gamma_{L_{2,3} \to MM}(\varphi)]_{as}$ and the $[\gamma_{L_{2,3} \to VV}(\varphi)]_{as}$ yields increase with φ more slowly than the experimental curve (fig. 47).

Better agreement could be reached taking into account the symmetrical collisions of recoils with target atoms. This is important if $\mu > 1$. This is the case in the experiment by Hou et al. (1986).

The Auger electron yield in symmetrical collisions is

$$[\gamma_{L_{2,3} \to VV}(\varphi)]_{sym} = \frac{Nd\omega_a P(E)}{\cos \varphi} \sum_i \sum_j \sum_k \int_{(i-1)d}^{x_m} \exp\left(-\frac{x}{\lambda}\right) R_{ijk}(\varphi, x) \, dx , \tag{204a}$$

$$[\gamma_{L_{2,3} \to MM}(\varphi)]_{sym} = \frac{N d\omega_a}{\cos \varphi} \sum_i \sum_j \sum_k \int_0^{-\infty} R_{ijk}(\varphi, x) \, dx \,, \tag{204b}$$

where $R_{ijk}(\varphi, x)$ is the symmetrical collision ionization rate of the kth atom located in the ith layer and hit by the jth recoil atom. $R_{ijk}(\varphi, x)$ equals

$$R_{ijk}(\varphi, x) = \int_0^{2\pi} d\psi' \int_{\theta_1'(\rho_m')}^{\theta_2'(\rho_m')} \frac{d\sigma(\theta)}{d\Omega(\theta)} \frac{d\sigma_{ion}(\theta')}{l^2 \, d\theta'}$$

$$\times W(\theta'(\theta), \psi', x) I(\theta, \psi; \theta', \psi') \, d\theta' \,, \tag{205}$$

where $I(\theta, \psi; \theta', \psi')$ is the Jacobian of the transition from (θ, ψ) to (θ', ψ'). $\Omega(\theta)$ is the solid angle and l is the distance between the atoms (fig. 46). The function $W(\theta'(\theta)\psi', x)$ is determined in analogy with eqs. (194)–(196), and the angles θ, ψ and θ', ψ' are the recoil and azimuthal angles of the primary recoil atom and the recoil atom ionized after the symmetrical collision, respectively. Integral (203) was calculated numerically.

The contribution of the symmetrical collisions $[\gamma_{L_{2,3}}(\varphi)]_{sym}$ is also displayed in fig. 47. It is about $\frac{1}{2}$ of the asymmetrical contribution, but increases very sharply with φ. Although it does not display any anisotropy, it affects the total yield to fit the experimental increase with increasing φ.

The total yield

$$[\gamma_{L_{2,3}}(\varphi)]_{th}$$

$$= [\gamma_{L_{2,3} \to MM}(\varphi) + \gamma_{L_{2,3} \to VV}(\varphi)]_{as} + [\gamma_{L_{2,3} \to MM}(\varphi) + \gamma_{L_{2,3} \to VV}(\varphi)]_{sym} \,,$$

calculated with use of eqs. (190)–(205), is displayed in fig. 47. As is seen, there is good agreement between the theoretical and experimental values.

The sharp increase of $\gamma_{L_{2,3} \to MM}(\varphi)$ for $\varphi > 35°$ is due to the fact that, at glancing bombardment, a considerable part of the ionized recoil atoms have a momentum directed toward the surface.

Thus, the Auger electron yield indeed displays an incident beam orientation anisotropy caused by atom shadowing. The band-like Auger spectrum being emitted in the bulk, while the atom-like Auger spectrum is emitted outside the solid, the sharp increase at glancing incidence testified the role of the recoil atoms in ion-excited Auger electron emission.

The ion-induced Auger electron yield versus the angle of ion beam incidence shows a sharp anisotropy which is caused by shading of the inner layers by the upper ones. It is expressed in the existence of minima and maxima on the curves $\gamma_{X \to YZ}(\varphi)$ which could be observed experimentally. It will be very interesting to record the ion-induced Auger electron yield anisotropy at rotating a single crystal around one of the low-index axes for different incidence energies, target temperatures and other parameters.

CHAPTER 12

Interaction of Multiple-Charged Ions with Surfaces

1. Introduction

When a multiple-charged ion approaches a solid surface and penetrates the solid, intense capture of electrons by the ion occurs. The energy expended in forming the ion, which can be comparable to or even higher than the kinetic energy of the ion, is dissipated.

Ion–electron emission from solids is affected both by the energy of the electrons of the solid in the ion field (potential emission) and the kinetic energy of the impinging ion (kinetic emission). Since the kinetic ejection of electrons has a definite velocity threshold ($v_{th} \leqslant 1.5 \times 10^7$ cm/s), the potential ejection could be observed alone below this threshold. It has been investigated in details for metals and other targets with single-charged ions and reported in some papers in the thirties–fifties and reviewed in several monographs (e.g., U.A. Arifov 1968, Kaminsky 1965).

This phenomenon is caused by Auger neutralization or Auger deexcitation of ions with previous charge exchange on solid surfaces (see fig. 1). The theoretical explanation of this phenomenon was first made by Shekhter (1937) and then a more detailed experimental study as well as phenomenological description were reported by Hagstrum (1954a,b).

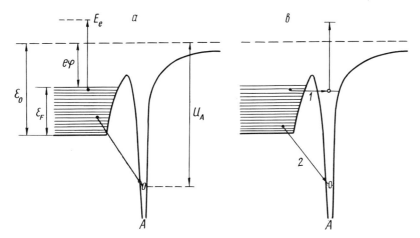

Fig. 1. Electron energy diagrams of the transition processes of an ion in front of a surface. (a) Auger neutralization (AN), (b) (1) resonance neutralization (RN) and (2) Auger deexcitation (AD).

In the part concerning the probability of electron escape from solids, this description was improved by Parilis (1962, 1969a). A simple formula for the description of the dependence of the coefficient of potential emission on both ion and metal parameters was derived by Kishinevsky (1973) using the Hagstrum's method (see section 2.2).

Parilis (1969b, 1970, 1980b, 1991) has pointed out that the ion emission and the sputtering of nonmetals under the influence of slow multiple-charged ions should be much more pronounced than the corresponding effects in the case of single-charged ions of the same mass and energy. The *Coulomb explosion* mechanism for the process has been proposed. According to this mechanism the ion is neutralized primarily by the Auger effect. With a high ion charge, the neutralization is a stepwise process (Hagstrum 1954b, U.A. Arifov et al. 1973b), and in each transition several Auger electrons are emitted from the lattice atoms. As a result, a charge greater than the initial ion charge is created near the surface. If the time required to neutralize this charge region is long enough, the ions which are formed can acquire (through Coulomb repulsion) a kinetic energy higher than the surface binding energy and can be emitted from the solid. In other words, sputtering below the ordinary s,plluttering threshold can occur.

2. Electron emission

2.1. Potential electron emission from metals

The first studies on the potential emission of electrons under multiple-charged ion bombardment were reported in the early fifties (Dunaev and Flaks 1953, Flaks 1955). The authors used the ions K^{q+}, Hg^{q+}, Zn^{q+} ($q \leqslant 3$) and Tl^{q+}, Pb^{q+} ($q \leqslant 4$) bombarding Pt. The potential emission for triple- and quadruple-charged ions was observed to increase sharply relative to the one from single-charged ions. It should be noted that the same increase of the number of emitted electrons for multiple-charged ions Ne^{q+} ($q \leqslant 4$) bombarding a gas target (Xe) has also been observed by Flaks et al. (1961). The theoretical description of the Auger ionization of atoms by multiple-charged ions was developed by Kishinevsky and Parilis (1968b).

The first detailed investigation of Auger emission of electrons from W under the bombardment by multiple-charged ions He^{2+}, $Ne^{2+,3+}$, $Ar^{2+,3+}$, $Kr^{(2-4)+}$, $Xe^{(2-5)+}$ with energies up to 1 keV was made by Hagstrum (1954b). The coefficients of emission $\gamma_p(E_0, q)$ were shown to increase sharply with increase of the ion charge q (and hence with total ionization energy increase) and then to decrease with the increase of ion energy E_0 (except He^{2+}), besides $|\partial\gamma_p/\partial E_0|$ increases with q increase (see fig. 2).

It was well established that for ions with a high ionization potential, U_q

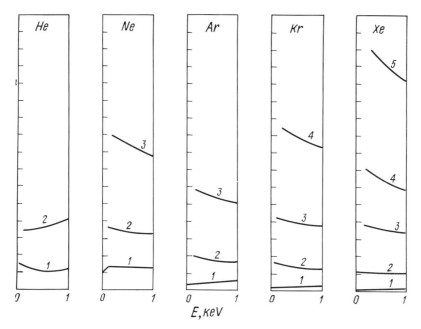

Fig. 2. Total yield versus ion kinetic energy for single and multiple-charged ions of noble gases incident on atomically clean tungsten. The charge of the ion is indicated at each curve (Hagstrum 1954b).

(helium, neon and other inert gases with $q \geqslant 3$), the ratios

$$\gamma_r = \gamma_p(q+1)/\gamma_p(q) ; \qquad U_r = U_{q+1}/U_q , \qquad (1)$$

are nearly equal, and the mean energy of emitted electrons is nearly independent of the ion charge. These facts as well as other peculiarities of the electron energy spectra obtained by Hagstrum (see fig. 3a,b) gave evidence of multistage neutralization of multiple-charged ions on metal surfaces.

The experimental study of the emission coefficients for Mo and W under Ar^{q+}, Kr^{q+} ($q \leqslant 6$), Xe^{q+} ($q \leqslant 8$) bombardment with ion energies 200–1500 eV (U.A. Arifov et al. 1971c, 1973a) was of great importance. These investigations allowed to develop the method of multiple-charged ion identification by their potential emission (section 2.3).

The experiments done by Schram et al. (1966), Perdrix et al. (1968), U.A. Arifov et al. (1971b), and Cano (1973) were reported by this time. The experiments by Hagstrum (1954b), U.A. Arifov et al. (1971c, 1973a) promoted the development of the theory of the phenomenon by Kishinevsky and Parilis (U.A. Arifov et al. 1973b, Kishinevsky and Parilis 1978) (see section 2.4). In the middle of the seventies, the experimental study per-

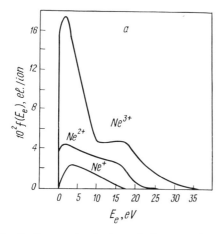
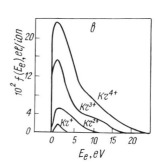

Fig. 3. Energy distributions of electrons ejected from tungsten by 200 eV (a) neon and (b) krypton ions (Hagstrum 1954b).

formed by U.A. Arifov et al. (1974) enabled to obtain the coefficients of electron emission from the Mo (110) and (111) faces bombarded by Ar^{q+} ($q \leq 5$) with energy 0.5–20 keV.

T.U. Arifov et al. (1976) obtained the coefficients of emission from Ta and the alloy CuBe, bombarded by Al^{q+} ($q \leq 5$), Cu^{q+} ($q \leq 6$), Re^{q+} ($q \leq 7$), In^{q+} and Bi^{q+} ($q \leq 9$) with energies from 0.5 to 28 keV, as well as from Ni, bombarded by Cr^{q+} ($q \leq 8$) with energies from 0.5 to 28 keV.

An intensive rise in the number of investigations of the phenomenon under consideration could be noted in the recent years. They are described in short in section 2.5. The investigations from nonmetals were considered in section 2.6.

2.2. Electron yield under single-charged ions

If the ionization energy of the incident particle exceeds the double work function ($E_i > 2\varphi$), then an Auger neutralization of the ion is possible resulting in the formation of a free electron whose emission is classified as *potential* electron emission. The electron release may also occur in another way, with a preliminary resonance neutralization and subsequent Auger relaxation. Both of these competing processes have a similar contribution to the emission (Hagstrum 1954a), hence in examining the complete coefficient of emission we shall not dwell at length on Auger relaxation.

Experimental studies of potential emission from a whole series of metals and metal films have shown that the coefficient of potential ion–electron emission, γ, depends to a great extent on the work function of the target (U.A. Arifov and Rakhimov 1960). However, there have been no special theoretical studies of the dependence of γ on individual parameters of the target and the ion, including the work function. The purpose of this section is to examine these dependences.

Fundamental research in potential electron emission falls into two groups of studies, consistently theoretic studies and semiphenomenological ones. The former demonstrates the possibility, in principle, of describing a phenomenon and also the practical impossibility of achieving good agreement with the experiment while remaining consistently within the framework of theory. These studies also show the ways and means for selecting the required parameters and functional dependences in the theories of the second kind. The latter strive to ensure agreement with the experiment through the utilization of a sufficient number of physically substantiated parameters.

The most detailed description on the basis of the semiphenomenological approach has been provided by Hagstrum (1954a). The method he had developed is applied in this section. Having assumed that the probability of a single Auger transition is representable as co-factors which separately take into consideration the angular and energy dependences and having substituted the co-factors which takes into account the dependence of the transition probability on the energy of the primary state level by a constant, which is equivalent to using the effective density of electron states $N_c(\varepsilon)$ within the conductivity zone, Hagstrum (1954a) derived the following ratio for the coefficient,

$$\gamma = \int_0^\infty N_o(E_K) \, dE_K = \int_{\varepsilon_0}^\infty N_i(\varepsilon_K) P_e(\varepsilon_K) \, d\varepsilon_K \tag{2}$$

Here $N_o(E_K)$, $N_i(\varepsilon_K)$ are the energy distributions of Auger electrons outside and inside the metal,

$$N_i(\varepsilon_K) = \frac{\frac{1}{2} N(\varepsilon_K) T\left[\dfrac{\varepsilon_K + \varepsilon_0 - E_i(s_m)}{2} \right]}{\displaystyle\int_{\varepsilon_F}^\infty \frac{1}{2} N(\varepsilon_K) T\left[\dfrac{\varepsilon_K + \varepsilon_0 - E_i(s_m)}{2} \right] d\varepsilon_K}, \qquad \varepsilon_K > \varepsilon_F,$$

$$= 0, \qquad\qquad\qquad\qquad \varepsilon_K < \varepsilon_F, \tag{3}$$

where ε_K, E_K are the energy of electrons, read-off correspondingly from the bottom of the conductivity zone and the vacuum level, $\varepsilon_0 = \varepsilon_F + \varphi$, and $E_i(s_m)$ is the ionization energy of the incident particle at distance s_m from the metal where Auger neutralization at the given velocity is most effective.

If the complete probability of Auger neutralization $R_t(s)$ changes with distance to the surface according to the law

$$R_t(s) = A \, e^{-as},$$

then

$$s_m = \frac{1}{a} \ln(A/av) .$$

$N(\varepsilon_K) \propto \sqrt{\varepsilon_K}$ is the state density over the Fermi level; $T(\varepsilon)$ is connected with the state density $N_c(\varepsilon)$ below the Fermi level by the ratio

$$T(\varepsilon) = \int_0^\varepsilon N_c(\varepsilon - \Delta) N_c(\varepsilon + \Delta) \, d\Delta , \qquad 0 < \varepsilon < \tfrac{1}{2}\varepsilon_F ,$$

$$= \int_0^{\varepsilon_F - \varepsilon} N_c(\varepsilon - \Delta) N_c(\varepsilon + \Delta) \, d\Delta , \quad \tfrac{1}{2}\varepsilon_F < \varepsilon < \varepsilon_F ,$$

$$= 0 , \qquad \varepsilon < 0, \; \varepsilon > \varepsilon_F ,$$

and turns out to be an Auger transformant; $P_e(\varepsilon_K)$,

$$P_e(\varepsilon_K) = 2\pi \int_0^{\theta_c} P(\varepsilon_K, \theta) \sin \theta \, d\theta ,$$

is the probability that the electron at energy ε_K will overcome the potential barrier on the surface of the metal.

$P_\Omega(\varepsilon_K, \theta) \, d\Omega$ is the probability that the momentum of the electron at energy ε_K is directed into the solid angle $d\Omega$.

Here and further we examine only the complete coefficient γ while the broadening of the energy spectrum of electrons is not taken into account.

If we do not limit ourselves to the assumption that the probability of Auger transition is independent of the energy of the primary state level, then the Auger transformant $T(\varepsilon)$ includes into the sub-integral expression a factor which is characteristic of this probability. Propst (1963), in particular, applying the quasiclassical investigation, has demonstrated that this factor diminishes exponentially with growing distance between the energy level of the electron and the Fermi level. Hagstrum et al. (1965) and Sternberg (1957) have arrived at similar conclusions.

According to Hagstrum's reasoning (Hagstrum 1954a), the angular distribution of electrons produced in the course of Auger neutralization is asymmetric. This is due to the fact that the matrix element for free electrons, i.e., those which are capable of escaping from the metal, is larger than for bound electrons. From a comparison with experiment, Hagstrum (1954a) established that

$$P_e(\varepsilon_K) = 0.5 \, \frac{1 - \sqrt{\varepsilon_0/\varepsilon_K}}{1 - \lambda\sqrt{\varepsilon_0/\varepsilon_K}} \equiv P_{Hag} , \tag{4}$$

where $\lambda = 1 - 1/f^2, f = 2.2$ is the ratio of matrix elements for free and bound electrons.

At $f = 1$, the distribution of $P_\Omega(\varepsilon_K, \theta)$ is spherically symmetric and

$$P_e(\varepsilon_K) = \tfrac{1}{2}(1 - \sqrt{\varepsilon_0/\varepsilon_K}) \equiv P_{sym}. \tag{5}$$

Contrary to Hagstrum (1954a), Propst (1963) believes that P_Ω is a spherically symmetric function of angle θ. In his view, the change in the form of $P_e(\varepsilon_K)$ is due to the contribution of secondary electrons.

The share of primary electrons which fail to be emitted from the metal makes up $N_i(\varepsilon_K)[1 - P_e(\varepsilon_K)]$. One electron at energy ε_K produces true secondary electrons whose number comprises the coefficient of true secondary emission

$$\delta = \frac{\sigma - r}{1 - r}.$$

Here, σ is the coefficient of secondary electron emission, and r is the coefficient of elastic reflection of electrons. These electrons have such an energy distribution $N_{sec}(\varepsilon_K, E)$, that $\int N_{sec}(\varepsilon_K, E)\,dE = \delta$. Hence, the complete number of electrons emitted at energy ε_K, i.e., the energy spectrum, is derived from the following,

$$N_o(\varepsilon_K) = N_i(\varepsilon_K)P_e(\varepsilon_K) + \int N_i(\varepsilon_K)[1 - P_e(\varepsilon_K')N_{sec}(\varepsilon_K', \varepsilon_K)]\,d\varepsilon_K'$$

$$= N_i(\varepsilon_K)P_e(\varepsilon_k) + N_{sec}(\varepsilon_K) \tag{6}$$

$$- N_i(P_e + N_{sec}/N_i) = N_i P_e^*(\varepsilon_K).$$

On the other hand, the complete number of emitted electrons during the formation of one primary electron at energy ε_K comprises

$$P_{sym}(\varepsilon_K) + \delta[1 - P_{sym}(\varepsilon_K)] = P_e^{**}(\varepsilon_K). \tag{7}$$

That is why the coefficient of potential emission can be written as

$$\gamma = \int N_i(\varepsilon_K)P_e^*(\varepsilon_K)\,d\varepsilon_K = \int N_i(\varepsilon_K)P_e^{**}(\varepsilon_K)\,d\varepsilon_K. \tag{8}$$

In the second case, the sub-integral expression, although it does not describe the energy spectrum of electrons, nevertheless takes the same form as given by Hagstrum (1954a).

In order to describe the difference between P_e^* and P_e^{**}, let us examine the form of the energy spectrum for true secondary electrons. If the energy of primary electrons is 15–20 eV, then the maximum energy of true secondary electrons does not exceed 4 eV (Bronstein and Freiman 1969). Hence, $P_e^* > P_e^{**}$ within the range of low energies while in the other energy ranges $P_e^* < P_e^{**}$.

The comparison used by Hagstrum for achieving a better agreement with experiment for the dependence $P_{Hag}(\varepsilon_K)$ with $P_e^{**}(\varepsilon_K)$ for tungsten (fig. 4) shows that within the low-energy range $P_e^{**} < P_{Hag}$, while in the range of $E_K > 15$ eV $P_e^{**} > P_{Hag}$. Taking into account the ratio between P_e^{**} and P_e^*, we arrive at the conclusion that P_{Hag} and P_e^* are close to each other, which apparently explains the good agreement of Hagstrum's calculations with experiment.

The coefficient of true secondary emission, δ, has been taken from the monograph by Bronstein and Freiman (1969) where it comprises $\delta/(1 - r)$. By comparing the coefficients of true secondary electron emission for various metals we arrive at the conclusion that they are very close to each other. Hence $P_e^{**}(\varepsilon_K)$ are also close. Taking this into account let us in examining the potential electron emission take only the dependence $P_e(\varepsilon_K)$ which is used by Hagstrum.

It will be noted that in examining P_e^{**} we did not take into account the asymmetry of the angular distribution of primary electrons, the contribution to the elastically reflected secondary electrons and the reverse reflection on the surface barrier of true secondary electrons. However, the last factor compensates the former two.

Parilis (1969a) has proposed another method of estimating the function $P_e(\varepsilon_K)$, allowing to keep the dependence on distance s from the metal surface. This method gives for the probability of electron escape

$$P_e(s, \varepsilon_K) = \frac{1}{2}\left\{1 - \left(\frac{\varepsilon_0}{\varepsilon}\right)^{0.5} \exp[-as(\varepsilon_K/\varepsilon_0 - 1)]\right\}. \tag{9}$$

P_e depends on s in such way that if $s \to \infty$ then $P_e \to \frac{1}{2}$, if $s = 0$, then $P_e \to P_{sym}$ and if as varies from 1 to 2 then P_e is close to P_{Hag}.

A good approximation for the function of the electron density within the filled part of the conduction band in the case of tungsten is either $N_c(\varepsilon) = K_1$

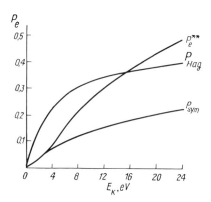

Fig. 4. Various types of dependence $P_e(\varepsilon_K)$.

or $N_c(\varepsilon) = K_2\varepsilon$ (Hagstrum 1954a). Further, we shall restrict ourselves to the first of these approximations and use it for other metals, too. With this

$$T(\varepsilon) = K_1^2\varepsilon , \qquad\qquad 0 < \varepsilon < \tfrac{1}{2}\varepsilon_F ,$$
$$\qquad = K_1^2(\varepsilon_F - \varepsilon) , \quad \tfrac{1}{2}\varepsilon_F < \varepsilon < \varepsilon_F , \qquad\qquad (10)$$
$$\qquad = 0 , \qquad\qquad \varepsilon < 0, \ \varepsilon > \varepsilon_F .$$

Taking into account eqs. (2) and (3), one derives

$$\gamma = \frac{\displaystyle\int_{\varepsilon_0}^{\infty} \sqrt{\varepsilon_K} P_e(\varepsilon_K) T[\tfrac{1}{2}(\varepsilon_K + \varepsilon_0 - E_i)]\, d\varepsilon_K}{\displaystyle\int_{\varepsilon_K}^{\infty} \sqrt{\varepsilon_K} T[\tfrac{1}{2}(\varepsilon_K + \varepsilon_0 - E_i)]\, d\varepsilon_K} = \frac{I_2}{I_1} . \qquad (11)$$

The ratio given in eq. (11) at the above-mentioned assumptions can be taken down in a clearly analytical but extremely cumbersome form. An attempt is, therefore, made to obtain simple dependences of γ on the individual parameters.

Coefficient γ depends on three parameters – the work function φ, the Fermi energy ε_F and the position of the vacant term E_i. However, if $T(\varepsilon)$ is a homogeneous function of its argument of the nth order ($n = 1$ at $N_c = K_1$; $n = 2$ at $N_c = K_2\sqrt{\varepsilon}$), then

$$I_j(\alpha\varphi, \alpha\varepsilon_F, \alpha E_i) = \alpha^{3/2+n} I_j(\varphi, \varepsilon_F, E_i) , \quad j = 1, 2 . \qquad (12)$$

Hence,

$$\gamma(\alpha\varphi, \alpha\varepsilon_F, \alpha E_i) = \gamma(\varphi, \varepsilon_F, E_i) .$$

It is sufficient to determine the behavior of γ at any arbitrary change of any two parameters to be able to judge the behavior of γ when the third parameter changes too.

Let us begin by examining the dependence of the coefficient γ on the Fermi energy ε_F at fixed values of φ and E_i. Curve γ_I in fig. 5 illustrates $\gamma_I = \gamma(\varepsilon_F)$ at $\varphi = 4$ eV, $E_i = 20$ eV in the interval of $4 < \varepsilon_F < 12$ eV, which with the increase of ε_F diminishes as const/ε_F. The function $\gamma_{\varphi,E_i}(\varepsilon_F)$ behaves analogically at other fixed values of φ, E_i [see curve γ_{II}, fig. 5, which corresponds to $\gamma_{II} = \gamma(\varepsilon_F)$ at $\varphi = 4.5$ eV; $E_i = 14$ eV]. For comparison, fig. 5 also illustrates sections of the hyperbolic line $1.6/\varepsilon_F$ and $0.33/\varepsilon_F$. They coincide with the rated curves γ_I, γ_{II} with a precision of $\pm 10\%$. Subsequently

$$\gamma_{\varphi,E_i}(\varepsilon_F) \approx \frac{\text{const}}{\varepsilon_F} , \qquad\qquad (13)$$

where the const depends on φ and E_i.

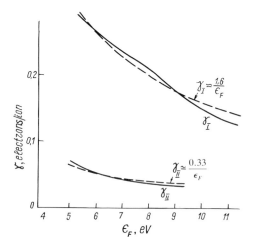

Fig. 5. Total yield versus Fermi energy ε_F at fixed values of φ and E_i.

At lower values of ε_F outside of the mentioned interval the growth of γ is slowed down with the reduction of ε_F ($\lim_{\varepsilon_F \to 0} \gamma_I = 0.42$; $\lim_{\varepsilon_F \to 0} \gamma_{II} = 0.345$), although this carries no practical interest since in the case of metals $\varepsilon_F \sim 5$ eV.

Let us now examine the dependence of the coefficient γ on the work function φ at fixed values of ε_F and E_i. As an example, fig. 6 illustrates $\gamma_{III} = \gamma(\varphi)$ at $\varepsilon_F = 6$ eV, $E_i = 14$ eV and $\gamma_{IV} = \gamma(\varphi)$ at $\varepsilon_F = 6$ eV, $E_i = 20$ eV at a change of φ from 0 to $\frac{1}{2}E_i$.

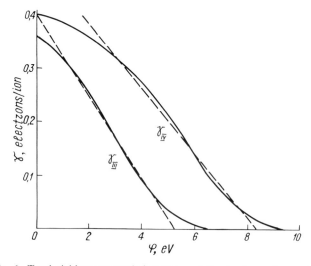

Fig. 6. Total yield versus work function φ at fixed values of ε_F and E_i.

As it appears from fig. 6, the central sections of these curves provide a satisfactory approximation by the linear dependences,

$$\gamma_{\text{III}} \approx 0.4 - 0.076\varphi , \qquad \gamma_{\text{IV}} \approx 0.5 - 0.060\varphi . \tag{14}$$

With the reduction of the work function within the range of small φ, the growth of $\gamma(\varphi)$ slows down. Since we have derived the dependence of γ on φ and ε_{F}, let us now derive the dependence $\gamma(E_i)$. From eqs. (12) and (13), it follows that

$$\gamma(\varphi, \varepsilon_{\text{F}}, E_i) = \frac{1}{\alpha} \gamma\left(\varphi, \frac{\varepsilon_{\text{F}}}{\alpha}, E_i\right) = \frac{1}{\alpha} \gamma(\alpha\varphi, \varepsilon_{\text{F}}, \alpha E_i) , \tag{15}$$

i.e., at fixed values of ε_{F},

$$\gamma_{\varepsilon_{\text{F}}}(\varphi, E_i) = \frac{1}{\alpha} \gamma_{\varepsilon_{\text{F}}}(\alpha\varphi, \alpha E_i) , \tag{16}$$

This ratio may be used in two ways. In the first place, by fixing α we can derive from the given curve $\gamma_{\varepsilon_{\text{F}}, E_i}(\varphi)$ new and analogous curves at other fixed values of E_i. Secondly, having fixed the product $\alpha\varphi = \Phi$ and knowing $\gamma_{\varepsilon_{\text{F}}, E_i}(\varphi)$, one derives $\gamma_{\varepsilon_{\text{F}}, \Phi}(E_i)$.

Taking eq. (14) in the form of

$$\gamma_{\varepsilon_{\text{F}}, E_i}(\varphi) = a - b\varphi , \tag{17}$$

one obtains

$$\gamma_{\varphi}(\alpha E_i) = \alpha(a - b\varphi) = \frac{(\alpha E_i)a}{E_i} - b\Phi . \tag{18}$$

Changing the symbols of the component values,

$$\Phi \rightarrow \varphi , \quad \alpha E_i \rightarrow E_i , \quad E_i \rightarrow E_i^* ,$$

and taking into account the dependence of γ on ε_{F}, we derive the final form,

$$\gamma(\varphi, \varepsilon_{\text{F}}, E_i) = \frac{6}{\varepsilon_{\text{F}}}\left(\frac{aE_i}{E_i^*} - b\varphi\right) = \frac{3b}{\varepsilon_{\text{F}}}\left(\frac{2aE_i}{bE_i^*} - 2\varphi\right) . \tag{19}$$

Substituting the numerical values of a, b, E_i^*, we derive two ratios,

$$\gamma_1 \approx \frac{0.23}{\varepsilon_{\text{F}}}(0.75E_i - 2\varphi) , \tag{20a}$$

$$\gamma_2 \approx \frac{0.18}{\varepsilon_{\text{F}}}(0.83E_i - 2\varphi) . \tag{20b}$$

These approximated expressions for $\gamma(\varphi, \varepsilon_{\text{F}}, E_i)$ are very close and differ by about $\pm 10\%$ at $E_i > 3\varphi$, i.e., they are as correct as ratios (13) and (14). Using the precise dependences $\gamma_{\varphi, E_i}(\varepsilon_{\text{F}})$ and $\gamma_{\varepsilon_{\text{F}}, E_i}(\varphi)$ we would have derived two coinciding ratios instead of eqs. (20a) and (20b). Hence, with an accuracy of $\pm 10\%$,

$$\gamma(\varphi, \varepsilon_F, E_i) \approx \frac{0.2}{\varepsilon_F} (0.8E_i - 2\varphi) . \tag{21}$$

Greater deviations from this dependence should be observed within the range of the potential emission threshold. As shown by the detailed analysis of ratios (4) and (11) at $E_i < 3\varphi$,

$$\gamma \approx \frac{0.093(E_i - 2\varphi)^{5/2}}{I_1} . \tag{22}$$

Deviations from ratio (21) should also be observed at higher values of E_i, e.g., for deep levels in multiple-charged ions. At $E_i > 2\varepsilon_0$, the coefficient γ changes in a nonlinear fashion and has a tendency to saturation,

$$\gamma \approx 0.5 - 0.16\sqrt{\frac{\varepsilon_0}{E_i - \varepsilon_0 - \varphi}} . \tag{23}$$

Within the intermediary ionization energy range ratio (21) is applicable and the ranges of application of eqs. (21)–(23) overlap.

The ratios derived provide a satisfactory explanation for the experimental results.

(a) In experiments by U.A. Arifov and Rakhimov (1960) a linear dependence of $\gamma(\varphi)$ has been observed during the bombardment of a number of metal targets by Ne^+ and Ar^+ ions. This is in agreement with ratio (21). Deviations of experimental points from the linear dependence in the above-mentioned work have been small. This is due to the fact that the Fermi energy, ε_F, changes only little during the transition from one metal to another. However, it is possible that the dependence of the coefficient γ on ε_F is weaker than follows the above-mentioned calculations.

(b) The approximate linear dependence $\gamma(E_i)$ during the bombardment of one metal by various inert gas ions is confirmed by experiments of U.A. Arifov (1968) and Hagstrum (1954b).

(c) At small values of the difference $E_i - 2\varphi$, the experimental values of $\gamma(E_i)$ (Hagstrum 1954b) and $\gamma(\varphi)$ (U.A. Arifov and Rakhimov 1960) deviate from the linear dependence in qualitative agreement with ratio (22).

(d) The results of this investigation were testified in the experiment of Oechsner (1976). The total yield γ of ion-induced electron emission has been measured for clean polycrystalline surfaces of 11 metals for normally incident Ar^+ ions of 1.05 keV. For Ti and Cu, the measurements have been extended to Kr^+ and Xe^+. The γ values for the transition metals display a periodic variation with the atomic number of the target. The results are discussed on the basis of the theory of Hagstrum (1954a) and, as has been noted by Oechsner, can be surprisingly well described by the interpolation formula of Kishinevsky (1973) derived from this theory [see table 1 (Oech-

Table 1
Comparison of the experimental electron yield γ_{exp} for normal incident 1.05 keV Ar^+ ions of with with the yield γ_{calc} is calculated by eq. (21).

Target	φ (eV)	ε_F (eV)	γ_{calc} (el/ion)	γ_{exp} (el/ion)
Ti	3.9	6.5	0.144	0.148
Ni	4.8	8.0	0.075	0.078
Cu	4.5	8.7	0.083	0.082
Zr	4.0	5.3	0.170	0.140
Nb	4.4	5.3	0.137	0.137
Mo	4.2	7.3	0.133	0.115
Pd	5.1	5.7	0.086	0.077
Ag	4.4	7.2	0.105	0.088
Ta	4.2	7.7	0.105	0.117
W	4.5	7.4	0.091	0.102
Au	5.0	9.1	0.057	0.062

sner 1976)]. The good agreement of expression (21) with the experimental data were repeatedly noted by many experimentators in papers of the last two decades.

2.3. Auger neutralization of multiple-charged ions on metals

U.A. Arifov et al. (1971c, 1973a) measured the emission of electrons under bombardment of multiple-charged Ar, Kr, Xe, S and Cu ions on Mo and W targets. The dependences γ_q on both ion energy and charge multiplicity were registrated. Besides the multiple-charged ion, single-charged ion abundance ratios were determined for many beams.

The dependence of γ_q on Ar, Kr and Xe ion energy for Mo targets are exhibited in figs. 7–9. It is seen that the coefficient of potential emission for multiple-charged ions depends on the velocity of the bombarding ion: the higher is the velocity, the less is the value of γ_q, as found in the experiments by Hagstrum (1954b). With the increase of ion velocity, the time of flight through the distance of highest Auger transition probability decreases. As a result, with increased velocity the multiple-charged ions penetrate the target, undergoing merely partial neutralization, which leads to γ_q decrease. The higher is the ion charge multiplicity, the steeper is the descending part of the curves.

The values of the coefficients of electron emission under slow multiple-charged ion bombardment (Ne, Ar, Kr and Xe), obtained experimentally, are plotted in fig. 10 using data by Hagstrum (1954a) and U.A. Arifov et al. (1971c). Along the x-axis the values $W_q = \Sigma_{i=1}^{q} U^{i+} - 2\varphi$ were plotted (U^{i+} being the ith ionization potential, φ being the metal work function). As it is

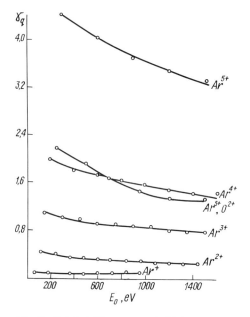

Fig. 7. Total yield γ_q versus Ar^{q+} ion energy.

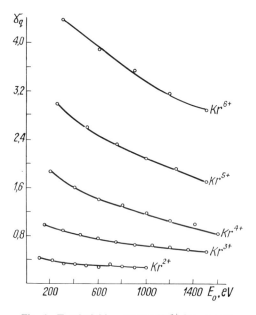

Fig. 8. Total yield γ_q versus Kr^{q+} ion energy.

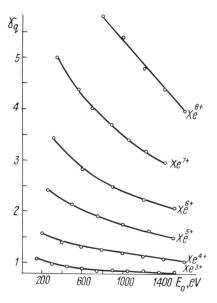

Fig. 9. Total yield γ_q versus Xe^{q+} ion energy.

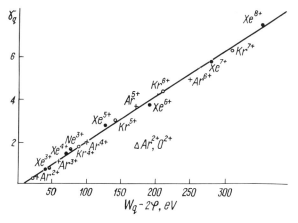

Fig. 10. Total yield γ_q versus $(W_q - 2\varphi)$.

seen, the experimental points do not differ much from the straight line

$$\gamma_q = K\left(\sum_{i=1}^{q} U^{i+} - 2\varphi\right) = K(W_q - 2\varphi), \tag{24}$$

where K is about $0.0183\,(\mathrm{eV})^{-1}$, and W_q is the sum of ionization potentials. The linear dependence γ_q on the total neutralization energy substantiates the assumption of stepwise behavior of the process and allows to predict the value γ_q for ions with higher charges q. For instance, the emission under slow 10-charged Zn ions should amount 20 electrons per ion and 11-charged Ca ions should amount 30 electrons per ion. Even the electron emission under slow multiply charged large clusters follows this rule reaching more than one hundred electrons per cluster (Mahoney et al. 1992).

2.4. Theory

The theoretical analysis of the emission of Auger electrons from metals under multiple-charged ion bombardment was carried out by U.A. Arifov et al. (1973b). As is seen from the experimental data obtained by U.A. Arifov et al. (1971c, 1973a), the coefficient of electron emission sharply increases with the increase of ion charge q. This is due to the fact that when a multiple-charged ion is approaching the metal it undergoes resonance and Auger neutralization into its highest excited states. By way of example, in fig. 11 the scheme of Ar^{6+} energy levels, corresponding to the successive transition of one (I), two (II) and more electrons from the metal into the ion is displayed. If the energy U^* of the level exceeds the double work

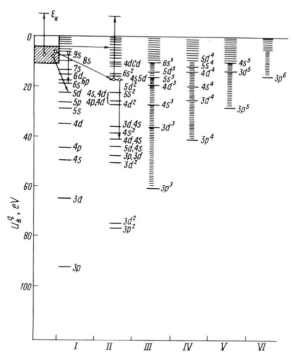

Fig. 11. Scheme of Ar^{6+} levels during the capture of one (I), two (II), and more electrons from metals.

function of the metal, 2φ, then the transition of an electron on this level gives an Auger process leading to the emission of an electron from the metal. In this case, the energy of ion neutralization $U^* - \varepsilon'$ is transferred to the electron being, e.g., on the ε'' level in the conduction band of the metal or on the corresponding resonance-excited ion level of the ion.

In the last case, the energy relations appeared to be the same, and the transition rate per time unit $W_t(s)$ is determined by the resonance charge transfer $W_n(s)$ as well as the Auger transition rate in the double-excited ion W_a.

The maximum energy of the emitted electrons is determined by the Auger transition between the most distant levels, which are not separated by any intermediate states.

For example, for Ne, Ar, Kr and Xe multiple-charged ions the shells $n + 1$ provide such levels (n is the main quantum number). The limiting energy is

$$E_{\lim} \approx U_q - 2|E^{(q-2)^{**}}|, \tag{25}$$

where U_q is qth ionization potential, and $E^{(q-2)^{**}}$ is the electron energy in the double-excited ion. The excitation energies can be determined following Slater (1930). In table 2, the limiting energies from eq. (25) are compared with the experimental ones as well as with the energy $U_q - 2\varphi$ corresponding to Auger neutralization of different ions W via direct transition onto the main level.

The relative rate of electron transition onto a given excited level of the multiple-charged ion is determined by the Auger neutralization rate per

Table 2
Limiting energies.

k	$E_{\lim}$ (eV)		$U_q - 2\varphi$ (eV)
	experiment[a]	theory[b]	
	$Ne^{q+} \rightarrow W$		
2	25	21	32
3	40	29	54
	$Ar^{q+} \rightarrow W$		
2	16	15	19
3	22	18	32
	$Kr^{q+} \rightarrow W$		
2	13	13	17
3	18	16.4	28
4	24	22	43
	$Xe^{q+} \rightarrow W$		
2	10	19	12
3	15	14.3	23
4	18	17.2	36
5	24	17	48

[a] Hagstrum (1954b). [b] Eq. (25).

time unit,

$$W = \frac{2\pi}{\hbar} |\langle \psi_{nl}^*(r_1)\psi_f^*(r_2)| \frac{e^2}{r_{12}} |\psi_{i'}(r_1)\psi_{i''}(r_2)\rangle|^2 = \frac{2\pi}{\hbar} |H_{if}|^2 , \qquad (26)$$

where ψ_{nl} is the electron wave function for the ion excited state, ψ_f is the Auger electron wave function, and $\psi_{i'}$ and $\psi_{i''}$ are electron wave functions on the ε' and ε'' levels in metal, respectively. They are written as (Sternberg 1957)

$$\psi_i = (\kappa L^{3/2})^{-1} \exp[i(k_{0x}x + k_{0y}y)][(\underline{k}_{0z} + k_{0z}) \exp(i\underline{k}_{0z}z)$$

$$+ (\underline{k}_{0z} - k_{0z}) \exp(-i\underline{k}_{0z}z) , \qquad \text{inside the metal} ,$$

$$\psi_i = (\kappa L^{3/2})^{-1} \exp[i(k_{0x}x + k_{0y}y)]2\underline{k}_{0z} \exp(ik_{0z}z) , \quad \text{outside the metal} .$$

$$(27)$$

Here L is the linear size of the box, in which the wave function was normalized,

$$\kappa = \left(\frac{2m}{\hbar^2} \varepsilon_0\right)^{1/2} , \qquad k_0^2 = \frac{2mE_k}{\hbar^2} , \qquad \underline{k}_{0z}^2 = k_{0z}^2 + \kappa^2 ,$$

$$E_k = \varepsilon_k - \varepsilon_0 , \qquad \underline{k}_0^2 = k_{0x}^2 + k_{0y}^2 + \underline{k}_{0z}^2 = k_0^2 + \kappa^2 . \qquad (28)$$

The line under the wave vector means that it was measured with respect to the bottom of the well with depth ε_0.

The matrix element H_{if} expresses the dependence of W on the transition energy $U^* - \varepsilon'$, which, as mentioned above, is determined by the overlapping of the wave functions $\psi_{i'}$, $\psi_{i''}$, ψ_f and ψ_{nl}. With increase of the level depth onto which Auger electron transfers, W sharply decreases. To display such decreases, the wave function of bound state was chosen in the form

$$\psi_{nl} = 1/\sqrt{\pi} \exp(-r/b) .$$

In this case, the energy U^* corresponding to this state amounts $-13.6/b^2$ eV, i.e., the less the value of the parameter b is, the deeper level.

For deep levels at large distances s between the ion and metal surface the matrix element H_{if} is written as

$$H_{if} = \frac{128i\pi^{3/2}\underline{k}_{0z}'k_{0z}''}{L^{9/2}\kappa^3\underline{k}_z} (\lambda + i\underline{k}_{0z}'') \frac{F}{(F^2 + k_{0z}'^2)^2}$$

$$\times \left[-k_{0z}' + \frac{k_{0x}'(k_{0x}'' - k_x)}{\lambda} + \frac{k_{0y}'(k_{0y}'' - k_y)}{\lambda}\right] \exp[i(k_{0z}' - \lambda)s]$$

$$\simeq \exp[i(k_{0z}' - \lambda)s][U^* + \varepsilon' - \varepsilon_0]^{-2} . \qquad (29)$$

Here

$$\lambda = [(k_x - k''_{0x})^2 + (k_y - k''_{0y})^2]^{1/2},$$

$$F = [b^{-2} + k'^2_{0x} + k'^2_{0y}]^{1/2}.$$

From eq. (29) follows that

$$W \propto (U^* + \varepsilon' - \varepsilon_0)^{-4}. \tag{30}$$

The strong dependence of the transition probability on the level depth allows to exclude practically the direct transitions to the ground state of the multiple-charged ion as well as transitions to the first excited state. The dependence W on ε_k points to transitions proceeding mainly to high excited states. On the other hand, the transitions to higher levels which are close to the bottom of the conduction band resulting in the formation of slow Auger electrons the energy of which is hardly sufficient for overcoming the surface potential barrier, seem to be unlikely. This follows from the behavior of the matrix element H_{if} in which the electron wave function ψ_f at $\varepsilon_k < \varepsilon_0$ exponentially decreases outside the metal, but at $\varepsilon_k \geq \varepsilon_0$ is small outside the metal due to internal reflection from the barrier. In fact, Auger transitions occur mainly to levels with energy depth 15–30 eV, being the optimum value for ε_k. In accordance with the theory of Auger neutralization and experimental data the corresponding value γ (per one transition) amounts 0.1–0.4 (Hagstrum 1954a).

Propst (1963) suggested that approximately half of the electrons escaping metal in the potential emission process are caused by tertiary electrons formed under those Auger electrons which were injected in the metal or reflected from the surface potential barrier. If this is taken into consideration, then the electron yield per one Auger relaxation event should be somewhat higher,

$$\gamma \approx \tfrac{1}{2}[\sigma(E_k)(1 + \sqrt{\varepsilon_0/\varepsilon_k}) + (1 - \sqrt{\varepsilon_0/\varepsilon_k})]. \tag{31}$$

Here $\sigma(E_k)$ is the coefficient of the secondary electron emission under electron bombardment. In deriving eq. (31) it was taken into account that, at electron energies $E_k < 100$ eV, $\sigma(E_k)$ is almost independent on the angle of incidence (Bruinning 1958). Then the value of the electron yield per one elementary event can, according to eq. (31), exceed $\tfrac{1}{2}$.

The theoretical and experimental values of electron yield increment $\Delta\gamma(A^{q+})$, corresponding to single-step partial neutralization $A^{q+} \to A^{(q-1)+}$ with unit decrease of the charge multiplicity are displayed in table 3.

Further development of the transitions takes the following course. The first electron drops down to one of the ion excited states lying below, transmitting the energy to a metal electron. The optimal energy of such transition also amounts 15–30 eV and, as a result, the second Auger electron

Table 3
The increment of electron yield.

$\Delta\gamma(A^{q+})$	Theory	Experiment
$\Delta\gamma(Ne^{2+})$	0.55	0.45
$\Delta\gamma(Ne^{3+})$	0.84	1.1
$\Delta\gamma(Ar^{2+})$	0.32	0.3
$\Delta\gamma(Ar^{3+})$	0.59	0.8
$\Delta\gamma(Kr^{2+})$	0.28	0.25
$\Delta\gamma(Kr^{3+})$	0.51	0.5
$\Delta\gamma(Kr^{4+})$	0.73	1.0
$\Delta\gamma(Kr^{5+})$	1.0	1.06
$\Delta\gamma(Kr^{6+})$	1.3	1.6
$\Delta\gamma(Kr^{7+})$	1.6	1.4
$\Delta\gamma(Xe^{2+})$	0.22	0.22
$\Delta\gamma(Xe^{3+})$	0.45	0.5
$\Delta\gamma(Xe^{4+})$	0.62	0.85
$\Delta\gamma(Xe^{5+})$	0.9	1.2
$\Delta\gamma(Xe^{6+})$	1.22	0.96
$\Delta\gamma(Xe^{7+})$	1.5	1.4
$\Delta\gamma(Xe^{8+})$	1.6	1.2

is ejected. Such descents can proceed up until filling the ground state, the energy steps of the mentioned value being the most probable ones.

Simultaneously, an ion can capture the second electron into one of the excited states. The energy scheme of such states for Ar is shown in fig. 11.

These states of double excitation provide an additional possibility for the emission of Auger electrons from the metal. The above discussion on the stepwise neutralization appears to be also applicable to them. The same is referred to the second, the third electron and so on. The number of possible level combinations onto which they could be consequently captured increases with increasing number of captured electrons. The levels of symmetric capture, when all electrons are captured onto one shell were shown in fig. 11. The levels corresponding to the capture onto different shells form in the intervals between them a set shown in fig. 11. The levels corresponding to the successive capture of n electrons form a dense set of levels, which provide practically any possibility for Auger transitions. Besides, the states with two and more excited electrons appear to be autoionizing ones. Such can decay with electron ejection, making additional contribution into the electron emission. Formation of autoionizing states at the capture of two electrons from a metal by He^{2+} ions has been observed by Hagstrum and Becker (1973). In accordance with the general theory of Auger effects (Parilis 1969a), the transition probability is less the larger the Auger electron energy is. It is well-known that in the multiple-excited atoms the Koster–Krönig transitions in the upper shells, giving low-energy Auger electrons, are the most probable ones.

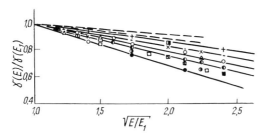

Fig. 12. $\gamma_p(E)/\gamma_p(E_1)$ versus $\sqrt{E/E_1}$ ($E_1 = 200$ eV). Broken curves: Ne^{3+}, $Ar^{2+,3+}$; $Kr^{2+,3+,4+}$, $Xe^{(2+)-(5+)}$ incident on W (Hagstrum 1954a). Points correspond to experimental data by U.A. Arifov et al. (1973a) for the multiple-charged ions Ar^{q+}, Kr^{q+} and Xe^{q+} (see figs. 7–9).

Both the experimental data by Hagstrum (1954b) and above considerations have shown that as a result of series of stepwise Auger transitions the multiple-charged ion is completely neutralized, and the total number of the emitted electrons should be proportional to the total ion neutralization energy, which explains the dependence $\gamma_q(W_q)$ (see fig. 10).

The values γ_q plotted in fig. 10 versus W_q refer to the low-velocity ions. As it is seen from the experimental data by U.A. Arifov et al. (1971c, 1973a), the emission of electrons under slow multiple-charged ion bombardment decreases with the increase of the bombarding ion velocity, this decrease being the faster the higher the ion charge is. The explanation of this dependence was brought forward in the theory given below.

The experimental data on the potential ion–electron emission under multiple-charged ion bombardment, obtained until the middle of the seventies, i.e., before the development of the theory (Hagstrum 1954b, U.A. Arifov et al. 1971b,c, 1973a, 1974, T.U. Arifov et al. 1976, Perdrix et al. 1968) cover both a large number of ion charges and a wide velocity range, but there was no investigation performed in any experiment for the energy range from $\simeq 100$ eV to several hundreds keV. The different experimental data correlate either qualitatively or semiqualitatively. In addition to the above-mentioned experimental data on potential emission decrease with the increase of ion energy, which is the faster the higher the ion charge is, it was shown that in the family $\gamma_p(E, q)/\gamma_p(E_1, q)$, where E_1 is the given energy, the curves are close to each other (see fig. 12). As far as, at mean velocities, the potential emission is observed against a background of kinetic emission, it was interesting to obtain the dependence $\gamma_p(E, q)$ theoretically (U.A. Arifov et al. 1973a, Kishinevsky and Parilis 1978). In the first of these papers, a simple theory concerning the ion velocity dependence of potential electron emission under multiple-charged ions has been given.

The filling of vacant levels in the outer shell occurs via a series of step-by-step Auger transitions with the mean step ε and the mean probability

$$W(s) = A\,e^{-as}, \quad s \geq 0,$$
$$\qquad = A, \qquad s < 0. \tag{32}$$

In eq. (32), it was taken into account that with decreasing distance the probability $W(s)$ increases not exponentially but tends to a constant as a result of the potential barrier diffusion, beginning with the surface of the metal $s = 0$. Apparently, ε should increase with increasing q, which was substantiated experimentally by recording the increase of the mean energy of the emitted electrons.

In view of the great variety of competing Auger transitions it is expedient to introduce a notion of the mean number of Auger transitions $\nu(q)$ which transfer a q-charged ion into a neutral atom,

$$\nu(q) = \sum_{k=1}^{q} U_k/\varepsilon = W_q/\varepsilon. \tag{33}$$

A peculiar consideration is needed for the filling of a vacancy in the inner shell for which the rate of transition, as a rule, is smaller but the released energy is rather larger.

In accord with section 2 of chapter 8, let us introduce P_i, the probability that $\nu-i$ transitions took place by time t, described by a system of equations,

$$dP_i/dt = -W_i P_i + W_{i+1} P_{i+1}, \quad i = 0, 1, \ldots, \nu, \tag{34}$$

where W_i is the Auger transition rate in the presence of i holes ($W_{\nu+1} = 0$). For the mean number of transitions $n(t)$ during the time t the following equation is written,

$$\frac{dn}{dt} = \frac{d}{dt}(\nu - i)P_i = W^*(t)(\nu - n), \tag{35}$$

where

$$W^*(t) = \sum_{i=1}^{\nu} W_i P_i \Big/ \sum_{i=1}^{\nu} iP_i = \sum_{i=1}^{\nu} W_i P_i/(\nu - n). \tag{36}$$

To calculate $W^*(t)$, the system of equations, eq. (34), has to be integrated analytically.

If Auger transitions are considered independent, then $W_i = iW(s)$, and $W^*(t) = W(s)$. In case of dependent events, W^* also sharply increases with s decrease and could be approximated by a formula of the type of eq. (32). Considering it, W^* is substituted by $W(s)$.

By integrating eq. (35), we get

$$n = \nu(1 - e^{-\tau}), \quad \tau = \int_{s}^{\infty} W(s)\,ds/v. \tag{37}$$

The coefficient of potential emission $\gamma_p(v, q)$ is connected with n by

$$\gamma_p = \int_{-\infty}^{\infty} \gamma_1(s)\, dn = v \int_{-\infty}^{\infty} \gamma_1(s) W(s) \exp\left[-\int W(\sigma)\, d\sigma/v\right] ds/v, \quad (38)$$

where $\gamma_1(s)$ is the mean electron yield per one transition at distance s from the surface.

Equations (37)–(38) are analogous to the corresponding formulas for both the transition probability during time t and the emission coefficient for the independent transitions, differing by a factor v only.

The dependence $\gamma_p(v)$ is determined mainly by the dependence $\gamma_1(s)$. The value $\gamma_1(s)$ decreases with decreasing s due to the level shift and broadening (Hagstrum 1954a), change of the shape of Auger electron angular distribution (Parilis 1969a) and different contributions from several channels (Kishinevsky et al. 1976). It could be described by an exponential dependence $(-\delta\gamma\, e^{-bs})$.

The decrease of $\gamma_1(s)$ inside the solid is connected with the escaping electron flux attenuation according to the exponential law $\exp[-|s|/\lambda]$.

Then

$$\gamma_1(s) = \gamma_0 - \delta\gamma \exp(-bs), \qquad s \geq 0,$$
$$= (\gamma_0 - \delta\gamma) \exp(-|s|/\lambda), \quad s < 0. \tag{39}$$

From eq. (38), taking into account eqs. (33), (37) and (39), we get

$$\gamma_p(v, q) = \gamma_p^0(q)\Bigg\{(1 - \delta\gamma/\gamma_0)(1 + v/a\lambda v_0)^{-1} \exp(-v_0/v)$$

$$+ \left[1 - \exp(-v_0/v) - \frac{\delta\gamma}{\gamma_0}\left(\frac{v}{v_0}\right)^{b/a} \int_0^{v_0/v} y^{b/a}\, e^{-y}\, dy\right]\Bigg\}, \tag{40}$$

where

$$v_0 = A/a, \qquad \gamma_p^0(q) = \gamma_p(v \to 0, q) = v\gamma_0 = W_q \gamma_0/\varepsilon. \tag{41}$$

According to eq. (40), the dependence of the coefficient of potential emission on q and v is described by several co-factors. As it is seen from eq. (41), $\gamma_p(v, q)$ is proportional to the total ionization energy. As mentioned above, indeed, the experimental values $\gamma_p^0(q) \propto (W_q - 2\varphi)$, i.e., at $q \geq 1$, when $W_q \gg 2\varphi$, they coincide with eq. (41). An additional term (-2φ) in eq. (41) permits to describe the cases where $q = 1$.

The linear dependence $\gamma_p(W_q)$ is confirmed by all available experimental data, obtained before the appearance of this theory as well as in recent publications [see eq. (24)]. The proportionality factor,

$$K = \gamma_p/W_q = \gamma_0/\varepsilon, \tag{42}$$

obtained in various investigations both for metals and for semiconductors and insulators is in the range 0.008–0.026. This follows from eq. (41), if consider $\gamma_0 = (0.2–0.4)$ and $\varepsilon = (15–30)$ eV. The dependence on the target parameters (the work function φ and Fermi energy ε_F) is included in γ_0 (subsection 2.2).

The increase of the metal work function (in case of insulators the depth of valence band) should lead to the decrease of γ_0. This explains the fact that for the Mo (111) face the emission is less than for the (110) face (U.A. Arifov et al. 1974). The dependence of the coefficient of the potential emission on work function (and the surface cleanness) should attenuate with the increase of ion charge, when the role of Auger transition with preliminary charge transfer increases, which was confirmed in experiments by T.U. Arifov et al. (1976).

If the impinging multiple-charged ion has not time enough to be neutralized completely at approaching the surface, it would lead to a decrease in the kinetic electron emission by a value $\Delta\gamma_k$.

Indeed, the cross section of ionization by a multiple-charged ion $\sigma_-(q)$ is less than at the collision of two neutral atoms by a value $\Delta\sigma(q)$, being equal to the stripping cross section.

Then, according to Parilis and Kishinevsky (1961),

$$\Delta\gamma_k = \int N\,\Delta\sigma(q(s))\exp(-|s|/\lambda)\,ds \;, \tag{43}$$

where $q(s)$ is the ion charge at depth s.

The charge $q(s)$ is connected with the number of transitions $n(s)$,

$$q(s) \leqslant \nu - n(s) = \nu \exp\left[-\int_s^\infty W\,ds/\nu\right], \tag{44}$$

where the sign $\leqslant$ appears because charge $q \leqslant \nu$.

For estimation, we take

$$\Delta\sigma = \sigma_{str}\,, \quad q(s) \geqslant 1\,,$$
$$= 0\,, \qquad q(s) < 1\,. \tag{45}$$

Then

$$\Delta\gamma_k = N\lambda\sigma_{str}(1 - \exp(-|s_n|/\lambda))\,, \tag{46}$$

where $|s_n|$,

$$|s_n| = 1/a[1 - (v/v_0)\ln\nu]\,, \tag{47}$$

is determined by the condition $q(s_n) = 1$ and means the path inside the solid, on which the multiple-charged ion is neutralized down to a single-charged one.

Considering that $\gamma_k = N\lambda\sigma_-$, we get

$$\Delta\gamma_k = \gamma_k[\sigma_{str}/\sigma_-][1 - \exp(-|s_n|/\lambda)] . \tag{48}$$

The ratio σ_{str}/σ_- slightly depends on ion velocity, being a function of atomic numbers of the colliding particles.

The obtained value $\Delta\gamma_k$ is overestimated, since $|s_n|$ was overestimated. The correction $\Delta\gamma_k$ is not important as long as $\gamma_k \ll \gamma_p$, i.e., with velocities $v \ll 10^8$ cm/s. With account of the correction $\Delta\gamma_k$, the total coefficient of electron emission under multiple-charged ion bombardment is written as

$$\gamma(v, q) = \gamma_p(v, q) + \gamma_k(v) - \Delta\gamma_k(v, q) , \tag{49}$$

The approximately linear decrease of γ_p with velocity increase at low v (see fig. 12), if $a = b$, follows from eq. (40). In this case,

$$\gamma_p = \gamma_p^0(q)\{1 - [v/v_0][\delta\gamma/\gamma_0][1 - \exp(-v_0/v)]$$
$$- (1 - \delta\gamma/\gamma_0)(1 + v/a\lambda v_0)^{-1} \exp(-v_0/v)\} . \tag{50}$$

Equation (50) allows to explain the experiments by Hagstrum (1954b), U.A. Arifov et al. (1971c, 1973a) and others.

1. Low velocities $(v \ll v_0)$. In this velocity range $\gamma_p(v, q)$ decreases linearly with increasing v,

$$\gamma_p \approx \gamma_p^0(q)(1 - v/v^*) , \quad v^* = v_0\gamma_0/(\delta\gamma) = A\gamma_0/(a\,\delta\gamma) . \tag{51}$$

To fit the data by Hagstrum (1954b), it is necessary to take $v^* = 3.3 \times 10^7$ cm/s (Ar^{q+}), $v^* = 2.2 \times 10^7$ cm/s (Kr^{q+}), $v^* = 1.65 \times 10^7$ cm/s (Xe^{q+}); and for the data by U.A. Arifov et al. (1971a, 1973a) $v^* = 2 \times 10^7$ cm/s (Ar^{q+}), $v^* = 1.1 \times 10^7$ cm/s (Kr^{q+}), $v^* = 9 \times 10^6$ cm/s (Xe^{q+}).

The parameter v^* could be estimated theoretically. For this purpose, let us estimate the rate of Auger ionization of a Coulomb center with charge q. Its wave function is $\psi = q^{3/2}/\sqrt{\pi}\, e^{-qr}$. In metals, the electron wave functions are plane waves, and the interaction between the electrons which causes the Auger transition with account of screening can be chosen as $r_{12}^{-1} \exp(-r_{12}/L_0)$ ($L_0 = 0.5$–0.8 Å being the Debye screening length).

Inside the metal (in atomic units),

$$W \equiv A \approx n^2/(2\pi)\overline{|V|^2 p_3} , \tag{52}$$

where

$$V = \frac{32\pi^{3/2}q^{5/2}}{(|\boldsymbol{p}_2 - \boldsymbol{p}_3|^2 + L_0^{-2})(|\boldsymbol{p}_1 + \boldsymbol{p}_2 - \boldsymbol{p}_3|^2 + q^2)^2} , \tag{53}$$

where n is electron density in the conduction band (for a number of metals $n \approx 0.1$ el/a_0^3), $\boldsymbol{p}_1$, $\boldsymbol{p}_2$ are the electron momenta in the initial state, $\boldsymbol{p}_3$ is

Auger electron momentum. The line means the averaging over electron momenta in the initial state.

With the changing level depth $U = \frac{1}{2}q^2$, the function A decreases by $\approx U^{-2}$ in the range $20 < U < 40$ eV, and further $A \approx U^{-3}$. For $\varepsilon = 15$–30 eV, $A = (1$–$2) \times 10^{15} \, \text{s}^{-1}$. Approximately the same value of A can be obtained at the description of the wave functions of a metal by orthogonalized plane waves.

Outside the metal surface, W decreases like the density of the conduction band electrons,

$$W = F(s) \exp(-2\sqrt{2\varepsilon_e} s) \approx A \, e^{-as} , \qquad (54)$$

where $a = 2\sqrt{2\varepsilon_e}$ a.u.

At $\varepsilon_e = 4$–10 eV, $a = (2$–$3) \times 10^8 \, \text{cm}^{-1}$, which agrees with the results by Hagstrum (1954b).

Using the obtained estimation for A, a and considering $\Delta\gamma/\gamma_0 = 0.2$–$0.3$, we get $v_0 = (0.3$–$1.0) \times 10^7$ cm/s, and $v^* = (1$–$5) \times 10^7$ cm/s, which agrees with the above-mentioned estimations of this value based on the results by Hagstrum (1954b), U.A. Arifov et al. (1971a, 1973a).

2. High velocities. At $v_0 < v < a\lambda v_0$,

$$\gamma_p = \gamma_p^0(q)(1 - \delta\gamma/\gamma_0) + O(v/v_0) , \qquad (55)$$

i.e., γ_p slightly depends on v, as was observed by T.U. Arifov (1976), and U.A. Arifov et al. (1971b, 1974). At $v > a\lambda v_0$, we have $\gamma_p \approx 1/v$.

Let us consider the dependence on v of the total emission coefficient $\gamma(v, q) = \gamma_p + \gamma_k - \Delta\gamma_k$.

In fig. 13, the theoretical curves of $\gamma(v)$ are shown for the following

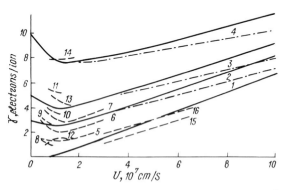

Fig. 13. Total yield γ versus velocity v. Theory: (1) γ_k, (2)–(4) γ by $\gamma_p^0(q) = 3$, 5 and 10, respectively. Full curves are without account of $\Delta\gamma_k$, chain curves are with account of $\Delta\gamma_k$. Experiment: (5)–(7) Ar^{3-}–Ar^{5+} incident on Mo (110) (U.A. Arifov et al. 1974), (8)–(11) Cu^{3+}–Cu^{6+}, (12)–(14) Al^{3+}–Al^{5+} incident on Ta (T.U. Arifov et al. 1976), (15) Ar, (16) Ar^{3+} incident on Mo (Perdrix et al. 1968).

parameters: $v_0 = 10^7$ cm/s, $\delta\gamma/\gamma_0 = 0.3$, and $a = 2 \times 10^8$ cm. The general behavior of $\gamma(v)$ is the following: a decrease at low velocities, a minimum nearby the threshold of kinetic emission, a displacement of the minimum to the right with increasing q, and a weak dependence of $\gamma(v)$ in the velocity range $(1-3) \times 10^7$ cm/s. These regularities agree with the above-mentioned data and manifest themselves more clearly with the account of the correction for the decrease of kinetic emission.

A very weak dependence or even independence of the total emission coefficient on energy in the energy range 1–10 keV has been observed for a wide variety of multiple-charged ions bombarding Au, Mo and CuBe in the experiments by Cano (1973). However, these experiments were carried out under poor vacuum conditions, using some gas-covered targets.

The shortcoming of eqs. (40) and (50), obtained phenomenologically, appears to be the great number of parameters induced to describe the $\gamma_p(v, q)$ decrease with velocity increase being influenced by the surface and bulk effects.

As is seen from fig. 14, an approximately identical decrease in a wide range of velocities could be achieved at different selection of values $\delta\gamma/\gamma_0$ and $a\lambda$.

Perhaps it seemed more preferable to use less parameters, considering the beam attenuation of the emitted electrons both inside and outside the target to be described by a united exponential factor, taking $\delta\gamma = 0$. In this case, we have to suppose the electron boundary of the metal to be shifted in the direction of the multiple-charged ion. Equation (50) is written as

$$\gamma_p(v, q) = \gamma_p^0(v, q)[1 - \exp(-v_0/v)/(1 + v_A/v)], \tag{56}$$

where $v_A = a\lambda v_0$.

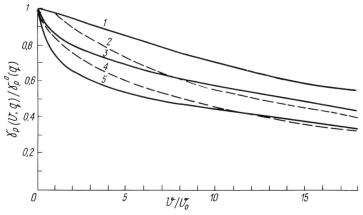

Fig. 14. $\gamma_p(v, q)/\gamma_p^0(q)$ versus v/v_0. (1), (2) $\delta\gamma/\gamma_0 = 0$,(3),(4) $\delta\gamma/\gamma_0 = 0.2$, (5) $\delta\gamma/\gamma_0 = 0.4$. For cases (1), (3), and (5), $a\lambda = 20$, for cases (2) and (4), $a\lambda = 10$.

At $v \gg v_0$, we get

$$\gamma_p(v, q) = \gamma_p^0(v, q)(1 + v_0/v_A)/(1 + v/v_A) . \tag{57}$$

However, at the consideration of the dependence of $\gamma_p(v, q)$ in a wide velocity range, which includes the low velocities ($\leqslant v$), eq. (50) must be used.

Recently, S.P. Apell and Monreal (1989) have proposed an analogous very simple model, which describes the $\gamma_p(v, q)$ decrease with increasing velocity. However, this decrease is connected only with the decrease of the number of Auger transitions which had time to take place before the multiple-charged ion approaches the solid boundary. The electron escape from the subsurface region inside the solid was neglected. Elsewhere, P. Apell (1988) has drawn the conclusion that the decrease of the potential emission under multiple-charged ion bombardment is caused both by the excitation of surface plasmons and the hard photon radiation competing with the Auger processes.

2.5. Recent investigations

2.5.1. Electron energy distribution

The electron energy distribution appear to be very informative; however, with increasing q the number of the processes which contribute to the spectrum increases sharply, which seldom fails to distinguish the elementary processes. In this respect, the experiments using the low-charged slow ions ($q = 2$) as well as the experiments using ions with a deep vacancy in the inner shells differ favorably.

In the experiments by Varga et al. (1982) the electron energy spectra of pure tungsten polycrystals bombarded by Ne^{2+}, $Ar^{2+,3+}$, Kr^{2+} and Xe^{2+} with energy 10–300 eV have been studied. At sufficiently low energies, 10–20 eV, in the spectra of electrons emitted by double-charged ions it is possible to distinguish approximately four components. However, with ion energies increasing up to 50–100 eV some striking peculiarities in the spectrum structure have disappeared (see fig. 15). The above-mentioned components the authors associated with the following processes:

$$X^{2+} \xrightarrow[\text{AN}]{\text{RN+AD}} X^{+*} \quad \text{(process 1)} ,$$

$$X^{+*} \xrightarrow{\text{AD}} X^{+} \quad \text{(process 2)} ,$$

$$X^{2+} \xrightarrow[\text{AN}]{\text{RN+AD}} X^{+} \quad \text{(process 3)} ,$$

$$X^{+} \xrightarrow{\text{AN}} X \quad \text{(process 4)} .$$

Later, Varga (1989) has distinguished the fifth spectrum component

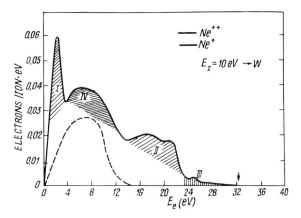

Fig. 15. Ejected electron energy distribution for impact of Ne^+/Ne^{2+} on clean tungsten (Varga et al. 1982). Hatched areas denote contributions from the following transitions: (I) first neutralization + deexcitation step of Ne^{2+} forming a low excited Ne^+ state, (II) deexcitation of this excited Ne^+ state resulting in one electron-like transition, (III) direct neutralization of Ne^{2+} into the Ne^+ ground state [by either Auger neutralization (AN) or resonance neutralization (RN) + Auger deexcitation (AD)], (IV) neutralization of the Ne^+ ground state following either steps (I) + (II) + (III).

connected with the autoionizing state formation. With increasing ion charge, the most drastic changes in the electron energy spectra were observed in the presence of a deep vacancy, e.g., a K-vacancy in N, O, Ne, and an L-vacancy in Ar, etc. The deexcitation of such a vacancy is connected with the release of a large amount of energy (several hundreds eV) carried out by a swift Auger electron in a single Auger transition. In the experiments by Zehner et al. (1986) and Meyer et al. (1987) on the interaction of N^{6+}, O^{7+}, Ar^{q+} ($q = 9$–12) with an Au target, K→LL electrons of nitrogen, oxygen, carbon (absorbed or inserted), L→MM electrons of argon, and $N_{6,7}$→VV electrons of Au were observed. The Auger electrons from C and Au gave evidence of the vacancy transfer from a multiple-charged ion to a target atom at their close approach, preceded by an Auger transition.

Similar experimental results were obtained by de Zwart (1987a,b) and de Zwart et al. (1989) for Ar^{q+}–W (polycrystal) interaction with $q = 9$–12. The electron peak at $E_{el} \approx 212$ eV for $q = 9$ was observed at low ion energy. At an ion energy of $E_0 = 2$ keV and angle of incidence $\theta = 45°$, the peak height halved. Above 5 keV a broad spectrum in the range 125–175 eV, which is typical for quasimolecular transitions, was observed. The analysis made by de Zwart (1987a,b) has indicated that the peak at 212 eV corresponded not to L→MM, but rather to L→MO transitions in Ar. The estimation of the swift electron intensity evidences in favor of a single electron emission per L vacancy. The observed Doppler shift of the peak corresponding to swift electrons testified that the filling of the deep vacancy in a slow ion ($\leqslant 1$ keV) occurred before its collision with the surface.

More swift ions have enough time to reach the surface having preserved the deep vacancies. The vacancies could be occupied with an electron from a deep shell of the target atom, which should lead to the ejection of a swift Auger electron from the target atom (with energies 125–175 eV). The total number of emitted electrons is reduced due to the comparatively large amount of energy carried out by the swift electrons.

2.5.2. Coefficient of swift electron emission

Recently, the coefficients of electron emission $\gamma(q, v)$ were measured for N^{q+} ($q \leqslant 6$), Ne^{q+} ($q \leqslant 7$), Ar^{q+} ($q \leqslant 12$), Ta^{q+} ($7 \leqslant q \leqslant 21$) on W and Au (Delauney et al. 1987, 1988, Fehringer et al. 1987) (see fig. 16).

As for low-charged ions, the decrease of $\gamma_p = \gamma - \gamma_k$ was observed with increasing ion velocity. For ions with low and middle charges, here the decrease of γ_p is less than the increment of γ_k, a minimum in the function $\gamma(v)$ was observed. With increasing q the decrease of γ_p exceeds of the increment of γ_k and the dependence of $\gamma(v)$ appears to be a decreasing one. The linear dependence $\gamma(W_q)$ was confirmed by us (see fig. 17). However, at $q = q^*$, if a deep vacancy appears, the occupation of which is connected with a swift electron emission, a break in the linear dependence $\gamma_p = kW_q$ was observed. Its appearance is understandable. Indeed, the contribution of swift electrons, Y_{swift}, to emission should be estimated apart. Thus, instead

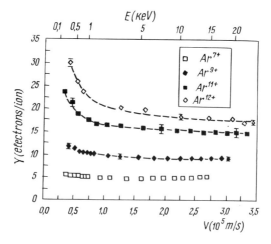

Fig. 16. Total electron yields γ for impact of Ar^{q+} on clean tungsten versus impact velocity v (Delauney et al. 1987).

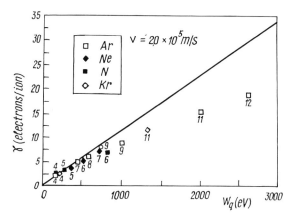

Fig. 17. Total electron yield γ for the impact of various ions on clean tungsten versus total ion
potential energy at a given impact velocity of 2×10^7 cm/s (Delauney et al. 1987).

of eq. (33), we get

$$\nu = \nu_{slow} + \nu_{swift} ,$$

$$\nu_{slow} = [W_q - (q - q^*)E^*]/\varepsilon , \tag{58}$$

$$\nu_{swift} = q - q^* .$$

Here, ν_{slow} and ν_{swift} are the numbers of transitions, E^* is the energy
expended for formation of a single swift electron. Equations (40), (50) and
(57) concern the slow electrons.

The total coefficient of the potential emission is given by

$$\gamma_p = (\gamma_p)_{slow} + (\gamma_p)_{swift} \approx (\gamma_p)_{slow} , \tag{59}$$

since $(\gamma_p)_{swift} < 0.5(q - q^*) \ll (\gamma_p)_{slow}$. Taking into account these relations,
we get

$$\gamma_p = kW_q , \qquad\qquad \text{at } q \leq q^* ,$$
$$= kW_q - (q - q^*)E^* , \quad \text{at } q > q^* . \tag{60}$$

which describes both the break and the linear dependence $\gamma_p(W_q)$ before
and after break.

Probably the decrease of the charge q^*, at which the break was observed,
with the increase of the ion energy is caused by a weak dependence of $\gamma_k(q)$
for metals, as was discussed in section 2.4.

Mirakhmedov and Parilis (1988) have studied theoretically the Auger
neutralization of highly charged ($q \leq 36$) and many-vacancy krypton near
the metal surface. It was shown that 95% of the electrons ejected during
Auger neutralization came from the transitions in the outer shells and the

fraction of swift electrons did not exceed 5%. In these transitions ≃40% of the total neutralization energy was released. Approximately 40% of the energy released during inner shell filling is carried away by X-rays. The dependence of the Auger electron yield on the energy W_q released at neutralization of Kr^{q+} on a metal surface is shown in fig. 18. This dependence is in agreement with eq. (60).

2.5.3. Other topics

In some investigations the question about the distance s_n at which the interaction of a multiple-charged ion with the surface became important was posed. According to the estimations made by Delauney et al. (1987), $s_n \simeq q$, but the estimation by P. Apell (1987), according to which the electron capture begins at the distance

$$d_n(\text{Å}) = 3q ,$$

seems to be overvaluated. The estimation by Snowdon (1988) appears to be more real,

$$d_n(\text{Å}) = q + 3.7 . \tag{61}$$

The role of the image forces and the level shift induced by them is discussed. The ion which suffers attraction in the field of this forces is accelerated and acquires an energy $(2–4)q$ eV (Varga 1989). This energy should be taken into account for slow ions. The level shift leads to a decrease of the binding energy by the value $(2q − 1)/4s$, which corresponds to the value 1–2 eV per transition.

It should be noted that the electrons are captured into some Rydberg states. Apparently, the Auger transitions are succeeded by the formation of a multiple-excited ion or atom. Auger relaxation of a system with low

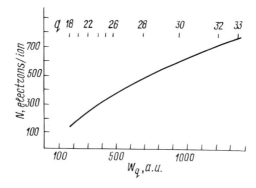

Fig. 18. Total Auger electron number N versus W_q for impact of Kr^{q+} ions on a metal surface.

charge takes place close to the surface, as follows from the above estimations.

2.6. Electron emission from nonmetals

Electron emission from alkali-halide crystals has been intensively investigated in the recent 15–20 years (U.A. Arifov et al. 1971a). This emission is known to consist of potential and kinetic emission, the first one strongly increasing with ion charge. Up to now, at least for metals, the second one was considered to be independent of ion charge. In contrast to the emission from metals, the kinetic electron emission from alkali-halide crystals is characterized by a low-energy threshold and high emission coefficients at energies as low as several keV. There was no theoretical interpretation of these facts. The electron emission from NaCl under multiple-charged Al^{q+} ($q \leqslant 10$) ion bombardment has been studied by Morozov et al. (1979) and a new phenomenon was discovered, namely: a strong dependence of the kinetic emission on ion charge.

Kishinevsky and Parilis (1982) theoretically described both the kinetic and potential electron emission from alkali-halide crystals as well as its dependence on ion charge. The comparison of the theory with the experiment and some new experiments proposed will be described in the subsections of this section.

2.6.1. Potential emission of electrons
Kishinevsky and Parilis (1978) have proposed a model of Auger neutralization by multiple-charged ions as a result of a series of stepped Auger transitions with a mean transition rate (see the section 2.4),

$$W = A \exp(-a|s|), \quad s \leqslant 0 \text{ out the solid},$$
$$= A, \quad s > 0,$$

(62)

and average number of transitions

$$\nu(q) = W_q/\varepsilon,$$

(63)

the transitions being considered as independent. In this case, the coefficient of potential emission is given by

$$\gamma_p = \nu \int \gamma_1(s)W(s) \exp\left[-\int_s^\infty W(\sigma)\,d\sigma/v\right] ds/v.$$

(64)

Here, $\gamma_1(s)$ is the mean emission yield per a transition. In case of an insulator, the Auger transition rates inside the solid $A \approx 10^{14}\,s^{-1}$ (Kishinevsky 1974a) and the change in $\gamma_p(v_0)$ with ion velocity v_0 is caused mainly by the attenuation of the escaping electron flux according to the law

$$\gamma_1(s) = \gamma_1(0) \exp(-s/\lambda) . \tag{65}$$

By integrating eq. (64), one can get approximately

$$\gamma_p(v_0, q) \approx \gamma_p^0(q)[1 + k\lambda/(m_b\{v_A + v_0\}^2) + 1/(a\lambda)](1 + v_0/v_A)^{-1} . \tag{66}$$

Here $\gamma_p^0(q) = \nu\gamma_1(0) = \gamma_p(v_0 \rightarrow 0, q)$, $v_A = \lambda A$. The value $1/a\lambda \sim 10^{-2}$ is connected with Auger neutralization outside the crystal, and the term $k\lambda/(m_b\{v_A + v_0\}^2) \sim 10^{-2}$ appears as a result of ion stripping inside the solid [see eq. (79) below]. However, some of these ν transitions have to be successive, i.e., dependent. So, another approach should be discussed. Let us take the filling of q lowest vacancies in the q-charged ion as independent events, i.e., q intricated events, q chains of events occurring with an average frequency W. If an intricated event is a result, e.g., of k transitions, the rate of each transition being W_k, then $1/W = \Sigma_{i=1}^k 1/W_i$. For the neutralization of low-charge ions, $k = 1$ and $W = W_1$.

Since $\gamma_p^0(q)$ but not ν appears in the final equation, eq. (66) remains valid. Nevertheless, the proposed approach seems to be more suitable for considering the dependence of the coefficient of kinetic emission, γ_k, on q as well as for describing the Auger neutralization in those cases where it would be completed in the formation of an ion, and not a neutral atom as, e.g. at Auger neutralization of Al^{q+}.

What consequences can the Auger neutralization at $q = 1$ have? If one consider all q low-lying vacancies to be equivalent, then the number of independent transitions equals q, but during the last transition the emission equals zero. Hence, from the total emission one should subtract the emission occurring during the last transition, but simultaneously taking into account the fact that the mean emission per transition increases by $q(q - 1)$ times.

The probability that $q - 1$ events took place from q independent ones is given by

$$P_1(q, s) = q(1 - e^{-\tau})^{q-1} = \sum_{i=1}^q (-1)^{i+1} i C_q^i e^{-i\tau} , \tag{67}$$

where $\tau = \int_{-\infty}^s W \, ds/v$. Thus,

$$\gamma_1 = 1/q\gamma_p^0(q) \int_{-\infty}^{\infty} W P_1 \exp(-s/\lambda) \, ds/v$$

$$\approx 1/q\gamma_p^0(q) \sum_{i=1}^q (-1)^{i+1} i C_q^i/(1 + \alpha/i) , \tag{68}$$

$$\gamma_p(v_0, q) \approx \gamma_p^0(q)q/(q-1)$$

$$\times [(1+\alpha)^{-1} - q^{-1} \sum_{i=1}^{q} (-1)^{i+1} C_q^i /(1+\alpha/i)]$$

$$\approx \gamma_p^0(q)/(1+\beta_q \alpha) . \tag{69}$$

Here,

$$\alpha = v_0/v_A , \qquad \beta_q = \left(q - \sum_{i=1}^{q} i^{-1}\right) \Big/ (q-1) . \tag{70}$$

In deriving eq. (69), the expansion in a series over the powers of α/i, the summation over i and then the approximate convolution of the obtained series were performed, valid with accuracy up to several percents even at $\alpha \approx 1$. Some minor corrections in eq. (66) were neglected. The obtained result is equivalent to the substitution of v_A in eq. (66) by v_A/β_q, i.e., to the increase of v_A in 2–1.27 times for $q = 2$–10. This increase is caused by a prompted Auger neutralization, since the rate of one vacancy neutralization in an i-vacancy atom is $W_i = iW_1$, and for aluminium $i > 1$.

2.6.2. Kinetic electron emission

The Firsov mechanism (Firsov 1959) of inelastic energy losses at violent atomic collisions promoted the development of the theory of kinetic electron emission from metals (Parilis and Kishinevsky 1961). However, this theory is not enough to describe the kinetic emission from alkali-halide crystals. The emission of electrons under ion (or neutral atom) bombardment of alkali-halide crystals is caused mainly by the electron detachment from anions rather than from cations. The cross sections σ_0 of electron ejection at negative-ion–neutral-atom collisions is determined by the interatomic distance R_c at which the electron term of the quasimolecule exists in continuum (Bydin and Dukelsky 1956, Smirnov 1968),

$$\sigma_0(E_0) = \pi R_c^2 \{1 - V(R_c)/E_{c.m.}\} = \sigma_m(1 - E_c/E_0) , \tag{71}$$

which is confirmed experimentally in the energy range up to the several keV. Here $V(R)$ is atom–ion interaction potential, $E_{c.m.}$ and E_0 are their energies in center-of-mass and laboratory systems, respectively, $\sigma_m = \pi R_c^2$, and $E_c = V(R_c)E_0/E_{c.m.}$.

The few experimental findings in the energy range up to 200 keV (Lichtenberg et al. 1980) corroborate the fact that the cross section σ_- of electron formation at negative-ion–atom collisions exceeds σ_0. This is caused by the contribution to the Firsov mechanism. Besides this mechanism is the only adequate for the electron emission from cations. With ion velocities v_0 exceeding $v_t \geqslant 10^7$ cm/s, according to Kishinevsky (1962) one can get

$$\sigma_F = \beta(v_0 - v_t) , \tag{72}$$

where

$$\beta = 1.1 \times 10^{-24}(N_1/U_1 + N_2/U_2)(\sqrt{Z_1} + \sqrt{Z_2})(\sqrt[6]{Z_1} + \sqrt[6]{Z_2})^3 \text{ cm s eV}.$$

(73)

N_i, Z_i and U_i (eV) ($i = 1, 2$) are the electron numbers, nucleus charges and ionization potentials of the colliding atoms and ions, respectively. In eq. (73), the sharing of the inelastically transmitted energy among the atoms was taken into account (Filippenko 1962). If one considers the alkali-halide crystal as a set of diatomic molecules, then

$$\sigma_-(E_0) = \sigma_0(E_0) + \sigma_F(E_0),$$

(74)

where σ_F includes the ionization of both anions and cations. Considering eq. (74), one can represent the coefficient of the kinetic emission as

$$\gamma_k = Nw \int_0^{s_c} \sigma_-(v) \exp(-s/\lambda)\, ds = \gamma_0 + \gamma_F.$$

(75)

Here, $N = \frac{1}{2}d^3$ is the number of molecules in unit volume, d is the lattice constant, λ is the effective range of the electron beam attenuation, and w is the probability of electron escape.

Using the Firsov elastic interaction potential $V = A/R^2$ (Firsov 1957), and the retarding law $E = E_0 - ks$, one can get

$$\gamma_0 = \gamma_m\{1 - \exp[(E_c - E_0)/E_1] - E_c/E_1 \exp(-E_0/E_1)$$
$$\times [Ei(E_0/E_1) - Ei(E_c/E_1)]\},$$

(76)

where

$$\gamma_m = Nw\lambda\sigma_m,$$

(77)

$$E_1 = k\lambda,$$

(78)

$$k = 3.1\pi Nm_b \sum_{i=1}^{2} A_i/(m_b + m_i),$$

(79)

$$A_i = 3.05 \times 10^{-16} Z_b Z_i(\sqrt{Z_b} + \sqrt{Z_i})^{-2/3} \text{ cm}^2 \text{ eV}.$$

(80)

$Ei(x) = \int_\varepsilon^x e^y\, dy/y$, Z_b, Z_1, Z_2, m_b, m_1, m_2 are the nuclei charges and masses of the projectiles and the anion and cation, respectively.

While considering the interaction of light atoms with alkali-halide crystals, the contribution of the inelastic energy losses U into k and E_1 have to be taken into consideration,

$$-dE/dx_{in} = k_{in} = N\sigma_- U.$$

(81)

At velocities v_0 close to v_t, coefficient $\gamma_F \ll 1$ is sharply increasing with increasing velocity and at $v_0 > 2v_t$,

$$\gamma_F \approx Nw\beta\lambda(v_0 - v_{\text{lim}}),\tag{82}$$

where $v_{\text{lim}}/v_t = 1.1\text{–}1.3$. Formulas (76)–(82) permit to discuss the general behavior of γ_0, γ_F and γ_k versus the velocity. According to Batanov (1961a,b, 1962), the energy E_t is rather low ($E_t \leqslant 50\text{–}300 \text{ eV}$). Beginning at energy $E_0 = (1.5\text{–}2)E_t$, $\gamma_0(E_0)$ increases approximately linear due to the increase of the beam penetration depth and then approaches the linear dependence on velocity v_0 up to $\gamma_0 = 0.7\gamma_m$. At $E_0 \gg E_1$, γ_0 asymptotically approaches γ_m. The Firsov mechanism holds at velocities $v_0 > v_t$, when γ_0 is already high. Near $v_0 \sim v_t$, the coefficient γ_F is exponentially small, and beginning with velocity $v_0 \approx 2v_t$ it increases approximately linearly with increasing velocity.

The dependence of the kinetic electron emission on ion charge is caused by the cross section σ_- and, consequently, γ_k dependence on the charge state of the colliding particles. The cross section of the electron formation depends on the depth s due to the charge distribution $P_i(s)$ inside the solid,

$$\sigma_-(q, s) = \sum_{i=0}^{q} P_i(s)\sigma_-(q = i).\tag{83}$$

The coefficient of the kinetic electron emission under multiple-charged ion bombardment can be written as

$$\gamma_k(q, E_0) = \gamma_k(q = 0, E_0) - \Delta\gamma_k(q, E_0),\tag{84}$$

where

$$\Delta\gamma_k = Nw \int_0^{s_0} \Delta\sigma(s, E_0) \exp(-s/\lambda)\, ds = \Delta\gamma_0 + \Delta\gamma_F,\tag{85}$$

$$\Delta\sigma = \Delta\sigma_0 + \Delta\sigma_F = \sigma_-(q = 0) - \sigma_-(q).\tag{86}$$

What a role does $\Delta\gamma_k$ play? If in the coefficient of electron emission,

$$\gamma(q, E_0) \equiv \gamma_q = \gamma_p + \gamma_k,$$

one considers only the decrease of $\gamma_p(E_0)$, then the curves γ_q should draw closer with the ion energy increase. The decrease of the kinetic electron emission could lead to the intersection of these curves. Three intersections in the energy range $0.8 < E_0 < 2$ keV have been observed by Morozov et al. (1979). At these energies the contribution of γ_F in γ_k is small or lacking. Thus, the intersections corroborate the intensive decrease of $\sigma_0(q)$ and $\gamma_0(q)$ with increasing q. It is interesting to verify the following assumptions,

(a) $\quad \sigma_0(q) = \sigma_0\delta_{0q}, \quad \sigma_0(q, s) = \sigma_0 P_0,\tag{87}$

i.e., detachment of electron is due to anion–neutral-atom collision, and

(b) $\sigma_0(q) = \sigma_0(\delta_{0q} + \delta_{1q})$, $\sigma_0(q, s) = \sigma_0(P_0 + P_1)$,   (88)

i.e., the detachment occurs during the collision with Al^0 and Al^+.

As it is shown below, the agreement with the experiment could be achieved for the second version. Nevertheless, the explanation is rather ambiguous. Equation (88) could be stipulated either by the fact that $\sigma_0(q = 0) \approx \sigma_0(q = 1)$ and the rate of Al^+ neutralization is of no importance, or that Al^0 inside the solid is unstable and the cross section σ_0 corresponds to Al^+–Cl^- interaction; another possibility: Al^+ captures an electron via charge exchange with the rate $W_{ex} \gg W$ and the cross section σ_0 corresponds to Al–Cl^- interaction. Since the first ionization potential of Al is small, apparently the second of the three considered assumptions is the most probable one. Considering

$$P_0(q, s) = [1 - \exp(-\tau)]^q,$$ (89)

and neglecting the change of $\sigma_0(q, s)$ and $\tau(s)$ with depth, one can get from eqs. (85)–(87) and (67)

$$\Delta\gamma_0 \approx \gamma_m(1 - E_c/E_0) \sum_{i=2}^{q} (-1)^i C_q^i / (1 + i/\alpha)^{-1}$$

$$= \gamma_m(1 - E_c/E_0)[-(q - 1) + (1 + \alpha) \sum_{i=2}^{q} (-1)^i C_q^i / (1 + \alpha/i)^{-1}].$$

(90)

The numerical integration with account of the dependence of $\sigma_0(q, s)$ and $\tau(s)$ on s has shown that $\Delta\gamma_0$ from eq. (90) has to be multiplied by a factor $(1 + 0.025q)^{-1}$. Taking into account eq. (69), one can get

$$\Delta\gamma_0 \approx \gamma_m\left(1 - \frac{E_c}{E_0}\right) \frac{q - 1}{1 + 0.025q} \frac{\alpha(1 - \beta_q)}{1 + \alpha\beta_q}.$$

(91)

It is necessary to estimate $\Delta\gamma_F$. The experiments on atom and ion stripping in gases (Fedorenko et al. 1960) as well as the theoretical estimations (Kishinevsky 1969) have shown that the stripping is considerable both at light-ion–heavy-atom collisions and in the proper gas. At heavy-ion–light-atom collisions the stripping is not important and the ratio $\delta = [\sigma_F(q) - \sigma_F(q = 0)]/\sigma_F(q = 0) = 0.05$–$0.1$. For Al^{q+}–$NaCl$ bombardment, apparently $\delta \ll 1$, but since $\Delta\gamma_F/\gamma_F < \delta$, $\Delta\gamma_F \ll \gamma_F$ being considerable at high velocities, while $\Delta\gamma_0$ is considerable already at low velocities and causes the intersection of γ_q curves at $q \leqslant 3$.

2.6.3. Auger neutralization rate
Unlike metals, the electrons of alkali-halide crystals are very localized on anions, the filled band being very narrow. The Auger neutralization involv-

ing two electrons of the same anion needs the energy exceeding 25 eV, i.e., the ionization energy of helium. Such process is possible with multiple-charged ions. Yet the Auger neutralization with the contribution of two electrons belonging to neighboring anions (Parilis 1972, Gaipov et al. 1973) is possible. The estimation of the rate of this process is given below (Kishinevsky 1974a).

In a first approximation of the perturbation theory, the Auger neutralization at distance s from the surface is

$$W_0(s) = 2\pi |V_{12}(s)|^2 g_2 . \tag{92}$$

Here and from here on atomic units are used. V_{12} is the matrix element for Auger transition from the initial state '1' to the final state '2', g_2 is the final state density.

For the wave functions of the valence electrons it is convenient to use the Wannier functions localized on single lattice points, which permits to consider the alkali-halide crystal as a set of ions.

The Auger neutralization occurs at approaching the projectile ion I^+ with one of the anions A^-. At such approach the wave functions of the ions are essentially overlapping and the quasimolecule $A^- + I^+ = B^0$ is formed in an excited state since the deep level of I^+ is vacant. In this case, the detachment of an electron from the neighboring A^- anion is analogous to the ionization in collision with an excited atom (Smirnov 1968). In case of a metastable excited state such ionization is called the Penning effect. The ionization cross section is the highest for resonance excitation. In the quasimolecule at least some of the six excited p electrons $(p\sigma)^2(p\pi)^4$ are resonantly excited. This allows to use the dipole–dipole interaction as the perturbation causing the electron detachment (Smirnov 1968),

$$V = L^{-3}[D_A D_B - 3(D_A n)(D_B n)] . \tag{93}$$

Here, L is the distance between the anions, $n = L/L$, $D_A = \Sigma r_A$ is the dipole momentum of the anion, and $D_B = \Sigma r_B$ is the dipole momentum of the quasimolecule.

By averaging the transition probability W_0 over the initial states and summarizing over the final ones one can get

$$W_0 = \frac{3c\sigma_{phot}(\omega)}{2\pi\omega^2 L^6} \sum_B f_B , \tag{94}$$

where c is the velocity of light, $\sigma_{phot}(\omega)$ is the cross section of the electron photodetachment from the anion by photons with frequency ω, f_B,

$$f_B = \tfrac{2}{3}\omega \langle r_B \rangle_{12}^2 , \tag{95}$$

is the oscillator strength for a quasimolecular transition from the final state into an excited one. The summation is carried out over different final states of the quasimolecule, ω is the energy released during a single-electron transition in the quasimolecule.

The Auger neutralization rate W_0 corresponds to the quasimolecular interaction with one of the neighboring anions. In alkali-halide crystals the number of neighboring anions on the surface $n_n = 8$, inside the crystal $n_n = 12$. Therefore, the total probability of Auger neutralization is $W = n_n W_0$.

The dependence of W on the distance to the surface (more precisely, on the quasimolecule size R) is mainly determined by $f_B(R)$. Let R_0 be the dimension of the atomic region, in which a significant overlapping of the atomic wave functions in the quasimolecule begins. At distance $R \gg R_0$, f_B is exponentially small and according to eq. (95) determined by the overlapping of the ground-state electron wave functions both of anion A^- and atom I, forming the quasimolecule. At $R \leqslant R_0$, f_B is of the order of the magnitude of the oscillator strength for the transitions in atoms, i.e., $\approx 10^{-1}$ (see fig. 19, curve 1). Due to the asymmetry at middle interatomic distance, the overlapping of the molecular wave functions is more significant than at $R \to 0$, i.e., within the united atom. Thus, f_B can pass through a maximum. This behavior of $f_B(R)$ is corroborated by the calculation of the oscillator strength for the transition in the molecule HeH^{2+} from the state $2p\sigma$ [$H(1s)$ at $R \to \infty$] into the state $1s\sigma$ [$He^+(1s)$ at $R \to \infty$] (Arthurs et al. 1957), (see fig. 19, curve 2). The cross sections of the photodetachment are known merely for certain negative ions (Cooper and Martin 1962). For negative ions

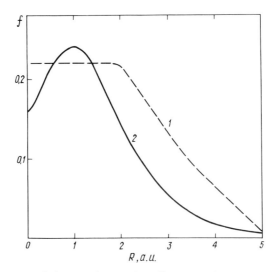

Fig. 19. Oscillator strength f_B versus internuclear distance R. Curve 1 calculation by Kishinev-sky (1974a); curve 2 calculation of $2p\sigma$–$1s\sigma$ in HeH^{2+} (Arthurs et al. 1957).

bound in a solid $\sigma_{phot}(\omega)$ is higher than for free ions. Indeed, at photo-detachment of the outer p-electrons of negative halogen ions, the sum rule is approximately fulfilled,

$$\frac{c}{2\pi^2} \int_{\omega_{min}}^{\infty} \sigma_{phot}(\omega)\, d\omega = 6 .$$

There is a wide forbidden band in the crystal, which diminishes the region of integration over ω, i.e., increases $\sigma_{phot}(\omega)$.

In the saturation region $(R < R_0)$, taking $\Sigma_B\, f_B = 0.2$, one can get

$$W_{surf} = \frac{3.1 \times 10^{35} \sigma_{phot}(\omega)}{\omega^2 d^6} , \qquad W_{ins} = \tfrac{3}{2} W_{surf} . \tag{96}$$

Here the energy ω is measured in eV, the cross section σ_{phot} in cm^2, the lattice constant d in Å, and the Auger neutralization rate in s^{-1}.

For an estimation, one can consider $E_e = 4\,eV$, since the energy of most electrons does not exceed this value (U.A. Arifov et al. 1966a). The calculated values W_{surf} and W_{ins} for four alkali-halide crystals are listed in table 4 along with the used values of lattice constant and photoionization cross sections.

At the decrease of the emitted electron energy, the function $W(\omega)$ increases as $\simeq \omega^2$ and exceeds the values listed in table 4. At $R > R_0$, the function $W(R)$ decreases exponentially due to the decrease of $f_B(R)$. Furthermore, with the increase of R the energies E_e and ω increase but $\sigma_{phot}(\omega)$ decreases, which leads to a more intensive decrease of $W(R)$.

It should be noted, in conclusion, that the considered processes of Auger neutralization, involving the neighboring anions, are analogous to the Auger ionization of a diatomic molecule (Parilis 1969c) and formula (94) could be used for the estimation of its rate.

2.6.4. Comparison with experiment

Atom–electron emission. The comparison of γ_k calculated using eqs. (76)–(82) with the experimental data (U.A. Arifov et al. 1971a) for LiF, NaCl and KBr bombarded with He, Ne, Ar and Kr neutral atoms has been made.

Table 4
Calculated values W_{surf} and W_{ins}.

crystal	d (Å)	ω (eV)	$\sigma_{phot} \times 10^{17}$ (cm^2)	W_{surf} (10^{14} s^{-1})	W_{ins} (10^{14} s^{-1})
KCl	3.14	12.2	5.0	1.13	1.70
NaCl	2.81	12.7	4.6	1.84	2.76
NaF	2.31	12.8	1.2	1.53	2.30
LiF	2.01	16	1.2	2.28	3.42

Considering that at energies exceeding 1 keV the emission coefficient for single-charged ions and neutral atoms are close, the calculations and the experiments by Baboux et al. (1971) are made for He^+, Ne^+, Ar^+ and Kr^+ ions bombarding NaCl within the energy range $1 \leqslant E_0 \leqslant 30$ keV. The typical dependences are shown in figs. 20 and 21. The values γ_m, E_1 and $d\gamma_F/dv_0$ calculated using eqs. (77), (78), (81) and (82) are listed in table 5. In this table the values of γ_m, E_1, $d\gamma_F/dv_0$, E_c and v_{lim} at which the agreement with experiments was achieved are also listed. In the calculations the following values of parameters were used: $d_{LiF} = 2.01$ Å, $d_{NaCl} = 2.81$ Å, $d_{KBr} = 3.29$ Å (Seitz 1940). The cross sections σ_m were estimated using experimental data by Bydin and Dukelsky (1956). The mean inelastic energy loss U for He^0 in all alkali-halide crystals was considered to amount 15 eV. For the atoms Ne, Ar and Kr the inelastic energy losses were ignored, since $k_{in} \ll k$. If one takes $w = 0.5$, then the agreement with the experiment by U.A. Arifov et al. (1971c) could be achieved at $\lambda_{LiF} = 40$ Å, $\lambda_{NaCl} = 60$ Å and $\lambda_{KBr} = 80$ Å. It should be noted that the mentioned experiments were performed at temperatures $T = 625$–675 K, and the data by Baboux et al. (1971) were obtained at $T = 450$ K. In accordance with theory (Dekker 1954),

$$\frac{\gamma(T_1)}{\gamma(T_2)} = \frac{\lambda(T_1)}{\lambda(T_2)} = \left(\frac{2n(T_2)+1}{2n(T_1)+1}\right)^{0.5}, \tag{97}$$

$$n(T) = [\exp(h\Omega/kT) - 1]^{-1}, \tag{98}$$

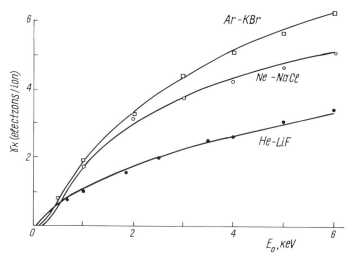

Fig. 20. Yield γ_k versus atom kinetic energy E_0 for impact of He, Ne, Ar atoms on LiF, NaCl, KBr, respectively. Full curve denotes theory. ($\bullet$), ($\bigcirc$), ($\triangle$) experiment by U.A. Arifov et al. (1971a).

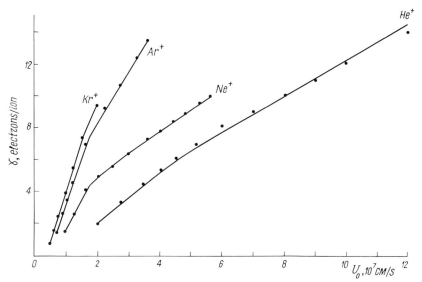

Fig. 21. Yield γ_k versus ion velocity v_0 for the impact of He$^+$, Ne$^+$, Ar$^+$ and Kr$^+$ on NaCl; full curves: theory; points: experiment by Baboux et al. (1971).

Table 5
Parameters for LiF, NaCl and KBr.

Atom	γ_m		$d\gamma_F/dv_0$ $(10^{-7}\,\text{s/cm})$		E_1 (keV)		E_c (keV)	v_{lim} $(10^7\,\text{cm/s})$
	theor.	exper.	theor.	exper.	theor.	exper.		
				LiF[a]				
He	3.7	1.0	0.69	0.72	0.25	0.27	0.05	2.05
Ne	6.2	6.2	1.56	1.63	1.6	1.6	0.05	1.65
Ar	5.9	5.0	2.3	1.9	3.2	3.0	0.12	1.4
Kr	7.6	7.0	4.8	–	6.4	6.4	0.2	–
				NaCl				
He[a]	2.8	2.5	0.75	1.1	0.20	0.25	0.05	2.0
He[b]	3.3	2.8	0.9	1.2	0.24	0.3	0.05	2.0
Ne[a]	3.9	4.0	1.42	1.55	1.25	0.84	0.17	1.5
Ne[b]	4.7	4.6	1.7	1.3	1.5	1.02	0.17	1.3
Ar[a]	5.1	5.7	2.2	2.3	2.9	1.93	0.15	1.3
Ar[b]	6.2	7.0	2.7	2.9	3.5	2.4	0.15	1.3
Kr[a]	5.4	5.7	3.7	–	6.5	4.0	0.2	1.1
Kr[b]	6.5	7.0	4.4	3.4	7.8	4.8	0.2	1.1
				KBr[a]				
He	2.2	2.2	1.16	1.05	0.13	0.13	0.05	1.9
Ne	3.0	1.63	2.0	2.0	1.1	0.6	0.3	1.2
Ar	4.5	5.5	2.4	2.0	2.9	1.6	0.1	1.2
Kr	5.9	6.3	4.2	–	7.2	4.0	0.12	–

[a] U.A. Arifov et al. (1971c).   [b] Baboux et al. (1971).

where Ω is the frequency of longitudinal optic oscillations. For NaCl, $\Omega = 0.79 \times 10^{13} \, \text{s}^{-1}$ (Baboux et al. 1971). Taking into account this dependence, it was considered that $\lambda_{\text{NaCl}} = 72 \, \text{Å}$.

It has been found that, for alkali-halide crystals, $\lambda > 100 \, \text{Å}$. However, according to the experiment by Afanasiev and Bronstein (1976) on the interaction of electrons with an energy $E_e < 30 \, \text{eV}$ with alkali-halide films, the main contribution ($>90\%$) to the coefficient of the secondary electron emission gives the layer thickness 50–$100 \, \text{Å}$, but the layer with thickness 100–$200 \, \text{Å}$ makes the main contribution to the proper secondary electron emission. If one considers the exponential dependence of the electron yield on the depth s, then a 90% yield is caused by the layer $\Delta s = \lambda \ln 10$, i.e., for slow electrons, $\lambda = 45$–$90 \, \text{Å}$. Since the spectrum of ion–electron emission involves mainly slow electrons, the obtained estimation corroborates the values of λ used in this calculation.

The results of the comparison of the theory with the experiment need to be discussed in detail. As a rule, the values of γ_m^{theor} and γ_m^{exper} are close, which testifies for the closeness of the cross sections σ_0 of electron detachment from a free negative ion and anion under collisions with a neutral atom. In two cases (He–LiF and Ne–KBr), the ratio $\gamma_m^{\text{theor}}/\gamma_m^{\text{exper}} \geqslant 2$. Perhaps, as with the collisions in gases, this is caused by the excitation of the initial state term into continuum (Bydin and Dukelsky 1956).

The parameter E_1 defines the energy loss on the path $s = \lambda$. The overvaluation of E_1^{theor} is apparently caused by the fact that the energy range $E_0 < 1 \, \text{keV}$, where γ_0 is the most sensitive to the value of E_1. The Firsov potential in its approximation $V = A/R^2$, and hence the energy losses, are overestimated. The overvaluation of E_1 was not found at He^0 bombardment, where the inelastic energy losses mainly contributed to E_1.

It should be noted that from comparison of the calculations with the experiment one cannot get the single parameters, but their products: $\sigma_m w \lambda$ from γ_m, $\beta w \lambda$ from $d\gamma_F/dv_0$ and $k\lambda = E_1$. Therefore, the achieved agreement with experiment could be attained with other values of parameters, which retain these products. So, for LiF, apparently $w < 0.5$ and $\lambda > 40 \, \text{Å}$, but if $w\lambda = 20 \, \text{Å}$, then γ_m and $d\gamma_F/dv_0$ do not change, and E_1^{theor} increases, and like the case of NaCl and KBr, exceeds E_1^{exp}.

The values of E_c listed in table 5 are close to those obtained via extrapolation of the experimental curves toward the threshold. The value of v_{lim} agrees with the above estimations $v_{\text{lim}} = (1.1$–$1.3)v_t > (1.1$–$1.3) \times 10^7 \, \text{cm/s}$; nevertheless, it should be noted that the accurate estimation of v_t from the experiment seems to be hampered, since near the threshold of γ_F the emission is intensive due to the detachment of weakly bound electrons. The error introduced by $\Delta\sigma_0$ was compensated by the changes of v_t and v_{lim} by $\Delta v \approx \Delta\sigma_0/\beta$. The blanks in the table 5 were caused by the fact that at Kr^0 bombardment (U.A. Arifov et al. 1971c) the velocity v_0 did not exceed 1.2×10^7, which is close to v_t.

At the comparison on the calculation results with the experimental data a rather good agreement was achieved with reasonable values of parameters.

Ion–electron emission. The calculations by Kishinevsky and Parilis (1982) can be compared with the experimental data (Morozov et al. 1979) for the total coefficient of electron emission in the collisions Al^{q+}–NaCl ($q = 1$–10). From the comparison of the experimental dependence $\gamma_i(E_0) = \gamma_k(q = 1, E_0)$ with eq. (76), the values of $E_c = 0.3$ keV, $E_1 = 1.25$ keV, $\gamma_m = 6$ were determined and according to the calculations $d\gamma_F/dv_0 = 2 \times 10^{-7}$ s/cm. It was considered that $v_{lim} = 1.2 \times 10^7$ cm/s. Three combinations were considered, namely:

(I) $\gamma_p(v_0)$ from eq. (66) and $\sigma_0(q)$ from eq. (87);
(II) $\gamma_p(v_0)$ from eq. (66) and $\sigma_0(q)$ from eq. (88);
(III) $\gamma_p(v_0)$ from eq. (69) and $\sigma_0(q)$ from eq. (88).

The best agreement with experiment of all families of curves γ_q was achieved with combination (III) with $v_A^{III} = 4 \times 10^7$ cm/s (see fig. 22), somewhat worse with combination (II) with $v_A^{II} = 5 \times 10^7$ cm/s and the worst with combination (I) with $v_A^I = 6 \times 10^7$ cm/s.

The location of the intersection points I ($\gamma_1 \times \gamma_2$), II ($\gamma_1 \times \gamma_3$) and III ($\gamma_2 \times \gamma_3$) appeared to be the most sensitive to the choice of parameters. The experimental and calculated values of energies E_I, E_{II} and E_{III} at which the intersections took place are listed in table 6.

The best agreement was achieved in combination (IIIa) with $v_A^{IIIa} = 3.4 \times 10^7$ cm/s. The bad agreement with experiment for combination (I) is caused by the fact that $\Delta\gamma_k$ was underestimated, which in this version corresponds to the assumption that $\sigma_0 \neq 0$ only for the collisions of Cl^- with neutral atoms Al^0. The weaker decrease of $\gamma_k(q, E_0)$ with the increase of E_0 in this version is compensated by the prompter decrease of $\gamma_p(E_0)$, hence $v_A^I > v_A^{II,III}$. Version (II) and (III) differ slightly by using $\gamma_p^{II}(v_0) = \gamma_p^0/(1 + \alpha)$

Table 6
The energies corresponding to the $\gamma_q(E_0)$ intersection points.

Intersection	Energy (keV)				
	experiment[b]	theory[a]			
		I	II	III	IIIa
I($\gamma_1 \times \gamma_2$)	0.8	0.8	0.75	0.8	0.75
II($\gamma_1 \times \gamma_3$)	1.2	2.0	1.4	1.35	1.2
III($\gamma_2 \times \gamma_3$)	2.0	6.0	3.0	2.6	2.0

[a] Kishinevski and Parilis (1982).   [b] Morozov et al. (1979).

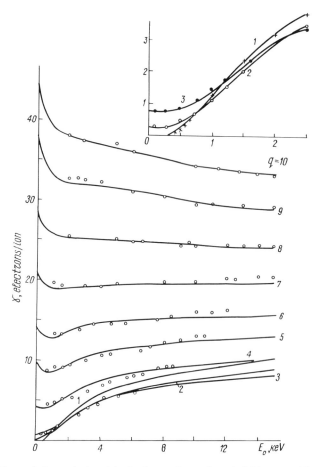

Fig. 22. (Full curve) theoretical and (points) experimental total yields versus kinetic energy for impact of Al^{q+} ($q = 1$–10) on NaCl.

and $\gamma_p^{III}(v_0) = \gamma_p^0 / (1 + \beta_q \alpha)$, which is equivalent to the substitution of v_A^{II} by v_A^{III}/β_q.

The alternation of $\gamma_k(E)$ at different q are shown in fig. 23. The strong decrease of $\gamma_k(q)$ with q increase which is notably considerable in the range of small q leads to the above-mentioned intersections of the curves $\gamma_q(E_0)$. Within the energy range $E_0 = 13$–14 keV, the intersection $\gamma_1 \times \gamma_4$ should be expected; however, $\gamma_1(E_0)$ was measured merely with the energies $E_0 \leqslant 2.5$ keV (Morozov et al. 1979). The measurement of $\gamma_1(E_0)$ within the energy range up to 15–20 keV should be of interest.

Since $v_A = A\lambda$, and $\lambda = 85$ Å for NaCl under the condition of the experiments by Morozov et al. (1979) ($T = 300$ K), the rate of Auger neutraliza-

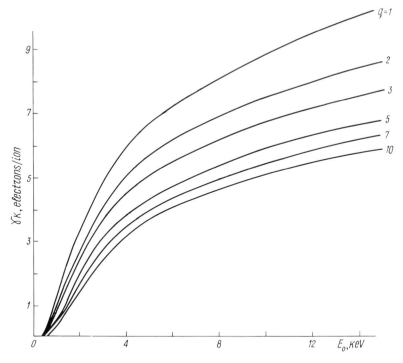

Fig. 23. Yield γ_k versus ion energy E_0 and charge q for impact of Al^{q+} ($q = 1$–10) on NaCl.

tion of a single vacancy amounts $0.5 \times 10^{14}\,s^{-1}$. This corroborates the theoretical estimation (subsection 2.6.3) and follows from the assumption that multiple-charged ions are neutralized inside the solid.

2.6.5. Conclusion

The above calculations permitted to explain a number of observed experimentally peculiarities of electron emission from alkali–halide crystals, namely: the low threshold and the high values of the coefficients of both kinetic and potential emission; as well as to estimate the Auger neutralization rates.

It should be noted in conclusion that, since the emission from alkali-halide crystals under bombardment with both atoms and single-charged ions of inert gases has been studied in considerable detail, the experimental study of emission from alkali–halide crystals under the bombardment with multiple-charged ions of these gases within a wide range of energies and atomic numbers should be performed.

Together with above emission from NaCl bombarded by ions Al^{q+}, Morozov et al. (1979) have obtained similar data for the ions Be^{q+} ($q \leqslant 4$),

V^{q+} ($q \leqslant 11$), Cu^{q+}, Sb^{q+} and Re^{q+} ($q \leqslant 8$) and for LiF and Si single crystals. It follows that the theoretical description developed above could be used for the potential electron emission from other semiconductors and insulators.

The dependence of the yield γ_p on energy, released at the neutralization of a multiple-charged ion, W'_q, is shown in fig. 24. Here,

$$W'_q = W_q - q(\Delta E + \chi), \tag{99}$$

ΔE is the width of the forbidden band of a semiconductor or insulator, χ is the electron affinity, and $(\Delta E + \chi)$ is the analog of the work function.

As it is seen from fig. 24 an approximately linear dependence of $\gamma_p(W'_q)$ is observed and consequently an approximately linear dependence $\gamma_p(W'_q)$, since $W'_q \simeq W_q$.

By comparing the dependence $\gamma_p(W_q)$ for metals, semiconductors and insulators one can conclude that the phenomenon of Auger emission of electrons from solids under multiple-charged ions has a universal character, since the dependence $\gamma_p(q)$ resolves the dependence $\gamma_p(W_q)$. The dependence on ion atomic number is displayed in the features of electron energy distributions, e.g., in the presence of deep vacancies.

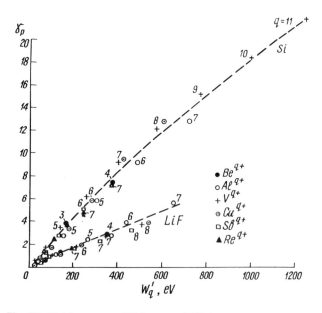

Fig. 24. Yield γ_p versus W'_q [see eq. (99)] (Morozov et al. 1979).

3. Sputtering of nonmetals under slow multiple-charged ions through Coulomb explosion

3.1. Introduction

The calculations by Parilis (1969b, 1970) refer to the sputtering of silicon under bombardment by slow, double-charged argon ions, and the results are compared with the only available experimental result by Wolsky and Zdanuk (1964) at the time, found by an indirect method.

The next experiments of the observation and study of this effect – ion emission from nonmetals under multiple-charged ion bombardment were reported by Radzhabov et al. (1976), and T.U. Arifov et al. (1976, 1978). The results show that the ion–emission yield does, in fact, increase with increasing ion charge, in contrast with the situation for metals, in which no charge dependence is observed (T.U. Arifov et al. 1976, Weathers et al. 1989).

Bitensky et al. (1979) has calculated on the basis of the Coulomb 'explosion' mechanism the sputtering of a nonmetal under slow multiple-charged ions and the results have been compared with the available experiments by T.U. Arifov et al. (1978) by that time.

In the experiments by Radzhabov and Rakhimov (1985) and de Zwart et al. (1986) the increase of the secondary ion emission with the ion charge increase has been confirmed.

The mass analysis of secondary ions has been carried out by Della-Negra et al. (1988b). These authors have shown that only the H^+ yield sharply increased with ion charge increase under 18 keV Ar^{q+} ($q = 1$–9) bombardment of CsI, Si and phenylalanine targets while the yield of Cs^+, Si^+ and M^+ did not depend on q.

As for the total sputtering yield, the experimental data were also contradictory. De Zwart et al. (1986) found the sputtering yield of Si under Ar^{q+} ($q = 1$–9) ions with an energy $E = 20$ keV to be independent of an ion charge q but Eccles et al. (1986) found the sputtering for different nonmetals (silicon, glass) under Ar^+ with $E = 10$ keV to exceed the sputtering under Ar atoms with the same energy by a factor 1.5–2.6. At the same time, the sputtering yield for Au did not depend on ion charge. The sputtering yield of Cs and Nb measured by Weathers et al. (1989) under Ar^{q+} ($q = 4, 8, 11$) bombardment of CsI, $LiNbO_3$ increased with ion charge increase.

3.2. Sputtering by Coulomb explosion

We consider the sputtering of nonmetals within the framework of the Coulomb 'explosion' mechanism following Parilis (1969b, 1970, 1980b, 1991), Bitensky et al. (1979, 1992c) and Bitensky and Parilis (1989).

Under low-energy bombardment, the neutralization of a multiple-charged ion occurs near the surface and the formed charged domain is a semisphere with radius R_0 (Bitensky et al. 1979). It could be calculated from the energy balance equation. The total neutralization energy W_q of a q-charged projectile is shared between the Coulomb repulsion energy W_C and the kinetic of $(N_q - q)$ Auger electrons E_k (N_q being the number of ions which appear at the Auger neutralization of q-charged ion). U.A. Arifov et al. (1973b) have assumed that $N_q \approx q^2$.

The number of Ions appearing per unit time in the charged domain is N_q/τ_i, where τ_i is effective time of the total Auger neutralization of a q-charged ion. Simultaneously, during the time τ_e they are neutralized by the conducting electrons. For an insulator, $\tau_e \gg \tau_i$; and for a metal, $\tau_e \ll \tau_i$. As a result, the charged domain of moment τ_i consists of N ions,

$$N = N_q(\tau_e/\tau_i)[1 - \exp(-\tau_i/\tau_e)] . \tag{100}$$

The electrostatic energy of an uniformly charged semisphere $W_c = 0.32\pi^2 n^2 e^2 R_0^5/\kappa$ enters into the energy balance equation used to determine N and, therefore, R_0 since $N = \frac{2}{3}n\pi R_0^3$,

$$W_q - qe\varphi = 0.55(\pi n)^{1/3}e^2 N^{5/3}/\kappa + (N_q - q)E_k , \tag{101}$$

where n is the number of atoms per unit volume, e is the electron charge, κ is the dielectric constant, and φ is the work function. For metals, $N = 0$ and the electron emission yield $\gamma = N_q \simeq (W_q - qe\varphi)$ in accordance with the paper by U.A. Arifov et al. (1973b). In case of nonmetals ($\tau_e > \tau_i$), the Coulomb energy of the charged domain should be taken into account both for secondary ion and electron emission. For high q, if the first term on the right-hand side of eq. (101) dominates, then N_q and, therefore, $\gamma \sim W_q^{3/5}$.

If an ion penetrates a solid, it is neutralized in a distance $H = v_i\tau_i$, where v_i is the ion velocity. The region in which the ionized atoms are produced can be regarded as a uniformly charged cylinder of height H and radius R with a hemispherical end.

First, we consider slow ions. Under the influence of Coulomb repulsion, the ionized atoms in this region acquire a certain kinetic energy which can be determined by solving the equation of motion. The velocity component normal to the surface, v_n, of a particle at the point $\{x, y, z\}$ is determined by

$$\frac{\mathrm{d}v_n}{\mathrm{d}t} = \frac{e\mathscr{E}_n\{x, y, z\}}{M} , \tag{102}$$

where $\mathscr{E}_n$ is the normal component of the Coulomb field, and M is the atomic mass. During the neutralization time, τ_e, a particle passes such a distance that $\mathscr{E}_n$ can be considered as a constant. Then, using eq. (102), we find an equation for the kinetic energy in the direction normal to the surface, E_n,

$$E_n = \frac{Mv_n^2}{2} = \frac{\mathscr{E}_n^2 e^2 \tau_e^2}{2M} . \tag{103}$$

A moving particle can escape from the surface if its energy exceeds the surface binding energy E_b, $E_n \geqslant E_b$. Using eq. (103), we can rewrite this inequality as

$$E_n \geqslant \sqrt{2ME_b}/(e\tau_e) . \tag{104}$$

The normal component of the field is

$$\mathscr{E}_n = \frac{ne}{\kappa} \int \frac{\cos \alpha \, dV_2}{r_{12}^2} . \tag{105}$$

At a surface point on a distance ρ from the center of the hemisphere, we have $\theta_1 = \frac{1}{2}\pi$ and $\cos \alpha = r_2 \cos \theta_2/r_{12}$. Evaluating the integral in eq. (105), and retaining terms of the first order in ρ/R_0, we find

$$\mathscr{E}_n(\rho) = \pi ne(R_0 - 0.49\rho)/\kappa . \tag{106}$$

The normal component of the field at a depth h $(h \ll R_0)$ below the surface of the hemisphere (on the symmetry axis) is found from eq. (105) by setting $\theta_1 = 0$ and $\cos \alpha = (h - r_2 \cos \theta_2)/r_{12}$,

$$\mathscr{E}_n(h) = \pi ne(R_0 - 2.7h)/\kappa . \tag{107}$$

Together with inequality (104) the values of ρ and h in eqs. (106) and (107) determine the region from which particles can be emitted. Approximating this region by a cone of radius ρ and height h, we find an equation for the sputtering yield,

$$S = \tfrac{1}{3}\pi\rho^2 hn = 0.49\pi n(R_0 - a)^3 , \tag{108}$$

where a is given by

$$a = \kappa(2ME_b)^{1/2}/(\pi e^2 n\tau_e) , \tag{109}$$

and R_0 can be expressed in terms of W_q using eq. (101).

The quantity a equals the effective thickness of a layer from which the particles cannot be ejected during time τ_e.

To determine how the sputtering yield varies with the velocity of the multiple-charged ion the electrostatic energy of the cylindrical region has been calculated.

Substituting the calculated value of W_c in the equation of energy balance, eq. (101), and evaluating $\mathscr{E}_n$ in eq. (105), we find that the ion yield can be approximated by a formula analogous to eq. (108),

$$K = 0.49\pi n\eta^+ (R - a)^3 , \tag{110}$$

where η^+ is ionization degree, and R is determined by

$$R = R_0^2/(R_0 + 0.37 v_i \tau_i) .$$ (111)

Equation (110) displays a threshold of ion emission from nonmetals under multiple-charged ions: for $R = a$, the yield $K = 0$. For a given solid, R decreases with primary energy increase. Besides, from eq. (110) it follows that the diagram $K^{1/3}$ versus R should be a straight line and one can determine the value a by the segment cut off from the axis R.

3.3. Comparison with experiments

In fig. 25, the dependence of $K^{1/3}$ on R is shown for different targets and different primary ions with different energies: (I) $E_0 = 0.1-1$ keV (Radzhabov et al. 1976) and extrapolated to $E_0 = 0$ (T.U. Arifov et al. 1978), (II) $E_0 = 1-8$ keV (ibid), and (III) $E_0 = 20$ keV (de Zwart et al. (1986). The yield K was calculated by subtracting the yield corresponding to $q = 1$ from the total yield.

The following values of parameters were used: for LiF, $n = 0.125 \text{ Å}^{-3}$, $\kappa = 2.4$, $e\varphi = 6$ eV, $E_k = 20$ eV, $\tau_i = 2 \times 10^{-15}$ s; and for Si, $n = 0.05 \text{ Å}^{-3}$, $\kappa = 4$, $e\varphi = 4$ eV, $E_k = 15$ eV, $\tau_i = 1.5 \times 10^{-15}$ s. As is seen from fig. 25, the points are well aligned along the straight lines $K^{1/3}$ versus R, i.e., eq. (110) satisfactorily describes the experimental data with the following fitting parameters: for LiF, $a = 0.7 \text{ Å}$, $\eta^+ = 3.7 \times 10^{-2}$; for Si (T.U. Arifov et al. 1978), $a = 2.2 \text{ Å}$, $\eta^+ = 9 \times 10^{-2}$; and for Si (de Zwart et al. 1986), $a = 3.4 \text{ Å}$ and $\eta^{+'} = 2.9 \times 10^{-2}$.

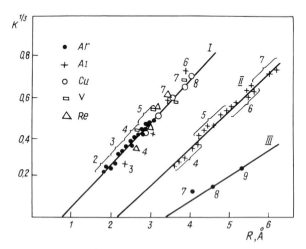

Fig. 25. $K^{1/3}$ versus R for (I) Ar^{q+} (Radzhabov et al. 1976) and metal ions (T.U. Arifov et al. 1978) on LiF; (II) Al^{q+} on Si (T.U. Arifov et al. 1978); and (III) Ar^{q+} on Si (de Zwart et al. 1986). The ion charges q are designated by numbers.

An estimation of the neutralization time τ_e for a positive q-charged defect in a solid has been obtained by Yunusov et al. (1976),

$$\tau_e = 0.25(2\kappa q^2 e^2 kT/m^2)^{-1/4} n_c^{-5/12}, \tag{112}$$

here m is the electron mass, k is the Boltzman constant, T is the temperature, and n_c is the current carrier concentration.

As follows from eqs. (109) and (112), $a \simeq T^{1/4} n_c^{5/12}$; therefore, with T and n_c increase, the lines $K^{1/3}(R)$ should be shifted to the right. Indeed, in the experiments by T.U. Arifov et al. (1978) and de Zwart et al. (1986) some samples with different concentration n_B of boron, which is the source of current carriers, have been used: Si wafer with $n_B = 10^{15}$–10^{16} atm/cm^3 corresponding to the specific resistance 25–2.5 kΩ/cm in the paper by de Zwart et al. (1986) and a high resistant silicon (50 kΩ/cm) with $n_B = 5 \times 10^{14}$ atom/cm in the paper by T.U. Arifov et al. (1978). Perhaps it explains the difference in a and η^+ for both experiments. It should be noted that for high-charged ions we get $R \simeq W_q^{1/3}$ and from eq. (110) $K \simeq W_q^{3/5}$.

The electron emission yield is given by

$$\gamma = p(N_q - q) = p(\tfrac{2}{3}\pi n R^3 - q), \tag{113}$$

where p is the probability of electron escape. De Zwart (1987b) has found that the dependence $\gamma(W_z)$ deviates from a linear one for Ar^{9+} ($E_0 = 20$ keV) impinging a tungsten polycrystal. This was explained by the emission of one fast Auger electron at filling the deep inner-shell vacancy, i.e., the assumption (U.A. Arifov et al. 1973b) of equal energy step Auger transitions became invalid. The Monte Carlo calculation (Mirakhmedov and Parilis 1987) of the cascade Auger neutralization of Kr^{q+} on a metal surface confirmed the deviation of $\gamma(W_q)$ from the linear dependence for high q. In the case of an insulator, this deviation is caused also by the transfer of part of the ion energy to electrostatic energy of the charged domain. In the paper by Morozov et al. (1979) the mentioned dependence indeed deviates from the linear one and as it is seen from fig. 24 satisfactorily agrees with eq. (113) for the same values of parameters as the ones used for calculation of the yield. The deviation from the linear dependence for high q has been observed recently by McDonald et al. (1992).

3.4. Discussion

The comparison of the theory with the experiment shows that the Coulomb explosion model describes satisfactorily the regularities of the dependence of ion emission from nonmetals on both ion charge and its velocity. As for the independence of total sputtering yield S on ion charge (de Zwart et al. 1986), it should be taken into account that we have assumed the additivity of Coulomb 'explosion' and collisional cascade contributions into sputtering, the former gives $S_{Coul} = K/\eta^+ \approx 0.6$, the second $S_{casc} = 1.7$ for $q = 9$, $E_0 = 20$ keV. This ratio means that in the charged domain the ions begin to move not only due to Coulomb repulsion, diminishing its effect. The dependence of K on q (ibid) can be explained by a rather high probability for the ejected ions to preserve their charge because they leave the domain which is not neutralized during the time $\tau_e \approx 10^{-13}$ s.

It is clear that for recording the proper Coulomb explosion sputtering, an experiment for lower energy should be carried out. Even at 10 keV Al→Si bombardment, $S_{casc} \approx 1$ (Andersen and Bay 1981) while according to eq. (111) $S_{Coul} \approx 3$, i.e., in this case the Coulomb explosion sputtering prevails.

In the experiment by de Zwart et al. (1986) a comparative measurement of the secondary ion energy distribution under Ar^{9+} and Ar^+ bombardment of an Si single crystal has drawn the authors to the conclusion that the enhancement of the ion yield occurs only due to the change of the electron state of the sputtered particles while the mechanisms of sputtering remains essentially of the cascade type. However, careful examination of these energy distributions shows that they are not similar: while the total yield ratio $K_0/K_1 \approx 3$, the ratio for different parts of spectrum has different values. For low secondary ion energy it equals 4, but in the high-energy part of the spectrum this ratio decreases to 2. If the enhancement of ion yield for $q = 9$ would result from particles with the same inner shell vacancies, then these ions should enhance just the high-energy part of the spectrum, while the low-energy ions are typical rather for Coulomb repulsion. Therefore, it seems that Coulomb explosion sputtering does contribute to secondary ion emission in the experiment by de Zwart et al. (1986).

Katin et al. (1990) have proposed a model of inelastic sputtering of insulators by slow multiple-charged ions. The dynamics of electron and ion temperature during the relaxation of the excited region due to the propagation of the ionization wave has been considered. It was obtained that the inelastic processes contribute to inelastic sputtering only at $W_q > Q_{th}$, which strongly depends on the width of the forbidden energy band and equals $Q_{th} = 5 \times 10^2 - 7 \times 10^4$ eV. This large value of Q_{th} cannot explain the results of the experiments by Weathers et al. (1989).

Schmieder and Bastasz (1989) have reported that the relative yield of ions released from a solid copper target by incident Xe^{q+} ions ($q = 10$–33) of kinetic energy 1.4 keV/q shows a significant increase with increasing projectile charge state. The possible explanation can be that this effect is connected with the structure on the metal target. It is known that the sputtering yield of fine-grained metal films under fast heavy-ion bombardment can exceed by three orders of magnitude the yield of a coarse-grained target (Baranov and Obnorskii 1983, Baranov et al. 1988). According to the model by Bitensky (1980) and Baranov et al. (1981), the weak conductivity of the grain boundaries results in grain charging. The escaping of electrons from a grain is hindered and it leads to a transfer of the energy from the electron system to the ion system. As a result, the whole grain (with a radius of about 100 Å) can be evaporated or even exploded. Martynenko and Yavlinsky (1988) have proposed the other mechanism of polycrystalline metal sputtering. The hot electrons transfer their energy to atoms located in the vicinity of grain boundaries leading to the temperature and pressure increase. The overpressure can result in small grain detachment.

3.5. Conclusion

Thus, theoretical analysis has shown that some high-charged ions with low energies (small v_i and S_{case}) bombarding an insulator or semiconductor with small conductivity at low temperature (large τ_e) should be used for recording the proper Coulomb explosion sputtering. It would be interesting to carry out a comparative investigation of sputtering and ion emission depending on the mentioned parameters.

The bombardment of thick biomolecule layers by slow high-charged ions (up to bare nuclei), e.g., Kr^{30+} or Xe^{50+}, the Auger neutralization of which leads to emission of about 10^2 electrons (Mirakhmedov and Parilis 1987) and accumulation of positive ions of the same quantity should result in a Coulomb explosion. It should be interesting to clear up experimentally whether it causes desorption of the intact biomolecules or only their fragmentation and sputtering.

4. Identification of multiple-charged ions

The investigation of the composition of multiple-charged ion mixed beams is of great interest because in ion-beam mass analysis the separation over A/q takes place (A and q are the mass and the charge of the ion, respectively) and the pairs of ions getting into one beam corresponds to equal or close values of this ratio ($^{40}Ar^{5+}$ and $^{16}O^{2+}$, $^{84}Kr^{3+}$ and $^{14}N^+$, $^{84}Kr^{7+}$ and $^{12}C^+$, $^{132}Xe^{11+}$ and $^{12}C^+$, etc).

The problem of quantitative discrimination of ions forming their pairs arose. For this purpose, a method of multiple-charged ion identification by potential electron emission from metals induced by them was proposed (Parilis 1967).

This method is based on the fact that the potential electron emission depends nonlinearly on the ion charge, being proportional to the total neutralization energy which in its turn is approximately proportional to q^2.

The problem of the identification of the multiple-charged ions consists in the following. For mixed beams it is necessary to determine the fractions N_q of the multiple-charged and N_1 of the single-charged ions in a beam ($N_1 + N_q = 1$).

In a pure beam, the determination of the coefficient of potential electron emission γ_q amounts to the measurement of the ion current on the target, I_t, and the secondary electron current on the collector, I_e, with

$$\gamma_q = (I_e/I_t)q . \tag{114}$$

For a mixed beam, the following formula is used,

$$\gamma_{meas} = \frac{N_q(\gamma_q - \gamma_1) + \gamma_1}{N_q(q - 1) + 1} , \tag{115}$$

where γ_{meas} is the measured value of the coefficient of the secondary emission, γ_q is the coefficient for the multiple-charged component, γ_1 is the coefficient for the single-charged component, and N_q is the fraction of multiple-charged ions in the beam. The quantities γ_{meas} and γ_1 could be determined experimentally, N_q is a unknown quantity, but γ_q could be determined either by the interpolation between the values γ_{q-1} and γ_{q+1}, or using the dependence of γ on the total energy of the multiple-charged ion neutralization. Then,

$$N_q = \frac{\gamma_{meas} - 1}{\gamma_q - \gamma_{meas}(q - 1) - \gamma_1} ; \tag{116}$$

if (i) $\gamma_{meas} = \gamma_1$, then $N_q = 0$, and if (ii) $\gamma_{meas} = \gamma_q/q$, then $N_q = 1$.

Multiple-charged ion identification by induced electron emission is possible, in principle, due to the fact that $\gamma_q \simeq q^2$.

The expected values γ_q for multiple-charged ions with velocity $v < 10^6$ cm/s and the ratio γ_q/γ_1 are listed in table 7.

As is seen from the table, for some pairs the difference in the emission under multiple-charged ions and single-charged ones is drastic. It should be noted that the above-mentioned values are rather rough and should be used merely by order of magnitude. However, a high value of the ratios γ_q/γ_1 indicates in principle the possibility of the separation of multiple-charged ions and single-charged ones by the emission of induced Auger electrons.

Table 7
The expected values γ_q and the ratio γ_q/γ_1.

Ion pairs	γ_q	γ_1	γ_q/γ_1
$^{84}Kr^{7+}$ & $^{12}C^+$	5.4	0.02	270
$^{84}Kr^{10+}$ & $^{16}O^{2+}$	15	0.3	50
$^{84}Kr^{6+}$ & $^{14}N^+$	4	0.1	40
$^{131}Xe^{10+}$ & $^{13}C^+$	15	0.02	750
$^{131}Xe^{15+}$ & $^{18}O^{2+}$	40	0.8	50
$^{40}Ar^{7+}$ & $^{17}O^{3+}$	5	2	2.5
$^{40}Ar^{8+}$ & $^{14}N^{3+}$	7	1.5	4.7
$^{238}U^{13+}$ & $^{18}O^+$	20	0.04	500

Apparently, it seems to be difficult to favor another phenomenon, which would depend to a great extent only on ion charge and be convenient for the identification of multiple-charged ions.

The dependence $\gamma_q = f(q)$ for a pure (homogeneous) beam Kr^{2+} up to Kr^{7+} as well as for mixed beams of $^{84}Kr^{3+}$ and $^{14}N_2^+$, $^{84}Kr^{6+}$ and $^{14}N^+$, $^{84}Kr^{7+}$ and $^{12}C^+$ with ion energy $E_0 = 400$ eV is shown in fig. 26. It is seen that for the pure beams the dependence $\gamma_q = f(q)$, as it follows from the theoretical estimation made by Parilis (1967), is nearly quadratic, but for mixed beams γ_{meas} is less than for the pure ones.

In this case, the coefficient γ_q was measured directly since a single peak corresponds to the ^{86}Kr isotope which is not contaminated. The coefficient γ_q for the impurity ions was estimated via direct measurement of the coefficient of the electron emission for C, N and O single- and double-charged ions with different ion energies (fig. 27).

The fraction N_q of multiple-charged ions for mixed beams was estimated using eq. (116).

For example, for the mixture $^{84}Kr^{7+}$ and $^{12}C^+$ the coefficients $\gamma_q = 5.35$, $\gamma_1 = 0.2$, and $\gamma_{meas} = 0.52$. Then for $q = 7$ and $N_q + N_1 = 1$, it was found that $N_q = 0.15$ and $N_1 = 0.85$, which means that the content of $^{84}Kr^{7+}$ is about 15% and that of $^{12}C^+$ is about 85%. For other mixed beams of Kr the following values N_1 and N_q were obtained:

for $^{84}Kr^{6+} + {}^{14}N^+$: $N_1 = 88\%$, $N_q = 12\%$;

for $^{84}Kr^{3+} + {}^{14}N_2^+$: $N_1 = 25\%$, $N_q = 75\%$.

The values N_1 and N_q estimated for the other ion pairs are listed in table 8.

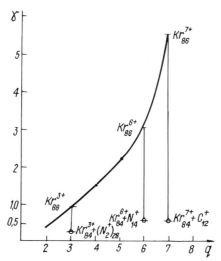

Fig. 26. Total yield γ_q versus charge q for pure and mixed beams of Kr ions with energy $E_0 = 400$ eV. (●) Kr ions without impurities, (○) mixed ion beams.

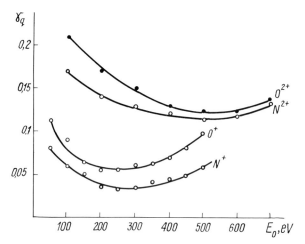

Fig. 27. Total yield γ_q versus ion energy for impurity ions.

Table 8
Values N_q and N_1.

Mixture	Multicharged ion	Impurity
	N_q (%)	N_1 (%)
$^{32}S^{2+}$ & $^{16}O^+$	71	29
$^{32}S^{4+}$ & $^{16}O^{2+}$	18	82
$^{32}S^{6+}$ & $^{16}O^{3+}$	11	89
$^{32}S^{7+}$ & $^{14}N^{3+}$	3	97
$^{64}Cu^{4+}$ & $^{16}O^+$	80	20
$^{64}Cu^{8+}$ & $^{40}Ar^{5+} + {}^{16}O^{2+a}$	2	98
$^{40}Ar^{5+}$ & $^{16}O^{2+}$	37	63
$^{84}Kr^{3+}$ & $^{14}N_2^+$	75	25
$^{84}Kr^{6+}$ & $^{14}N^+$	12	88
$^{84}Kr^{7+}$ & $^{12}C^+$	15	85

The graph in fig. 10 can be used directly for the determination of the fraction of multiple-charged ions in a mixed beam. To do so, the measured values γ_{meas} for a mixed beam and values γ_q and γ_1 obtained using the graph, in those cases where they were not measured experimentally, have to be substituted in eq. (116). Of course, this method is less precise than with direct measurements of γ_q and γ_1.

For example, the value γ_{meas} for a mixed beam containing $^{40}Ar^{5+}$ ions (47%) and $^{16}O^{2+}$ ions (53%) are shown in the graph fig. 10.

From eq. (116) it follows that the error in the determination of the fraction of multiple-charged ions N_q and impurity ions N_1 in the mixed beams depends on the accuracy of measuring the coefficients γ_q, γ_{meas} and γ_1.

An incomplete neutralization of the multiple-charged ions leads to the increase of the potential emission, which affects the accuracy of determination γ_q. To minimize the error, the measurements should be made with the lowest available ion energy. Low energy is also necessary to eliminate the kinetic emission which appears starting with ion velocity $(5-7) \times 10^6$ cm/s and masks the emission of Auger electrons.

CHAPTER 13

Application of Ion Scattering to Surface Diagnostics

One of the major goals in the investigation of ion scattering from the surface of a solid is the establishment of the relation between the characteristics of the scattered beam and the elemental composition, structure, and state of the surface.

The development of various models and simulations of ion scattering from point defects and extended defects in the crystal surface structure, as well as on adsorbed atoms and molecules has laid a theoretical foundation for using ion scattering as a tool in surface analysis and diagnostics.

Apart from the method involving Rutherford backscattering of high-energy ions, an approach is being developed to study surfaces by means of low- and medium-energy ion scattering which is called ion-scattering spectroscopy. In studies of the crystal structure of surfaces it is called ion-beam crystallography.

A large number of reviews deal with the various applications of low- and medium-energy ($\sim$0.5–50 keV) ion scattering from the surface of solids (Mashkova and Molchanov 1985, Petrov and Abroyan 1977, Aono 1984, Morrison 1977, Czanderna 1975, Fiermans et al. 1978, Treitz 1977, Poate and Buck 1976, Niehus and Bauer 1975, Taglauer 1985, Goff 1973, Buck 1975, Taglauer and Heiland 1980, Niehus and Comsa 1984, Van den Berg et al. 1980, Engelmann et al. 1985, Woodruff and Delchar 1986). Patents were obtained for the use of ion scattering techniques in the analysis of surface composition and structure (Smith 1969, Mckinney and Goff 1977, Heiland et al. 1978a, Brongersma 1978, Mckinney and Thomas 1978, Erickson et al. 1975, Andrä 1979, Hartmann et al. 1982). A distinctive feature of this method is its combination with a mathematical simulation of the scattering trajectories required for the interpretation of the angular diagrams and energy spectra of scattered ions. The various effects revealed in the scattering of medium-energy ions from the surface of a solid are used to analyze its elemental composition and crystalline structure. The studies devoted to the application of the low- and medium-energy ion spectroscopy to the determination of composition, number and position of impurity atoms on metal surfaces (Smith 1967, Taglauer and Heiland 1975, Prigge et al. 1977, Englert et al. 1982, 1979, Brongersma and Theeten 1976) and of sputtering rates (Ball et al. 1972), to the investigation of film epitaxy (S.H.A. Begemann et al. 1972) and surface topography (Czanderna 1975, Engelmann et al. 1985, Brongersma and Mul 1973, Heiland and Taglauer 1972), as well as of the structure of nonideal surfaces (Suurmeijer and Boers 1973, Czanderna 1975,

Heiland and Taglauer 1973b), which had been published before 1980, were reviewed in detail by Mashkova and Molchanov (1980), Petrov and Abroyan (1977), Czanderna (1975), and Fiermans et al. (1978). Therefore, we will confine ourselves here only to mentioning briefly the most typical of them while focusing major attention on recent publications. First, the main features which make ion-scattering spectroscopy different from other spectroscopic methods of surface analysis should be pointed out. This is primarily the high surface selectivity of the method, namely, it is based on the detection of the ions which were scattered from one or two surface layers, because the contribution of the subsequent layers is small, the depth from which the scattered and detected ions escape being 3–10 Å. Ion-scattering spectroscopy and secondary ion mass spectrometry are the only methods permitting analysis of the topmost monolayer. Another specific feature is the possibility of using the shadowing and multiple-scattering effects to obtain information on the structure, e.g., on the mutual arrangement of atoms on the surface (Brongersma and Mul 1973, Heiland and Taglauer 1972) and the presence of atomic steps and vacancies on the surface (Suurmeijer and Boers 1973, Heiland et al. 1973a, Eckstein et al. 1974), on the orientation of adsorbed molecules (Smith 1967) and even on such parameters as vibrational modes (phonon spectrum) of the surface of a solid (Walker and Martin 1982). Such information can be obtained only by combining a real experiment with computer experiment, i.e., if one has good computer programs for scattering calculations (Czanderna 1975, Fiermans et al. 1978, Heiland 1982). The sensitivity of this method is fairly high. Indeed, the minimum detectable concentration of adsorbed atoms is $\sim 10^{-2}$ monolayer for light elements on a high-Z substrate, and $\sim 10^{-4}$ monolayer for metals on a low-Z substrate (Heiland 1973, Ball et al. 1972, Taglauer and Heiland 1974, 1975, Brongersma and Mul 1973, Feldman and Mayer 1986).

Since the projectile interacts primarily with the nearest surface atom, this method is essentially local, and this determines its high sensitivity to atomic-scale inhomogeneities of the surface. However, an overall quantitative calculation of the sensitivity and, accordingly, determination of calibration graphs for various atoms are complicated at present by the absence of precise data on the interaction potential and probability of neutralization. Considerable progress in this direction comes with the possibility of detecting by time-of-flight (Buck et al. 1975) and stripping cell (Kasi and Rabalais 1990, Chicherov 1972, Eckstein et al. 1978, Luitjens et al. 1980c) techniques besides the ionic component also the total number of reflected particles, including neutrals. An alternative approach consists of the use of ions of alkali metals, which possess a low neutralization probability and are backscattered primarily in the ionic state (U.A. Arifov 1968, Terzic et al. 1979a, Niehus and Preuss 1982, Von dem Hagen et al. 1982, Taglauer et al. 1980), besides the use of noble-gas ions. This permits one to base the interpretation of the angular and energy distributions of scattered particles entirely on the scattering cross sections while excluding from the consideration the process of neutralization and charge exchange which are still inadequately well studied.

Detection of all scattered particles or the use of alkali ions enables the operation with very small ion doses, which reduces substantially the degree of concomitant surface damage. Using in an analysis all scattered particles (in the case of noble-gas ions) yields an additional possibility of quantitative comparison of the measured and calculated distributions based only on the scattering cross sections and excluding from the consideration the processes of neutralization and charge exchange. In the analysis of the composition and structure of crystal surfaces by ion-scattering spectroscopy one usually confines oneself to $\theta < 90°$, i.e., to forward scattering. In the case of noble-gas ions the possibilities of quantitative analysis are limited by the current lack of data on their neutralization, and in the case of alkali-metal ions by the need of knowing the exact interaction potential and employing complex computer programs to determine the contribution of multiple scattering to the experimentally measured angular and energy distributions.

The modification of ion-scattering spectroscopy (Aono 1984, Aono et al. 1982a,b, Souda et al. 1983, Oshima et al. 1981, Fauster 1988, Orrman-Rossiter et al. 1990), based on using large-angle ion backscattering ($\theta \approx 180°$) and the properties of the shadow cone behind the target atom (the so-called impact-collision ion-scattering spectroscopy) permits one to avoid these difficulties and to obtain directly information on the atomic positions of a single-crystal surface, its relaxation and reconstruction. By using alkali-metal ions in this modification (Niehus and Comsa 1984, Niehus et al. 1990), one gains an additional possibility of excluding neutralization from the consideration. The scattering angles close to 180° reduce the background due to multiple scattering to a minimum. Hence, arbitrariness in the selection of the interaction potential reveals itself in this case least of all.

This method can be used in surface structure analysis if it is combined with studying the distributions of directly recoiled atoms and the orientational sensitivity of the degree of scattered ion neutralization (Nizhnaya et al. 1979c, 1982a, Overbosch et al. 1980, Hou et al. 1978, Shultz et al. 1986, Möller et al. 1984, O'Connor et al. 1986, Volkov et al. 1982). While in contrast to secondary ion mass spectrometry, the ion-scattering method is not in principle destructive; nevertheless, some damage of the surface and a change in its properties caused by ion bombardment should be taken into account. They can be reduced to a minimum by decreasing the ion-beam intensity and use of scanning. The need of high vacuum and low mass resolution limit partially the applicability of the method; however, these conditions are usually met in setups designed for molecular beam epitaxy, ion doping, etching, decoration and ion beam lithography. Since the use of the ion-scattering method requires proper calibration, in principle it should be considered as semi-quantitative. The possibility of quantitative analysis can be broadened by employing calibration against reference samples and by carrying out a comprehensive theoretical and experimental investigation of the physical processes involved in the scattering of ions, the formation of their charge state on the atomic structures and various elements of topography of real single-crystal surfaces.

1. Analysis of surface composition

The possibility of elemental analysis of the surface composition is validated by the predominantly binary character of collisions of low- and medium-energy ions with individual atoms on the surface of an irradiated solid, which was established by U.A. Arifov and Ayukhanov (1951) and Eremeev (1951). These early studies revealed that the position of the main peak in the energy distribution of reflected ions on the energy axis approximately corresponds to the energy calculated by eq. (4) of chapter 1. As follows from this expression, the energy of the ion which suffered a single collision with a target atom depends on the scattering angle, the mass ratio of the collision partners and the initial ion energy, but does not depend on the interaction potential affecting only the probability of scattering at a given angle. Therefore, the position of this peak on the energy axis can be used for unambiguous determination of the mass of the particles responsible for the scattering from a surface and, hence, of its elemental composition.

A spectacular implementation of this possibility was the installation on Surveyor-5 and 6 spacecraft of energy analyzers of α-particles (i.e., helium ions) emitted by a radioactive source and reflected from lunar soil at a certain angle. The analysis of the elemental composition of the lunar soil carried out in this way revealed the presence in its rocks of oxygen, silicon, aluminium and other light elements which are abundant also in the Earth's crust. The presence of heavier elements could not be established reliably because of the small mass of helium. This was the first time in history that a direct analysis of the elemental composition of lunar soil could be made.

Smith (1967) was the first to apply ion scattering to analyze the elemental composition of metal surfaces. He used 0.5–3 keV He^+, Ne^+, Ar^+ ions and polycrystalline Ni and Mo targets. The sharp peaks in the energy spectrum revealed by him corresponded to single scattering from atoms of the target material and adsorbed oxygen and carbon atoms. The scattering angle θ was chosen 90°. Equation (4) of chapter 1 can in this case be simplified to

$$\frac{E_1}{E_0} = \frac{m_2 - m_1}{m_2 + m_1}. \tag{1}$$

Solving it for m_2, we obtain

$$m_2 = m_1 \frac{1 + E_1/E_0}{1 - E_1/E_0} . \tag{2}$$

Here, E_1/E_0 corresponds to the position of the single scattering peak in the spectrum, m_1 is the projectile ion mass, m_2 is the mass of the target or adsorbed atom to be found.

The intensity of single scattering I_i in an element of solid angle $\Delta\Omega$ determining the sensitivity of the method is proportional to the initial beam intensity I_0, the concentration of scattering centers N_i (cm^{-2}), neutralization probability P_i, differential scattering cross section $\mathrm{d}\sigma_i(\theta, \varphi)/\mathrm{d}\Omega$ cm^2/sr and the shadowing coefficient α_i for the ith component on the surface,

$$I_i \approx I_0 N_i P_i \alpha_i \frac{\mathrm{d}\sigma_i(\theta, \varphi)}{\mathrm{d}\Omega} \Delta\Omega . \tag{3}$$

Equations (4) of chapter 1 and (3) form a basis for the application of the ion scattering to elemental analysis of the surface of a solid (Mashkova and Molchanov 1980).

Figure 1a shows the energy distribution of 1 keV He$^+$ ions scattered to $\theta = 90°$ from a Cu/Ni film deposited on an Al substrate (Taglauer and Heiland 1976). One clearly sees distinct peaks due to single scattering from the various elements present on the surface, their positions and heights being determined by relations (4) of chapter 1 and (3).

The shadowing of atoms on the surface by other atoms causes a change in the relative intensity of single-scattering peaks (Goff 1973, Brongersma and Mul 1972, 1973) and even a total disappearance of the peak due to target atoms in the case of adsorption (Niehus and Bauer 1975, Brongersma and Mul 1972, Novacek et al. 1990). Indeed, adsorption of Br on Si (111) at a Br vapor pressure of $\sim 10^{-4}$ Pa suppressed completely the single-scattering peak due to Si atoms in the energy spectrum (Brongersma and Mul 1972). While not complete, such shadowing was observed to occur also in the adsorption of oxygen on tungsten (Niehus and Bauer 1975).

Brongersma et al. (1978) studied by ion-scattering spectroscopy the equilibrium segregation of copper on the surface of a polycrystalline Cu/Ni alloy with different Cu contents. Figure 1b shows the energy distributions of $E_0 = 3$ keV Ne$^+$ ions scattered to $\theta = 142°$ from a Cu/Ni alloy containing 52% copper. At $T = 25°C$, the surface concentration of Cu does not differ from the bulk concentration, the single scattering peaks due to Cu and Ni atoms being practically equal in height. With the temperature raised to $T = 510°C$, the pattern changes dramatically, namely, the surface becomes enriched in copper which manifests itself in a virtually complete disappearance of the Ni peak and a strong growth of the Cu peak. Removal of a part ($\sim 20\%$) of the outermost monolayer results in a linear growth of the Ni

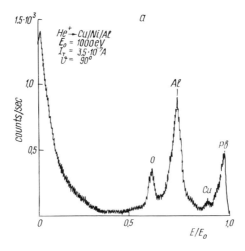

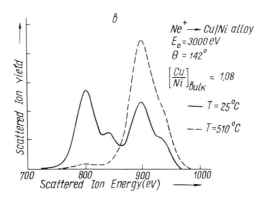

Fig. 1. Energy spectra of (a) He$^+$ ions scattered from a metal film deposited on an Al substrate, and (b) Ne$^+$ ions scattered from a Cu/Ni alloy with 52% Cu. Full curve: $T = 25°C$, broken curve $T = 510°C$.

atom signal. Hence the second layer is enriched in copper in a lesser degree than the topmost one.

Taglauer (1982) employed ion-scattering and Auger-electron spectroscopy to study the surface changes of multicomponent compounds caused by preferential sputtering of their constituents. Both methods yield qualitatively similar results; however, absolute measurements would require to find appropriate standards.

The relative magnitude of the scattering signals from different surface components provides information not only on their content but on their mutual positions as well. For a surface coated by an adsorbate, relation (3) takes on the form

$$I_s(\gamma) \sim I_0[N_s - \alpha_i(\gamma)N_a] \frac{d\sigma_s}{d\Omega} \Delta\Omega P_s(\gamma) , \tag{4}$$

where the indexes s and a refer to the substrate and adsorbate, respectively, and γ is the coverage. As seen from eq. (4), the single-scattering peak intensity is determined not only by the surface concentration of atoms but on the shadowing coefficient α_i as well, which depends on their mutual arrangement.

2. Structural analysis of crystal surfaces

Practically all the above-mentioned orientational effects which manifest themselves in ion scattering from the surfaces of solids can be used to study the structure of surfaces and thin films (Mashkova and Molchanov 1980, Petrov and Abroyan 1977). Indeed, the shadowing and double-scattering phenomena can be employed in determining the spacing between two neighboring atoms in the ion incidence plane, as well as the presence on the surface of vacancy-type and step defects (U.A. Arifov 1968, Mashkova and Molchanov 1980, Petrov and Abroyan 1977, Algra et al. 1980b, Overbury 1981), and the atom row and multiple-scattering effects can be employed in studies of the regular atom arrangement on the crystal surface which provides information on the mean length of undamaged atom rows on the irradiated surface and, hence, on the degree of its imperfection (Kaminsky 1965, Mashkova and Molchanov 1980, Petrov and Abroyan 1977, Taglauer et al. 1988).

Investigation of the energy spectra obtained with use of scattering data from surfaces coated with adsorbed atoms and molecules, i.e., measurement of signals from the substrate and the adsorbate, provides information (at least, qualitative) on the positions occupied by adatoms on a surface. Quantitative data on the mutual location of adatoms and substrate atoms can be derived with impact-collision ion-scattering spectroscopy.

2.1. Ion-scattering spectroscopy

Ion-scattering spectroscopy as a tool in surface studies has attractive assets as being relatively simple, operating with large scattering cross sections (10^{-2} Å^2/sr) and being extremely sensitive to surface characteristics. This method is used to advantage in studying the structure of the atomic layers involved in catalysis, ordered adsorption systems on metal surfaces, restructuring and disordering of surfaces caused by defects and thermal motion. The structural analysis of surfaces by this method is based on a comparison of experimentally measured energy and angular distributions of scattered particles with calculations (most frequently done on a computer) based on a chosen surface model. The results of the comparison can be used to refine the model parameters and, accordingly, the surface structure.

Algra et al. (1980c,d, 1982a,b,c) and Visscher et al. (1986) employed the ion-scattering technique to investigate the structure and relaxation of Cu (410) and determined the position of the edge atom on this stepped surface.

Landnyt (1984) used this method to investigate stepped Ni (111) surfaces and showed most peaks in the ion-scattering spectra to be due to single and double scattering.

A reconstructed Au (110) surface was studied by ion-scattering spectroscopy by Overbury et al. (1981).

The position of edge atoms on a stepped Cu (410) surface was probed by 21 keV H^+ ions (Algra et al. 1980c). The bombardment was performed under the conditions favorable for channeling ($\psi = 45° \rightarrow \langle 110 \rangle$) in the downward and upward step directions (fig. 2). The scattered-ion intensity is determined by the number of surface atoms seen by the detector, which in the case of a downward step does not depend on the scattering (θ) or escape (β) angles. In the case of an upward step the ion intensity will be determined not only by the surface atoms but by the θ- or β-dependent contribution of subsurface atoms as well. Thus, the essence of the method consists in finding the minimum of the ratio $I_{up}/I_{down} \approx 1$ for a certain β_{min} near $\beta = 45°$ by properly varying the angle θ (or β). For an ideal lattice and under specular scattering ($\psi = 45°$ and $\beta = 45°$), the contribution of subsurface atoms to scattering should be zero. The experimentally found value $\beta_{min} = 42.0 \pm 1.0°$ suggests that the edge atom is depressed into the lattice by an amount $\Delta a/a = 5.0 \pm 1.5\%$, where a is the Cu lattice constant. The absolute value of the displacement is $\Delta a = 0.18 \pm 0.05$ Å. The depression of the step edge atom found in this way is correct provided the underlying atom is not relaxed. The absence of relaxation of the subsurface atom for Cu (001) was established by Ma et al. (1978).

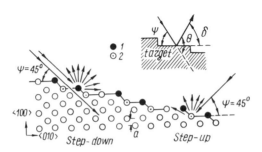

Fig. 2. Side view of Cu (410) face (left: motion down the step, right: motion upward. In the figure are shown (1) an edge atom, and (2) an atom below the edge one.

Algra et al. (1980d) used multiple scattering of Ne^+, Ar^+, and Kr^+ ions to determine the structure of the stepped Cu (410) surface, namely, the edge separation d for this face and the (100) terrace width l_0. Studies of stepped surfaces have gained recently particular significance in searching for a basic relation between the catalytic activity of such surfaces and their stepped structure (Wagner 1979).

To reduce to a minimum the radiation damage due to the primary beam, the target was maintained at a temperature of 540 K (Poelsema et al. 1977b). The possibility of structural analysis is based on the fact that for an energy E_0, grazing angle ψ, scattering angle θ, and azimuthal target rotation angle ξ the ions scattered from different surface features enter the detector on having suffered different numbers of collisions and with different final energies.

Determination of d was based on the fact that a particle undergoing strong single scattering from an atom suffers also a weak deflection from neighboring atoms both before and after the main collision (the so-called quasisingle scattering). The extent of this deflection and, hence, the energy E of the scattered particle depend on the spacing between the neighboring atoms and the grazing angle ψ. Evaluation of the inelastic losses capable of affecting the precision of the method showed them to be small in this case and to be independent on ψ for $\psi \approx \frac{1}{2}\theta$. The scattered ion energies were measured as a function of ψ for $\theta = 50°$ and $\xi = 0$, 18.4, +90, −90°. For $\xi = 90°$, the ion beam is directed up the step ($\langle 1\bar{3}0 \rangle$–$\langle 1\bar{5}0 \rangle$); and for $\xi = -90°$, down the step ($\langle 1\bar{3}0 \rangle$)–$\langle 150 \rangle$). The unknown interatomic distances in these directions were determined by comparing the experimental $E(\psi)$ relations with the computer simulation data obtained for $d = 3.6$ and 5.7 Å, respectively.

The ion energy E was found to increase in all cases with decreasing ψ, starting from $\psi = \frac{1}{2}\theta$, which is due to an increasing effect of weak deflection from the atoms which precede the atom responsible for the main scattering. The $E(\psi)$ relations obtained for the upward and downward steps differ from one another. Calculation of the scattering process within the three-atom semianalytic model of Poelsema and Boers (1978) using the Molière potential and including thermal vibrations describes well the experimental data. The model of an alternating $\langle 1\bar{3}0 \rangle$–$\langle 1\bar{5}0 \rangle$ surface sequence finds experimental confirmation. Therefore, the structure of the Cu (410) face determined earlier by LEED is supported by ion-scattering spectroscopy, namely, the step height d was found to be equal to one interplanar spacing 1.8 Å, and the (100) terrace width l_0, to four atom rows (7.2 Å).

The scattered-ion energy depends also on the length of the plateau connecting the edges of two neighboring steps in a perpendicular plane. The effective plateau length is $l = l_0/\sin \xi = 7.2$ Å$/\sin \xi$. For small ψ and large l, the ions are scattered through an angle θ along two different trajectories: (1)

part of the ions undergo single scattering from an edge atom and retain their energy, $E = E_s$; and (2) the other ions, while also becoming scattered in a close collision with a step atom, undergo preliminarily multiple specular scattering from plateau atoms, with $E_p > E_s$. The energy E_p depends on plateau length l, and decreases with increasing l. The corresponding analysis led to the conclusion that the dependence of the relative scattered-ion energy difference $(E_p - E_s)/E_0$ on plateau length can be used to determine l in the range 15–60 Å to within ~5 Å.

Luitjens et al. (1980d) showed that the results of these experiments can be used to study the dependence of radiation-induced surface damage on radiation dose. The measured high-energy peak of the Ar$^+$ ions back-scattered from Cu (100) is comparable with the above-mentioned peak due to a plateau; note that as the dose and, accordingly, the number of defect structures (isolated atoms and steps) increases, the intensity of the peak increases, and it is displaced toward higher energies. This displacement can be attributed to increased surface damage at the atomic level and, hence, to a smaller average length of the plateau sections.

As has been shown in chapter 6, Boers et al. (1980) estimated the number of atomic steps formed on the Cu (100) face under bombardment by Ar$^+$ ions by measuring the energy spectra of these ions on scattering from the surface as a function of grazing angle. Dzhurakhalov et al. (1988b, 1990a) used these energy spectra to estimate the dimensions (length) of these steps, the spacing between them and their relative abundance on the irradiated surface. This was done by computer simulating the trajectories of the particles scattered from both the ordered areas of the surface and mono-atomic steps of different length, one atomic layer high. The trajectories of scattered particles were calculated in the binary collision approximation using the universal interaction potential of Biersack–Ziegler, eq. (16) of chapter 1, with inclusion of the time integral. The elastic and inelastic energy losses were summed along the trajectories. The inelastic energy losses were calculated by Kishinevskii's relation, eq. (38) of chapter 1, and included into the scattering kinematics. If one takes into account thermal vibrations, the target atoms were assumed to vibrate independently of one another with their displacements from the equilibrium positions obeying the Gaussian distribution.

Figure 3 shows the energy distributions of the total number (ions and neutrals) (points) of argon particles scattered from Cu (100) in the ⟨100⟩ direction for grazing angles $\psi = 7°$, and a constant scattering angle $\theta = 30°$, obtained by Boers et al. (1980). The straight lines identify the energies of the particles scattered along the various trajectories. The energies of the particles on trajectories 1 and 2 are seen to coincide with the experimentally determined positions of peaks 1 and 2 in the spectrum (full curves in fig. 3). At the same time, the broad maximum 3 between peaks 1 and 2 could not

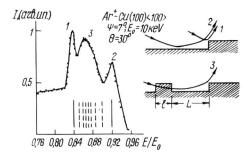

Fig. 3. Energy distribution of (points) the total flux of Ar$^+$ ions scattered at an angle $\theta = 30°$ from Cu (100) ⟨100⟩ for $\psi = 7°$ (Boers et al. 1980). The full and broken vertical curves on the spectra specify the calculated scattered ion energies.

be attributed to scattering along type 1 and 2 trajectories on an ordered surface even when thermal vibrations of its atoms are included. Figure 4 presents histograms of the angular distributions of scattered Ar particles for $\psi = 7°$ and target temperatures (a) $T = 0$ and (b) $T = 300$ K. We see that scattering at $\theta = 30°$ at $\psi = 7°$ is not observed to occur on an ordered surface even when thermal vibrations are included. To identify the origin of maximum 3 in the spectrum, we calculated the trajectories of the ions scattered from a surface bearing a number of isolated steps of different length l which are separated by plateaus, i.e., areas of ordered surface of length L (fig. 3, lower panel).

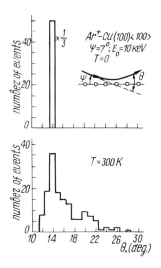

Fig. 4. Angular distribution histograms of 10 keV Ar$^+$ ions scattered from an ordered Cu (100) ⟨100⟩ face at a grazing angle $\psi = 7°$ for (a) $T = 0$ and (b) $T = 300$ K.

As shown by calculations, the broad maximum 3 in the spectrum is due to particles with trajectories 3 determined by two neighboring steps, a process in which the particles pass under the first of the steps and are then scattered from the edge atom of the second step. The trajectories of quasidouble scattering from edge atoms of the neighboring steps likewise contribute to the broad maximum. The calculations reveal that ion bombardment performed under the experimental conditions of Boers et al. (1980) forms on the surface monoatomic steps consisting of several (two to five) atoms, their length not exceeding ~15 Å. Thus, by varying the number of atoms in steps and, hence, their dimensions, as well as the separation between them, one can construct the trajectories of ions with energies extending throughout the broad maximum 3 in the spectrum (broken curves). The spacing between the steps, L, is varied from the minimum value equal to two lattice constants up to ~45 Å. The most probable turn out to be steps consisting of 2 and 3 atoms separated by areas of ordered surface (plateaus) of length $L = 15$–45 Å. Note that the calculated path length of particles scattered along the ridge of a row and contributing to peak 2 was found to be ~30 Å. This correlates with the values of L derived from an analysis of trajectory 3 and quasidouble scattering. Thus, the presence of peak 2 in the spectrum confirms the existence of step spacings in excess of 30 Å. Quasidouble scattering trajectories were found to be only weakly sensitive to the separation between steps; however, their probability falls off sharply with increasing L.

Similar calculations were carried out also for the $\psi = 23°$ spectrum, the disappearance of peaks 2 and 3 in the ionic component of the scattered flux being in accord with the pattern of trajectories 2 and 3. To estimate the effect of the actual interaction potential chosen, the trajectories of the scattered particles were calculated using the matched potential, eq. (20) of chapter 1. The conclusion, using step length and separations, drawn based on the matched potential turned out to be in good agreement with similar data obtained with the Biersack–Ziegler potential.

Thus, a comparison of experimental data with computer simulation suggests that intense bombardment of the Cu (100) face in the $\langle 100 \rangle$ direction at $T = 300$ K by 10 keV Ar$^+$ ions at a current density $J = 10^{-6}$–10^{-8} A/cm^2 results in the formation of monoatomic steps containing two to five atoms and separated by areas of ordered surface (plateaus) ~15–45 Å long. An estimate of the number of atomic steps made based on the proposed model of surface damage is in agreement with the value 2×10^{14} cm^{-2} obtained by Boers et al. (1980).

De Wit et al. (1979) used ion-scattering spectroscopy to determine the positions of adsorbed oxygen atoms on Cu (110). The method is based on a qualitative analysis of the azimuthal distribution of ions backscattered from a clean and an oxygen-coated surface, as well as on the use of the shadow cone model. By this model (Martynenko 1973), if a uniform ion beam falls on a target atom a shadow region may form behind it where no scattered ions can penetrate. At the same time, the scattered particle distribution will exhibit a strongly enhanced flux at the shadow boundary (rainbow effect). Depending on the actual experimental geometry (polar and azimuthal incidence and escape angles), the target atom next to the one causing the shadow may turn out to lie either within the shadow cone, at its boundary, or, finally, outside it. In the case where the atom lies inside the shadow cone, scattering from it at a large angle is impossible. If it lies at the boundary, it will face the strongly enhanced ion flux. The enhanced flux of the particles scattered from a shadow boundary atom (whether it is a target atom or an adatom) permits determination of its relative position.

To reduce the effects of blocking and neutralization, only ions with large escape angles to the surface were analyzed. Figure 5 shows azimuthal distributions of 4 keV Ne^+ ions single-scattered from Cu atoms on a clean (curve a) and a partially oxygen-coated (curve b) Cu (110) face for $\psi = 6°$. Presented on the right is a schematic presentation of the two top surface layers on Cu (110). The dips in the curves at the angles $\varphi = 0$, 55 and 90° correspond to scattering along the closely packed directions $\langle 110 \rangle$, $\langle 112 \rangle$ and $\langle 001 \rangle$ on Cu (110). Because of the small grazing angle, the atoms A, B, and C lie in these cases inside the shadow cone produced by atom M, making single scattering to $\theta = 45°$ in these directions impossible, since each subsequent atom is located within the shadow cone of the preceding one. As the grazing angle increases, the boundary of the shadow cone moves toward

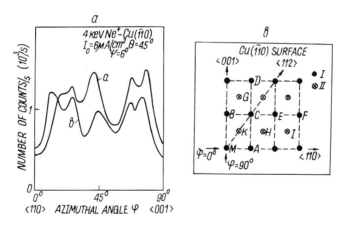

Fig. 5. (a) Azimuthal distributions of Ne^+ ions scattered from Cu atoms on a (curve a) clean, and (curve b) partially oxygen-coated Cu ($\bar{1}$10) face, and (b) schematic representation of the first two layers of the Cu ($\bar{1}$10) face: (I) Cu atoms in the topmost, and (II) in the second layer.

atom B, so that at $\psi = 16°$ it will lie already at the shadow cone boundary. This is indicated by a sharp maximum in the azimuthal distribution of ions scattered in the $\langle 001 \rangle$ direction at $\psi = 16°$.

The azimuthal distributions $I(\varphi)$ computed for different faces with the C/r^2 potential are qualitatively similar to those obtained experimentally.

Curve b in fig. 5 is similar to curve a, with the exception of an additional peak close to the $\langle 110 \rangle$ direction. The behavior of this peak with variation of the grazing angle ψ (it disappeared for $\psi > 13°$) permitted the conclusion that it is due to a shadow cone produced by an oxygen atom. Comparison of curves a and b suggests that copper atom displacements in the plane of the oxygen-coated Cu $(1\bar{1}0)$ face, as well as perpendicular to it, do not exceed 4% of the interatomic separation at room temperature.

To determine the position of chemisorbed oxygen atoms, azimuthal distributions of 3 keV Ne$^+$ ions single-scattered from oxygen atoms on Cu $(\bar{1}10)$ were measured as a function of grazing angle ψ. An analysis of the distributions shows the oxygen atom to be located inside the shadow cone produced by the copper atom in the $\langle 001 \rangle$ direction. From the symmetry of the distribution with respect to the $\langle 110 \rangle$ direction and the existence of a limiting angle of Ne$^+$ ion scattering from oxygen atoms, one may conclude that the oxygen atom lies midway between two copper atoms in the $\langle 001 \rangle$ row under the topmost layer. The distance from the center of the oxygen atom to the surface plane was found to be $\sim 0.6 \pm 0.1$ Å. Experiments show that ion bombardment does not affect the measured oxygen-atom position in the topmost layer.

Bronckers and de Wit (1981a,b,c) used the technique developed by de Wit et al. (1979) to investigate the structure of two top layers of Cu (110) both with and without adsorbed oxygen atoms. To separate structural phenomena from charge exchange effects, the surface was bombarded with Ne$^+$ and H$_2$O$^+$ ions, and a comparison was made of the intensities of scattered Ne$^+$ and O$^-$ ions. The similarity between the anisotropies in the azimuthal Ne$^+$ and O$^-$ distributions was attributed to the blocking and ion focusing effects, the differences being assumed to be due to charge exchange. Surface layer structure analysis is made possible by a dependence of the above effects and processes on the mutual atom arrangement which manifests itself in the anisotropy in the intensity of scattered atoms.

Initially, the measurements were carried out in such a way that only atoms of the topmost layer could contribute to the structure of the azimuthal scattered ion distributions (Bronckers and de Wit 1981a). Figure 6 presents azimuthal distributions of 4 keV Ne$^+$ ions single scattered at $\theta = 45°$ from copper atoms on Cu (110) for (a) $\psi = 8°$, (b) 23°, and (c) 36°. The dips in the curves correspond to the principal crystallographic directions, and their halfwidth and depth are related to interatomic spacing, the dip depth and halfwidth increasing with atom row packing density. While the structures of

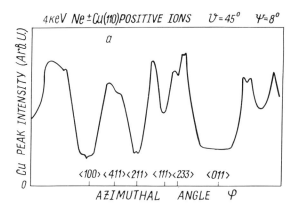

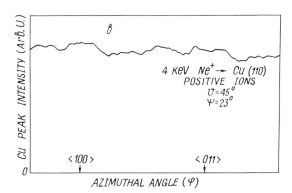

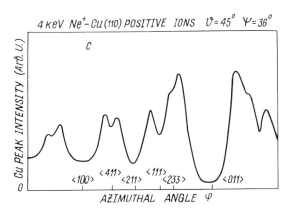

Fig. 6. Azimuthal distributions of 4 keV Ne$^+$ ions scattered at $\theta = 45°$ from a Cu atom on Cu (110) for (a) $\psi = 8°$, (b) $\psi = 23°$, and (c) $\psi = 36°$.

the distributions for $\psi = 36°$ and $8°$ are similar, the ion intensity for the larger ψ is substantially lower. At near-specular scattering ($\psi = 23°$) there is no more mutual atom screening in the topmost layer, just as there is no corresponding structure in the distributions. Because of strong neutralization, the ions backscattered from the second layer do not contribute to the distribution, the variation of the neutralization probability with φ being small. The pattern of the angular distributions can be used to determine the structure of the surface layers. The width of the shadowing dip in the azimuthal distributions ($\Delta\varphi$) depends not only on E_0 and d, but on the grazing angle ψ as well. Then we will have

$$\Delta\varphi^2 = \Omega^2 - \psi^2 , \tag{5}$$

where Ω is the angular dimension of the shadow cone, which for the C/r^2 potential is $\Omega = \frac{3}{2}(\pi C/E_0 l^2)^{1/3}$, l is the distance to the shadowing atom. The precision with which the relative position of atoms in a row can be determined depends on the accuracy of the angular positions of the shadowing dips and is estimated as 0.1 Å.

Bronckers and de Wit (1981b) changed the experimental geometry in such a way as to make possible observation of Ne^+ and O^- ion scattering from atoms in the second layer. The angular distributions were complicated here by the distorted shape of the shadow cone and the effect of focusing of the ions incident on the second atom layer.

The distortion of the shadow cone shape may result both in a decrease and in an increase of the reflected ion yield. In the first case, the ions pass approximately midway between two target atoms and become focused at a point close to the intersection of the two undistorted shadow cones produced by the target atoms (wedge focusing). The ion yield increases when the second atom is located inside the shadow cone produced by the first atom. This occurs in the case where the ion beam propagates approximately parallel to one of the principal crystallographic directions. Atoms in the rows along these directions form a shadow similar in shape to the enveloping surface of the shadow cones due to the individual atoms. For small interatomic distances it resembles the shadow of a string and is determined by a continuous potential $U(r')$ depending on atom density in the string and distance r' from the string axis.

Calculation of the string shadow shape can be simplified by turning to the projection of ion trajectory on the plane perpendicular to the string axis. The shape of the shadow in the transverse plane is similar to that of an individual atom provided one uses the continuous potential and the energy component $E_\perp$ in the transverse plane,

$$E_\perp = E_0(\sin^2\psi + \cos^2\psi \sin^2\varphi) . \tag{6}$$

Thus, the shape of the string shadow depends not only on E_0 and distance to the atom responsible for it, but on the angles ψ and φ as well, which requires numerical procedures for its calculation.

A comparison of azimuthal distributions for the Ne^+ and O^- ions showed the peaks of the distributions around the principal crystallographic directions to be more clearly pronounced in the O^- case, which can be accounted for by the focusing of the O^- ions by atoms of the topmost layer onto the second, and by increased yield of scattering on the second layer. Similar focusing is observed to occur also with Ne^+ ions which, however, become primarily neutralized in scattering from the second layer. The scattered ion flux is most intense in the directions for which the atom of the second layer lies at the edge of the shadow produced by atoms of the first layer. Thus, by comparing the intensities of the ions scattered at different angles φ, one can evaluate the relative probability of neutralization of the ions scattered from the layer and obtain information on target-atom positions in the first two layers of the crystal.

Bronckers and de Wit (1981c) applied Ne^+ and H_2O^+ ion scattering to study the reconstruction of the $Cu\,(110)$ surface caused by oxygen adsorption, and determined the position of adsorbed oxygen. This was done by measuring the $\psi-\varphi$ diagrams of scattered Ne^+ and O^- ions, as well as of positively and negatively charged oxygen recoils. Reconstruction is described with a 'missing-row' model, by which all $\langle 100 \rangle$ atom rows adjacent to oxygen-containing $\langle 100 \rangle$ rows disappear from the surface. Chemisorbed oxygen atoms occupy bridging positions in the $\langle 100 \rangle$ rows, the copper atoms residing at the same sites as on a clean surface.

Figure 7 presents azimuthal distributions of Ne^+ ions backscattered from an oxygen-coated $Cu\,(110)$ face. A comparison with distributions obtained for a clean surface (see fig. 6a) shows that in the case of an oxygen-coated

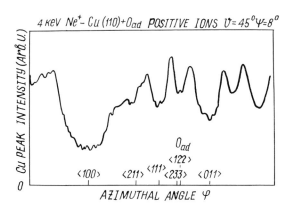

Fig. 7. Azimuthal distribution of 4 keV, $\psi = 8°$, Ne^+ ions scattered from a Cu atom at $\theta = 45°$ on an oxygen-coated Cu (110) surface.

surface the dips in the $\langle 011 \rangle$ and $\langle 211 \rangle$ directions are much smaller implying a substantial increase of the interatomic separations in these directions. As for the $\langle 100 \rangle$ dip, it broadens due to the additional shadowing and blocking caused by the oxygen adsorbed in this row. The $\langle 111 \rangle$ dip does not change. These observations are in accord with the preceding conclusion that every second row disappears completely from the surface under oxygen adsorption, and that the oxygen atoms occupy bridging positions in the $\langle 100 \rangle$ rows, 0.6 ± 0.2 Å below the first layer. The polar distributions measured for a clean and an oxygen-coated Cu (110) face also indicate clearly an increase in the interatomic separation in the $\langle 211 \rangle$ and $\langle 011 \rangle$ directions. The distance between two chemisorbed oxygen atoms in the $\langle 110 \rangle$ direction is ~5 Å, and the maximum oxygen coverage on the Cu (110) face is 0.5 monolayer, which is in agreement with the results of Poate and Buck (1976) and Niehus and Bauer (1975).

Thus the surface sensitivity of the LEIS method originates not only from the high probability of neutralization in scattering from atoms in deeper layers, but from the shadowing and blocking of these atoms as well.

Thermal vibrations of atoms in a solid affect considerably the characteristics of scattered particles. Since the time during which particles interact in scattering is much shorter than the time of thermal motion of the surface atoms, the disordering of the surface caused by the ions may be considered quasistatic and studied by ion-scattering spectroscopy, just as surface defects.

Poelsema et al. (1983) studied the temperature dependence of the quasisingle scattering peak, and derived from it the Debye temperature for the Cu (100) face bombarded by 10 keV Kr^+ ions.

Martin (1980) considered correlated thermal vibrations but did not draw any conclusions from the experiment concerning θ_D and the character of these vibrations.

Engelmann et al. (1985, 1986) investigated the temperature dependence of the scattering characteristics of 1 keV Ne^+ and Na^+ ions incident on Cu (110) in the range of 100–600 K. From a comparison of experiment with calculations a conclusion is drawn on the amplitude of surface atom vibrations. A comparison is made of the experimental Na^+ ion spectra ($\theta = 60°$, $\psi = 30°$, $\varphi = 0° \rightarrow \langle 110 \rangle$) with the spectra obtained by computer simulation. Computer calculations carried out for the Debye temperature component perpendicular to the surface, $\theta_{D_\perp} = 150$ K, and for $\theta_{D_\parallel}$ in the plane of the surface which is equal to that in the bulk, i.e., $\theta_{D_\parallel} = 343$ K, agree with experiment. The contribution of different trajectories to the energy spectrum is discussed. The difference in QS scattering from the first and second atomic layers comes from the stronger thermal disordering of the topmost layer which suppresses the QS peak caused by it. The quasidouble peak due to the first layer is shifted toward lower energies. This is explained by the high probability of large deviations of the microscopic grazing angle ψ_a from the macroscopic angle ψ (because of the large thermal vibrations), which results in a large asymmetry in scattering (i.e., $\theta_{tot} \geqslant \theta_i$; $\theta_{tot} = \theta_1 + \theta_2$). By analyzing the energy spectrum of scattered ions at $\varphi = 32°$, which does not coincide with the low-indexed direction, one can evaluate more accurately the in-plane thermal atomic vibrations since the spectrum exhibits in this case features associated with the trajectories which leave the scattering plane, the so-called zigzag collisions. A comparison of the calculated with experimental spectrum leads to the conclusion that the differences originate from undervaluation of the magnitude of in-plane thermal vibrations, namely, $\theta_{D_\parallel}$ should be somewhat lower on the surface than in the bulk, i.e., less than the value $\theta_{D_\parallel} = 343$ K accepted in the calculation.

Engelmann et al. (1985) studied also the temperature dependence of the single-scattering intensity of 1 keV Ne$^+$ ions for $\varphi = 0° \rightarrow \langle 110 \rangle$ Cu (110), $\theta = 60°$ and for various grazing angles. The experiments are analyzed in terms of the double-scattering model and compared with computer calculations. The shadow cone concept is invoked to explain the temperature dependence of the single-scattering peak intensity. Figure 8 shows three different cases of target atom position with respect to the shadow cone produced by the neighboring atom. In case a, the atom is located far from the shadow cone ($\psi > \psi_{cr}$, $S_0 > 0$). In this case, large thermal vibrations are required for the shadow cone to reach the atom. As the amplitude of thermal vibrations increases (with increasing T), the atom becomes partially

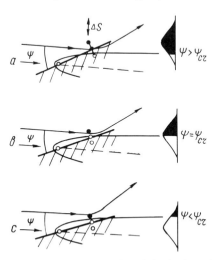

Fig. 8. Two-atom model illustrating the probability of shadowing of the second atom by the first one as a function of ion incidence angle (the shadowed part in the distributions specifies the probability for the second atom to move outside the shadow cone because of thermal vibrations).

shadowed, and the scattering intensity decreases. In case b, the thermal vibration distribution is divided by the shadow cone boundary in half, so that the scattering intensity should not depend on temperature. In case c, only the atoms that leave the shadow cone because of thermal vibrations contribute to scattering, and, therefore, one should observe an increase in intensity with temperature.

In the harmonic approximation and neglecting correlation, the probability for the second atom to be outside the shadow cone because of thermal vibrations can be written as

$$W(\overline{\Delta S^2}) = \frac{1}{2} + \frac{1}{2}\,\phi\!\left(\frac{S_0^2}{\overline{\Delta S^2}}\right), \quad \text{for } S_0 < 0,$$

$$= \frac{1}{2} - \frac{1}{2}\,\phi\!\left(\frac{S_0^2}{\overline{\Delta S^2}}\right), \quad \text{for } S_0 > 0, \tag{7}$$

where S_0 is the distance along the normal from the atom to the edge of the shadow cone; ϕ is the error function, and $\overline{\Delta S^2}$ is the squared amplitude of thermal displacement perpendicular to the surface (Leibfried 1955),

$$\overline{\Delta S^2} = \frac{1}{3}\,\frac{18\hbar^2 T}{m_2 k \theta_{\mathrm{D}}^2} \quad \text{for } T \geqslant \tfrac{1}{4}\theta_{\mathrm{D}} . \tag{8}$$

In fig. 8, the probability for an atom to be located outside the shadow cone is proportional to the ratio of the hatched part of the thermal distribution to its total area. Thus, depending on its initial position, the probability for an atom to be outside the shadow cone and contribute to scattering may decrease or increase with increasing temperature. Engelmann et al. (1985) measured the dependence of the single-scattering peak intensity I on temperature for different ψ. It was shown that for $\psi > \psi_{\mathrm{cr}} = 21°$, $I(T)$ falls off; and, for $\psi < \psi_{\mathrm{cr}}$, grows, in full accord with the model, the value $\theta_{\mathrm{D}_\perp} = 150 \pm 30\,\mathrm{K}$ being in agreement with experiment. Three-dimensional computer calculations support this conclusion.

The investigations of Engelmann et al. (1985, 1986) and Poelsema et al. (1983) present a good illustration of the application of ion-scattering spectroscopy to studying thermal vibrations of surface atoms. This method yields information on the mutual arrangement of surface atoms, namely, whether they lie one under another or along a simple scattering trajectory. A decrease of the substrate signal in the presence of an adsorbate depends not only on its concentration, but on the degree of shadowing as well, which opens up new possibilities in structural analysis.

Figure 9 shows an energy spectrum of 580 eV He$^+$ ions backscattered from Cu (110) (Taglauer 1985) for two different cases: equal coverage (0.3 monolayer) by oxygen and by CO molecules. We readily see that the oxygen peak is noticeably higher when CO molecules are adsorbed, and is

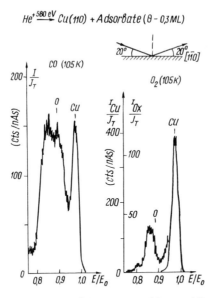

Fig. 9. Energy spectrum of 580 eV He$^+$ ions scattered by $\theta = 40°$ at $\psi = 20°$ from a Cu (110) surface covered by $\frac{1}{3}$ monolayer of (left) CO or (right) O.

about equal to the peak due to shadowed Cu. This means that in the CO case the O atom protrudes higher above the Cu (110) face. Such considerations are being actively discussed and already used in ion-scattering spectroscopy to investigate adsorbed layers.

Many studies deal with the shadowing of a substrate atom by adsorbed atoms at monolayer or lower coverages (Taglauer 1985, Taglauer and Heiland 1980, Smith 1967, Prigge et al. 1977, Englert et al. 1979, Brongersma and Theeten 1976, Heiland and Taglauer 1972, Oshima et al. 1981, de Wit et al. 1979, Heiland et al. 1978b, 1975, Godfrey and Woodruff 1981b, Saitoh et al. 1980, Onsgaard et al. 1980, Higginbottom et al. 1985, Priggemeyer et al. 1990). The sensitivity of ion-scattering spectroscopy to the atomic layer structure permits the determination of depth profiles within a few layers. The ion-scattering spectroscopy method is used to study stimulated catalysis (Taglauer 1985, Shelef et al. 1975) and surface segregation (Brongersma et al. 1978, Overbury and Millins 1989, Hinch et al. 1989).

Taglauer (1985) describes an ion-scattering study with 500 eV He$^+$ ions of the surface of an Ni–Mo catalyst deposited on an Al$_2$O$_3$ substrate, at target temperatures of 570–870 K. At $T = 770$ K, in the beginning of bombardment, the peak due to Ni is much higher than that due to Mo, i.e., the signal ratio I_{Ni}/I_{Mo} is large. As bombardment progresses, the upper layers are sputtered off, the Mo signal increasing compared with the signal due to Ni. This means that at this temperature Ni is on the surface and shadows Mo, while the bombardment-induced sputtering of the upper layers expose Mo. At $T = 870$ K, the Ni atoms move into the bulk of the target which is evidenced by the dependence of the ratio I_{Ni}/I_{Mo} on bombardment duration. Thus, measurements of the relative magnitude of the I_{Ni}/I_{Mo} signal complemented by other techniques permitted Taglauer (1985) to draw a conclusion on the structure of the surface.

Apparently at temperatures up to $T = 770$ K, molybdenum resides initially in spots on aluminum, Ni covering primarily the uppermost layers. At $T = 870$ K, Mo is distributed uniformly over the surface (the initial I_{Mo}/I_{Al} ratio is very high), the Ni profile extending over several atomic layers into the aluminum substrate. Experiments show that the shadowing effect can be used to study ordered structures on single-crystal surfaces as a complement to LEED.

By rotating the target azimuthally, one can determine the adatom position on the surface in a given crystallographic direction, and, by varying the grazing angle, measure the height of adatom position to within a few tenths of Å. In this way the adsorption of O on Ni (110) was investigated (Van den Berg et al. 1980, Heiland and Taglauer 1972, Verhey et al. 1979); of O, S on Ni (100) (Brongersma and Theeten 1976, Godfrey and Woodruff 1979), O on Ag (110) (Heiland et al. 1975); of Co on Ni (111) and W (110) (Heiland et al. 1978b); O on Cu (110) (de Wit et al. 1979); O on Cu (100) (Godfrey and Woodruff 1981b); Ag on Si (111) (Saitoh et al. 1980); Co on Si (111) (Onsgaard et al. 1980); and O on W (110) (Prigge et al. 1977).

Heiland and Taglauer (1972) used ion-scattering spectroscopy to obtain spectra on the scattering of 600 eV He$^+$ ions at $\theta = 60°$ from the oxygen-coated Ni (110) face for different grazing angles ψ, and $\langle 110 \rangle$ and $\langle 100 \rangle$ azimuthal directions. In the $\langle 110 \rangle$ direction, the Ni signal at $\psi = 10°$ is one half that for $\psi = 22.5°$, whereas the oxygen signal practically does not change, i.e., the I_{Ni}/I_O ratio falls off with decreasing ψ, which implies that oxygen shadows Ni atoms in the $\langle 110 \rangle$ direction and does not shadow them along $\langle 100 \rangle$. This correlates with the surface model where each second $\langle 100 \rangle$ row is absent, with oxygen atoms localized in place of these rows. The dependence of I_{Ni}/I_O on ψ permits the conclusion that oxygen protrudes 0.4–0.8 Å above the surface. Note that in the $\langle 110 \rangle$ direction oxygen atoms can shadow (for certain ψ) Ni atoms in the first layer, and in the $\langle 100 \rangle$ direction, Ni atoms located under them in the second layer. The I_{Ni}/I_O signal ratio in the $\langle 110 \rangle$ direction is larger than that along $\langle 100 \rangle$, which implies that atoms in the second layer contribute substantially to scattering.

Adsorption of O on Ag (110) was investigated by ion-scattering spectroscopy together with LEED and Auger electron spectroscopy (Heiland et al. 1975). The authors measured energy spectra of 600 eV He$^+$ ions back-scattered at $\theta = 60°$ at $\psi = 30°$ in the $\langle 110 \rangle$ and $\langle 100 \rangle$ directions from Ag (110). A comparison of the spectra showed the intensity of He$^+$ ions scattered from Ag atoms along $\langle 100 \rangle$ to be one half of that in the $\langle 110 \rangle$ direction. Scattering from oxygen in the $\langle 100 \rangle$ direction was barely seen.

An analysis based on the shadowing effect and calculations of the scattering cross section for Ag and O atoms using the Bohr and Born–Mayer potentials led to the conclusion that the oxygen adatom occupies bridging positions between neighboring Ag atoms in the $\langle 001 \rangle$ direction. Since the grazing angle in the experiment was fixed, and no O signal was detected at $\psi = 30°$, the height of the oxygen atom with respect to the surface could not be deduced one could only conclude that at $\psi = 30°$ the O atom is located in the shadow of the Ag atoms, i.e., this height cannot be larger than 0.74 Å (for the Born–Mayer potential) or -0.36 Å (for the Bohr potential).

Information on the orientation of adsorbed molecules can be obtained from a comparison of signals from the components of the molecule, e.g., for the CO molecule on Mo (Taglauer 1985), and CO on Ni (111), Ni (110) and W (100) (Heiland et al. 1978b). The results in these cases were similar, i.e., the signal from oxygen is many times that from carbon. Hence, the CO molecule is adsorbed in the vertical position, with oxygen on top.

Ion-scattering spectroscopy is a useful tool to probe surface kinetics. Adsorption of CO on W (100) is a typical surface reaction. At room temperature the I_O/I_C ratio for W is the same as for Ni. After annealing at $T = 700$ K this ratio reduces to one fourth its room temperature value for W. This can be due to the fact that after heat treatment the bulk of the CO molecules transforms into the β-phase, i.e., the molecules are arranged parallel to the surface and may even be dissociated. After heating to 1100 K, carbon diffuses under the first layer which is evidenced by a strong increase of the I_O/I_C ratio.

Another example of surface reactions is the dissociation by electron impact (Heiland et al. 1978b). After surface bombardment by 2 keV electrons, the carbon signal increases.

An experimental study of surface reactions by ion-scattering spectroscopy was carried out by Taglauer and Heiland (1980) who investigated copper film oxidation by oxygen isotopes (^{18}O).

Verhey et al. (1979) studied the structure of the oxidized Ni (110) surface. Measurement of the energy distributions of 6 keV Ar^+ and Ne^+ ions backscattered from Ni (100) at $\theta = 30°$ led to the conclusion that as a result of oxygen adsorption the interatomic separation in the $\langle \bar{1}10 \rangle$ direction increases by a factor two, reaching 5 ± 0.4 Å, while decreasing along $\langle 00\bar{1} \rangle$, which implies that the initial stages of oxidation of the Ni (110) surface are accompanied by its reconstruction. The oxygen atoms reside either in the $\langle 00\bar{1} \rangle$ rows or close to them. The experimental data were compared with qualitative predictions basing on various models of the surface (Germer and McRae 1962, Demuth et al. 1973).

The effects of ion focusing and surface hyperchanneling (Mashkova and Molchanov 1985, Karpuzov et al. 1978) were also used for ion beam control. Kumakhov (1979a,b) put forward the idea of the possibility of turning the beam by scattering it off a bent single crystal and made the corresponding estimates. Akkerman and Chubisov (1981) proposed to use the latter effect for turning a charged particle beam with a system of single crystal surfaces. Numerical calculations of the scattering of Ne^+ ions with $E_0 = 3$, 10 and 30 keV from the Si (001) face along the $\langle 110 \rangle$ and $\langle \bar{1}10 \rangle$ directions corresponding to shallow and deep surface semichannels revealed the possibility of using crystal surfaces for turning charged particle beams through small angles.

2.2. *Impact-collision ion-scattering spectroscopy*

A modification of the ion-scattering spectroscopy, proposed by Aono (1984), Aono et al. (1982a,b), Souda et al. (1983), and Oshima et al. (1981), greatly simplifies surface analysis and makes it a quantitative tool. It is based on measuring the intensity of single scattering at large angles ($\theta \approx 180°$) as a function of grazing angle. The analysis is based on the concept of a shadow cone behind the target atom. The ion flux at the shadow cone boundary increases compared with the incident flux because of small-angle scattering. Depending on the grazing angle, one atom (atom B) may be totally in the shadow of the preceding atom (inside the shadow cone), enter the focussed flux at the cone boundary, or be outside the cone far from its boundary (fig. 10a). If the atom is inside the shadow cone, there can be no direct scattering from it at angles $\theta \approx 180°$.

As the grazing angle is increased, the incident ion flux reaches first the totally shadowed atom, increases to a maximum at the shadow cone boundary (the scattering intensity will in this case also be the highest) and falls off down to the initial flux when the atom is far from the shadow cone.

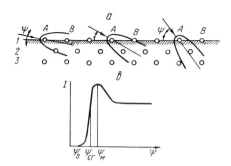

Fig. 10. (a) Position of atom B relative to the shadow cone produced by atom A as a function of ψ, and (b) schematic profile of the intensity I of ions scattered at $\theta \approx 180°$ versus grazing angle ψ.

This dependence of scattered-ion intensity on grazing angle, $I(\psi)$, is shown schematically in fig. 10b. For $\psi < \psi_0$, the target atom lies completely within the shadow cone, for $\psi = \psi_M$, scattering from this atom produces an enhanced flux at the cone boundary, and for $\psi > \psi_M$ the flux reaches the initial level. In the absence of thermal vibrations the drop of $I(\psi)$ from ψ_M to ψ_0 should be exactly vertical. In reality, an atom vibrates about its equilibrium position. The equilibrium position of an atom exactly at the edge of the shadow cone will be determined by the angle ψ_{cr} which, in the absence of focusing, would correspond to one half the height of the $I(\psi)$ maximum but is shifted towards ψ_M because of increased intensity. Computer calculations show ψ_{cr} to correspond to about 70% of the intensity maximum (Aono 1984). Very essential is the selection of the scattering angle $\theta \approx 180°$, since in this case, in contrast to smaller angles, the beam sees a small region around the center of each atom. As shown in fig. 10a, while at $\theta \neq 180°$ atom B may be located inside the shadow cone, nevertheless, scattering from it will be observed since the impact parameters in this case are not equal to zero. At $\theta \approx 180°$, when atom A has started to shadow atom B, the edge of the shadow cone will very nearly pass through the center of atom B. All this makes the method quantitative, since it permits one to consider the contributions to scattering coming from individual atoms.

Very essential in this analysis is the shape of the shadow cone. It can be calculated theoretically and measured experimentally, as was done by Aono et al. (1982b). By measuring ψ_{cr} for different interatomic separations, one can calculate the shadow cone radius $R_{cr} = d \sin \psi_{cr}$ (fig. 11). By choosing

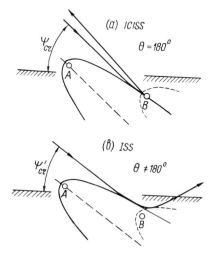

Fig. 11. Schematic illustration of the difference between (a) impact-collision, and (b) conventional ion-scattering spectroscopies.

different crystallographic orientations, one can measure R_{cr} for different d and, hence, for different distances $L = d \cos \psi_{cr}$ and, thus, derive the shape of the shadow cone. The shape of the shadow cone from a Ti atom bombarded by 1 keV He$^+$ ions (Aono et al. 1982b) was found to agree well with that calculated using the Molière potential. Niehus and Comsa (1984), Aono et al. (1982b), and Souda et al. (1983) calculated the shadow cone shapes for a number of collision partners and initial energies, the results being in agreement with experiment.

This method is based on the following approximations: (1) scattering at $\theta \approx 180°$ is assumed to be essentially single; (2) the ion flux producing the shadow cone behind the target atom in small-angle scattering is considered as the flux of the initial energy.

This method is widely used in studying clean surfaces of single crystals (Aono 1984, Czanderna 1975, Niehus and Comsa 1984, Yarmoff and Williams 1986, Niehus 1986, Möller et al. 1986, C.S. Chang et al. 1989, Spitzl et al. 1990), single-crystal alloys (Aono 1984), adsorbate-coated surfaces (Oshima et al. 1981, Rabalais et al. 1989), and the kinetics of variation of the crystalline surface structure (Aono 1984). As an illustration, consider the application of this method to investigate the structure of a TiC (001) surface. Quantitative structure analysis consists usually in measuring the relaxation of the surface components, i.e., of the relative displacements of the two atomic species making up the surface. This quantity for the TiC (001) surface was measured by Aono (1984) in the following way. The experimental dependence of the intensity of 1 keV He$^+$ ions backscattered from surface carbon atoms was investigated as a function of angle ψ in the $\langle 100 \rangle$ direction where the carbon and titanium atoms alternate (fig. 12a). The edge of the shadow cone from the Ti surface atom (fig. 12b) passed through the neighboring C atom at $\psi_{cr} = 24°$. Knowing the shape of the shadow cone from Ti for 1 keV He$^+$ ions, one could establish the position of the C atom relative to the Ti surface atom. The broken curve in fig. 12b specifies the shadow cone from Ti along two opposite directions, $\langle 100 \rangle$ and $\langle 010 \rangle$. Since in both directions $\psi_{cr} = 24°$ and the edge of the shadow cone passed through the center of atom C, one can easily determine the position of the surface atom C as the intersection of the edges of the two cones. The position of the carbon atom calculated in this way turned out to be slightly (about 0.1 Å) below the surface.

This method provides information on impurity localization on the surface, on surface defects and surface atom kinetics under various treatments. Aono et al. (1983) studied the structure, chemical activity and atomic vacancy concentration on TiC (001). The surface was first annealed at $T = 1600°C$ to remove surface defects, after which the $I(\psi)$ dependence was measured with 1 keV He$^+$ ions backscattered from Ti atoms in the $\langle 100 \rangle$ direction. Figure 13 shows the drop of intensity in the region of $\psi_{cr} = 22°$ (curve a). All

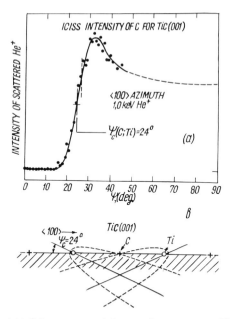

Fig. 12. (a) Intensity of He$^+$ ions scattered from carbon atoms on TiC (001) versus grazing angle, and (b) C atom position on this surface in the $\langle 100 \rangle$ direction derived from this dependence.

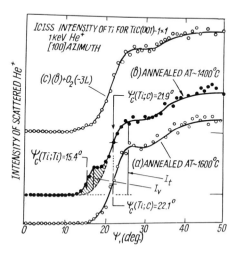

Fig. 13. Intensity of H$^+$ ions scattered from Ti atoms on TiC (001) in the $\langle 100 \rangle$ direction versus grazing angle ψ. Curve a: annealed at ~1600°C; curve b: annealed at ~1400°C, and curve c annealed at ~1400°C + O$_2$ (~3 L).

surface Ti atoms are shadowed by C atoms. After ion bombardment (current density ~0.3 mA/cm², $t = 10$ min) and annealing at $T \approx 1000°C$, the $I(\psi)$ dependences were measured again; we see (curve b) that the shadowing effect is observed, as before, at $\psi_{cr} = 22°$; however, after this the intensity, rather than dropping to zero as is the case with curve a, flattens out to a shoulder with $\psi_{cr} = 15.4°$, thus evidencing the formation of vacancies at carbon sites which results from the preferential sputtering of carbon. Now the titanium atoms will be shadowed not by the neighboring C atoms but by the more distant Ti atoms adjacent to the vacancies. An analysis shows that the Ti atom adjacent to a vacancy is not displaced from its position in the perfect crystal within measurement precision. From the relative magnitude of the shoulder one can derive the number of the vacancies, although this is not a simple problem since the magnitude of the shoulder depends both on E_0 and the degree of neutralization of He⁺ ions. Aono et al. (1983) found out from curve b in fig. 13 that under the given experimental conditions the carbon vacancies make up $10 \pm 2\%$. A conclusion on the chemical activity of the surface could also be drawn. The vacancy-free TiC surface is known to be very inert; indeed, for a monolayer of oxygen to form on it, the TiC should be maintained in oxygen environment for more than 1000 L (1 L = 10^{-6} Torr s). To study the chemical activity of TiC with vacancies, the sample characterized by curve b in fig. 13 was exposed to oxygen. Curve c in this figure displays the results obtained after a short (~3 L) exposure to oxygen. Since there is no more shoulder in the $I(\psi)$ dependence observed in curve b, one may conclude that the carbon vacancies captured oxygen atoms into the vacancy sites, which evidences the high chemical activity of the surface.

This method was further improved by Yarmoff et al. (1986), who investigated in detail the flux distribution at the boundary of the shadow cone with inclusion of thermal vibrations of the surface atoms.

Souda et al. (1983) calculated the shape of the distribution. However, the calculation was based on a Monte Carlo simulation using the atom row model. While the results agree well with experiment, they require a lot of computer time, since one has to follow each direction of incidence.

Yarmoff et al. (1986) proposed an original technique of calculating the flux at the edge of each shadow cone and blocking cone from the results of a Monte Carlo simulation on a single-atom model. This simple model requires little computer time but is capable of yielding the flux distribution at the boundary of the shadow and blocking cones as a function of vibration amplitude and interaction potential. Since only single scattering at $\theta \approx 180°$ is considered, the scattering cross section should be constant for given E_0, and the intensity should depend only on the distribution of the incident flux produced by the shadowing atom. A polar $I(\psi)$ scan can be constructed simply by summing up the flux distributions for each atom pair contributing to scattering. A series of scattering trajectories in a plane for an atom producing a shadow cone was considered for a number of impact parameters measured from the equilibrium positions of the atom. The atoms were assumed to vibrate independently of one another in two mutually perpendicular directions in the scattering plane. Since thermal displacements were prescribed only for one, shadowing or blocking, atom, the magnitude of the rms displacement was taken twice that expected for the given temperature. The results of the simulation (initial impact parameters p and scattering angles θ) were stored in the memory. Next, the trajectories of an ion passing through the scattering atom were considered. As seen from fig. 14a, the incidence angle ψ_s is related to the interatomic separation d, impact parameter p and angle θ of scattering from the shadowing atom in the following way

$$\psi_s = \theta + \arcsin[\cos \theta \, (p + \Delta \tan \theta)/d] \, . \tag{9}$$

The total flux incident on the scattering atom as a function of ψ_s was determined by the number of 1°-spaced trajectories in the histogram. To calculate the flux distribution in the case of blocking, the trajectories corresponding to the angle ψ_b (fig. 14b) between the line connecting the atom centers and the direction to the detector were taken from the memory,

$$\psi_b = \theta + \arcsin(p/d) \, . \tag{10}$$

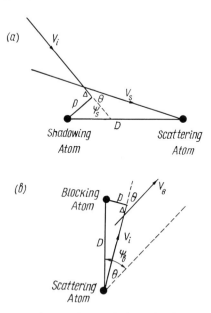

Fig. 14. Geometric representation used to calculate the flux enhancement caused by (a) shadowing, and (b) blocking.

The scattering atom acted in this case as a source of trajectories with an energy equal to that of single scattering from the scattering atom. The flux from the blocking atom as a function of ψ_b was likewise found by summing the 1°-spaced trajectories in the histogram. For a given model of the surface the distribution of the flux produced by each atom pair is obtained by summation of histograms from individual flux distributions. If a pair is arranged parallel to the surface, then ψ_b is simply the grazing angle, while in the opposite case one should add to ψ_s the angle between the surface and the line connecting the atom pair. If the detector-to-source angle is δ, then $\psi = \psi_b - \delta$. The $I(\psi)$ dependence obtained in this way is compared with experiment, and the structure parameters are refined until a reasonable agreement is reached.

Yarmoff et al. (1986) studied in a similar way the state of clean Cu (110) and Cu (110) + (2 × 1)-O surface coated with oxygen. It was found that the clean surface can be represented by a model where the first and second atomic layers are closer by $10 \pm 5\%$ compared with the structure in the bulk, and the thermal vibration amplitude in the topmost layers exceeds that in the bulk by a factor 1.5.

The results obtained for the Cu (110) face exposed to 200 L oxygen are in agreement with the 'missing-row model' and the Cu (110) + (2 × 1)-O structure where each second ⟨100⟩ atom row is missing. The separation between the first and second layers is increased by $25 \pm 10\%$, and that between the second and third row reduced by $10 \pm 5\%$ compared with the interlayer distance in the bulk. Higher oxygen exposures result in the surface copper layer becoming disordered. The position of adsorbed oxygen could not be determined. No direct signal from O was observed, since with a sample bombarded by Na^+ ions, which are heavier than oxygen, no scattering at 180° on O occurs because of the existence of θ_{lim} for $\mu < 1$. No reduction of the Cu signal due to shadowing and blocking by oxygen was seen in the $I(\psi)$ scan because of the low cross section of scattering from oxygen and of the large amplitude of thermal vibrations of the latter.

2.3. Ion-scattering spectroscopy combined with recoil-atom spectroscopy and orientational sensitivity of the degree of scattered-ion neutralization

As already pointed out, lack of knowledge of the degree of ionization is an obstacle to apply the ion scattering spectroscopy to surface diagnostics. On the other hand, the ion spectra of noble gases are more sensitive to the surface layer structure; thus since the charge composition of atoms is sensitive to orientation (Nizhnaya et al. 1979c, 1982a, Parilis and Verleger 1985d, Hou et al. 1978), it carries information on the surface structure. Indeed, the energy spectra obtained by Hou et al. (1978) in the scattering of 3 keV Ne^+ ions from Ni (110) revealed, besides the main Ni peak, additional lower-energy peaks which are believed to originate from light adsorbed atoms on the surface. While being barely distinguishable in the ionic spectra, these maxima are seen clearly in the dependence of the degree of ionization on energy, $\eta(E)$, at different values of the azimuthal scattering angle φ. Such an additional maximum at $\varphi = 52°$, corresponding to the $\langle 112 \rangle$ direction, has a double-peak structure. It is natural to assume that under these experimental conditions the most abundant impurities are the nitrogen molecules leaking from the stripping chamber. However, the number of these molecules was so small that they were not revealed by Auger electron spectroscopy. Indeed, calculations (Nizhnaya et al. 1982a) showed the energy of this additional maximum to coincide with that of Ne^+ single scattering from N atoms, while the double-peak structure at $\varphi = 52°$ is due to Ne^+ scattering from N_2 molecules. This implies that the N_2 molecule under the experimental conditions of Hou et al. (1978) lies along the $\langle 112 \rangle$ semichannel axis. If the N_2 molecule were above the row, this would yield for the low-energy peak a higher degree of ionization than for the main one, which disagrees with experiment . If the N_2 molecule were incorporated in the Ni row, it would be blocked by the atoms in the row and, hence, would not participate in scattering.

To determine the correct position of the molecule on single crystal surface, calculations were carried out of η^+ in scattering from the N_2 molecule located in three different positions on the axis of the $\langle 112 \rangle$ semichannel (fig. 15). The degree of ionization was calculated under the assumption that the electron capture and loss processes occur in the interaction of the projectile particle with individual surface atoms. In each position, the height of the molecule above the surface, h, was varied from 0 to $0.4a$, with a $0.05a$ step (a is the lattice constant). The $\eta^+(E)$ dependences calculated for all φ directions were found to agree with experiment only for position 3 at the height $h = 0.5$ Å. Thus, calculation of $\eta^+(E)$ for different φ and comparison with the experiment of Hou et al. (1978) made possible the

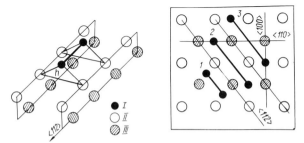

Fig. 15. Arrangement of adsorbed N_2 molecule on Ni (110). (1), (2), (3) possible locations of (I) N_2 atoms relative to Ni atoms of the (II) topmost and (III) second layer.

determination of the positions of the N_2 molecule under conditions where the amount of the impurity present is at a minimum and cannot be measured by other techniques.

Möller et al. (1984) applied the ion-scattering spectroscopy to studying the weakly chemisorbed N_2 molecule on Ni (110) at coverages less than one half of a monolayer. Energy spectra of 600 eV He^+ ions scattered from Ni (110) are considered for different nitrogen coverages. The Ni signal was shown to decrease with increasing coverage because of the shadowing of atoms of the substrate by those of the adsorbate. Infrared absorption analysis established that at coverages <0.5 monolayer the N_2 molecule is arranged perpendicular to the surface; however, this analysis is not capable of revealing the actual position of this molecule, namely, whether it is at the ridge or between the $\langle 110 \rangle$ atom rows. The reduction of scattering from Ni due to N_2 adsorption (at coverages where the shadow cones produced by individual molecules do not overlap) is used by Möller et al. (1984) to argue that adsorption occurs most probably over the ridge of atomic rows rather than between them. As follows from fig. 16b (view from above), the shadow cone produced by 600 eV He^+ ions from an individual molecule is too narrow to shadow the Ni atoms in the adjacent rows, so that if the N_2 molecule lies between the rows one could not expect a substantial decrease of the Ni signal compared with a clean target. If, however, the molecule is located above the ridge of the $\langle 110 \rangle$ row as shown in fig. 16a (view from the left), then the outermost nitrogen atom will shadow the two Ni atoms denoted by 'b'. Apart from this, the He^+ ions scattered from the atoms labelled by 'n' will undergo preferential neutralization since they pass through the electron cloud of the N_2 molecule. Thus, in the given geometry, the N_2 molecules should reduce the ion scattering from Ni by three or four

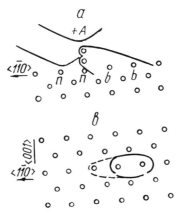

Fig. 16. Geometric model of scattering: (a) side view, and (b) view from above.

times, which is in accord with experiment. As we see, the ion-scattering spectroscopy combined with the IR absorption technique and even a qualitative inclusion of neutralization offers the possibility of localizing an adsorbed molecule. The difference between the conclusions drawn by Nizhnaya et al. (1982a) and Möller et al. (1984) on the position of N_2 on Ni (110) can apparently be attributed to different concentrations of N_2 molecules on the Ni surface.

By analyzing the intensity of scattered particles with inclusion of neutralization, Möller et al. (1984) showed that ion-scattering spectroscopy is capable of detecting less than 0.015 monolayer of N_2 on Ni (110).

Shultz et al. (1986) studied the spectra of scattered ions and neutrals from clean, oxidized and OH-coated Mg surfaces bombarded by Ne^+, Ar^+ and Ar^{2+} ions. The scattered neutrals were detected by the time-of-flight technique. High-energy VUV photons were also analyzed. Progress was achieved by employing recoil atoms of the quantitative measurement of the surface concentration of light impurities.

Shultz et al. (1986) showed the degree of ionization of scattered particles and recoils to depend strongly on the structure of the surface and provide a key to its analysis. Indeed, the degree of ionization of knocked out Mg recoils increases dramatically with increasing exposure to oxygen up to coverages in excess of 0.7–0.8 monolayer. This is apparently associated with a change in the surface electronic structure and may indicate that oxygen transfers from the chemisorbed to the free state, its islands covering the Mg surface. A small peak of hydrogen recoils (due to an impurity present) knocked out directly from clean Mg grows as it becomes coated by O_2 molecules, which implies that oxygen adsorption favors the hydrogen release from the surface.

After exposure of Mg to 30 L of H_2O, the peak of hydrogen recoils becomes comparable with that of Ne^+ ions scattered from Mg, which shifts toward lower energies because of multiple collisions while the Mg recoil peak disappears. This evidences the formation on the surface of $Mg(OH)_2$ molecules shadowing the clean Mg, whereas Ne^+ ions interact primarily with atoms of oxygen and hydrogen, which results in the disappearance of the recoil peak and the absence of single scattering from Mg.

This investigation of the spectra and degree of ionization of the directly knocked-out recoils and scattered particles may yield information on the kinetics of a surface under adsorption and on its electronic structure.

O'Connor et al. (1986) studied oxygen adsorption on Ni (110) by 2 keV H^+ ion scattering combined with the measurement of directly knocked-out negative O^- recoils. They measured the energy spectra of hydrogen ions scattered at $\theta = 30°$ for different grazing angles ψ from a clean and an oxygen-exposed Ni (110) surface. For $\psi = 45°$, the beam is oriented along

⟨010⟩, and the detector along ⟨100⟩. The scattering plane is parallel to the ⟨110⟩ direction. The energy spectrum represents in this case a surface peak against suppressed background. Adsorption of oxygen at 0.4 L results in a considerable lowering of the surface peak. This means that the oxygen adatom is located either at half the interatomic ⟨110⟩ separation above the outermost atomic layer (S), or at half the ⟨100⟩ separation and shadows the second layer (L) (fig. 17). To decide between these two possibilities, one can compare the O⁻ recoil signals for the ⟨001⟩ and ⟨110⟩ directions.

The ion recoil spectra obtained for $\theta = 60°$ with ψ varied from 0 to 60° under bombardment by 2 keV Ne⁺ ions showed the recoil signal in the ⟨110⟩ direction to exceed by far that along ⟨100⟩. This immediately invalidates the assumption of the oxygen being located at the center, since in this case the ⟨001⟩ and ⟨110⟩ signals should be comparable in magnitude. The S (short-bridging) position likewise is invalid since it should result in a higher signal in the ⟨001⟩ direction. Only the L position between atoms in the ⟨110⟩ direction agrees with experiment. For a given experimental geometry, the projectile ion interacts only weakly with the O atom in the ⟨001⟩ direction, since the latter lies inside the shadow cone produced by the neighboring Ni atoms. The dimensions of the shadow cone calculated using the Molière potential permit the following conclusion to be drawn; for the recoil yield in the ⟨110⟩ direction to be comparable with that along ⟨001⟩ under the given experimental conditions, the oxygen atom should lie 0.42 Å

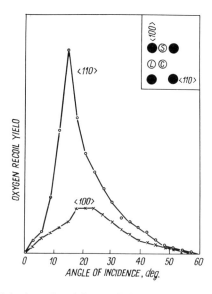

Fig. 17. Dependence of the intensity of O⁻ recoils knocked out at an angle $\theta = 60°$ on grazing angle ψ for the ⟨100⟩ and ⟨110⟩ direction.

above the Ni surface. The recoil atom yields in these directions are different. O'Connor et al. (1986) believe that the oxygen atom is located 0.2 ± 0.1 Å above the surface which corresponds to the Ni–O bond length (1.77 Å).

Using negatively charged recoil ions expands the possibilities of ion-scattering spectroscopy. They can be easily separated from the scattered-ion spectrum of noble gases which only rarely form negative ions.

Recoil atoms and scattered particles can be used to determine the mutual positions of the adatom and the substrate atom, as well as the impurity concentration on the surface (Teplov 1990).

In all these cases, experiments should be combined with computer simulation of the processes under study.

2.4. Determination of adatom species and position on single-crystal surfaces by ions channeling through thin films

Determination of the equilibrium positions of atoms adsorbed on crystalline surfaces is a fundamental problem in surface physics. Stensgaard and Jakobson (1985) were the first to propose for the determination of the position of D atoms adsorbed on the rear (100) surface of a thin Ni single crystal, 800 keV He$^+$ ions which channel through the sample. The adatoms were detected by monitoring the yield of the D(^{3}He, p)^{4}He reaction.

Nizhnaya et al. (1988) proposed to use for this purpose double scattering of medium-energy ions, with the first scattering from a host lattice atom, and the second from an adsorbed atom, provided that $M > m_1 > m_2$. Here M, m_1, m_2 are the masses of the host lattice atom, ion, and adatom, respectively. For $m_1 > m_2$, there exists a limiting angle $\theta_{\text{lim}} = \sin^{-1}(m_2/m_1)$ which leads to the rainbow effect in scattering, i.e., enhanced reflection that results in the appearance in the angular and energy distributions of characteristic peaks useful for diagnostic purposes.

Dzhurakhalov et al. (1989d, 1990c) proposed for the determination of the position and kind of light adatoms on the rear surface of a thin crystal to use interaction with these adatoms of medium-energy ions of mass m_1 greater than the adatom mass m_2, which channel through the crystal. To improve the usefulness of the angular and energy distributions of transmitted particles for diagnostic purposes, it can be recommended to employ, besides the rainbow effect, also the flux peaking phenomenon, i.e., a spatial redistribution of the channeling in the transverse channel plane which produces an increase of the flux in the central part of the channel and its sharp decrease on its periphery (Kumakhov 1975). The species and position of the adatoms are determined from the rainbow effect in the angular and energy distributions of transmitted particles caused by their interaction with adatoms above the rear surface of the crystal. The trajectories of 1–5 keV He$^+$ and

Ne$^+$ ions channeling in thin ($\Delta z = 100$–500 Å) Ni (100) films were computer simulated in the binary collision approximation using the universal Biersack–Ziegler potential, eq. (16) of chapter 1, and with the inclusion of elastic and inelastic energy losses.

Successive multiple scattering of ions from atoms in the rows lying along the principal crystallographic axes is followed in a special search procedure, with the impact parameters for all target atoms forming the walls of a channel calculated for each layer in the crystallite. The possible simultaneity of the ion interaction with atoms in the row was included by the procedure proposed by M.T. Robinson and Torrens (1974).

Thermal vibrations of the crystal atoms were assumed to be uncorrelated with the Gaussian distribution of displacements, eq. (4) of chapter 2.

For ions emerging out of the rear side of the crystal, one calculates the coordinates of the emergence points projected on a plane perpendicular to the channel axis, the total scattering angle (the angle between the initial and final ion directions), and the final ion energy. Next, one calculates the interaction of the transmitted ions with the adsorbed atoms residing in various positions relative to the channel axis and at given heights h with respect to the rear surface of the crystal. The adatom species and positions are determined from a comparison of the angular and energy distributions of the transmitted ions with or without adatoms present in the crystal (fig. 18a,b).

Figure 18a illustrates the possibility of using the channelling effect to detect the adatom position above the rear crystal surface along the principal

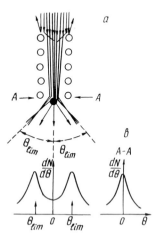

Fig. 18. (a) The possibility of using the channeling effect to detect the adatom position above the rear crystal surface along the principal axes and (b) the changes in the angular distribution of the redistributed transmitted flux caused by the ion interaction with the adatom.

axes. Figure 18b shows schematically the changes in the angular distribution of the redistributed transmitted flux caused by the ion interaction with the adatom. The right side of fig. 18b demonstrates the angular distribution in the absence of adatoms. We readily see that the presence of a peak in the angular distribution at the limiting scattering angle permits the identification of the adatom species. Thus, the present method is efficient only when $m_2 < m_1$, i.e., when the adatoms are lighter than the channeling ions.

Figure 19 shows the distributions of the ion emergence points in projection on the plane perpendicular to the channel axis for 5 keV Ne^+ ions channeling through single crystal Ni (100) films in the $\langle 100 \rangle$ direction, and the corresponding angular distributions after interaction with oxygen adatoms located at the channel center at $h = 0.9$ Å above the rear crystal surface, for film thicknesses Δz of (a) 100 Å, (b) 300 Å, and (c) 500 Å. The

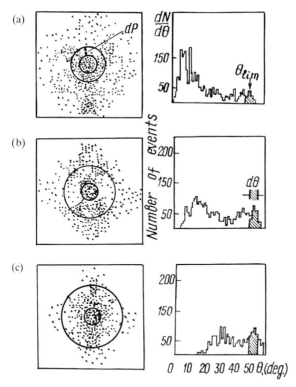

Fig. 19. Distributions of the ion emergence points in projection on the plane perpendicular to the channel axis for 5 keV Ne^+ ions channeling through single crystal Ni (100) films in the $\langle 100 \rangle$ direction and the corresponding angular distributions after interaction with oxygen adatoms: (a) $\Delta z = 100$ Å; (b) $\Delta z = 300$ Å, and (c) $\Delta z = 500$ Å.

Ne$^+$ ions with emergence point coordinates confined between the two circles possess impact parameters of collision with an O adatom in the interval of p to $p \pm \Delta p$, and are scattered after interaction with it at angles ranging from θ_{lim} to $\theta_{lim} \pm \Delta\theta$, where $\Delta\theta \cong 2.5°$. We see that as the ion flux becomes redistributed over the channel cross section with increasing film thickness, the impact parameter range corresponding to ion scattering from an adatom at angles $\theta \cong \theta_{lim}$ increases too. In the angular distributions this manifests itself in an increase of the area bounded by the peaks corresponding to scattering at angles $\theta_{lim} \pm \Delta\theta$, as well as in a decrease and, eventually, disappearance with increasing Δz of the peaks in the low scattering angle region. Thus, by choosing properly the ion energy and crystal thickness, one can obtain angular distributions with the main peak corresponding to the limiting scattering angle. This is illustrated in fig. 20, showing angular distributions of 1 keV Ne$^+$ ions channeling through thin ($\Delta z = 100$ Å) Ni (100) films along $\langle 100 \rangle$ in the (a) absence and (b) presence of adatoms of carbon and (c) oxygen, located at the channel center at a height $h = 0.9$ Å above the rear crystal surface, which were calculated for $T = 0$ K. The presence of adatoms affects noticeably the distributions in that they shift

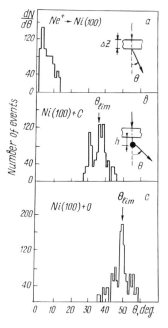

Fig. 20. Angular distributions of 1 keV Ne$^+$ ions channeling through thin Ni (100) $\langle 100 \rangle$ films in (a) the absence and (b) presence of adatoms of carbon and (c) oxygen.

toward scattering angles $\theta \cong \theta_{\text{lim}}$, their main peak corresponding to rainbow scattering ($\theta_{\text{lim}} \approx 35°$ for carbon, and $\theta_{\text{lim}} \approx 50°$ for oxygen). An essential advantage of this method lies in its capability of detecting very light adatoms, such as hydrogen and deuterium. Figures 21 and 22 show the angular distributions of 1 keV He^{+} ions passing by channeling through $\Delta z = 100$ Å thick Ni (100) films along $\langle 100 \rangle$ and through $\Delta z = 200$ Å thick Ni (111) films along $\langle 110 \rangle$, in (a) the absence and (b) presence of hydrogen and deuterium adatoms located at channel center at heights $h = 0.5$ and 0.8 Å, respectively. As follows from a comparison of figs. 21 and 22, this method allows discrimination between the isotopes of an adsorbed element.

Indeed, in the case of hydrogen, the limiting scattering angle, with the inelastic energy losses taken into account, is $\theta_{\text{lim}} \cong 13°$, and for deuterium it is $\cong 29°$. Similar arguments are valid also for close isotopes of carbon and oxygen. Thus, when $m_1 > m_2$ the angular distributions of transmitted ions

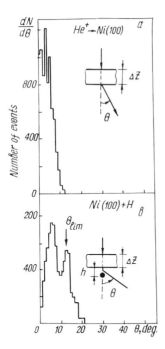

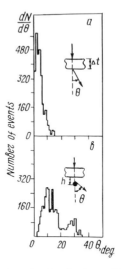

Fig. 21. Angular distributions of 1 keV He^{+} ions passing by channeling through Ni (100) $\langle 100 \rangle$ films of thickness $\Delta z = 100$ Å in (a) the absence and (b) presence of hydrogen adatoms.

Fig. 22. Angular distributions of 1 keV He^{+} ions passing by channeling through $\Delta z = 200$ Å thick Ni (111) $\langle 110 \rangle$ films in (a) the absence and (b) presence of deuterium adatoms.

allow a correct determination of the adatom species. The film thickness is
not a crucial factor in this technique, and it can be taken thicker if one
increases correspondingly the initial ion energy. Taking into account the
target atom and adatom thermal vibrations results in a washing out of the
distributions and a lowering of the peak at $\theta \simeq \theta_{lim}$. The rms displacements
for target atoms were calculated by eq. (4) of chapter 2 for $T = 300$ K. For
the hydrogen and deuterium adatoms, they were taken equal to ~0.24 Å
parallel to the surface (two dimensions) and ~0.12 Å perpendicular to it
(one dimension) (Stensgaard and Jakobson 1985). As seen from fig. 23, the
magnitude of the main maximum at $\theta \cong \theta_{lim}$ depends on the height h of the
adatom above the crystal surface. For a given crystal thickness and initial
ion energy, the peak reaches a maximum at a certain height h. Indeed, for
the case Ne$^+ \to$ Ni (100) + O, film thickness $\Delta z = 100$ Å, and $E_0 = 1$ keV, it
occurs at $h = 0.9$ Å, which is in agreement with the experimental value
(Demuth et al. 1973).

Figure 24 follows the evolution of the angular distributions of 5 keV Ne$^+$

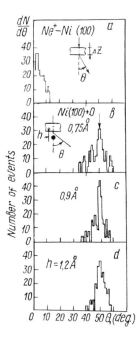

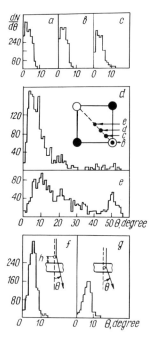

Fig. 23. The magnitude of the main max-
imum at $\theta \cong \theta_{lim}$ in angular distributions versus the height h of the adatom above the
crystal surface.

Fig. 24. Evolution of the angular distributions of 5 keV Ne$^+$ ions passing through a
200 Å thick Ni (100) $\langle 100 \rangle$ film as a function of adatom position relative to the crystal unit cell in the transverse plane of the
channel.

ions passing through a 200 Å thick Ni (100) film along $\langle 100 \rangle$ with oxygen adatoms present above the rear surface at $h = 0.9$ Å as a function of adatom position relative to the crystal unit cell in the transverse plane of the channel (b,c,d,e). Figure 24a corresponds to a clean target. Figure 24f shows the angular distribution corresponding to fig. 24e but with inclusion of the target atom thermal vibrations. When an adatom resides above a lattice site $[b(\tau, 0, \frac{1}{2}\tau)]$ or close to it $[c(\frac{5}{6}\tau, \frac{1}{6}\tau, \frac{1}{2}\tau)]$, it is shadowed for the incident ion beam, and, therefore, the angular distributions practically do not change. As the adatom approaches the channel center, it leaves the shadow region $[d(\frac{2}{3}\tau, \frac{1}{3}\tau, \frac{1}{2}\tau)]$, which reveals itself in a broadening of the angular distribution up to the limiting scattering angle. With the adatom at the channel center $[e(\frac{1}{2}\tau, \frac{1}{2}\tau, \frac{1}{2}\tau)]$, a clearly pronounced maximum appears at $\theta = \theta_{lim}$. Here, τ is one half the lattice constant.

The sensitivity of the angular distributions of transmitted ions to shadowed adatom positions which do not correspond to the channel center in the given direction can be revealed by varying the direction of particle channeling. The accuracy, with which the adatom position with respect to the unit cell in the transverse plane of the channel can be determined, is evaluated as ~0.1 Å.

Rainbow scattering of channeled ions from an adatom results in a substantial change not only of their angular, but of energy distributions as well (fig. 25). (a) and (b) in fig. 25 correspond to spectra obtained without and with oxygen adatoms present at the $\langle 100 \rangle$ channel center at a height of 0.9 Å above the rear Ni (100) surface. In the absence of adatoms, the total energy distribution of transmitted ions has a conventional pattern with peaks corresponding to axial (for high energies) and planar channeling in the scattering angle range $0 \leqslant \theta \leqslant 10°$. Rainbow scattering from an adatom shifts the distribution strongly toward low energies and large scattering angles with a peak corresponding to scattering at $\theta = \theta_{lim}$.

It thus follows that the presence in the angular and energy distributions of transmitted channeling ions of rainbow peaks produced by scattering at adatoms permits the determination of the kind of adatoms and their position on the crystal surface; also, that by properly varying the direction of particle channelling one can reveal the sensitivity of these distributions to adatom positions outside the central region of the channel, when viewed in the given direction. The channeling of ions of medium energy and mass through thin single-crystal metal films can be used to determine the species and position of light atoms adsorbed on a clean rear surface provided the mass of the ions is greater than that of the adatoms to be detected. The presence, position, and magnitude of the rainbow peaks in the angular and energy distributions of the transmitted ions are determined by the species of the adatom and its position relative to the unit cell (the height above the surface and the displacement with respect to the channel center in the transverse plane).

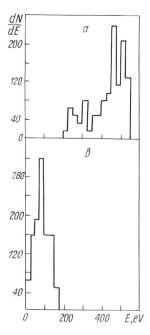

Fig. 25. Energy distribution histograms of 5 keV Ne⁺ ions channeling through thin Ni (100) ⟨100⟩ films (a) without and (b) with oxygen adatoms.

An experimental test of this technique would permit a realistic evaluation of its capabilities.

High-precision experiments, combined with detailed computer calculations using various surface models, permit one already to carry out both qualitative and, in some cases, also quantitative analysis by ion-scattering spectroscopy of the composition and structure of surfaces and films.

3. Adatom identification and location

3.1. Introduction

The properties of a metal surface covered with some adsorbate depend on the structure of the outermost layer. Therefore, the precise location of an adatom on a single-crystal surface is very important for the determination of its catalytic characteristics. Information on the position of the adsorbates is also necessary for the preparation of the substances with the given properties and for thin-film technology.

Low-energy ion-scattering, spectroscopy, referred to as ion-scattering spectroscopy, has been used over the past two decades to a variety of problems of surface analysis. Among the successful applications there are studies of the arrangement of atomic layers on supported catalysts, ordered adsorption systems on metal surfaces, and surface reconstruction and dis-ordering (Petrov and Abroyan 1977, Nizhnaya et al. 1982a,b, Aono 1984, Taglauer 1985, Yarmoff et al. 1986, Rabalais 1988). This method is one of the most powerful techniques for the analysis of the topmost atomic layers due to its simple principles and its extreme surface sensitivity.

This conceptually simple method is limited by the fact that interaction potentials and charge-exchange processes are only approximately known. Progress has been made by applying a new form of this method – impact-collision ion-scattering spectroscopy with a special geometry. The basis of the technique is to monitor the intensity of the quasisingle scattered ions as a function of the glancing angle for ions scattered at an angle θ close to 180°. In the case $\theta \approx 180°$, the ion 'sees' the close vicinity of the center of each atom, which makes this method quantitative (Aono 1984).

However, this is not a suitable method for diagnostics of light impurities (m_2) on heavy substrates (M) by ions (m_1) heavier than the adatoms (m_1, m_2, M are the masses of the colliding particles) due to single scattering from the adatoms over an angle $\theta \approx 180°$ for the inverse mass ratio ($\nu = m_1/m_2 \geqslant 1$) is not possible. The limiting scattering angle for $\nu > 1$ is $\theta_L = \arcsin(m_1/m_2)$. The peaks of single scattering on a matrix atom also hardly give any information about the adatom location because the effects of blocking and shadowing are not well exhibited in this case.

In this case, only a method of direct recoil spectrometry can provide information about the adatom location (Rabalais 1988). Nizhnaya et al. (1988) have shown that for $M > m_1 > m_2$ the characteristic peaks suitable for the diagnostic purposes should appear in the energy and angular distributions of the scattered ions as well as adatom recoils.

3.2. Analytical estimations

3.2.1. Scattered ions
We have considered the normal incidence of an ion (m_1) with energy E_0 on a surface row of atoms (M) above which an adatom (m_2) is located ($M > m_1 > m_2$). The position of the adatom is determined by coordinates x and y (fig. 26).

As has been mentioned above, the single scattering of primary ions by adatoms over the angle $\theta > \theta_L$ is not possible and cannot contribute to the spectra.

Let us consider the trajectory of the double scattering contribution to the spectra at the given escaping angle ψ (fig. 26). This double scattering consists of, first, the violent scattering over an angle θ_1 by the substrate atom and a second scattering over an angle θ_2 by the adatom, the adatom being scattered over the angle θ_a.

We have considered the regularities of such scattering with the following assumptions (Nizhnaya et al. 1988):

(i) atom M is a point source of the diverging beam of the scattered particles (m_1) because, at E_0 in the keV region and $\theta_1 \geqslant \frac{1}{2}\pi$, the distance of the closest approach with the substrate atom $R(\theta_1) \ll d$ $[d = (x^2 + y^2)^{1/2}]$;

(ii) the diverging beam of particles m_1 is an isotropic one as, at $\theta_1 \geqslant \frac{1}{2}\pi$ the cross section of the single scattering $\sigma(\theta_1)$ is a slowly changing function of θ_1.

With such approximations the cross section of the double scattering over the angle $\theta = \theta_1 + \theta_2$ is proportional to the cross section of the divergent isotropic ion beam scattering over the angle θ_2. Kurnaev et al. (1985) have obtained a simple analytical expression for $\sigma(\theta_2)$,

$$\sigma(\theta_2) = \{[p^{-1} \sin \theta_2 (1 - p^2/d^2)^{1/2} + \cos \theta_2 \, d^{-1}]$$
$$\times [(1 - p^2/d^2)^{1/2} \, d\theta_2/dp + d^{-1}]\}^{-1} . \tag{11}$$

Here, $p = p(\theta_2)$ is the ion impact parameter with the adsorbed atom.

The function $p(\theta_2)$ depends on the interaction potential of the colliding particles. At E_0 in the keV region, one could use the potential $U(R) = A/R^2$ and obtain

$$p(\theta_2) = R_0(\pi - \chi)(2\pi\chi - \chi^2)^{-1/2} , \tag{12}$$

where $R_0 = (A/E_1)^{1/2}$ is the distance of closest approach in a head-on collision, $E_1 = E_1(E_0, \theta_1)$ and χ are the ion energy after the first scattering and scattering angle in the center-of-mass system, respectively. It is known (Landau and Lifshitz 1960) that the equation $\tan \theta_2 = \sin \chi \, (\chi + \cos \chi)^{-1}$ at $\nu = m_1/m_2 > 1$ has two solutions,

$$\chi_1 = \theta_2 + \arcsin(\nu \sin \theta_2)$$
$$\chi_2 = \pi + \theta_2 - \arcsin(\nu \sin \theta_2) . \tag{13}$$

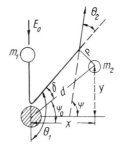

Fig. 26. Scattering diagram.

Hence, at $\nu > 1$, the function $p(\theta_2)$ will be two-fold. This is schematically shown in fig. 27.

Singular peaks of the rainbow scattering in the angular distribution should be observed for those angles $\theta_2 = \theta^*$ where $\sigma(\theta^*) = \infty$. From eq. (11) it follows that $\sigma(\theta^*) = \infty$ if

$$\frac{\mathrm{d}p}{\mathrm{d}\theta_2}\bigg|_{\theta_2 = \theta^*} = -[d^2 - p^2(\theta_2)]^{1/2}, \tag{14}$$

where

$$\frac{\mathrm{d}p}{\mathrm{d}\theta_2} = \frac{\mathrm{d}p}{\mathrm{d}\chi}\frac{\mathrm{d}\chi}{\mathrm{d}\theta_2}.$$

Taking into account eqs. (12)–(14), we obtain the following equations for the determination of θ^*,

$$\pi^2(2\chi_1 - \chi_1^2)^{-3/2}[1 + \cos\theta_2(1 - \nu^2\sin^2\theta_2)^{-1/2}] = (d^2 - p^2)^{1/2}/R_0. \tag{15}$$

The graphical solution of eq. (15) is schematically shown in fig. 28, where the left-hand part of eq. (15), which is independent on energy, and the right-hand part are denoted as $f_1(\theta_2)$ and $f_2(E_0, \theta_2)$, respectively.

From fig. 28 one could draw the following conclusions:

(i) The solution of eq. (15) and the corresponding singular peaks in the angular distribution, eq. (11), are observed only at initial energies exceeding a given energy E_t, which considerably depends on both the coordinates of the adsorbate atom and the masses of the colliding particles. In the case of ion scattering from a (100) Au surface covered by oxygen adatoms located at height $y = 1.5\ \text{Å}$ in the middle between the Au atoms of a [100] row, the values $E_t = 64\ \text{eV}$ for Ne^+, $E_t = 2.6\ \text{keV}$ for Ar^+ and $E_t = 700\ \text{keV}$ for Xe^+

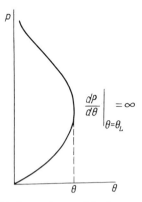

Fig. 27. Dependence of the impact parameter on the scattering angle in the laboratory system for the case $m_1 > m_2$.

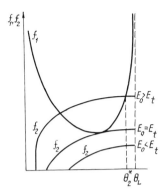

Fig. 28. Graphical solution of eq. (15).

primary ions have been obtained. If $m_1 > m_2$ and $m_1 \ll M$, then a simple analytical expression can be obtained,

$$E_t \approx 4A\pi^{-2}d^{-2} .$$ (16)

From eq. (16) it follows that if the constant of potential A is known, then the measurement of E_t gives the possibility to determine the distance d between the adsorbate atom and the target one. However, it should be taken into account that due to neutralization the peak in the ion spectrum could appear at higher energies.

(ii) At $E_0 > E_t$, eq. (15) has two solutions: θ_1^* and θ_2^*. With E_0 increasing, the angle θ_2^* rapidly tends to θ_L^* and $\theta_1^* \to 0$. The solution θ_1^* is connected with the nonphysical divergence of the cross section $\sigma(\theta_2)$ at $\theta_2 = 0$ and the solution θ_2^* gives the singular peak as a result of scattering concentration in the region $\theta_2 \approx \theta_L$.

3.2.2. Adatom recoils

The angular distribution of the adatom recoils also has been considered in the approximation of a point source. In this approximation, the adatom scattering cross-section is determined by Kurnaev et al. (1985)

$$\sigma(\theta_a) = d^2/\{[(d^2 - p^2)^{1/2}\sin\theta_a/p - \cos\theta_a][(d^2 - p^2)^{1/2}\, d\theta_a/dp - 1]\} .$$ (17)

Singular peaks in the spectrum of adatom recoils would be observed for those scattering angles θ_a and the corresponding escape angles ψ_a for which $\sigma(\theta_a)$ becomes infinite. The equation $\sigma(\theta_a) = \infty$ with account of eq. (17), leads also to two equations,

$$p(\theta_a) = d\sin\theta_a ,$$ (18a)

$$dp/d\theta_a = (d^2 - p^2)^{1/2} .$$ (18b)

As is seen from fig. 29, the escape angle of the adatom recoil is

$$\psi_a = \psi_0 - \arcsin(p/d) + \theta_a .$$ (19)

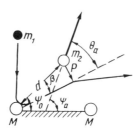

Fig. 29. Diagram of the knocking-out of an adatom.

From eqs. (18a) and (19), one could get the position of the first singular peak,

$$\psi_{a,1}^* = \psi_0 \ . \tag{20}$$

As follows from the analysis of eq. (18a), for one escape angle ψ_1 there are two scattering angles. The first one corresponds to head-on collisions and does not depend on the kind of potential,

$$\theta_{a,1}^* = 0 \ . \tag{21a}$$

The second one is the solution of eq. (18a) where $p(\theta_a)$ is determined by the kind of potential. For the potential $U = A/R^2$,

$$\theta_{a,2}^* = \tfrac{1}{2}\pi\sqrt{1 - R_0^2/d^2} \ . \tag{21b}$$

Thus, the energy spectrum of the adatoms having escaped over the angle $\psi_{a,1}^*$ should contain two peaks.

Using the same potential one could get the following solution of eq. (18b),

$$\theta_{a,3}^* = \tfrac{1}{2}\pi\sqrt{1 - [2R_0/(\pi d)]^{2/3}} \ , \tag{22}$$

which gives the position of the second singular peak $\psi_{a,2}^*$,

$$\psi_{a,2}^* = \psi_0 - \arcsin\frac{p(\theta_{a,3}^*)}{d} + \theta_{a,3}^* \ . \tag{23}$$

By measuring the angles $\psi_{a,1}^*$ and $\psi_{a,2}^*$, one could define the coordinates x, y of an adatom on a single-crystal surface.

3.3. Computer simulation

The qualitative estimations performed allow us to determine only the existence of rainbow peaks. To define the diagnostics feasibilities of the proposed method, a more detailed analysis of the angular and energy distributions is necessary. First of all, one should know to what extent the rainbow peaks are distinguished on the background of other spectrum peculiarities. For this purpose the following computer simulation was made. We considered Ar^+ ion scattering with $E_0 = 1 - 30$ keV on a linear [100] chain of Au single crystal above which the adsorbate oxygen atoms were located in the middle between Au atoms at height y. Oxygen atoms were located uniformly on the surface with a coverage corresponding to one adatom per four Au atoms. The calculations were made with a power potential (Vajda 1964), using the method described by Nizhnaya et al. (1979a).

The results of the calculation are presented in figs. 30–34. The computer simulation has shown that the energy spectra contain two groups of well-

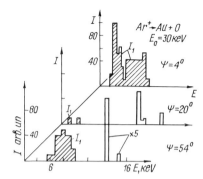

Fig. 30. Energy spectra of the scattered Ar$^+$ ions with $E_0 = 30$ keV for different escape angles.

distinguished peaks (fig. 30). High-energy peaks correspond to the quasisingle and quasidouble scattering of Ar$^+$ only by Au atoms, while low-energy peaks result from Ar$^+$ scattering by Au and oxygen atoms. At $\theta_2 = \theta_L$, the low-energy peaks become more intensive and much broader than for other scattering angles. The appropriate characteristics of these diagnostic peaks is the area under the peak.

Figure 31 shows the area under the low-energy peak $I_1(\psi)$ versus the

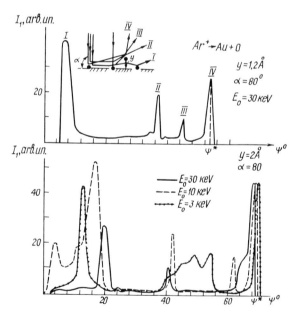

Fig. 31. Dependence of the intensity I_1 of the low-energy scattered Ar$^+$ ions on the escape angle ψ. At the top is shown $I_1(\psi)$ for $y = 1.2$ Å, $E_0 = 30$ keV. At the bottom, $y = 2$ Å, $E_0 = 3$, 10 and 30 keV.

escape angle ψ. The angular distribution $I_1(\psi)$ for glancing angle $\alpha = 80°$ contains four peaks corresponding to trajectories of rainbow scattering. These trajectories are displayed on the top of fig. 31. The common feature of these trajectories is first violent scattering by an Au atom and second scattering by an adatom over the angle $\theta_2 \approx \theta_L$. In the angle region corresponding to $\theta = \theta_1 + \theta_L$, sharp maxima were observed. Figure 31 shows the dependence $I_1(\psi)$ for two different adatom heights $y = 1.2$ Å ($E_0 = 30$ keV) and $y = 2$ Å ($E_0 = 3$, 10 and 30 keV). It is seen from the figure that the position of the peak IV in the angular distributions for a given y slightly depends on the primary ion energy in a wide range of $E_0 = 3$–30 keV. However, it is rather sensitive to the adatom height. In addition, peak IV is situated on the right edge of the angular distribution $I_1(\psi)$, which could simplify its measuring. Hence peak IV is the most suitable for diagnostics.

Figures 32–34 show the energy and angular distributions of the directly knocked-out adatoms. It is seen (fig. 32) that in the energy spectra of adatom recoils escaped over an angle ψ_a close to $\psi_{a,1}$ both a rather intensive peak corresponding to $\theta_a = \theta^*_{a,2}$ with lower energy and a low-intensity peak corresponding to the head-on collision with the highest energy exist. However, the peak A at $\psi_a \approx \psi^*_{a,1}$ in the angle distribution (fig. 33) has a large angular extension, hence it is not suitable for diagnostics. A sharp peak B at $\psi_a \approx \psi^*_{a,2}$ is also observed. Its position is very sensitive to the adatom location. The peak position is shifted over an angle $\Delta\psi \approx 1°$ with a height change by $\Delta y \approx 0.1$ Å (fig. 34).

In fact, the calculations have been made for a planar geometry and one atomic row. However, ion scattering by the row ridge mainly contributes to the energy spectra (Nizhnaya et al. 1979a). In addition within the chosen geometry (adatom located in the middle of the [100] row on the (100) plane,

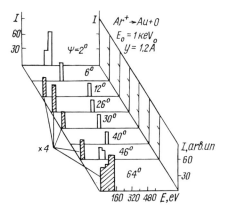

Fig. 32. Energy spectra of the knocked-out oxygen atoms for the case $Ar^+ \rightarrow Au$ (100) + O, $E_0 = 1$ keV, $\alpha = 90°$.

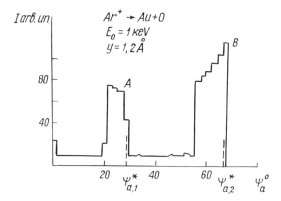

Fig. 33. Angular distribution of the adatom recoils.

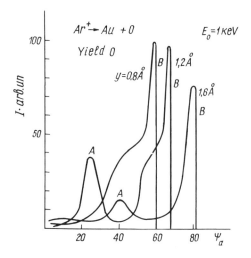

Fig. 34. Angular distributions of the adatom recoils for different adatom locations on the surface.

just above the second layer atom) the ions scattered by the second layer do not contribute to the characteristic peak. At other adatom locations the contribution of the second layer must be taken into account, but it should not exceed 10% of the first layer contribution, mainly due to preferable neutralization of the particles, emerged from the bulk (Matschke et al. 1977, Nizhnaya et al. 1979b,c).

Thus the computer simulation has shown that in the case $M > m_1 > m_2$ the rainbow peaks in both the spectra of scattered primary ions and adatom recoils were clearly distinguished and the position of these peaks in the angular distributions were related to the coordinates of adsorbate atoms.

Comparing the experimental distributions with computer calculations for different models of submonolayer coverage shows that it is possible to determine the adatom location.

It should be noted that the calculations of both angular and energy distributions of ions scattered by a surface with a submonolayer covering were made for total flux (ions and neutrals) and without the consideration of thermal vibrations.

Since the measurement of the neutral component is difficult and the most commonly experiments are carried out at room temperature, it is necessary to calculate the ion spectra as well as to investigate the influence of thermal vibrations on the characteristic peak widths and, therefore, the possibility to resolve them at $T > 0$.

3.4. Ion spectra

Nizhnaya et al. (1991) have shown that in the angular distribution of scattered ions $I_1(\psi)$ for $\alpha = 90°$ only two maxima I and II, corresponding to the quasidouble trajectories of ions which underwent additional scattering on an adatom, were observed. Peak I in the small-angle region comprises also multiple scattering on matrix atoms, hence its location changes with the change of the primary energy E_0. The peak II (analogous to peak IV at $\alpha = 80°$) corresponds to the double-scattering trajectory, shown in fig. 26. Its location is practically unaffected by change of E_0, and its right side is built by ions which underwent the last scattering at the limiting angle θ_L. This allows to determine unambiguously the distance of the adatom from the surface using only the scattering kinematics.

To calculate the ion energy spectra for a single-crystal surface with submonolayer covering, the charge-state formation model, based on the concepts of electron capture and loss along the scattering trajectory, was used (Nizhnaya et al. 1982a,b). The violent collisions in which the transferred inelastic energy appeared to be larger than ionization energy were considered to be ionizing, the rest was taken as neutralizing ones.

The electron capture occurs during the Auger neutralization, the rate of which was a three-parameter function (Kishinevsky et al. 1976),

$$\omega_c(R) = B, \qquad\qquad\qquad \text{at } R \leqslant R_a,$$
$$\quad = B \exp[-\beta(R - R_a)], \quad \text{at } R > R_a. \qquad (24)$$

The parameter R_a is approximately the sum of radii of the colliding atoms, for a typical case, $R_a = 1.5$–2 Å. The parameter $\beta \approx 1$–2 Å. The parameter B was chosen to be proportional to the number of electron pairs participating in the Auger transition,

$$B = B_0 C_n^2 = \tfrac{1}{2} B_0 n(n-1), \qquad\qquad (25)$$

where n is the number of electrons in the outer shells, B_0 was chosen from a comparison of the calculated with the experimental data by Buck et al. (1975). For $Ar^+ \rightarrow Au$, it equals $B_0 = 10^{13}\,s^{-1}$.

The choice of $\omega_c(R)$ in the form of eq. (24) permitted to get a good agreement with numerous experimental data for different pairs of the colliding particles (Nizhnaya et al. 1979b,c, 1982a,b).

The calculation of both angular and energy distributions of single-charged Ar^+ ions scattered from an Au (100) surface covered by oxygen atoms (0.2 monolayer located at height 1.2 Å above the row [100] in the middle of the Au atoms) was made by computer simulation using this model.

The ion and atom spectra turned out to be similar (see fig. 35). At small escaping angles ψ the particles are scattered mostly as neutrals. At large ψ the ionization degree η^+ increases and, moreover, is higher for particles double scattered on Au and O than for ones scattered only on Au, although in the last case the particles have a larger energy, therefore, the ion fraction in the low-energy peak is higher than in high-energy one. Therefore, at scattering on an oxygen-covered surface η^+ increases in comparison with scattering on a clean surface, which agrees with the experimental data.

The distributions $I_1(\psi)$ calculated both for the total flux of the scattered particles in the low-energy peak and its ion component are shown in fig. 36.

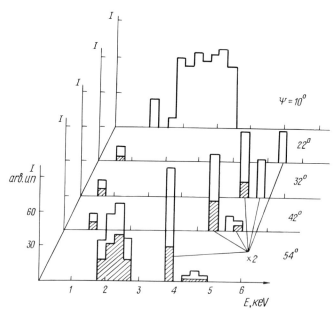

Fig. 35. The energy spectra of scattered Ar particles for different escaping angles ψ $(Ar^+ \rightarrow Au + O,\ E_0 = 10\,keV,\ y = 1.2\,Å,\ \alpha = 90°)$. The ion fraction of scattered particles is shown by the shadowing.

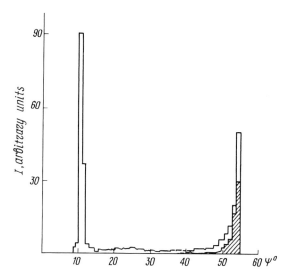

Fig. 36. The angular distribution $I_1(\psi)$ and $I_1^+(\psi)$ of scattered Au$^+$ particles for the case Ar$^+ \rightarrow$ Au + O, $E_0 = 10$ keV, $y = 1.2$ Å, $\alpha = 90°$. The area under $I_1^+(\psi)$ is shadowed.

At normal incidence, the angular distribution $I_1(\psi)$ contains two peaks, the peak II located in the large scattering angle region being the most informative one. This peak remains the sole one in the ion distribution $I_1^+(\psi)$. The ionization degree $\eta^+ \approx 50\%$ in the peak II, which allows to obtain reasonably intensive peaks with low density at bombarding beams. All this renders the ion spectra to be the most convenient for getting the diagnostic information.

It should be noted that the change of parameters R_A and β within reasonable limits change the magnitude of η^+, but does not influence the spectrum shape.

3.5. Thermal vibrations

The characteristic peak, II, in the angular distribution is observed for the escaping angle (see fig. 26).

$$\psi^* = \psi_0 + \theta^* + \arcsin[\, p(\theta^*)/d\,]\,, \tag{26}$$

where ψ_0 and D are the polar coordinates, $\psi_0 = \arctan(y/x)$, $d = (x^2 + y^2)^{1/2}$, and θ^* is the angle of scattering on adatom, which is close to the limiting angle for the inverse mass ratio $\theta^* \approx \theta_L$, $p(\theta^*)$ is the impact parameter corresponding to the angle θ^*. This simple formula allows to determine the height of the adatom location y if the coordinate x is determined from experiments (e.g., with LEED).

The most informative characteristic peak, peak II, is located in the extreme right region of the angular distribution $I_1(\psi)$, which simplifies its identification. Besides, the angular location of this peak practically does not change with changing E_0 and depends on the location of the adatom only.

For example, in case of scattering of Ar^+ with an energy E_0 from 5 to 30 keV along the plane [100] of the Au (100) surface, covered by oxygen (0.2 monolayer) located at the height $y = 1.2$ Å, the computer simulation has shown that $\psi^* = 54°$ at the analyzer width 1°. In this case, $\psi_0 = 30.9°$, $\theta_L = 23.5°$ and $\psi^* \approx \psi_0 + \theta_L$, i.e., the angle $\delta = \arcsin[p(\theta^*)/D] \leqslant 1°$. This means that the indeterminacy introduced by a potential into the calculation of ψ^* could not essentially influence the measurement of the adatom height y, when $p(\theta_L) \leqslant d$.

To take into account the influence of thermal vibrations on the angular distributions of the scattered particles, simple analytical estimations were performed.

One needs to consider the trajectory of double scattering by a target atom with mass M and an adatom with mass m_2. The target atom is considered to be a stationary point source of the bombarding particles with mass m_1 (Nizhnaya et al. 1988) and the amplitude of the adatom vibration is doubled.

The adatom vibrations could be considered as two independent displacements:

(i) one along the axis connecting the adatom with the target atom; and

(ii) one normal to this axis (see fig. 26).

Then the width of the characteristic peak is given by

$$\Delta\psi = \left|\frac{\partial\psi}{\partial\psi_0}\right|\Delta\psi_0 + \left|\frac{\partial\psi}{\partial d}\right|\Delta d . \tag{27}$$

The angle ψ^*, at which the singular peak II could be observed, is determined by formula (26). Then

$$\frac{\partial\psi}{\partial\psi_0} = 1 , \qquad \frac{\partial\psi}{\partial d} = -\frac{p(\theta^*)}{d[d^2 - p^2(\theta^*)]^{1/2}} , \tag{28}$$

where $\theta^* \approx \theta_L = m_2/m_1$.

At $m_2 \ll m_1$, for the potential $U = A/R^2$,

$$p(\theta^*) = \frac{R_0(\frac{1}{2}\pi - \theta^*)}{[(\frac{1}{2}\pi + \theta^*)(\frac{3}{2}\pi - \theta^*)]^{1/2}} , \tag{29}$$

where $R_0 = (A/E_0)^{1/2}$ is the distance of closest approach of atoms at a head-on collision.

Since $\theta^* \ll \pi$, it follows from formula (29) that $p(\theta^*) < R_0\sqrt{3}$. Then, if

$d \gg p$, i.e., $E_0 \gg A/(3d^2)$,

$$\frac{\partial \psi}{\partial d} < \frac{R_0}{\sqrt{3}d^2} . \tag{30}$$

The displacement of an adatom normal to the axis d could be given by

$$\Delta \psi = \arctan(\Delta d/d) \approx \Delta d/d , \tag{31}$$

where Δd is root-mean-square displacement from the equilibrium state, $\Delta d = [\hbar^2 T/(2m_2 K\theta Q_D^2)]^{1/2}$. Here, Q_D is the Debye temperature of the surface, T is temperature, K, $\hbar$ are Boltzmann and Planck constants, respectively.

Finally, the broadening of the characteristic peak could be written as

$$\Delta \psi \leq \frac{\Delta d}{d}\left(1 + \frac{R_0}{\sqrt{3}d}\right) . \tag{32}$$

Under conditions $Ar^+ \rightarrow Au + O$, $E_0 = 10$ keV, $T \approx 300$ K the width $\Delta\psi \ll 3°$, which allows to resolve the characteristic peaks in angular distribution.

However, to obtain the exact shape of both angular and energy distributions it is necessary to take into account the thermal vibrations.

3.6. Discussion

Getting diagnostic information from the comparison of the spectra observed experimentally with computer calculations is a usual procedure in the ISS technique. In addition to the computer time consumption, the major disadvantage of such a technique is the ambiguity of the interpretation of experimental results. Our computer simulation has shown that the rainbow peak in the angular distribution of the scattered particles (peak II in fig. 35) and the peak in the distribution of adatom recoils (peak B in fig. 32) are the most suitable for the diagnostics of the submonolayer coverage. Peak II corresponds to the most simple trajectory of double scattering which was analytically estimated. It is observed at the largest scattering angle and is hardly sensitive to changes in the initial energy E_0 because its angular position is determined by nothing but the adatom position. Such properties of peak II allows us to determine the height of the adatom location by a simple formula,

$$\psi = \psi_0 + \delta + \theta_L$$

where $\psi_0 = \operatorname{arctanh} y/x$ and $\delta = \arcsin[\,p(\theta_L)/(x^2 + y^2)^{1/2}]$ (see fig. 26).

Let us consider several examples showing what additional information on the characteristics of the submonolayer coverage can be obtained with the use of the peaks II or IV and B.

(i) Let only one adsorbed state of the adatom with coordinates x, y be on the surface at the initial moment. When varying the experimental conditions (temperature, coverage degree), an adatom transition into another adsorbed state (x_1, y_1) occurs. In the angular distribution of scattered ions and adatom recoils this transition will be observed as the appearance of peaks II and B corresponding to new adatom coordinates.

(ii) Let N_0 be the number of adsorbate atoms in a given adsorbed state on the surface at the initial moment. During the bombardment the number of these atoms will decrease with time according to the law

$$\frac{dN}{dt} = -N(t)\sigma_i j^+/e , \tag{33}$$

where j^+ is the density of the incident ion current, e is the electron charge, and σ_i is the cross section describing the processes leading to the reduction of the adatoms on the surface due to ion bombardment. If one neglects the adatom penetration inside the target, $\sigma_i = \sigma_{isd}$ is the cross section of the ion-induced desorption. Then

$$N(t)/N_0 = \exp(-\sigma_i j^+ t/e) . \tag{34}$$

It is clear that $N(\tau)/N_0 = S(\tau)/S_0$, where $S(\tau)$ and S_0 are the areas under the characteristic peak at the moment $t = \tau$ and $t = 0$, respectively. Therefore, by measuring the rate of $S(t)$ drop, it is possible to define the value σ_i for the given adsorbed state.

(iii) Let the flux of the slow atoms of the adsorbed coverage with the flux density q impinge on the surface simultaneously with the ion flux. Then,

$$\frac{dN}{dt} = -N(t)\sigma_i j^+/e + (1 - N)q\sigma_s , \tag{35}$$

where σ_s is the cross section of adatom sticking on the surface in the given adsorbed state, $\sigma_s = P/N_0'$, where P is the sticking coefficient, and N_0' is the number of adsorbate atoms per $1\,cm^2$ of the surface with a monolayer coverage. In this case, an equilibrium coverage of the surface, N_e, is established and the area under peak II will not change with time,

$$N_e = \frac{q\sigma_s e}{\sigma_i j^+ + \sigma_s qe} = N_0 \frac{S_e}{S_0} . \tag{36}$$

By measuring S_e with known σ_i it is possible to define the cross section of sticking to a given adsorbed state.

(iv) Let an electron beam with density j^- impinge on the surface simultaneously with an ion beam and adatom flux. Then,

$$\frac{dN}{dt} = -N(t)(\sigma_i j^+ + \sigma_{esd} j^-)/e + (1 - N(t))q\sigma_s , \tag{37}$$

where σ_{esd} is the cross section of electron-induced desorption from the given adsorbed state. The effect of the electron beam will lead to the establishment of a new value of the equilibrium coverage $N'_e < N_e$ and the area under the peak $S'_e \sim N'_e$,

$$N_e = \frac{q\sigma_s e}{\sigma_i j^+ + \sigma_s q e + \sigma_{esd} j^-} = N_0 \frac{S'_e}{S_0}. \tag{38}$$

From relations (36) and (38), it follows that

$$\frac{\sigma_{esd} j^-}{\sigma_i j^+ + \sigma_s q e} = \frac{S_e - S'_e}{S'_e}. \tag{39}$$

It is possible to define σ_{esd} from eq. (39) by measuring $\Delta S = S_e - S'_e$ using the cross sections σ_i, and σ_s obtained earlier. The cross section σ_{esd} is determined in such a way for all charged states of the desorbed particles, while σ_{esd} is usually measured for desorbed ions only (Ageev 1982).

Actually, the proposed applications of this method concern a dynamical situation on a surface under ion bombardment.

However, first, it is possible to diminish the number of defects by decreasing the flux density. Secondly, the problems of ion technology (e.g., the investigation of the features of epitaxial growth of films stimulated by ion bombardment) require knowledge about the cross section of the considered process exactly under the conditions of the ion bombardment of a surface.

Thus, the diverse information on the properties of submonolayer coverage could be obtained by measuring the characteristics of rainbow peaks of double scattering of primary ions and adatom recoils.

Conclusion

The physics of ion bombardment is experiencing a rejuvenation as is the case now in many other fields of atomic physics. Classical investigations performed in the twenties and thirties when the principal attention was drawn to atomic physics were aimed at the elucidation of the general quantum laws of the microcosm rather than at the study of the phenomena proper used in these investigations. There had been neither time nor means for detailed studies. With the beginning of the atomic era, attention was focused once more on these seemingly forgotten phenomena in the atomic physics and electronics.

The studies were resumed on the base of both new understanding of the physics of the phenomena involved and new experimental and computer techniques. No drastic change in our knowledge of the basic physical laws were brought by the wide development of the ion-beam–solid-surface interaction investigations in the sixties and up to nowadays. However, some unexpected and unpredicted phenomena of correlated plural collisions, including blocking, channeling, the *carambole*-type double scattering, interaction of highly charged ions and very large biomolecules and clusters with solid surface on one hand, and their influence on the well-known inelastic electronic processes on the other hand, gave new views on all processes and new tools for applications in the rapidly developing microelectronics.

APPENDIX

List of symbols

The following list contains the symbols most frequently used in the book.

Symbol	Quantity
a	Screening length for atomic potentials
a_B	Bohr screening length
a_F	Firsov screening length
a_L	Lindhard screening length
a_0	Bohr radius, $\hbar^2/(m_e e^2) = 0.529\ \text{Å}$
d	distance between atoms in a row
$d\sigma$	differential cross section
$d\Omega$	element of solid angle
e	electron charge
E_1	energy of reflected particles
E_2	recoil energy
E_0	energy of projectile
E_d	displacement energy
E_f	focussing energy
E_r	energy of relative motion
$E_\perp$	energy in the transverse plane
E	inelastic energy loss in an elementary collision event
F	focal length
I	reflected particle intensity
k	Boltzmann constant
K_N	total particle reflection coefficient
l_0, m_0, n_0	the projections of the unit vector $L_0(l_0, m_0, n_0)$ on the coordinate axes
m_1, M_1	mass of the projectile; mass number of the projectile
m_2, M_2	mass of the target atom; mass number of the target atom
N	number of atoms per unit volume
p	impact parameter
q	charge of ion
r	distance between interacting particles; distance between the projectile and the string
r_0	apsis of a collision
r_c	cut-off radius
R	distance between the projectile and the string atom
R_s	shadow radius
S	sputtering yield
t, τ	time integral for a binary collision
T	temperature

T_m	melting temperature
$\overline{u_1^2}$	mean square amplitude of one-dimensional thermal vibrations
$[uvw]$	direction of an axis
v	velocity of moving particle
v_B	Bohr velocity, $v_B = e^2/\hbar = 2.2 \times 10^8$ cm/s
$V(R), V(r)$	two-body interaction potentials
w	probability for an excited electron to escape from the solid
W	rate of transition
(x, y, z)	coordinates of the aiming point
Y	molecule desorption yield
Z_1, Z_2	charge numbers of the projectile and target atom
β, δ	emergence (ejection) angle of the reflected particle or recoil atom in the plane of incidence
γ	coefficient of ion–electron emission
γ_k	coefficient of kinetic ion–electron emission
$\gamma_p(q)$	coefficient of potential ion–electron emission for q-charged ion
ε	the analyser resolution
η^+	ionization degree
$\eta_\perp$	incidence angle in the transverse plane
θ	total scattering angle
θ_1	polar scattering angle in the laboratory system
θ_2	recoil observation angle
θ_D	Debye temperature
θ_{lim}	limiting scattering angle
$\theta_\perp$	scattering angle in the transverse plane
λ	effective range
λ_0	the effective depth for the ion backscattering
μ	reduced mass, $\mu = m_2/m_1$
ξ	azimuthal angle of target rotation
φ	azimuthal scattering angle
φ	work function of metal
φ_0	collision integral
φ_1	the azimuthal angle after the first scattering
$\varphi_\perp$	reflection angle in the transverse plane
$\Phi(\theta_D/T)$	the Debye function
χ	scattering angle in centre-of-mass system
$\chi(r/a)$	screening function
ψ	glancing angle of primary particles
ψ_1, ψ_2	critical angle for channeling at high and low energies

References

Aarset, B., R.W. Cloud and J.G. Trump, 1954, J. Appl. Phys. **25**, 1365.

Abel, F., M. Bruneaux, C. Cohen, G. Della Mia, A.V. Drigo, S. Lorusso and P. Mazzoldi, 1976, Nucl. Instrum. Methods **132**, 197.

Aberdam, D., R. Baudoing, E. Blanc and C. Gaubert, 1978, Surf. Sci. **71**, 279.

Abrahamson, A.A., 1963, Phys. Rev. **130**, 693.

Abrahamson, A.A., 1969, Phys. Rev. **178**, 76.

Abroyan, I.A., 1961, Fiz. Tverd. Tela (Leningrad) **3**, 588.

Abroyan, I.A., V.P. Lavrov and A.I. Titov, 1963, Fiz. Tverd. Tela (Leningrad) **7**, 3660.

Abroyan, I.A., M.A. Eremeev and N.N. Petrov, 1967, Usp. Fiz. Nauk **92**, 105.

Adamson, A.W., 1976, Physical Chemistry of Surfaces (Wiley Interscience, New York).

Afanasaiev, A.G., and I.M. Bronstein, 1976, Radiotekh. i Elektron. **21**, 2247.

Afrosimov, V.V., Yu.S. Gordeev, A.N. Panov and N.V. Fedorenko, 1964a, Zh. Tekh. Fiz. **34**, 1613.

Afrosimov, V.V., Yu.S. Gordeev, A.N. Panov and N.V. Fedorenko, 1964b, Zh. Tekh. Fiz. **34**, 1624.

Afrosimov, V.V., Yu.S. Gordeev, A.N. Panov and N.V. Fedorenko, 1964c, Zh. Tekh. Fiz. **34**, 1637.

Afrosimov, V.V., Yu.S. Gordeev, A.N. Zinovjev and D.Kh. Rasulev, 1976, Zh. Eksp. Teor. Fiz. Pis'ma **24**, 33.

Afrosimov, V.V., G.G. Meskhi, N.N. Tsarev and A.P. Shergin, 1983, Zh. Eksp. Teor. Fiz. **84**, 454.

Ageev, V.N., 1982, Poverkhnost'. Fiz. Khim. Mekh. **4**, 1.

Agranovich, V.M., and V.V. Kirsanov, 1976, Usp. Fiz. Nauk **118**, 1.

Agranovich, V.M., and D.D. Odintsov, 1965, Dokl. Akad. Nauk SSSR **162**, 778.

Agranovich, V.M., and V.A. Ryabov, 1970, Fiz. Tverd. Tela (Leningrad) **12**, 260.

Akazawa, H., and Y. Murata, 1988, Surf. Sci. **207**, L971.

Akishin, A.I., and S.S. Vasilev, 1959, Fiz. Tverd. Tela (Leningrad) **1**, 5.

Akkerman, A.F., and M.A. Chubisov, 1981, Zh. Tekh. Fiz. **51**, 2151.

Akkerman, A.F., Yu.M. Nikitushev and V.A. Botvin, 1972, Problems on Fast Electrons Channelling in Substance in Monte-Carlo Simulation (Nauka, Alma-Ata) (in Russian).

Akoev, I.G., S.S. Yurov and B.I. Akoev, 1985, Adv. Space Res. **1**, 75.

Alberts, A., K. Wien, P. Duck, W. Treu and H. Voit, 1982, Nucl. Instrum. Methods **198**, 69.

Algra, A.J., P.P. Maaskant, S.B. Luitjens, E.P.Th.M. Suurmeijer and A.L. Boers, 1980a, J. Phys. D **13**, 2363.

Algra, A.J., S.B. Luitjens, E.P.Th.M. Suurmeijer and A.L. Boers, 1980b, Inst. Phys. Conf. Ser. No. 54: ch. 4, p. 119.

Algra, A.J., S.B. Luitjens, E.P.Th.M. Suurmeijer and A.L. Boers, 1980c, Phys. Lett. A **75**, 496.

Algra, A.J., S.B. Luitjens, E.P.Th.M. Suurmeijer and A.L. Boers, 1980d, Surf. Sci. **100**, 329.

Algra, A.J., S.B. Luitjens, H. Borggreve, E.P.Th.M. Suurmeijer and A.L. Boers, 1982a, Radiat. Eff. **62**, 7.

Algra, A.J., E. Van Loenen, E.P.Th.M. Suurmeijer and A.L. Boers, 1982b, Radiat. Eff. **60**, 173.

Algra, A.J., S.B. Luitjens, E.P.Th.M. Suurmeijer and A.L. Boers, 1982c, Appl. Surf. Sci. **10**, 273.

Aliev, A.A., S.L. Nizhnaya, F.F. Umarov and N.N. Flyants, 1991, Poverkhnost'. Fiz. Khim. Mekh. **3**, 60.

Alimov, Sh.A., B.E. Baklitzky, V.Ya. Kanarov and E.S. Parilis, 1990, Izv. Akad. Nauk SSSR Ser. Fiz. **54**, 1237.

Allaby, J.V., A.N. Diddens, R.J. Glauber, A. Kloving, O. Kofoed-Hansen, E.J. Sacharidis, K. Schlupmann, A.M. Thorndike and A.M. Wetherell, 1969, Phys. Lett. B **30**, 549.

Allen, J.S., 1939, Phys. Rev. **55**, 336.

Alonso, E.V., R.A. Baragiola, J. Ferron, M.M. Jakas and A. Oliva-Florio, 1980, Phys. Rev. B **22**, 80.

Andersen, H.H., and H.L. Bay, 1973, Radiat. Eff. **19**, 139.

Andersen, H.H., and H.L. Bay, 1981, Sputtering yield measurements, in: Sputtering by Particle Bombardment I, ed. R. Behrisch (Springer, Berlin) p. 145.

Andersen, H.H., and P. Sigmund, 1965, Risø Report No. **103**, 1.

Andersen, H.H., and P. Sigmund, 1966, K. Dan. Vidensk. Selsk. Mat. Fys. Medd. **34**, 15.

Andrä, H.J., 1979, DE Patent 2650455.

Anno, J.N., 1962, J. Appl. Phys. **33**, 1678.

Anno, J.N., 1963, J. Appl. Phys. **34**, 3495.

Aono, M., 1984, Nucl. Instrum. Meth. Phys. Res. B **12**, 374.

Aono, M., C. Oshima, S. Zaima, S. Otani and Y. Ishizawa, 1981, J. Appl. Phys. **20**, L829.

Aono, M., Y. Hou, C. Oshima and Y. Ishizawa, 1982a, Phys. Rev. Lett. **49**, 567.

Aono, M., Y. Hou, R. Souda, C. Oshima, S. Otani, Y. Ishizawa, K. Matsuda and R. Shimizi, 1982b, Jpn. J. Appl. Phys. **21**, L670.

Aono, M., Y. Hou, R. Souda, C. Oshima, S. Otani and Y. Ishizawa, 1983, Phys. Rev. Lett. **50**, 1293.

Apell, P., 1987, Nucl. Instrum. Meth. Phys. Res. B **23**, 242.

Apell, P., 1988, J. Phys. B **21**, 2665.

Apell, S.P., and R. Monreal, 1989, Nucl. Instrum. Meth. Phys. Res. B **42**, 171.

Arifov, T.U., E.K. Vasileva, D.D. Gruich, S.F. Kovalenko and S.N. Morozov, 1976, Izv. Akad. Nauk SSSR Ser. Fiz. **40**, 2621.

Arifov, T.U., D.D. Gruich and S.N. Morozov, 1978, Ion emission from nonmetals at multiply-charged ion bombardment, in: Proc. 5th All-Union Conf. on Interaction of Atomic Particles with Solids (Minsk, USSR) Vol. 1, p. 200.

Arifov, U.A., 1961, Atomic Particles Interaction with Metal Surface (Fan, Tashkent) (in Russian) [1963, Engl. transl., AEC-tr-6089, Washington].

Arifov, U.A., 1968, Interaction of Atomic Particles with Solid Surface (Nauka, Moscow) (in Russian) [1969, Engl. transl., Plenum, New York].

Arifov, U.A., and A.A. Aliev, 1967, Dokl. Akad. Nauk SSSR **172**, 65.

Arifov, U.A., and A.A. Aliev, 1968a, Dokl. Akad. Nauk SSSR **183**, 60.

Arifov, U.A., and A.A. Aliev, 1968b, Dokl. Akad. Nauk SSSR **183**, 1041.

Arifov, U.A., and A.A. Aliev, 1974, Angular Dependence of Atomic Particles Interaction with Solids (Fan, Tashkent) (in Russian).

Arifov, U.A., and A.Kh. Ayukhanov, 1951, Dokl. Akad. Nauk Uzb. SSR **4**, 12.

Arifov, U.A., and R.R. Rakhimov, 1960, Izv. Akad. Nauk SSSR Ser. Fiz. **24**, 657.

Arifov, U.A., N.N. Flyants and A.Kh. Ayukhanov, 1961a, Dokl. Akad. Nauk Uzb. SSR **10**, 10.

Arifov, U.A., Kh.Kh. Khadjimukhamedov, E.S. Parilis and L.M. Kishinevsky, 1961b, Izv. Akad. Nauk Uzb. SSR, Ser. Fiz. Mat. Nauk **6**, 50.

Arifov, U.A., A.A. Aliev and A.Kh. Ayukhanov, 1962a, Izv. Akad. Nauk SSSR Ser. Fiz. **26**, 1440.

Arifov, U.A., R.R. Rakhimov and O.V. Khozinsky, 1962b, Izv. Akad. Nauk SSSR Ser. Fiz. **26**, 1398.

Arifov, U.A., R.R. Rakhimov, M. Abdullaeva and S. Gaipov, 1962c, Izv. Akad. Nauk SSSR Ser. Fiz. **26**, 1403.

Arifov, U.A., R.R. Rakhimov and Kh. Dzhurakulov, 1962d, Dokl. Akad. Nauk SSSR **143**, 309.

Arifov, U.A., R.R. Rakhimov and Kh. Dzhurakulov, 1963a, Radiotekh. i Elektron. **8**, 299.

Arifov, U.A., G.A. Tashkhanova and R.R. Rakhimov, 1963b, Dokl. Akad. Nauk Uzb. SSR **10**, 5.

Arifov, U.A., N.N. Flyants and R.R. Rakhimov, 1964, Dokl. Akad. Nauk Uzb. SSR **10**, 15.

Arifov, U.A., S. Gaipov, M. Ikramova and R.R. Rakhimov, 1966a, Izv. Akad. Nauk SSSR Ser. Fiz. **30**, 896.

Arifov, U.A., Kh. Dzhurakulov and R.R. Rakhimov, 1966b, Izv. Akad. Nauk Uzb. SSR Ser. Fiz. Mat. Nauk **6**, 33.

Arifov, U.A., R.R. Rakhimov and S. Gaipov, 1971a, Izv. Akad. Nauk SSSR Ser. Fiz. **35**, 562.

Arifov, U.A., R.R. Rakhimov, Sh.S. Radjabov and A.N. Mavlyanov, 1971b, Dokl. Akad. Nauk Uzb. SSR **6**, 19.

Arifov, U.A., E.S. Mukhamadiev, E.S. Parilis and A.S. Pasjuk, 1971c, Preprint by Joint Inst. on Nucl. Res. P7-6165, Dubna.

Arifov, U.A., D.D. Gruich, G.E. Ermakov and Z. Khalmirzaev, 1972, Slow alkali metalic ions reflection from solid surface, in: Atomic Collisions on Solid Surface (Fan, Tashkent) (in Russian) p. 28.

Arifov, U.A., E.S. Mukhamadiev, E.S. Parilis and A.S. Pasjuk, 1973a, Zh. Tekh. Fiz. **43**, 375.

Arifov, U.A., L.M. Kishinevsky, E.S. Mukhamadiev and E.S. Parilis, 1973b, Zh. Tekh. Fiz. 43, 181 [1973, Sov. Phys. Tech. Phys. **18**, 118].

Arifov, U.A., Sh.S. Radjabov and R.R. Rakhimov, 1974, Electron emission from (100) and (111) faces of Mo single crystal at bombardment by multicharged Ar$^+$ ions, in: Proc. 3rd All-Union Conf. on Atomic Particles Interaction with Solids (Kiev) Part 2, p. 46.

Arifov, U.A., N.N. Flyants and A.A. Zibitsker, 1976, Izv. Akad. Nauk SSSR Ser. Fiz. **40**, 2548.

Arthurs, A.M., R.A. Bond and J. Hylop, 1957, Proc. Phys. Soc. London Sect. A **70**, 617.

Askaryan, G.A., 1957, At. Enegr. **3**, 152.

Avakyantz, G.M., 1951a, Dokl. Akad. Nauk Uzb. SSR **6**, 3.

Avakyantz, G.M., 1951b, Dokl. Akad. Nauk Uzb. SSR **9**, 9.

Babanskaya, L.N., M.A. Vasiliev and S.D. Gorodetsky, 1983, Metallofizika **5**, 38.

Baboux, J.C., M. Perdrix, R. Goutte and C. Guillaud, 1971, J. Phys. D **4**, 1617.

Baklitzky, B.E., and E.S. Parilis, 1972, Radiat. Eff. **12**, 137.

Baklitzky, B.E., and E.S. Parilis, 1976, Izv. Akad. Nauk SSSR Ser. Fiz. **40**, 2538.

Baklitzky, B.E., and E.S. Parilis, 1982, Poverkhnost'. Fiz. Khim. Mekh. **8**, 92.

Baklitzky, B.E., and E.S. Parilis, 1986, Zh. Tekh. Fiz. **56**, 27.

Baklitzky, B.E., E.S. Parilis and V.K. Verleger, 1973, Radiat. Eff. **19**, 155.

Baklitzky, B.E., E.S. Parilis and V.K. Verleger, 1975, Zh. Tekh. Fiz.**15**, 2427.

Balashova, L.L., E.S. Mashkova and V.A. Molchanov, 1977a, Phys. Lett. A **61**, 193.

Balashova, L.L., E.S. Mashkova and V.A. Molchanov, 1977b, Zh. Tekh. Fiz. Pis'ma **3**, 339.

Balashova, L.L., A.M. Borisov, E.S. Mashkova and V.A. Molchanov, 1977c, Radiat. Eff. **31**, 249.

Balashova, L.L., E.S. Mashkova and V.A. Molchanov, 1978, Surf. Sci. **76**, L590.

Balashova, L.L., I.N. Evdokimov, E.S. Mashkova and V.A. Molchanov, 1979, Izv. Akad. Nauk SSSR Ser. Fiz. **43**, 1887 [1979, Bull. Acad. Sci. USSR Phys. Ser. **43**(9), 86].

Balashova, L.L., Sh.N. Garin, A.I. Dodonov, E.S. Mashkova and V.A. Molchanov, 1982a, Surf. Sci. **119**, L378.

Balashova, L.L., Sh.N. Garin and A.I. Dodonov, 1982b, Vestn. Mosk. Univ. Ser. Fiz. **3**, 85 [1982, Moscow Univ. Phys. Bull. **37**(3), 102].

Balashova, L.L., A.I. Dodonov, O.B. Firsov, Sh.N. Garin, E.S. Mashkova and V.A. Molchanov, 1983, Radiat. Eff. **77**, 67.

Ball, D.J., T.M. Buck, D. Macnair and G.H. Wheatley, 1972, Surf. Sci. **30**, 69.

Baragiola, R.A., E.V. Alonso, J. Ferron and A. Oliva-Florio, 1979, Surf. Sci. **90**, 240.

Baranov, I.A., and V.V. Obnorskii, 1983, Radiat. Eff. **79**, 1.

Baranov, I.A., A.S. Krivokhatskii and V.V. Obnorskii, 1981, Zh. Tekh. Fiz. **51**, 2457.

Baranov, I.A., Yu.V. Martynenko, S.O. Tsepelevich and Yu.N. Yavlinsky, 1988, Usp. Fiz. Nauk **156**, 447.

Barker, J.A., and D.J. Auerbach, 1984, Surf. Sci. **4**, 1.

Barker, J.A., A.W. Kleyn and D.J. Auerbach, 1983, J. Chem. Phys. Lett. **97**, 9.

Barret, J.H., 1971, Phys. Rev. B **3**, 1527.

Barret, J.H., and D.P. Jackson, 1980, Nucl. Instrum. Methods **170**, 115.

Bastasz, R.J., T.E. Felter and W.P. Ellis, 1989, Phys. Rev. Lett. **63**, 558.

Batanov, G.M., 1960, Fiz. Tverd. Tela (Leningrad) **2**, 2048.

Batanov, G.M., 1961a, Fiz. Tverd. Tela (Leningrad) **3**, 558.

Batanov, G.M., 1961b, Fiz. Tverd. Tela (Leningrad) **3**, 642.

Batanov, G.M., 1962, Fiz. Tverd. Tela (Leningrad) **4**, 1778.

Baun, W.L., 1978, Surf. Sci. **72**, 536.

Bay, H.L., H.H. Andersen, W.O. Hofer and O. Nielsen, 1976, Nucl. Instrum. Methods **132**, 301.

Becker, O., S. Della-Negra, Y. Le Beyec and K. Wien, 1986, Nucl. Instrum. Meth. Phys. Res. B **16**, 321.

Beeler Jr, J.R., 1966, Phys. Rev. **150**, 470.

Beeler Jr, J.R., 1970, Adv. Mater. Res. **4**, 295.

Beeler Jr, J.R., and D.G. Besco, 1963, J. Appl. Phys. **34**, 2873.

Begemann, S.H.A., and A.L. Boers, 1972, Surf. Sci. **30**, 134.

Begemann, S.H.A., J.A. Van den Berg and A.L. Boers, 1972, Surf. Sci. **32**, 595.

Begemann, W., K.H. Meiwes-Broer and H.O. Lutz, 1986, Phys. Rev. **56**, 2248.

Beining, P., J. Scher, E. Nieschler, B. Nees and H. Voit, 1988, Phys. Rev. B **37**, 9197.

Belenky, A.Ya., 1981, Usp. Fiz. Nauk **134**, 125.

Bellina Jr, J.J., and H.H. Farnsworth, 1972, J. Vac. Sci. Technol. **9**, 616.

Benazeth, N., 1982, Nucl. Instrum. Methods **194**, 405.

Benazeth, N., C. Leonard, C. Benazeth, L. Viel and M. Negre, 1980, Surf. Sci. **97**, 171.

Bernstein, R.B., 1966, Phys. Rev. Lett. **16**, 385.

Berry, H.W., 1958, J. Appl. Phys. **29**, 1219.

Berry, H.W., 1961, Phys. Rev. **121**, 1714.

Bertrand, P., 1980, Nucl. Instrum. Methods **170**, 489.

Besenbacher, F., J. Bottiger and O. Graversen, 1981, Nucl. Instrum. Methods **191**, 221.

Bess, L., 1957, Phys. Rev. **105**, 1469.

Betz, H.D., and L. Grodzins, 1970, Phys. Rev. Lett. **25**, 211.

Bhalla, C.P., and J.N. Bradford, 1968, Phys. Lett. A **27**, 318.

Bhattacharya, R.S., W. Eckstein and H. Verbeek, 1980, Surf. Sci. **93**, 563.

Bitensky, I.S., 1980, At. Energ. **49**, 232.

Bitensky, I.S., and E.S. Parilis, 1978a, Dissociationless scattering of fast heteroatomic molecules by single crystal surface, in: Proc. 5th All-Union Conf. on Atomic Particles Interaction with Solids (Minsk, USSR) p. 105.

Bitensky, I.S., and E.S. Parilis, 1978b, Zh. Tekh. Fiz. **48**, 1941 [1978b, Sov. Phys. Tech. Phys. **23**, 1104].

Bitensky, I.S., and E.S. Parilis, 1980, Scattering of swift molecules and knocking-out of high energy clusters at correlated collisions on solid surface, in: Proc. Symp. on Sputtering (Perchtoldsdorf/ Vienna, Austria) p. 688.

Bitensky, I.S., and E.S. Parilis, 1981a, Zh. Tekh. Fiz. **51**, 1798 [1981a, Sov. Phys. Tech. Phys. **26**, 1042].

Bitensky, I.S., and E.S. Parilis, 1981b, Zh. Tekh. Fiz. **51**, 2134 [1981, Sov. Phys. Tech. Phys. **26**, 1247].

Bitensky, I.S., and E.S. Parilis, 1982, Zh. Tekh. Fiz. Pis'ma **8**, 419 [1982, Sov. Phys. Tech. Phys. Lett. **8**, 183].

Bitensky, I.S., and E.S. Parilis, 1984, Nucl. Instrum. Meth. Phys. Res. B **2**, 384.

Bitensky, I.S., and E.S. Parilis, 1985, Surf. Sci. **161**, L565.

Bitensky, I.S., and E.S. Parilis, 1987, Nucl. Instrum. Meth. Phys. Res. B **21**, 26.

Bitensky, I.S., and E.S. Parilis, 1989, J. Phys. (Paris) **50**, 227.

Bitensky, I.S., and E.S. Parilis, 1990, Gordon Conf. on Interaction of Particles with Solids, unpublished.

Bitensky, I.S., M.N. Mirakhmedov and E.S. Parilis, 1979, Zh. Tekh. Fiz. **49**, 1042 [1979, Sov. Phys. Tech. Phys. **24**, 618].

Bitensky, I.S., Ya.S. Gilenko and E.S. Parilis, 1985, Izv. Akad. Nauk SSSR Ser. Fiz. **49**, 1788.

Bitensky, I.S., Ya.S. Gilenko and E.S. Parilis, 1987, Zh. Tekh. Fiz. **57**, 1692 [1987, Sov. Phys. Tech. Phys. **32**, 9].

Bitensky, I.S., Ya.S. Gilenko and E.S. Parilis, 1988, Zh. Eksp. Teor. Fiz. **94**, 66 [1988, Sov. Phys. JETP **67**, 470].

Bitensky, I.S., A.M. Goldenberg and E.S. Parilis, 1989a, J. Phys. (Paris) **50**, 213.

Bitensky, I.S., A.M. Goldenberg and E.S. Parilis, 1989b, Khim. Vys. Energ. **23**, 141.

Bitensky, I.S., A.M. Goldenberg and E.S. Parilis, 1989c, in: Proc. Int. Conf. on Ion Formation from Organic Solids (IFSOV), Lovanger, Sweden, eds A. Hedin, B.U.R. Sundqvist and A. Beninghoven (Wiley, New York) p. 205.

Bitensky, I.S., E.S. Parilis and I.A. Wojciekhovski, 1990, Nucl. Instrum. Meth. Phys. Res. B **47**, 243.

Bitensky, I.S., E.S. Parilis and I.A. Wojciechowski, 1992a, Nucl. Instrum. Meth. Phys. Res. B **67**, 359.

Bitensky, I.S., E.S. Parilis and I.A. Wojciechowski, 1992b, Nucl. Instrum. Meth. Phys. Res. B **67**, 595.

Bitensky, I.S., E.S. Parilis, S. Della-Negra and Y. Le Beyec, 1992c, Nucl. Instrum. Meth. Phys. Res. B, to be published.

Bitensky, I.S., E.S. Parilis and I.A. Wojciechowski, 1992d, Nucl. Instrum. Meth. Phys. Res. B, to be published.

Black, I.E., B. Zaks and D.L. Mills, 1980, Phys. Rev. B **22**, 1818.

Boardman, A.D., B.V. Paranjape and R. Teshima, 1975, Surf. Sci. **49**, 275.

Boers, A.L., 1977, Surf. Sci. **63**, 475.

Boers, A.L., and L.K. Verhey, 1973, The influence of thermal lattice vibrations on multiple ion scattering, in: Proc. 5th Int. Conf. on Atomic Collisions in Solids (Gatlinburg) p. 583.

Boers, A.L., S.B. Luitjens and E.P.Th.M. Suurmeijer, 1980, Surf. Sci. **100**, 315.

Bohdansky, J., J. Roth, M.K. Sinha and W. Ottenberger, 1976, J. Nucl. Mater. **63**, 115.

Bohr, N., 1948, K. Dan. Vidensk. Selsk. Mat. Fys. Medd. **18**, 8.

Bolbach, G., S. Della-Negra, C. Deprun, Y. Le Beyec and K.G. Standing, 1987, Rapid. Commun. Mass Spectrom. **1**, 22.

Bolbach, G., R. Beavis, S. Della-Negra, C. Deprun, W. Ens, Y. Le Beyec, D.E. Main, B. Schueler and K.G. Standing, 1988, Nucl. Instrum. Meth. Phys. Res. B **30**, 74.

Bolshov, L.A., A.P. Napartovich, A.G. Naumovets and A.G. Fedorus, 1977, Usp. Fiz. Nauk **122**, 125.

Born, M., and J.E. Mayer, 1932, Z. Phys. **75**, 1.

Bosch, S.H., and G. Kuskevics, 1964, Phys. Rev. A **134**, 1356.

Bosch, S.H., and G. Kuskevics, 1965, Phys. Rev. A **137**, 255.

Bourne, H.C., 1952, Thesis, Massachusetts Technical Institute.

Brandt, W., 1975, in: Atomic Collisions in Solids, Vol. 1, eds S. Datz, B.R. Appleton and C.D. Moak (Plenum Press, New York) p. 261.

Brice, D.K., 1972, Phys. Rev. A **6**, 1971.

Briggs, J.S., and A.P. Pathak, 1974, J. Phys. C **7**, 1929.

Briggs, J.S., and K. Taulbjerg, 1975, J. Phys. B **8**, 1909.

Brinkman, J.A., 1954, Appl. Phys. **25**, 961.

Bronckers, R.P.N., and A.G.J. De Wit, 1981a, Surf. Sci. **104**, 384.

Bronckers, R.P.N., and A.G.J. De Wit, 1981b, Surf. Sci. **112**, 111.

Bronckers, R.P.N., and A.G.J. De Wit, 1981c, Surf. Sci. **112**, 133.

Brongersma, H.H., 1978, Apparatus for surface layer analysis by ion scattering, US Patent No. 4100409.

Brongersma, H.H., and P.M. Mul, 1972, Chem. Phys. Lett. **14**, 380.

Brongersma, H.H., and P.M. Mul, 1973, Surf. Sci. **35**, 393.

Brongersma, H.H., and J.B. Theeten, 1976, Surf. Sci. **54**, 519.

Brongersma, H.H., M.I. Sparnay and T.M. Buck, 1978, Surf. Sci. **71**, 657.

Bronstein, I.M., and B.S. Freiman, 1969, Secondary Electron Emission (Nauka, Moscow) (in Russian).

Brown, A.B., C.W. Snyder, W.A. Fowler and C.C. Lauritsen, 1951, Phys. Rev. **82**, 159.

Brown, W.L., W.M. Augustiniak, E. Brody and B. Cooper, 1980, Nucl. Instrum. Methods **170**, 321.

Bruinning, G., 1958, Physics and Secondary Electron Emission Application (Sov. Radio, Moscow) (in Russian).

Brünne, C., 1957, Z. Phys. **147**, 161.

Brusilovsky, B.A., 1985, Vacuum **35**, 595.

Brusilovsky, B.A., 1990a, Kinetic Ion-Induced Electron Emission (Energoatomizdat, Moscow) (in Russian).

Brusilovsky, B.A., 1990b, Appl. Phys. A **50**, 111.

Buck, T.M., 1975, Low energy ions scattering spectroscopy, in: Methods of Surface Analysis, ed. A.W. Czanderna (Elsevier, Amsterdam) p. 102.

Buck, T.M., Y.S. Chen, G.H. Wheatley and W.F. Van der Weg, 1975, Surf. Sci. **47**, 244.

Buck, T.M., G.H. Wheatley, D.P. Jackson, A.L. Boers, S.B. Luitjens, E. Van Loenen, A.J. Algra, W. Eckstein and H. Verbeek, 1982, Nucl. Instrum. Methods **194**, 649.

Büker, M., 1977, in: Proc. Eur. Symp. on Life Sci. Res. in Space (Frankfurt am Main) p. 23.

Burrekoven, H., 1956, Dissertation, University of Köln.

Bydin, Yu.F., and V.M. Dukelsky, 1956, Zh. Eksp. Teor. Fiz. **31**, 569.

Campbell, B.L., and R.N. Whittem, 1961, J. Sci. Instrum. **38**, 516.

Cano, G.L., 1973, J. Appl. Phys. **44**, 5293.

Carlston, C.E., G.D. Magnuson, P. Mahadevan and D.E. Harrison Jr, 1965, Phys. Rev. A **139**, 729.

Carter, G., 1979, Radiat. Eff. Lett. **43**, 193.

Carter, G., 1983, Nucl. Instrum. Methods **209/210**, 1.

Carter, G., and J.S. Colligon, 1968, Ion Bobardment of Solids (Heinemann, London).

Cawthron, E.R., 1971, Aust. J. Phys. **24**, 859.

Chang, C.-C., 1971, Surf. Sci. **25**, 53.

Chang, C.-C., L.A. De Louise, N. Winograd and B.J. Garrison, 1985, Surf. Sci. **154**, 22.

Chang, C.S., T.L. Poster and I.S.T. Tsong, 1989, Vacuum **39**, 1195.

Cheshire, J., G. Dearneley and J.P. Poate, 1968, Phys. Lett. A **27**, 304.

Chicherov, V.M., 1968, Zh. Eksp. Teor. Fiz. **55**, 25 [1969, Sov. Phys. JETP **28**, 13].

Chicherov, V.M., 1972, Zh. Eksp. Teor. Fiz. Pis'ma **16**, 328 [1972, JETP Lett. **16**, 231].

Christensen, C.H., J.O. Jensen and K. Lefman, 1986, Nucl. Instrum. Meth. Phys. Res. B **13**, 230.

Cocconi, G., A.N. Diddens and E. Lillethun, 1962, Phys. Rev. **126**, 277.

Colombie, N., C. Benazeth, L. Mishler and L. Viel, 1973, Radiat. Eff. **18**, 251.

Colton, R.J., M.M. Roos and D.A. Kidwell, 1986, Nucl. Instrum. Meth. Phys. Res. B **13**, 259.

Cooper, J.W., and J.B. Martin, 1962, Phys. Rev. **126**, 1482.

Czanderna, A.W., ed., 1975, Methods of Surface Analysis (Elsevier, Amsterdam).

Dahl, P., and N. Sandager, 1969, Surf. Sci. **14**, 305.

Datz, S., and C. Snoek, 1964, Phys. Rev. A **134**, 347.

Davis, H.L., and J.R. Noonan, 1983, Surf. Sci. **126**, 245.

De Jonge, R., J. Los and A.E. De Vries, 1988, Nucl. Instrum. Meth. Phys. Res. B **30**, 159.

De Wit, A.G.J., I.A. Van der Shootbrugge and J.M. Fluit, 1975, Surf. Sci. **47**, 258.

De Wit, A.G.J., R.P.N. Bronckers and J.M. Fluit, 1979, Surf. Sci. **82**, 177.

De Zwart, S.T., 1987a, Nucl. Instrum. Meth. Phys. Res. B **23**, 238.

De Zwart, S.T., 1987b, Thesis, Groningen.

De Zwart, S.T., T. Fried, U. Jellen, A.L. Boers and A.G. Drentje, 1985, J. Phys. B **18**, L623.

De Zwart, S.T., T. Fried, D.O. Boerma, R. Moekstra, A.G. Drentje and A.L. Boers, 1986, Surf. Sci. **177**, L939.

De Zwart, S.T., A.G. Drentje, A.L. Boers and R. Morgenstern, 1989, Surf. Sci. **217**, 298.

Debevec, L., E. Heldt, K.H. Krebs and J. Marsel, 1966, Proc. 7th Int. Conf. on Phenomena in Ionized Gases (Belgrade, 1965).

Dekker, A.J., 1954, Phys. Rev. **94**, 1179.

Delauney, M., M. Fehringer, R. Geller, D. Hitz, P. Varga and H. Winter, 1987, Phys. Rev. B **35**, 4232.

Delauney, M., R. Geller, J. Debernardi, P. Ludwig and P. Sortais, 1988, Surf. Sci. **197**, L273.

Della-Negra, S., Y. Le Beyec, B. Monart, K.G. Standing and K. Wien, 1988a, Nucl. Instrum. Meth. Phys. Res. B **32**, 360.

Della-Negra, S., J. Depauw, H. Joret, Y. Le Beyec and E.A. Schweikert, 1988b, Phys. Rev. Lett. **60**, 948.

Della-Negra, S., J. Depauw, H. Joret, Y. Le Beyec, I.S. Bitensky, G. Bolbach, R. Galera and K. Wien, 1990, Nucl. Instrum. Meth. Phys. Res. B **52**, 121.

Demkov, Yu.N., 1980, Zh. Tekh. Fiz. Pis'ma **6**, 933.

Demkov, Yu.N., 1981, Zh. Eksp. Teor. Fiz. **80**, 127 [1981, Sov. Phys. JETP **53**, 63].

Demkov, Yu.N., 1988, The kinematics of change reactions at fast collisions, in: Problems in Theoretical Physics (Leningrad State University) p. 48.

Demuth, J.E., D.W. Jepsen and P.M. Marcus, 1973, Phys. Rev. Lett. **31**, 540.

Derks, H., A. Narmann and W. Heiland, 1989, Nucl. Instrum. Meth. Phys. Res. B **44**, 125.

Diehl, R.D., M. Lindroos, A. Kearsley, C.J. Barnes and D.A. King, 1985, J. Phys. C **18**, 4069.

Dodonov, A.I., Sh.N. Garin, E.S. Mashkova and V.A. Molchanov, 1984, Surf. Sci. **140**, L244.

Dodonov, A.I., E.S. Mashkova and V.A. Molchanov, 1988, Radiat. Eff. **107**, 27.

Dodonov, A.I., E.S. Mashkova and V.A. Molchanov, 1989, Radiat. Eff. and Defects in Solids **110**, 227.

Domeij, B., and K. Bjorquist, 1965, Phys. Lett. **14**, 127.

Dorozhkin, A.A., 1967, Tr. Leningr. Politekh. Inst. Fiz. Elektron. 277, p. 89.

Dorozhkin, A.A., 1985, Thesis, Leningrad.

Dorozhkin, A.A., and N.N. Petrov, 1965, Fiz. Tverd. Tela (Leningrad) **7**, 118.

Dorozhkin, A.A., and N.N. Petrov, 1983, Ionic Auger-Spectroscopy (Leningrad Pedagogical Institute Press, Leningrad) (in Russian).

Dorozhkin, A.A., A.N. Mishin and N.N. Petrov, 1974, Izv. Akad. Nauk SSSR Ser. Fiz. **38**, 249.

Dorozhkin, A.A., G.A. Krysov, A.A. Petrov and N.N. Petrov, 1978, Zh. Tekh. Fiz. **48**, 526.

Drentje, S.A., 1967, Phys. Lett. A **24**, 12.

Drozdova, N.V., and Yu.V. Martynenko, 1980, Voprosy Atom. Nauki i Tekhn., Ser. Termoyad. Sintez **1**(5), 111.

Du Commun, J.P., M. Cantagel and M. Marchal, 1974, J. Mater. Sci. **9**, 725.

Duck, P., W. Treu, H. Fröhlich, W. Galster and H. Voit, 1980, Surf. Sci. **95**, 603.

Duke, K.B., 1974, Perspectives of surface science, in: Surface Science: Recent Progress and Perspectives, eds Y.T.S. Jayadevaiah and R. Vanselov (CRC Press, Cleveland, OH) p. 344.

Dunaev, Yu.A., and I.P. Flaks, 1953, Dokl. Akad. Nauk SSSR **91**, 1.

Dunlap, B.I., J.E. Campana, B.N. Green and R.H. Beteman, 1983, J. Vac. Sci. Technol. A **1**, 432.

Dzhemilev, N.Kh., 1980, Metal surface sputtering in form of clusters under ion bombardment, in: Proc. All-Union Conf. - Symp. on Surface Diagnostics by Ion Beams (Donetsk, USSR) p. 88.

Dzhemilev, N.Kh., and R.T. Kurbanov, 1979, Izv. Akad. Nauk SSSR Ser. Fiz. **43**, 606.

Dzhemilev, N.Kh., and R.T. Kurbanov, 1984, Poverkhn. Fiz. Khim. Mekh. **1**, 56.

Dzhemilev, N.Kh., U.Kh. Rasulev and S.V. Verkhoturov, 1987, Nucl. Instrum. Meth. Phys Res. B **29**, 531.

Dzhurakhalov, A.A., E.S. Parilis and F.F. Umarov, 1987a, Zh. Tekh. Fiz. **57**, 372.

Dzhurakhalov, A.A., E.S. Parilis, N.Yu. Turaev and F.F. Umarov, 1987b, Radiotekh. & Elektron. **32**, 1019.

Dzhurakhalov, A.A., E.S. Parilis and F.F. Umarov, 1988a, Modernized algorythm and computer simulation program of ion scattering processes and recoil atoms forming on a single crystal surface, Deposited in All-Union Institute of Scientific and Technical Information (VINITI), Dep. No. 2363-B88.

Dzhurakhalov, A.A., E.S. Parilis and F.F. Umarov, 1988b, Ion scattering by atomic steps on a single crystal surface, in: Proc. All-Union Conf. - Symp. on Young Scientists (Donetsk, USSR) p. 96.

Dzhurakhalov, A.A., E.S. Parilis and F.F. Umarov, 1988c, Anomalous energy losses of ion at sliding scattering in surface semichannels, in: Proc. 5th All-Union Conf. on Secondary Ion and Ionphoton Emission (Kharkov, USSR) p. 80.

Dzhurakhalov, A.A., E.S. Parilis and F.F. Umarov, 1989a, Elastic and inelastic energy losses and trajectory peculiarity under sliding ion scattering by a single crystal surface, Preprint No. 17 of Arifov Institute of Electronics (Tashkent).

Dzhurakhalov, A.A., E.S. Parilis, N.Yu. Turaev, F.F. Umarov and I.D. Yadgarov, 1989b, Sliding ion scattering by the surface of single crystal of compound composition, in: All-Union Arifov Symp. on Atomic Particles Interaction with Solids (Fan, Tashkent) p. 32.

Dzhurakhalov, A.A., N.Yu. Turaev, F.F. Umarov and I.D. Yadgarov, 1989c, Sliding ion scattering by single crystal surface of compound structure, in: Proc. 9th All-Union Conf. on Atomic Particles Interaction with Solids (Moscow) Vol. 1, Part. 1, p. 6.

Dzhurakhalov, A.A., A.M. Rasulov and F.F. Umarov, 1989d, Determination of adatom sort and their place on single crystal surface by shooting the thin films under channeling conditions, in: Proc. 9th All-Union Conf. on Atomic Particles Interaction with Solids (Moscow) Vol. 1, Part 1, p. 49.

Dzhurakhalov, A.A., E.S. Parilis and F.F. Umarov, 1990a, Poverkhn. Fiz. Khim. Mekh. **7**, 56.

Dzhurakhalov, A.A., N.Yu. Turaev, F.F. Umarov and I.D. Yadgarov, 1990b, Izv. Akad. Nauk SSSR Ser. Fiz. **54**, 1250.

Dzhurakhalov, A.A., E.S. Parilis, A.M. Rasulov and F.F. Umarov, 1990c, Poverkhn. Fiz. Khim. Mekh. **6**, 148.

Eccles, A.J., J.A. Van den Berg, A. Brown and J.C. Vickerman, 1986, Appl. Phys. Lett. **49**, 188.

Eckstein, W., 1981, Charge fractions of reflected particles, in: Inelastic Particle-Surface Collisions, eds E. Taglauer and W. Heiland (Springer, Berlin) p. 157.

Eckstein, W., and J.P. Biersack, 1983, Z. Phys. A **310**, 1.

Eckstein, W., and F.E.P. Matschke, 1976, Phys. Rev. B **14**, 3231.

Eckstein, W., and H. Verbeek, 1978, J. Nucl. Mater. **76–77**, 365.

Eckstein, W., H.G. Schaffler and H. Verbeek, 1974, Scattering of Rare Gas Ions from Metal Surface, JPP 9/16 (Max-Planck-Institut für Plasmaphysik, Garching).

Eckstein, W., H. Verbeek and S. Datz, 1975, Appl. Phys. Lett. **27**, 527.

Eckstein, W., F.E.P. Matschke and H. Verbeek, 1976, J. Nucl. Mater. **63**, 199.

Eckstein, W., V.A. Molchanov and H. Verbeek, 1978, Nucl. Instrum. Meth. **149**, 599.

Eckstein, W., H. Verbeek and J.P. Biersack, 1980, J. Appl. Phys. **51**, 1194.

Edge, R.D., R. Sizmann and C. Varelas, 1974, Radiat. Eff. **22**, 79.

Edwards, A.K., and M.E. Rudd, 1968, Phys. Rev. **170**, 140.

Eguiluz, A., and I.I. Quinn, 1975, Phys. Lett. A **53**, 151.

Eisen, F.H., 1968, Can. J. Phys. **46**, 561.

Elber, R., and B.B. Gerber, 1983, Chem. Phys. Lett. **97**, 4.

Engelmann, G., E. Taglauer and D.P. Jackson, 1985, Surf. Sci. **162**, 921.

Engelmann, G., E. Taglauer and D.P. Jackson, 1986, Nucl. Instrum. Meth. Phys. Res. B **13**, 240.

Engelmann, G., W. Englert, S. Kato and E. Taglauer, 1987, Nucl. Instrum. Meth. Phys. Res. B **26**, 522.

Englert, W., W. Heiland, E. Taglauer and D. Menzel, 1979, Surf. Sci. **83**, 243.

Englert, W., E. Taglauer and W. Heiland, 1982, Nucl. Instrum. Methods **194**, 663.

Ens, W., B.U.R. Sundqvist, P. Håkansson, D. Fenyo, A. Hedin and G. Johnson, 1989, J. Phys. (Paris) **50**, 9.

Eremeev, M.A., 1951, Dokl. Akad. Nauk SSSR **79**, 775.

Eremeev, M.A., and M.V. Zubchaninov, 1942, Zh. Eksp. Teor. Fiz. **12**, 358.

Erginsoy, C., 1965, Phys. Rev. Lett. **15**, 360.

Erickson, R.L., and D.P. Smith, 1975, Phys. Rev. Lett. **34**, 297.

Erickson, R.L., F. Minn and D.P. Smith, 1975, Ion scattering spectrometer, in which charge exchange processes are used, US Patent No. 3920989.

Escovitz, W.H., T.R. Fox and R. Levi-Setti, 1979, IEEE Trans. Nucl. Sci. **NS-26**, 1147.

Evdokimov, I.N., 1976, Abnormal energy losses in ion scattering, in: Proc. 4th All-Union Conf. on Atomic Particles Interaction with Solids (Kharkov, USSR) Part 1, p. 89.

Evdokimov, I.N., 1982, Poverkhn. Fiz. Khim. Mekh. **7**, 48.

Evdokimov, I.N., 1983, Poverkhn. Fiz. Khim. Mekh. **3**, 74.

Evdokimov, I.N., 1984, Poverkhn. Fiz. Khim. Mekh. **8**, 31.

Evdokimov, I.N., 1985, Mechanisms of small angle ion scattering, in: Proc. All-Union Conf. - Symp. on Surface Diagnostic by Ion Beams (Uzhgorod, USSR) p. 238.

Evdokimov, I.N., 1987, Ion stimulated secondary effects in near surface region of solid, in: Proc. XXth All-Union Conf. on Emission Electronics (Kiev) Part 2, p. 163.

Evdokimov, I.N., and V.A. Molchanov, 1969, Izv. Akad. Nauk SSSR Ser. Fiz. **33**, 763.

Evdokimov, I.N., V.A. Molchanov, D.D. Odintsov and V.M. Chicherov, 1966, Fiz. Tverd. Tela (Leningrad) **8**, 2939.

Evdokimov, I.N., E.S. Mashkova and V.A. Molchanov, 1967a, Phys. Lett. A **25**, 619.

Evdokimov, I.N., E.S. Mashkova, V.A. Molchanov and D.D. Odintsov, 1967b, Fiz. Tverd. Tela (Leningrad) **9**, 407.

Evdokimov, I.N., E.S. Mashkova and V.A. Molchanov, 1967c, Fiz. Tverd. Tela (Leningrad) **9**, 6.

Evdokimov, I.N., E.S. Mashkova and V.A. Molchanov, 1969, Dokl. Akad. Nauk SSSR **186**, 549 [1969, Sov. Phys. Dokl. **14**, 467].

Evdokimov, I.N., R.P. Webb, D.G. Armour and D.S. Karpuzov, 1979, Radiat. Eff. **42**, 83.

Evdokimov, I.N., R.P. Webb and D.G. Armour, 1981, Radiat. Eff. Lett. **58**, 59.

Everhart, E., and Q.C. Kessel, 1966, Phys. Rev. **146**, 27.

Everhart, E., G. Stone and R.J. Carbone, 1955, Phys. Rev. **99**, 1287.

Evstigneev, S.A., S.N. Zvonkov and V.M. Chicherov, 1982, Zh. Eksp. Teor. Fiz. **82**, 1091 [1982, Sov. Phys. JETP **55**, 640].

Fagot, B., and C. Fert, 1964, C.R. Acad. Sci. **258**, 1180.

Fagot, B., N. Colombie and C. Fert, 1964, C.R. Acad. Sci. **258**, 4670.

Fagot, B., N. Colombie, R. Thirty and C. Fert, 1966, C.R. Acad. Sci. **262**, 173.

Fano, U., and W. Lichten, 1965, Phys. Rev. Lett. **14**, 627.

Farnsworth, M.E., and K. Hayek, 1967, Surf. Sci. **8**, 35.

Fastrup, B., G. Hermann and K.J. Smith, 1971, Phys. Rev. A **3**, 1591.

Fauster, Th., 1988, Vacuum **38**, 129.

Fedorenko, N.V., 1959, Usp. Fiz. Nauk **68**, 481.

Fedorenko, N.V., I.P. Flaks and L.G. Filippenko, 1960, Zh. Eksp. Teor. Fiz. **38**, 719.

Fehn, U., 1974, Int. J. Mass. Spectrom. Ion Phys. **15**, 391.

Fehringer, M., M. Delauney, R. Geller, P. Varga and H. Winter, 1987, Nucl. Instrum. Meth. Phys. Res. B **23**, 245.

Feijen, H.H.W., 1975, Thesis, Groningen.

Feldman, L.C., and J.W. Mayer, 1986, Fundamental of Surface and thin Film Analysis (North-Holland, Amsterdam).

Fenyö, D., B.U.R. Sundqvist, B. Karlsson and R.E. Johnson, 1989, J. Phys. (Paris) **50**, 33.

Fermi, E., 1928, Z. Phys. **48**, 73.

Ferron, J., E.V. Alonso, R.A. Baragiola and A. Oliva-Florio, 1980, J. Phys. D **14**, 1707.

Fert, C., N. Colombie, B. Fagot and Phan Van Chuong, 1962, Le bombardment ionique (Editions CNRS, Paris).

Fiermans, L., J. Vannik and V. Dekeyser, eds, 1978, Electron and Ion Spectroscopy of Solids (Plenum, New York).

Filippenko, L.G., 1962, Zh. Tekh. Fiz. **32**, 356.

Firsov, O.B., 1953, Zh. Eksp. Teor. Fiz. **24**, 280.

Firsov, O.B., 1957, Zh. Eksp. Teor. Fiz. **33**, 696 [1958, Sov. Phys. JETP **6**, 534].

Firsov, O.B., 1958, Zh. Eksp. Teor. Fiz. **34**, 447 [1958, Sov. Phys. JETP **7**, 308].

Firsov, O.B., 1959, Zh. Eksp. Teor. Fiz. **36**, 1517 [1959, Sov. Phys. JETP **9**, 1076].

Firsov, O.B., 1970, Zh. Tekh. Fiz. **40**, 83.

Firsov, O.B., E.S. Mashkova and V.A. Molchanov, 1973, Radiat. Eff. **18**, 257.

Firsov, O.B., E.S. Mashkova, V.A. Molchanov and V.A. Snisar, 1976, Nucl. Instrum. Methods **132**, 695.

Flaks, I.P., 1955, Zh. Tekh. Fiz. **25**, 2463.

Flaks, I.P., G.N. Ogurtsov and N.V. Fedorenko, 1961, Zh. Eksp. Teor. Fiz. **41**, 1438.

Fleischmann, H.H., R.C. Dehmel and S.K. Lee, 1972, Phys. Rev. A **5**, 5.

Fluit, J.M., P.K. Rol and J. Kistemaker, 1963, Physika **34**, 690.

Fogel, I.M., R.P. Slabospitsky and A.V. Rastropin, 1960a, Zh. Tekh. Fiz. **30**, 63.

Fogel, I.M., R.P. Slabospitsky and I.M. Karnaukhov, 1960b, Zh. Tekh. Fiz. **30**, 824.

Forstmann, F., and H. Stenschke, 1978, Phys. Rev. B **17**, 1489.

Forsythe, G.E., M.A. Malcolm and C.B. Moler, 1977, Computer Methods for Mathematical Computations (Prentice-Hall, Englewood Cliffs, NY) 07632.

Foster, S., and F.W. Saris, 1973, X-rays from $Ar–Cl_2$ (Caramboles), in: Proc. 7th Int. Conf. on Physics of Electronic and Atomic Collisions (Belgrade) Vol. 2, p. 716.

Francke, B., 1982, Ann. Phys. Soc. **4**, 111.

Frenkel, Ya.I., 1941, Zh. Eksp. Teor. Fiz. **11**, 706.

Fröhlich, H., 1955, Proc. Phys. Soc. B **68**, 9.

Gaipov, S., L.M. Kishinevsky, E. Koshchanov and R.R. Rakhimov, 1973, On the potential electron emission of alkali-halid crystals, in: Proc. XV All-Union Conf. on Emission Electronics (Kiev) Part 2, p. 192.

Garcia, J.D., and R.J. Fortner, 1973, J. Phys. B **6**, L174.

Gay, W.L., and D.E. Harrison Jr, 1964, Phys. Rev. A **135**, 1780.

Gaydaenko, V.I., and V.K. Nikulin, 1970, Chem. Phys. Lett. **7**, 296.

Gemmel, D., 1974, Rev. Mod. Phys. **46**, 129.

Gerasimenko, V.I., and Yu.D. Oksyuk, 1965, Zh. Eksp. Teor. Fiz. **48**, 499 [1965, Sov. Phys. JETP **21**, 333].

Gerhard, W., 1975, Z. Phys. B **22**, 31.

Gerhard, W., and H. Oechsner, 1975, Z. Phys. B **22**, 41.

Germer, L.A., and A.U. McRae, 1962, J. Appl. Phys. **33**, 2923.

Ghosh, S.N., and S.P. Khare, 1962, Phys. Rev. **125**, 1254.

Giber, J., and W.O. Hofer, 1980, Postionization experiments on cluster sputtering of solids, in: Proc. Symp. on Sputtering (Vienna, Austria) p. 697.

Gibson, J.B., A.N. Goland, M. Milgram and G.H. Vineyard, 1960, Phys. Rev. **120**, 1229.

Glauber, R.J., 1955, Phys. Rev. **100**, 242.

Glauber, R.J., 1971, Usp. Fiz. Nauk **109**, 137.

Gnasser, H., and W.O. Hofer, 1989, Appl. Phys. A **48**, 261.

Godfrey, D.J., and D.P. Woodruff, 1979, Surf. Sci. **89**, 76.

Godfrey, D.J., and D.P. Woodruff, 1981a, Surf. Sci. **105**, 438.

Godfrey, D.J., and D.P. Woodruff, 1981b, Surf. Sci. **105**, 459.

Goff, R.F., 1973, J. Vac. Sci. Technol. **10**, 335.

Goldansky, V.I., E.Ya. Lantzburg and P.A. Yampolsky, 1975, Zh. Eksp. Teor. Fiz. Pis'ma **21**, 365.

Goldstein, H., 1959, Classical Mechanics (Addison-Wesley, Reading, MA).

Gombas, P., 1949, Die Statistische Theorie des Atoms und ihre Anwendungen (Springer, London).

Gomoyunova, M.V., 1982, Usp. Fiz. Nauk **136**, 105.

Gomoyunova, M.V., S.L. Zaslavsky and I.I. Pronin, 1978a, Fiz. Tverd. Tela (Leningrad) **20**, 1586.

Gomoyunova, M.V., S.L. Zaslavsky and I.I. Pronin, 1978b, Fiz. Tverd. Tela (Leningrad) **20**, 3645.

Goodman, F.O., and H.Y. Wachman, 1976, Dynamics of Gas-Surface Scattering (Academic Press, New York).

Goodstein, D.M., R.L. McEachern and B.H. Cooper, 1989, Phys. Rev. B **39**, 13129.

Gott, Yu.V., and Yu.N. Yavlinsky, 1973, Interaction of Slow Particles with Matter and Plasma Diagnostics (Atomizdat, Moscow).

Green, T.A., 1970, Phys. Rev. **1**, 1416.

Grishin, A.N., and Yu.G. Skripka, 1975, Differential cross-section of ion scattering by atom chain, in: Proc. 6th All-Union Conf. on Physics of Interaction of Charged Particles with Single Crystals (Moscow) p. 85.

Grozdanov, T.P., and R.K. Janev, 1977, Phys. Lett. **65**, 396.

Gruich, D.D., N. Rakhimbaeva, G. Ikramov and T.U. Arifov, 1964, Izv. Akad. Nauk Uzb. SSR Ser. Fiz. Mat. Nauk **1**, 53.

Gruich, D.D., G.E. Ermakov and V.S. Grigoriev, 1975, Study of energetic accomodation coefficient under slow ions bombardment of metal, in: Ion Bombardment as a Method of Research of Feature of the Surface (Fan, Tashkent) (in Russian) p. 15.

Gruich, D.D., A.A. Dzhurakhalov, M.A. Tukhtaev and F.F. Umarov, 1990, Poverkhn. Fiz. Khim. Mekh. **5**, 42.

Guinan, M., 1974, J. Nucl. Mater. **53**, 171.

Günther, K., 1964, Ann. Phys. **14**, 296.

Gurney, R.W., 1928, Phys. Rev. **32**, 467.

Gurtovoj, M.E., 1940, Zh. Eksp. Teor. Fiz. **10**, 483.

Gvozdover, R.S., V.M. Efremenkova, L.B. Shelyakin and V.E. Yurasova, 1976, Radiat. Eff. **27**, 237.

Hagstrum, H.D., 1953a, Phys. Rev. **89**, 244.

Hagstrum, H.D., 1953b, Phys. Rev. **91**, 543.

Hagstrum, H.D., 1954a, Phys. Rev. **96**, 336.

Hagstrum, H.D., 1954b, Phys. Rev. **96**, 325.

Hagstrum, H.D., and G.E. Becker, 1973, Phys. Rev. B **8**, 107.

Hagstrum, H.D., Y. Takeshi and D.D. Pretzer, 1965, Phys. Rev. A **139**, 526.

Håkansson, P., and B. Sundqvist, 1982, Radiat. Eff. **61**, 179.

Håkansson, P., E. Jayasinghe, A. Johansson, I. Kamensky and B. Sundqvist, 1981, Phys. Rev. Lett. **47**, 1227.

Håkansson, P., I. Kamensky and B. Sundqvist, 1982, Surf. Sci. **116**, 302.

Håkansson, P., I. Kamensky, M. Salehpour, B. Sundqvist and S. Widdiyasekera, 1984, Radiat. Eff. **80**, 141.

Harrison Jr, D.E., and C.B. Delaplain, 1976, J. Appl. Phys. **47**, 2252.

Harrison Jr, D.E., C.E. Carlston and G.D. Magnuson, 1965, Phys. Rev. A **139**, 737.

Harrison Jr, D.E., W.L. Gay and H.M. Effron, 1969, J. Math. Phys. **100**, 1179.

Harrower, G.A., 1956, Phys. Rev. **102**, 340.

Hart, G., and C.B. Cooper, 1979, Surf. Sci. **82**, L283.

Hartmann, H., H.H. Brongersma and J. Walinga, 1982, Analysevorrichtung für eine Oberflachen-schicht, DE Patent No. 2458025.

Hasselkamp, D., 1988, Comments At. Mol. Phys. **21**, 241.

Hauffe, W., 1971, Phys. Status Solidi **4**, 111.

Hayashiuchi, Y., Y. Kitazoe, T. Sekiya and Y. Yamamura, 1977, J. Nucl. Mater. **71**, 181.

Hedin, A., P. Håkansson, B. Sundqvist and R.E. Johnson, 1985, Phys. Rev. B **31**, 1780.

Hedin, A., P. Håkansson and B. Sundqvist, 1987a, Nucl. Instrum. Meth. Phys. Res. B **22**, 491.

Hedin, A., P. Håkansson, M. Salehpour and B. Sundqvist, 1987b, Phys. Rev. B **35**, 7377.

Heiland, W., 1973, IPP. Rep. **9**(11), 1.

Heiland, W., 1982, Appl. Surf. Sci. **13**, 282.

Heiland, W., and E. Taglauer, 1972, J. Vac. Sci. Technol. **9**, 620.

Heiland, W., and E. Taglauer, 1973a, Bombardment induced surface damage in a nickel single crystal observed by ion scattering and LEED, in: Ion Surface Interaction, Sputtering and Related Phenomena (Gordon and Breach, London) p. 117.

Heiland, W., and E. Taglauer, 1973b, Radiat. Eff. **19**, 1.

Heiland, W., H.G. Schaffler and E. Taglauer, 1973a, Surf. Sci. **35**, 381.

Heiland, W., H.G. Schaffler and E. Taglauer, 1973b, Surface scattering of low energy ions, in: Atomic Collisions in Solids, Vol. 2 (Plenum Press, New York) p. 599.

Heiland, W., F. Iberl, E. Taglauer and D. Menzel, 1975, Surf. Sci. **53**, 383.

Heiland, W., E. Taglauer and M.T. Robinson, 1976, Nucl. Instrum. Methods **132**, 655.

Heiland, W., E. Taglauer and M. Grundner, 1978a, Ion-optical apparatus for sample surface investigation by ion bombardment and by analysis of ions, emitted from irradiated part of surface, DE Patent No. 2414221.

Heiland, W., W. Englert and E. Taglauer, 1978b, J. Vac. Sci. Technol. **15**, 419.

Heiland, W., U. Beitat and E. Taglauer, 1979, Phys. Rev. B **19**, 1677.

Heiland, W., U. Imke, S. Schubert and K.J. Snowdon, 1987, Nucl. Instrum. Meth. Phys. Res. B **27**, 167.

Helbig, H.F., M.W. Linder, G.A. Morris and S.A. Steward, 1982, Surf. Sci. **114**, 251.

Hennequin, S.-F., and P. Viaris de Lesegno, 1974, Surf. Sci. **42**, 50.

Herzberg, G., 1950, Molecular Spectra and Molecular Structure, Vol. 1, Spectra of Diatomic Molecules (Van Nostrand Reinold, New York).

Herzog, R.F.K., W.P. Poschenruder and F.G. Satkievich, 1973, Radiat. Eff. **18**, 199.

Higatsberger, M.J., H.L. Domorest and O.A. Nier, 1954, J. Appl. Phys. **25**, 883.

Higginbottom, P.R., J. Homer, D.J. O'Connor and R.J. MacDonald, 1985, Appl. Surf. Sci. **22/23**, 100.

Hill, A.G., W.W. Buechner, J.S. Clark and J.B. Fisk, 1939, Phys. Rev. **55**, 463.

Hinch, B.J., A. Lock, J.P. Toennies and G. Zhang, 1989, J. Vac. Sci. Technol. B **7**, 1260.

Hofer, W.O., 1980, Nucl. Instrum. Methods **170**, 275.

Hofer, W.O., 1990, Ion-induced Electron Emission from Solids, in: Scanning Microscopy Suppl. **4**, 1.

Hogberg, G., 1971, Phys. Status Solidi B **48**, 829.

Holloway, S., and J.W. Gadzuk, 1985, J. Chem. Phys. **82**, 5203.

Holmen, G., B. Svensson, J. Schou and P. Sigmund, 1979, Phys. Rev. B **20**, 2247.

Holmen, G., B. Svensson and A. Buren, 1981, Nucl. Instrum. Methods **185**, 523.

Hoogerbrugge, R., and P.G. Kistemaker, 1987, Nucl. Instrum. Meth. Phys. Res. B **27**, 37.

Hou, M., 1989, Vacuum **39**, 309.

Hou, M., and M.T. Robinson, 1976, Nucl. Instrum. Methods **132**, 641.

Hou, M., W. Eckstein and H. Verbeek, 1978, Radiat. Eff. **39**, 107.

Hou, M., C. Benazeth, N. Benazeth and C. Mayoral, 1986, Nucl. Instrum. Meth. Phys. Res. B **13**, 645.

Houston, J.E., G.E. Laramore and R.L. Park, 1973, Surf. Sci. **34**, 477.

Hren, J., and S. Ranganathan, eds, 1968, Field-Ion Microscopy (Plenum, New York).

Hutchence, D.K., and S. Hontzeas, 1974, Nucl Instrum. Methods **116**, 217.

Iferov, G.A., A.G. Kadmensky and A.F. Tulinov, 1972, Particles scattering by atomic chain, in: Proc. 3rd All-Union Conf. on Charged Particles Interaction with Single Crystals (Moscow State University Press, Moscow) p. 24.

Imke, U., K.J. Snowdon and W. Heiland, 1986a, Phys. Rev. B **34**, 41.

Imke, U., K.J. Snowdon and W. Heiland, 1986b, Phys. Rev. B **34**, 48.

Inghram, M.G., and R.J. Hayden, 1954, Mass Spectroscopy (Washington).

Izmajlov, S.V., 1958a, Uch. Zap. Leningr. Gos. Pedagog. Inst. **166**, 309.

Izmajlov, S.V., 1958b, Zh. Tekh. Fiz. **28**, 2209.

Izmajlov, S.V., 1959a, Fiz. Tverd. Tela (Leningrad) **1**, 1546.

Izmajlov, S.V., 1959b, Fiz. Tverd. Tela (Leningrad) **1**, 1557.

Izmajlov, S.V., 1961, Fiz. Tverd. Tela (Leningrad) **3**, 2804.

Jackson, D.P., 1974, Surf. Sci. **43**, 431.

Jackson, D.P., 1975, Computer simulation of atomic collisions, in: Atomic Collisions in Solids, Vol. 1, eds S. Datz, B.R. Appleton and C.D. Moak (Plenum, New York) p. 185.

Jackson, D.P., W. Heiland and E. Taglauer, 1981, Phys. Rev. B **24**, 4198.

Jacobs, S.D.L., and R.D. Macfarlane, 1988, Energy distribution of organic molecules desorbed under MeV ions, in: Proc. 36th ASMS Conf. on Mass Spectrometry and Allied Topics (San Francisco) p. 1205.

Jacobson, R.L., and G.K. Wehner, 1965, J. Appl. Phys. **36**, 2674.

Jahrreiss, H., 1964, Ann. Phys. **14**, 325.

Jakas, M.M., and D.E. Harrison Jr, 1985, Surf. Sci. **149**, 500.

Jarger, L.L., and J.N. Anno, 1966, J. Appl. Phys. **37**, 2929.

Jayadelvaiah, T.S., and R. Vauselow, eds, 1974, Surface Science: Recent Progress and Perspectives (CRC Press, Cleveland, OH).

Jo, Y.-S., A. Shultz, T.R. Schuler and J.W. Rabalais, 1985, J. Phys. Chem. **89**, 2113.

Johar, S.S., and D.A. Thompson, 1979, Surf. Sci. **90**, 319.

Johnson, R.E., 1988, Int. J. Mass Spectrom. and Ion Phys. **78**, 357.

Johnson, R.E., 1989, J. Phys. (Paris) **50**, 251.

Johnson, R.E., B.U.R. Sundqvist, A. Hedin and D. Fenyö, 1989, Phys. Rev. **40**, 49.

Joyes, P., 1971, J. Phys. B **22**, 31.

Kadmensky, A.G., and V.V. Samarin, 1982, Channeled particles scattering by atomic chain, in: Proc. 11th All-Union Conf. on Physics of Charged Particles Interaction with Crystals (Moscow State University Press, Moscow) p. 141.

Kadmensky, A.G., and V.V. Samarin, 1983, Temperature dependence calculation under axial channelling description and simulation, in: Proc. 12th All-Union Conf. on Physics of Charged Particles Interaction with Crystals (Moscow University Press, Moscow) p. 12.

Kadmensky, A.G., and A.F. Tulinov, 1973, Fast charged particles scattering by atomic chain, in: Proc. 4th All-Union Conf. on Physics of Charged Particles Interaction with Crystals (Moscow State University Press, Moscow) p. 46.

Kadmensky, A.G., G.A. Iferov and A.F. Tulinov, 1974, Calculation of ring-like angular distributions of particles scattered by thin single crystals, in: Proc. 5th All-Union Conf. on Physics of Charged Particles Interaction with Crystals (Moscow State University Press, Moscow) p. 45.

Kalmykov, B.M., Yu.A. Ryzhov, D.S. Strizhenov and I.I. Shkarban, 1972, Computer simulation of 10^3 to 10^4 eV energy ions scattering by polycrystal surface, in: Proc. 2nd All-Union Symp. on Atomic Particles Interaction with Solids (Moscow) p. 84.

Kamensky, I., P. Håkansson, B. Sundqvist, G.J. McNeal and R. Macfarlane, 1982, Nucl. Instrum. Methods **198**, 65.

Kaminsky, M., 1965, Atomic and Ionic Impact Phenomena on Metal Surfaces (Springer, Berlin).

Kapitza, P.L., 1923, Philos. Mag. **45**, 989.

Karashima, S., 1982, Jpn. J. Appl. Phys. **21**, 434.

Karplus, R., and Y. Yamaguchi, 1967, Nuovo Cimento **22**, 588.

Karpuzov, D.S., and V.E. Yurasova, 1971a, Izv. Akad. Nauk SSSR Ser. Fiz. **35**, 393.

Karpuzov, D.S., and V.E. Yurasova, 1971b, Phys. Status Solidi **47**, 41.

Karpuzov, D.S., V.A. Eltekov and V.E. Yurasova, 1966, Fiz. Tverd. Tela (Leningrad) **8**, 2173.

Karpuzov, D.S., V.I. Shulga and V.E. Yurasova, 1969a, Izv. Akad. Nauk SSSR Ser. Fiz. **33**, 819.

Karpuzov, D.S., V.I. Shulga and V.E. Yurasova, 1969b, Some pecularities of ion reflection from a single crystal, in: Proc. IX Intern. Conf. on Phen. in Ionized Gases (Inst. of Physics, Acad. SR Romania, Bucharest) p. 103.

Karpuzov, D.S., I.N. Evdokimov, D.G. Armour and G. Carter, 1978, Phys. Lett. A **68**, 485.

Karpuzov, D.S., E.V. Alonso, R.P. Walker, D.G. Armour and D.G. Martin, 1983, Surf. Sci. **129**, L271.

Kasi, S.R., and J.W. Rabalais, 1990, Rad. Eff. and Defects in Solids **112**, 119.

Kasi, S.R., H. Kang, C.S. Sass and J.W. Rabalais, 1989, Surf. Sci. Rep. **10**(1/2).

Katakuze, I., T. Ichihara, Y. Fujuta, T. Matsuo, T. Sakurai and H. Matsuda, 1985, Int. J. Mass Spectrom. Ion Phys. **67**, 229.

Katin, V.V., Yu.V. Martynenko and Yu.N. Yavlinsky, 1989, Zh. Tekh. Fiz. **59**, 88.

Katin, V.V., Yu.V. Martynenko and Yu.N. Yavlinsky, 1990, Poverkhnost'. Fiz. Khim. Mekh. **4**, 51.

Katz, R., 1978, Nucl. Track Detection **2**, 1.

Katz, R., and P. Kobetich, 1968, Phys. Rev. **170**, 401.

Kesmodel, L.L., and G.A. Somorjai, 1975, Phys. Rev. B **11**, 630.

Kesselmann, V.S., 1971, Zh. Tekh. Fiz. **41**, 1708.

Kimura, K., and M. Mannami, 1984, J. Phys. Soc. Jpn. **53**, 3372.

King, B.V., I.S.T. Tsong and S.H. Lin, 1987, Int. J. Mass Spectrom. Ion Phys. **78**, 341.

Kishinevsky, L.M., 1962, Izv. Akad. Nauk SSSR Ser. Fiz. **26**, 1410.

Kishinevsky, L.M., 1969, Thesis, Tashkent.

Kishinevsky, L.M., 1970, Zh. Tekh. Fiz. **40**, 2585.

Kishinevsky, L.M., 1973, Radiat. Eff. **19**, 23.

Kishinevsky, L.M., 1974a, On estimation of noble gases ion Auger-neutralization rate on alkali-haloid crystals surfaces, in: Proc. 3rd All-Union Conf. on Atomic Particles Interaction with Solids (Kiev) p. 24.

Kishinevsky, L.M., 1974b, Izv. Akad. Nauk SSSR Ser. Fiz. **38**, 392.

Kishinevsky, L.M., and E.S. Parilis, 1962, Izv. Akad. Nauk SSSR Ser. Fiz. **26**, 1409.

Kishinevsky, L.M., and E.S. Parilis, 1966, Dokl. Akad. Nauk Uzb. SSR **9**, 10.

Kishinevsky, L.M., and E.S. Parilis, 1968a, Zh. Tekh. Fiz. **38**, 760.

Kishinevsky, L.M., and E.S. Parilis, 1968b, Zh. Eksp. Teor. Fiz. **55**, 1932.

Kishinevsky, L.M., and E.S. Parilis, 1978, Potential electronic emission under Auger-neutralization of fast multiple-charged ions, in: Proc. 5th All-Union Conf. on Atomic Particles Interaction with Solids (Minsk, USSR) Part 1, p. 217.

Kishinevsky, L.M., and E.S. Parilis, 1982, Zh. Tekh. Fiz. **52**, 1290.

Kishinevsky, L.M., E.S. Parilis and V.K. Verleger, 1974, Orientational effects in charge state of atoms scattered by surface, in: Interaction of Atomic Particles with Solids (Kiev, USSR) Part 1, p. 19.

Kishinevsky, L.M., E.S. Parilis and V.K. Verleger, 1975, Orientational effects in charge state of atoms scattered by crystals, in: Proc. VIIth All-Union Conf. on Physics of Interaction of Charged Particles with Single Crystals (Moscow State University, Moscow) p. 20.

Kishinevsky, L.M., E.S. Parilis and V.K. Verleger, 1976, Radiat. Eff. **29**, 215.

Kishinevsky, L.M., B.G. Krakov and E.S. Parilis, 1983, Zh. Tekh. Fiz. **53**, 1456.

Kislovsky, L.D., ed., 1976, Solid Surface Decorating (Nauka, Moscow) (in Russian).

Kitagawa, M., and Y.-H. Ohtsuki, 1976, Phys. Rev. B **13**, 4682.

Kitazoe, Y., and Y. Yamamura, 1980, Radiat. Eff. Lett. **50**, 39.

Kitazoe, Y., N. Hiroaka and Y. Yamamura, 1981, Surf. Sci. **111**, 381.

Kitov, V.Yu., and E.S. Parilis, 1979, Zh. Tekh. Fiz. Pis'ma **5**, 1297.

Kitov, V.Yu., and E.S. Parilis, 1981a, Zh. Tekh. Fiz. Pis'ma **7**, 818.

Kitov, V.Yu., and E.S. Parilis, 1981b, Surf. Sci. **107**, 363.

Kitov, V.Yu., and E.S. Parilis, 1982a, Zh. Tekh. Fiz. Pis'ma **8**, 725.

Kitov, V.Yu., and E.S. Parilis, 1982b, Poverkhnost'. Fiz. Khim. Mekh. **6**, 19.

Kitov, V.Yu., and E.S. Parilis, 1984a, Poverkhnost'. Fiz. Khim. Mekh. **3**, 22.

Kitov, V.Yu., and E.S. Parilis, 1984b, Surf. Sci. **138**, 203.

Kivilis, V.M., E.S. Parilis and N.Yu. Turaev, 1967, Dokl. Akad. Nauk. SSSR **173**, 805 [1967, Sov. Phys. Dokl. **12**, 328].

Kivilis, V.M., E.S. Parilis and N.Yu. Turaev, 1970, Dokl. Akad. Nauk. SSSR **192**, 1259 [1970, Sov. Phys. Dokl. **15**, 587].

Kleyn, A.W., A.C. Luntz and D.J. Auerbach, 1981, Phys. Rev. Lett. **47**, 1169.

Kleyn, A.W., A.C. Luntz and D.J. Auerbach, 1982, Surf. Sci. **117**, 33.

Klinger, M.I., Ch.B. Luschik and T.V. Mashovets, 1985, Usp. Fiz. Nauk **147**, 523.

Knake, O., and I.N. Stransky, 1959, Usp. Fiz. Nauk. **68**, 261.

Kochboch, L., 1976, J. Phys. B **9**, 2269.

Kolybasov, V.M., and M.S. Marinov, 1973, Usp. Fiz. Nauk **109**, 137.

Komarov, F.F., and M.A. Kumakhov, 1973, Phys. Status Solidi B **58**, 387.

Komarov, F.F., and M.M. Temkin, 1976, Phys. Lett. **9**, L255.

Können, G.P., A. Tip and A.E. De Vries, 1975, Radiat. Eff. **26**, 23.

Konoplev, V.M., 1986, Radiat. Eff. **87**, 207.

Kopetsky, Ch.V., 1979, Vestn. Akad. Nauk SSSR **9**, 3.

Kovalyov, E.E., and O.D. Brill, 1984, Effects of fast heavy ions on biological substances, in: Issues of Biological Effects and Dosimetry of Heavy Charged Particles and High Energy Hadrons, ed. S. Yurov (Pushchino, USSR) (in Russian).

Kovalyova, E.A., and E.T. Shipatov, 1979, Izv. Vuzov., Ser. Fizika **6**, 124, Deposited in All-Union Institute of Scientific and Technical Information (VINITI), Dep. No. 1516-79.

Kovalyova, E.A., and E.T. Shipatov, 1980a, Izv. Vuzov, Ser. Fizika **3**, 141, Deposited in All-Union Institute of Scientific and Technical Information (VINITI), Dep. No. 141-80.

Kovalyova, E.A., and E.T. Shipatov, 1980b, Investigation of distribution of H^+ ion scattering by binary atomic chain from KF^+ crystal surface layer including atomic displacements, in: Proc. 10th All-Union Conf. on Physics of Charged Particles Interaction with a Single Crystal (Moscow State University Press, Moscow) p. 230.

Kovalyova, E.A., and E.T. Shipatov, 1983, Ion hyperchannelling in crystals, in: Proc. 12th All-Union Conf. on Physics of Charged Particles Interaction with Crystals (Moscow State University Press, Moscow) p. 30.

Krebs, K.H., 1968, Fortschr. Phys. **16**, 419.

Krebs, K.H., 1983, Vacuum **33**, 555.

Krueger, F.R., 1979, Surf. Sci. **86**, 246.

Kumakhov, M.A., 1975, Usp. Fiz. Nauk **115**, 427.

Kumakhov, M.A., 1979a, Zh. Tekh. Fiz. **5**, 1530.

Kumakhov, M.A., 1979b, Zh. Tekh. Fiz. **5**, 689.

Kumakhov, M.A., and F.F. Komarov, 1979, Energy Loss and Ion Ranges in Solids (Belorussuian State University Press, Minsk) (in Russian).

Kumakhov, M.A., and G. Shirmer, 1980, Atomic Collisions in Crystals (Atomizdat, Moscow) (in Russian).

Kumar, R., M.H. Mintz and J.W. Rabalais, 1984, Surf. Sci. **147**, 15.

Kurnaev, V.A., E.S. Mashkova and V.A. Molchanov, 1985, Light Ion Reflection from Solid Surface (Energoatomizdat, Moscow) (in Russian).

Kvlividze, V.A., V.A. Molchanov and E.S. Mashkova, 1964, Izv. Akad. Nauk SSSR Ser. Fiz. **28**, 1409.

Landau, L.D., and E.M. Lifshitz, 1960, Mechanics (Pergamon Press, Oxford).

Landau, L.D., and E.M. Lifshitz, 1964, Statistical Physics (Nauka, Moscow) (in Russian).

Landman, U., R.M. Hill and M. Mostoller, 1988, Phys. Rev. B **21**, 448.

Landnyt, J.P., 1984, Acad. anal. **46**, 57.

Large, L.N., 1963, Proc. Phys. Soc. London **81**, 1101.

Lee, H.-W., and Th.F. George, 1985, Surf. Sci. **159**, 214.

Lee, S.-L., and R.R. Lucchese, 1988, Surf. Sci. **193**, 486.

Legg, K.O., W.A. Metz and E.W. Thomas, 1980, Nucl. Instrum. Methods **170**, 561.

Lehmann, C., 1977, Interaction of Radiation with Solids and Elementary Defect Production (North-Holland, Amsterdam).

Leibfried, G., 1955, Physical Book Manual (Springer, Berlin) (in German).

Lenchenko, V.M., and Yu.Z. Akilov, 1972, Dokl. Akad. Nauk SSSR **203**, 315.

Lichten, W., 1980, J. Phys. Chem. **84**, 2102.

Lichtenberg, W.J., K. Bethe and H. Schmidt-Bocking, 1980, J. Phys. B **13**, 343.

Lichtenstein, V.Kh., and I.I. Tankov, 1975, Zh. Tekh. Fiz. Pis'ma **1**, 982.

Lichtenstein, V.Kh., A.E. Shabelnikova and K.L. Yasnopolsky, 1978, Radiotekh. i Elektron. **23**, 1251.

Lindhard, J., 1965, K. Dan. Vidensk. Selsk. Mat. Fys. Medd. **34**(14).

Lindhard, J., and M. Scharff, 1961, Phys. Rev. **124**, 128.

Lindhard, J., and A. Winter, 1964, K. Dan. Vidensk. Selsk. Mat. Fys. Medd. **32**, 34.

Lindhard, J., M. Scharff and H.E. Schiott, 1963, K. Dan. Vidensk. Selsk. Mat. Fys. Medd. **33**(14).

Linford, L.H., 1935, Phys. Rev. **47**, 279.

Little, F.P., 1956, Handbuch der Physik **Bd.XXI** S. 574.

Los, J., and J.J.C. Geerlings, 1990, Phys. Rep. **190**, 133.

Louchet, F., L. Viel, C. Benazeth, B. Fagot and N. Colombie, 1972, Radiat. Eff. **14**, 123.

Lucas, A.A., 1979, Phys. Rev. Lett. **43**, 1350.

Lucchese, R.R., 1987, J. Chem. Phys. **86**, 443.

Luitjens, S.B., H.R. Verbeek, A.J. Algra and A.L. Boers, 1978, Surf. Sci. **76**, 609.

Luitjens, S.B., A.J. Algra and A.L. Boers, 1979a, Surf. Sci. **80**, 566.

Luitjens, S.B., A.J. Algra and A.L. Boers, 1979b, Radiat. Eff. **42**, 259.

Luitjens, S.B., A.J. Algra and A.L. Boers, 1979c, Radiat. Eff. **42**, 253.

Luitjens, S.B., A.J. Algra, E.P.Th.M. Suurmeijer and A.L. Boers, 1980a, Surf. Sci. **99**, 631.

Luitjens, S.B., A.J. Algra, E.P.Th.M. Suurmeijer and A.L. Boers, 1980b, Surf. Sci. **99**, 652.

Luitjens, S.B., A.J. Algra, E.P.Th.M. Suurmeijer and A.L. Boers, 1980c, J. Phys. E **13**, 665.

Luitjens, S.B., A.J. Algra, E.P.Th.M. Suurmeijer and A.L. Boers, 1980d, Appl. Phys. **21**, 205.

Luitjens, S.B., A.J. Algra, E.P.Th.M. Suurmeijer and A.L. Boers, 1980e, Surf. Sci. **100**, 315.

Lukyanov, S.Yu., and V.M. Chicherov, 1973, Zh. Eksp. Teor. Fiz. Pis'ma **17**, 360.

Lyamshev, L.P., ed., 1987, Radiational Acoustics (Nauka, Moscow) (in Russian).

Ma, S.K.S., F.W. Wette and G.P. Alldredge, 1978, Surf. Sci. **78**, 598.

MacDonald, R.J., and P.J. Martin, 1981, Surf. Sci. **111**, L739.

MacDonald, R.J., and D.J. O'Connor, 1983, Surf. Sci. **124**, 433.

MacDonald, R.J., and D.J. O'Connor, 1984, Aust. J. Phys. **37**, 389.

MacDonald, R.J., L.C. Feldman and P.J. Silverman, 1985, Surf. Sci. **157**, L335.

Magnuson, G.D., and C.E. Carlston, 1963, Phys. Rev. **129**, 2409.

Magnuson, G.D., and C.E. Carlston, 1965, Phys. Rev. A **137**, 739.

Mahoney, J.F., E.S. Parilis, J. Perel and S.A. Ruatta, 1992, Nucl. Instrum. Meth. Phys. Res. B, to be published.

Mannami, M., K. Kimura, K. Nakanishi and A. Nishimura, 1986, Nucl. Instrum. Meth. Phys. Res. B **13**, 587.

Maradudin, A.A., E.W. Montroll and G.H. Weiss, 1963, Theory of Lattice Dynamics in the Harmonic Approximation (Academic Press, New York).

Marchenko, S.D., E.S. Parilis and N.Yu. Turaev, 1972, Dokl. Akad. Nauk SSSR **206**, 313.

Marchenko, S.D., E.S. Parilis and N.Yu. Turaev, 1973, Radiat. Eff. **19**, 147.

Martin, D.J., 1980, Surf. Sci. **97**, 586.

Martynenko, Yu.V., 1964, Fiz. Tverd. Tela (Leningrad) **6**, 2003 [1965, Sov. Phys. Solid State **6**, 1581].

Martynenko, Yu.V., 1966a, Fiz. Tverd. Tela (Leningrad) **8**, 637.

Martynenko, Yu.V., 1966b, Phys. Status Solidi **15**, 767.

Martynenko, Yu.V., 1973, Radiat. Eff. **20**, 211.

Martynenko, Yu.V., and Yu.N. Yavlinsky, 1988, Poverkhnost'. Fiz. Khim. Mekh. **6**, 5.

Marwick, A.D., M.W. Thompson, B.W. Farmery and J.S. Harbinson, 1972, Radiat. Eff. **15**, 195.

Mashkova, E.S., 1979, Fiz. Plazmy. **5**, 1385.

Mashkova, E.S., and V.B. Flerov, 1983a, Neon ions reflection from the Cu target at a sliding falling, in: Proc. 4th All-Union Conf. on Secondary Ion and Ion-photon Emissions (Kharkov, USSR) Part 1, p. 233.

Mashkova, E.S., and V.B. Flerov, 1983b, Poverkhnost'. Fiz. Khim. Mekh. **3**, 41.

Mashkova, E.S., and V.A. Molchanov, 1962, Dokl. Akad. Nauk SSSR **146**, 585.

Mashkova, E.S., and V.A. Molchanov, 1963, Fiz. Tverd. Tela (Leningrad) **5**, 2383.

Mashkova, E.S., and V.A. Molchanov, 1964a, Fiz. Tverd. Tela (Leningrad) **6**, 3486.

Mashkova, E.S., and V.A. Molchanov, 1964b, Fiz. Tverd. Tela (Leningrad) **6**, 3704.

Mashkova, E.S., and V.A. Molchanov, 1965, Zh. Tekh. Fiz. **35**, 575.

Mashkova, E.S., and V.A. Molchanov, 1966, Fiz. Tverd. Tela (Leningrad) **8**, 1517 [1966, Sov. Phys. Solid State **8**, 1206].

Mashkova, E.S., and V.A. Molchanov, 1967, Dokl. Akad. Nauk SSSR **172**, 813 [1967, Sov. Phys. Dokl. **112**, 133].

Mashkova, E.S., and V.A. Molchanov, 1972a, Radiat. Eff. **13**, 131.

Mashkova, E.S., and V.A. Molchanov, 1972b, Radiat. Eff. **16**, 143.

Mashkova, E.S., and V.A. Molchanov, 1972c, Radiat. Eff. **13**, 183.

Mashkova, E.S., and V.A. Molchanov, 1973, Izv. Akad. Nauk SSSR Ser. Fiz. **37**, 2568.

Mashkova, E.S., and V.A. Molchanov, 1974, Radiat. Eff. **23**, 215.

Mashkova, E.S., and V.A. Molchanov, 1975, Radiat. Eff. **25**, 33.

Mashkova, E.S., and V.A. Molchanov, 1980, Medium-energy Ion Scattering by Solid Surfaces (Atomizdat, Moscow) (in Russian).

Mashkova, E.S., and V.A. Molchanov, 1985, Scattering of Medium Energy Ions by Solid Surfaces (Elsevier, Amsterdam).

Mashkova, E.S., V.A. Molchanov and D.D. Odintsov, 1963a, Dokl. Akad. Nauk SSSR **151**, 1074.

Mashkova, E.S., V.A. Molchanov and D.D. Odintsov, 1963b, Fiz. Tverd. Tela (Leningrad) **5**, 3426.

Mashkova, E.S., V.A. Molchanov, E.S. Parilis and N.Yu. Turaev, 1965, Phys. Lett. **18**, 7.

Mashkova, E.S., V.A. Molchanov, E.S. Parilis and N.Yu. Turaev, 1966, Dokl. Akad. Nauk SSSR **166**, 330 [1966, Sov. Phys. Dokl. **11**, 52].

Mashkova, E.S., V.A. Molchanov and W. Soszka, 1967, Phys. Status Solidi **19**, 425.

Mashkova, E.S., V.A. Molchanov and Yu.G. Skripka, 1969, Phys. Lett. A **29**, 645.

Mashkova, E.S., V.A. Molchanov and Yu.G. Skripka, 1970a, Dokl. Akad. Nauk SSSR **190**, 73 [1970, Sov. Phys. Dokl. **15**, 26].

Mashkova, E.S., V.A. Molchanov and Yu.G. Skripka, 1970b, Phys. Lett. A **33**, 373.

Mashkova, E.S., V.A. Molchanov and Yu.G. Skripka, 1971, Dokl. Nauk SSSR **198**, 809 [1971, Sov. Phys. Dokl. **16**, 440].

Mashkova, E.S., V.A. Molchanov and V.I. Shulga, 1982, Zh. Tekh. Fiz. **52**, 532.

Massey, H.S.W., and E.H.S. Burhop, 1952, Electronic and Ionic Impact Phenomena (Clarendon Press, Oxford).

Matschke, F.E.P., W. Eckstein and H. Verbeek, 1977, Charge state of hydrogen backscattered from a gold single crystal, in: Proc. VII Int. Conf. on Atomic Collisions in Solids, Vol. 2 (Moscow State University Publishing House) p. 98.

Matzumura, H., and S. Furukawa, 1973, Jpn. J. Appl. Phys. **14**, 1783.

Mayer, J.W., L. Eriksson and J.A. Davies, 1970, Ion Implantation in Semiconductors (Academic Press, New York).

Maynard, G., and C. Deutsch, 1989, Radiat. Eff. and Def. in Solids **110**, 157.

McCraken, G.M., 1975, Rep. Prog. Phys. **38**, 241.

McCraken, G.M., and N.J. Freeman, 1969, J. Phys. Rev. B **2**, 661.

McDonald, J.W., D. Schneider, M.W. Clark and D. Dewitt, 1992, Phys. Rev. Lett. **68**, 2297.

McGuire, E.J., 1971, Phys. Rev. A **3**, 1801.

McGuire, E.J., 1972, Phys. Rev. A **5**, 1043.

Mckinney, J.T., and R.F. Goff, 1977, Ion-scattering spectrometer with two analysers settled in tandem, US patent No. 4058724.

Mckinney, J.T., and W. Thomas, 1978, Ion-scattering spectrometer with modified displacement, US patent No. 4107526.

McRae, A.U., 1964, Surf. Sci. **2**, 522.

Meckbach, W., G. Braunstein and N. Arista, 1975, J. Phys. B **8**, L344.

Medved, P.B., 1963, J. Appl. Phys. **34**, 3142.

Medved, P.B., and Y.E. Strausser, 1965, Kinetic ejection of electrons from solids, Adv. Electron. Electron. Phys. **21**, 101.

Medved, P.B., P. Mahadevan and J.K. Layton, 1963, Phys. Rev. **129**, 2086.

Merkle, K.L., and W. Jager, 1981, Philos. Mag. **44**, 741.

Meyer, F.W., C.C. Havener, S.H. Overbury, K.J. Snowdon and D.M. Zehner, 1987, Nucl. Instrum. Meth. Phys. Res. B **23**, 234.

Migdal, A.V., 1939, Zh. Eksp. Teor. Fiz. **9**, 1163.

Mirakhmedov, M.N., and E.S. Parilis, 1987, Emission of Auger electrons by neutralization of slow multicharged ions on metal surface, in: Proc. XX Soviet Conf. on Emission Electrons (Kiev, USSR) p. 151.

Mirakhmedov, M.N., and E.S. Parilis, 1988, Neutralization of multiple charged ions on metal surface, in: Atomic Collisions in Gases and on the Solid Surface (Fan, Tashkent) (in Russian) p. 144.

Mischler, J., M. Negre, N. Benazeth, D. Spanjaard, C. Gaubert and D. Aberdam, 1979, Surf. Sci. **82**, 453.

Mishin, A.N., and N.N. Petrov, 1975, Tr. Leningr. Politekh. Inst. Fiz. Elektron. 345, p. 3.

Miterev, A.M., I.G. Kaplan and E.A. Borisov, 1974, Khim. Vys. Energ. **8**, 537.

Moe, D.E., and O.H. Petsh, 1958, Phys. Rev. **110**, 1358.

Molchanov, V.A., and W. Soszka, 1964, Dokl. Akad. Nauk SSSR **155**, 70 [1964, Sov. Phys. Dokl. **9**, 209].

Molchanov, V.A., and W. Soszka, 1965, Zh. Eksp. Teor. Fiz. **35**, 963.

Molchanov, V.A., V.A. Snisar and V.M. Chicherov, 1969, Dokl. Akad. Nauk SSSR **184**, 74 [1969, Sov. Phys. Dokl. **14**, 43].

Molière, G., 1949, Z. Naturforsch. **2**a, 133.

Möller, I., W. Heiland and W. Unertl, 1984, Nucl. Instrum. Meth. Phys. Res. B **2**, 366.

Möller, I., H. Niehus and W. Heiland, 1986, Surf. Sci. **166**, L111.

Monreal, R., F. Flores, A. Naermann, W. Heiland and S. Schubert, 1989, Radiat. Eff. and Defects in Solids **109**, 75.

Morgan, D.V., and D. Van Vliet, 1968, Can. J. Phys. **46**, 503.

Morgan, G.H., and E. Everhart, 1962, Phys. Rev. **128**, 667.

Morgulis, N.D., 1934, Zh. Eksp. Teor. Fiz. **4**, 499.

Morgulis, N.D., 1939, Zh. Eksp. Teor. Fiz. **9**, 1484.

Morozov, S.N., D.D. Gruich and T.U. Arifov, 1979, Izv. Akad. Nauk SSSR Ser. Fiz. **43**, 612.

Morrison, S.R., 1977, The Chemical Physics of Surfaces (Plenum, New York).

Moshammer, R., R. Matthaus, K. Wien, Y. Le Beyec and G. Bolbach, 1989, Energy and angular distributions of secondary ions ejected from organic solids by MeV ions, in: Proc. Int. Conf. on Ion Formation from Organic Solids (IFOSV), Lovanger, Sweden, eds A. Hedin, B. Sundqvist and A. Benninghoven (Wiley, New York) p. 16.

Mosunov, A.S., 1983, Computer simulation of low energy ion scattering by polycrystalic surface (Preprint of Sci. Res. Calc. Centre of Acad. of Sci., USSR, Pushchino) (in Russian).

Mosunov, A.S., I.I. Mosunova, L.B. Shelyakin and V.E. Yurasova, 1978, Computer simulation of ion scattering by polycrystals, in: Proc. 5th All-Union Conf. on Atomic Particles Interaction with Solids (Minsk, USSR) Part 1, p. 144.

Mott, N.F., and H.S.W. Massey, 1965, The Theory of Atomic Collisions, 3rd Ed. (Clarendon Press, Oxford).

Mros, S., and A. Mros, 1981, Surf. Sci. **109**, 444.

Mukhamadiev, E.S., and R.R. Rakhimov, 1966, Izv. Akad. Nauk SSSR Ser. Fiz. **30**, 892.

Mukhamedov, A.A., and E.S. Parilis, 1975, Multiple scattering of heavy ions on two-atomic molecules, in: Proc. 6th All-Union Conf. on Physics of Electronic and Atomic Collisions (Tbilisi, USSR) p. 248.

Mukhamedov, A.A., and E.S. Parilis, 1981, Zh. Tekh. Fiz. Pis'ma **7**, 1474 [1981, Sov. Phys. Tech. Phys. Lett. **32**(9)].

Mukhamedov, A.A., and E.S. Parilis, 1982, Multiple heavy atom scattering on two-atomic molecule, Deposited in All-Union Institute of Scientific and Technical Information (VINITI), Dep. No. 967-82.

Mukhamedov, A.A., E.S. Parilis, N.Yu. Turaev and F.F. Umarov, 1973, Izv. Akad. Nauk SSSR Ser. Fiz. **37**, 2562.

Mukhamedov, A.A., E.S. Parilis, N.Yu. Turaev and F.F. Umarov, 1976, Izv. Akad. Nauk SSSR Ser. Fiz. **40**, 2533.

Mukhamedov, A.A., E.S. Parilis and F.F. Umarov, 1979, Izv. Akad. Nauk SSSR Ser. Fiz. **43**, 630.

Müller-Jahreis, U., 1970, Dokl. Akad. Nauk SSSR **91**, 801 [1970, Sov. Phys. Dokl. **15**, 353].

Müller-Jahreis, U., 1977, Vestn. Mosk. Univ. Ser. Fiz. **18**, 88.

Murdock, J., and G. Miller, 1955, AEC Rep. ISC 652.

Navinsek, B., 1972, Investigation of surfaces changes induced by ion bombardment, in: Proc. Int. Conf. on Physics of Ionized Gases (Belgrade, Yugoslavia) p. 221.

Nees, B., E. Nieschler, N. Bischof, H. Fröhlich, K. Riemer, W. Tiereth and H. Voit, 1984, Surf. Sci. **145**, 197.

Nelson, R.S., M.W. Thompson and H. Montgomery, 1962, Philos. Mag. **7**, 1385.

Niehus, H., 1986, Surf. Sci. **166**, L107.

Niehus, H., and E. Bauer, 1975, Surf. Sci. **47**, 222.

Niehus, H., and G. Comsa, 1984, Surf. Sci. **140**, 18.

Niehus, H., and E. Preuss, 1982, Surf. Sci. **119**, 349.

Niehus, H., W. Raunan, K. Besocke, R. Spitzl and G. Comsa, 1990, Surf. Sci. **225**, L8.

Nieschler, E., B. Ness, N. Bischof, H. Fröhlich, W. Tiereth and H. Voit, 1984a, Surf. Sci. **145**, 294.

Nieschler, E., B. Ness, N. Bischof, H. Fröhlich, W. Tiereth and H. Voit, 1984b, Radiat. Eff. **83**, 294.

Nikolaev, V.S., and I.S. Dmitriev, 1968, Phys. Lett. A **28**, 277.

Nizhnaya, S.L., and F.F. Umarov, 1985, Azimuthal anisotropies in ion scattering by semichannels on a single crystal surface, in: Proc. All-Union Conf.-Symp. on Surface Diagnostics by Ion Beams (Uzhgorod, USSR) p. 243.

Nizhnaya, S.L., E.S. Parilis and N.Yu. Turaev, 1976, Ion scattring by atomic chains on a single crystal surface, in: Proc. 9th All-Union Conf. on Atomic Particles Interaction with Solids (Kharkov, USSR) Part 1, p. 22.

Nizhnaya, S.L., E.S. Parilis and N.Yu. Turaev, 1979a, Radiat. Eff. **40**, 43.

Nizhnaya, S.L., E.S. Parilis and V.K. Verleger, 1979b, Izv. Akad. Nauk SSSR Ser. Fiz. **43**, 1906.

Nizhnaya, S.L., E.S. Parilis and V.K. Verleger, 1979c, Radiat. Eff. **40**, 23.

Nizhnaya, S.L., E.S. Parilis and N.Yu. Turaev, 1979d, Scattering of atoms by surface of well regulating single crystal alloys, in: Proc. All-Union Arifov Symp. on Atomic Particles Interaction with Solids (Fan, Tashkent) p. 40.

Nizhnaya, S.L., E.S. Parilis and V.K. Verleger, 1982a, Radiat. Eff. **62**, 173.

Nizhnaya, S.L., E.S. Parilis and V.K. Verleger, 1982b, Poverkhnost'. Fiz. Khim. Mekh. **4**, 72.

Nizhnaya, S.L., E.S. Parilis and F.F. Umarov, 1984, Azimuthal orientational effect in intensity of atoms scattering by semichannels on a single crystal surface, in: Proc. 7th All-Union Conf. on Atomic Particles Interaction with Solids (Minsk, USSR) p. 36.

Nizhnaya, S.L., U.Kh. Rasulev and V.K. Verleger, 1988, Poverkhnost'. Fiz. Khim. Mekh. **12**, 100.

Nizhnaya, S.L., L.M. Nepomnyashchy, U.Kh. Rasulev and V.K. Verleger, 1991, Poverkhnost'. Fiz. Khim. Mekh. **9**, 137.

Nobes, M.J., J.S. Colligon and G. Carter, 1969, J. Mater. Sci. **4**, 730.

Northcliffe, L.C., and R.F. Schilling, 1970, Nucl. Data Tables **7**, 233.

Novacek, P., A. Liegl, M. Borrell, W. Hofer and P. Varga, 1990, Vacuum **40**, 113.

O'Connor, D.J., and J.P. Biersack, 1986, Nucl. Instrum. Meth. Phys. Res. B **15**, 14.

O'Connor, D.J., and R.J. MacDonald, 1980a, Radiat. Eff. **45**, 205.

O'Connor, D.J., and R.J. MacDonald, 1980b, Nucl. Instrum. Methods **170**, 495.

O'Connor, D.J., R.J. MacDonald, W. Eckstein and P.R. Higginbottom, 1986, Nucl. Instrum. Meth. Phys. Res. B **13**, 235.

O'Connor, D.J., I.G. Shen, I.M. Wilson and R.J. MacDonald, 1988, Surf. Sci. **197**, 277.

Odintsov, D.D., 1963, Fiz. Tverd. Tela (Leningrad) **5**, 1114.

Odintsov, D.D., 1964, Izv. Akad. Nauk SSSR Ser. Fiz. **28**, 1427.

Oechsner, H., 1976, Phys. Rev. B **17**, 1052.

Oen, O.S., 1965, Phys. Lett. **19**, 358.

Oen, O.S., 1982, Nucl. Instrum. Methods **194**, 87.

Oen, O.S., and M.T. Robinson, 1976a, J. Nucl. Mater. **63**, 210.

Oen, O.S., and M.T. Robinson, 1976b, Nucl. Instrum. Methods **132**, 647.

Oen, O.S., and M.T. Robinson, 1976c, Nucl. Instrum. Methods **132**, 672.

Oen, O.S., D.K. Holmes and M.T. Robinson, 1963, J. Appl. Phys. **34**, 302.

Ogurtsov, G.N., 1972, Rev. Mod. Phys. **44**, 1.

Ohtsuki, Y.-H., 1983, Charged Beam Interaction with Solids (Taylor and Francis, London).

Okorokov, V.V., 1965, Pis'ma Zh. Eksp. Teor. Fiz. **2**, 175.

Ollerhead, R.W., J. Bottiger and J.A. Davies, 1981, Radiat. Eff. **49**, 203.

Onsgaard, J., W. Heiland and E. Taglauer, 1980, Surf. Sci. **99**, 112.

Orrman-Rossiter, K.G., A.H. Al-Bayati and D.G. Armour, 1990, Surf. Sci. **225**, 341.

Oshima, C., M. Aono, T. Tanaka, S. Kawai, S. Zaima and S. Shibata, 1981, Surf. Sci. **102**, 312.

Oura, K., H. Ugawa and T. Hanawa, 1988, J. Appl. Phys. **64**, 1795.

Overbosch, E.G., B. Rasser, A.D. Tenner and J. Los, 1980, Surf. Sci. **92**, 310.

Overbury, S.H., 1981, Surf. Sci. **112**, 23.

Overbury, S.H., and D.R. Millins, 1989, J. Vac. Sci. Technol. A **7**, 1942.

Overbury, S.H., W. Heiland, D.M. Zehner, S. Datz and R.S. Thoe, 1981, Surf. Sci. **109**, 239.

Pabst, H.J., 1975, Radiat. Eff. **24**, 233.

Palmer, G.H., 1959, Adv. Mass Spectrosc. **1**, 96 (London).

Panin, B.V., 1961, Zh. Eksp. Teor. Fiz. **41**, 3.

Panin, B.V., 1962, Zh. Eksp. Teor. Fiz. **42**, 313 [1962, Sov. Phys. JETP **14**, 1].

Parilis, E.S., 1962, Radiotekh. i Electron. **7**, 1979.

Parilis, E.S., 1963, Thesis, Tashkent.

Parilis, E.S., 1965, On the theory of reflection of ions from solid surfaces, in: Proc. VII Int. Conf. on Phen. in Ionized Gases (Belgrade) p. 129.

Parilis, E.S., 1967, Preprint by Joint Institute on Nuclear Res., Dubna P7-3355.

Parilis, E.S., 1968, Kinetic electron emission and fast ion bombardment, in: A Survey of Phenomena in Ionized Gases. Invited Papers (IAEA, Vienna) p. 309.

Parilis, E.S., 1969a, Auger Effect (Fan, Tashkent) (in Russian) [English translation ORNL-tr-1814].

Parilis, E.S., 1969b, A mechanism for sputtering of nonmetals by slow multiple charged, in: Proc. IX Int. Conf. on Phen. in Ionized Gases (Bucharest) p. 94.

Parilis, E.S., 1969c, Multiple ion scattering on molecular atoms, in: Proc. 4th All-Union Conf. on Physics of Electronic and Atomic Collisions (Riga, USSR) p. 103.

Parilis, E.S., 1970, A mechnism for sputtering of non-metals by slow multiply charged ions, in: Atomic Collisions in Solids, ed. D.W. Palmer (North-Holland, Amsterdam) p. 324.

Parilis, E.S., 1972, On the theory of interaction of multiply charged ions with matter, in: Atomic Particles Interaction in Matter and Solid Surface (Fan, Tashkent) (in Russian) p. 35.

Parilis, E.S., 1973, Izv. Akad. Nauk SSSR Ser. Fiz. **37**, 2565.

Parilis, E.S., 1980a, Charge state of light ions scattered by solid surface, in: Proc. VII Int. Conf. on Atomic Collissions in Solids, Vol. 2 (Moscow, 1977), p. 96.

Parilis, E.S., 1980b, Interaction of multiply charged ions with solid surface, in: Proc. Symp. on Sputtering (Perchtoldsdorf, Vienna) p. 668.

Parilis, E.S., 1991, Z. Phys. D Suppl. **21**, S127.

Parilis, E.S., and L.M. Kishinevsky, 1961, Fiz. Tverd. Tela (Leningrad) **3**, 1219.

Parilis, E.S., and N.Yu. Turaev, 1964, Dokl. Akad. Nauk USSR **12**, 16.

Parilis, E.S., and N.Yu. Turaev, 1965, Dokl. Akad. Nauk SSSR **161**, 84.

Parilis, E.S., and N.Yu. Turaev, 1966, Izv. Akad. Nauk SSSR Ser. Fiz. **30**, 1983.

Parilis, E.S., and N.Yu. Turaev, 1972, On theory of ion scattering by a single crystal, in: Atomic Particles Interaction in Substance and on Solid Surface (Fan, Tashkent) (in Russian) p. 21.

Parilis, E.S., and V.K. Verleger, 1973, Charge state of ions scattered by metal surface, in: Proc. XV All-Union Conf. on Emission Electronics, Part 2 (Kiev, USSR) p. 131.

Parilis, E.S., and V.K. Verleger, 1979a, Izv. Akad. Nauk SSSR Ser. Fiz. **43**, 560.

Parilis, E.S., and V.K. Verleger, 1979b, Captures and losses of electrons out of metal surface under fast particles scattering, in: Proc. All-Union Arifov Symp. on Atomic Particles Interaction with Solids (Tashkent) p. 46.

Parilis, E.S., and V.K. Verleger, 1980, J. Nucl. Mater. **93/94**, 512.

Parilis, E.S., and V.K. Verleger, 1983, Connection of reflected atoms charge state with scattering characteristic and solid surface structure, in: Proc. Rep. Symp. on Surface Diagnostics by Ion Beams (Zaporozhje, USSR), p. 112.

Parilis, E.S., and V.K. Verleger, 1985a, Poverkhnost'. Fiz. Khim. Mekh. **6**, 17.

Parilis, E.S., and V.K. Verleger, 1985b, Poverkhnost'. Fiz. Khim. Mekh. **6**, 26.

Parilis, E.S., and V.K. Verleger, 1985c, Poverkhnost'. Fiz. Khim. Mekh. **7**, 13.

Parilis, E.S., and V.K. Verleger, 1985d, Poverkhnost'. Fiz. Khim. Mekh. **7**, 21.

Parilis, E.S., N.Yu. Turaev and V.M. Kivilis, 1967, Influence of multiple scattering and the thermal vibrations upon the reflection of ions from single crystal surface, in: Proc. 8th Int. Conf. on Phen. in Ionized Gases (Vienna) p. 47.

Parilis, E.S., N.Yu. Turaev and V.M. Kivilis, 1969, Simulation of ion scattering by surfaces semichannels of a single crystal, in: Proc. IX Int. Conf. on Phen. in Ionized Gases (Bucharest) p. 105.

Parilis, E.S., N.Yu. Turaev and V.M. Kivilis, 1970, Radiotekh. i Elektron. **15**, 214.

Parilis, E.S., N.Yu. Turaev and F.F. Umarov, 1975, Radiat. Eff. **24**, 207.

Parilis, E.S., N.Yu. Turaev, F.F. Umarov and S.L. Nizhnaya, 1987, The Theory of Medium Energy Atoms Scattering by Solid Surface (Fan, Tashkent) (in Russian).

Parilis, E.S., L.M. Kishinevsky, I.S. Bitensky, B.E. Baklitzky, I.A. Wojciekhovsky, Ya.S. Gilenko, T.Z. Kalanov, V.Yu. Kitov, M.N. Mizakhmedov, S.L. Nizhnaya, F.F. Umarov and V.K. Verleger, 1988, Atomic Collisions in Gases and on Solid Surface (Fan, Tashkent) (in Russian).

Parilis, E.S., F.F. Umarov, A.A. Dzhurakhalov and A.M. Rasulov, 1989a, Chain effect for inverse mass ratio of colliding particles, in: Proc. All-Union Arifov Symp. (Fan, Tashkent) p. 36.

Parilis, E.S., L.M. Kishinevsky, V.I. Matveev and B.G. Krakov, 1989b, Auger Processes at Atomic Collisions (Fan, Tashkent) (in Russian).

Partensky, M.B., 1979, Usp. Fiz. Nauk **128**, 69.

Pathak, A.P., 1974, J. Phys. F **4**, 1883.

Perdrix, M., S. Poletto, R. Goutte and C. Guillaud, 1968, Phys. Lett. A **28**, 534.

Petrov, N.N., 1960a, Fiz. Tverd. Tela (Leningrad) **2**, 949.

Petrov, N.N., 1960b, Fiz. Tverd. Tela (Leningrad) **2**, 1300.

Petrov, N.N., 1962, Izv. Akad. Nauk SSSR Ser. Fiz. **26**, 1327.

Petrov, N.N., and I.A. Abroyan, 1977, Surface Diagnostics Using Ion Beams (Leningrad State University Press, Leningrad).

Philbert, G., 1953, C.R. Acad. Sci. **237**, 882.

Pincherle, L., 1955, Proc. Phys. Soc. London Sect. B **68**, 319.

Pleshivtsev, N.V., 1968, Cathode Sputtering (Atomizdat, Moscow) (in Russian).

Ploch, W., 1951, Z. Phys. **130**, 174.

Poate, J.M., and T.M. Buck, 1976, Ion-Scattering Spectroscopy in Experimental Methods in Catalitical Research, Vol. 3 (Academic Press, New York) p. 175.

Poelsema, B., and A.L. Boers, 1978, Radiat. Eff. **41**, 229.

Poelsema, B., L.K. Verhey and A.L. Boers, 1975, Surf. Sci. **47**, 256.

Poelsema, B., L.K. Verhey and A.L. Boers, 1976a, Surf. Sci. **56**, 445.

Poelsema, B., L.K. Verhey and A.L. Boers, 1976b, Surf. Sci. **60**, 485.

Poelsema, B., L.K. Verhey and A.L. Boers, 1977a, Surf. Sci. **64**, 554.

Poelsema, B., L.K. Verhey and A.L. Boers, 1977b, Surf. Sci. **64**, 537.

Poelsema, B., L.K. Verhey and A.L. Boers, 1983, Surf. Sci. **133**, 344.

Powell, R.A., 1978, J. Vac. Sci. Technol. **15**, 1797.

Pradal, F., and R. Simon, 1958, C.R. Acad. Sci. **247**, 438.

Predvoditelev, A.A., and V.N. Opekunov, 1977, Fiz. Khim. Obrab. Mater. **5**, 44.

Preuss, E., 1978, Radiat. Eff. **38**, 151.

Prigge, S., H. Niehus and E. Bauer, 1977, Surf. Sci. **65**, 141.

Priggemeyer, S., A. Brockmeyer, H. Dotsch, H. Koschmieder, D.J. O'Connor and W. Heiland, 1990, Appl. Surf. Sci. **44**, 255.

Pronko, P.P., B.R. Appleton, O.W. Holland and S.R. Wilson, 1979, Phys. Rev. Lett. **43**, 779.

Propst, P.M., 1963, Phys. Rev. **129**, 7.

Pugachova, T.S., 1978, Investigation of different multiplacy collisions contribution in ion scattering by polycrystal surface, in: Proc. 17th All-Union Conf. on Emission Electronics (Leningrad) p. 479.

Pugachova, T.S., and Z.M. Khakimov, 1979, Researches in ion scattering by polycrystal by statistic trial method, in: Proc. All-Union Arifov Symp. (Fan, Tashkent) p. 24.

Pugachova, T.S., E.I. Levina and I.A. Yanovsky, 1975, Izv. Akad. Nauk Uzb. SSR Ser. Fiz. Mat. Nauk **4**, 40.

Rabalais, J.W., 1988, CRC Crystal Reviews in Solid State and Material Sciences **14**, 319.

Rabalais, J.W., O. Grizzi, M. Shi and H. Bu, 1989, Phys. Rev. Lett. **63**, 51.

Radzhabov, Sh.S., and R.R. Rakhimov, 1985, Izv. Akad. Nauk SSSR Ser. Fiz. **49**, 1812.

Radzhabov, Sh.S., R.R. Rakhimov and D. Abdusalamov, 1976, Izv. Akad. Nauk SSSR Ser. Fiz. **40**, 2543.

Rasser, B., and M. Remy, 1980, Surf. Sci. **93**, 223.

Remizovich, V.S., and Sh.A. Shakhmametyev, 1983, Poverkhnost'. Fiz. Khim. Mekh. **7**, 12.

Remizovich, V.S., M.I. Ryazanov and I.S. Tilinin, 1980a, Dokl. Akad. Nauk SSSR **251**, 848.

Remizovich, V.S., M.I. Ryazanov and I.S. Tilinin, 1980b, Dokl. Akad. Nauk SSSR **254**, 616.

Remizovich, V.S., M.I. Ryazanov and I.S. Tilinin, 1980c, Zh. Eksp. Teor. Fiz. **79**, 448.

Robinson, J.E., K.K. Kwok and D.A. Thompson, 1976, Nucl. Instrum. Methods **132**, 667.

Robinson, M.T., 1963, Tables of classical scattering integrals for Bohr, Born–Mayer and Thomas–Fermi potentials, Report ORNL-3493, Oak Ridge National Laboratory.

Robinson, M.T., 1981, Theoretical aspects of monocrystal sputtering, in: Sputtering by particle bombardment, Vol. I, ed. R. Behrish (Springer, Berlin) p. 73.

Robinson, M.T., and O.S. Oen, 1963, Phys. Rev. **132**, 2385.

Robinson, M.T., and I.M. Torrens, 1974, Phys. Rev. B **9**, 5008.

Rol, P.K., J.M. Fluit and J. Kistemaker, 1960, Physika **26**, 1009.

Ronchi, C., 1973, Appl. Phys. **44**, 3575.

Rostagni, A., 1934, Nuovo Cimento **8, 11**, 99.

Rudd, M.E., T.Jr. Jorquensen and D.J. Volz, 1966, Phys. Rev. **151**, 28.

Russek, A., 1961, Phys. Rev. **132**, 246.

Russek, A., and M. Thomas, 1958, Phys. Rev. **109**, 2015.

Rutherford, E., 1905, Philos. Mag. **10**, 193.

Ryabov, V.A., 1968, Fiz. Tverd. Tela (Leningrad) **10**, 3436.

Ryazanov, M.I., and I.S. Tilinin, 1985, Surface Structure Analysis Using Backscattering of Particles (Energoatomizdat, Moscow) (in Russian).

Ryzhik, J.M., and J.M. Gradstein, 1951, Tables of Integrals, Rows and Sums and Products (GITTL, Moscow-Leningrad) (in Russian).

Saitoh, M., F. Shoji, K. Oura and T. Hanawa, 1980, Jpn. J. Appl. Phys. **19**, L421.

Salehpour, M., P. Håkansson and B. Sundqvist, 1984, Nucl. Instrum. Meth. Phys. Res. B **2**, 752.

Salehpour, M., P. Håkansson, B. Sundqvist and S. Widdiyasekera, 1986, Nucl. Instrum. Meth. Phys. Res. B **13**, 278.

Sass, C.S., and J.W. Rabalais, 1988, J. Chem. Phys. **89**, 3870.

Säve, G., P. Håkansson and B. Sundqvist, 1987a, Nucl. Instrum. Meth. Phys. Res. B **26**, 259.

Säve, G., P. Håkansson, B. Sundqvist, R.E. Johnson, E. Soderstrom, S.-E. Lunddqvist and J. Berg, 1987b, Appl. Phys. Lett. B **51**, 1379.

Säve, G., P. Håkansson, B. Sundqvist, E. Soderstrom, S.-E. Lunddqvist and J. Berg, 1987c, Int. J. Mass Spectrom. Ion Phys. **78**, 259.

Scharkert, P., 1966, Z. Phys. **197**, 32.

Schmieder, R.W., and R.J. Bastasz, 1989, Nucl. Instrum. Meth. Phys. Res. B **43**, 318.

Schneider, P.J., W. Eckstein and H. Verbeek, 1982, J. Nucl. Mater. **111-112**, 795.

Schou, J., 1980a, Phys. Rev. B **22**, 2141.

Schou, J., 1980b, Nucl. Instrum. Methods **170**, 317.

Schram, B.L., A.J.H. Boerboom, W. Klein and J. Kistemaker, 1966, Physica **32**, 749.

Schroeer, J.M., T.N. Rhodin and R.C. Bradley, 1973, Surf. Sci. **34**, 571.

Schubert, S., J. Neumann, U. Imke, K.J. Shouldon, P. Varga and W. Heiland, 1986, Surf. Sci. **171**, L375.

Schubert, S., U. Imke and W. Heiland, 1989, Surf. Sci. **219**, L576.

Schuster, M., and C. Varelas, 1985, Nucl. Instrum. Meth. Phys. Res. B **9**, 145.

Seiberling, L.E., J.E. Griffith and T.A. Tombrello, 1980, Radiat. Eff. **52**, 201.

Seitz, F., 1940, The Modern Theory on Solids (McGraw-Hill, New York).

Seitz, F., 1949, Discuss. Faraday Soc. **5**, 271.

Shamir, N., D.A. Baldwin, T. Darko, J.W. Rabalais and P. Hochmann, 1982, J. Chem. Phys. **76**, 6417.

Sharma, S.P., 1975, Surf. Sci. **49**, 106.

Sharma, S.P., and T.M. Buck, 1975, J. Vac. Sci. Technol. **12**, 468.

Shekhter, Sh.Sh., 1937, Zh. Eksp. Teor. Fiz. **7**, 750.

Shelef, M., M.A.Z. Wheeler and H.C. Yao, 1975, Surf. Sci. **47**, 697.

Shi, M., J.W. Rabalais and V.A. Esaulov, 1989, Radiat. Eff. **109**, 81.

Shkarban, I.I., 1978, Angular dependences of energy accomodation coefficient under noble gas interaction with metal surface, in: Proc. 5th All-Union Conf. on Atomic Interaction with Solids (Minsk, USSR) p. 148.

Shoji, F., K. Kashihara, K. Oura and T. Hanawa, 1989, Surf. Sci. **220**, L719.

Shulga, V.I., 1975, Radiat. Eff. **26**, 61.

Shulga, V.I., 1976a, Ion beam focusing by pair of atomic chains and such effect appearance under ion scattering by crystal, in: Proc. 7th All-Union Conf. on Charged Particles Interaction with a Single Crystal (Moscow State University Press, Moscow) p. 69.

Shulga, V.I., 1976b, Two-atomic focusing of ion beam under their scattering by single crystal surface, in: Proc. 4th All-Union Conf. on Atomic Particles Interaction with Solids (Kharkov, USSR) Part 1, p. 73.

Shulga, V.I., 1977, The angular distribution of ions scattered by single crystal surface semichannels, in: Proc. VII Int. Conf. on Atomic Collisions in Solids (Moscow) p. 128.

Shulga, V.I., 1978, Radiat. Eff. **37**, 1.

Shulga, V.I., 1980, Radiat Eff. **51**, 1.

Shulga, V.I., 1982, Zh. Tekh. Fiz. **52**, 534.

Shulga, V.I., 1983a, Poverkhnost'. Fiz. Khim. Mekh. **3**, 74.

Shulga, V.I., 1983b, Poverkhnost'. Fiz. Khim. Mekh. **9**, 40.

Shulga, V.I., 1985, Zh. Tekh. Fiz. **55**, 2027.

Shulga, V.I., 1986a, Phys. Status Solidi **134**, 87.

Shulga, V.I., 1986b, Radiat. Eff. **100**, 71.

Shulga, V.I., 1990a, Poverkhnost'. Fiz. Khim. Mekh. **1**, 108.

Shulga, V.I., 1990b, Poverkhnost'. Fiz. Khim. Mekh. **7**, 33.

Shulga, V.I., and V.E. Yurasova, 1971, Fiz. Tverd. Tela (Leningrad) **13**, 2224 [1972, Sov. Phys. Solid State **13**, 1867].

Shulga, V.I., I.G. Bunin, V.E. Yurasova, V.V. Andreev and B.M. Mamaev, 1971, Phys. Lett. A **37**, 181.

Shulga, V.I., M. Vicanek and P. Sigmund, 1989, Phys. Rev. A **39**, 3360.

Shultz, I.A., C.R. Blakley, M.H. Mintz and J.W. Rabalais, 1986, Nucl. Instrum. Meth. Phys. Res. B **14**, 500.

Sigmund, P., 1969, Phys. Rev. **184**, 383.

Sigmund, P., 1972a, Rev. Roum. Phys. **17**, 823.

Sigmund, P., 1972b, Rev. Roum. Phys. **17**, 969.

Sigmund, P., 1972c, Rev. Roum. Phys. **17**, 1079.

Sigmund, P., 1974, Appl. Phys. Lett. **25**, 169.

Sigmund, P., 1987, Nucl. Instrum. Methods Phys. Res. B **27**, 1.

Sigmund, P., 1989, J. Vac. Sci. Technol. A **7**, 585.

Sigmund, P., and P. Vajda, 1964, Risø Report No. 83.

Sizmann, R., and C. Varelas, 1976, Nucl. Instrum. Methods **132**, 633.

Slater, J.C., 1930, Phys. Rev. **36**, 57.

Slodzian, G., 1958, C.R. Acad. Sci. **246**, 3631.

Smirnov, B.M., 1968, Atomic Collisions and Elementary Processes in Plasma (Atomizdat, Moscow) (in Russian).

Smith, D.P., 1967, Appl. Phys. **38**, 340.

Smith, D.P., 1969, Low energy ion scattering apparatus and method for analysing the surface of a
     solid, US Patent N 3480774.
Snoek, C., W.F. Van der Weg, R. Geballe and P.K. Rol, 1966, Study of collison phenomena at Copper
     surfaces, in: Proc. 7th Int. Conf. on Phen. in Ionized Gases, Vol. 1 (Belgrade, 1965) p. 145.
Snowdon, K.J., 1988, Nucl. Instrum. Meth. Phys. Res. B **34**, 309.
Snowdon, K.J., D.J. O'Connor and R.J. MacDonald, 1989a, Radiat. Eff. and Defects in Solids **109**,
     25.
Snowdon, K.J., D.J. O'Connor and R.J. MacDonald, 1989b, Radiat. Eff and Defects in Solids **109**,
     33.
Snowdon, K.J., D.J. O'Connor and R.J. MacDonald, 1989c, Surf. Sci. **221**, 465.
Snowdon, K.J., D.J. O'Connor, M. Kato and R.J. MacDonald, 1989d, Surf. Sci. **222**, L871.
Sommeria, J., 1954, J. Phys. Radiat. **15**, 1126.
Sommermeyer, K., 1936, Ann. Phys. **25**, 481.
Sonthern, A.L., W.R. Willis and M.T. Robinson, 1963, J. Appl. Phys. **34**, 153.
Sosnovsky, L., 1957, Izv. Akad. Nauk SSSR Ser. Fiz. **21**, 70.
Soszka, W., 1981, Phys. Lett. A **83**, 364.
Souda, R., M. Aono, C. Oshima, S. Otani and Y. Ishizawa, 1983, Surf. Sci. **128**, L236.
Spitzl, R., H. Niehus and G. Comsa, 1990, Rev. Sci. Instrum. **61**, 760.
Standing, K.G., B.F. Chait and W. Ens, 1982, Nucl. Instrum. Methods **198**, 33.
Stanyukovich, K.P., ed., 1975, Physics of Explosion (Nauka, Moscow) (in Russian).
Staudenmaier, G., 1972, Radiat. Eff. **13**, 87.
Stensgaard, I., 1983, Surf. Sci. **128**, 281.
Stensgaard, I., and F. Jakobson, 1985, Phys. Rev. Lett. **54**, 711.
Sternberg, D., 1957, Thesis, Dept. of Phys. (Columbia University).
Sternglass, E.I., 1957, Phys. Rev. **108**, 1.
Strizhenov, D.S., Yu.A. Ryzhov and B.M. Kalmykov, 1971, Izv. Akad. Nauk SSSR Ser. Fiz. **35**, 398.
Sundqvist, B., I. Kamensky, P. Håkansson and J. Kjellberg, 1984, Biomed. Mass Spectrom. **11**, 242.
Sundqvist, B., A. Hedin, P. Håkansson, M. Salehpour and G. Säve, 1986, Nucl. Instrum. Meth. Phys.
     Res. B **14**, 429.
Suurmeijer, E.P.Th.M., and A.L. Boers, 1973, Surf. Sci. **43**, 309.
Svensson, B., and G. Holmen, 1981, J. Appl. Phys. **52**, 6928.
Svensson, B., and G. Holmen, 1982, Phys. Rev. B **25**, 3056.
Svensson, B., G. Holmen and A. Buren, 1981, Phys. Rev. B **24**, 3749.
Tabor, D., and J.M. Wilson, 1972, J. Vac. Sci. Technol. **9**, 695.
Taglauer, E., 1982, Surf. Sci. **13**, 80.
Taglauer, E., 1985, Appl. Phys. **38**, 161.
Taglauer, E., and W. Heiland, 1972, Surf. Sci. **33**, 23.
Taglauer, E., and W. Heiland, 1974, Appl. Phys. Lett. **24**, 437.
Taglauer, E., and W. Heiland, 1975, Surf. Sci. **47**, 234.
Taglauer, E., and W. Heiland, 1976, Appl. Phys. **9**, 261.
Taglauer, E., and W. Heiland, 1980, Ion scattering spectroscopy, in: Applied Surface Analysis, eds
     T.L. Barr and L.E. Davis (American Society of Testing and Materials, New York) p. 111.
Taglauer, E., W. Englert and W. Heiland, 1980, Phys. Rev. Lett. **45**, 740.
Taglauer, E., M. Beckschulte, R. Margraf and D. Mehl, 1988, Nucl. Instrum. Methods Phys. Res. B
     35, 404.
Takeuchi, W., and Y. Yamamura, 1985, Surf. Sci. **154**, 1.
Telkovsky, V.G., 1956, Izv. Akad. Nauk SSSR, Ser. Fiz. **20**, 1179.
Teplov, S.V., 1990, Poverkhnost'. Fiz. Khim. Mekh. **8**, 31.
Teplova, Ya.A., V.S. Nikolaev, I.S. Dmitriev and L.N. Fateeva, 1962, Zh. Eksp. Teor. Fiz. **24**, 44.
Terzic, I., D. Ciric and B. Perovic, 1976, Zh. Tekh. Fiz. Pis'ma **24**, 492.
Terzic, I., D. Ciric and B. Perovic, 1979a, Surf. Sci. **85**, 149.
Terzic, I., N. Neskovic and D. Ciric, 1979b, Surf. Sci. **88**, L71.

Thomas, E.W., 1984, Vacuum **34**, 1031.

Thomas, L.H., 1927, Proc. Cambridge Philos. Soc. **23**, 542.

Thompson, D.A., 1981, Radiat. Eff. **56**, 105.

Thompson, D.A., and S.S. Johar, 1979, Appl. Phys. Lett. **34**, 342.

Thompson, M.W., 1968, Contemp. Phys. **9**, 375.

Thompson, M.W., 1969, Defects and Radiation Damage in Metals (Cambridge University Press, Cambridge).

Thompson, M.W., and H.J. Pabst, 1978, Radiat. Eff. **37**, 105.

Thum, F., and W.O. Hofer, 1979, Surf. Sci. **90**, 331.

Tilinin, I.S., 1983, Poverkhnost'. Fiz. Khim. Mekh. **3**, 10.

Toennies, J.P., 1974, Appl. Phys. **3**, 91.

Tongson, L.L., and C.B. Cooper, 1975, Surf. Sci. **52**, 263.

Torgerson, D.F., R.P. Skowronsky and R.D. Macfarlane, 1974, Biochem. Biophys. Res. Commun. **60**, 616.

Torrens, I.M., 1972, Interaction Potentials (Academic Press, New York).

Tracy, J., 1972, Electron Emission Spectra (Boston, MA).

Treitz, N., 1977, J. Phys. E **10**, 573.

Trubnikov, B.A., and Yu.N. Yavlinsky, 1967, Zh. Eksp. Teor. Fiz. **52**, 1638.

Tulinov, A.F., V.S. Kulikauskas and M.M. Malov, 1965a, Dokl. Akad. Nauk SSSR **162**, 546.

Tulinov, A.F., V.S. Kulikauskas and M.M. Malov, 1965b, Phys. Lett. **18**, 304.

Turkenburg, W.C., W. Soszka, F.W. Saris, H.H. Kersten and B.G. Colenbrander, 1976, Nucl. Instrum. Methods **132**, 587.

Umarov, F.F., 1975, Thesis (Moscow).

Urbassek, H.M., 1988, Nucl. Instrum. Meth. Phys. Res. B **31**, 541.

Urbassek, H.M., 1989, Radiat. Eff. and Def. in Solids **109**, 293.

Vajda, P., 1964, Risø Report **84**.

Van den Berg, J.A., L.K. Verhey and D.G. Armour, 1980, Surf. Sci. **91**, 218.

Van den Hoek, P.J., T.C.M. Horn and A.W. Kleyn, 1988, Surf. Sci. **198**, L335.

Van der Weg, W.F., and D.J. Bierman, 1968, Physica **38**, 406.

Van der Weg, W.F., and D.J. Bierman, 1969, Physica **44**, 177.

Van der Weg, W.F., D.J. Bierman and B. Onderdelinden, 1969, Physica **44**, 161.

Van Leerdam, G.C., K.-M.H. Lenssen and H.H. Brongersma, 1990, Nucl. Instrum. Meth. Phys. Res. B **45**, 390.

Van Veen, A., and J. Hoack, 1972, Phys. Lett. A **40**, 398.

Van Zoest, J.M., C.E. Van der Meij and J.M. Fluit, 1984, Nucl. Instrum. Meth. Phys. Res. B **2**, 406.

Varelas, C., K. Günther and R. Sizmann, 1977, Influence of surface contamination on energy distribution of surface channeled ions, in: Proc. VIIth Int. Conf. on Atomic Collisions in Solids (Moscow) p. 116.

Varga, P., 1989, Comments At. Mol. Phys. **23**, 111.

Varga, P., W. Hofer and H. Winter, 1982, Surf. Sci. **117**, 142.

Veksler, V.I., 1962, Zh. Eksp. Teor. Fiz. **42**, 325.

Veksler, V.I., 1978, Secondary Ionic Emission of Metals (Nauka, Moscow) (in Russian).

Veksler, V.I., and V.V. Evstifeev, 1973, Izv. Akad. Nauk SSSR Ser. Fiz. **37**, 2570.

Verbeek, H., W. Eckstein and R.S. Bhattacharya, 1980, Nucl. Instrum. Methods **170**, 539.

Verhey, L.K., B. Poelsema and A.L. Boers, 1975, Phys. Lett. A **53**, 381.

Verhey, L.K., J.A. Van den Berg and D.G. Armour, 1979, Surf. Sci. **84**, 408.

Verhey, L.K., E. Van Loenen, J.A. Van den Berg and D.G. Armour, 1980, Nucl. Instrum. Methods **168**, 595.

Viaris de Lesegno, P., and S.-F. Hennequin, 1981, Surf. Sci. **103**, 257.

Viaris de Lesegno, P., G. Rivais and S.-F. Hennequin, 1974, Phys. Lett. A **49**, 265.

Viel, L., C. Benazeth, B. Fagot, F. Louchet and N. Colombie, 1971, C.R. Acad. Sci. B **273**, 30.

Viel, L., C. Benazeth and N. Benazeth, 1976, Surf. Sci. **54**, 635.

Villard, M.P., 1899, J. Phys. (Paris) **8**, 5.

Vinokurov, Ya.A., L.M. Kishinevsky and E.S. Parilis, 1976, Izv. Akad. Nauk SSSR Ser. Fiz. **40**, 1745.

Visscher, B., A.L. Boers, L.K. Verhey and B. Poelsema, 1986, Appl. Surf. Sci. **26**, 121.

Vlasov, V.P., S.A. Evstigneev, S.N. Zvonkov, S.Yu. Lukyanov and V.M. Chicherov, 1982, Zh. Eksp. Teor. Fiz. Pis'ma **35**, 508.

Volkov, S.S., S.B. Rutkovsky and A.V. Tolstoguzov, 1982, Elektron. Promyshl. **10/11**, 41.

Von dem Hagen, T., M. Hou and E. Bauer, 1982, Surf. Sci. **117**, 134.

Von Rock, C.V., 1970, Phys. Lett. **25**, 792.

Von Roos, O., 1957, Z. Phys. **147**, 210.

Vorobjova, I.V., Ya.E. Geguzin and V.S. Monastyrenko, 1986, Poverkhnost'. Fiz. Khim. Mekh. **4**, 141.

Vrakking, J.J., and A. Kroes, 1979, Surf. Sci. **84**, 153.

Wagner, H., 1979, Physical and chemical properties of stepped surfaces, in: Solid Surface Physics, eds J. Holzl, F.K. Shulte and H. Wagner, Vol. 85 (Springer, Berlin) p. 151.

Walker, R.P., and D.J. Martin, 1982, Surf. Sci. **118**, 659.

Waters, P.M., 1958, Phys. Rev. **111**, 1053.

Way, K.R., and J. Stwalley, 1973, J. Chem. Phys. **59**, 5298.

Weathers, D.L., T.A. Tombrello, M.H. Prior, R.G. Stokstad and R.E. Tribble, 1989, Nucl. Instrum. Meth. Phys. Res. B **42**, 307.

Weber, K., and F. Bell, 1976, Phys. Rev. A **16**, 1075.

Weizel, W., and O. Beeck, 1932, Z. Phys. **76**, 250.

Whitton, J., and G. Carter, 1980, The development of surface topography by heavy ion sputtering, in: Proc. Symp. on Sputtering (Perchtoldsdarf, Vienna, Austria) p. 552.

Widdiyasekera, S.D., P. Håkansson and B.U.R. Sundqvist, 1988, Nucl. Instrum. Meth. Phys. Res. B **33**, 836.

Wien, K., 1989, Radiat. Eff. and Def. in Solids **109**, 137.

Wien, K., O. Becker, W. Guthier, S. Della-Negra, Y. Le Beyec, B. Monart, K.G. Standing, G. Maynard and C. Deutsch, 1987, Int. J. Mass Spectrom. Ion Phys. **78**, 243.

Willerding, B., 1984, Thesis (Osnabruck).

Willerding, B., W. Heiland and K.J. Snowdon, 1984a, Phys. Rev. Lett. **53**, 2031.

Willerding, B., H. Steninger, K.J. Snowdon and W. Heiland, 1984b, Nucl. Instrum. Meth. Phys. Res. B **2**, 453.

Willerding, B., K.J. Snowdon, U. Imke and W. Heiland, 1986, Nucl. Instrum. Meth. Phys. Res. B **13**, 614.

Williams, P., and B. Sundqvist, 1987, Phys. Rev. Lett. **58**, 1031.

Williams, R.S., M. Kato, R.S. Daley and M. Aono, 1990, Surf. Sci. **225**, 355.

Winterbon, K.B., 1980, Radiat. Eff. Lett. **57**, 89.

Winterbon, K.B., P. Sigmund and J.B. Sanders, 1970, K. Dan. Vidensk. Selsk. Mat. Fys. Medd. **37**(14).

Witcomb, M.J., 1975, Appl. Phys. **46**, 5053.

Wittmaack, K., 1979b, Phys. Lett. A **74**, 197.

Wittmaack, K., 1979a, Phys. Lett. A **69**, 322.

Wittmaack, K., 1980, Nucl. Instrum. Methods **170**, 565.

Wolff, P.A., 1954, Phys. Rev. **95**, 56.

Wolsky, S.P., and E.J. Zdanuk, 1964, Phys. Rev. **121**, 374.

Woodruff, D.P., 1982, Surf. Sci. **116**, L219.

Woodruff, D.P., and T.A. Delchar, 1986, Modern Techniques of Surface Science (Cambridge University Press, Cambridge).

Xu, M.L., and S.Y. Tong, 1985, Phys. Rev. B **31**, 6332.

Xu, N.S., and J.L. Sullivan, 1990, Vacuum **39**, 1201.

Yamamura, Y., and W. Takeuchi, 1979, Phys. Lett. A **70**, 215.

Yamamura, Y., and W. Takeuchi, 1980, Radiat. Eff. **49**, 251.

Yamamura, Y., and W. Takeuchi, 1983, Phys. Lett. A **94**, 109.

Yamamura, Y., and W. Takeuchi, 1984, Radiat. Eff. **82**, 73.

Yanushkevich, V.A., 1979, Fiz. Khim. Obrab. Mater. **2**, 47.

Yarkulov, U., 1970, Izv. Akad. Nauk Uzb. SSR Ser. Fiz. Mat. Nauk **2**, 69.

Yarlagadda, B.S., J.E. Robinson and W. Brandt, 1978, Phys. Rev. B **17**, 3473.

Yarmoff, J.A., and R.S. Williams, 1986, Surf. Sci. **165**, L73.

Yarmoff, J.A., D.M. Cyr, J.H. Huang, S. Kim and R.S. Williams, 1986, Phys. Rev. B **33**, 3856.

Yavlinsky, Yu.N., B.A. Trubnikov and V.F. Elesin, 1966, Izv. Akad. Nauk SSSR Ser. Fiz. **30**, 1917.

Yunusov, M.S., M.A. Zaikovskaya, B.L. Oksengendler and K.R. Tokhirov, 1976, Phys. Status Solidi A **35**, K145.

Yurasova, V.E., 1976, Surface and bulk phenomena in single crystal sputtering, in: Proc. 8th Symp. on the Physics of Ionized Gases (Dubrovnik, Yugoslavia) p. 493.

Yurasova, V.E., and D.S. Karpuzov, 1967, Fiz. Tverd. Tela (Leningrad) **9**, 2508.

Yurasova, V.E., G.V. Spivak and A.I. Krokhina, 1965, Cathod sputtering as prepairing method in electron microscopy (Sovrem. Elektr. Mikrosk., Scientific and technical society of radiotechnic and electric connection, Moscow) (in Russian).

Yurasova, V.E., V.I. Shulga and D.S. Karpuzov, 1967, Reflection of oblique incident ions from a single crystal, in: Proc. VIII Int. Conf. on Phenom. in Ionized Gases (Vienna) p. 50.

Yurasova, V.E., V.I. Shulga and D.S. Karpuzov, 1968, Can. J. Phys. **46**, 759.

Yurasova, V.E., I.G. Bunin, V.I. Shulga and B.M. Mamaev, 1972, Radiat. Eff. **12**, 175.

Zaikov, V.P., E.A. Kralkina, V.S. Nikolaev, Yu.A. Fainberg and N.F. Vorobjov, 1986, Nucl. Instrum. Meth. Phys. Res. B **17**, 97.

Zaikov, V.P., E.A. Kralkina, V.S. Nikolaev and E.I. Sirotin, 1988, Nucl. Instrum. Meth. Phys. Res. B **33**, 202.

Zehner, D.M., S.H. Overbury, C.C. Havener and F.W. Meyer, 1986, Surf. Sci. **178**, 359.

Zel'dovich, Ya.B., and Yu.P. Raizer, 1966, Physics of Shock Waves and High Temperature Hydrodynamic Phenomena (Nauka, Moscow) (in Russian).

Zhabrev, G.I., 1979, Thesis (Moscow).

Zhabrev, G.I., V.A. Kurnaev and V.G. Telkovsky, 1974, Zh. Tekh. Fiz. **44**, 1560.

Zhukov, V.P., and A.A. Boldin, 1987, At. Energ. **63**, 325.

Zscheile, H., 1964, Phys. Status Solidi **6**, K87.

Author Index

Subject Index